INTRODUCTION TO ORGANIC GEOCHEMISTRY

Introduction to Organic Geochemistry

SECOND EDITION

Stephen Killops and Vanessa Killops

Blackwell Publishing

© 2005 Blackwell Science Ltd, a Blackwell Publishing company

BLACKWELL PUBLISHING
350 Main Street, Malden, MA 02148-5020, USA
108 Cowley Road, Oxford OX4 1JF, UK
550 Swanston Street, Carlton, Victoria 3053, Australia

The right of Stephen Killops and Vanessa Killops to be identified as the Authors of this Work has been asserted in accordance with the UK Copyright, Designs, and Patents Act 1988.

All rights reserved. No part of this publication may be reproduced, stored in a retrieval system, or transmitted, in any form or by any means, electronic, mechanical, photocopying, recording or otherwise, except as permitted by the UK Copyright, Designs, and Patents Act 1988, without the prior permission of the publisher.

First edition published 1993 by Longman Scientific and Technical
Second edition published 2005 by Blackwell Publishing Ltd

Library of Congress Cataloging-in-Publication Data

Killops, S. D. (Stephen Douglas), 1953–
Introduction to organic geochemistry / Stephen Killops and Vanessa Killops.–2nd ed.
  p. cm.
Includes bibliographical references and index.
ISBN 0-632-06504-4 (pbk. : alk. paper)
1. Organic geochemistry. I. Killops, V. J. (Vanessa Jane), 1955– II. Title.

QE516.5K55 2005
553.2–dc22
2004003107

A catalogue record for this title is available from the British Library.

Set in $9\frac{1}{2}$ on 11 pt Bembo
by SNP Best-set Typesetter Ltd., Hong Kong
Printed and bound in the United Kingdom
by TJ International, Padstow, Cornwall

The publisher's policy is to use permanent paper from mills that operate a sustainable forestry policy, and which has been manufactured from pulp processed using acid-free and elementary chlorine-free practices. Furthermore, the publisher ensures that the text paper and cover board used have met acceptable environmental accreditation standards.

For further information on
Blackwell Publishing, visit our website:
www.blackwellpublishing.com

Contents

Preface vii
Acknowledgements ix

1 Carbon, the Earth and life **1**
1.1 Carbon and the basic requirements of life 1
1.2 Chemical elements, simple compounds and their origins 2
1.3 The origin of life 5
1.4 Evolution of life and the atmosphere 15
1.5 Major contributors to sedimentary organic matter 23

2 Chemical composition of organic matter **30**
2.1 Structure of natural products 30
2.2 Carbohydrates 35
2.3 Amino acids and proteins 40
2.4 Lipids 43
2.5 Lignins, tannins and related compounds 62
2.6 Nucleotides and nucleic acids 67
2.7 Geochemical implications of compositional variation 69

3 Production, preservation and degradation of organic matter **71**
3.1 How and why organic-rich deposits form 71
3.2 Controls on primary production 71
3.3 Preservation and degradation of organic matter 89
3.4 Depositional environments associated with accumulation of organic matter 109

4 Long-term fate of organic matter in the geosphere **117**
4.1 Diagenesis 117
4.2 Humic material 119
4.3 Coal 122
4.4 Kerogen 132
4.5 Catagenesis and metagenesis 144
4.6 Temporal and geographical distribution of fossil organic carbon 162

5 Chemical stratigraphic concepts and tools **166**
5.1 Biologically mediated transformations 166
5.2 Examples of source indicators in Recent sediments 169
5.3 Diagenesis at the molecular level 174
5.4 Source and environmental indicators in ancient sediments and oil 196
5.5 Thermal maturity and molecular transformations 207
5.6 Palaeotemperature and age measurement 212
5.7 Maturity of ancient sedimentary organic matter 221
5.8 Isotopic palaeontology 234

6 The carbon cycle and climate **246**
6.1 Global carbon cycle 246
6.2 Changes in carbon reservoirs over geological time 254
6.3 Palaeoclimatic variations 263
6.4 Isotopic excursions at period boundaries 280
6.5 Human influence on the carbon cycle 285

7 Anthropogenic carbon and the environment **295**
7.1 Introduction 295
7.2 Halocarbons 295
7.3 Hydrocarbon pollution in aquatic environments 298
7.4 Endocrine-disrupting chemicals 304
7.5 Environmental behaviour of selected xenobiotic compounds 308

7.6 Factors affecting the fate of
anthropogenic components 317

Appendix 1 SI units used in this book 322
Appendix 2 SI unit prefixes 323
Appendix 3 Geological time scale 324

References 325

Index 363

Preface

To begin with, a brief statement of what constitutes organic geochemistry is probably called for. It is the study of the transformation undergone by organic matter of all types, whether of biological or manmade origin, in the Earth System. The transformations involved vary from those mediated by biological processes involved in the production of living tissue and the operation of food-chains, to those controlled by temperature and pressure at depth in the crust. Photochemical processes in the atmosphere and hydrosphere can also be important in controlling the environmental behaviour of some organic compounds.

Our knowledge of organic geochemistry has been expanding at such a great rate that a comprehensive text on the subject would fill many books of this size. To a newcomer, the bulk of information and the terminology adopted from a range of disciplines, such as chemistry, geology, ecology, biochemistry, botany and oceanography, can be quite daunting. However, to those not readily deterred, the fascination of the subject soon becomes apparent. If only the basics of organic geochemistry could be found readily at hand rather than scattered through textbooks and journals of a number of disciplines! These were our thoughts when we first came to the subject in the 1980s, and they subsequently provided the stimulus for this book when one of us (SDK) began teaching the subject to undergraduates and postgraduates.

This book is an attempt to present a readily accessible, up-to-date and integrated introduction to organic geochemistry, at a reasonable price. It does not assume any particular specialist knowledge, and explanatory boxes are used to provide essential information about a topic or technique. Technical terms are also highlighted and explained at their first appearance in the text. SI units are presented in Appendix 1, prefixes used to denote exponents in Appendix 2 and a geological time scale in Appendix 3. A comprehensive reference list is provided for those wishing to explore the original sources of the concepts and case studies covered, which concentrates on articles in the most readily available journals.

The text is intended to serve undergraduate and postgraduate courses in which organic geochemistry is an important component. It may also be found a useful companion by experienced scientists from other disciplines who may be moving into the subject for the first time. Whereas the first edition of this book concentrated on organic-rich deposits, of particular interest to those involved in petroleum exploration, this edition considers the fate of organic matter in general. The importance of environmental geochemistry has not been overlooked, and consideration has been given to environmental change at various times during Earth's history in order to provide a background for assessing modern changes and how the carbon cycle works. Naturally, in a book of this size it is impossible to cover everything; for example, it is not possible to do justice to the wide range of analytical techniques used in organic geochemistry, which draw on all aspects of separation science and spectrometry. However, there are many texts on these topics, some of which are listed below. We hope the topics we have selected for this edition stimulate the reader to continue studying organic geochemistry.

Further reading

Harwood L.M., Claridge T.D.W. (1997) *Introduction to Organic Spectroscopy*. Oxford: Oxford Science Publications.

Lewis C.A. (1997) Analytical techniques in organic chemistry. In *Modern Analytical Geochemistry* (ed. Gill R.), 243–72. Harlow: Longman.

Peters K.E., Moldowan J.M. (1993) *The Biomarker Guide: Interpreting Molecular Fossils in Petroleum and Ancient Sediments*. Englewood Cliffs, NJ: Prentice Hall.

Poole C. (2002) *The Essence of Chromatography*. Amsterdam: Elsevier Health Sciences.

Settle F. (ed.) (1997) *Instrumental Techniques for Analytical Chemistry*. Englewood Cliffs, NJ: Prentice Hall.

Silverstein R.M., Bassler C.G., Morrill T.C. (1991) *Spectrometric Identification of Organic Compounds*. New York: Wiley.

Acknowledgements

The authors wish to thank their colleagues, for advice and helpful comments during the preparation of this book, and the library at Plymouth University. We are particularly grateful to those who volunteered their time and expertise in reviewing the manuscript: Dr Geoff Abbott (NRG in Fossil Fuels and Environmental Geochemistry, University of Newcastle-upon-Tyne); Prof. Carrine Blank (Dept of Earth and Planetary Sciences, Washington University); Dr Paul Finch (Centre for Chemical Science, Royal Holloway, University of London); Prof. Andy Fleet (Dept of Mineralogy, Natural History Museum, London); Dr Anthony Lewis (School of Environmental Sciences, Plymouth University); Dr Dave McKirdy (School of Earth and Environmental Sciences, University of Adelaide); Prof. Phil Meyers (Dept of Geological Sciences, University of Michigan); Dr Fred Prahl (College of Oceanic and Atmospheric Sciences, Oregon State University); Dr Lloyd Snowdon (Geological Survey of Canada, Calgary).

1 | Carbon, the Earth and life

1.1 Carbon and the basic requirements of life

In its broadest sense, organic geochemistry concerns the fate of carbon, in all its variety of chemical forms, in the Earth system. Although one major form of carbon is strictly inorganic, carbon dioxide, it is readily converted by photosynthesis into the stuff of life, organic compounds (see Box 1.9), and so must be included in our consideration of organic geochemistry. From chiefly biological origins, organic compounds can be incorporated into sedimentary rocks (Box 1.1) and preserved for tens of millions of years, but they are ultimately returned to the Earth's surface, by either natural processes or human action, where they can participate again in biological systems. This cycle involves various biochemical and geochemical transformations, which form the central part of the following account of organic geochemistry. To understand these transformations and the types of organic compounds involved we must first consider the origins and evolution of life and the role played by carbon.

Growth and reproduction are among the most obvious characteristics of life, and require the basic chemicals from which to build new cellular material, some form of energy to drive the processes and a means of harnessing and distributing this energy. There is an immense range of compounds involved in these processes. For example, energy is potentially dangerous; the sudden release of the energy available from complete oxidation of a single molecule of glucose is large when considered at a cellular level. Therefore, a range of compounds is involved in bringing about this reaction safely by a sequence of partial oxidations, and in the storage and transport to other sites in the cell of the more moderate amounts of energy released at each step. We look at the geochemically important compounds involved in life processes in Chapter 2.

What makes carbon such an important element is its ability to form an immense variety of compounds—

Box 1.1 | Sediments and sedimentary rocks

Sediment is the solid material, inorganic or organic, that settles out of suspension from a fluid phase (normally water, ice or air) in which it has been transported. Over time, under the right conditions, it can undergo lithification (i.e. conversion into a solid body of rock). Various processes can be involved in lithification: compaction, cementation, crystallization and desiccation.

Inorganic sediment is supplied by erosion of material from exposed areas of high relief, and can be transported a considerable distance to the area of deposition. The composition of this **detrital** (or **clastic**) material varies, but aluminosilicate minerals are usually important. There are also biogenic sediments, resulting from the remains of organisms (e.g. calcareous and siliceous tests, peat) and **chemical sediments** formed by precipitation of minerals from solution (e.g. evaporites, some limestones and authigenic infills of pores by quartz and calcite cements).

The nature of the sediments accumulating in a particular location can change over time, allowing the recognition of different bodies of sedimentary rock. Such a body is termed a **facies**, and it displays a set of characteristic attributes that distinguish it from vertically adjacent bodies. Various distinguishing attributes include sedimentary structures, mineral content and fossil assemblages. Organofacies can also be recognized, based on compositional differences in the organic material present (Jones 1987; Tyson 1995).

primarily with the elements hydrogen, oxygen, sulphur and nitrogen, as far as natural products are concerned—with an equally wide range of properties; this is unparalleled by other elements. This variety of properties allows carbon compounds to play the major role in the creation and maintenance of life. The strength of the chemical bonds in organic compounds is sufficiently high to permit stability, which is essential in supportive tissue, for example, but low enough not to impose prohibitive energy costs to an organism in synthesizing and transforming compounds.

Another prerequisite for life is liquid water, the medium in which biochemical reactions take place and usually the main constituent of organisms. Although bacteria, and even some simple animals, like the tardigrade, can survive in a dormant state without water, the processes that we associate with life can only take place in its presence. This requirement obviously imposes temperature limits on environments that can be considered suitable for life; hence one of the criteria in the search for life on other planets is evidence for the existence of liquid water at some stage of a planet's life.

1.2 Chemical elements, simple compounds and their origins

1.2.1 Origin of elements

Carbon is the twelfth most abundant element in the Earth's crust, although it accounts for only $c.0.08\%$ of the combined lithosphere (see Box 1.2), hydrosphere and atmosphere. Carbon-rich deposits are of great importance to humans, and comprise diamond and graphite (the native forms of carbon), calcium and magnesium carbonates (calcite, limestone, dolomite, marble and chalk) and fossil fuels (gas, oil and coal). Most of these deposits are formed in sedimentary environments, although the native forms of C require high temperature and pressure, associated with deep burial and metamorphism.

Where did the carbon come from? The universe is primarily composed of hydrogen, with lesser amounts of helium, and comparatively little of the heavier elements (which are collectively termed metals by astronomers). The synthesis of elements from the primordial hydrogen, which was formed from the fundamental particles upon the initial stages of cooling after the Big Bang some 15 Gyr ago, is accomplished by nuclear fusion, which requires the high temperatures and pressures within the cores of stars. Our Sun is relatively small in stellar terms, with a mass of $c.2 \times 10^{30}$ kg, and is

---

Box 1.2 | Earth's structure

Temperature and pressure both increase with depth in the Earth and control the composition and properties of the material present at various depths. The Earth comprises a number of layers, the boundaries between which are marked by relatively abrupt compositional and density changes (Fig. 1.1). The inner core is an iron–nickel alloy, which is solid under the prevailing pressure and temperature ranges. In contrast, the outer core is molten and comprises an iron alloy, the convection currents within which are believed to drive the Earth's magnetic field. The core–mantle boundary lies at $c.$ 2900 km depth and marks the transition to rocky material above. The **mantle** can be divided into upper and lower parts, although the boundary is quite a broad transitional zone ($c.$1000–400 km depth). It behaves in a plastic, ductile fashion and supports convection cells. The upper mantle layer from $c.$100 to 400 km depth is called the **asthenosphere**, and its convection system carries the drifting continental plates.

With decreasing temperature towards the surface, the top part of the mantle is sufficiently cool that it behaves as a strong, rigid solid. The cold, relatively thin, layer of solid rock above the mantle is the **crust**, which is $c.$5–7 km thick under the oceans but $c.$30–70 km thick on the continents. The topmost mantle and crust are often considered together as **lithosphere**. Under excessive strain, such as during earthquakes, the lithosphere undergoes brittle failure, in contrast to the ductile deformation that occurs within the asthenosphere.

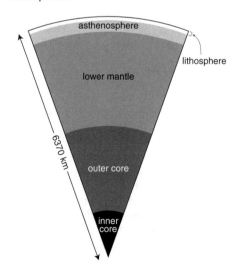

Fig. 1.1 Simplified layering within the Earth.

capable of hydrogen fusion, which involves the following reactions:

$$^{1}H + ^{1}H \rightarrow {}^{2}H + energy \qquad [Eqn\ 1.1]$$

$$^{2}H + ^{1}H \rightarrow {}^{3}He + energy \qquad [Eqn\ 1.2]$$

$$^{3}He + ^{3}He \rightarrow {}^{4}He + ^{1}H + ^{1}H + energy \qquad [Eqn\ 1.3]$$

(where ^{2}H can also be written as D, or deuterium, and the superscript numbers represent the mass numbers as described in Box 1.3). Because of the extremely high temperatures and pressures, electrons are stripped off atoms to form a plasma and it is the remaining nuclei that undergo fusion reactions. Ultimately, when enough helium has been produced, helium fusion can then begin. This process is just possible in stars of the mass of our Sun, and results in the creation of carbon first and then oxygen:

$$^{4}He + ^{4}He + ^{4}He \rightarrow {}^{12}C + energy \qquad [Eqn\ 1.4]$$

$$^{12}C + ^{4}He \rightarrow {}^{16}O + energy \qquad [Eqn\ 1.5]$$

There is still usually plenty of hydrogen left in a star when helium fusion starts in the core. If the products of helium fusion mix with the outer layers of the star it is possible for other elements to be formed. The CNO cycle is an important fusion pathway (Fig. 1.2), which primarily effects the conversion of H to He. However, the cycle can be broken, resulting in the formation of heavier elements; for example, by the fusion reaction shown in Eqn 1.5.

Only more massive stars can attain the higher temperatures needed for the synthesis of heavier elements. For example, magnesium can be produced by fusion of carbon nuclei and sulphur by fusion of oxygen nuclei. Fusion of this type can continue up to ^{56}Fe, and ideal conditions are produced in novae and supernovae explosions. Heavier elements still are synthesized primarily by neutron capture.

Our Sun is too young to have produced carbon and heavier elements. These elements in the nebula from which the Solar System was formed $c.4.6$ Gyr ago, together with the complex organic molecules in our bodies, owe their existence to an earlier generation of stars.

1.2.2 The first organic compounds

Away from the nuclear furnaces of the stars elements can exist as the atoms we are familiar with, which in turn can form simple compounds if their concentrations are sufficiently great that atomic encounters can occur. The highest concentrations are found in interstellar clouds, and in particular in molecular clouds, where densities of 10^{9}–10^{12} particles per m^3 can exist. This is still a very low density, and the most common constituents of these clouds are H (atomic hydrogen), H_2 (molecular hydrogen) and He, which can be ionized by bombardment with high-energy particles, originating from phenomena like supernovae, and can then take part in ion–molecule reactions, such as:

$$H_3^+ + CO \rightarrow HCO^+ + H_2 \qquad [Eqn\ 1.8]$$

$$H_3^+ + N_2 \rightarrow N_2H^+ + H_2 \qquad [Eqn\ 1.9]$$

$$H_3^+ + O \rightarrow OH^+ + H_2 \qquad [Eqn\ 1.10]$$

$$H_3^+ + C_2 \rightarrow C_2H^+ + H_2 \qquad [Eqn\ 1.11]$$

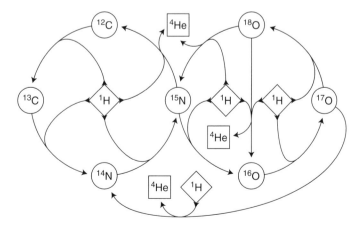

Fig. 1.2 Hydrogen fusion via the CNO cycle.

Box 1.3 | Stable isotopes

Isotopes are atoms of the same element that contain the same numbers of protons and electrons, so are chemically identical, but contain different numbers of neutrons, so their masses are different. Each element has an individual atomic number, equal to the number of electrons (or protons) in an atom (six for carbon). Electrons carry a unit negative charge but very little mass. The negative charge of the electrons in an atom is offset by an equal number of positively charged particles, protons, which have masses considerably greater than the electron. The protons exist in a nucleus, around which the electrons orbit. Also in the nucleus are uncharged particles called neutrons, with similar masses to the protons. Isotopes of an element differ in the number of neutrons in their nuclei and, therefore, in their atomic mass, which is the sum of the protons and neutrons (12 and 13 for the stable isotopes of carbon). So, in general, we can represent an isotope by ^m_nE, where m is the mass number and n the atomic number of the element E, but often the atomic number is omitted for simplicity (e.g. ^{13}C instead of $^{13}_6C$).

Carbon is a mixture of two stable isotopes, ^{12}C and ^{13}C. In the Earth as a whole the relative abundances of ^{12}C and ^{13}C are 98.894% and 1.106%, respectively. Carbon compounds of biological origin are relatively enriched in the lighter isotope, while the heavier isotope is retained in the main forms of inorganic carbon (e.g. carbonate, bicarbonate and carbon dioxide). Biogenic substances usually contain more of the lighter isotope than exists in the substrate from which the element was sequestered, a process termed **isotopic fractionation**. This is because, in the main assimilatory pathways and, to a lesser extent, the ensuing metabolic processes, the reactions involving isotopically lighter molecules of a compound, such as in the primary carbon fixation reaction of photosynthesis, occur slightly faster, a phenomenon termed the **kinetic isotope effect**. Isotopic fractionation can also take place during diffusion of a gas across a cell membrane—e.g. the uptake of carbon dioxide by unicellular algae—because the slightly smaller molecules of lighter isotopic composition diffuse at a faster rate (see Box 3.8).

The ratio ^{13}C to ^{12}C in a geological sample is measured by mass spectrometry after converting the carbon to CO_2. To minimize inaccuracies in measuring the absolute amounts of $^{12}CO_2$ and $^{13}CO_2$ the ratio of the two in a sample is compared with that in a standard analysed at the same time. The isotopic ratio of a sample is normally expressed by δ values (with units of permil, or ‰) relative to the standard, and its general form can be represented by:

$$\delta^m E(‰) = [R_{sample}/R_{standard} - 1] \times 10^3 \quad [Eqn\ 1.6]$$

where m = mass number of the heavier isotope, E = the element and R = the abundance ratio of a heavier to the lightest, most abundant isotope. So for carbon we have:

$$\delta^{13}C(‰) = [(^{13}C/^{12}C)_{sample}/(^{13}C/^{12}C)_{standard} - 1] \times 10^3 \quad [Eqn\ 1.7]$$

Other biogeochemically important elements have a range of stable isotopes, as shown in Table 1.1, and the isotopic ratios are expressed using the general formula in Eqn 1.6. A different standard is used for each element, and the standard can also vary depending upon the form of the element (e.g. oxygen in Table 1.1). By definition, the $\delta^m E$ value of a standard is 0‰, so negative values for a sample indicate depletion in the heavier isotope compared with the standard and positive values indicate enrichment in the heavier isotope (for PDB $^{13}C/^{12}C = 0.011237$).

Some elements have unstable isotopes, which undergo radioactive decay, such as ^{14}C (see Box 5.5). Those of ^{238}U, ^{235}U, ^{232}Th and ^{40}K are responsible for the heat production in the Earth's crust.

Table 1.1 Stable isotope abundances of biogeochemically important elements and their associated standards (after Hoefs 1997)

element	stable isotopes	(% relative abundance)	common reference standard
hydrogen	^1H (99.9844)	^2H or D (0.0156)	Vienna standard mean ocean water (V-SMOW)
carbon	^{12}C (98.89)	^{13}C (1.11)	Cretaceous Peedee formation belemnite (PDB)
nitrogen	^{14}N (99.64)	^{15}N (0.36)	atmospheric N_2 (air)
oxygen*	^{16}O (99.763)	^{18}O (0.1995)	PDB for low-temperature carbonates, otherwise standard mean ocean water (SMOW)
sulphur*	^{32}S (95.02)	^{34}S (4.21)	Canyon Diablo meteorite troilite (CDT)

*The above stable isotopes are those commonly used in geochemistry, but others exist for oxygen (^{17}O (0.0375%)) and sulphur (^{33}S (0.75%), ^{36}S (0.02%)).

Among the eventual products of these reactions are methanal (HCHO, also known as formaldehyde), ammonia, water and various simple organic molecules, respectively. Just a few examples of the types of simple molecules that have been detected in interstellar space and also in comets (see Box 1.4) are given in Table 1.2. These compounds are all gases when in the interstellar medium, but are solids when accreted on to dust particles (formed *inter alia* from carbonaceous grains, and oxides of magnesium and aluminium). Interestingly, carbon dioxide has been detected in comets but not in molecular clouds, and it is likely that the more intimate associations of molecules in comets can lead to different products and perhaps more complex organic molecules. One source of energy to fuel such reactions is ultraviolet (UV) radiation from the Sun.

1.3 The origin of life

1.3.1 The young Earth

It is likely that conditions on the newly accreted Earth were not favourable for life: hence the naming of the Era from 4.6 to 3.8 Ga as the Hadean (see Appendix 3 for geological time scale). The Earth's primary atmosphere, immediately after its formation, would have probably reflected the composition of the nebula from which the Solar System formed. It would have contained mainly hydrogen and helium, which would have tended to escape the gravitational field of the Earth, but would, in any event, have been stripped away by the violent solar winds during the early T-Tauri stage of the Sun's evolution (Hunten 1993). The collision of the Earth with another body that ejected material to form the Moon before 4.5 Ga, shortly after the core and mantle had differentiated (Halliday 2000), would also have had a major influence on the atmospheric composition. The Earth's secondary atmosphere owes its existence to juvenile volatiles outgassing from the interior of the planet (although a proportion of the water may have been acquired subsequently from meteorites). In view of the composition of volcanic emissions today these volatiles probably comprised mainly water vapour, nitrogen, carbon dioxide, carbon monoxide, sulphur dioxide and hydrogen chloride, although opinions vary over the importance of reducing gases (see Box 1.5) such as methane, ammonia and hydrogen. Whether methane and ammonia could have been present depends upon whether the oxidation state of the mantle has varied, and the amount of time it took to reach its current degree of oxidation. It is believed that no free oxygen was present

---

Box 1.4 | Comets, asteroids and meteorites

Comets are mostly aggregates of interstellar dust, ice (H_2O, CO and CO_2) and some organic molecules. They originate from two regions in the Solar System. The most distant is the Oort cloud, which is up to 10^5 AU from the Sun, well outside the orbit of Pluto (1 AU = Astronomical Unit, the mean orbital distance of the Earth from the Sun), and which is probably the source of the long-period comets (e.g. Hale–Bopp). The nearer is the Kuiper Belt, which lies between *c.*30 AU (just beyond Neptune) and 100 AU, and is the likely source of short-period comets (e.g. Swift–Tuttle). Comets are ejected from these source regions by gravitational perturbations, resulting in the usually very eccentric orbits we are familiar with. Meteor showers are associated with the Earth crossing the orbit of short-period comets (e.g. Swift–Tuttle is responsible for the Perseids). Cometary composition is believed to reflect the primordial material from which the Solar System formed.

Asteroids originate from a belt between Mars and Jupiter (*c.*2–4 AU), and seem to represent primordial Solar System material that failed to aggregate into a planet. As for comets, gravitational perturbations can destabilize orbits, sometimes resulting in collisions that eject fragments (meteoroids). Some asteroids have Earth-crossing orbits.

Meteorites are the grains of meteoroids or meteors that survive the journey through the Earth's atmosphere and reach the surface. Some are almost pure iron–nickel alloy, whereas others contain silicates and sulphides, and yet others (the **carbonaceous chondrites**) contain organic compounds.

Table 1.2 Some simple molecules detected in both interstellar space and comets

organic	inorganic
CH_4	H_2O
H_2CO	CO
H_3COH	NH_3
$HCOOH$	HCN
CH_3CH_2OH	H_2S
$HC\equiv CCN$	SO_2
H_2CS	OCS

> **Box 1.5 | Oxidation and reduction**
>
> The most obvious definition of **oxidation** is the gain of oxygen by a chemical species, as in the burning of methane:
>
> $$CH_4 + 2O_2 \rightarrow CO_2 + 2H_2O \quad \text{[Eqn 1.12]}$$
>
> A further example is provided by the oxidation of ferrous ions (iron(II)) to ferric (iron(III)) during the sedimentary deposition of iron oxide:
>
> $$4Fe^{2+} + O_2 + 4H_2O \rightarrow 2Fe_2O_3 + 8H^+ \quad \text{[Eqn 1.13]}$$
>
> Oxidation can also be defined as the loss of hydrogen, as occurs with methane above (Eqn 1.12). A further definition of oxidation is the loss of electrons. This is the net process undergone by iron in the above oxidation of iron(II) to iron(III), and can be represented by:
>
> $$Fe^{2+} \rightarrow Fe^{3+} + e^- \quad \text{[Eqn 1.14]}$$
>
> All three definitions of oxidation are encountered in geochemistry, and **reduction** is the opposite of oxidation. Oxidation and reduction occur in unison, because the oxidation of one chemical species results in the reduction of another, and the combination is termed a **redox** reaction.
>
> Oxidizing conditions in sedimentary environments are termed **oxic** and are related to free oxygen being available for oxidative reactions to take place. In **anoxic** conditions there is no such available oxygen and conditions are described as reducing. In water (whether in water bodies or in sedimentary pore waters) dissolved oxygen levels of >0.5‰ (parts per thousand, or per mil) correspond to oxic conditions, while those of <0.1‰ correspond to anoxic conditions. Conditions related to intermediate values of oxygen concentration are generally described as **suboxic**.
>
> Slightly different terms are used for zones with differing oxygen availability when biological activity is being described. Well oxygenated conditions and associated metabolic processes are described as **aerobic**, oxygen-starved conditions are **anaerobic** and intermediate conditions may be called **dysaerobic**.

near the surface for c.2.5 Gyr, although some may have been formed high up in the atmosphere by the photodissociation of CO_2 and H_2O (Kasting 1993). Without significant amounts of oxygen there would not have been an effective ozone layer, so life at the surface would have been exposed to damaging UV radiation (at wavelengths < c.300 nm, see Box 7.1).

It is thought that the formation of significant amounts of continental crust did not begin until the start of the Archaean, when heat production had fallen to levels permitting the initiation of crustal differentiation processes. Continental crust production seems to have reached a maximum in the late Archaean (3.0–2.5 Ga), although continental crust formation in the style observed at modern subduction zones did not commence until heat production had fallen still further, after c.2.5 Ga (Martin 1986). Not only would a lot of heat have been inherited from the accretionary processes of the Earth's formation, but the infant Earth had hardly begun to deplete its store of heat-generating radiogenic isotopes (^{40}K, ^{232}Th, ^{235}U, ^{238}U; see Box 1.3 for an explanation of isotopes) and continued to experience bombardment from the remaining unaccreted debris (adding heat to the system) until the Solar System settled into its current, relatively stable, state. Cratering evidence on the Moon suggests that there was a period of particularly heavy bombardment from 4.0 to 3.8 Ga (the **Late Heavy Bombardment**; Cohen et al. 2000), which seems not to have been limited to the Earth–Moon system (Ash et al. 1996). These are hardly conditions conducive to the establishment of life, yet there appears to be microstructural and carbon isotopic evidence for the existence of life around 3.8 Ga on Earth (Mojzsis et al. 1996). If this evidence is not misleading, and the result of abiotic processes, life appeared very soon after the heavy bombardment ceased, or it appeared earlier and survived in some refuge; or it may even have originated more than once, only to be wiped out by large impacts.

Initially the surface temperature would have been too high for liquid water to exist. Until recently it was assumed that abundant liquid water was not likely to be present much before 3.8 Ga, but oxygen isotopic evidence from zircons suggests that liquid water was present at 4.4 Ga (Valley et al. 2002). The $\delta^{18}O$ values (see Box 1.3) suggest a temperature of at most 200°C, whereas a temperature of at least 374°C would be required for a hydrosphere of the present-day size to have existed entirely in the vapour phase. So the Hadean may not have been quite so hell-like after all. Unfortunately, it is not possible to tell how hot it may have become during the Late Heavy Bombardment because of the lack of zircon evidence.

The earliest known sedimentary rocks deposited under water date from the beginning of the Archaean Era (3.8 Ga), and by 3.5 Ga oceans seem to have been widespread, based on the occurrence of sedimentary rocks and pillow lavas. The evidence for life before c.2.7 Ga is not conclusive. In the oldest rocks (c.3.8 Ga metamorphic rocks from Greenland) it relies entirely upon light carbon isotopic values, but the carbon may not always be of biogenic origin (van Zuilen et al. 2002; see Box 1.3). The earliest fossil evidence, in the form of **stromatolite**-like structures (layered domes similar to those produced today by cyanobacterial mats), is from c.3.5 Ga (e.g. Apex chert, northwestern Australia; Schopf 1993). However, the microstructural features in the stromatolite-like bodies, together with the bulk compositional (e.g. from Laser–Raman analysis) and isotopic characteristics of the graphitic carbon they contain, could have abiotic origins, resulting from hydrothermal activity (Schopf et al. 2002; Brasier et al. 2002). Fischer–Tropsch-type reactions (involving reduction of CO_2) in hydrothermal settings are potential sources of hydrocarbons with light carbon isotopic signatures (Lancet & Anders 1970; Horita & Berndt 1999; Holm & Charlou 2001). By 2.7 Ga there is clear evidence from chemical fossils for the presence of life, as we see in Section 1.4.1.

1.3.2 The raw material for life

The story of the evolution of life on Earth is relatively straightforward in comparison to the problem of how it first arose. The abiotic synthesis on Earth of the organic compounds necessary for the creation of life appears to require a reducing atmosphere (see Box 1.5). Early experiments with electrical discharges in mixtures of methane, ammonia and hydrogen sulphide in the presence of water created various organic compounds essential for life, including many amino acids (e.g. Miller & Urey 1959), which have also been found in meteorites. However, as mentioned in the previous section, the Earth's early secondary atmosphere was probably less reducing, with a somewhat different composition. Due to the major environmental and time constraints on the abiotic synthesis of the basic compounds of life and their subsequent assembly into a viable cell (Maher & Stevenson 1988), other theories have been propounded. It has been suggested that a suitable supply of organic material could have originated from space via comets etc. (Chyba et al. 1990; Greenberg 1997). Other theories have postulated that life originated elsewhere and travelled to Earth (panspermia; Wickramasinghe et al. 1997), although the proposition that bacteria-like organisms travelled from Mars to Earth protected within cometary material (McKay et al. 1996) remains to be proven, because the evidence for biological activity is equivocal (Grady 1999).

Estimations of the amount of organic compounds produced abiotically on Earth (**endogenous**) or delivered directly by extraterrestrial bodies (**exogenous**) during the heavy bombardment 4 Gyr ago are shown in Table 1.3 (Chyba & Sagan 1992). Delivery of intact organic compounds is favoured by the gentle deceleration in the atmosphere experienced by interplanetary dust particles and intermediate-sized meteorites. In comparison, small meteorites tend to undergo complete ablation and large meteorites are often heated sufficiently to pyrolyse the organic material and render it useless for life. The estimated amount of exogenous organic material is probably effectively independent of the degree to which the atmosphere was reducing. The energy needed to produce organic compounds at the Earth's surface can be generated by impact shocks, UV light or electrical discharges, and the amounts of organic products are strongly influenced by the atmospheric

Table 1.3 Estimated endogenous and exogenous supply of organic material to Earth 4 Gyr ago (after Chyba & Sagan 1992)

source	organic production rate (kg yr^{-1})	
	reducing atmosphere	neutral atmosphere*
lightning	3×10^9	3×10^7
coronal discharge	2×10^8	2×10^6
UV (H$_2$S absorption <270 nm)	2×10^{11}	–
UV (CO$_2$ absorption <230 nm)	–	3×10^8
UV (H$_2$O absorption <200 nm)	3×10^9	–
atmospheric shock (meteor)	1×10^9	3×10^1
atmospheric shock (post-impact plume)	2×10^{10}	4×10^2
interplanetary dust particles	6×10^7	6×10^7

*Neutral atmosphere has H$_2$/CO$_2$ concentration ratio of 0.1.

composition. A reducing atmosphere (mostly methane and water) has an abiotic synthesis potential at least two orders of magnitude greater than a neutral atmosphere. Long-wavelength UV may have been the dominant energy source, but the influence of shock-wave energy is not easy to quantify because it is highly dependent upon atmospheric chemistry. In a neutral atmosphere, electrical discharges, UV, shock-waves and interplanetary dust particles could have made equally significant contributions to the Earth's inventory of abiotic organics.

Meteorites (see Box 1.4), and in particular carbonaceous chondrites, provide a glimpse of the chemical composition of the Solar System, because their elemental composition is very like that observed spectroscopically in the Sun and in other nearby stars. Carbonaceous chondrites contain significant amounts of organic compounds (up to 5% by weight) as well as plentiful water, but relatively few (c.36) have been collected soon after arrival, and there is always the danger of contamination of the indigenous organic matter in the meteorites with terrestrial material upon impact. One of the most thoroughly studied examples is the Murchison meteorite, which fell in Australia in 1969 (e.g. Engel & Macko 2001). Among the amino acids in the meteorite (Table 1.4), several that are common to biological systems (e.g. serine and threonine) were found in only trace amounts or were below detection limits, suggesting minimal contamination subsequent to impact (Engel & Nagy 1982).

Amino acids are important because they are the building blocks of proteins, which are responsible for many cellular functions (see Section 2.3). They can exist as optical isomers (see Section 2.1.3), and life on Earth has adopted the L form rather than the D, for reasons that are as yet unclear. The pair of optical isomers for alanine is shown in Fig. 1.3. Abiotic synthesis of amino acids in the laboratory results in equal amounts of the two optical isomers (a racemic mixture), and so it is assumed that extraterrestrial abiotic synthesis would proceed in a similar fashion. The isomeric composition of amino acids in meteorites should, therefore, provide information on their extraterrestrial origin, although few such analyses have been undertaken to date. One analysis is shown in Table 1.4, and the D/L ratio in the initial water extract shows a dominance of the L isomers. When the remaining rock was ground up and digested with acid, the amino acids that were liberated had an even greater dominance of the L isomers. Because the latter amino acids had been more protected from contamination within the rock matrix than those in the initial extract, it would seem unlikely that the L predominance is attributable to contamination by terrestrial sources after impact; instead, it probably represents the extraterrestrial source.

Further support for a non-terrestrial origin for the excess of the L isomers of amino acids in the Murchison meteorite is provided by the stable-isotope compositions, as described in Box 1.6. The origin of the L isomer

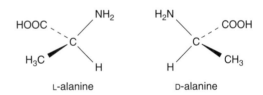

Fig. 1.3 The optical isomers of alanine.

Table 1.4 Amino acids detected in hydrolysed aqueous extracts of Murchison meteorite (after Engel & Nagy 1982)

	concentration (nmol g^{-1})*	D/L ratio	
		pre acid digestion*	post acid digestion†
common amino acids			
glutamic acid	18.2	0.30	0.18
aspartic acid	8.5	0.30	0.13
proline	13.5	0.30	0.11
glycine	45.8		
β-alanine	13.1		
leucine	1.9	0.17	0.03
sarcosine	4.7		
alanine	15.3	0.60	0.31
valine	8.6		
exotic amino acids			
α-aminobutyric acid	107.8		
isovaline	23.6		

*Initial hydrolysed water extract; †second hydrolysed water extract after acid digestion of rock.

Box 1.6 | Stable isotopic composition of amino acids in the Murchison meteorite

If the amino acids in the Murchison meteorite have an abiotic origin, the D and L isomers would be expected to have identical $\delta^{13}C$ and $\delta^{15}N$ values. Similarly, racemization of the dominant L form of biotic origin on Earth would result in mixtures of D and L isomers with identical isotopic compositions because none of the C and N atoms are exchanged with external sources during racemization (Engel & Macko 1986). However, bulk $\delta^{13}C$ and $\delta^{15}N$ values for the Murchison amino acids are higher than commonly found in biological materials on Earth (Epstein et al. 1987), so it should be possible to discern whether terrestrial contamination has given rise to the excess L isomers, because those L isomers would be depleted in ^{13}C and ^{15}N compared to the D isomers. The isotopic compositions shown in Table 1.5 are all enriched in ^{13}C and ^{15}N, and D and L isomeric pairs have virtually identical $\delta^{13}C$ and $\delta^{15}N$ values, apparently confirming an indigenous, extraterrestrial origin.

It is worth commenting upon the variations in isotopic signatures between different amino acids, because it is not necessarily obvious why there should be any differences. Abiotic synthesis generally involves an apparent kinetic isotope effect (Box 1.3), with ^{13}C becoming progressively depleted as carbon chain length increases. The Murchison $\delta^{13}C$ data in Table 1.5 are broadly consistent with this trend, suggesting the longer-chained acids are formed from shorter precursors. The $\delta^{15}N$ data show something different: α-aminoisobutyric acid, and to a lesser extent sarcosine, are enriched in ^{15}N, whereas glycine is depleted; the other amino acids (apart from proline) have a value of c.60‰, suggesting a common source. All of these amino acids have a single N atom, so there is no kinetic isotope effect of the kind attributed to account for the $\delta^{13}C$ variations. There appear to be no reported $\delta^{15}N$ data for abiotic synthesis, such as the Strecker reaction in Eqn 1.15, which could occur in aqueous fluids on the parent meteorite body.

Table 1.5 Stable C and N isotope values of amino acids in Murchison meteorite

amino acid	$\delta^{13}C$ (‰)*	$\delta^{15}N$ (‰)†
α-aminobutyric acid	5	184
isovaline	17	66
sarcosine		129
glycine	22	37
β-alanine		61
D-glutamic acid		60
L-glutamic acid	6	58
D-alanine	27.7	60
L-alanine	26.1	57
L-leucine		60
D,L-proline		50
D,L-aspartic acid		61

*Unhydrolysed water extract (after Engel et al. 1990); †hydrolysed water extract (after Engel & Macko 1997).

So it is not possible to determine whether the Murchison values are consistent with an abiotic source as far as N isotopes are concerned. However, the N sources of α-aminoisobutyric acid and sarcosine are clearly distinct. Glycine is a common decomposition product of other amino acids, so its ^{15}N-depletion may reflect a kinetic isotope effect during decomposition of precursor amino acids. Alternatively, gylcine, which is a relatively simple compound, may have formed in interstellar space (Snyder 1997) from isotopically distinct precursor(s) prior to the synthesis of more complex amino acids on the Murchison parent body. At present there are insufficient data to conclude whether the stable isotopic composition of amino acids in the Murchison meteorite reflects the range associated with biosynthetic processes on Earth.

$$\underset{R'}{\overset{R}{\diagdown}}C=O + NH_3 + HCN \longrightarrow \underset{\underset{H_2N}{|}}{\overset{R}{\underset{|}{R'--C}}}-C\underset{OH}{\overset{O}{\diagup}} + \underset{\underset{R'}{|}}{\overset{R}{\underset{|}{H_2N--C}}}-C\underset{OH}{\overset{O}{\diagup}} \quad [Eqn\ 1.15]$$

L   D

excess among the meteorite amino acids remains to be explained. Could abiotic synthesis have involved unknown stereoselective pathways? Or could a racemic mixture have been subjected to alteration processes that led to preferential destruction of the D isomers, such as exposure to circularly polarized light from neutron stars (Bailey et al. 1998)? A further possibility is that the Murchison meteorite contains the residue of a once living system, with the amino acids having undergone partial racemization after death, as occurs on Earth. It is not unprecedented for racemization to be incomplete after a long period.

There are other chemical questions surrounding the emergence of life. For example, membranes are needed to confine the contents of cells and provide a stable, controlled environment for biochemical reactions. Some fairly simple organic molecules (**amphipathic** molecules), in which one end has an affinity for water (i.e. it is **hydrophilic**) but the other does not (i.e. it is **hydrophobic**, like oil), can naturally form membrane-like structures, which, in the presence of water, form cell-like spheres (see Box 1.7). Self-replication is another vital factor, which needs to be accomplished with minimal error. The information necessary for con-

---

Box 1.7 | Membrane formation

Life on Earth is based upon cells, the membranes of which keep the essential biochemical apparatus together and in a suitable environment. The formation of cell-like vesicles, effectively impermeable to water, was likely to have been a key step in the appearance and evolution of life. In modern organisms cellular membranes have inner and outer surfaces comprising hydrophilic (water-loving) groups attached to a core of aligned hydrophobic (water-repelling) hydrocarbon chains. The compounds responsible, called phospholipids (see Section 2.4.1b), have a hydrophilic head containing a phosphate group bonded to a pair of hydrophobic hydrocarbon chains via a glycerol unit. In eukaryotic membranes there are two layers of these compounds, which assemble with the hydrophobic tails pointing to the interior of the membrane and the hydrophilic heads forming the inner and outer surfaces. Compounds with hydrophilic and hydrophobic ends are often termed amphipathic (or **amphiphilic**). To help to stiffen the membrane other molecules, which must conform to precise dimensional constraints, are slotted between some of the phospholipids. In eukaryotes cholesterol is common, and lines up with its polar alcohol end in the phosphate group layer (Fig. 1.4a).

In the organisms apparently most closely related to the common ancestor of all life, the archaebacteria (see Section 1.3.3), rigidifying molecules are not needed because the hydrocarbon chains have regular methyl branches (every fourth carbon in the chain), which appear to confer sufficient stability to the membrane on their own. These chains are terpenoidal, formed from enzymatic condensation of C_5 isoprene units (Section 2.4.3). However, isopentenol can be produced abiotically from simple compounds likely to have been present on the young Earth; for example, by the acid-catalysed reaction of formaldehyde (methanal) and *iso*butene (Eqn 1.16). Isopentenol can undergo abiotic acid-catalysed condensation to form polyprenol chains like those in archaebacterial membranes, which could have been readily phosphorylated by pyrophosphates or volcanically produced phosphorus pentoxide, again on catalytic mineral surfaces, with the chains eventually breaking off and assembling into a vesicle (Fig. 1.4b). This may be how the first cellular membranes formed (Ourisson & Nakatani 1994).

$$\text{isobutene} + CH_2O \xrightarrow{\text{acid catalysed}} \text{isoprenol} \xrightarrow{\text{acid catalysed condensation}} \xrightarrow{\text{further condensations}} \text{polyprenols } (C_{20}-C_{35})$$

[Eqn 1.16]

Continued

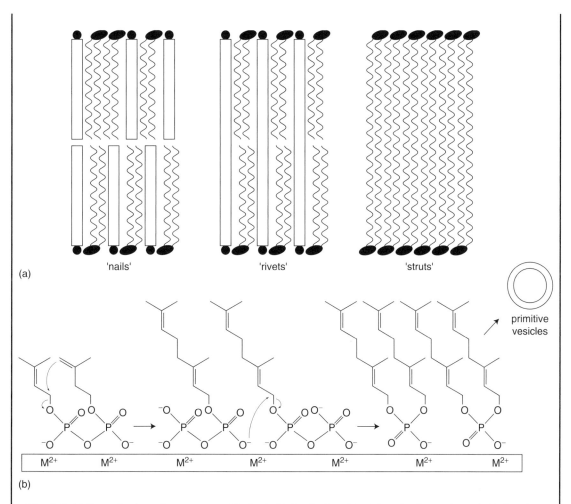

Fig. 1.4 (a) Generalized membrane structure. Hydrophobic heads of amphipathic molecules are shown in black and hydrophilic tails as wavy lines. Membrane reinforcers are shown as boxes and can be likened to 'nails' spanning half the membrane (e.g. sterols and hopanoids) or 'rivets' spanning the entire membrane (e.g. carotenoids). Where they are absent, in archaebacterial thermophiles, the amphipathic molecules span the entire membrane and act as 'struts'. (b) Possible steps in the self-assembly of primitive membranes (M^{2+} are metal ions).

structing an organism is carried in its genes, encoded in deoxyribonucleic acid (DNA) sequences (see Section 2.6.2). The information in DNA strands is translated into the amino acid sequences of proteins using enzymes (which are themselves proteins) and ribonucleic acids (RNAs). But which came first, the code for protein formation, DNA, or the protein catalysts (see Box 2.6) needed to make DNA work? It is believed that the early Earth was an RNA world, because RNA has been found to have some catalytic activity of the type now provided by proteins, and has the ability to carry genetic information (Horgan 1991). Today, only RNA viruses do not use DNA as their genetic material. It is difficult to explain how pyrimidines, a basic component of nucleic acids, could have arisen abiotically, and also how all the components of molecules as complex as nucleic acids could have been assembled. As Fred Hoyle has commented, the random assemblage of such a molecule is as likely as a Boeing 747 airliner being assembled by the passage of a tornado through a junk yard. Various theories have been propounded, such as the necessary concentration of the starting materials and

Box 1.8 | Classification of organisms

Organisms (excluding viruses) can be broadly separated into **prokaryotes** and **eukaryotes** on the basis of cellular structure. Prokaryotic cells contain no nuclear membranes and their DNA is not arranged in chromosomes. Eukaryotic cells always contain a nuclear membrane and their DNA is usually in chromosomal form. There are other differences between the two cell types but these are the most useful general distinguishing features. The prokaryotes comprise all bacteria (including cyanobacteria) and are therefore unicellular (also termed the Monera). All other organisms are eukaryotes, either unicellular (the Protista) or multicellular.

The eukaryotes are often divided into two kingdoms: animals (Animalia) and plants (Plantae). This distinction is blurred in the case of flagellates, which can fall into either kingdom, and some can be classified in both. Further categorization results in the main **taxonomic ranks** shown in Table 1.6, in which the classification of humans is given as an example.

It is possible to use subcategories of the main ranks. The rank below kingdom is generally termed division for plants but phylum for animals. For the plant kingdom the following divisions are generally recognized: algae (Phycophyta), fungi (Mycophyta), lichens (Lichenes), bryophytes (Bryophyta), pteridophytes (Pteridophyta) and spermatophytes (Spermatophyta). The last two divisions comprise the vascular plants, while the Bryophyta include mosses, liverworts and hornworts. Some organisms are difficult to classify, particularly unicellular forms, and changes in classification are inevitable as our knowledge of these organisms grows.

There is flexibility in the application of classification ranks (e.g. Holmes 1983), which can confuse the inexperienced reader. For example, some classes of algae are considered to be sufficiently distinct to warrant divisional status and so the algae as a whole become a subkingdom (even a kingdom in some classifications), and the cyanobacteria can be treated as a division of this subkingdom (the Cyanophyta). The prokaryotes can be treated as a division (Schizophyta), with the bacteria and cyanobacteria as constituent classes (Schizomycetes and Schizophyceae, respectively). The fungi are sometimes considered to be a kingdom in their own right, with two main divisions comprising the slime moulds (Myxomycota) and the true fungi (Eumycota).

In the latest taxonomic system based on rRNA sequencing, three domains have been recognized: Archaea (the archaebacteria), Bacteria (the eubacteria) and Eukarya (the eukaryotes). When this system is used, bacteria with a lower-case 'b' is generally assumed to refer to all bacteria (archaebacteria + eubacteria).

Table 1.6 Major taxonomic ranks, using humans as an example

taxonomic rank	classification for humans
kingdom	Animalia
division or phylum	Chordata
class	Mammalia
order	Primates
family	Hominidae
genus	*Homo*
species	*sapiens*

---

their catalytic reorganization on clay or iron pyrite particles, but there is no evidence of any system that could accomplish the synthesis of RNA or even other types of perhaps simpler self-replicating molecules that may have preceded RNA.

Perhaps life originated near the surface at an early stage of Earth's life and avoided complete obliteration during the heavy bombardment period by migrating to deep environments, and possibly remaining dormant for tens of millions of years (an ability demonstrated by some bacteria; Parkes 2000). It is safe to conclude that we do not know how or where life originated, and an extraterrestrial origin may be considered as plausible as any other theory. More recently it has been suggested that the cradle of life may have been hydrothermal vents (e.g. Corliss 1990), which would have provided a ready supply of energy and raw materials for chemoautotrophes. However, on a geological time-scale an individual vent is an ephemeral feature, so would there have been enough time for life to evolve? In addition, superheated water temperatures can reach 300°C, and it is questionable whether the complex organic molecules

Box 1.9 | Autotrophes and heterotrophes

All organisms use organic material as an energy source and for growth and reproduction. How they get the necessary substrates determines their trophic status.

Autotrophes are organisms that can manufacture their organic materials from inorganic sources, and do not rely directly upon other organisms for any of their organic chemical or energy requirements. If they obtain the energy needed for the primary C-fixation step from sunlight (i.e. photosynthesis) they are termed **phototrophes** (or photoautoptrophes). If instead they obtain it from chemical energy (i.e. the energy liberated by inorganic redox reactions; Box 1.5) they are called **chemotrophes** (or chemoautotrophes) (see Box 1.10). Because the chemoautotrophes use inorganic species as terminal electron acceptors they are commonly also termed **lithotrophes**. Overall, chemosynthesis of organic matter is minor compared with photosynthesis. The amount of carbon fixed into new growth by autotrophes is often termed **primary production**.

All other organisms gain their energy supplies and organic substrates by feeding, directly or indirectly, upon autotrophes, and are termed **heterotrophes**. They comprise herbivores, saprophytes, carnivores and parasites. In geochemical terms the first two groups are the most important. Energy is passed from the autotrophes to the various heterotrophes along food chains and ultimately reaches the top carnivores.

Herbivores represent the first link in the **grazing food chain**. Dead plant matter together with the faecal material of animals and their remains upon death are collectively termed **detritus** and pass to organisms of the **detrital food chain** in soil, water and sediments. Invertebrate animals form one group of detrital-feeding organisms, the **detritivores**, which are often termed deposit feeders if they obtain their sustenance from soil and sediments, or filter (or suspension) feeders if they scavenge it from the water column. Detritivores may, therefore, be herbivores and/or carnivores. The saprophytes obtain their organic sustenance in dissolved form by the use of extracellular enzymes, and comprise fungi and a variety of bacteria, often collectively termed **decomposers**. They may feed directly on the remains of dead plants or animals, but usually decomposition proceeds more rapidly via the comminuted products of the detritivores, due to both the greater surface area presented and the extensive disruption of protective tissues.

---

needed for life would be stable under such conditions. More favourable conditions may be found at the boundary between vent waters and normal seawater, where there are rapid redox and pH changes. It has been proposed that iron sulphide can form membrane-like bubbles at this interface, the surfaces of which also contain nickel sulphide and can catalyse organic synthesis and facilitate protometabolic processes (Russell & Hall 1997; Martin & Russell 2003). Subsequently, the organosulphide polymers could have taken over the membrane role, while iron, sulphur and phosphate continued to be involved in metabolic processes in the proto-organisms.

1.3.3 The common ancestor

If life originated on Earth, it has been postulated that the earliest forms may have been anaerobic prokaryotic heterotrophes (see Boxes 1.8 and 1.9), eking out a living from the simple abiogenic organic compounds at the Earth's surface. It seems most likely that the early stages in the evolution of life were confined to aquatic environments, because water is an essential requirement for all life. A diffuse abiotic food supply would have limited the proliferation of life and, as mentioned above, the heavy bombardment by meteorites etc. (bolides) would have hampered its development at the surface.

The search for a common ancestor of life involves tracking the divergences of related groups of organisms back through time. How different groups of organisms are related to one another has traditionally been based upon physiological characteristics and the sequence of gradual changes assumed to result from random mutation. When combined with the effects of Darwinian selection, the result is the 'tree of life', with the ancestral organism of all life on Earth at its root. However, it is

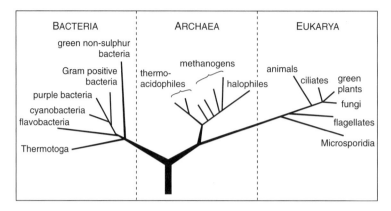

Fig. 1.5 Evolutionary relationship of organisms based on rRNA sequence comparisons, with division of the 'tree of life' into three fundamental domains: Bacteria (or eubacteria), Archaea (or archaebacteria) and Eukarya (or eukaryotes) (after Woese et al. 1990).

based primarily on multicellular (eukaryotic) organisms that appeared significantly later in the fossil record than the prokaryotes. Our knowledge of the microbial world, although expanding dramatically at present, is still very limited, and it is clear that microbial morphological characteristics are too simple to enable a reliable, universal phylogenetic taxonomy to be established. A new tool, molecular phylogenetics, for examining the arrangement for the tree of life has recently become available that potentially overcomes this problem. It is based on the sequencing of nucleotides (see Section 2.6) in genes, and the assumption that random mutation occurs at a constant rate, permitting an estimation of interrelationships and the relative timing of divergences. The first studies were conducted on genes that encode ribosomal RNAs (rRNAs; Woese et al. 1990), which were chosen because they are part of a basic cell activity, protein synthesis, and so were thought unlikely to have undergone radical change and so might act as a slow, regular clock. A simplified version of the resulting tree of life is shown in Fig. 1.5, in which there are three kingdoms (or domains): the Archaea (or archaebacteria), the Bacteria (or eubacteria) and the Eukarya (or eukaryotes). This classification demonstrates the importance of bacteria, which, by this taxonomic system, are divided into the **archaebacteria** (methanogens, halophiles and thermoacidophiles; Woese & Wolfe 1985) and true bacteria, the **eubacteria**. The eukaryotes occupy a much less significant position than in older classifications.

Archaebacteria are considered the most ancient form of cellular life; they are closest to the root of the tree and so are the closest extant relatives of the universal ancestor. They occur in extreme environments that are hostile to other forms of life—such as hydrothermal vents, fumaroles and soda lakes—and they can all tolerate high temperatures. Those that can grow at temperatures >80°C are termed **hyperthermophiles**. The highest growth temperatures so far are 113°C for *Pyrolobus fumarii* (Blöchl et al. 1997) and 121°C for an Fe(III)-reducing strain (see Box 1.13 and Section 3.3.2b) from a hydrothermal vent in the north-eastern Pacific Ocean (Kashefi & Lovley 2003). The stability of archaebacterial enzymes at high temperatures is the key to modern DNA sequencing using the polymerase chain reaction (PCR) to amplify DNA, which relies on an enzyme from *Thermus aquaticus*, a bacterium that thrives at 70°C in hot spring communities in Yellowstone Park. The position of hyperthermophilic archaebacteria nearest the root of the tree might be considered support for an origin of life at high temperature. In the previous section we touched upon the potential role of hydrothermal vents as sources of all the ingredients needed by chemoautotrophes. Another possibility is that the ancestral archaebacteria could have originated at depth within the Earth, although migration to the surface becomes problematical as depth increases because growth rates, and hence migration rates, decrease, leading to entrapment and burial (Parkes et al. 1999). The subsurface bacterial biomass is extremely large, probably exceeding that of all surface-dwelling and marine organisms (Whitman et al. 1998), although activity is low at depth. The concept of a deep hot biosphere (originally proposed by Gold in 1992) is now recognized, but the depth/temperature limit has yet to be established.

The initial optimism surrounding the belief that a universal tree of life could be assembled has been tempered by the discovery that the sequences of other genomes (e.g. genes encoding for various proteins) present a different picture (Pennisi 1998). The conclusion is that genes have not all evolved at the same rate or in the same way. Some eukaryotic genomes contain what appears to be a mixture of DNA with some components associated with an archaebacterial ancestry and others of eubacterial legacy. This could represent early organisms acquiring genes via food or even swapping DNA with neighbours. Instead of a tree branching out from a single trunk, there may have been considerable merging and splitting of lines before the offshoot of modern kingdoms. Clearly, rapid gene swapping (lateral transfer; Doolittle 1999) would have enabled early organisms to adapt to new environmental conditions, an advantage that may have ensured that the DNA code became universal, because organisms not able to read DNA-based genes would not have had such an adaptive capability. Lateral transfer complicates the task of determining phylogenetic relationships sufficiently that the universal ancestor (if there was just one) may not be traceable. At present it is not clear whether the three kingdoms in Fig. 1.5 are truly representative; it has recently been proposed that eukaryotic-like cells may have predated prokaryotes (Penny & Poole 1999).

It is not only DNA that seems to have been shared by organisms; there is good reason to believe that mitochondria (the energy generators of cells) and chloroplasts (the sites of photosynthesis in plant cells) may once have been prokaryotes that were ingested by, or entered into a symbiotic relationship with, other prokaryotes (Doolittle 1998).

1.4 Evolution of life and the atmosphere

1.4.1 Atmospheric oxygen and photosynthesis

Important stages in the evolution of the Earth's surface are presented in Fig. 1.6; one of the most dramatic changes has been the development of an oxygenated atmosphere. Reducing, or at least non-oxidizing (see Box 1.5), conditions prevailed during the Archaean, and so this time belonged to the anaerobic prokaryotes. Not until the advent of oxygen-producing (oxygenic) photosynthesis could oxidizing conditions begin to develop at the surface; and only with the subsequent availability of free oxygen was it possible for multicellular, eukaryotic organisms to develop and diversify. The composition of the atmosphere has both affected and been affected by the development of life on the planet.

As mentioned in the previous section, it is possible that the first organisms were heterotrophes, obtaining their energy requirements from simple, abiotic organic compounds by fermentation. Ancestors of methanogens can be envisaged as one such group of prokaryotes, utilizing simple organic compounds like methanoic acid ($HCOOH$, also known as formic acid) and ethanoic acid (CH_3COOH, also known as acetic acid):

$$CH_3COOH \rightarrow CH_4 + CO_2 \quad [\text{Eqn 1.17}]$$

Modern methanogens can all reduce carbon dioxide with molecular hydrogen, while synthesizing carbohydrates (which can be represented by their empirical, i.e. simplest, formula CH_2O):

$$2CO_2 + 6H_2 \rightarrow CH_2O + CH_4 + 3H_2O \quad [\text{Eqn 1.18}]$$

Carbohydrates are used in the biosynthesis of other organic compounds and to provide an energy store for the performance of normal cellular functions. In the absence of free oxygen, fermentation would have provided the energy-releasing step for anaerobic organisms, but compared to oxidation involving oxygen, it is a relatively inefficient process in terms of the amount of energy it releases. The overall fermentation of simple carbohydrates by anaerobes can be represented by:

$$2CH_2O \rightarrow CH_4 + CO_2 \quad [\text{Eqn 1.19}]$$

The initial step is actually the formation of pyruvic acid with the liberation of free hydrogen atoms, which can then be combined with other groups:

$$\underset{\text{glucose}}{C_6H_{12}O_6} \rightarrow \underset{\text{pyruvic acid}}{2C_3H_4O_3} + 4H \quad [\text{Eqn 1.20}]$$

Fermentation is one of a number of anaerobic respiration (i.e. energy-liberating) processes that are discussed more fully in Section 3.3.2b.

Other plausible candidates for the first prokaryotes are the ancestors of chemotrophes colonizing hydrothermal vents. Today they use the oxygen dissolved in ocean water to obtain energy from the sulphides emanating from the hot springs in chemosynthetic reactions (see Box 1.10) that can be summarized by:

$$CO_2 + H_2S + O_2 + H_2O \rightarrow CH_2O + 2H^+ + SO_4^{2-} \quad [\text{Eqn 1.21}]$$

Clearly, anaerobic ancestors would have had to utilize a chemical energy source other than oxygen, but there is no clear evidence that anaerobic chemotrophes existed.

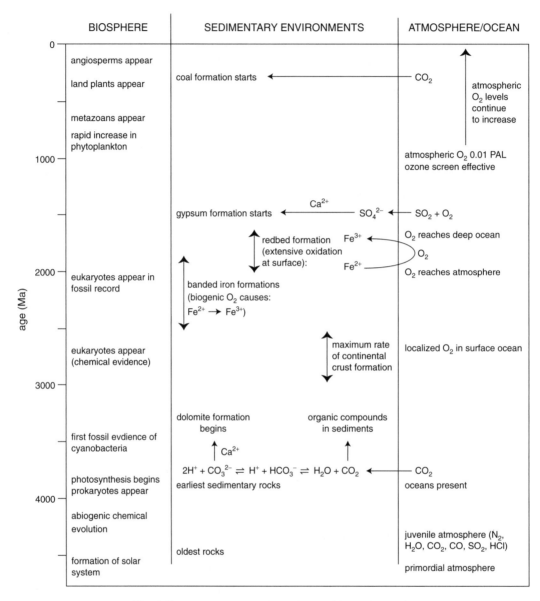

Fig. 1.6 Important events in the evolution of the Earth and life.

The first autotrophes to use the Sun's energy (i.e. photosynthesizers) would also have been prokaryotic anaerobes living in aquatic environments, and they used hydrogen sulphide as a source of the hydrogen needed for carbohydrate synthesis, yielding sulphur as a by-product, not oxygen:

$$CO_2 + 2H_2S \rightarrow CH_2O + 2S + H_2O \quad \text{[Eqn 1.22]}$$

The nearest modern relatives of these organisms are the purple and green sulphur bacteria, which are all anaerobes.

As mentioned in Section 1.3.1, the first indirect (carbon isotopic) evidence for life appears in rocks aged some 3.8 Ga, and there are also structures dating from 3.5 Ga that have been interpreted as the counterparts of modern microbial mats, primarily formed by large

Box 1.10 | Photosynthesis and chemosynthesis

Chlorophyll-*a* is the primary pigment of **photosynthesis** and absorbs photons of light energy in order to convert carbon dioxide into carbohydrates. Water is essential, as it provides hydrogen, while oxygen is expelled into the environment as a by-product. Photosynthesis can be divided into a light (photochemical) stage and a dark (chemical) stage. The former requires light but is unaffected by temperature, while the latter does not require light and proceeds more rapidly with increasing temperature. These stages are presented in Fig. 1.7 and can be summarized as follows.

Light stage
1 Generation of energy: an electron (e^-) is liberated from chlorophyll-*a* upon adsorption of light energy.
2 Storage of energy: the high-energy electron can recombine with positively charged chlorophyll and the excess energy is used to convert ADP (adenosine diphosphate) into ATP (adenosine triphosphate) by the addition of inorganic phosphate.
3 Storage of reducing power: an electron from chlorophyll may be captured by a hydrogen ion (H^+), produced from the self-ionization of water, to yield a hydrogen atom, which is immediately taken up by NADP (nicotinamide adenine dinucleotide phosphate), storing the reducing power in the form of NADPH. The chlorophyll ion can regain an electron from a hydroxyl ion (OH^-), which is also formed during the self-ionization of water, and the resulting hydroxyl radical ($OH\cdot$, which has no charge) combines with others to form oxygen and water.

Dark stage
Overall, the hydrogen stored in NADPH is used to reduce CO_2 to carbohydrate units (CH_2O). This is not a direct reaction because the CO_2 is first combined with a C_5 compound, ribulose diphosphate (RDP), which then spontaneously splits into two identical C_3 molecules, phosphoglyceric acid (PGA). Most of the PGA is used to synthesize further RDP but some is reduced by NADPH, using energy supplied by the ATP/ADP system, to give triose phosphate, which in turn is converted into the glucose phosphate from which various carbohydrates are synthesized. This assimilatory path is known as the **Calvin cycle** and is involved in all autotrophic carbon fixation, whether photosynthetic or chemosynthetic.

Plants that use the Calvin cycle alone in carbon fix-

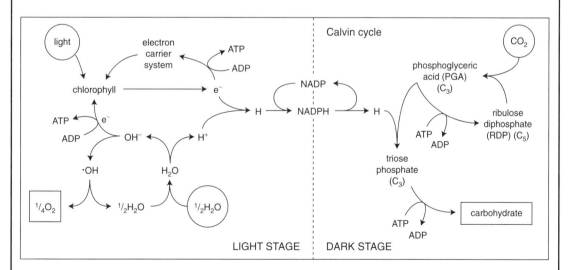

Fig. 1.7 Summary of the chemical processes involved in oxygenic photosynthesis. Net inputs are shown in circles and products in rectangles. ADP/ATP = adenosine di/triphosphate; NADP/NADPH = nicotinamide adenine dinucleotide phosphate and its reduced form, respectively.

Continued

ation are termed **C₃-plants** because of the involvement of PGA, which contains three carbon atoms. Most plants and the cyanobacteria use the C₃ path. However, under conditions of high temperature and low CO_2 levels much energy can be lost by C₃-plants due to photorespiraton. Two smaller groups of plants overcome this by using an additional biochemical pathway that fixes CO_2 at night. The carbon dioxide is released again within the plant tissue during the day for incorporation into the Calvin cycle. In this way stomatal pores can be closed during the day to reduce photorespiration without cutting off essential supplies of CO_2. These two groups are named after their additional mechanisms: **C₄-plants** and **CAM-plants**.

As well as the oxygenic phototrophic cyanobacteria, there are also **anoxygenic photosynthetic bacteria**, which are anaerobes (although some are O_2-tolerant), generally use simple organic compounds as a C supply, never use water as a source of hydrogen and do not produce oxygen. These anoxygenic phototrophes comprise the purple sulphur bacteria, purple non-sulphur bacteria, green sulphur bacteria, green non-sulphur bacteria and *Heliobacter*. There are also some anoxygenic aerobic photoheterotrophes (belonging to the proteobacteria), which are obligate aerobes, metabolizing organic substrates in the water column when available by O_2-dependent respiration, but otherwise performing anoxygenic photosynthesis. They are probably related to the purple non-sulphur bacteria (*Rhodospirillum*; Fenchel 2001).

Chemosynthesis is performed by some types of bacteria, most of which are obligate aerobes. The initial energy-generating process does not involve light or water, but utilizes the energy stored in chemicals, mainly the reduced forms of simple inorganic species. An enzyme, dehydrogenase, is used to liberate the energy (in the form of electrons) and reducing power (in the form of protons) from a chemical species such as H_2S. While the necessary compounds for chemosynthesis are generally found throughout the water column they are in highest concentrations in anoxic waters beneath areas of high productivity, resulting from the breakdown of organic constituents in detritus. Chemosynthetic bacteria are, therefore, found at the oxic-anoxic boundary (see Box 1.5).

colonies of cyanobacteria in shallow water (Schopf & Packer 1987). **Cyanobacteria** (previously classified as blue-green algae) perform oxygenic photosynthesis (i.e. oxygen is liberated), during which a complex series of processes occurs, as summarized in Box 1.10. Light energy in the visible part of the spectrum emitted by the Sun is adsorbed by a green pigment, chlorophyll. This results in the transfer of hydrogen atoms from water to carbon dioxide molecules (i.e. reduction of CO_2) to build up carbohydrate units, while oxygen is liberated from the water molecules. The overall reaction for the formation of a carbohydrate such as glucose (which contains six of the basic CH_2O units) can be simplified to:

$$6CO_2 + 6H_2O \rightarrow C_6H_{12}O_6 + 6O_2 \quad [\text{Eqn 1.23}]$$

Around 1500 extant species of cyanobacteria have been identified, some of which are colonial mat formers, while others are non-colonial and inhabit open water. Whether the anaerobic ancestral forms were oxygenic photosynthesizers is not clear; although there are many proponents of the idea that the early Archaean stromatolites represent oxygenic photosynthesizers (e.g. Rothschild & Mancinelli 1990), others have suggested that the organisms responsible for the structures may have been non-oxygenic phototrophes or even chemotrophes (e.g. Lasaga & Ohmoto 2002). One thing is clear, however: oxygenic photosynthesis must have evolved during the Archaean. In addition to the stromatolitic evidence (Buick 1992), there is compelling molecular evidence, in the form of 2α-methylhopanes (Brocks et al. 1999; see Sections 2.4.3e and 5.4.2e), for the presence of cyanobacteria, and by inference for oxygenic photosynthesis, dating back to 2.7 Ga, although recent phylogenetic studies suggest a somewhat later appearance of cyanobacteria, c.2.4–2.2 Ga (Blank 2004).

1.4.2 Geological record of oxygen levels

Oxygen liberated by early oxygenic photosynthesizers would have been rapidly immobilized in the generally reducing environment, primarily by the oxidation of iron from Fe(II) to Fe(III) and reducing gases emitted from volcanoes (see Box 1.5). Exactly when oxygen reached the atmosphere and the rate at which levels subsequently rose is extremely difficult to determine, and relies on indirect evidence of minimum and maximum oxygen concentrations, as used in the construction of Fig. 1.8. The amount of oxygen in the atmosphere can be represented in various ways, such as its partial pressure (pO_2; Box 1.11) and the amount relative to the present atmospheric level (**PAL**).

Variations between $\delta^{33}S$ and $\delta^{34}S$ values (see Box 1.3) in Archaean sulphides and sulphates have been attributed to photodissociation of atmospheric SO_2, a process that could only occur if oxygen was at most a trace component (due to the adsorbtion of the required UV radiation by O_2 and O_3), so the Archaean atmosphere seems to have been virtually devoid of oxygen (Farquhar et al. 2002; Wiechert 2002). Among other parameters that restrict oxygen concentrations are the conditions under which redbeds, banded iron formations and detrital uraninites could form, and for various aerobic organisms to survive (Kasting 1993).

For common metals with more than one oxidation state, like Fe(II) and Fe(III), the higher oxidation state is much less water soluble, so aqueous transport of iron occurs in the form of Fe^{2+} ions, whereas Fe^{3+} ions (as in Fe_2O_3) are immobile. Banded iron formations (**BIFs**) may represent localized oxidation of Fe(II) involving the free oxygen liberated during photosynthesis. They are mostly confined to an age range of 2.7–1.9 Ga (early Proterozoic), although some are found among the oldest known sedimentary rocks dating back to 3.8 Ga in Isua (Greenland), and a few others appear to date to as recently as 1.85 Ga (Kasting 1993). The Fe(II) could have originated from hydrothermal emissions near spreading ridges in anoxic deep water and have been transported to shallow marginal basins containing slightly oxic waters, leading to the precipitation of Fe(III) (Morris & Horwitz 1983). The Fe(III)-rich bands alternate with chert, and it has been suggested that this varve-like banding could arise from the transport of Fe(II) in oxygenic phototrophes (in which the iron behaves as a micronutrient; see Box 3.7). Upwelling of nutrient-rich waters in nearshore areas, as occurs in some areas of the ocean today (see Section 3.2.5), could have resulted in seasonal blooms and the development of minor oxygen oases in surface waters. The large quantities of detritus from the phototrophes would then lead to oxidation of the iron and deposition of Fe(III)-rich bands, interspersed with iron-poor bands deposited

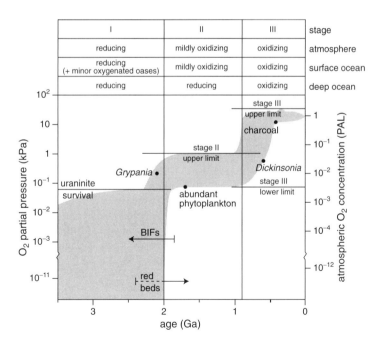

Fig. 1.8 Estimated atmospheric concentration of oxygen throughout the Earth's life (shaded area; after various sources, including Kasting 1993; Canfield 1998). Key upper and lower limits are shown; dashed line for redbeds denotes disputed data. See text for *Grypania* and *Dickinsonia*.

> **Box 1.11** | **Partial pressure**
>
> The behaviour of gases under varying conditions of temperature, pressure and volume can be predicted to a first approximation by the ideal gas equation:
>
> $$pV = nRT \quad \text{[Eqn 1.24]}$$
>
> where p = pressure, V = volume, n = number of moles of gas, R = universal gas constant and T = temperature.
>
> An ideal gas is one whose molecules are infinitely small compared to the size of the container housing the gas, behave perfectly elastically in all collisions (i.e. there is no loss of kinetic energy) and do not react with or attract each other. One further property is assumed, that the duration of collisions is negligible (so that the effective number of molecules in motion remains constant).
>
> From the ideal gas equation it can be seen that, at constant temperature and volume, the pressure exerted by any gas is proportional to the number of molecules (or atoms, if the gas is monatomic, such as helium and argon). This is because pressure is a consequence of molecules colliding with the container walls, so the more molecules, the more collisions. For a mixture of gases, the total pressure is related to the total number of molecules present. To put this another way, the total pressure of the mixture is equal to the sum of the pressures that each gas would exert if it alone occupied the container. This is Dalton's Law of **partial pressures**. So atmospheric pressure is approximately given by:
>
> $$P_{atmospheric} \approx pN_2 + pO_2 + pAr + pCO_2 \quad \text{[Eqn 1.25]}$$

at other times (Cloud 1973). More recently it has been suggested that iron-oxidizing bacteria could account for the deposition of BIFs (Konhauser et al. 2002).

An upper limit of atmospheric O_2 based on BIF deposition is c.0.08 PAL. The earliest postulated eukaryotes in the fossil record, *Grypania spiralis* (believed to be an alga), have an age of 2.1 Gyr (Han & Runnegar 1992), and so overlap with BIFs. It is assumed they required at least 0.01 PAL of dissolved O_2 (Runnegar 1991), while most eukaryotic microbes require at least 0.05 PAL (Jahnke & Klein 1979). Although the fossil record of eukaryotes dates back to 2.1 Ga, there is molecular evidence, in the form of steranes (derived from sterols; see Sections 2.4.3f and 5.3.3d), for the emergence of eukaryotes at least 2.7 Gyr ago (Brocks et al. 1999), in the form of unicellular algae, which are phototrophes and members of the phytoplankton (Box 1.12).

In contrast to BIFs, **redbeds** are formed under oxidizing, subaerial conditions, and are characterized by quartz grains coated with haematite (Fe_2O_3). They are generally absent from the geological record before c.2 Ga, although there are some disputed redbeds that date back to c.2.4 Ga, which presents the possibility of some overlap with BIFs (Kasting 1993).

Unlike iron and other common metals, the higher oxidation state of uranium, U(VI), is water soluble, and its lower oxidation state, U(IV), is insoluble. So during the deposition of detrital uraninites, which are mostly older than 2 Gyr, the level of atmospheric oxygen must have been low, <0.01 PAL (Kasting 1993; Rasmussen & Buick 1999). Large-scale deposition of detrital pyrite (FeS_2) similarly mostly occurred before 2 Ga.

The iron content of **palaeosols** (fossil soils) can also provide some information on atmospheric oxygen levels, although the interpretation is not straightforward and unanimous agreement upon the evidence has yet to be reached. The identification of palaeosols is not always easy, and assessment of the losses of iron (as Fe(II)) by leaching depends on a knowledge of the parent rock from which the soil formed and various environmental factors, such as the $O_2:CO_2$ ratio in the groundwater responsible for the leaching. The palaeosol record appears to indicate that atmospheric O_2 levels were ≤ 0.004 PAL prior to 2.25 Ga, and rose to ≥ 0.15 PAL after 1.92 Ga (Rye & Holland 1998).

Another major increase in atmospheric O_2 may have occurred towards the end of the Proterozoic (Knoll et al. 1986; Derry et al. 1992; Des Marais et al. 1992), providing the opportunity for more complex, multicellular eukaryotes to evolve. Further evidence for such an increase is provided by the isotopic record of sedimentary marine sulphides (Canfield 1998; see Box 1.13). From c.2.3 to between 1.05 and 0.64 Ga the sulphur isotopic fractionation was $\leq 4‰$ with respect to marine sulphate, suggesting low sulphate levels, consistent with limited oxygenation of the atmosphere and surface waters, sufficient to allow enough oxidation of sulphide to sulphate for the use of sulphate-reducing bacteria in deeper anoxic environments. Enough sulphide may have been produced by these bacteria to remove dissolved iron(II) as pyrite. So for a large part of the Proterozoic bottom waters remained anoxic and sulphidic

Box 1.12 | Plankton classification

Plankton are organisms living primarily in the upper part of the water column and although often capable of some motion, particularly vertical migration in zooplankton, are unable to maintain their overall lateral position and drift with the oceanic currents. Buoyancy aids such as oil bodies are sometimes present. The plankton can be divided into phytoplankton and zooplankton.

Phytoplankton are photosynthesizing microorganisms, usually dominated by unicellular algae. However, free-floating cyanobacteria (unicellular forms rather than the colonial mat formers) and photosynthetic bacteria should also be strictly included by this definition. **Zooplankton** are animals ranging from unicellular microorganisms (**protozoa**) to multicellular organisms (**metazoa**). Many of the smaller zooplankton are herbivores, feeding on phytoplankton, and are, in turn, food for larger carnivorous zooplankton.

Classification of the plankton as a whole is often made on the basis of size (Fig. 1.9). The **ultrananoplankton** (or picoplankton) is composed almost entirely of bacteria, the **nanoplankton** of algae (phytoplankton), and the macro- and megaplankton of animals (zooplankton), mainly invertebrates.

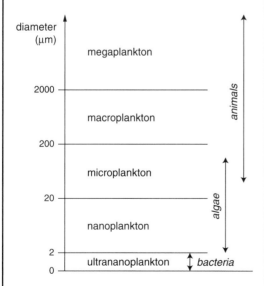

Fig. 1.9 Size classification of plankton (the 0.2 to 2 μm size range is often termed the picoplankton).

(**euxinic**), while surface waters were oxygenated but contained only modest amounts of sulphate (Canfield 1998; Shen et al. 2003). Subsequently, the S isotopic fractionation recorded by pyrite was much greater (51 ± 10‰), suggesting that higher sulphate concentrations were available to the sulphate reducers and that some bacterial sulphide oxidation must have contributed to the fractionation (Box 1.13). At this point oxygen concentration must have increased above 0.05–0.18 PAL (Canfield & Teske 1996; Canfield 1998).

Piecing together all the above indicators leaves us with the envelope of possible atmospheric O_2 content represented by the shaded area in Fig. 1.8. There seems to be reasonable agreement among the indicators of redox conditions that a major increase in atmospheric oxygen occurred around 2 Ga. It is possible that the mantle had effectively been oxidized by then (its iron is in the form of silicates and oxides) as a result of the mixing and exposure to surface conditions caused by tectonic activity. In addition, tectonic activity decreased in the early Proterozoic (from 2.5 Ga), leading to smaller inputs of hydrothermal Fe(II) and reducing gases, which would have aided the increase in levels of atmospheric oxygen. A second major increase in atmospheric oxygen appears to have occurred at c.0.9 Ga, triggering the evolution of aerobic organisms, and particularly the metazoa. Three stages are shown in Fig. 1.8, the first corresponding to a predominantly reducing environment, with minor oxygen oases in the surface waters of the oceans and an upper atmospheric O_2 limit of c.0.08 PAL. The third stage corresponds to oxygenated conditions, for which O_2 levels could not have fallen below 0.002 PAL or the deep ocean would have remained anoxic (on the basis of mass balance considerations; Kasting 1993). During the intermediate stage (II), the deep waters of the ocean appear to have been anoxic, but the surface environment was partially oxidizing.

It has been estimated that atmospheric O_2 would have to have reached 0.01–0.1 PAL in order for an effective ozone layer to develop (Kasting 1987). The increase in abundance of phytoplankton, the dominant phototrophes of surface waters, noted during the early Proterozoic in Fig. 1.8 may reflect the improved growth conditions afforded by an ozone screen, protecting the delicate photosynthesizing organisms from the harmful effects of UV radiation (see Box 7.1). The earliest phototrophes may have had some protection from UV by the photolysis products of methane, given the likely activity of methanogens (Kasting 1991). The development of oxidizing conditions in the atmosphere and ocean would have put severe ecological pressure on the prokaryotes that had originated and evolved in reducing

Box 1.13 | Isotopic fractionation in sulphides and its relationship to oxygen levels

The burial/weathering relationship between oxygen and pyrite can be simplified to the following overall reaction (Holland 1984):

$$2Fe_2O_3 + 8SO_4^{2-} + 16H^+ \underset{\text{weathering}}{\overset{\text{burial}}{\rightleftharpoons}} 4FeS_2 + 8H_2O + 15O_2 \quad \text{[Eqn 1.26]}$$

The conversion of sulphate into sulphide in the burial reaction results from the activity of sulphate-reducing bacteria, which strip the oxygen from sulphate to use in the oxidation of organic matter, thereby obtaining energy (see Box 3.10). The major source of sulphate for sulphate-reducing bacteria, which appear to have emerged $c.2.4$ Gyr ago (Blank 2004), is oceanic water, but the levels would have been very low prior to the development of an oxidizing atmosphere and the large-scale oxidation of sulphide to sulphate. As for all biochemical reactions, the assimilation of S involves isotopic fractionation favouring the lighter isotope (^{32}S; Box 1.3). However, at low sulphate concentrations of <100 mg l^{-1}, bacterial sulphate reduction (BSR) results in negligible isotopic fractionation, so that the sulphide produced is ≤4‰ lighter than the precursor oceanic sulphate (Canfield & Teske 1996). At higher concentrations, BSR is observed to result in fractionations of 4–46‰.

Isotopic fractionations >46‰ appear to require additional fractionation processes to be involved, such as bacterial sulphide oxidation, which would afford a supply of sulphate with a lighter isotopic signature than that of average oceanic water (Canfield & Thamdrup 1994; Canfield & Teske 1996). Sulphur-oxidizing bacteria are confined to the zone of steep oxygen and sulphide gradients between oxic and anoxic environments, so their presence requires atmospheric oxygen to be $c.5$–18% of its present concentration. These bacteria appeared some 0.76 ± 0.32 Gyr ago (Canfield & Teske 1996). Since the late Proterozoic, bacterial processes have been important in the interconversion of sulphate and sulphide (and also of iron(II) and iron(III)), as represented in Fig. 1.10.

---

conditions. Modern descendants of these organisms, the archaebacteria, have been forced into extreme environments, and it is likely that competition for ecological niches from newly evolving organisms would have played an important part. Aerobic heterotrophic prokaryotes, in contrast, became widespread and important in the recycling of organic material.

With free oxygen in the atmosphere, the advantages of aerobic respiration could be exploited, a process that can be represented by the reverse of Eqn 1.23 (i.e. the converse of oxygenic photosynthesis):

$$C_6H_{12}O_6 + 6O_2 \rightarrow 6CO_2 + 6H_2O \quad \text{[Eqn 1.27]}$$

Aerobic respiration is a much more efficient process than fermentation, releasing about 18 times more energy, and so it is no coincidence that the proliferation of the eukaryotes began $c.2$ Gyr ago. The soft-bodied Ediacaran fauna (e.g. the worm-like *Dickinsonia*) that dates from $c.580$ Myr ago, towards the end of the Proterozoic, probably required O$_2$ levels of at least 0.03 PAL (Runnegar 1991), and the subsequent diversification of eukaryotes in the Cambrian is arguably testament to the importance of oxygen.

During the Phanerozoic, atmospheric O$_2$ concentrations have probably remained fairly constant, within $c.0.7$–1.7 PAL (i.e. 15–35% by vol., cf. present-day 21%; Berner & Canfield 1989). The charcoal lower limit in Fig. 1.8 corresponds to 0.6 PAL (12% by vol.), the level at which vegetation fires can occur forming charcoal, which is found in the fossil record since the rise of woody plants $c.425$ Ma (Chaloner 1989). The dramatic increase in photosynthesis and coal formation during the Carboniferous is responsible for the modelled increase in O$_2$ at $c.300$ Ma to possibly as much as 1.7 PAL (35% by vol.), although at such elevated levels runaway vegetation fires become a likelihood, which would limit the O$_2$ levels (hence the upper limit for stage III in Fig. 1.8). Although it has been questioned whether O$_2$ levels could have risen much above 1.2 PAL (25% by vol.; Lenton & Watson 2000), a variety of adaptations has been noted that could be associated with elevated oxygen levels around the Carboniferous–Permian boundary, such as giantism in insects and invasion of the land by vertebrates (Graham et al. 1995; Dudley 1998). The 1.2 PAL limit was based on studies of the spontaneous combustion of paper (Watson et al. 1978), but the lignified tissues of trees contain more

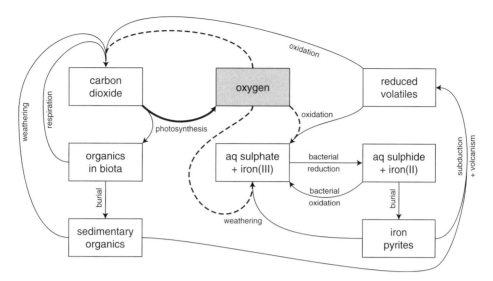

Fig. 1.10 A summary of the main processes involved in the regulation of atmospheric oxygen levels (aq = aqueous; reduced volatiles include CO and SO_2). Oxygen removal processes are shown by broken lines, and generation by the bold line.

water and the charring of the outer layers resists complete combustion, so atmospheric oxygen levels could have approached 2 PAL before runaway wild fires were induced (Wildman et al. 2004). In fact, extensive fires produce abundant charcoal, which resists degradation, leading to greater C burial and hence potentially favouring higher atmospheric oxygen concentrations.

For oxygen to build up in the atmosphere the rate of oxygenic photosynthesis must exceed that of O_2 consumption by sinks. A summary of interactions between the main sinks and sources of oxygen is shown in Fig. 1.10. Major processes consuming oxygen are the oxidative weathering of sedimentary organic matter and reduced minerals (primarily iron(II) and sulphide from pyrite), the atmospheric oxidation of reduced volatiles in volcanic emissions and the conversion of biomass back into CO_2 during respiration (Petsch & Berner 1998; Berner et al. 2003).

The amount of oxygen presently in the surface environment gives us an estimation of the amount of phototrophic biomass that has ever been produced and has been locked away from aerobic respiration by burial in sedimentary rocks. It is a minimum value because of the removal of oxygen into other sinks. Molecular oxygen (O_2) is not particularly soluble in water, so most of it resides in the atmosphere (1.2×10^{18} kg) rather than in the oceans (7.8×10^{15} kg). From Eqn 1.23 we can estimate the corresponding amount of phototrophic material that must have been produced as $c.1.1 \times 10^{18}$ kg, or $c.4.5 \times 10^5$ Gt of C (see Appendices 1 and 2 for units of mass). However, there is only $c.600$ Gt of C in living biomass today, which would only have generated an atmospheric oxygen concentration of $c.0.001$ PAL. Clearly, a large amount of organic material must be locked away from oxidation. The long-term burial of organic matter in sedimentary rocks is the key to the build-up and maintenance of high levels of oxygen in the atmosphere (Van Valen 1971).

If oxygen is produced by the reduction of carbon dioxide, then atmospheric CO_2 levels must have declined over the life of the Earth, and there is evidence that this has indeed occurred. The climatic implications of changes in atmospheric CO_2 levels and how the global carbon cycle operates are addressed in Chapter 6, after the various elements of the carbon cycle have been introduced in the intervening chapters. For the moment, we continue with a more detailed look at the evolution of life in terms of the organisms that are most important by way of their contributions to the organic matter that is eventually buried in sedimentary rocks.

1.5 Major contributors to sedimentary organic matter

1.5.1 Major present-day contributors

Autotrophes, and particularly phototrophes, provide the energy needed by all other organisms, the heterotrophes. The energy stored in the form of reduced carbon—the organic material in the tissues of organisms—is transferred along a food chain by the feeding of heterotrophes. There can be several links, or **trophic levels**, in a food chain, but energy transfer between adjacent trophic levels is not very efficient. Of the organic matter consumed at a particular trophic level some is lost in excreted material, and of the energy that is assimilated only part is available for growth and reproduction (the elements that constitute net production). Heterotrophes at higher trophic levels (i.e. carnivores) often expend a considerable proportion of assimilated energy on movement in obtaining food and the resulting high respiration levels generally lead to low net production. The effect can be seen in the low numbers of top carnivores, like the large cats in Africa, compared to the herbivores upon which they prey.

The efficiency of energy transfer between adjacent trophic levels can be measured from the ratio of the net production at the higher level to that at the lower level. Values of this ratio, known as the **transfer efficiency**, in terrestrial and marine ecosystems are typically only $c.10\%$. Net production is, therefore, significantly higher for the main primary producers than for organisms at higher trophic levels. This explains why phytoplankton (see Box 1.12) are the major contributors to marine and lacustrine sedimentary organic matter, and higher plant remains dominate in peat mires (although a considerable amount of higher plant material is also transported to sedimentary environments in coastal areas and lakes). Today, marine and terrestrial primary production are of similar size and have been so since about the Cretaceous, but earlier in the Earth's history marine production was dominant, and was, indeed, the only form of photoautotrophy before plants began to colonize the land, possibly in the Ordovician, but certainly by the Silurian.

The transfer efficiency from phytoplankton to zooplankton, the first step in the marine grazing chain, appears to be a little higher than average at $c.20\%$. This is partly due to the greater digestibility of phytoplankton compared with terrestrial plants, particularly the more woody species. Zooplankton can, therefore, be a significant source of organic matter for sediments, whereas other animals at similar trophic levels do not appear to be important in this respect. As a result of the predator–prey link, concentrations of zooplankton tend to be greatest where phytoplanktonic production is high. However, the transfer efficiency to herbivorous zooplankton is often lower in areas of highest productivity and there is a consequential increase in the proportion of phytoplanktonic remains reaching the sediment and passing to the detrital food chain (see Box 1.9) or undergoing preservation.

There is one further major contributor to sedimentary organic matter, bacteria. A large proportion of the energy flow in ecosystems can pass through the detrital food chain, in which heterotrophic bacteria are prominent participants. Heterotrophic bacteria are important in all sedimentary environments, and although they consume organic detritus they supplement the organic matter with their own remains. In some environments autotrophic bacteria may also be important (e.g. the Black Sea; see Section 3.4.3c).

Although bacterial biomass may be relatively small, bacterial productivity can be very high in aquatic environments. For example, in the Caspian Sea the biomass and productivity of heterotrophic bacteria have been estimated to be about half the corresponding values of the phytoplankton (Bordovskiy 1965), while in the Black Sea bacterial production (autotrophic and heterotrophic) may be an order of magnitude greater than that of the phytoplanktonic algae. Consequently, the organic-rich remains of bacteria may make significant contributions to most, if not all, sedimentary organic matter.

There are, therefore, four major sources of sedimentary organic matter, in general order of importance: phytoplankton, higher plants, bacteria and zooplankton. Fungi, perhaps surprisingly, do not appear to make significant contributions to sedimentary organic matter. Although they are important organisms in terrestrial environments they are much less so in marine environments, where decomposition is primarily carried out by bacteria.

A much simplified summary of the flow of energy through an ecosystem is shown in Fig. 1.11 (n.b. the chemical energy supply utilized by chemoautotrophes is omitted). Eventually the solar energy originally captured by photoautotrophes is released to the environment by respiration, either via the grazing (herbivore–carnivore) food chain or via the detrital food chain. In the latter, the action of detritivores (see Box 1.9) in comminuting the larger particles of organic matter facilitates bacterial decomposition.

1.5.2 The fossil record of major contributors

The fossil record can be used to gain information on the relative importance of organisms throughout geological time but it has limitations, such as the inherent tendency to give undue weight to those organisms that have recognizable preservable parts, while organisms comprised of only soft tissue are grossly underrepresented

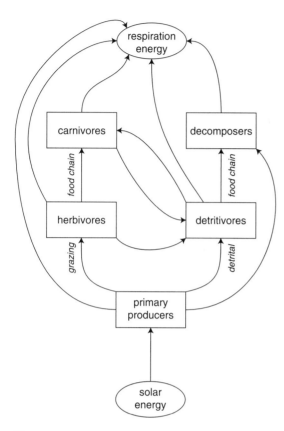

Fig. 1.11 Simplified representation of energy flow through an ecosystem (a proportion of herbivore and detritivore remains is recycled directly by decomposers).

(e.g. Briggs & Crowther 2001). This means that organism numbers and productivity may not be accurately reflected. For example, the predominant phytoplankton of the Palaeozoic (acritarchs, green algae and cyanobacteria) had cell walls composed of organic material and so were less likely to leave evidence of their existence than the phytoplankton secreting calcareous or siliceous **tests** (skeletal parts) that became dominant in more recent times.

Formation of dark, organic-rich, marine shales appeared to be widespread in the early Palaeozoic but was generally less common after the Silurian. This change coincides with an increase in abundance of herbivorous zooplanktonic remains in the fossil record and so could be interpreted as reflecting increasing importance of the grazing food chain at that time, resulting in less detritus reaching marine sediments in general. However, as is seen in Chapter 3, other factors are important in determining whether organic matter undergoes long-term preservation, and it is possible that herbivores were no less abundant during the early Palaeozoic than later but may have lacked identifiable preservable parts.

Because of the problems associated with selective preservation in interpreting the fossil record, the representation of the evolutionary trends in the groups of organisms shown in Fig. 1.12 should be treated with caution; the width of bars is only intended as a guide to trends in the relative abundance within each group of organisms, and even this limited assessment is extremely difficult for bacteria and fungi. In the absence of preserved hard parts, chemical evidence in the form of specific compounds may provide information on contributing organisms, but again this approach is limited with respect to fungi. Preservation potential, whether applied to hard parts or specific molecules, is often referred to as **taphonomy**.

1.5.3 Evolution of marine life

Throughout the Archaean, prokaryotes, in the form of photosynthetic bacteria and cyanobacteria, were the main producers of organic carbon. The resting stages (cysts) of early unicellular eukaryotic organisms, probably representing planktonic algae and referred to as **acritarchs**, first appeared in the fossil record at $c.1.85$ Ga and became abundant from $c.1.0$ Ga.

Stromatolite abundance is seen to decline from $c.1.0$ Ga, and at an increasing rate from 850 Ma. This could reflect competition from the emerging groups of benthonic algae for the favourable shallow-water environments (the term **benthonic** is generally applied to organisms that live on top of or within sediments). It might also reflect the effects of grazing by herbivores, although the earliest fossil evidence of what appear to be animals, the soft-bodied multicellular Ediacaran faunas, dates from the end of the Proterozoic (the Vendian Period, $c.610–540$ Ma). It is unknown whether the Ediacaran organisms are the ancestors of the succeeding Cambrian faunas, despite their pronounced physiological differences, or whether they represent an evolutionary dead end (Fortey 1997). A large range of invertebrate phyla appeared explosively in the fossil record at the beginning of the Cambrian, and we have a good indication of their feeding habits from the remarkable preservation of their soft tissues in the Burgess Shale (Gould 1991; Fortey 1997). The grazing and detrital food chains were clearly well developed by that time. It seems an inescapable conclusion that the organisms associated with the Cambrian 'explosion' did not appear out of thin air but must have had Precambrian ancestors, which were perhaps too small to leave a visi-

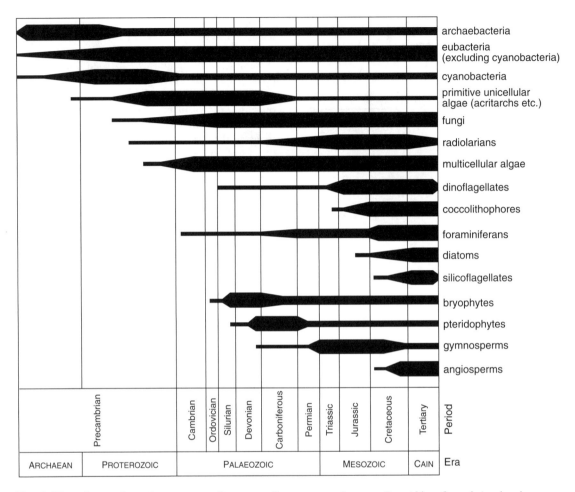

Fig. 1.12 Evolution of some important contributors to sedimentary organic matter. Bar widths reflect relative abundance trends within each group of organisms. Early origins of cyanobacteria are equivocal (see text).

ble fossil record (as is true for many members of the modern zooplankton; Fortey 1997).

As noted above, the early phytoplanktonic communities had organic cell walls (e.g. **dinoflagellates**), but during the Mesozoic, planktonic organisms appeared that secreted calcium carbonate (i.e. calcareous) tests. Hard coatings would have offered a degree of protection from predation, until predators developed effective means to breach the shells (like rasps). Among these calcareous organisms nanoplankton, particularly **coccolithophores**, dominated primary production (see Box 1.9), and zooplanktonic **foraminiferans** also made a significant contribution to biogenic sediments. Siliceous phytoplankton (i.e. secreting silica tests), especially **diatoms** and **silicoflagellates**, became impor-

tant in the Late Cretaceous and Cainozoic. The main primary producers within the phytoplankton today — in terms of net production — are the diatoms, but also important are the dinoflagellates, coccolithophores, silicoflagellates, cyanobacteria and prochlorophytes. In terms of numbers of cells, the **prochlorophytes** (the smallest of the picoplankton and related to the cyanobacteria; Partensky et al. 1999) are the most abundant, followed by the cyanobacteria.

The abundance of the fossilized remains of herbivorous zooplankton (see Boxes 1.9 and 1.12) is relatively low for the early Palaeozoic but increases subsequently for orders such as the **radiolarians** and foraminiferans (both protozoa of the class Rhizopoda). Grazing of phytoplankton by zooplankton greatly reduces the direct contribution of the former to sedimentary organic matter. However, there is still an indirect input via detrital material from zooplankton, comprising both their faecal pellets and their remains upon death, and in this respect the most important zooplankton today are **copepods** (small crustaceans) and foraminiferans.

Multicellular (macroscopic or macrophytic) algae appeared during the late Proterozoic and increased in numbers during the early Palaeozoic. These organisms were generally non-planktonic, being attached to the substrate, and so can be considered as benthonic organisms. They can be classified according to colour: green (Chlorophyta), red (Rhodophyta) and brown (Phaeophyta). The Devonian saw significant changes in the evolution of the algae, with many forms dying out and others replacing them. Some forms were different from earlier and later types, and among them may be representatives of the first plants to attempt colonization of the land.

Fungi appear to have evolved alongside the algae, and there are unicellular (yeasts) as well as colonial forms. Their evolutionary record is, however, difficult to piece together due to the lack of preservable parts. Fossil remains have been found in Archaean marine sediments and today fungi are highly successful in terrestrial environments. Fungi probably invaded the land at the same time as plants.

1.5.4 Evolution of terrestrial life

The last important group of photosynthesizing organisms to appear were the terrestrial **vascular plants**, probably evolving from ancestral green algae. They are commonly termed the **higher plants** in order to differentiate them from the lower plants such as algae. The bryophytes (mosses, hornworts and liverworts) probably also originated from ancestral algae; they can be important contributors to peat (e.g. *Sphagnum*), although their overall contribution to sedimentary organic material is relatively minor.

The colonization of the land by plants began in the Ordovician, based on an age of $c.475$ Ma for the earliest known fossil spores, which appear to have derived from liverwort-like plants (Wellman et al. 2003). By this time there was a well developed ozone (O_3) layer providing protection against harmful UV radiation. Aquatic plants are supported by the density of water, do not require protection from desiccation and have their nutrient requirements delivered to all parts by the water they inhabit. Like aquatic plants, bryophytes do not have a vascular system, and so are restricted to areas of plentiful water supply. In order to spread across the continents in the relatively dry and low density medium of air, land plants had to develop structural tissues to provide support, cuticles to protect against dehydration and a vascular system to deliver water and nutrients to all parts. The early vascular plants were prostrate forms, like the bryophytes, minimizing the need for support. They reproduced by spore formation, again like the bryophytes, and areas with standing water or subject to flooding (e.g. tidal mud-flats) were ideal primary habitats, transporting spores and gametes, as well as supplying all the water needed by the plants. *Cooksonia* is generally accepted as the earliest erect pteridophyte-like plant (Briggs & Crowther 2001). Diversity, distribution and size of land plants increased greatly during the Devonian, the major vascular plant types belonging to the psilophytes, a class of the pteridophytes (see Box 1.8). As plants increased in size their photosynthetic requirements increased, so a larger surface area was required for the capture of light as well as CO_2 (levels of which in the atmosphere were declining, although still high). The solution to this problem was the development of leaves from modified branches (Stewart & Rothwell 1993).

Dense vegetation stands grew during the Late Devonian and Carboniferous, when the clubmosses (lycopsids) and horsetails (sphenopsids) reached their peak of development, forming the great coal forests of the period. These plants grew to considerable size, with heights of 45 m or so for the largest clubmoss, *Lepidodendron*. When the climate changed and the swamps dried out at the end of the Carboniferous some of the larger of the pteridophytes died out, but survivors remain today among the ferns and clubmosses, and one horsetail, *Equisetum*. The first large woody trees, which were important members of early Late Devonian forests, were the progymnosperms (e.g. *Archaeopteris*), which reproduced by spores and had a fern-like foliage.

Seed-bearing plants (spermatophytes), in the form of the **gymnosperms**, emerged during the Late Devonian from pteridophyte ancestors. Important orders of the gymnosperms were cycads, conifers and ginkgoes, and they dominated the terrestrial flora until another group of seed-bearing plants, the **angiosperms**, rose to prominence by the mid-Cretaceous. Although of diminished importance, members of the gymnosperms exist today, most notably the conifers of the temperate regions. The angiosperms, or flowering plants, emerged in the Early Cretaceous, and are generally characterized by broad and veined leaves, in contrast to the needles of gymnosperms. They are highly versatile and successful, and their productivity is reflected in the large coal deposits of the Cretaceous and Tertiary. Grasslands are a relatively recent feature and, although not important contributors to sedimentary organic matter, they serve as a useful example of the evolution of the photosynthetic pathway (Box 1.10). Grasses (*Poaceae*) first appeared during the Oligocene, utilizing the C_3 pathway of earlier organisms. C_4 grasses seem to have emerged later in the Miocene (Cerling et al. 1993), and today dominate at latitudes less than 40°, where the ability to keep stomata closed during the hottest times of the day provides a competitive advantage. In contrast, C_3 grasses dominate in regions with cool growing seasons.

1.5.5 Ecosystem variations

The broad picture painted in the preceding sections of the temporal variations in the Earth's biota is the result of the interplay of many factors, including gradual evolutionary changes as well as relatively rapid (in geological terms) and catastrophic events. The overall trend is one of increasing diversity, as organisms adapted to the plethora of habitats available. So the explosive radiation at the beginning of the Cambrian is not unexpected, but can diversity continue to increase indefinitely, as suggested in the overall trends in Fig. 1.13? A basic tenet of ecology is that if two species occupy precisely the same ecological niche (i.e. the same habitat, environmental conditions, food source etc.), one will be outcompeted by the other and become excluded from the niche (a potential mechanism for extinction); this is the principle of competitive exclusion (e.g. Begon & Mortimer 1986). The great diversity of life on modern coral reefs shows how large the range of ecological niches can be, and it would seem that the diversity ceiling has yet to be reached for the Earth as a whole.

Estimating diversity is far from easy, and depends upon how taxonomic distinctions are made on the basis of the preserved physiology, which contributes to the different appearance of the trends (a) and (b) in Fig. 1.13 (Benton 1995). Nevertheless, the fossil record is not one of uninterrupted diversity increase; the trend is punctuated by sharp declines, marking mass extinctions, near several period boundaries (Hallam & Wignall 1997). The five largest mass extinctions occurred in the Late Ordovician, Late Devonian, Late Permian, Late Triassic and end-Cretaceous (Fig. 1.13a). Three major, distinct, faunal groups have been recognized among Phanerozoic marine organisms with skeletal parts that are generally well preserved: the Cambrian, Palaeozoic and Modern faunas. Each of these groups has a characteristic assemblage of taxa that remains relatively stable over a significant time period, as shown in Fig. 1.13a (Sepkoski 1984). Although the Cambrian faunal diversity was decreasing throughout the Ordovician, while

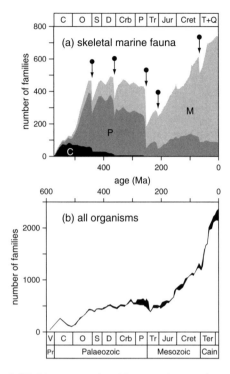

Fig. 1.13 Diversity trends in (a) marine fauna with preserved skeletal hard parts (C = Cambrian fauna, P = Palaeozoic fauna, M = Modern fauna, • = major extinctions; after Sepkoski 1984); (b) all organisms (variation in line thickness reflects uncertainty in taxonomic differentiation; after Benton 1995).

the Palaeozoic faunal diversity was increasing, the first three of the major extinction events clearly had a major impact on the Cambrian fauna, and similarly the Late Permian extinction brought about a dramatic decline in the Palaeozoic faunal diversity. As life recovered from the extinction events, new families and species appeared relatively rapidly, filling the vacated niches.

Perhaps the best known of the mass extinctions is that associated with the demise of the dinosaurs at the Cretaceous–Tertiary boundary, which has been attributed to a meteorite impact (Alvarez et al. 1980; Kruge et al. 1994; Skelton et al. 2003), although natural climate change may also have been a factor. Planktonic organisms (particularly foraminiferans, calcareous phytoplankton and radiolarians) were also severely affected. On land, trees seem to have been devastated, with the recovery period marked by the proliferation of ferns. The pollen record is dominated by fern spores, and hence known as the fern spike, which is observed throughout the world (Nichols et al. 1986; Vajda et al. 2001).

However, the mass extinction with the greatest impact occurred in the Late Permian, with >60% of marine and terrestrial families disappearing (Erwin 1993, 1994; Benton 1995). Over a period of <1 Myr, 85% of the main species were wiped out (Bowring et al. 1998). The extinction event also brought an end to the *Glossopteris* flora, which dominated the forests at high southern latitudes, and their northern equivalents, the *Cordaites* flora of Siberia (Retallack 1995). There was a corresponding proliferation of fungal spores (Eshet et al. 1995; Visscher et al. 1996). Lycopsids dominated for c.5 Myr in the tropics (e.g. Europe), and conifers did not proliferate until the mid-Triassic (Looy et al. 1999). The cause(s) of the extinction event are a matter for speculation due to the lack of sufficiently precise dating of events. Some of the factors that may have been instrumental in affecting ecosystems on a global scale are listed in Table 1.7.

Global changes in ecosystems clearly have the potential to affect the type and amount of organic material deposited in sediments and, as we have seen, can be driven by extinction events as well as more gradual evolution resulting from random mutation. Less dramatic but equally important influences on ecosystem composition and evolution are major climatic changes. The Earth has experienced ice ages at various times—during the late Proterozoic, the Late Carboniferous–Early Permian and the Quaternary—which we examine in Chapter 6. The arrangement of continents can play a key role in modifying climate as well as having a direct influence on the variety of habitats. The continents have been assembled in one large mass at least three times in the Earth's history, at c.250 Ma (Pangaea), c.600 Ma (Vendian), c.1.0 Ga (Rodinia) and possibly also c.1.9 Ga (Rogers & Santosh 2002). During such periods climate and ecosystems are uniform over large areas, potentially restricting diversity (Valentine & Moores 1970). For example, during the Devonian and Carboniferous, as Pangaea was assembling, large areas of the northern hemisphere land mass (Laurasia) were subjected to an almost uniformly warm and moist climate, as in the present-day tropics. Under these conditions a uniform *Lepidodendron* flora of the Laurasian coal swamps flourished, while a large area of southern hemisphere land mass (Gondwana) was covered by ice sheets. During the Permian, the *Glossopteris* flora (named after the dominant sporophyte) produced significant coal deposits at high southern latitudes across what is now Antarctica, Australia, South America and India, and there was an equivalent flora at high northern latitudes, but arid conditions prevailed over a wide expanse of Pangaea at lower latitudes. As Pangaea split up and dispersed during the Mesozoic, the distinctions between habitats and their associated plant communities became more pronounced.

While the spread of marine organisms throughout the oceans is largely unhindered by topographical obstacles, a uniform distribution does not always result even after sufficient time has elapsed for extensive colonization to occur by a newly evolved species. Coastal and shelf environments are more extensive during periods of continent dispersion, and there is more opportunity for local populations of marine organisms to become isolated and undergo divergent evolution. The arrangement of continents can also have a marked effect on oceanic currents and climate. Today, there are latitudinal variations in distributions of both abundance and species of phytoplankton related to climate, oceanic circulation and nutrient distribution patterns, and there is no reason to believe that such controls did not operate in the past. Such variations and the factors that control microorganism distributions in saline and freshwater environments are discussed in Chapter 3. But first, in the next chapter, we review the organic composition of organisms in order to understand why the composition of sedimentary organic matter varies.

Table 1.7 Some factors that may have influenced ecosystems on a global scale

---

bolide impacts
bursts of cosmic radiation, e.g. nearby supernovae
magnetic reversals—loss of protection from charged particle fluxes
dispersal of continents
sea-level changes—particularly their effect on area of shelf seas
climatic change—particularly rapid periods of change (e.g. volcanic CO_2 emissions)

2 | *Chemical composition of organic matter*

2.1 Structure of natural products

2.1.1 Introduction

Having established the major contributors to sedimentary organic matter in Chapter 1, it is important to consider their chemical composition in order to understand the changes that result in the fossil forms of carbon preserved in ancient sediments. This also helps us to understand the behaviour of man-made compounds in the environment. A comprehensive study of natural products would require many volumes the size of this book (e.g. Barton & Nakanishi 1999), so we confine our review to those of major geochemical importance. All organisms are composed of the same basic chemical classes, the most important of which, geochemically, are carbohydrates, proteins and lipids. In addition, higher plants contain significant quantities of lignin, a major component of their supportive tissue. Lipids are believed to play a dominant role in petroleum formation and so organisms relatively rich in lipids, such as the plankton (c.10% lipids, dry wt), are quantitatively very important contributors to oil source rocks.

In the following sections we review the compositions of the main chemical classes and their biochemical functions. These functions are related to the structure and shape of the basic carbon skeleton of a molecule and the functional groups (see Box 2.1) attached to it.

2.1.2 Bonding in organic compounds

Atoms within organic molecules are held together by **covalent bonds**, which are formed by adjacent atoms sharing pairs of electrons (usually each atom donates one of its outermost, or valency, electrons to the bond). Single, double and even triple bonds can be formed, in which one, two and three electron pairs are shared, respectively, although triple bonds are rare among natural products. Compounds where all the carbon atoms are joined together by single bonds are called **saturated**. Saturated hydrocarbons are termed **alkanes** and can be acyclic or cyclic. The simplest acyclic alkanes are the straight-chain compounds called normal alkanes (n-alkanes).

Compounds that contain one or more pairs of adjacent C atoms joined together by a double bond (C=C) are termed **unsaturated**. An unsaturated hydrocarbon is called an **alkene**. Double bonds are stronger than single bonds, but are also reactive and can undergo addition reactions to form saturated compounds. Particularly stable arrangements of unsaturation are formed where several C=C bonds are present and alternate with C–C bonds producing the pattern: ~C–C=C–C=C–C=C~. This arrangement of the double bonds is described as **conjugated** and is frequently encountered in polyunsaturated natural products. Conjugation is possible in rings of atoms, and when it involves $2n + 1$ (where n = integer) double bonds it results in **aromatic** compounds, which have enhanced stability. The simplest aromatic compound is the hydrocarbon benzene (containing three double bonds; Table 2.2), but it is possible for a number of aromatic rings to be fused into polycyclic structures. Organic compounds are conventionally termed either aromatic or aliphatic; the simplest definition of **aliphatic** is a compound that is not aromatic. In both aliphatic and aromatic compounds C atoms can be replaced by other atoms (heteroatoms), most commonly O, N and S. Atoms such as N and O are able to form additional single bonds in which they donate both of the electrons involved (a pair of valency electrons), in what is termed a dative (or coordinate) bond (e.g. two of the bonds between N atoms and magnesium in the chlorophylls in Fig. 2.27).

As well as strong covalent bonds, there are weaker, electrostatic interactions between molecules that influence their properties, one of the most important of

Box 2.1 | Geochemically important functional groups

An organic compound contains a basic skeleton of carbon atoms, which can be arranged as a simple straight chain, a branched chain, one or more rings, or a combination of these. Many natural products involve cyclic structures, with rings of six carbon atoms being particularly common. The simplest organic compounds contain only hydrogen atoms bonded to the basic carbon skeleton and are called **hydrocarbons**. Atoms of other elements (**heteroatoms**) can be incorporated in the basic hydrocarbon structure, often in the form of peripheral **functional groups**. The most common heteroatoms in natural products are oxygen, nitrogen and sulphur. A functional group confers characteristic chemical properties on a compound, so that compounds are often considered collectively under the name of the functional group they contain, e.g. carboxylic acids for the compounds bearing a carboxyl group. The main types of functional groups encountered in natural products are listed in Table 2.1.

More than one type of functional group may be present in a molecule, as in amino acids, which contain both carboxylic acid and amino groups. Similarly, amides can be considered to comprise amino and carbonyl groups, and carboxylic acids to comprise carbonyl and hydroxyl groups, although the behaviour of each group is modified by the neighbouring group. The double bond of an alkene can also be thought of as a functional group.

Heteroatoms can be incorporated into cyclic systems, and some of the geochemically important are shown in Table 2.2.

Table 2.1 Geochemically important functional groups

formula*	group/compound name
R—OH	hydroxyl: alcohol (R = alkyl group)
	phenol (R = phenyl group)
—C=O | R	carbonyl: aldehyde (R = H) ketone (R = alkyl or phenyl)
—C=O | OR	carboxyl: carboxylic acid (R = H) ester (R = alkyl group)
—O—	ether
—NH$_2$	amine
—C=O | NH$_2$	amide
—SH	thiol (or mercaptan)
—S—	sulphide
>C=C<	alkene

* R is used to represent aliphatic chains (alkyl groups) or aromatic rings. The latter can also be called aryl groups, and are sometimes represented by Ar.

Table 2.2 Geochemically important cyclic units

formula	unit name
(1,2 and 1,4 quinone structures)	quinones (1,2 and 1,4) (i.e. 2 carbonyl groups within a cyclic system)
(5-membered ring with Z)	cyclopentadiene (Z = CH$_2$) furan (Z = O) pyrrole (Z = NH) thiophene (Z = S)
(6-membered ring with Z)	benzene (phenyl group) (Z = CH) pyridine (Z = N)
(6-membered ring with O)	pyran

---

which is the **hydrogen bond**. It is 10–20 times weaker than a normal single covalent bond, and involves an electrostatic attraction between a slightly positively charged H atom and a slightly negatively charged heteroatom. The slight positive charge on the H atom results from partial withdrawal of electron density by the electronegative heteroatom, usually O or N, to which it is directly covalently bonded. The hydrogen is attracted to another slightly negatively charged heteroatom in a different molecule or elsewhere in the same molecule. Consequently, hydrogen bonding can affect the shape adopted by molecules and their interactions with other molecules (e.g. Fig. 2.33). Throughout this book hydrogen bonds are represented by dotted lines.

It is important to remember that organic compounds are generally not planar, but for simplicity they are drawn as if they were. The geometry of bonds at each carbon atom can be tetrahedral, trigonal or linear, depending on whether the atom is bonded to four, three or two substituents, respectively. This variation in bond geometry for carbon arises from the different spatial arrangements possible for the four bonds generally formed by carbon. If there are four single bonds a tetrahedral arrangement results; if there are two single and one double bond a trigonal geometry results; and if there are two double bonds or one triple and one single, a linear geometry arises. Conventions for representing the three-dimensional nature of structures are given in Box 2.2.

2.1.3 Stereoisomerism

The possibility for different spatial arrangements of atoms or groups in molecules gives rise to the phenomenon of **stereoisomerism**. There are two forms of stereoisomerism, configurational and conformational.

Configurational isomerism can occur where the possibility exists for two spatially different arrangements of bonding at one or more atoms. The simplest example is a carbon atom bonded to four different atoms or groups: two arrangements are possible for the tetrahedral geometry, which are non-superimposable mirror images, e.g. (+)- and (−)-glyceraldehyde in Fig. 2.2a. Such stereoisomers are called optical isomers or **enantiomers** and most of their physical properties are identical, but they rotate the plane of polarized light in opposite directions (and the compounds can therefore be termed chiral). The carbon atom around which the four different groups are arranged is termed a **stereogenic centre**. If two of the groups were identical, there would no longer be a stereogenic centre because it would be possible to superimpose the structures. Stereogenic centres can also exist as part of a cyclic system, as in carvone (Fig. 2.2a). At first sight it may appear that the indicated carbon is not surrounded by four different groups because it is directly bonded to two identical -CH_2- units in the cyclic system. However, proceeding around this cyclic system in the clockwise direction we find a -CH_2- unit followed by a -CH=, while in the anticlockwise direction a -CH_2- unit is followed by -C(O)-. The cyclic system, then, does not appear the same when viewed from each direction of attachment to the stereogenic carbon and, effectively, this atom is surrounded by four different groupings.

Natural products can have several stereogenic centres. If there are n centres, there are 2^n possible stereoisomers, comprising 2^{n-1} pairs of enantiomers. In any particular pair of enantiomers the configuration at each centre in one isomer is opposite to that of the corresponding centre in the other isomer. An enantiomer that rotates the plane of polarized light in a clockwise direction is termed dextrorotatory and is labelled (+). The other enantiomer is laevorotatory, is labelled (−) and rotates plane-polarized light by the same amount in an anticlockwise direction. This property provides a means of identifying the relative configuration of a molecule. However, there is no simple relationship between the sign and magnitude of optical rotation and molecular structure. The true spatial orientation of groups at each stereogenic centre in a molecule, the **absolute configuration**, can be represented by the labels R and S. An older system with D and L labels is commonly used to distinguish between enantiomers of amino acids and sugars (see Box 2.3). A 50:50 mixture of the enantiomers of a compound is optically inactive and is termed **racemic**.

All stereoisomers that are not enantiomers are termed **diastereomers**. For example, in a molecule containing two or more stereogenic centres any pair of stereoisomers that are not enantiomers are diastereomers. In other words, a pair of diastereomers share the same configuration at one or more, but not all, of their stereogenic centres and, unlike enantiomers, they have different physical properties. Two diastereomers that differ in configuration at only one of a number of stereogenic centres are called **epimers**, and their interconversion **epimerization**.

Diastereomers are also encountered in unsaturated acyclic compounds. When two C atoms are joined together by a double bond, all the remaining four single bonds to the two C atoms lie in the same plane as the C=C bond. If each of these two carbon atoms is bonded to a H atom and a hydrocarbon (alkyl) chain, the alkyl chains can be either on the same side of the C=C bond as each other or on opposite sides, and the resulting diastereomers (which used to be known as geometric isomers), shown in Fig. 2.2b, are termed *cis* and *trans*, respectively. Again, these diastereomers have different physical properties (see also Box 2.3). Optical isomerism is not possible about a C=C bond (the mirror images are superimposable).

Although complex natural products may have a number of stereogenic centres, they tend to exhibit only a limited number of stereoisomeric variations (e.g. see Finar & Finar 1998). This is because biosynthesis results in the selective formation of one configuration at many of the centres, as can be seen in subsequent sections. The overall shape of a molecule can be very

Box 2.2 | Structural representation of organic compounds

Bonds between atoms are represented by lines, with the number of lines indicating the number of bonds. While it is possible to label each carbon and hydrogen atom by C and H, the overall structure of complex molecules is often clearer if H atoms are omitted and the carbon skeleton is shown as lines representing bonds between C atoms. For example, alternative representations of n-hexane and phenol are given in Fig. 2.1a. The presence of a carbon atom is inferred at each change in angle of the line drawing and at each end of a chain. Heteroatoms are represented by the normal chemical symbol, as in phenol (Fig. 2.1a). Sometimes the conjugated bond system of aromatic rings is represented by a circle, again as shown for phenol (Fig. 2.1a). This reflects the delocalization of bonding that occurs in conjugated systems, such that bonds between all the C atoms involved are approximately identical and of intermediate strength (and length) between single and double bonds.

Where the relative three-dimensional spatial arrangement of atoms in a molecule is important, symbols exist to represent bonds projecting above and below the plane of the paper. Thickened lines show bonds projecting above the paper, dotted lines show bonds projecting below it. It is sometimes more convenient in cyclic structures to show the relative orientation of a hydrogen atom rather than that of the adjacent C–C bonds. In the convention adopted in this book an open ring around the C atom to which the H atom is bonded indicates that the H atom lies below the plane of the paper (α configuration), while a filled-in circle denotes that the H atom is above the paper plane (β configuration). An example is shown in Fig. 2.1b and the conventions are summarized in Table 2.3. It should be noted in the structure on the right side of Fig. 2.1b that it is not necessary to show the configuration for the hydrogen atom bonded to the bottom C atom in the left-hand ring: because of the tetrahedral arrangement about the C atom, if two bonds are drawn in the plane of the paper and the methyl projects above it, the H atom must lie below the plane.

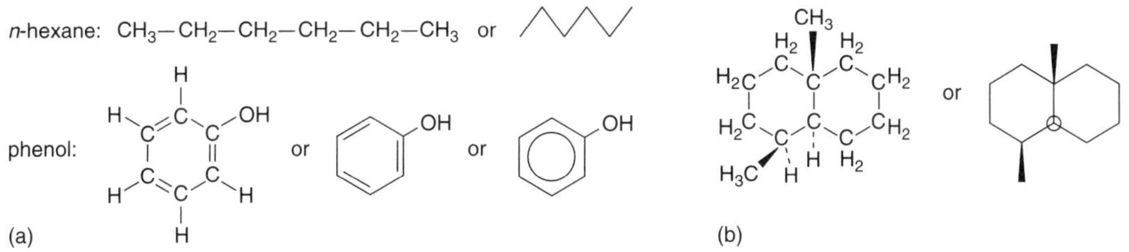

Fig. 2.1 Representation of (a) carbon skeletons in molecules and (b) spatial arrangement of atoms and groups.

Table 2.3 Symbols representing the spatial arrangement of bonds

C–C (or C–H) bond projects above plane of page	C–H bond projects above plane of page (β-configuration)
C–C (or C–H) bond projects below plane of page*	C–H bond projects below plane of page (α-configuration)
C–C (or C–H) bond lies in plane of page	unspecified C–C or C–H bond configuration (either possibility exists)

(* The following symbol can also be used).

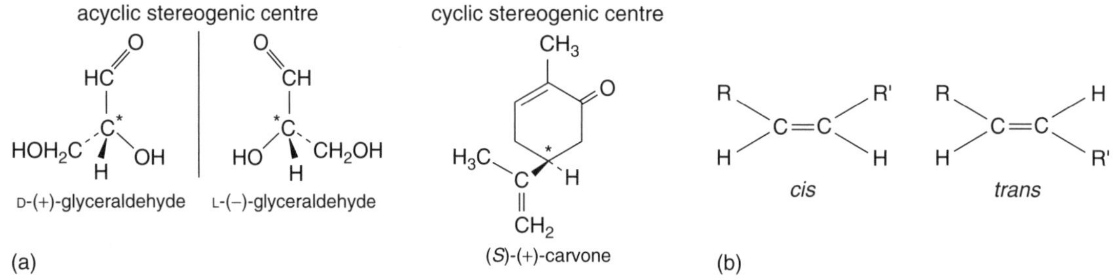

Fig. 2.2 Examples of stereoisomerism: (a) optical isomerism involving acyclic and cyclic carbon atoms (*indicates stereogenic centre); (b) *cis* and *trans* diastereomers in unsaturated compounds (R and R′ = aliphatic or aromatic groups).

Box 2.3 | Systems for denoting absolute stereochemical configuration

Prior to the advent of anomalous dispersion X-ray crystallography the absolute configuration of optically active molecules was not known. However, it was still possible to compare the absolute configurations of different molecules by relating them to an optically active reference compound, glyceraldehyde (see Fig. 2.2a), which is biosynthetically related to a variety of other natural products. The absolute configuration of the (+) enantiomer of glyceraldehyde was arbitrarily assigned the D label, and that of the (−) enantiomer the L label (see Fig. 2.2a). A compound that can be converted into D-glyceraldehyde by reactions that conserve the configuration of the stereogenic centre in the compound is also given the D label, while one that can be converted to L-glyceraldehyde is given the L label. It is important to remember that the D and L labels relate to the absolute configuration and not the direction of rotation of plane-polarized light. Although the D enantiomer of glyceraldehyde is dextrorotatory, for another compound the D enantiomer may be laevorotatory. For monosaccharides, although more than one stereogenic centre may exist, a single D/L label is used, denoting configuration at the highest-numbered centre (e.g. C-5 in hexoses; see Section 2.2.1).

A more recent and universal convention for representing absolute configuration is the *R/S* system. It is based on a series of rules defining priority among the substituents at a stereogenic centre, which can be found in undergraduate organic chemistry texts (e.g. Finar & Finar 1998; Smith & March 2000; Morrison & Boyd 2001) and at the IUPAC website (http://www.chem.qmul.ac.uk/iupac/stereo/). In determining which assignment is given, the centre is viewed with the bond to the lowest priority group, usually a hydrogen atom in the compounds we are concerned with, pointing away from the viewer. The other three bonds then appear to form a trigonal arrangement. If the priority of these groups decreases in a clockwise direction the *R* configuration is assigned. The *S* configuration is given to an anticlockwise decrease in priority. An example is given in Fig. 2.3, in which only alkyl groups are involved and increasing priority equates with increasing mass.

The *R/S* convention can be used for all configurational isomers, but sometimes an α/β convention is used, particularly for atoms that are part of a cyclic system (e.g. sugars and steroids). When applied to steroids, α indicates that the labelled atom or group projects below the plane of the paper (and cyclic structure) and β indicates that it projects above (as noted for H atoms in Box 2.2). The principles of the *R/S* system can also be applied to describing the configuration about a C=C bond (formerly termed geometric isomerism), but two new symbols, *E* and *Z*, are substituted. Where the groups with highest priority on both C atoms lie on the same side of the C=C bond (e.g. the *cis* isomer in Fig. 2.2b) the *Z* label is used. This system is most useful when either or both of the C atoms involved in the C=C bond have two alkyl substituents because the *cis* and *trans* system then becomes difficult to apply.

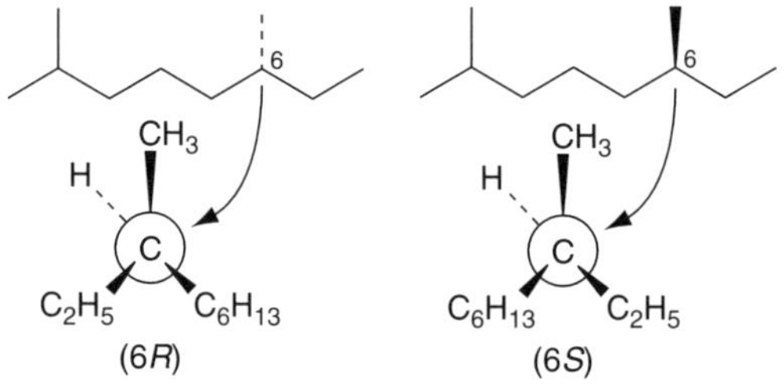

Fig. 2.3 *R* and *S* configurational assignment at C-6 in 2,6-dimethyloctane.

important in controlling its biochemical functions, and depends upon both types of stereoisomerism.

Conformational isomerism involves stereoisomers that can be interconverted by rotation of groups around single bonds. The possibilities for a simple acyclic compound, *n*-butane, are shown in Fig. 2.4. The staggered arrangement of groups is preferred, with the lowest energy arrangement being that in which the methyl groups are furthest apart (far left, Fig. 2.4). The bulkier the groups the more energy is required to overcome the strain involved in allowing the largest groups to rotate through the intermediate eclipsed arrangement (i.e. the three groups on one C atom are in direct line with those on the next).

The conformation in a ring system is critical in minimizing the strain on bonds, an important factor being potential interactions between bulky substituent groups. Maintaining the tetrahedral angle required by C–C bonds in a C_6 ring results in a chair-shaped conformation being the most stable. Two possible chair conformers exist for such a ring, as shown in Fig. 2.5 for *trans*-1,2-dimethylcyclohexane, the interchange between which involves rotation about single C–C bonds in the ring, but with somewhat higher energy barriers than in acyclic compounds. The preferred conformer in Fig. 2.5 is that in which both the bulky methyl groups are **equatorial** (i.e. bonds to them are parallel to the plane of the main part of the ring system) rather than **axial** (bonds perpendicular to the main ring plane), because the diaxial conformer involves greater crowding between the methyl groups and the axial hydrogen atoms.

2.2 Carbohydrates

2.2.1 Composition

The term **carbohydrate** derives from the fact that many members of this group of compounds have the general formula $C_n(H_2O)_n$, i.e. they contain only carbon, hydrogen and oxygen, with the H and O atoms being in the same ratio as in water. They are polyhydroxy-substituted carbonyl (i.e. aldehyde or ketone) compounds. The simplest molecules are **monosaccharides**, which are named according to the number of carbon atoms present; for example, tetroses, pentoses, hexoses and heptoses contain four, five, six and seven carbons, respectively. Both aldehyde and ketone derivatives of these units exist, called respectively aldoses and ketoses. Hence a C_6 monosaccharide may be an aldohexose (e.g. glucose) or a ketohexose (e.g. fructose). Most naturally occurring monosaccharides are hexoses or pentoses, some examples of which are given in Fig. 2.6a. They mainly exist as cyclic systems, with those forming five-membered rings being called furanoses and those forming six-membered rings pyranoses, after the simplest parent compounds furan and pyran (Table 2.2). Another way of representing these cyclic systems is by the Haworth structure, which is shown for glucose in Fig. 2.7.

Some modified monosaccharides exist, such as D-glucosamine and D-galactosamine, in which the hydroxy group at C-2 has been replaced by an amino group (Fig. 2.6a). Other modifications include deoxygenation, e.g. the absence of oxygen at C-6, as in L-rhamnose and L-fucose (Fig. 2.6a; which can also be called 6-deoxy-L-mannose and 6-deoxy-L-galactose, respectively), and the presence of carboxylic acid groups (which can complex a variety of cations), giving rise to the **uronic acids** (e.g. glucuronic and galacturonic acids, Fig. 2.6a).

We have already considered the photosynthetic formation of carbohydrates (Box 1.10) and have seen that the basic building blocks are C_3 compounds—triose phosphates—D-glyceraldehyde-3-phosphate and dihydroxyacetone phosphate (Fig. 2.6b), the latter being formed from the former by enzymatic isomerization.

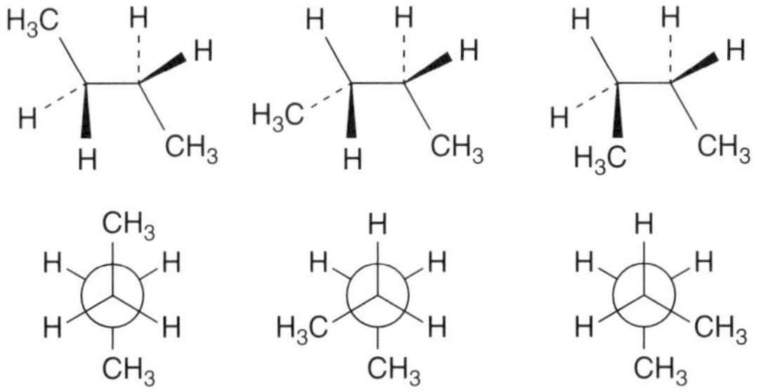

Fig. 2.4 Conformational isomers of *n*-butane (lower row shows projection along horizontal C–C bond in top row).

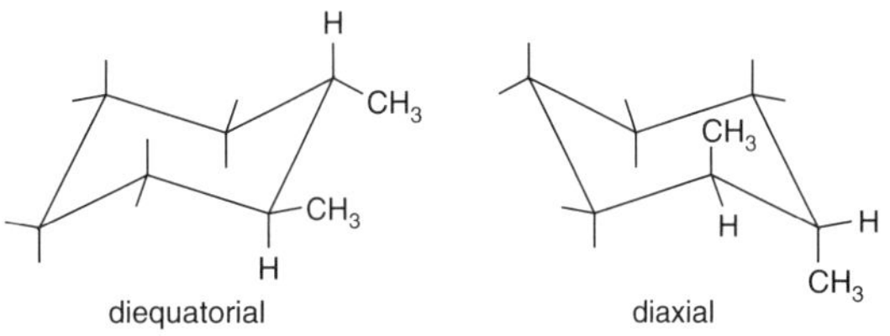

Fig. 2.5 Chair conformations of *trans*-1,2-dimethylcyclohexane (only hydrogen atoms adjacent to methyl groups are shown for clarity).

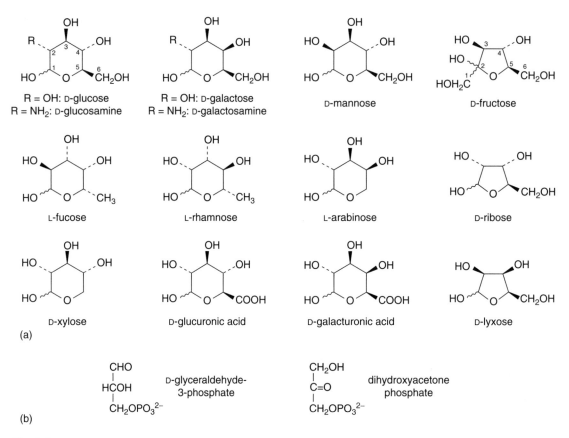

Fig. 2.6 (a) Geochemically important monosaccharides (carbon numbering convention shown for D-glucose and D-fructose) and (b) their biological precursors.

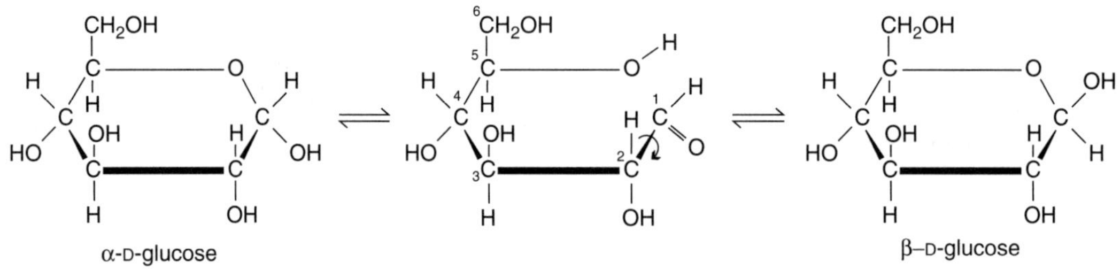

Fig. 2.7 Mutarotation of D-glucose (the arrow in intermediate, open-chain structure denotes bond rotation).

The formal carbon numbering scheme for glucose is shown in Fig. 2.6a. At the C-1 position there is a stereogenic centre, the two possible stereoisomers being termed α where the OH group lies on the opposite side of the ring plane from the CH$_2$OH group on C-5 and β where it lies on the same side (see Fig. 2.7). In solution the α and β isomers of the cyclic monosaccharides interconvert to give equilibrium mixtures of both isomers, a process that is discussed in more detail in Box 2.4.

The four other carbon atoms in the ring system of aldohexoses (C-2 to C-5) are also stereogenic centres,

Box 2.4 | Ring formation in monosaccharides

Isomerism at C-1 occurs via ring opening by cleavage of the bond between the O and C-1 atoms in the ring, yielding the open-chain form with a carbonyl group: an aldehyde group for the aldoses and ketone group for the ketoses. In the open-chain form C-1 of the free carbonyl group is no longer stereogenic. Rotation about the bond joining the carbonyl group to the rest of the monosaccharide allows either diastereomer (referred to as **anomers**) to be formed upon ring closure. This process of anomerism (or mutarotation) is shown for D-glucose in Fig. 2.7. It occurs upon the breakdown of polysaccharides by hydrolysis into monosaccharide units (e.g. during compositional analysis); configurational specificity is unavoidably lost at C-1, the carbon linking monosaccharide units together. In polysaccharides only the terminal monosaccharide units can exist in open-chain form, the rest of the polymer comprising cyclic (pyranose and furanose) forms.

The reaction of the carbonyl group with the alcohol group at C-5 is a typical reaction of a carbonyl with an alcohol, yielding hemiacetals from aldehydes (as shown in Fig. 2.8) and hemiketals from ketones. Reaction with the C-5 OH group is favoured because a six-membered ring is formed, imposing minimal strain on the bonds forming the ring. In aqueous solution <0.1% of D-glucose exists in the intermediate open-chain form, while the most abundant form is the β anomer (63%), which has the more stable orientation of the C-1 OH group.

The forms of glucose in aqueous solution are quite complex and small proportions exist as furanose structures. These arise from the reaction between the OH group at C-4 and the aldehyde group in aldoses. The five-membered ring system is quite stable but pyranose forms are favoured over furanose forms by most monosaccharides; notable exceptions are the ketohexose fructose and the aldopentose ribose. Too much strain would be involved in reducing bond angles to produce smaller rings, so trioses and tetroses do not form intermolecular cyclic systems.

so there is a total of 2^4 or 16 possible stereoisomers. Therefore, there are eight diastereomeric forms (i.e. enantiomer pairs) of the aldohexoses, each having a different name and possessing D or L configuration at C-5 and α or β stereoisomers at C-1.

A final consideration in describing the structure of a monosaccharide is its conformation. Most aldopyranoses adopt a 'chair' conformation, the most stable form being that in which as many as possible of the bulkier OH and CH_2OH groups are equatorial, as shown for glucose in Fig. 2.8. A corresponding 'envelope' conformation is the most stable for furanoses (as for fructose, Fig. 2.8).

Monosaccharide units can be linked together by **condensation** (a reaction that joins together two molecules with the elimination of a simple molecule, water in this instance). The resulting linkage is called the **glycosidic bond**, and condensation usually involves the OH group on C-1 of one unit and an OH group on a C atom other than C-1 (often C-4) in the second unit. Condensation of two monosaccharides yields a **disaccharide**, such as sucrose, which contains a glucose and

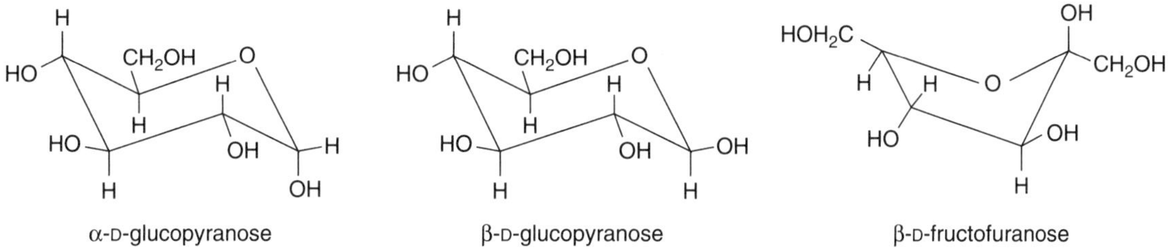

α-D-glucopyranose   β-D-glucopyranose   β-D-fructofuranose

Fig. 2.8 Preferred conformations adopted by cyclic monosaccharides: chair form for six-membered rings (e.g. glucose) and envelope form for five-membered rings (e.g. fructose). Each carbon atom in the cyclic system possesses one axial and one equatorial substituent; the axial bonds are represented by vertical lines. All OH groups and the CH_2OH group are equatorial in β-D-glucose.

a fructose unit (Fig. 2.9). Further units can be linked to give tri- and tetrasaccharides, etc. Those formed from two to ten monosaccharide units are generally termed **oligosaccharides**, while those with more units, such as cellulose (Fig. 2.9), are called **polysaccharides**. The mono- and disaccharides are commonly termed **sugars**, and polysaccharides can be termed **glycans**.

The number of possible monosaccharide units and their order and orientation in polysaccharides lead to an immense variety of possible structures (e.g. Kennedy 1988). For example, amylose and cellulose are both glucose polymers but differ in the configuration of the bridging C-1 atom (Fig. 2.9). Branching of the polysaccharide chain may also occur, as in amylopectin (Fig. 2.9), leading to further structural variety. Polysaccharides formed from only one type of monosaccharide, like cellulose and amylose, are termed **homopolysaccharides**, whereas those formed from different types of monosaccharides are **heteropolysaccharides**. Homopolysaccharides can be named after their constituent monosaccharides, such as glucans (after glucose), mannans (after mannose) and galacturonans (after galacturonic acid). Similarly, heteropolysaccharides comprising primarily one major and one less abundant monosaccharide can be named after those units, such as arabinogalactans (where galactose is the dominant unit and arabinose the subordinate). The configuration can also be included in these names (e.g. β-D-xylans).

2.2.2 Occurrence and function

Carbohydrates can function as energy reserves, structural material and antidesiccants (MacGregor & Greenwood 1980; Kennedy 1988). Polysaccharides are major components in most **cell walls**, which provide a rigid, reinforcing layer around the cell membranes in plants, bacteria and fungi. D-Glucose is by far the most abundant monosaccharide. It is important as an energy source and it is the basic unit of the polysaccharide **cellulose** (a β-D-glucan), the main structural building material of plants, a molecule of which contains $c.10^4$ glucose units. Cellulose is the most abundant natural organic compound, with higher plants containing the largest amounts, whereas some algae appear to have none.

As an energy reserve D-glucose is stored in the form of polysaccharides: starch in plants and glycogen in animals. Starch normally comprises 80% amylopectin and 20% amylose (an α-D-glucan; Fig. 2.9), although some compositional variations exist. Its structure, therefore, differs from that of cellulose in containing a C-6 branch in the amylopectin units, which occurs about every 20–25 glucose units, and also in the configuration at C-1. Glycogen has a similar structure to amylopectin but with more frequent branching. Utilization of these reserves initially involves breaking down the polysaccharides into glucose units, which then undergo **glycolysis**, a respiration process that yields pyruvic acid and some energy. Glycolysis is common to all organisms, aerobes and anaerobes alike. In aerobes the pyruvic acid is converted to acetyl coenzyme A, which then enters the **citric acid cycle**, where it can be oxidized to CO_2 with the release of energy in the form of ATP (the process of oxidative phosphorylation). The citric acid cycle, which is also known as the Krebs cycle and the tricarboxylic acid (TCA) cycle, derives its name from the initial combining of acetyl coenzyme A with oxaloacetic acid ($HOOC-CO-CH_2-COOH$) to form citric acid [$HOOC-CH_2-C(OH)(COOH)-CH_2-COOH$], which then undergoes a sequence of energy-liberating reactions to yield oxaloacetic acid, completing the cycle. A summary of the aerobic respiration of carbohydrates is given in Fig. 2.10. Although glycolysis contributes only $c.5\%$ of the energy available to aerobes from the complete oxidation of glucose, it is an important source of energy for some anaerobes (anaerobic respiration is considered in more detail in Section 3.3.2b).

D-Fructose is the most important ketose and is found in all living plant tissue. In the form of homopolysaccharides (fructans) it serves as a short-term energy reserve in plants, and similar homopolysaccharides are found in bacteria.

After cellulose the next most abundant group of carbohydrates in plants is the **hemicelluloses**, compounds that form a matrix surrounding the cellulose fibres in plant cell walls (i.e. secondary cell wall material). They are a complex mixture of polysaccharides, mostly containing 50–2000 monosaccharide units, the most abundant of which are D-xylose, D-mannose and D-galactose, with lesser amounts of others (e.g. L-arabinose, D-glucose). Hemicelluloses contain some homopolysaccharides but heteropolysaccharides predominate. Also present in higher plant primary (i.e. non-woody) cell walls and in intercellular layers are pectins. They are present in only minor amounts in woody (i.e. secondary cell wall) tissues but are more abundant in fruits. **Pectins** are complex mixtures of mainly heteropolysaccharides in which polygalacturonic acid structures predominate (Fig. 2.9), and among the lesser amounts of other monosaccharides are galactose and arabinose side-chains.

Cellulose is replaced as a structural material in most fungi, some algae, arthropods (e.g. insects and crustaceans) and molluscs by **chitin** (Fig. 2.9), a

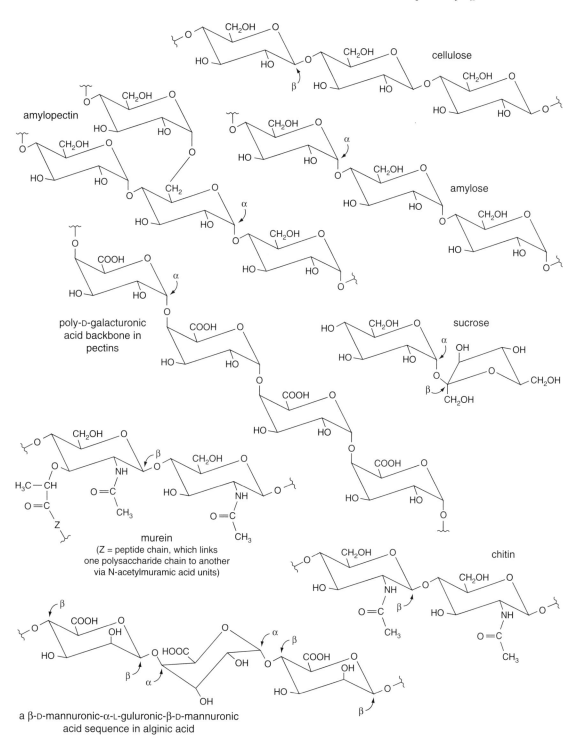

Fig. 2.9 Some important carbohydrates (showing configuration at C-1 in monosaccharide units).

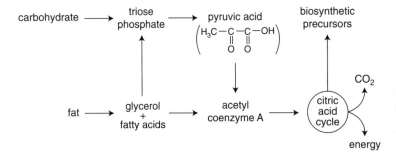

Fig. 2.10 Summary of aerobic respiration of carbohydrates and fats. (The biosynthetic precursors are used in the synthesis of fatty acids, amino acids, terpenoids etc.)

homopolysaccharide of N-acetyl-D-glucosamine, which probably rivals lignin for the position of second most abundant organic substance on Earth. All eubacterial cell walls contain **murein**, which comprises polysaccharide chains of alternating N-acetyl-D-glucosamine and N-acetylmuramic acid units that are cross-linked by chains of amino acids (Fig. 2.9). Because of its formation from peptide and carbohydrate units this material is often referred to as a **peptidoglycan**, and it can account for up to 75% (dry wt) of bacterial biomass. Eubacteria can be classified by a stain test as either **Gram-positive** (e.g *Clostridium*, *Bacillus*, acetomycetes) or **Gram-negative** (e.g. *Pseudomonas*, *Methylomonas*), reflecting differences in cell wall architecture. A major difference is the presence in Gram-negative bacteria of an outer membrane, covering the murein layer, in which **lipopolysaccharides** (compounds in which lipids are bound to polysaccharides, see Section 2.4.1b) are major constituents. Again, N-acetyl-D-glucosamine is an important unit in these polysaccharides. Gram-positive bacteria have a capsule of polysaccharides outside their cell walls in which D-glucose and D-galactose are major components.

Archaebacterial cell wall composition can vary markedly between species. The cell walls of some species contain only protein, whereas others comprise glycoprotein, polysaccharide or a type of peptidoglycan, but murein is never present. The Gram stain test can also be applied to archaebacteria (e.g. *Halobacterium* are Gram-negative).

Fungal polysaccharides are mainly homopolysaccharides of D-glucose, D-galactose or D-mannose. D-Glucose and D-galactose are also important constituents of algal polysaccharides, such as agar. The algae provide examples of how certain carbohydrates can be characteristic of, although not exclusive to, various groups of organisms. Among marine algae the Chlorophyta have relatively large amounts of L-rhamnose and D-ribose. Some species of brown algae contain up to 40% (dry wt) of alginic acid, a copolymer of D-mannuronic and L-guluronic acids (Fig. 2.9), the proportions and order of linking of which vary between species (Gacesa 1988). In contrast, freshwater algae and higher aquatic plants generally contain significant amounts of L-arabinose and D-xylose.

Carbohydrates are important in the production of fats and proteins. As well as generating energy, the citric acid cycle (Fig. 2.10) provides precursors for the biosynthesis of fatty acids, amino acids and terpenoids, which are discussed in the following sections.

2.3 Amino acids and proteins

2.3.1 Composition

Proteins account for most of the nitrogen present in organisms. They are polymers (polypeptides) of **α-amino acids** (i.e. the amino (NH_2) and carboxylic acid (COOH) groups are attached to the same carbon atom). In the general structure in Fig. 2.11 the α-carbon atom can be seen to be a stereogenic centre for all amino acids except glycine (in which the C atom is bonded to two H atoms and so cannot be stereogenic). This reflects the stereospecificity of enzymes involved in the formation (and utilization) of these compounds; in fact, proteins contain only L-amino acids. However, some D-amino acids (e.g. D-alanine and D-glutamic acid) occur naturally, e.g. in the peptidoglycans of bacterial cell walls. The existence of acid and amine groups in the same molecule leads to zwitterion formation, i.e. the transfer of the acidic proton to the amine group. The zwitterion

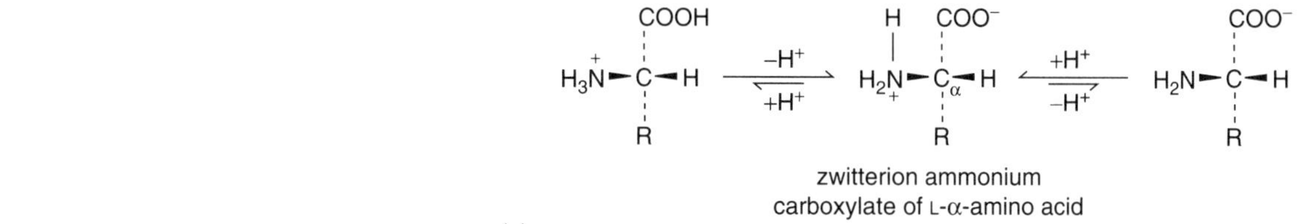

Chemical composition of organic matter | 41

(a)

$$H_3\overset{+}{N}-\underset{R}{\overset{COOH}{\underset{|}{C}}}-H \underset{+H^+}{\overset{-H^+}{\rightleftharpoons}} H_2\overset{+}{N}-\underset{R}{\overset{H\ \ COO^-}{\underset{|}{C_\alpha}}}-H \underset{-H^+}{\overset{+H^+}{\rightleftharpoons}} H_2N-\underset{R}{\overset{COO^-}{\underset{|}{C}}}-H$$

zwitterion ammonium
carboxylate of L-α-amino acid

(b)

R group	name	R group	name
—H	glycine	—CH$_2$CNH$_2$ (with C=O)	asparagine
—CH$_3$	alanine	—CH$_2$CH$_2$CNH$_2$ (with C=O)	glutamine
—CH(CH$_3$)$_2$	valine	—(CH$_2$)$_4$NH$_2$	lysine
—CH$_2$CH(CH$_3$)$_2$	leucine	—(CH$_2$)$_3$NHCNH$_2$ (with C=NH)	arginine
—CHCH$_2$CH$_3$ | CH$_3$	isoleucine	—CH$_2$–(indole)	tryptophane
—CH$_2$–(phenyl)	phenylalanine		
—CH$_2$OH	serine		
—CHOH | CH$_3$	threonine	—CH$_2$–(imidazole)	histidine
—CH$_2$–(phenyl)–OH	tyrosine		
—CH$_2$OOH	aspartic acid	—CH$_2$SH	cysteine
—CH$_2$CH$_2$OOH	glutamic acid	—CH$_2$CH$_2$SCH$_3$	methionine

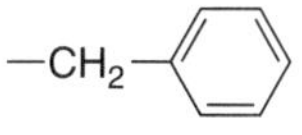

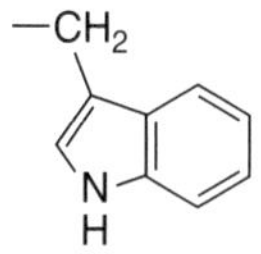

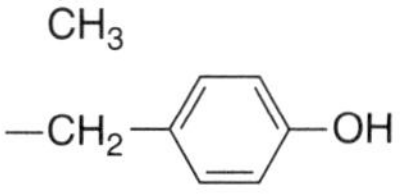

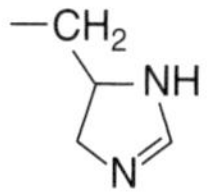

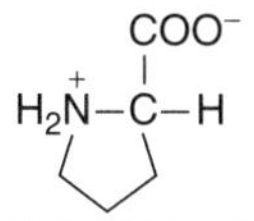

proline: $H_2\overset{+}{N}-\overset{COO^-}{\underset{|}{C}}-H$ (in pyrrolidine ring)

Fig. 2.11 General structure of L-α-amino acids and (b) important members of this compound class. (R refers to group in general structure, but for proline the complete structure is shown.)

ammonium carboxylate is the predominant form at physiological pH values (4–9), as shown in Fig. 2.11a.

The different types of amino acids can be broadly classified as neutral, acidic or basic (see Box 2.5), depending on whether the number of carboxylic acid groups in a molecule equals, is greater than or is less than the number of amine groups respectively. Proteins are made up from some 20 different amino acids, shown in Fig 2.11b. Sulphur is an important component in some amino acids (e.g. cysteine). In plants amino acids are generally synthesized from glutamic acid (Fig. 2.11b) by transfer of the amino group to other carbon skeletons (**transamination**). Animals cannot synthesize all the amino acids they need for protein formation and so must obtain these essential amino acids (there are nine for humans) directly or indirectly from plants.

An acid group on one amino acid molecule can undergo a condensation reaction with an amine group on another, with the elimination of water. The resulting amide group (see Box 2.5) that joins the two amino acids together to form a dipeptide is usually called the **peptide linkage**:

Box 2.5 | Acids and bases

In the context of organic geochemistry the most appropriate definition of an **acid** is a proton donor, i.e. it contains a hydrogen atom that can be transferred to another chemical species as H⁺ (a proton, which is a hydrogen atom less an electron). In addition to traditional inorganic acids (e.g. H_2SO_4), certain organic compounds contain groups that possess acidic hydrogens, such as the –COOH group of carboxylic acids. **Bases** are compounds that can accept a proton, and include amines. This behaviour of acids and bases is demonstrated in eqns 2.1 and 2.2:

$$CH_3COOH \rightarrow CH_3COO^- + H^+ \quad \text{[Eqn 2.1]}$$
$$\text{acid} \quad\quad\quad \text{base}$$

$$C_2H_5NH_2 + H^+ \rightarrow C_2H_5NH_3^+ \quad \text{[Eqn 2.2]}$$
$$\text{base} \quad\quad\quad\quad \text{acid}$$

A carboxylic acid can react with an amine to form an amide, and water is liberated:

$$CH_3COOH + C_2H_5NH_2 \rightarrow CH_3CONHC_2H_5 + H_2O \quad \text{[Eqn 2.3]}$$
$$\text{acetic acid} \quad \text{ethylamine} \quad\quad \text{N-ethylacetamide} \quad \text{water}$$

An important reaction in natural product chemistry is that of organic acids with alcohols to form esters, like ethyl acetate, with the liberation of water:

$$CH_3COOH + C_2H_5OH \rightarrow CH_3COOC_2H_5 + H_2O \quad \text{[Eqn 2.4]}$$
$$\text{acetic acid} \quad \text{ethanol} \quad\quad \text{ethyl acetate} \quad \text{water}$$

The reverse of the reactions in Eqns 2.3 and 2.4 is achieved by **hydrolysis**, so named because it involves the splitting of the molecule with the addition of water. Hydrolysis reactions can be speeded up (catalysed, see Box 2.6) by the presence of an acid or base (e.g. the alkalis NaOH and KOH). The alkaline hydrolysis of fatty esters is often termed **saponification**.

$$\begin{array}{c} R \ \ O \\ | \ \ \ || \\ HN-C-C-OH \\ | \ \ | \\ H \ \ H \end{array} + \begin{array}{c} R \ \ O \\ | \ \ \ || \\ H-N-C-C-OH \\ | \ \ | \\ H \ \ H \end{array} \rightarrow \begin{array}{c} R \ \ O \ \ \ \ \ \ R \ \ O \\ | \ \ \ || \ \ \ \ \ \ \ \ | \ \ \ || \\ HN-C-C-N-C-C-OH \\ | \ \ | \ \ | \ \ | \\ H \ \ H \ \ H \ \ H \\ \text{peptide} \\ \text{linkage} \end{array} + H_2O \quad \text{[Eqn 2.5]}$$

Condensation reactions can continue until large molecules, **polypeptides** (which can also be called polyamides), have been built up. Proteins are large polypeptides and can contain >8000 amino acid units with molecular weights >10^6. In such large molecules the types of amino acids incorporated and their order allow a multitude of different structural possibilities. The overall shape (conformation) of a protein molecule is a key factor in its biochemical function and is governed by the sequence of amino acids (primary structure), the rigidity of the amide bond, hydrogen bonding (see Section 2.1.2) and the formation of S–S bonds between the SH groups of cysteine residues.

2.3.2 Occurrence and function

Proteins comprise the greater part of the nitrogen-containing organic material in organisms, although lesser amounts of free amino acids and peptides are found. In fact, proteins can be a sizable fraction of the bulk organic material in an organism. The polypeptide chains often fold into regularly repeating structures (secondary structure), especially when the primary structure (the amino acid sequence) is dominated by a few amino acids or has a repetitive sequence. A particularly important secondary structure is the α-helix, which is stabilized by hydrogen bonding (e.g. Stryer 1988). Several helical strands can intertwine to form bundles of fibres, which are suitable for structural roles in organisms because the multiple intermolecular forces make the protein insoluble. These fibrous proteins serve as supportive tissues in animals, e.g. in skin and bone (collagen), hooves and claws (keratin), silk and sponge. This is in contrast to plants, where cellulose and lignin (see Section 2.5.1) perform the structural role, although they can be associated with a collagen-like protein. There are also globular proteins, which also contain α-helices, but they are folded up into compact globules. They generally perform important regulatory functions and include enzymes (the biochemical catalysts; see Box 2.6), hormones (e.g. insulin, which regulates metabolism), antibodies (which are glycoproteins, i.e. they contain carbohydrates) and transport and storage units (e.g. haemoglobin for oxygen transfer and cytochromes for electron transfer).

2.4 Lipids

Lipids can be defined as all the substances produced by organisms that are effectively insoluble in water but extractable by solvents that dissolve fats (e.g. chloroform, hexane, toluene and acetone). This broad definition is suitable for our purposes and encompasses a wide variety of compound classes, including photosynthetic

---

Box 2.6 | Catalysis

A **catalyst** is a substance that increases the rate of a reaction but is not itself consumed by the reaction. We can represent the energy changes taking place during the course of a simple reaction by Fig. 2.12. Two substances that are to react must be brought close together and certain bonds need to be weakened to allow the transfer of atoms and formation of new bonds that lead to the generation of products. This requires energy to be supplied. As we move along the reaction path in Fig. 2.12 we reach a maximum in the energy curve at some intermediate state; thereafter energy decreases as products are formed and move apart. Most reactions result in the overall evolution of excess energy (ΔH) as heat, as in Fig. 2.12. The minimum amount of energy needed for the reaction to overcome the high-energy intermediate state, the **activation energy** (E_{act}), can be reduced by a catalyst. The catalyst partially bonds to the reactants, weakening one or more of the bonds that have to be broken, which means less external energy is required. When the products are formed the catalyst reverts to its original state. The overall energy change in the reaction (ΔH) is not affected by the action of the catalyst. Although there is no net energy gain, the benefits of a catalyst are that the reaction can proceed faster and less initial energy (e.g. in the form of ATP in cells) is required to supply the necessary activation energy.

Catalysts are used throughout the chemical industry and **enzymes** are Nature's catalysts. In addition to reducing the energy needed to break bonds, enzymes also act as templates to hold the reactants in the best orientation for the desired reaction to occur, often at a specific site in a molecule (**regiospecific**), or to produce a specific configuration (**stereoselective**). Mineral surfaces (e.g. clays) can also behave as catalysts in geochemical reactions.

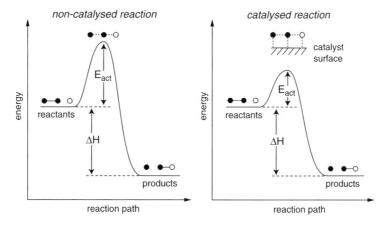

Fig. 2.12 Effect of a catalyst on the energetics of a reaction.

pigments. However, application of the term lipid can vary, sometimes being restricted to fats, waxes, steroids and phospholipids, and sometimes to fats alone. Simple organic compounds like aliphatic carboxylic acids and alcohols can be found among the lipids, but most lipids exist as combinations of these simple molecules with one another (e.g. wax esters, triglycerides, steryl esters and phospholipids) or with other compound classes such as carbohydrates (glycolipids) and proteins (lipoproteins). We review the quantitatively and geochemically most important classes of these lipids.

2.4.1 Glycerides

Glycerides are esters of the alcohol glycerol. A glycerol molecule contains three hydroxyl groups and so it can react with one, two or three carboxylic acid molecules, forming mono-, di- and triglycerides, respectively. Among the important types of glycerides are the fats and phospholipids.

(a) Fats

Fats are triglycerides, formed from straight-chain, aliphatic carboxylic acids, called **fatty acids**. Each of the fatty acids in a triglyceride molecule can be different:

$$\begin{array}{l} H_2C-OH \\ HC-OH \\ H_2C-OH \\ \text{glycerol} \end{array} + \begin{array}{l} RCOOH \\ R'COOH \\ R''COOH \\ \text{fatty acids} \end{array} \rightarrow \begin{array}{l} H_2C-OOCR \\ HC-OOCR' \\ H_2C-OOCR'' \\ \text{triglyceride} \end{array} + 3H_2O \quad \text{[Eqn 2.6]}$$

The fatty acids are typically of C_{12} to C_{36} chain length and in animals they are predominantly saturated (called alkanoic acids), whereas in plants more unsaturated (alkenoic acids) and polyunsaturated acids are present.

For the same chain length an unsaturated acid has a lower melting point than a saturated fatty acid and so, at ambient temperatures, plant-derived fats are often oils, while animal fats are solids. In animals C_{16} and C_{18} saturated fatty acids predominate, whereas the major fatty acids in plants are the C_{18} mono-, di- and triunsaturated forms. Polyunsaturated fatty acids are more common in algae than higher plants. The number of double bonds and their geometric configuration are important factors in the function of these compounds. Most unsaturated fatty acids adopt the *cis* configuration (Fig. 2.2b). The common and systematic names and the structures of some fatty acids are given in Table 2.4.

As can be seen from Table 2.4, systematic names and formulae can become cumbersome, while trivial names require memorizing. These problems can be overcome by the use of shorthand notations for fatty acids, and two of the more common are described in Box 2.7.

Fatty acids have predominantly even numbers of carbon atoms because they are effectively formed from acetyl (C_2) units, which are derived from glucose in the presence of various enzymes, coenzymes and carrier proteins. An overall scheme for saturated fatty acid biosynthesis is presented in Fig. 2.13, in which it can be seen that the first step is the formation of acetyl coenzyme A (often abbreviated to acetyl-CoA). One molecule of acetyl-CoA undergoes addition of CO_2 to form malonyl-CoA, while the acetyl group on another molecule is transferred to an enzyme (fatty acid synthase). The malonyl unit (C_3) is added to the enzyme-bound acetyl unit, which produces a butyryl group following loss of CO_2, dehydration and reduction. Six further steps of combined malonyl addition, decarboxylation, dehydration and reduction occur to yield palmitate (C_{16}). Higher acids are built from palmitate in a similar

Table 2.4 Common fatty acids

common name	systematic name	structure
lauric acid	dodecanoic acid	$CH_3(CH_2)_{10}COOH$
myristic acid	tetradecanoic acid	$CH_3(CH_2)_{12}COOH$
palmitic acid	hexadecanoic acid	$CH_3(CH_2)_{14}COOH$
stearic acid	octadecanoic acid	$CH_3(CH_2)_{16}COOH$
arachidic acid	eicosanoic acid	$CH_3(CH_2)_{18}COOH$
palmitoleic acid	hexadec-9-enoic acid	$CH_3(CH_2)_5CH=CH(CH_2)_7COOH$
oleic acid	octadec-9-enoic acid	$CH_3(CH_2)_7CH=CH(CH_2)_7COOH$
linoleic acid	octadec-9,12-dienoic acid	$CH_3(CH_2)_4CH=CHCH_2CH=CH(CH_2)_7COOH$
linolenic acid	octadec-9,12,15-trienoic acid	$CH_3CH_2CH=CHCH_2CH=CHCH_2CH=CH(CH_2)_7COOH$
arachidonic acid	eicos-5,8,11,14-tetraenoic acid	$CH_3(CH_2)_4CH=CHCH_2CH=CHCH_2CH=CHCH_2CH=CH(CH_2)_3COOH$

Box 2.7 | Simple notation schemes for fatty acids

The important attributes of a fatty acid are its carbon chain length, the number of double bonds present and their positions, which can be represented by a simple notation scheme. For example, oleic acid can be represented by *cis*-18:1ω9, where *cis* refers to the stereochemistry about the C=C bond (see Fig. 2.2b and Box 2.3), 18 is the number of C atoms, the number of double bonds (1) is given after the colon, and the number following ω is the position of the double bond from the opposite end to the acid group. As double bonds in polyunsaturated acids are usually conjugated (see Section 2.1.2), it is only necessary to give the position of the first double bond because all others follow on alternate carbon atoms. Hence arachidonic acid is 20:4ω6, in which the first C=C bond occurs between C-6 and C-7, numbering from the opposite end to the acid group, and the other three C=C bonds are between C-8 and C-9, C-10 and C-11, and C-12 and C-13.

An alternative convention is also often used in which the position of the first double bond is numbered from the end of the molecule bearing the functional group, the ω symbol being replaced by Δ. Thus vaccenic acid, which is most commonly found as the *trans* isomer, can be represented as *trans*-18:1ω7 or *trans*-18:1Δ11. The alternative *E/Z* system for labelling *cis/trans* isomers (see Box 2.3) is not commonly applied to fatty acids.

---

way but using different enzymes. Enzymatic desaturation of these acids (i.e. dehydrogenation) can occur, resulting in unsaturated products.

Fats are used as energy stores by animals and plants. On a weight-for-weight basis, during complete oxidation fats liberate just over twice as much energy as carbohydrates because they contain less oxygen to start with. Fats, therefore, are particularly useful where a compact energy source is required (e.g. in seeds and fruits). Aerobic respiration of fats proceeds by hydrolysis of

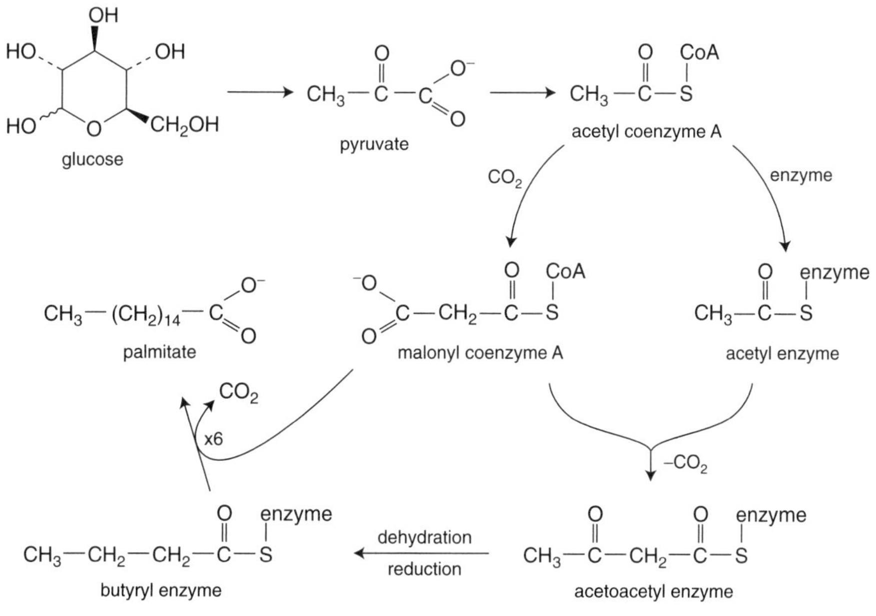

Fig. 2.13 Biosynthesis of saturated fatty acids in plants and animals. Palmitate is formed by successive additions of malonyl coenzyme A to the enzyme-bound chain, with CO_2 being lost at each addition. This results in chain elongation by a $(CH_2)_2$ unit at each step. Details of the formation of butyryl (C_4) from acetyl (C_2) are shown, while the subsequent six further additions, terminating in palmitate, proceed similarly.

triglycerides to release fatty acids, which then undergo successive loss of C_2 units, which in turn, in the form of acetyl-CoA, are oxidized to CO_2 in the citric acid cycle (Fig. 2.10) with the release of energy as ATP. The glycerol part of fats also contributes to energy generation in the citric acid cycle after conversion to pyruvic acid (Fig. 2.10). This mechanism of fatty acid oxidation is termed **β-oxidation** and occurs in all aerobes. However, aerobic oxidation can also occur via enzymatic removal of single carbon atoms from fatty acids, termed **α-oxidation**. It operates in bacteria, and in plants it is often the more important of the two oxidation pathways.

(b) Phospholipids, glycolipids and ether lipids

All these lipids are important constituents of the membranes that isolate the contents of cells from the surrounding environment in various organisms. **Cell membranes** (or plasma membranes) chiefly comprise lipids and proteins, with phospholipids being the main type of membrane lipid (up to 65% in plants; e.g. Finean et al. 1984).

Phospholipids (or phosphatides) are triglycerides containing one phosphoric acid and two fatty acid units. The phosphate group is often linked to a nitrogen base, such as choline in the phospholipid lecithin, which is found in all animals and plants (Fig. 2.14a). The phospholipids are arranged in a bilayer, with the nonpolar (hydrophobic) alkyl chains of the fatty acids directed towards the interior of the bilayer and the polar (hydrophilic) phosphate ends lying on each surface of the membrane (Fig. 2.14b). Cell membrane proteins are of two types: integral, which bridge the membrane and are involved in transfer processes across it; and peripheral, which are located on a surface of the membrane and are bonded to integral proteins (Fig. 2.14b).

The term **glycolipid** can be applied to any compound in which lipids and carbohydrates are combined.

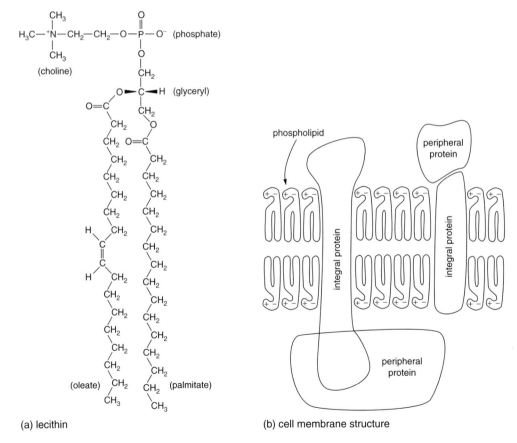

Fig. 2.14 (a) Molecular structure of a phospholipid, lecithin (n.b. adopted stereochemistry of glyceryl unit), and (b) phospholipid arrangement in cell membranes.

Glycolipids in which the phosphate group of a phospholipid has been replaced by a sugar are found in plant cell membranes (accounting for up to 20% of membrane lipids) and are important components in the cell membranes of Gram-positive bacteria. They are also major components of the membranes surrounding chloroplasts (the photosynthesizing organelles) in higher plants, algae and cyanobacteria (e.g. diacylgalactosylglycerol, Fig. 2.15a). Teichoic acids are another type of glycolipid and are major components in the cell membranes and walls of Gram-positive bacteria. Glycerol teichoic acids (polymers of glycerol phosphate in which the OH groups of the glycerol units are substituted by D-glucose, D-alanine and fatty acids; Fig. 2.15a) are present in both cell walls and membranes, while ribitol teichoic acids (in which the open-chain form of ribose replaces glycerol; Fig. 2.15a) are found in cell walls only, where they are bonded to murein (see Section 2.2.2). In contrast, the outer membrane of Gram-negative bacteria is composed of lipopolysaccharides, comprising polysaccharide chains linked to glycolipids containing sugar, fatty acid and phosphate units.

Ether lipids are glycerides, but instead of being formed from fatty acids they are formed from fatty alcohols (*n*-alkanols), leading to ether rather than ester linkages (see Table 2.1). These lipids can contain phosphate and sulphate units and also sugar residues. The cell membranes of anaerobic eubacteria contain large amounts of plasmalogens, which have both ester and ether linkages (Fig. 2.15a). Archaebacterial cell membranes differ from those of other organisms in being formed from ether lipids containing only phytanyl chains (i.e. formed from the saturated counterpart of the alcohol phytol; Fig. 2.17c). These glycerol ether lipids comprise diphytanyl diethers (the diphytanylglycerol unit shown in Fig. 2.15a is also known as archaeol) and biphytanyl tetraethers, in which the biphytanyl groups (formed from linking two phytyl units tail-to-tail) link two glyceride units (Fig. 2.15b; e.g. Bu'lock et al. 1981). The cell membranes of halophiles and methanogens contain mixed di- and tetraethers, whereas those of thermoacidophiles contain mainly tetraethers (including some with varying numbers of cyclopentanyl groups within the biphytanyl chains). One species of methanogen has been found to synthesize a cyclic biphytanyl diether (Fig. 2.15b). Archaebacterial cell walls lack the murein (Fig. 2.9) that provides rigidity and strength in eubacterial cell walls. However, these functions may be performed in archaebacteria by the biphytanyl tetraethers that span the cell membrane (Fig. 2.15b), locking the two halves of the lipid bilayer together, providing enhanced rigidity and strength to the membrane.

2.4.2 Waxes and related compounds

(a) Waxes

Waxes mainly function as protective coatings, such as those found on leaf cuticles. They are mixtures of many constituents with high melting points, important members being esters of fatty acids with straight-chain saturated alcohols (fatty alcohols). The fatty acids and alcohols in these **wax esters** have similar chain lengths, mainly in the range C_{24} to C_{28}. They have predominantly an even number of carbon atoms because the alcohols are biosynthesized from fatty acids by enzymatic reduction (Eqn 2.7). Lesser amounts of ketones, branched alkanes and aldehydes are present.

$$CH_3(CH_2)_nCH_2COOH \underset{acid}{} \rightarrow CH_3(CH_2)_nCH_2CHO \underset{aldehyde}{}$$
$$\rightarrow CH_3(CH_2)_nCH_2CH_2OH \underset{alcohol}{} \quad \text{[Eqn 2.7]}$$

(where n = odd number)

Plant waxes also contain hydrocarbons, mainly long-chain *n*-alkanes. In contrast to the fatty acids and alcohols, the *n*-alkanes contain chiefly odd numbers of C atoms, mostly in the range C_{23} to C_{35}, with C_{27}, C_{29} and C_{31} generally predominating. This results from biosynthesis of the alkanes from acids by enzymatic decarboxylation:

$$CH_3(CH_2)_nCH_2COOH \underset{acid}{} \rightarrow CH_3(CH_2)_nCH_3 \underset{alkane}{} + CO_2 \quad \text{[Eqn 2.8]}$$

(where n = odd number)

Fungal cell walls contain *n*-alkanes similar to those in the higher plants and possibly fulfilling a similar role. In contrast, waxes are absent in most bacteria. However, the mycobacteria and the related nocardiae and corynebacteria, which contain greater amounts of lipids than other types of bacteria, contain high-molecular-weight waxy molecules in their cell walls (Harwood & Russell 1984). These molecules comprise various mycolic acids, such as β-mycolic acid (Fig. 2.15a), which are bonded to polysaccharides and, via phosphate groups, to murein (Fig. 2.9).

A particular type of wax ester, a **steryl ester**, is produced where the alcohol unit to which the fatty acid is joined is a sterol. These are also waxy solids, an example being lanosteryl palmitate in lanolin, which is found on sheep's wool.

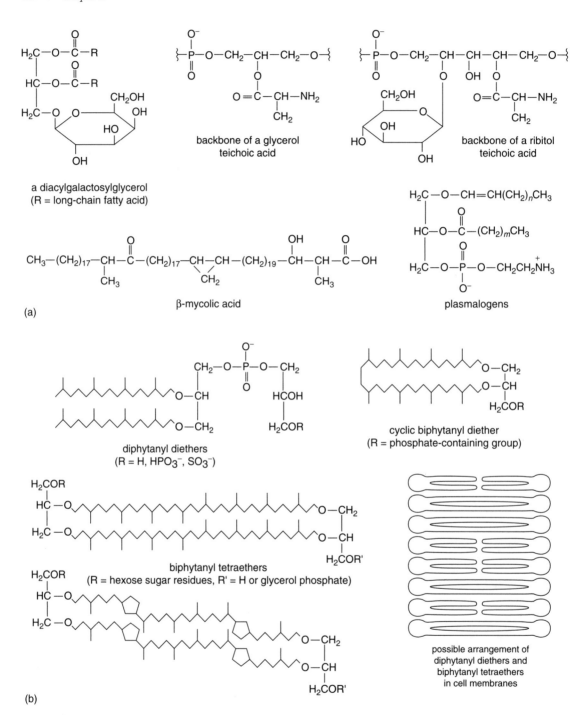

Fig. 2.15 (a) Examples of lipids in the membranes and cell walls of eubacteria. (b) Phytanyl ether lipids in archaebacterial cell membranes.

(b) Cutin, suberin and related polymers

Related to the plant waxes are **cutin** and **suberin**, which also form protective coatings for plant tissue. They are polymerized and cross-linked structures of hydroxy fatty acids, are resistant to oxidation and to microbial and enzymatic attack, and help to prevent excessive evaporative loss of water. Cutin is mainly a polymer of hydroxy (often dihydroxy) fatty acids and forms a layer on the outer surface of exposed plant tissue (i.e. the cuticle). Suberin is mainly associated with underground parts of plants and with protecting wounds. Its constituents are mainly α,ω-diacids (i.e. fatty acids with a COOH group at each end) and ω-hydroxy fatty acids (i.e. an OH group at the opposite end of the molecule to the COOH group). The units from which cutin and suberin are composed chiefly contain an even number of carbon atoms in the C_{16} to C_{26} range, with C_{16} and C_{18} dominating (Holloway 1982).

In addition to the polyester cutin, plant cuticles may contain minor amounts of a highly aliphatic polymer, lacking ester linkages, formed predominantly from chains of CH_2 units. This polymethylenic material may be cross-linked by ether bonds and has been termed **cutan**. The equivalent counterpart of suberin is termed **suberan**. The related chemically resistant material in algae is called **algaenan**. All these polymethylenic materials are discussed in more detail in Section 4.4.1a.

2.4.3 Terpenoids

Terpenoids are a class of lipids displaying a great diversity of structures and functions, ranging from the small, volatile molecules in sex pheromones to the very large molecules in natural rubber. There is, however, a unifying theme, in that they are all formally constructed from C_5 **isoprene** units, the number of which can be used to classify terpenoids (Table 2.5).

Most naturally occurring terpenoids contain oxygen, commonly in alcohol, aldehyde, ketone and carboxylic acid groups. The suffix '-ene' is sometimes used to denote a whole class of compounds or just the alkenes within a class. In order to avoid such confusion we shall apply the general suffix '-oid' to all classes of compounds within a group (e.g. diterpenoids), '-ane' to denote alkanes within a group (e.g. diterpanes) and '-ene' to specify alkenes (e.g. diterpenes). As seen below, many terpenoids form cyclic systems but there are also non-cyclic terpenoids, which are often referred to as **acyclic isoprenoids**.

Terpenoids do not necessarily contain exact multiples of five carbons and allowance has to be made for the loss or addition of one or more fragments and possible molecular rearrangements during biosynthesis. In reality the terpenoids are biosynthesized from acetate units derived from the primary metabolism of fatty acids, carbohydrates and some amino acids (see Fig. 2.10). Acetate has been shown to be the sole primary precursor of the terpenoid cholesterol. The major route for terpenoid biosynthesis, the mevalonate pathway, is summarized in Fig. 2.16. Acetyl-CoA is involved in the generation of the C_6 mevalonate unit, a process that involves reduction by NADPH. Subsequent decarboxylation during phosphorylation (i.e. addition of phosphate) in the presence of ATP yields the fundamental isoprenoid unit, isopentenyl pyrophosphate (IPP), from which the terpenoids are synthesized by enzymatic condensation reactions. Recently, an alternative pathway has been discovered for the formation of IPP in various eubacteria and plants, which involves the condensation of glyceraldehyde 3-phosphate and pyruvate to form the intermediate 1-deoxy-D-xylulose 5-phosphate (Fig. 2.16; e.g. Eisenreich et al. 1998). We consider some of the more common examples of the main classes of terpenoids below.

(a) Monoterpenoids

Monoterpenoids, nominally containing two isoprene units, are particularly abundant in higher plants and algae. Because of their volatility they are important as attractants (e.g. insect pheromones), but are probably best known as components of the essential oils of plants (e.g. menthol in peppermint oil). Esters of chrysanthemic acid, found in pyrethrum flower heads, are natural insecticides. Some examples are shown in Fig. 2.17a.

(b) Sesquiterpenoids

Some sesquiterpenoids function as essential oils in plants, while others act as fungal antibiotics. The acyclic compound farnesol is widely distributed in nature, being found in many plants and in the chlorophyll of some bacteria (see Section 2.4.4). Mono- and dicyclic sesquiterpenoids are common in plants. Some examples are given in Fig. 2.17b.

Table 2.5 Terpenoid classification

classification	number of isoprene units
monoterpenoids	2
sesquiterpenoids	3
diterpenoids	4
sesterterpenoids	5
triterpenoids	6
tetraterpenoids	8
polyterpenoids	>8

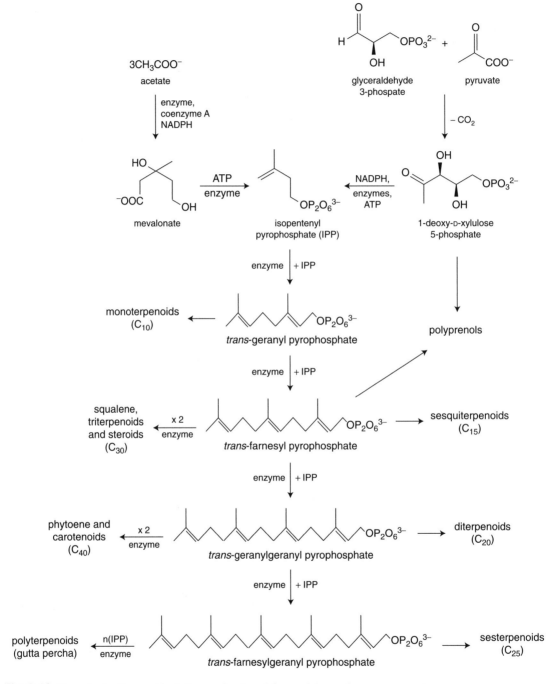

Fig. 2.16 Biosynthesis of terpenoids via the mevalonate and deoxyxylulose pathways.

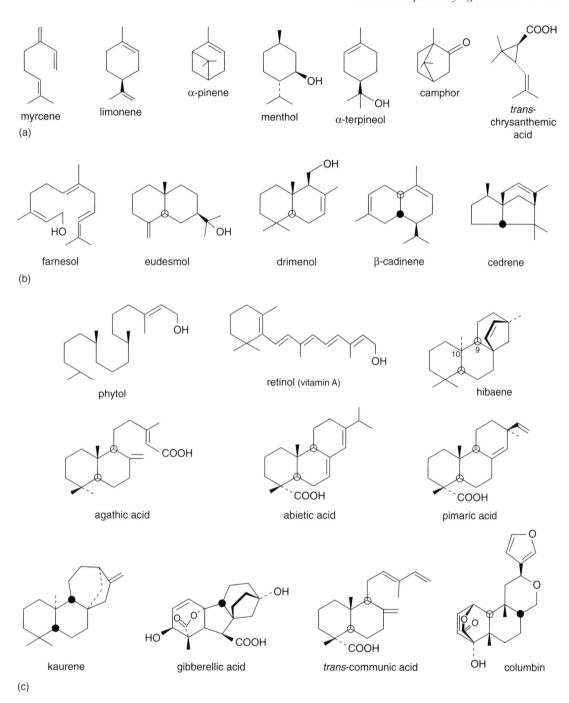

Fig. 2.17 Examples of (a) monoterpenoids, (b) sesquiterpenoids and (c) diterpenoids.

(c) Diterpenoids

The most important acyclic diterpenoid is **phytol**. It forms part of the chlorophyll-*a* molecule and is present in many other chlorophylls (see Section 2.4.4). The saturated analogue of phytol, dihydrophytol (or phytanol), is present in a variety of bacterial glyceride ether lipids (Section 2.4.1b).

Most diterpenoids are di- and tricyclic compounds, and are especially common in higher plants. In particular, the resins of gymnosperms (e.g. conifers) are characterized by diterpenoids such as agathic, abietic, communic and pimaric acids, and alkenes, such as kaurene and hibaene. Most of these resin diterpenoids exist as isomers differing in configuration at a stereogenic centre or in C=C bond position, which gives rise to a large number of components. They act as generally protective agents, sealing wounds and discouraging insect and animal attack. Other diterpenoids include compounds that give the bitter taste in plants (bitter principles, e.g. columbin). Important, but quantitatively minor, constituents of plants are the growth-regulating gibberellins (e.g. gibberellic acid). Some representative structures are shown in Fig. 2.17c.

(d) Sesterterpenoids

Sesterterpenoids are not well represented among the geochemically important terpenoids, other than the highly branched alkenes found in planktonic (Belt et al. 2001) and benthonic (Volkman et al. 1994) diatoms. These C_{25} acyclic alkenes have been found to contain three, four or five double bonds, some examples of which are shown in Fig. 2.18.

(e) Triterpenoids

All triterpenoids appear to be derived from the acyclic isoprenoid squalene ($C_{30}H_{50}$, see Fig. 2.21), which is a ubiquitous component in organisms (e.g. shark oil, vegetable oils, fungi). Most triterpenoids are either pentacyclic or tetracyclic. The latter belong primarily to the important class of compounds, the **steroids**, which are considered separately. Most C_{30} pentacyclic triterpenoids with a six-membered E ring (see Box 2.8) are of higher plant origin, commonly occurring as resin constituents. Three major series can be distinguished among these higher plant triterpenoids: the oleanoid (e.g. β-amyrin), ursanoid (e.g. α-amyrin) and lupanoid (e.g. lupeol) series (Fig. 2.19). In contrast, the **hopanoids** have a five-membered E ring and are often called the bacteriohopanoids because they are common components of cell membranes in eubacteria, particularly diploptene, diplopterol, bacteriohopanetetrol and aminobacteriohopanetriol (Ourisson et al. 1979, 1987; Rohmer et al. 1992; Kannenberg & Poralla 1999). However, as can be seen from Fig. 2.19, this distinction between sources based on E-ring size is not absolute; for example, lupeol is synthesized by angiosperms and diploptene is found in ferns as well as bacteria. The C_{35}

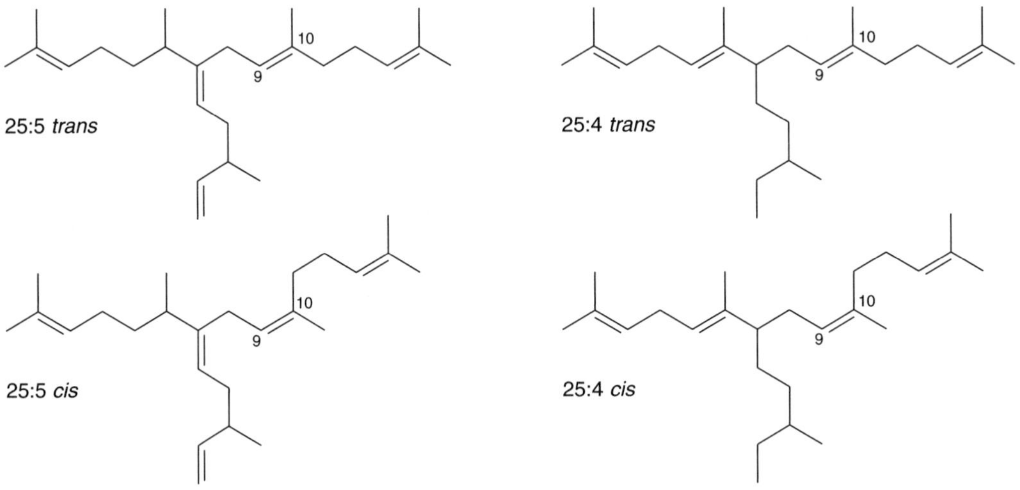

Fig. 2.18 Examples of highly branched sesterterpenes isolated from planktonic diatom cultures (*trans* and *cis* refer to configuration at double bond between C-9 and C-10; after Belt et al. 2001).

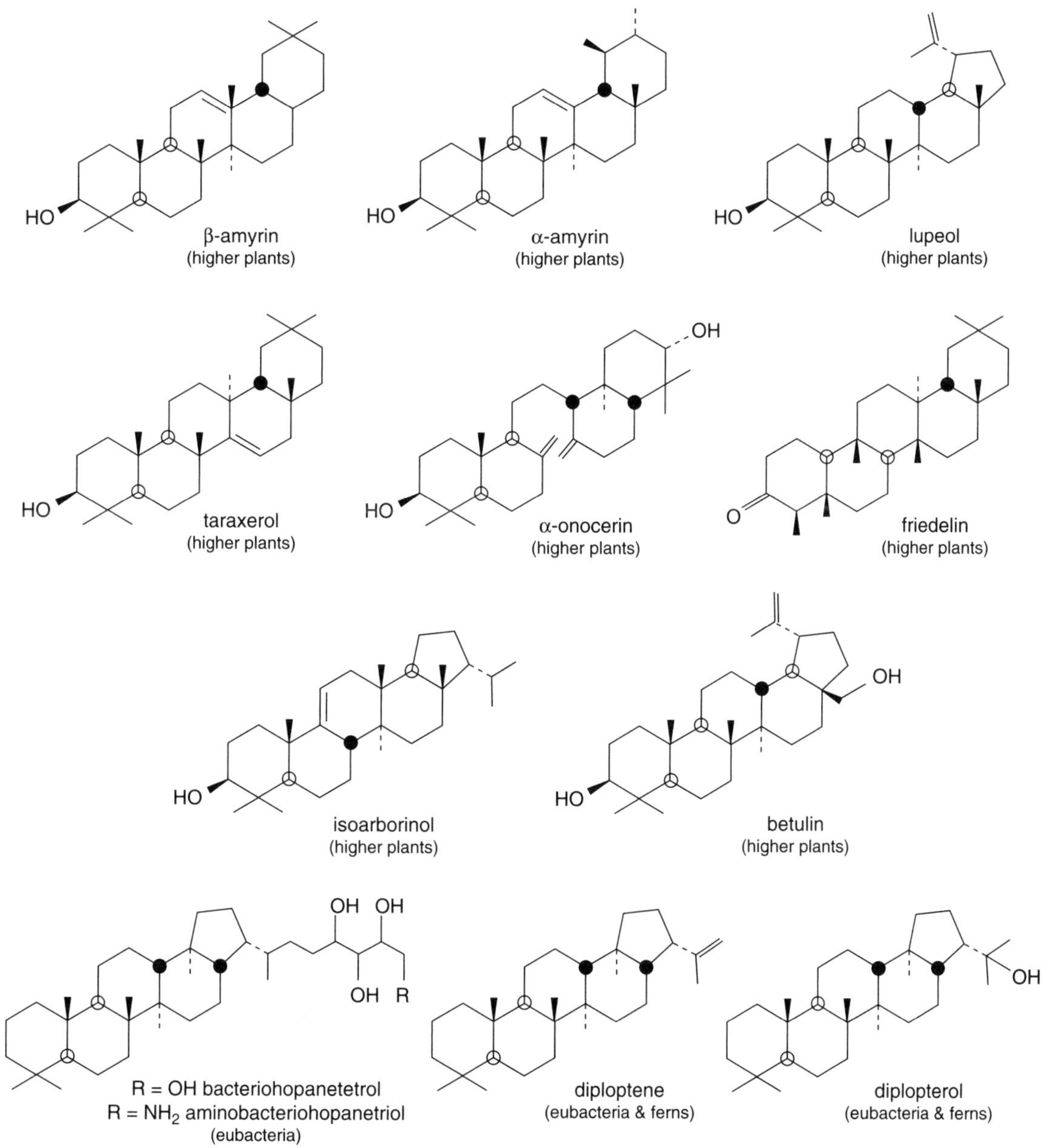

Fig. 2.19 Some geochemically important polycyclic triterpenoids and their major sources.

bacteriohopanepolyols are biosynthesized by addition of ribose (a C_5 sugar) to diplopterol (Rohmer et al. 1989).

(f) Steroids

Formation of steroids results from enzymatic oxidation of squalene followed by cyclization. This produces either cycloartenol, the precursor of most plant steroids, or lanosterol, the precursor of animal and fungal steroids, and also some plant steroids (Fig. 2.21). Enzymatic oxidation and decarboxylation converts lanosterol (C_{30}) into cholesterol (C_{27}), the precursor of all other animal steroids. Most of the cholesterol in animals and the related sterols in plants is bound in cell

Box 2.8 | Carbon numbering sequence for steroids and triterpenoids

The method of identifying particular carbon atoms in these polycyclic compounds is by a systematic numbering sequence. Individual rings are also denoted alphabetically. The conventions for steroids and hopanoids are shown in Fig. 2.20a. Numbering for higher plant triterpenoids follows a similar sequence to that of hopanoids, with C-22 being incorporated in the six-membered E ring, and C-29 and C-30 being the substituents on the E ring at C-19 and/or C-20.

As in carbohydrates the six-membered rings adopt predominantly the chair conformation and five-membered rings the envelope conformation (Templeton 1969; see Section 2.1.3). Ring junctions are generally *trans*, in which methyl and hydrogen atoms at ring junctions occupy axial positions as

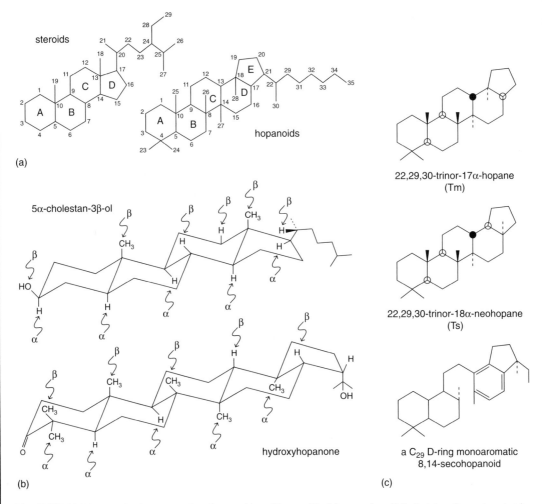

Fig. 2.20 (a) Ring numbering conventions for steroids and hopanoids; (b) examples of 'all-chair' conformations (with *trans* ring junctions) for steroids and hopanoids; (c) examples of application of hopanoidal nomenclature system. (Ts may also be called 17-methyl-18,22,29,30-tetranor-17α,18α-hopane.)

Continued

shown in Fig. 2.20b for 5α-cholestan-3β-ol and hydroxyhopanone. Where there is no ambiguity about the substituent to which an α or β label refers, it is strictly unnecessary to denote that group. For example, the 5α of 5α-cholestan-3β-ol clearly refers to the single substituent (an H atom) at the C-5 ring junction (Fig. 2.20b), and there is no need to write 5α(H)-cholestan-3β-ol. A shorthand notation is sometimes used for the position of C=C bonds in steroids and triterpenoids: a Δ followed, in superscript, by the position numbers of the carbon atoms involved, as described above.

Steroids can be formally named using the basic C_{27} cholestane structure (i.e. the sterane formed by reduction of cholesterol), e.g. cholesterol can be called cholest-5-en-3β-ol. The position of the first (lowest numbered) carbon atom involved in the double bond (denoted by 'en') and the C atom bonded to the hydroxy group are indicated. The second (higher numbered) carbon atom of a C=C bond is usually only given (in parentheses after the first number) where more than one possibility exists. Where alkyl groups additional to the cholestane skeleton are present the position of the carbons to which they are attached is noted, and the configuration at various positions (where the stereochemistry is known) can be given, e.g. stigmasterol can be called (22E)-24β-ethylcholesta-5,22-dien-3β-ol (see Box 2.3 for explanation of E configuration). The absence of a methyl group from the basic cholestane structure is indicated by the position number of the missing carbon atom followed by 'nor'. Diasteroids share the same numbering scheme as the regular steroids, with the 'rearranged' methyl groups C-10 and C-19 being attached to the cyclic system at C-14 and C-5 respectively, rather than at C-13 and C-10.

In hopanoids, nomenclature is based on the C_{30} compound hopane. The presence of additional carbon atoms in the alkyl chain at C-21 is indicated by their position number followed by 'homo', e.g. the C_{35} hopanoidal alkane in Fig. 2.20a is called 31,32,33,34,35-pentahomohopane. However, additional methyl groups elsewhere are often denoted as substituents by their attachment position (as for steroids above, e.g. 3-methylhopane). As with steroids, the absence of carbon atoms is indicated by the prefix 'nor' with the relevant position numbers. For example, the absence of the alkyl chain attached to the E ring of hopane results in the C_{27} compound 22,29,30-trinorhopane (also known as Tm; Fig. 2.20c). There are also variations on the regular hopanoidal structure that give rise to other series of hopanoids, such as the neohopanoids (the C_{27} member is known as Ts; Fig. 2.20c), in which the C-28 methyl is attached to C-17 rather than C-18 at the D,E-ring junction. The absence of a C–C bond that would normally form part of the ring system is indicated by the position numbers involved followed by 'seco', e.g. 8,14-secohopanoids (Fig. 2.20c).

membranes and in lipoproteins. **Lipoproteins** are the main means by which hydrophobic lipids like sterols are transported within organisms. In eukaryotic cell membranes sterols act as rigidifiers, inserted between the adjacent fatty acid chains on phospholipids to give the precise geometry required of the membrane (molecular size and conformation are critical factors). The equivalent role in eubacteria is performed by the hopanoids, which can be formed via a similar cyclization of squalene (Rohmer et al. 1980). Steroids are rare in heterotrophic bacteria, their limited occurrence seemingly reflecting ingestion (i.e. heterotrophic uptake) rather than bacterial biosynthesis. Although sterols have been reported in cyanobacteria (e.g. Volkman et al. 1998), it is doubtful that these organisms biosynthesize sterols (Ourisson et al. 1987; see Section 5.2.2), and most use bacteriohopanetetrol as membrane rigidifiers (Rohmer et al. 1989).

Sterols of geochemical significance are mainly C_{27} to C_{30} compounds with a β-hydroxy group at C-3, a C=C bond within the ring system (usually at the 5,6 position, i.e. Δ^5) and a side chain on the D ring (at C-17) with some branching and often some unsaturation. Some examples are shown in Fig. 2.22. The term **sterol** is commonly used to denote steroidal alcohols, which may or may not be unsaturated. More specific nomenclature can be used to differentiate the saturated alcohols, **stanols**, and unsaturated alcohols, **stenols**. The main sterols in photosynthetic organisms are sometimes referred to as **phytosterols** and include the main higher plant sterols, stigmasterol and β-sitosterol, as well as phytoplankton sterols like epibrassicasterol. Of less importance, geochemically, are the animal bile acids (e.g. cholic acid) and hormone steroids (e.g. testosterone, cortisone, aldosterone). There are also plant steroids that can have medicinal uses, such as digitonin (a cardiac

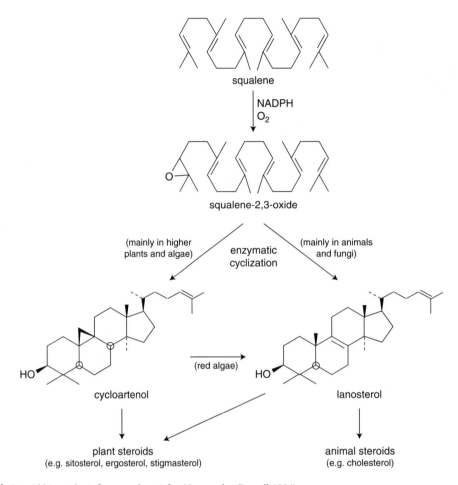

Fig. 2.21 Steroid biosynthesis from squalene (after Harwood & Russell 1984).

stimulant) and conessine (which is active against amoebic dysentery).

(g) Tetraterpenoids

The most important members of this group are the **carotenoid** pigments, which are widely distributed compounds with a variety of functional groups and a broad range of reactivities. They are subdivided into hydrocarbons, the **carotenes**, and oxygen-containing compounds, the **xanthophylls**. They are highly unsaturated and the configuration of C=C bonds (see Section 2.2.1) enables carotenoids to absorb light of a particular wavelength, endowing them with a yellow/red colouration. All carotenoids are derived from phytoene (see Fig. 2.16), and some representative structures are presented in Fig. 2.23.

Carotenoids are found in most organisms, including non-photosynthetic bacteria, fungi and mammals (in which they are essential for vitamin-A production). They are abundant in higher plants and algae, sometimes constituting >5% of the organic carbon of algae, and are also important in photosynthetic bacteria. Carotenoids are present in all marine phytoplankton and are responsible for the colour of the 'red tides' encountered during periods of high dinoflagellate productivity (blooms). The major role of carotenoids is as accessory pigments to chlorophyll-*a* in the light-harvesting part of photosynthesis (see Box 2.9). Lutein is one such carotenoid (Fig. 2.23), which usually constitutes 30–60% of total

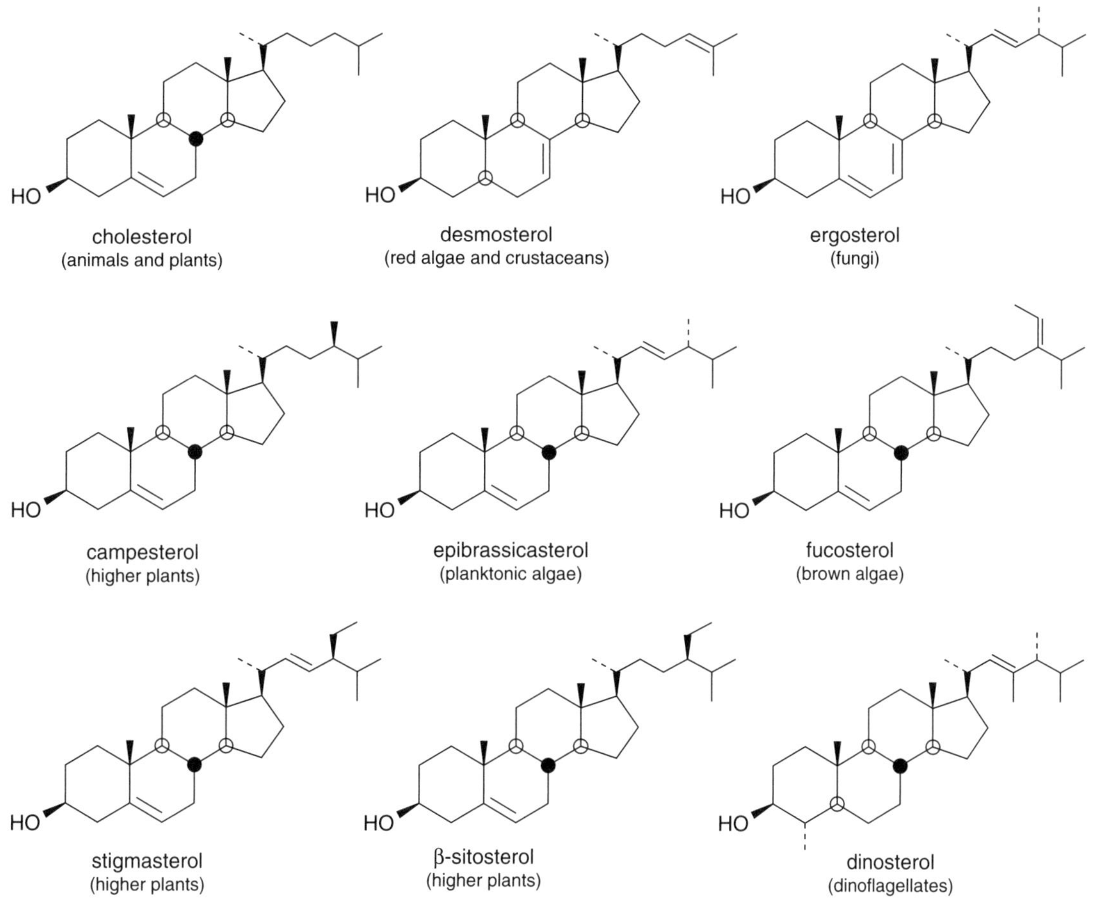

Fig. 2.22 Some geochemically important sterols and their major sources.

xanthophylls in higher plants (Bungard et al. 1999). Although most of the 300 or so naturally occurring carotenoids are tetraterpenoids, there are also some C_{30} and C_{50} compounds unique to non-photosynthetic bacteria.

Carotenoids also provide protection against free radicals (see Box 2.10), which undergo damaging reactions with cellular chemicals. Phototrophes are particularly susceptible to the oxidizing effects of too large an accumulation of photosynthetic by-products when light intensity is high, but they make use of the oxygen-capturing ability of xanthophylls, forming epoxides. Two major **xanthophyll cycles** exist, as shown in Fig. 2.25 (Müller et al. 2001). That involving zeaxanthin, antheraxanthin and violaxanthin is found in higher plants and green and brown algae, whereas the cycle comprising diatoxanthin and diadinoxanthin dominates in most other algae (Lohr & Wilhelm 1999).

Distributions of carotenoids can be characteristic for various groups of photosynthetic organisms. Fucoxanthin is characteristic of diatoms (Bacillariophyceae) and peridinin is found in many dinoflagellates (Dinophyceae; Fig. 2.23). In contrast, diatoxanthin and diadinoxanthin occur in many phytoplanktonic classes due to their xanthophyll cycle role (Fig. 2.25). Although it is a less specific marker compound, β-carotene is abundant in cyanobacteria. Photosynthetic bacteria produce acyclic and aromatic carotenoids (e.g. lycopene and okenone, respectively; Fig. 2.23). Astaxanthin and its esters are major constituents of marine zooplankton (Fig. 2.23).

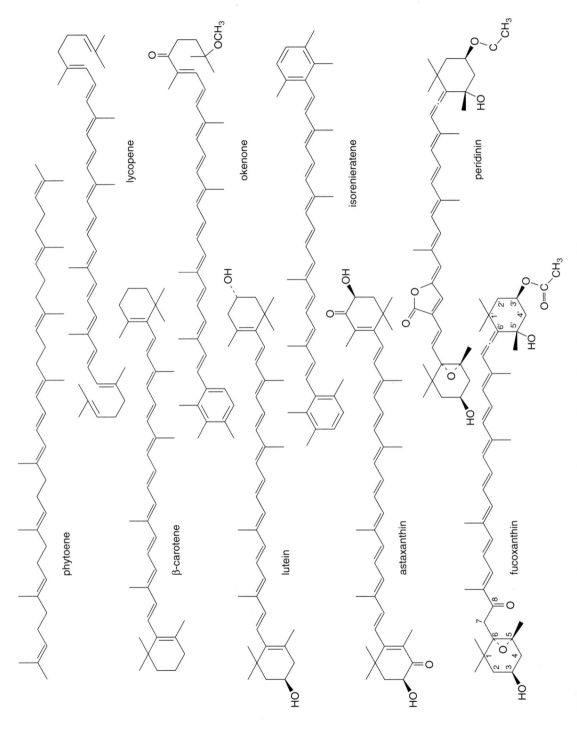

Fig. 2.23 Some geochemically important carotenoids (ring numbering scheme shown for fucoxanthin).

Box 2.9 | Accessory pigments

Chlorophyll-*a* does not absorb light uniformly over all the visible region of the electromagnetic spectrum (400–700 nm wavelength). Maximum absorption occurs at the red and blue ends of the spectrum (Fig. 2.24), resulting in the green colour of most photosynthesizing organisms (particularly terrestrial plants). **Accessory pigments** include carotenoids and a variety of chlorophylls and related tetrapyrroles (see Section 2.4.4), all of which enable more light of shorter wavelengths to be absorbed (Fig. 2.24), increasing the ability of organisms to photosynthesize at lower light levels. This is especially useful where daylight hours may be restricted (e.g. winter at high latitudes) or light intensity may be reduced (e.g. a low angle of incidence, again at high latitudes). Light intensity falls off quite rapidly with increasing water depth, due to absorption processes, so there is a limit to the depth at which photosynthesis can occur. This limit is about 200 m in clear water, which is much greater than would be possible with cholorphyll-*a* alone and is due to light 'scavenging' by accessory pigments. Light at the blue end of the spectrum penetrates to greater depths and is absorbed by accessory pigments, particularly carotenoids (which consequently appear red-brown). The utilization of light energy during photosynthesis is represented by **action spectra**, which show the amount of light absorbed at different wavelengths by the pigments present. The action spectrum in Fig. 2.24 demonstrates the effectiveness of accessory pigments.

Accessory pigments do not perform any function other than capturing light energy in photosynthesis and must pass on the absorbed light energy to a special form of chlorophyll-*a*. The colours of some accessory pigments tend to mask the green of chlorophyll-*a* and often give rise to the names of certain classes of multicellular, generally benthonic algae, the Rhodophyta (red algae) and the Phaeophyta (brown algae e.g. fucoids and kelps). In contrast, chlorophylls-*a* and -*b* dominate in the Chlorophyta (green algae), a large group, including both unicellular phytoplankton and multicellular macrophytes (e.g. *Enteromorpha* and *Ulva*), which do not tend to live at such great depths as the red and brown algae.

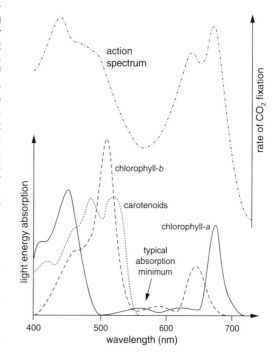

Fig. 2.24 Light absorption characteristics of some photosynthetic pigments and their relationship with utilization of light energy during photosynthesis (action spectrum).

Sporopollenin, the polymeric material forming the protective coatings of spores and pollens, may contain carotenoid and/or carotenoid ester structures such as astaxanthin dipalmitate. However, the major part of sporopollenin is synthesized from unbranched aliphatic chains, probably derived from fatty acids. Polyhydroxybenzene units similar to lignin precursors (see Section 2.5.1) have also been identified. Sporopollenin is found in the pollen grains of angiosperms and gymnosperms, in the spores of algae, fungi and lycopods and in fossil spores (e.g. tasmanites).

(h) Polyterpenoids

Some examples of polyterpenoids from higher plants are shown in Fig. 2.26, and they all serve protective functions (sealing wounds). Perhaps the best known is rubber, secreted in latex (which also contains polycyclic

> **Box 2.10 | Free radicals and ions**
>
> As noted in Section 2.1.2, covalent bonds involve the sharing of pairs of the outermost (valency) electrons. When a single bond is broken, the pair of electrons may be equally divided between the two atoms that they formerly joined. This is termed **homolytic fission**, and the products are **free radicals**, each possessing an unpaired electron. In **heterolytic fission**, one of the atoms gets both of the electrons, forming a negatively charged ion (**anion**), and the other a positively charged ion (**cation**). Where heterolytic fission of a C–C bond occurs in an organic compound and the charges are located largely on each of the C atoms, the anion is called a **carbanion**, and the cation a **carbocation**.
>
> How the electrons are distributed upon fission of a C–C bond depends upon the ability of various groups attached to each of the C atoms to push electrons towards or pull them away from the site of fission (termed the **inductive effect**). For C atoms in virtually identical environments, such as in an *n*-alkyl chain, homolytic fission is most likely. However, electron-withdrawing groups (e.g. COOH, CHO, halogens) help to stabilize the formation of a carbanion by the C atom to which they are bonded. Conversely, electron-releasing groups (e.g. OH, NH_2, OCH_3, CH_3) favour formation of a carbocation.
>
> Free radicals are extremely reactive. The reactivity of carbanions and carbocations depends upon the extent to which the charge is dissipated by the inductive effect of other groups, but they are generally highly reactive. A carbocation is termed an **electrophile** because it reacts with a site in another molecule that can provide electrons. The delocalized electrons of an aromatic ring are one such suitable source of electrons. Conversely, carbanions are called **nucleophiles** because they tend to react with sites that are deficient in electrons. Electrophiles and nucleophiles can also be neutral molecules, possessing groups with inductive effects that leave a slight residual charge on one of their C atoms.

triterpenoids), which comprises *cis*-linked isoprene units (Fig. 2.26). The number of isoprene units involved varies with plant species (which mainly belong to the Euphorbia); the greatest have been found in *Hevea brasiliensis*, averaging *c*.7000 (Archer et al. 1963). In contrast, gutta-percha has *trans*-linked isoprene units (Schaeffer 1972), an arrangement that appears to result in slightly lower degrees of polymerization (Duch & Grant 1970). Few species of plant appear to synthesize both polymers.

Woody higher plants can also exude resins. For example, dammar resin from certain angiosperms (e.g. dipterocarps) contains a polycadinene (Fig. 2.26), based on a Δ^5-cadinene monomer (van Aarssen et al. 1990). Gymnosperm resin polymers are based on labdatriene (diterpenoid) monomers such as communic acid (Fig. 2.26; Mills et al. 1984/5; Hatcher & Clifford 1997), and some angiosperm resins also contain these polymers. In contrast to polycadinene formation, polymerization of labdatrienes is thought to occur after exudation.

Polyprenols comprise long chains of unsaturated isoprene units (up to *c*.30), terminating in an alcohol group (Fig. 2.26). They are widespread, being most abundant in plants, but also occurring in lesser amounts in fungi, bacteria and animals. Those in gymnosperms have two *trans* isoprene units at the ω end (the opposite end from the hydroxyl group), whereas those from angiosperms have three such units (Ibata et al. 1984). These units are followed by a variable number of *cis* units, including that bearing the terminal alcohol.

2.4.4 Tetrapyrrole pigments

These compounds contain four pyrrole units (see Table 2.2) linked together by =CH- units, either in the form of an open chain or as a large ring (macrocyclic) system. They are all involved in photosynthesis, as primary or accessory pigments (see Box 2.9). Chlorophyll-*a* is just one of a variety of **chlorophylls** comprising the macrocyclic group of **tetrapyrrole pigments**, some of which are shown in Fig. 2.27. The general name **porphyrin** is often applied to this cyclic tetrapyrrole structure. Attached to the tetrapyrrole ring of many chlorophylls is the isoprenoidal alcohol phytol.

Chlorophyll-*a* is quantitatively by far the most important chlorophyll and is present in all algae, higher plants and the cyanobacteria. In addition, all higher plants contain chlorophyll-*b*. The prochlorophytes, a group of oxygenic photosynthesizers related to the cyanobacteria, possess divinyl derivatives of chlorophyll-*a* and *b*, often referred to as chlorophyll-a_2 and b_2 respectively. Photosynthetic bacteria contain **bacteriochlorophylls** but, as in plants and cyanobacteria (see Box 2.9), the '*a*' form is the primary photosynthetic pigment and the other forms act as accessory pigments,

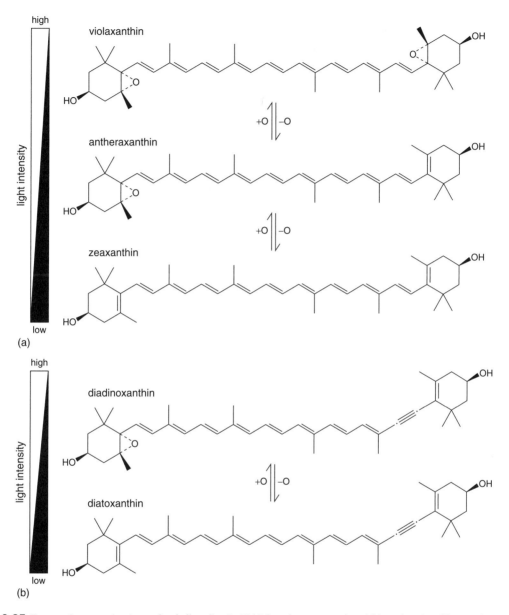

Fig. 2.25 Enzymatic conversion in xanthophyll cycling in (a) higher plants, green algae (chlorophytes) and brown algae (phaeophytes), (b) most other algae (e.g. diatoms, dinoflagellates, chrysophytes and haptophytes; see Table 3.6).

passing captured light energy on to the '*a*' form. Bacteriochlorophyll-*b* is present together with the '*a*' form in most species of photosynthetic bacteria. Species of the green photosynthetic bacteria *Chlorobium* also contain some unique chlorophylls, including bacteriochlorophyll-*e* (Fig. 2.27), which is characterized by a methyl group at C-20 (sometimes called the δ position).

Oxygen transport in animals is performed by related pigments such as **haem** (Fig. 2.28), which is associated with a protein, globin, in haemoglobin. The metal atom in haems (iron) and chlorophylls (magnesium) is impor-

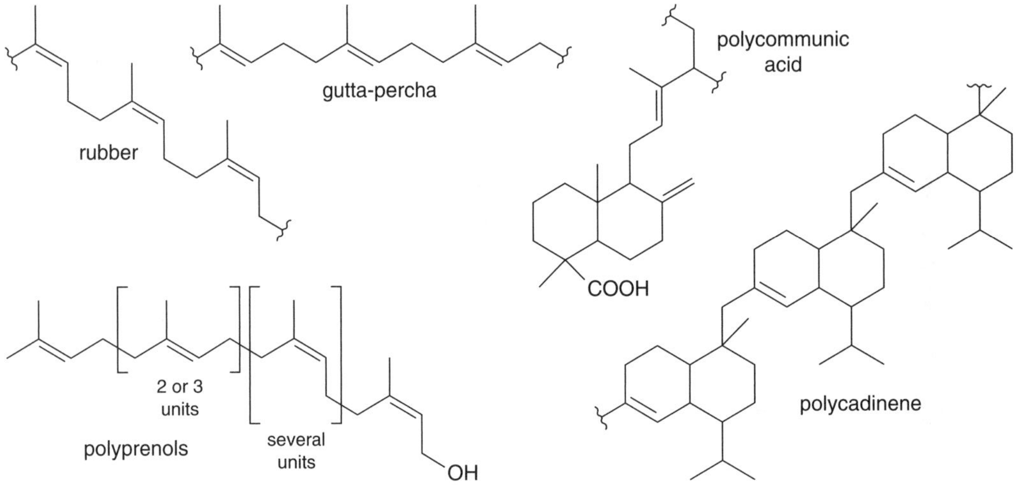

Fig. 2.26 Examples of polyterpenoids.

tant in electron transfer processes that occur during respiration and photosynthesis, respectively. Haems are also found in cytochromes, which perform electron transfer in energy-releasing processes in most organisms (but not, for example, in methanogens).

Among the accessory pigments are non-cyclized, non-metallated tetrapyrroles. Phycobilins are such compounds, the most important examples of which are phycocyanobilin, found in cyanobacteria, and phycoerythrobilin, found in certain red algae (Fig. 2.28). In these pigments the tetrapyrrole system is bonded to a protein (molecular weight $c.2 \times 10^4$) through a cysteine residue. A related blue-green photochromic pigment, phytochrome, is found in all higher plants and it is also bound to a protein (molecular weight $c.12 \times 10^4$). It controls various growth-related functions, such as germination, and a wide range of developmental and metabolic processes.

2.5 Lignins, tannins and related compounds

These higher plant components are characterized by phenolic (hydroxy-aromatic) structures. Such structures, which derive originally from monosaccharides, are common in plant but not animal tissue.

2.5.1 Lignin

Lignin is second only to cellulose as the most abundant biopolymer. It is found mostly in cell walls, where it is intimately associated with hemicelluloses (see Section 2.2.2), forming a network around cellulose fibres in maturing xylem (the channelled, woody core of terrestrial plants) and fulfilling an important supportive function. Cellulose is also an important component of wood (40–60%) but lignin comprises most of the remaining material (20–30% of vascular tissue). Lignin is a high molecular weight, polyphenolic compound, formed by condensation reactions (involving dehydrogenation and dehydration) between three main building blocks: coumaryl, coniferyl and sinapyl alcohols (Adler 1977). These compounds are biosynthesized from glucose (under enzymatic control) by the much simplified reaction scheme in Fig. 2.29, in which some of the intermediates are shown. One intermediate, shikimic acid, occurs widely in plants and is transformed into phenylalanine by a series of reactions, including NH_3 transfer from an amino acid (transamination). Phenylalanine then undergoes deamination and hydroxylation to form 4-hydroxycinnamate (which can also be called *p*-coumarate), from which the three main lignin precursors are formed by reactions that include carboxyl reduction, ring hydroxylation and partial methylation of hydroxyl groups.

The types of structural elements in lignin that are likely to be formed from the three precursor phenolic

Chemical composition of organic matter

chlorophyll	structure	R^1	R^2	R^3	R^4
chlorophyll-a	I	–CH$_2$=CH$_2$	–CH$_3$	–CH$_2$–CH$_3$	phytyl
chlorophyll-b	I	–CH$_2$=CH$_2$	–C(=O)H	–CH$_2$–CH$_3$	phytyl
chlorophyll-a_2	I	–CH$_2$=CH$_2$	–CH$_3$	–CH$_2$=CH$_2$	phytyl
chlorophyll-b_2	I	–CH$_2$=CH$_2$	–C(=O)H	–CH$_2$=CH$_2$	phytyl
chlorophyll-c_1	II	–CH$_2$=CH$_2$	–CH$_3$	–CH$_2$–CH$_3$	H
chlorophyll-c_2	II	–CH$_2$=CH$_2$	–CH$_3$	–CH=CH$_2$	H
chlorophyll-d	I	–C(=O)H	–CH$_3$	–CH$_2$–CH$_3$	phytyl
bacteriochlorophyll-a	III	–C(=O)CH$_3$	–CH$_3$	–CH$_2$–CH$_3$	phytyl, farnesyl or geranylgeranyl
bacteriochlorophyll-b	III	–C(=O)CH$_3$	–CH$_3$	=CH–CH$_3$	phytyl, farnesyl or geranylgeranyl
bacteriochlorophyll-e	IV	–C(H)(OH)CH$_3$	–C(=O)H	–CH$_2$–CH$_3$, –CH$_2$–CH$_2$–CH$_3$ or –CH$_2$–CH(CH$_3$)–CH$_3$	farnesyl

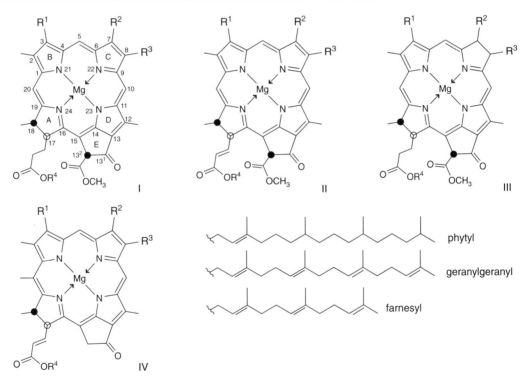

Fig. 2.27 Some geochemically important chlorophylls (ring numbering scheme shown). The arrows from N atoms to Mg^{2+} ions denote dative bonds (see Section 2.1.2).

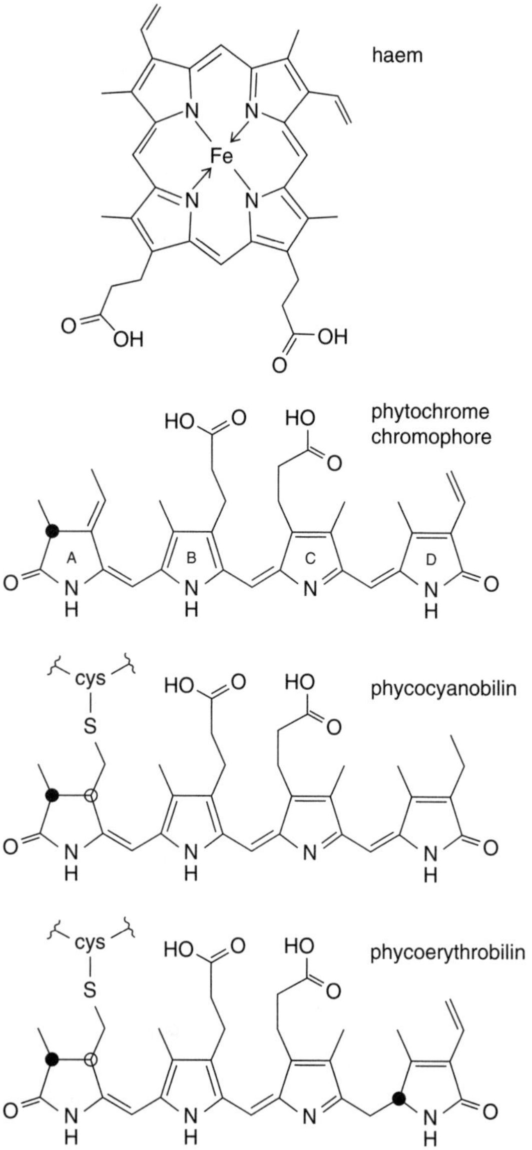

Fig. 2.28 Some non-chlorophyll tetrapyrrole pigments. The A ring of the phytochrome chromophore binds to a protein via the S atom of cysteine in the same way as shown for the two phycobilins (the three chromophore chains differ only in their D-ring substitution patterns).

compounds are shown in Fig. 2.30. The coumaryl alcohol precursor gives rise to *p*-hydroxyphenyl units, coniferyl alcohol gives vanillyl units (also called guaiacyl units) and sinapyl alcohol yields syringyl units. Lignins vary in composition according to plant type: vanillyl units predominate in gymnosperm lignins, whereas angiosperm lignins contain approximately equal amounts of vanillyl and syringyl units (as in Fig. 2.30), and both plant types contain only small amounts of *p*-hydroxyphenyl units. More than ten different types of linkage can be formed between the phenylpropane units during the polymerization process, but the dominant linkage (>50%) is between the β-C atom of the propenyl chain and the phenolic O atom (at C-4), which can be abbreviated to β-O-4 (see Fig. 2.30). Various other secondary reactions take place during polymerization, resulting in cross-linking between lignin and hemicelluloses.

It had been thought that the structure of lignin is quite random (Adler 1977), unlike the uniform structural arrangements in polysaccharides and proteins that we have seen so far. Such a structure is shown in Fig. 2.30. However, the morphological preservation observed in wood during its early stages of transformation into coal suggests a more ordered structure (Hatcher & Clifford 1997), as is considered further in Section 4.3.2. A helical conformation of lignin chains has been proposed, which can arise if the monomers are joined through β-O-4 linkages; the other types of linkages observed in lignins could be accommodated in links between adjacent helices (Faulon et al. 1994). The helical arrangement would require stereoselectivity at the α and β C atoms (Fig. 2.30) to allow the molecule to adopt the necessary shape, and there is precedence for such stereoselectivity during biosynthesis of the monomer units (Davin et al. 1997).

Although fungi are important in the degradation of lignins, no lignins have been found in fungi. However, some fungi have partially developed the phenylpropanoid-acetate pathway to cinnamic and *p*-coumaric acids, which they can condense with acetate-derived components to produce styrylpyrones such as hispidin (Fig. 2.31). Dimers formed from various combinations of the three lignin precursors give rise to the lignans, which can have diverse compositions, although the 8–8′ linkage is most widespread, as in guaiaretic acid (Fig. 2.31).

2.5.2 Tannins and other hydroxy-aromatic pigments

Tannins are widespread in nature but are quantitatively and geochemically less important than lignin. They are important in making plants less palatable to herbivores and, when extracted by steeping plant material, particularly bark and leaves, in water, they can be used for tanning leather. The structural units of tannins are polyhydroxy aromatic acids, such as gallic acid and

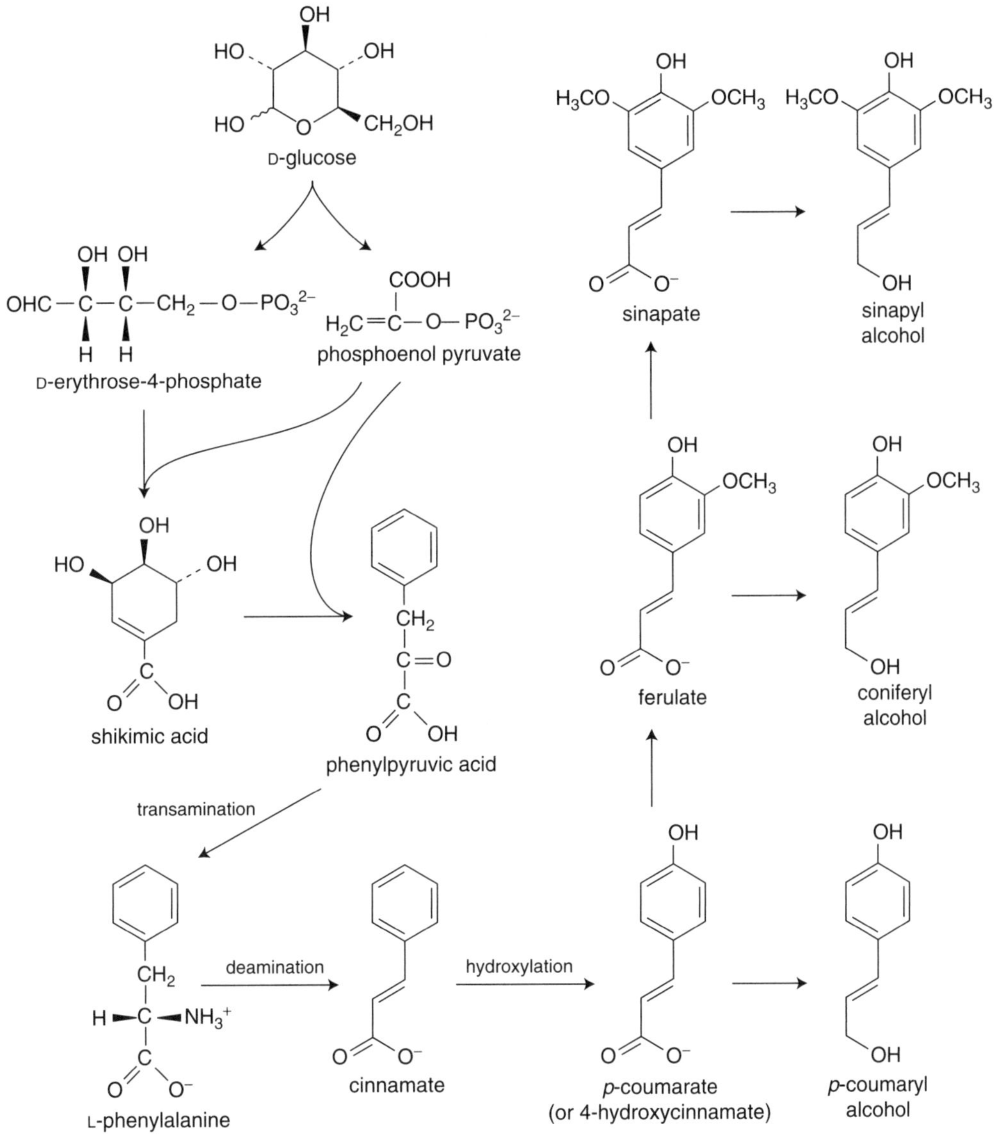

Fig. 2.29 Biosynthesis of major lignin precursors sinapyl, coniferyl and coumaryl alcohols (after MacGregor & Greenwod 1980).

ellagic acid (Fig. 2.31), which are produced by similar biosynthetic routes to lignin precursors. The condensed, polymeric structures of tannins generally incorporate glucose esters on residual carboxylic acid groups and have molecular weights of 500–3000.

Polyhydroxyflavonol units have also been recognized in some tannins. These units are directly related to the **flavonoid** higher plant pigments, which include anthocyanins, flavones and flavonols (e.g. Britton 1983). All flavonoids share the same basic carbon skeleton, as shown for cyanidin chloride in Fig. 2.31, but differ in the number and position of hydroxy and methoxy substituents (e.g. coumestan, Fig. 2.31; coumestrol and isoflavone, Fig. 7.11). They can be formed from the

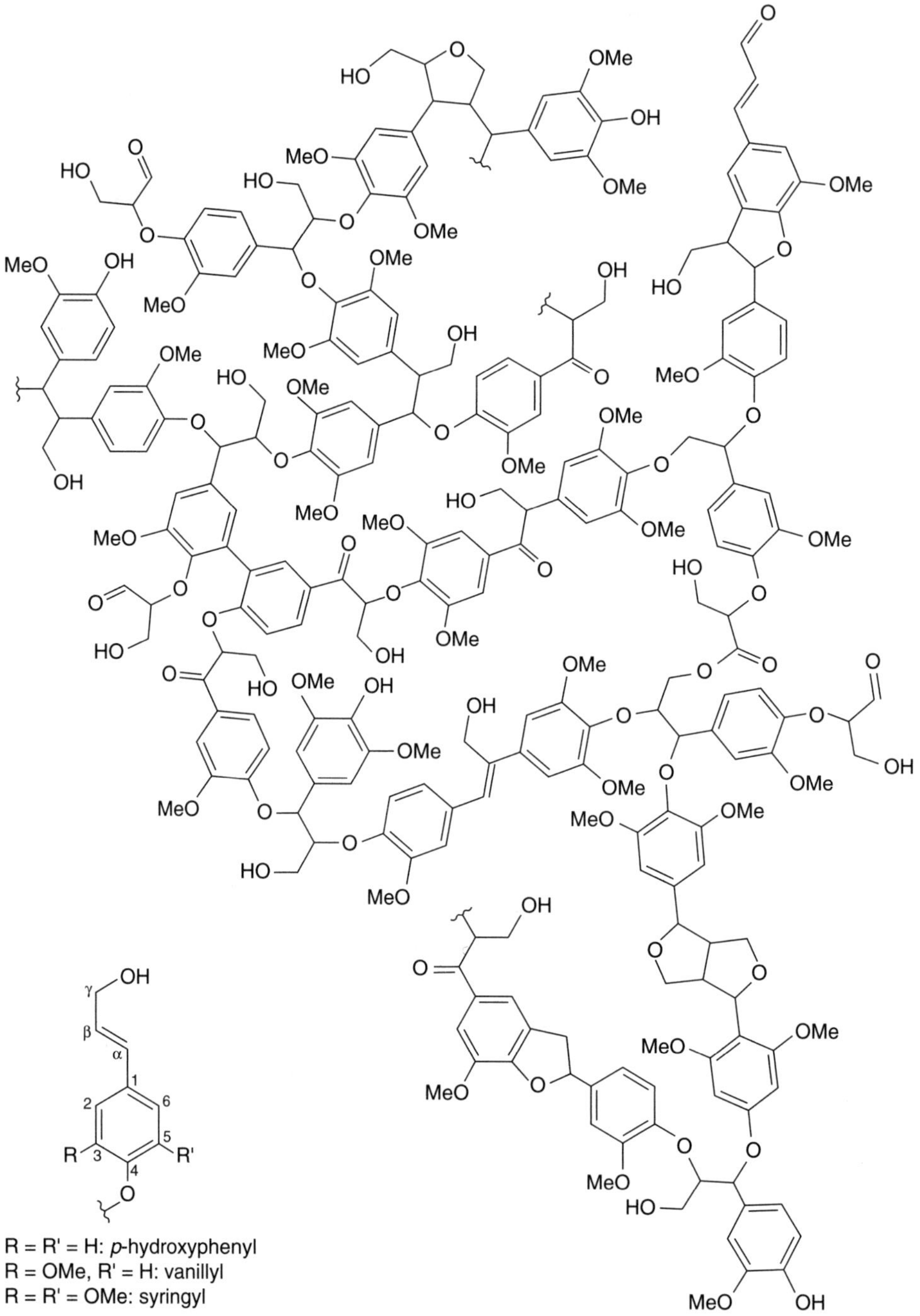

R = R' = H: *p*-hydroxyphenyl
R = OMe, R' = H: vanillyl
R = R' = OMe: syringyl

Fig. 2.30 A partial, random, structure of beechwood lignin, illustrating the types of condensation linkages present (after Nimz 1974; Me = CH$_3$), together with unit names and C numbering system.

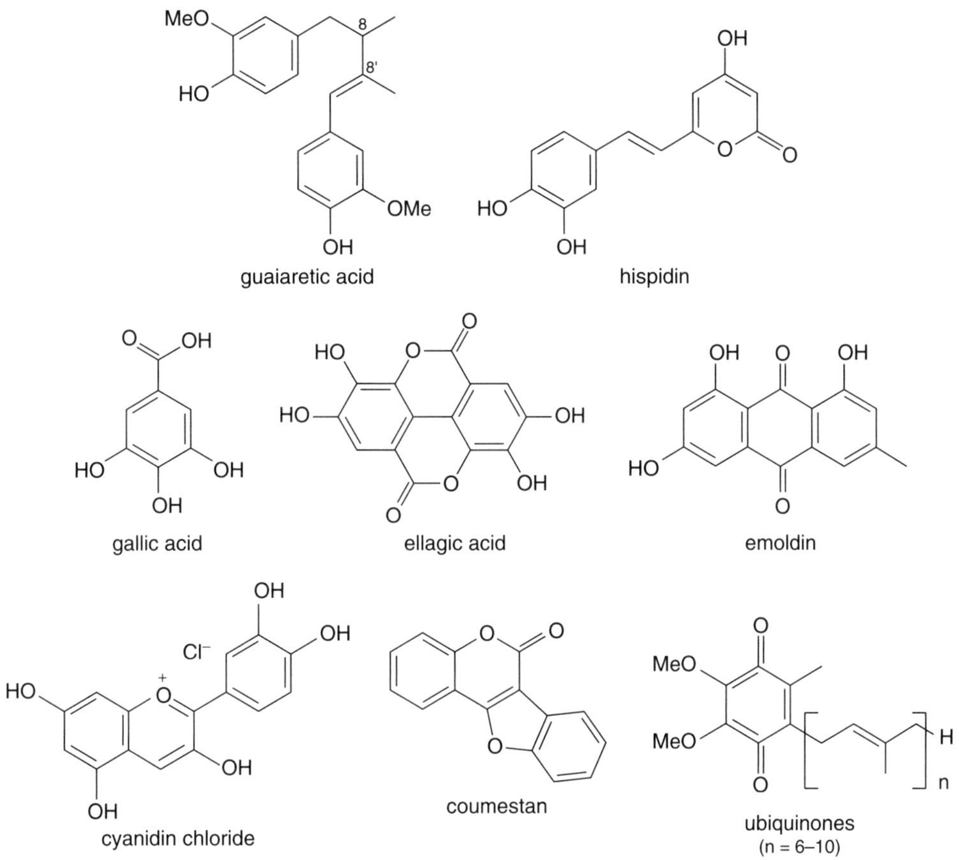

Fig. 2.31 Some hydroxyaromatic plant pigments and quinones.

Co-A ester of *p*-coumaric acid (Fig. 2.29), as well as via an acetate pathway. Flavonoids are often found as **glycosides** (i.e. they are complexed with sugars in a similar way to tannins) and appear to offer protection against UV radiation and microbial attack, as well as providing the colour in petals and autumn leaves.

Anthraquinones, another group of hydroxyaromatic pigments, are found as glycosides in higher plant tissues, particularly bark, heartwood and roots. Although not generally associated with tannin formation, these compounds occur in a range of organisms: fungi, lichens and insects, as well as vascular plants. Emoldin is perhaps the most widely distributed (Fig. 2.31). Other quinones based on benzene and naphthalene units also occur widely in nature, such as the ubiquinones (also called coenzymes Q) involved in electron transport during respiration (Fig. 2.31).

2.6 Nucleotides and nucleic acids

2.6.1 Nucleotides

Proteins are not the only nitrogen-containing compounds: there are others that are present in low amounts relative to proteins but perform some essential functions. The alkaloids are one such large and diverse group, synthesized by plants and fungi, which can have pharmacological activity but are geochemically unimportant. Another group is the **nucleotides**, which comprise a phosphate, a pentose sugar (D-ribose or 2-deoxy-D-ribose) and a nitrogen-containing organic base. The base is either a purine or a pyrimidine derivative (Fig. 2.32), generally containing additional amino or carbonyl groups. We have already met two important nucleotides, ATP and NADP$^+$ (Fig. 2.32),

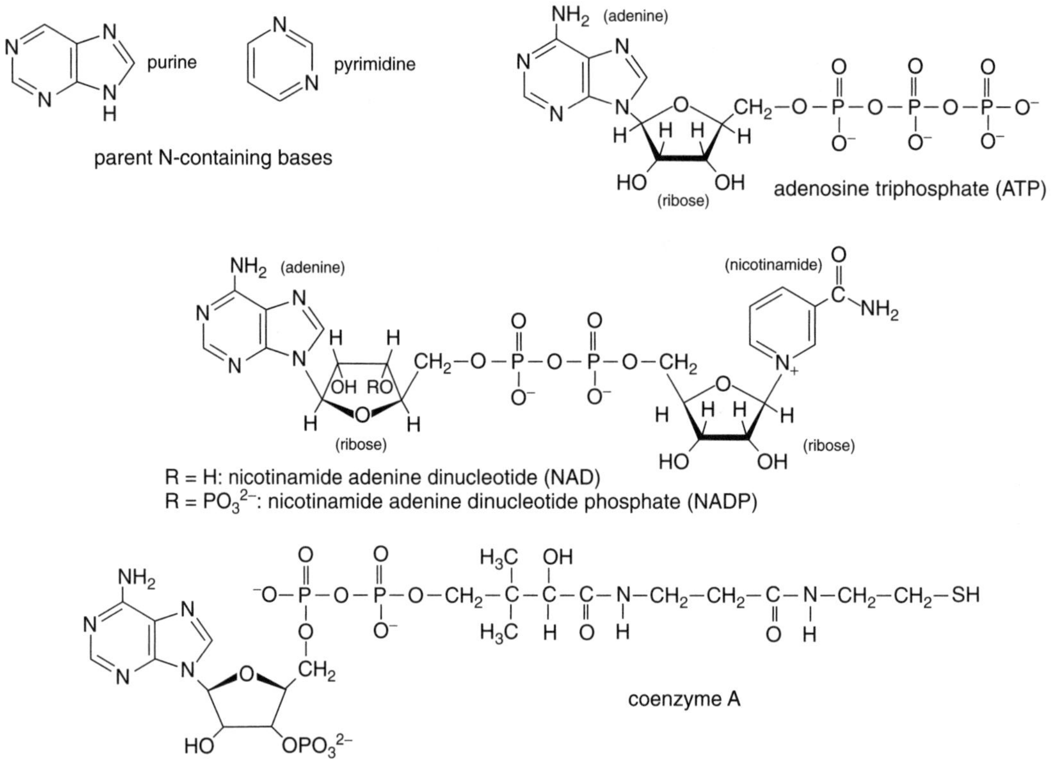

Fig. 2.32 Parent nitrogen-containing bases and structures of some important nucleotides.

which contain the purine derivative adenine and are involved in many biochemical reactions in addition to photosynthesis. Another important nucleotide, closely related to NADP⁺ and used by all organisms in oxidation and reduction processes, is NAD⁺ (Fig. 2.32). Adenine forms part of coenzyme A, yet another important compound in many biochemical reactions (Fig. 2.32).

2.6.2 Nucleic acids

Nucleic acids are polymers of nucleotides: ribonucleic acid (RNA) and deoxyribonucleic acid (DNA). Although nucleic acids do not survive for geological periods in sediments (the maximum appears to be $c.50$ kyr under favourable conditions) they are very important; they control the self-replication of organisms (and hence provide information on evolutionary relationships) and act as the templates for protein biosynthesis. There are four nitrogen-containing bases in DNA: the pyrimidines cytosine and thymine, and the purines adenine and guanine, which are often abbreviated to their initials C, T, A and G respectively. In RNA thymine is replaced by another pyrimidine, uracil (U). In both RNA and DNA there are long chains of alternating phosphate and sugar groups, and the nitrogen-containing bases are bonded to the sugars at C-1. In DNA there are often two strands, forming a helix, that are linked together by hydrogen bonding between the opposing base groups. Because of the spatial constraints on this form of bonding, thymine always hydrogen bonds with adenine and cytosine with guanine (Fig. 2.33).

During cell division the DNA strands separate and act as templates for the construction of new strands. Protein synthesis makes use of various forms of RNA to translate the DNA code into a sequence of amino acids, with each amino acid being coded for by groups of three consecutive bases (e.g. GGA codes for glycine).

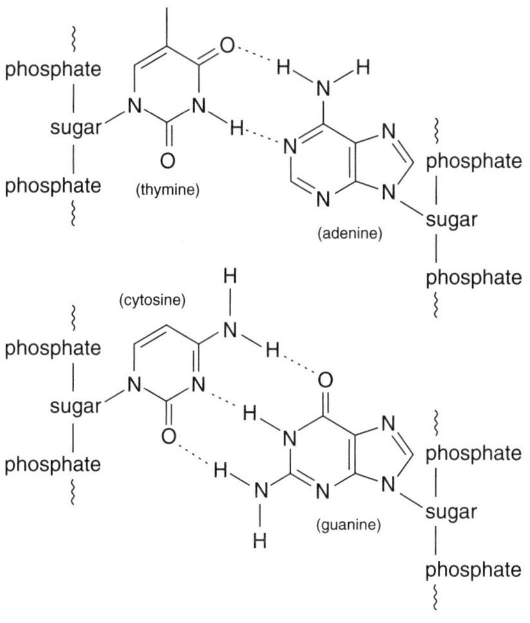

Fig. 2.33 Hydrogen bonding between bases in DNA.

2.7 Geochemical implications of compositional variation

2.7.1 Compositional variation of organisms

Despite the fact that all organisms contain broadly the same groups of chemical classes performing generally the same biochemical functions, there can be considerable variation in the relative amounts of each class of compound between different groups of organisms, reflecting varying physiological and metabolic requirements. An obvious contrast is the composition of higher plants, in which cellulose and lignin can account for up to 75% of the organic material, and phytoplankton, which do not contain these structural components. Diatoms and dinoflagellates contain $c.$25–50% protein, $c.$5–25% lipids and up to $c.$40% carbohydrates (dry wt; Raymont 1983). Bacteria are quite variable in their chemical constitution, but as a guide their composition can be considered as similar to that of planktonic algae. Higher plants contain $c.$5% protein, $c.$30–50% carbohydrate (mainly cellulose) and $c.$15–25% lignin. The lipid content of higher plants is relatively low and chiefly concentrated in fruiting bodies and leaf cuticles.

We have also seen examples of how compound distributions within a given chemical class vary between different groups of organisms. We examine the use of individual compounds in sedimentary organic matter as indicators of contributing organisms in Chapter 5. However, there are some complicating factors that require consideration. Small organisms are particularly sensitive to changes in factors such as temperature and salinity because they have high surface area to volume ratios. As a result, variation in environmental conditions may bring about greater changes in lipid distributions within a species than exist between species of the same genera or even between more distantly related groups of planktonic organisms. For example, in copepods the abundance of lipids in general, and wax esters (which are used as energy stores) in particular, depends on factors such as activity, nutritional status and apparently also water temperature. It has been found that copepod species that live in cold waters (i.e. permanently below 250 m depth or in high latitude surface waters) generally have more lipids and wax esters than those from warmer waters (Raymont 1983). An indication of the abundance range of total lipids, wax esters and triglycerides observed for copepods and their latitudinal variations is given in Table 2.6.

It can now be appreciated that the nature of the organic material deposited in sediments does not just depend on the type of organisms contributing to the sediments, although higher plant lignins are the main source of aromatic compounds in contemporary sediments, while plankton and bacteria contribute primarily aliphatic material. Inputs of sedimentary organic matter are classified as **autochthonous** if they originate at or close to the site of deposition, or **allochthonous** if transported from another environment. Autochthonous inputs to most aquatic environments include the remains of phytoplankton and organisms that feed directly or indirectly on phytoplankton (e.g. zooplankton and bacteria), and that live within the water column and upper layers of sediment. Allochthonous organic material mostly derives from higher plants, usually

Table 2.6 Abundance range of copepod lipids (after Lee et al. 1971)

lipid class	subtropical		polar	
	min	max	min	max
total lipids (% of dry wt)	3	37	31	73
triglycerides (% of lipids)	1	42	2	11
wax esters (% of lipids)	0	46	61	92

transported by water from adjacent areas of land to the deposition site. However, in peat swamps higher plants make a large autochthonous contribution. The composition of sedimentary organic matter depends to a large extent upon the relative contributions from the various autochthonous and allochthonous inputs and their chemical compositions.

2.7.2 Variations through geological time

The evolution of organisms throughout geological time means that we do not necessarily see the same types of organic matter in what are otherwise similar sedimentary environments. For example, stromatolites are more frequently found in sediments deposited before the cyanobacterial communities had to cope with predation by herbivores and compete for habitats with other photosynthetic organisms. Higher plant inputs are not seen prior to land colonization in the Silurian. The 'fingerprint' of organic compounds preserved in any particular environmental setting will reflect the stage of evolution of life-forms and, especially since the Devonian, the increasing diversity of species.

The geographical distribution of organisms is also an important consideration for terrestrial ecosystems. In Chapter 1 we noted how regional variations in flora occurred as Pangaea split up in the Carboniferous and became more pronounced as the continental fragments drifted apart and greater regional variation in climates and environments developed. The marine environment is more uniform than the terrestrial and there is less restriction to the spread of organisms. However, marine habitats have been significantly affected by variation in tectonic activity and the related changes in land mass distributions over geological time, as we see in Chapter 3 when examining the conditions necessary for the production and preservation of organic matter in sediments.

3 | *Production, preservation and degradation of organic matter*

3.1 How and why organic-rich deposits form

Having considered the main biological sources of sedimentary organic matter and its general chemical composition, it is time to examine the physical, chemical and geological processes that control the production of organic matter and its subsequent degradation or preservation within sediments. The organic carbon content of sedimentary deposits can range from zero up to almost 100% in coals, and very occasionally has reached 20–30% in some marine settings (e.g. Holocene sapropel in the Black Sea, and Neogene–Quaternary and Cretaceous sapropels in the Mediterranean; see Section 6.3.4a). In freshly deposited marine sediments today organic carbon content rarely exceeds 2% (it very occasionally reaches up to $c.15\%$ locally) and over most of the ocean floor it is <0.25% (Pedersen & Calvert 1990). Organic-rich sediments have not been laid down continuously throughout geological time, and to understand the reasons for their deposition it is necessary to examine the production and fate of organic material in present-day environments.

There are some general conditions that favour the formation of organic-rich sediments. First, a sufficiently large amount of organic material is needed, which, as we have seen in Section 1.5, is predominantly derived (directly or indirectly) from the main primary producers: higher plants on land and phytoplankton in aquatic environments. High primary productivity is, therefore, an important factor, providing at least the opportunity for relatively large amounts of organic matter to reach and be incorporated into the sediment, rather than being completely recycled within the water column. Second, a low-energy depositional environment (e.g. low water current velocities and limited wave action) is needed if relatively low-density organic material is to be deposited and accumulate over time, rather than being eroded periodically. Third, inputs of inorganic mineral matter should not overwhelm the organic matter and dilute it significantly. Finally, conditions must favour preservation of organic matter within the sediment, rather than degradation by detritivores and decomposers (see Section 1.5.1), and this appears generally to be favoured by the development of anoxicity, which in turn is aided by high accumulation rates. The following sections are concerned with the major factors controlling the production and the immediate fate of organic matter in contemporary sediments within various depositional environments.

3.2 Controls on primary production

3.2.1 General controls on photosynthesis

Photosynthesis is thought to have been the most important form of primary production for at least the past 3.5 Gyr (Schidlowski 1988; Section 1.4.1) and involves the conversion of carbon dioxide into cellular organic material by the utilization of light energy. Chemosynthesis (which uses chemical rather than light energy for carbon fixation; see Boxes 1.9 and 1.10) is globally far less important, although it may be locally significant (Sorokin 1966). Light, then, is a critical factor in primary production. Water is vital for all life and provides a ready source of the hydrogen needed for aerobic photosynthesis (Fig. 1.7). Most photosynthesizing organisms are aerobes: vascular plants, macroscopic algae (seaweeds), unicellular algae (phytoplankton), cyanobacteria and prochlorophytes. Water availability is obviously not a problem for aquatic organisms but it can be an important factor controlling terrestrial primary production. Some of the most productive areas, such as the tropical rain-forests (net annual primary production $c.15.3\,\text{Gt}\,\text{C}$; biomass $c.340\,\text{Gt}\,\text{C}$), have a more than ample water supply, and it is likely that the great coal-forming forests of the past grew in swampy areas maintained by high rainfall.

For a long period of the year at high latitudes water is present as ice and is, therefore, unavailable for plant uptake. This is one reason for high terrestrial productivity being confined to the middle and lower latitudes under current climatic conditions. As well as this important indirect effect, temperature directly affects photosynthesis because the dark reactions (see Box 1.10) are restricted to a certain temperature range. Within this range the rate of dark reactions tends to increase with increasing temperature. However, towards the top end of this range net production in C_3 vascular plants tends to decline because of elevated photorespiration (see Box 1.10) and because stomata are constricted to limit water loss by evapotranspiration, which also restricts CO_2 uptake. Excessively high temperatures inhibit photosynthesis by destroying or altering enzymes and cellular components.

Temperature variations are less extreme in the marine environment (owing to the high thermal capacity of water compared to air) and probably affect species diversity but not necessarily productivity (unless they affect other important factors, such as causing a reduction in light intensity under opaque ice). This is because organisms adapted to high or low temperatures are likely to have less competition from other species and may consequently be more productive. Although the solubility of carbon dioxide in water decreases with increasing temperature, the concentration of dissolved CO_2 is usually high enough not to be a limiting factor in aquatic primary production (but see Section 5.8.2).

The salinity of water bodies (Box 3.1) has an effect on the composition of primary producer communities. Fresh water and seawater in typical open marine environments contain the greatest numbers (diversity) of species. However, relatively few organisms can tolerate large fluctuations in salinity (e.g. where fresh water meets seawater in estuaries) and hypersaline conditions. In hypersaline conditions (Box 3.1) phytoplanktonic diversity is much reduced but the species adapted to these environments can produce large amounts of organic material. In addition, herbivore abundance may be low, so much of the net primary production may be available for incorporation into sediments. Cyanobacterial mat communities tend to be successful in the shallow areas of such environments, and productivity can reach $8-12\,g\,C_{org}\,m^{-2}\,day^{-1}$ (Schidlowski 1988).

Table 3.2 compares the productivities of various environments. The most prolific ecosystems are the oceanic surface waters (dominated by phytoplankton)

Box 3.1 | Salinity of seawater

Eight major ions account for almost 99% of the mass of salts present in seawater, as shown in Table 3.1. The dominant ions are chloride (i.e. the anion Cl^-) and sodium (the cation Na^+). Virtually all known elements have been detected in seawater, but most are at extremely low concentrations (below the parts per million, or ppm, level).

The **salinity** of natural waters is a measure of the total mass of ions present. On average there are c.35 g of salts in 1 kg of seawater, or 35 per-mil by weight, so the salinity is expressed as 35‰. **Fresh water** is usually defined as having a salinity of <0.5‰, and **hypersaline** water a salinity of >40‰. The salinity of oceanic surface-waters varies considerably, because it depends upon the balance between the removal of fresh water by evaporation and its supply, mainly by precipitation. It is highest at a latitude of c.20° (high evaporation and low precipitation) and decrea-ses towards higher latitudes (primarily due to low evaporation and high precipitation) and towards the equator (primarily due to high precipitation; see Figs 3.3a, 3.12). Salinity can be variable near the poles, again owing to the addition or removal of fresh water, but via the balance between the melting and formation of sea-ice, rather than precipitation/evaporation (see Box 3.2).

Table 3.1 Major ionic constituents of seawater (after Harvey 1982 and other sources)

ion	symbol	% by wt	concentration (g l^{-1})
chloride	Cl^-	55.08	19.35
sodium	Na^+	30.62	10.76
sulphate	SO_4^{2-}	7.72	2.67
magnesium	Mg^{2+}	3.68	1.29
calcium	Ca^{2+}	1.18	0.41
potassium	K^+	1.10	0.39
bicarbonate	HCO_3^-	0.40	0.14
bromide	Br^-	0.19	0.07
total		99.97	35.08

Table 3.2 Estimated primary productivity and biomass (dry wt) for various natural environments (after several sources, including Whittaker & Likens 1975; De Vooys 1979)

ecosystem type	area (A) (10^6 km^2)	plant biomass (Gt)	annual net primary production (P) (Gt yr^{-1})	P/A (g m^{-2} yr^{-1})
tropical rain-forest	17.0	765	37.4	2200
seasonal rain-forest	7.5	260	12.0	1600
temperate forest	12.0	385	14.9	1200
boreal forest	12.0	240	9.6	800
wood/shrub land	8.5	50	6.0	700
savannah	15.0	60	13.5	900
temperate grassland	9.0	14	5.4	600
tundra and alpine	8.0	5	1.1	140
desert scrub	18.0	13	1.6	90
swamp and marsh	2.0	30	6.0	3000
lake and stream	2.0	0.05	0.8	400
open ocean	332.0	1.0	41.5	130
upwelling zones	0.4	0.008	0.2	500
continental shelf	26.6	0.27	9.6	360
algal bed and reef	0.6	1.2	1.6	2700
estuary (open water)	1.4	1.4	2.1	1500

Table 3.3 Estimated aquatic primary productivity (after De Vooys 1979)

type	annual net primary production		major organisms
	(Gt C yr^{-1})	(% total)	
kelps	0.02	0.04	laminarians
other weeds	0.01	0.02	fucoids
angiosperms	0.49	1.1	grasses, mangroves
estuaries	0.92	2.0	algae
ocean	43.5	94.9	phytoplankton
coral reefs	0.30	0.65	coraline algae
fresh water	0.58	1.3	phytoplankton

and tropical rain-forests (dominated by woody higher plants). Some other ecosystems are more productive by unit area, such as swamps and coral reefs, but are not particularly extensive on a global scale.

The estimated primary production (in terms of C content) for various aquatic ecosystems is shown in Table 3.3. Freshwater primary production, in lakes and streams, amounts to a little over 1% of total aquatic primary production. Phytoplankton account for $c.95\%$ of marine primary production, which totals $c.40$ Gt C yr^{-1}, whereas coastal ecosystems make relatively minor contributions. Important macrophytes in intertidal zones include *Rhizophora* in mangrove swamps, turtle grass (*Thalassia*) in tropical tidal flats and eelgrass (*Zostera*) and cordgrass (*Spartina*) in salt marshes. Mangrove swamps are characteristic of warm, wet climates, and salt marshes can be considered their counterparts in cooler and/or drier coastal areas. In these coastal ecosystems algae can be important primary producers, such as benthonic diatoms and epiphytes growing on the vascular plants. In deeper water, kelp forests are important in cool temperate regions (water $<c.20°C$). Coral reefs, although extremely productive on a unit area basis, account for <1% of total aquatic, primary production, and are confined to latitudes <30° (water $>c.20°C$).

As well as CO_2 and H_2O, primary producers, and

indeed all organisms, need a variety of elements other than C, H and O (e.g. N, P, S and certain metals), as can be seen in Chapter 2. Bacteria play an important role in the cycling of these essential elements, or nutrients (see Section 3.2.4), in both terrestrial and marine environments. Light and nutrient availability are probably the main factors controlling primary production in aquatic environments and they are considered in more detail in Sections 3.2.3 and 3.2.4. However, the operation of these factors is affected by stratification of the water column and so we examine this phenomenon first.

3.2.2 Stratification of the water column

In water bodies of sufficient depth, stratification can occur as a result of density differences related to temperature and/or salinity. Stable thermal stratification arises when water warmed by solar radiation (insolation) overlies colder, denser water. At the boundary between the two there is a layer of water in which temperature changes rapidly, termed a **thermocline**. The associated sharp change in density in the thermocline effectively isolates the warm layer of water from the cold body beneath (Fig. 3.1).

(a) Oceanic stratification

As a consequence of the general deep circulation of the ocean (Box 3.2), a permanent thermocline is present in temperate and tropical oceans (at depths of $c.300$ and $100\,m$, respectively). The permanent thermocline persists throughout the year at middle and low latitudes but is absent at latitudes above $c.60°$ because the cooler climate and reduced insolation do not cause sufficient heating of the surface water to produce an adequately steep thermal gradient (Fig. 3.1a).

Above the permanent thermocline at middle and low latitudes smaller diurnal (i.e. daily) and seasonal thermoclines can develop, caused by the warming of surface waters by insolation on a daily or seasonal basis. For example, a **seasonal thermocline** can be established at depths of $c.50–100\,m$ in temperate latitudes during spring and summer, when winds are light, but it is destroyed by general cooling of surface waters and the turbulent effects of strong winds (the effects of which can extend to $200\,m$ depth) as winter approaches, resulting in mixing of the water column down to the permanent thermocline. There is insufficient seasonal contrast in insolation in the tropics for a seasonal thermocline to form, and consequently the permanent thermocline is nearer the surface (see Fig. 3.1). However, a shallow diurnal thermocline can develop in calm tropical waters.

When stratification arises from large differences in salinity, the resulting boundary layer of high salinity gradient is termed the **halocline**. Because temperature and salinity affect density, a rapid change in density is associated with thermoclines and haloclines and is termed a pycnocline. Solid material suspended in the water can also create density gradients, leading to pycnocline formation, and is important in glacial lakes and lakes fed by rivers with large suspended sediment loads.

(b) Lacustrine stratification

Thermal stratification can occur in lakes as well as the ocean. Lakes can be classified according to their mixing regimes, as described in Box 3.3. The warm, oxygenated and less dense layer above the thermocline (or

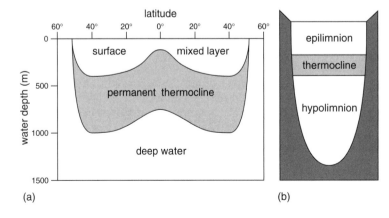

Fig. 3.1 (a) Generalized thermal stratification of the oceans; (b) corresponding nomenclature for lakes (in which the thermocline can be termed the metalimnion).

Box 3.2 | General oceanic circulation

Deep currents and the permanent thermocline

The water in the oceans behaves as though it is made up of discrete bodies that lie at depths dictated by their densities. In the North Atlantic and off Antarctica surface waters become sufficiently dense to sink to the bottom of the ocean. This is achieved by a combination of wind-induced cooling and an increase in salinity brought about by evaporation and sea-ice formation (salt is left behind in solution when water molecules crystallize as ice, a process termed brine rejection). These bottom-water masses flow over the floor of the ocean, as depicted in Fig. 3.2a, under the influence of the Earth's rotation. The general deep-water flow is from north to south Atlantic, and finally to the north Pacific via the Indian Ocean (e.g. Chester 2000). Waters of intermediate density are also formed (e.g. Antarctic Intermediate Water and the warm but high-salinity Mediterranean water, which enters the mid Atlantic at $c.1000$ m depth; Fig. 3.2b). Freshly formed dense bottom water displaces older bottom water, which has incorporated some overlying, less dense water. The result is a general upward percolation of relatively cold water, offsetting the convective downward transfer of heat from warm surface waters at middle and low latitudes (where insolation is greatest). This creates thermal stratification, involving a layer of water, termed the main or **permanent thermocline**, in which temperature rapidly decreases with depth (often exceeding 5°C per 100 m over a range of $c.15$ to 5°C). As the name implies, the permanent thermocline is a constant feature at latitudes below $c.60°$ (Fig. 3.1) and beneath it water temperature decreases steadily down to 2°C or less. The whole system of water recycling in the world's oceans is termed the **thermohaline circulation**, owing to the dominant mode of deep-water formation by cooling of waters with elevated salinity.

Surface winds and currents

Surface currents in the oceans mostly result from the frictional stress imposed by wind action and, because of the inertia of water, there is a relatively constant circulation pattern broadly reflecting the main prevailing winds. These winds include the Trade Winds, predominantly north-easterlies in the northern hemisphere and south-easterlies in the southern hemisphere. They blow from the subtropical high-pressure belts at $c.30°$ in both hemispheres towards the equatorial low-pressure system (Doldrums or Intertropical Convergence Zone (**ITCZ**)). The westerlies are another important set of winds, related to air movement from the subtropical highs towards the subpolar lows at $c.60°$ in both hemispheres (Fig. 3.3a). At still higher latitudes are the polar easterlies, which result from the deflection of air moving from the polar highs to the subpolar lows.

The different relative movements of water bodies cause various convergences and divergences at the surface. At a **divergence** surface waters in adjacent bodies move away from each other and are replaced

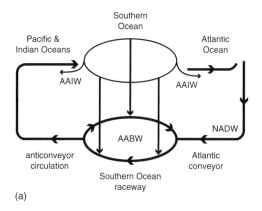
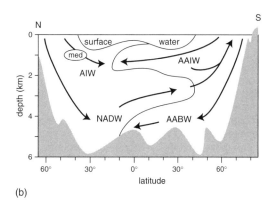

Fig. 3.2 (a) Simplified, schematic representation of oceanic deep water circulation (after Broecker 1997). (b) Water masses in the Atlantic (after Stowe 1979). AABW = Antarctic Bottom Water; AAIW = Antarctic Intermediate Water; AIW = Atlantic Intermediate Water; med = Mediterranean Water; NADW = North Atlantic Deep Water.

Continued

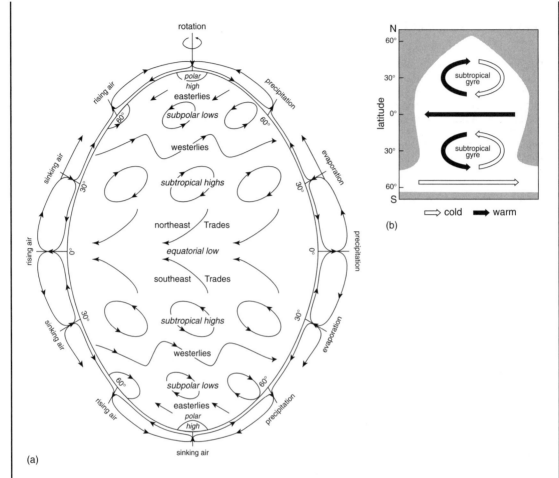

Fig. 3.3 (a) Generalized atmospheric pressure, wind and precipitation patterns on an ocean-covered Earth (n.b. sphere is elongated to aid clarity of wind patterns). (b) Idealized major oceanic surface current systems in the Pacific and Atlantic oceans. The Coriolis effect is responsible for disruption of the meridional circulation cells between 30 and 60° in (a).

by water welling up from depth. The converse occurs at a **convergence**: the more dense of the two colliding water bodies sinks. The permanent thermocline is affected in these areas, being nearer the surface at divergences and deeper (e.g. 300–400 m) at convergences (see Fig. 3.1).

The movement of water and air masses over the Earth is subject to the **Coriolis effect**, which results from the fact that the velocity of the Earth's surface in the direction of spin (eastwards) is greatest at the equator and decreases to zero at the poles. Consequently, as water moves away from the equator at the western margins of the main oceans (where it is piled up by the action of the Trade Winds), it moves progressively faster than the Earth's surface in an easterly direction. In the northern hemisphere its motion is turned progressively clockwise, while in the southern hemisphere the imposed direction of motion is anticlockwise. With continental masses acting as barriers to the east and west the result is the formation of two large, roughly circular, surface circulation cells, or **gyres**, in both the Atlantic and Pacific Oceans between $c.10°$ and 50° latitude (Fig. 3.3b). Only a single gyre is present in the Indian Ocean because it lies mostly within the southern hemisphere. The northern subtropical gyres have a clockwise circulation and the southern subtropical gyres an anticlockwise rotation, reflecting the movement of air around centres of high pressure (anticyclones) that generally overlie these regions. The Coriolis effect is also responsible for cyclonic and anticyclonic air circulation at mid latitudes (which disrupts the meridional circulation cells, or Hadley cells, shown in Fig. 3.3a).

> ### Box 3.3 | Lake classification
>
> The following classification of lakes is based on their mixing patterns (after Hutchinson & Löffler 1956).
>
> **Amictic lakes**
> Never mix, are usually ice covered year round, and so are found in polar regions or high mountains where the temperature is mostly below freezing.
>
> **Meromictic lakes**
> Mix incompletely, once or more annually (e.g. they can be dimictic). Even when the temperature is uniform throughout, mixing occurs only in the upper layer, because the large density contrast between it and the deep layer resists mixing. Over time the deep water layer reaches very high density by accumulating dissolved material, and the thickness of this layer also increases.
>
> **Holomictic lakes**
> Mix completely at least once annually, and there are four types:
> 1 Oligomictic. Relatively rare, and mostly in the tropics. The poor degree of mixing is irregular or sporadic, and usually of short duration. These lakes are generally warm throughout, but the surface waters are even warmer, so stratification can develop. The opportunity for mixing is only provided by infrequent cooling of surface waters.
> 2 Polymictic. Many mixing periods, with some lakes mixing nearly continuously all year. They are often small and shallow, and located in the tropics or at high altitude. Diurnal variations in surface temperature are often more important than seasonal changes. Circulation in some lakes occurs mostly at night by the action of convection currents rather than wind.
> 3 Monomictic. One regular period of mixing annually in cold (cold monomictic) or warm (warm monomictic) climates. Cold monomictic lakes are generally at high latitudes; they freeze over during the long winter but are ice-free in summer. Because the surface water does not warm much above 4 °C in summer, frequent mixing may occur during that relatively short season. These lakes would be defined as amictic in years of continuous ice cover. Warm monomictic lakes are typically subtropical and have the opposite cycle to cold monomictic lakes: they have a long summer and short winter, and stratification exists most of the year but for a short period in winter it breaks down and mixing can occur.
> 4 Dimictic lakes. Represent the average temperate lake. They have two mixing periods, spring and autumn, when the temperature (and hence density) of the water column becomes uniform throughout, enabling strong winds to effect an overturn.

metalimnion) is called the **epilimnion** (Fig. 3.1b). The lower layer is termed the **hypolimnion** and contains colder denser water, which may become depleted in oxygen. Circulation within each layer is independent; it is likely to be more sluggish in the hypolimnion because it is not affected by wind action.

When the water column reaches a uniform temperature (and hence density) from top to bottom it is susceptible to mixing (overturn), which is usually caused by wind action. The unusual density behaviour of water is a controlling factor. In contrast to normal seawater, which exhibits a continuous increase in density with decreasing temperature because of the salt content, the density of fresh water reaches a maximum at 4 °C (i.e. while it is still a liquid), which causes ice to float. We can examine the consequences of this property by considering a typical temperate dimictic lake (one that mixes twice a year). In winter, under ice, the water temperature generally increases with depth, from 0 °C just under the ice to 4 °C within 1–2 m, and stays much the same to the bottom. In spring, the surface water warms until it reaches its maximum density at 4 °C and the entire water column is at about the same temperature and density. Strong winds can then mix the lake from top to bottom in all but very deep lakes. As the season proceeds the winds become lighter. Most of the Sun's energy is absorbed in the top few cm, so the surface water warms above 4 °C and becomes less dense than the insulated water below, initiating stratification. Eventually the epilimnion is mixed to a depth of c.5–6 m. By early summer the lake is stratified and the epilimnion continues to mix in strong winds, to a depth of c.5–6 m. The well developed thermocline, which is c.2 m thick, prevents deep mixing. In the hypolimnion temperature falls slowly, possibly down to c.4 °C at the bottom. As autumn approaches nights become cooler and days shorter. The surface water cools and becomes more dense than the underlying warmer water, so it sinks, forming convection currents. Cooling of the epilimnion continues until the entire water column again attains a uniform temperature and density. The autumn overturn can occur without wind, but strong winds accelerate the process and cause overturn in just a few hours. Mixing ceases once the lake surface has become covered with ice.

Under abnormal conditions there are times when there is not enough wind to force overturn in a dimictic lake, or moderate winds cause incomplete or partial overturns. A lake with a relatively small fetch (small or sheltered surface area) requires strong winds to produce currents extending to the bottom, and often only the upper waters are mixed. The proportion of water in the hypolimnion depends upon the lake depth; a lake only a few metres deep would not develop a thermocline and hypolimnion, so some shallow lakes can mix entirely during the summer whenever winds are sufficiently strong.

The density of fresh water does not change uniformly with temperature; it decreases progressively more rapidly as temperature rises. For a given temperature difference between epilimnion and hypolimnion, more energy is required to mix the water masses in a stratified lake at higher temperatures because the density difference is greater. Consequently, stratification is more readily achieved in tropical than in temperate lakes.

3.2.3 Light

The quantity (number of photons) and quality (wavelength) of light affects the rate of photosynthesis and, therefore, primary production. The rate of photosynthetic production varies with the intensity of light and this is demonstrated in Fig. 3.4, which shows three distinct phases: a linear phase, a saturation plateau and a photo-inhibition phase. Each phase has different limiting factors: the linear phase is probably limited by insufficient light intensity; the saturation plateau is probably limited by the rate of dark reactions (see Box 1.10); and during the inhibition stage the high light intensity may limit photosynthesis by damaging enzymes and possibly also cell structure. Most phytoplankton live just beneath the water surface to avoid cell damage and destruction of chlorophyll by excessive exposure to UV radiation. The variation in efficiency of light utilization with wavelength for different classes of phytoplankton is shown in Fig. 3.4.

A certain amount of photosynthetic production is needed to offset phytoplanktonic respiration before

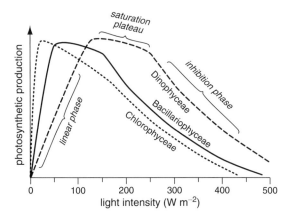

Fig. 3.4 Variation in relative photosynthetic production with light intensity for three classes of phytoplankton: green algae (dotted line), diatoms (solid line) and dinoflagellates (broken line) (after Parsons et al. 1977).

there is any surplus energy that can be channelled towards growth and reproduction. The level of light at which the photosynthetic rate is balanced by the rate of respiration is called the compensation light intensity.

In terrestrial ecosystems the amount of light available for photosynthesis is limited by factors such as day length, the angle of incidence of sunlight (both functions of latitude and season) and shading of individual leaves by the rest of the canopy. The importance of light availability is demonstrated by the short growing season and relatively low productivity at high latitudes.

There are similar restrictions on aquatic primary production (including self-shading when phytoplanktonic numbers are very high), but there is the additional effect of water depth and clarity on light penetration. The water depth at which the rate of oxygen consumed by phytoplanktonic respiration equals that produced by photosynthesis is known as the **compensation depth**, and the depth at which the amount of respiration in the water column above is equivalent to the gross primary production is the **critical depth** (Fig. 3.5). The surface layer of water in which light intensity is sufficient for photosynthesis to be possible is termed the **euphotic zone** (or photic zone). Water clarity is dependent on the amount of suspended and dissolved material in the water column, which can adsorb or scatter light, so reducing the depth of the euphotic zone. In exceptionally clear oceanic water the euphotic zone can extend down to c.200 m, while in some estuaries light may only penetrate a few centimetres because of the large amount of suspended material. Typical average depths for the

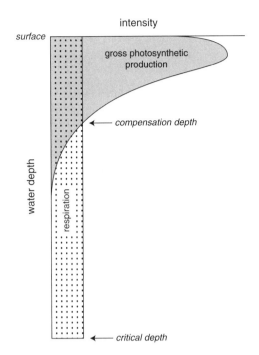

Fig. 3.5 Relationships between phytoplanktonic production, respiration and depth (after Lalli & Parsons 1997). The compensation depth is where the rate of respiration equals that of production. The critical depth is where the total respiration in the water column above equals the gross primary production (i.e. the areas under the two shaded curves are equal).

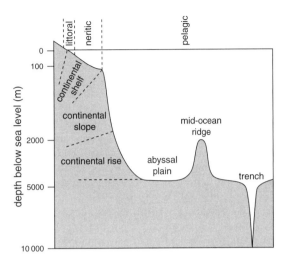

Fig. 3.6 Major regions of ocean basin floors and associated water column zones (note non-linear depth scale and exaggerated gradients).

euphotic zone in the open ocean and in coastal areas are 150 m and 30 m, respectively. In general, then, the euphotic zone lies above the permanent and seasonal thermoclines.

The maximum depth of the euphotic zone would be significantly shallower if only chlorophyll-a was involved in light adsorption. Accessory pigments, such as carotenoids, enable organisms to gain maximum use of the available light at greater depths (see Box 2.9). Much of the ocean floor lies at a depth of $c.$3700 m (the abyssal plains, Fig. 3.6) and so is below the euphotic zone, but in coastal regions where light can penetrate down to the sediment surface, benthonic plant communities can be significant primary producers on a localized basis. For example, benthonic diatoms can be prolific producers in estuaries, while productive macrophyte communities include seagrass (*Thalassia*) meadows in shallow coastal waters (down to $c.$15 m) and forests of kelp (e.g. *Laminaria*, *Macrocystis*, *Ascophyllum*) in deeper coastal waters ($c.$20–40 m depth).

3.2.4 Nutrients

The rate of growth and reproduction of organisms depends not only upon the availability of carbon, water and energy but also upon a variety of essential mineral **nutrients**. In Chapter 2 we saw that a number of elements are important, such as N in chlorophyll and amino acids, P in ATP and phospholipids, Si in diatom tests and Ca in coccoliths. Some of these essential elements (e.g. N, P, Ca and Si) are generally abundant, and so can be termed macronutrients, whereas others (e.g. Fe and Mg) are required by organisms in only trace amounts and are called micronutrients.

Nitrogen, phosphorus or silicon availability can limit primary production in the ocean. These three elements are, therefore, termed **biolimiting** (Table 3.4). More recently, iron availability has been postulated as an important control on production in certain parts of the oceans (Behrenfeld et al. 1996), so iron can be added to the list of biolimiting nutrients. The biolimiting elements, their main assimilable forms in the environment and some of their important biochemical roles are listed in Table 3.5. Phosphorus limitation has been demonstrated in freshwater lakes, with any nitrogen deficiency being made up by nitrogen-fixing cyanobacteria (Schindler 1976; Hecky & Kilham 1988), whereas terrestrial primary production is mostly controlled by the availability of nitrogen, and to a lesser extent by phosphorus. Phosphorus is utilized by phototrophes as phosphate, silicon as silicate and nitrogen mainly as nitrate but also sometimes as ammonium (by some terrestrial plants) and nitrite. In terrestrial ecosystems these and other mineral nutrients are obtained in solution from the soil via root systems, whereas in aquatic ecosystems they are absorbed directly from solution into the bodies of phytoplankton. The availability of nutrients can depend on factors like pH and E_h (Box 3.4), which affect the chemical form and solubility of nutrients. Although phosphate and silicate are the major, commonly occurring forms of P and Si, nitrate can be converted to volatile ammonia or oxides of nitrogen, which are readily lost from terrestrial ecosystems to the atmosphere (see Box 3.9). In both terrestrial and aquatic ecosystems recycling of mineral nutrients, involving the microbial decomposers, is highly important (see Fig. 1.11) and is considered in more detail in Section 3.3.2. Inputs of fresh nutrients are derived from weathering of continental crust and by nitrogen-fixing microorganisms.

In aquatic systems stratification of the water column has significant effects on nutrient availability and the timing of bursts of primary productivity. It is convenient to consider these effects in the oceans in terms of the latitudinal variations in stratification. Overall, due to the sinking of detritus out of the euphotic zone and the deep circulation pattern (Fig. 3.2), there is a flux of nutrients towards the deep Pacific.

(a) Low-latitude oceans

At low latitudes there is a permanent thermocline, but because a seasonal thermocline does not develop the surface and nutrient-rich deeper waters remain unmixed. Consequently, the euphotic zone, which lies above the permanent thermocline, becomes depleted in nutrients. Continued primary production is dependent upon the release of nutrients from detritus in surface waters. Productivity is, therefore, relatively low in tropical oceans but it is fairly constant, due to a year-round warm climate and ample sunlight (Fig. 3.7a).

(b) Mid-latitude oceans

Increasing light intensity during the spring allows phytoplankton to assimilate the available nutrients, resulting in a marked increase in productivity, giving rise to a spring phytoplanktonic bloom. As the seasonal thermocline becomes fully established, at depths approximately corresponding to the base of the euphotic zone ($c.50–100\,m$), mixing between the water layers above and below it ceases. One or more biolimiting nutrients become depleted in the euphotic zone and phytoplanktonic production falls to a level dictated by the rate of nutrient recycling within the surface layer. While most organic material is recycled within the euphotic zone some can sink below the seasonal thermocline, in the form of dead organisms and faecal material from zooplankton, which exacerbates the shortage of nutrients. Nutrients are, however, abundant beneath the euphotic zone because they are released during bacterial decomposition of detritus, but their assimilation rates are relatively low in the absence of phototrophes.

In the autumn the surface water begins to cool, reducing the strength of the seasonal thermocline and, therefore, the density difference between the water layers on either side of it. Nutrients are once again brought

Table 3.4 Classification and availability of essential elements in the ocean

category	element	ambient concentration
biolimiting elements	P, N, Si, Fe	can be almost totally depleted in surface waters
biointermediate elements	Ba, Ca, C, Ra	partially depleted in surface waters
non-limiting elements	B, Br, Cs, Cl, F, Mg, K, Rb, Na, Sr, S	show no measurable depletion in surface waters

Table 3.5 Biolimiting nutrients

element	major substrates	important biochemical forms
N	nitrate (NO_3^-); nitrite (NO_2^-); ammonium (NH_4^+)	amino acids, proteins, nucleotides, chlorophylls
P	phosphate (mostly orthophosphate HPO_4^{2-})	phospholipids, ATP/ADP, NAD/NADPH, co-enzymes
Si	silicate (mostly orthosilicic acid H_4SiO_4)	silicate tests (diatoms, radiolarians, silicoflagellates)
Fe	hydroxyoxides (FeOOH); organic-Fe(III) complexes	ferredoxins, cytochromes, nitrogenase

> **Box 3.4 | E_h and pH**
>
> **Redox potential (E_h)**
> Oxidation and reduction can be considered solely in terms of electron transfer (see Box 1.5). The electrons liberated during the oxidation of one chemical species must be taken up by another species, resulting in its reduction. Therefore, oxidation and reduction must occur together in what is termed a redox reaction (see Box 1.5). The chemical species that is oxidized gives up electrons and so is termed a reducing agent. Conversely, the species that is reduced accepts electrons and so is an oxidizing agent. This relationship can be seen in the oxidation of Fe(II) by oxygen, which is the sum of two half-reactions:
>
> $$4Fe^{2+} \rightarrow 4Fe^{3+} + 4e^- \quad \text{[Eqn 3.1a]}$$
> $$4H^+ + O_2 + 4e^- \rightarrow 2H_2O \quad \text{[Eqn 3.1b]}$$
> $$\overline{4H^+ + 4Fe^{2+} + O_2 \rightarrow 4Fe^{3+} + 2H_2O} \quad \text{[Eqn 3.1c]}$$
>
> In this reaction Fe(II) loses electrons and is oxidized, while oxygen gains electrons and is reduced. Oxidation can also be defined as the addition of oxygen or the loss of hydrogen, although the above electron transfer still operates. For example, when organic matter is oxidized by oxygen in aerobic environments the carbon is oxidized to CO_2, while the oxygen is reduced to H_2O.
>
> The tendency for a redox reaction to occur is reflected by the tendency for electrons to flow and produce an electrical current, which can be measured as a voltage, the redox potential (E_h). At the oxic/anoxic boundary in sediments $E_h = 0$, and conditions become more oxidizing (i.e. organic matter is more vigorously attacked by oxygen) as E_h becomes increasingly positive. Conversely, conditions become more reducing as E_h values increase in negative value. The value of E_h depends on temperature (T), the number of electrons involved (n) and the concentrations of oxidants and reductants, according to the Nernst equation:
>
> $$E_h = E_0 + \frac{RT}{nF} \log_e \left(\frac{[\text{oxidized species}]}{[\text{reduced species}]} \right) \quad \text{[Eqn 3.2]}$$
>
> (where E_0 = standard electrode potential, R = universal gas constant, F = Faraday constant and square brackets represent concentration).
>
> **Acidity (pH)**
> In Box 2.5 an acid was defined as a substance that donates H^+ ions. The greater the concentration of hydrogen ions in a solution, the more acidic it is. The acidity of a solution can be expressed in terms of its pH, which is defined in terms of the hydrogen ion concentration ($[H^+]$) as:
>
> $$pH = -\log_{10}[H^+] \quad \text{[Eqn 3.3]}$$
>
> Strictly speaking, hydrogen ion activity should be substituted for $[H^+]$, but effectively the latter is sufficiently accurate. Water is considered to be neutral, but it is partially ionized into H^+ and OH^- ions. The concentration of H^+ ions in pure water is 10^{-7} mol dm^{-3}, so a neutral solution has pH = 7. Below this value a solution becomes increasingly acidic as pH falls, while above it conditions become increasingly alkaline as pH rises.

into the euphotic zone as the seasonal thermocline breaks down and circulation occurs to greater depth in surface waters, giving rise to a second bloom. This bloom is usually smaller than that in the spring because the light level and nutrient supply are slightly lower. During the winter, rough conditions result in complete disappearance of the seasonal thermocline and mixing down to the permanent thermocline is re-established, renewing the nutrient supply in surface waters ready for the next spring bloom.

In the northern Atlantic the two phytoplanktonic blooms are reflected in two peaks in biomass because there appears to be a lag between the onset of rapid primary production and increase in herbivorous zooplanktonic numbers (Fig. 3.7b). During the summer the zooplankton graze down the phytoplankton and their numbers decline before the second algal bloom occurs, whereupon zooplanktonic reproduction again follows the bloom (Barnes & Hughes 1988). In contrast, in the northern Pacific phytoplanktonic biomass appears relatively constant, although there is the same overall pattern of estimated primary production as in the Atlantic (Fig. 3.7c). The absence of a spring maximum in phytoplanktonic biomass seems to arise because of a different strategy employed by the major Pacific herbivorous zooplankton (copepods), which use reserves of food stored over the winter to enable them to breed before the spring bloom. Consequently, large numbers of

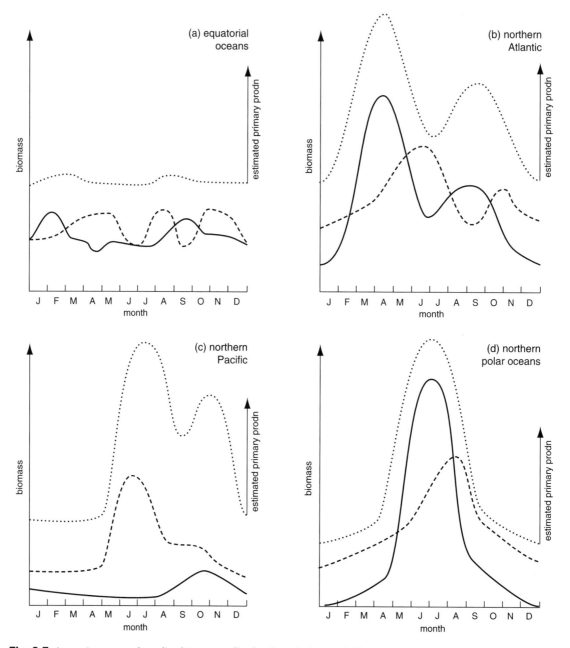

Fig. 3.7 Approximate annual standing biomass profiles for phytoplankton (solid lines) and zooplankton (broken lines) in various regions of the oceans. Estimated phytoplanktonic production is indicated by dotted lines (after Heinrich 1962; Parsons et al. 1977).

zooplankton are present and effectively graze down the phytoplankton as they appear (Barnes & Hughes 1988). Zooplanktonic numbers decline somewhat in the late summer with the exhaustion of the initial phytoplanktonic bloom, resulting in the autumn bloom showing up as a slight increase in phytoplanktonic biomass.

(c) High-latitude oceans

In polar regions sufficient light intensity for photosynthesis is only available for five or six months of the year. There is no permanent thermocline at latitudes above 60° because there is insufficient insolation to create a significant vertical temperature gradient. As a result, nutrients are constantly being replenished in the euphotic zone, leading to a single large algal bloom during the summer (Fig. 3.7d).

(d) Stratified lakes

In temperate lakes phytoplanktonic blooms occur in response to the seasonal availability of nutrients, which is controlled by stratification of the water column (see Section 3.2.2b). The situation is, therefore, similar to seasonal thermocline development and decline in temperate oceans. During summer stratification nutrients in the epilimnion become depleted. During the spring and autumn overturns, nutrient-rich (and oxygen-deficient) waters of the hypolimnion mix with the epilimnion and fuel phytoplanktonic blooms, primarily in the spring.

We have seen in Section 3.2.2b that, when fully developed, thermal stratification is very stable in tropical lakes. Consequently, overturning of the water column depends upon wind strength. Overturn is a rare event in oligomictic lakes (Box 3.3), so nutrient levels in the epilimnion are low, productivity is low and the lakes can be desribed as **oligotrophic**. In contrast, winter winds are sufficiently strong to cause overturn in warm monomictic lakes (Box 3.3), and nutrient levels are replenished in surface waters.

In very deep tropical and temperate lakes, mixing seldom occurs to the very bottom. Lake Tanganyika (East Africa) and Lake Baikal (Siberia) are examples of such meromictic lakes (Box 3.3). Lake Baikal is the deepest (1620 m) and largest freshwater lake in the world, containing >10% of the total standing fresh water, and it is well oxygenated to at least 500 m depth. In contrast, the waters below *c.*200 m in Lake Tanganyika, which is only slightly less deep at 1400 m, are permanently anoxic. The great depth of these lakes arises from their position over extensional axes in continental crust (see Box 3.11).

In shallow lakes at mid and low latitudes it is not possible for a steep density gradient to become established and so if any stratification occurs it is weak and readily broken down, sometimes on a diurnal basis, during nightly cooling. In such polymictic lakes (Box 3.3) nutrient supply is virtually constant, so other factors, such as light and temperature, control any variations in phytoplanktonic production. Primary production can be higher in these lakes than in those where nutrients are immobilized within the hypolimnion for long periods.

3.2.5 Spatial variation in marine primary production

The previous sections have given a general idea of aquatic primary productivity. However, production is not uniform over the surface of the oceans because of the effect of upwellings and downwellings on nutrient availability (see Box 3.5), which leads to some areas being particularly productive while others are the oceanic equivalent of deserts. The general pattern can be seen in Fig. 3.8, and it has important implications for the likely location of the deposition of organic-rich sediments. Estimates of global marine primary production can be achieved by satellite monitoring of light absorption by chlorophyll-*a*.

The centres of the main subtropical gyres (see Fig. 3.3b) correspond to convergences (Box 3.2), where nutrient supply is limited and primary productivity extremely low, as can be seen in Fig. 3.8. Such nutrient-depleted areas are termed oligotrophic (as for lakes). Conversely, in areas of divergence, phytoplanktonic production is fed by nutrient-rich waters welling up from depth. These **eutrophic** areas are extremely productive, as can be seen from Table 3.1, although their total area is limited. Among the most productive of these areas are the upwellings associated with the eastern boundary currents of the subtropical gyres (Box 3.5).

It can be seen that the shape of the ocean basins and the distribution of continental masses have an important effect on currents and the location of high productivity zones. The degree of flooding of continental margins, which is controlled by the relative volumes of ocean basins and seawater, is also important in terms of the area of productive coastal waters. Nutrient supply from land, via rivers, can contribute to the productivity of shelf seas.

3.2.6 Variation in phytoplanktonic populations

The main classes of phytoplankton are given in Table 3.6 and most are represented in both seawater and

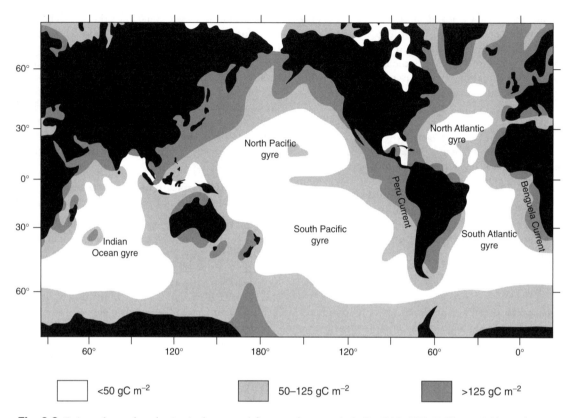

Fig. 3.8 Estimated annual production in the oceans (after several sources, including FAO 1972; Koblenz-Mishke et al. 1970).

fresh water, although species vary between the two environments. In the oceans, the prochlorophytes and cyanobacteria (picoplankton) are numerically very important. In addition, aerobic photoheterotrophic bacteria (see Box 1.10) may account for 5–10% of all marine bacteria (Kolber et al. 2001). Among the nanoplanktonic classes, most of the Haptophyceae are marine, whereas the Chrysophyceae are mainly freshwater. The Euglenophyceae are also predominantly found in fresh water.

There are latitudinal variations in the major families of phytoplankton responsible for marine primary production. Diatoms appear particularly important at high latitudes and along the equatorial belt, but in intermediate latitudes dinoflagellates and coccolithophores are more important. Cyanobacteria and the related prochlorophytes appear to make a major contribution within 40° of the equator. There can also be variations in phytoplanktonic communities with depth. For example, in the subtropical mid-Pacific diatoms are responsible for c.80–90% of total algal production in the surface waters (where their production is probably nutrient-limited), but in deeper waters (although still above the thermocline at c.85–180 m) coccolithophores are the largest group of primary producers (their activity is probably light-limited; Venrick 1982). The prochlorophytes congregate at depths of 100–200 m, where the wavelength range of penetrating sunlight is efficiently harvested by their divinylchlorophylls (chlorophylls a_2 and b_2, Fig. 2.27; Partensky et al. 1999).

Planktonic cyanobacteria can be very important in eutrophic lakes, particularly in the tropics. Eutrophic lakes and tropical lakes in general tend to contain mainly larger species (e.g. Euglenophyceae and Chlorophyceae), whereas cold lakes and oligotrophic lakes tend to be dominated by small flagellates (e.g. Chrysophyceae).

In estuaries phytoplankton are generally less important than benthonic and littoral (shoreline; Fig. 3.6) communities, but benthonic diatoms can be important. The larger plants (macrophytes) of salt marshes and mangrove swamps may host a significant community of

Box 3.5 | Oceanic upwelling and downwelling

The subtropical gyres (Fig. 3.3b) tend to draw warm surface water towards their centres, resulting in a convergence that depresses the main thermocline to as deep as 400 m (Fig. 3.1) and prevents nutrients from deep water coming to the surface (see Box 3.2). In contrast, there are also smaller subpolar gyres in the northern hemisphere, underlying the subpolar low-pressure belt, in which surface water moves outwards, forming a divergence.

Large divergent or convergent frontal zones can form in continental boundary currents. For example, where the subtropical gyral currents travel towards the equator along roughly north–south trending western coasts — such as the Peru Current in the Pacific and the Benguela Current in the Atlantic (Fig. 3.8) — the prevailing winds in both hemispheres cause a net transport of warm surface water offshore (a divergence; Fig. 3.9a), due to the influence of the Coriolis effect (Box 3.2). This causes upwelling of deep, cool, nutrient-rich water (from up to $c.500$ m depth; Fig. 3.9b). Another example is the Antarctic Divergence, which supports vast stocks of krill. On the western boundary currents of the subtropical gyres, where water flows away from the equator, the Coriolis effect causes a convergence, in which warm, nutrient-poor water accumulates (major coral reefs are often associated with such convergencies).

Other large frontal zones can form where two water masses meet. Divergence at the front between surface currents at the equator results in an upwelling that supports the high productivity tongue seen extending westward across the Pacific from the Americas in Fig. 3.8. A major convergence between surface waters occurs at the Antarctic Polar Front (at $c.58°$ S), which is the source of Antarctic Intermediate Water (Fig. 3.2b).

Frontal systems can also develop at shelf breaks, where shallowing and the associated increase in current velocity can result in upwelling and high primary productivity. Smaller frontal zones can be produced in the wake of islands, resulting in upwelling and elevated productivity (e.g. Hawaii and Isles of Scilly).

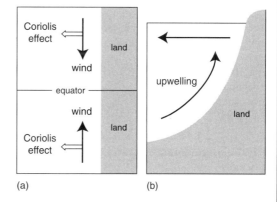

Fig. 3.9 Upwelling induced by divergence at eastern boundary currents of subtropical gyres. (a) Plan view showing forces involved; (b) cross section showing movement in water column.

Table 3.6 Phytoplankton classes

class	description and occurrence*
Bacillariophyceae	diatoms [M + F]
Bangiophyceae	subclass of Rhodophyceae [M]
Chlorophyceae	green algae [M + F]
Chrysophyceae	chrysomonads, or golden algae (include silicoflagellates) [M + F]
Cryptophyceae	cryptomonads [M + F]
Cyanophyceae†	cyanobacteria [M + F]
Dinophyceae‡	dinoflagellates [mostly M]
Euglenophyceae	green flagellates [F + B]
Eustigmatophyceae	similar to xanthophytes [mostly F]
Haptophyceae	yellow-brown algae (e.g. prymnesiophyte coccolithophores) [mostly M]
Prasinophyceae	prasinomonads, similar to chlorophytes [M + F]
Prochlorophyceae	picoplankton related to cyanobacteria [M]
Raphidophyceae	chloromonads, similar to dinoflagellates [M]
Xanthophyceae	yellow-green algae [mostly F]

*M = marine, F = freshwater, B = brackish. †Although bacterial, this class is sometimes included among the algae due to its previous classification as blue-green algae. ‡Only some dinoflagellates are strictly autotrophic, $c.50\%$ possess no chloroplasts and are strict heterotrophes, and some are capable of both forms of production. Aerobic photoheterotrophic bacteria are not included in this table (see Box 1.8).

attached microalgae (**epiphytes**). The flow rate of rivers is generally too great to support major phytoplanktonic communities.

During phytoplanktonic blooms in seasonally stratified water bodies the major families of algae present change as the season progresses, a process known as **succession**. In dimictic lakes the first to appear are the faster-growing diatoms, which later decline as silicate becomes depleted. They are followed by other algal classes that require a different balance of nutrients, and finally, if stratification persists, nitrogen-fixing cyanobacteria may form dense blooms (Reynolds 1984). A simpler succession occurs in temperate oceans, with diatoms being followed by dinoflagellates. The succession probably results from varying requirements for light, temperature and specific nutrients between each family and also from grazing pressure. Within each family a succession of species can also be observed. Such successions are influenced by the presence of trace chemicals released by preceding species, some of which (e.g. vitamins) act as promoters, while others inhibit the growth of potentially competing species (Barnes & Hughes 1988).

3.2.7 Biolimiting nutrient variations in the oceans

The question of which nutrient controls the overall primary productivity of the oceans is vigorously debated. Geochemists have tended to consider that phosphate is the main biolimiting nutrient over geological periods (Box 3.6), not least because its primary source is the weathering of continental rocks, whereas nitrogen can be fixed by a few types of cyanobacteria (especially *Trichodesmium*) as well as being transported from the land as nitrate in river outflows. However, biologists have demonstrated that N is biolimiting in large areas of the ocean. This can be seen from the atomic N:P ratio (known as the **Redfield ratio**) of 16:1 in plankton (and also in early degradation products), whereas deep waters exhibit a ratio closer to 15:1 (Falkowski 1997). The apparent paradox may be explained by the interaction of communities of nitrogen-fixing cyanobacteria and major non-N-fixers, such as diatoms. Under conditions of abundant nitrate the cyanobacteria are at a competitive disadvantage because nitrogen fixation is more energy intensive (see Box 3.9): the numbers of cyanobacteria fall and levels of available nitrate effectively decline until equilibrium between N and P is re-established. If phosphate increases in concentration relative to nitrate, the proportion of N-fixers increases, again re-establishing the N:P equilibrium.

So, although N may be the proximate limiter of primary production, P is likely to be the ultimate limiter (Tyrrell 1999). The dependence of nitrogen fixation on iron availability introduces a complicating factor (Wu et al. 2000).

Three oceanic regions have been distinguished with regard to macronutrient status and apparent productivity (Dugdale & Wilkerson 1992):
- HNHC: high nutrient, high chlorophyll (i.e. high productivity, eutrophic).
- LNLC: low nutrient, low chlorophyll (oligotrophic).
- HNLC: high nutrient, low chlorophyll.

The HNLC areas were enigmatic until the discovery of the role of iron in phytoplankton production. In the three main HNLC areas—the equatorial Pacific, subarctic Pacific and Southern Ocean—the phytoplanktonic communities are generally dominated by picoplankton. There are also periodic diatom blooms, from which a large proportion of the detritus sinks out of the euphotic zone, so silicate levels are also often low (Dugdale et al. 1995). Iron concentrations are very low, which in the subarctic Pacific and Southern Ocean may be due to the low aeolian contribution in the form of continental dust (Martin et al. 1991).

The biolimiting effects of iron (Box 3.7) have been demonstrated by the large-scale addition of soluble iron to areas of surface water in the equatorial Pacific (IronExI, Martin et al. 1994; IronExII, Coale et al. 1996) and the Southern Ocean (Boyd et al. 2000). Addition of iron, whether artificially or via natural processes such as upwellings, causes a shift in the dominant phytoplanktonic members, from picoplankton to the larger types of diatoms (Chavez et al. 1991). This is believed to be because the picoplankton have a larger surface area to volume ratio, and so have a competitive advantage in nutrient-poor waters. The main grazers are protozoans, which can respond rapidly to changes in picoplanktonic production, so organic carbon is efficiently recycled in the HNLC regions via grazing. However, a transient flux of biolimiting nutrient favours large diatoms, which can store nutrients in vacuoles for later use (Chavez et al. 1991). The resulting blooms can produce a significant amount of detritus that sinks out of the euphotic zone. This detritus mostly escapes the attention of the grazers because the larger zooplankton involved have larval growth stages, so there is a lag of a few days between the onset of the bloom and the appearance of large numbers of the grazing forms (Falkowski et al. 1998). The HNLC areas are, therefore, probably the result of the interplay of at least three factors—grazing, iron limitation and nutrient export in sinking particulates—which control primary produc-

Box 3.6 | Phosphorus as the ultimate oceanic biolimiter

If nitrate were the biolimiting nutrient, there is no reason why oceanic primary production should not vary significantly over geological periods, depending upon changes in the supply of nitrate from the continents and *in situ* nitrogen fixation by cyanobacteria (see Box 3.9). Large increases in nitrate supply would lead to enhanced primary production and increases in atmospheric oxygen concentrations. Conversely, the periodic exposure and oxidation of large deposits of pyrite and organic-rich sediments would lower atmospheric oxygen concentrations. Without obvious stabilizing feedback mechanisms, oxygen levels could vary dramatically, but throughout the Phanerozoic they seem to have remained close to present-day levels (Fig. 1.8). This is thought to be due to the stabilizing mechanism of a negative feedback involving phosphate bioavailability (Fig. 3.10; van Cappellen & Ingall 1996).

Although the main source of phosphate is the weathering of continental rocks, the efficiency of phosphate recycling within the oceans plays a key part in the availability of phosphate to phytoplankton. As primary production increases and atmospheric oxygen builds up, the oceanic water column (and most importantly bottom water) becomes more oxidizing, causing an increase in the amount of Fe(III) hydroxyoxides, which readily complex with the phosphate and precipitate. The resulting decline in available phosphate gradually (over the time-scale of oceanic turnover) provides a negative feedback on primary productivity, so atmospheric oxygen levels reach a maximum and then decline, as respiration and other oxygen-depleting reactions outweigh photosynthetic generation. As the oxygen content of bottom waters falls, aerobic respiration can bring about reducing conditions more readily, resulting in the liberation of phosphate from reactive sediments, eventually stimulating higher levels of primary production and causing atmospheric oxygen levels to rise once more. This model demonstrates the coupling of the carbon, phosphorus, oxygen and iron cycles.

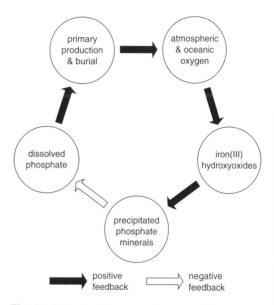

Fig. 3.10 Feedback between redox-controlled phosphate recycling in the oceans and atmospheric oxygen levels.

tivity and the composition of the phytoplanktonic community (Chisholm & Morel 1991).

3.2.8 Variations in higher plant populations

Because light levels (which control temperature, total productivity and the duration of seasonal growth outside the tropics) and rainfall exhibit latitudinal variations (Fig. 3.12), it is not surprising that the distribution of the major higher plant associations on land (**biomes**) are arranged in broadly latitudinal belts, as shown in Fig. 3.13. The general trends are modified by factors such as altitude, mountain rain-shadows and the thermal influence of oceanic currents. For example, the southern limits of boreal and temperate forest in North America occur at lower latitude than in western Europe because of the warming effect of the North Atlantic Drift. Superimposed on the generalized regional pattern in Fig. 3.13, factors such as soil condition (and the related nutrient availability) play an important role in determining the local composition of plant assemblages.

The biome pattern is somewhat more complex than the obvious climatic tropical, arid, temperate and polar zonation. Another representation is given in Fig. 3.14, which shows the influence of mean annual precipitation and temperature, and can be used to predict

Box 3.7 | Iron as a biolimiting element

Iron is a vital component of the proteins involved in photosynthesis, respiration and nitrogen fixation, but under oxic conditions it is predominantly found in its extremely low solubility Fe(III) form and so is not readily available for uptake by organisms. Unlike anoxic environments, in which soluble Fe(II) is generally abundant, the well oxygenated surface waters of the open oceans can contain as little as $10^{-8}\,g\,l^{-1}$ of

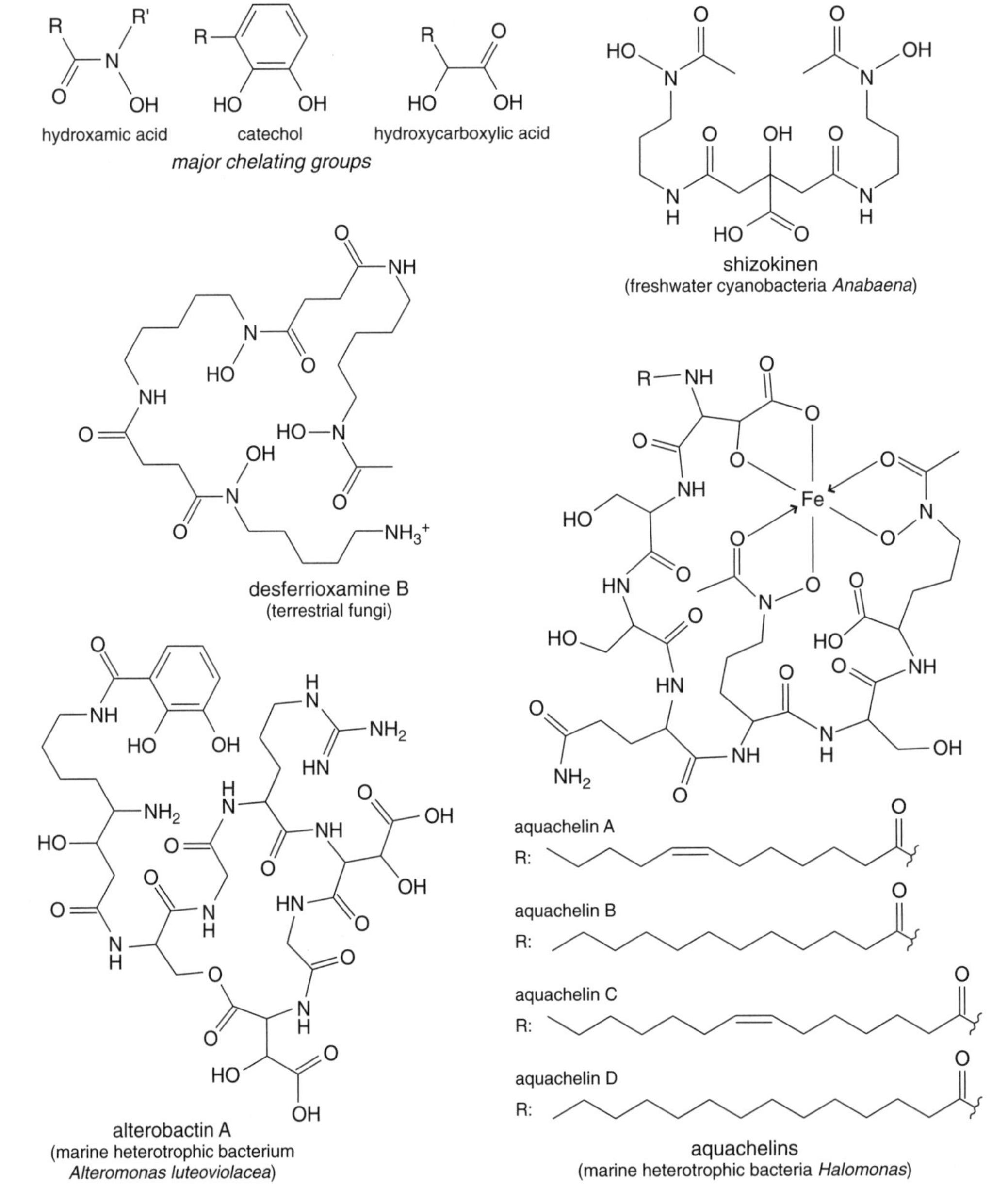

Fig. 3.11 Major iron(III)-chelating groups in siderophores and structures of some siderophores.

Continued

dissolved iron. In oceanic surface waters, inorganic Fe(III) is present as hydroxyoxides, which, with increasing age, undergo dehydration and crystallization, decreasing the bioavailability of the iron. However, >99% of the element in these waters is bound to organic compounds, which are as yet poorly characterized (Hutchins et al. 1999), but are known to bind particularly strongly with Fe(III), maintaining its solubility and potential bioavailability.

Microbes and higher plants release a range of organic compounds (**siderophores**) into aerated soils to sequester Fe(III); an example of a fungal siderophore is shown in Fig. 3.11. Siderophores can be termed **chelators**, because each molecule surrounds one ferric ion and forms several bonds to it, mostly via hydroxamate, catecholate and hydroxycarboxylate units. Marine heterotrophic bacteria and cyanobacteria have been found to synthesize siderophores (see Fig. 3.11 for examples), although production is only activated under conditions of extreme iron depletion, due to the high energy requirements and the potential loss of siderophores in the open ocean. Iron complexes with siderophores bearing an alkyl chain, like the aquachelins (Fig. 3.11), are amphiphilic and can spontaneously form membrane-like vesicles (Martinez et al. 2000; see Box 1.7). The iron–siderophore complexes are transported into bacterial cells where they undergo reductive decoupling to release Fe(II).

No conclusive evidence has been found for siderophore synthesis by algae (i.e. eukaryotic phytoplankton), although they appear to be able to obtain iron from chelated forms via a non-specific ferrireductase system on the exterior of the cell membrane (Hutchins et al. 1999). This system is more effective for iron porphyrins (possibly originating from zooplanktonic detritus), in which the square-planar chelated iron is more accessible than in the octahedrally chelated siderophores (compare Figs 2.28 and 3.11).

All phytoplankton can assimilate the more labile forms of dissolved inorganic Fe(II) and Fe(III), and this is the main assimilation mode for bacteria under iron-replete conditions (Granger & Price 1999). Extracellular bioreduction is an important route to these labile forms, but the photoreductive dissociation of Fe(III) chelates in oceanic surface waters may play a major role in sustaining phytoplanktonic productivity (Sunda & Huntsman 1995), as has been proposed for the aquachelins (Barbeau et al. 2001).

The differences between prokaryotic and eukaryotic phytoplanktonic uptake of iron must affect competition and hence the composition of primary producer communities (Hutchins et al. 1999). The rate of transport of iron into cells depends upon the number of receptors on the membrane surface, so low iron concentrations favour growth of the picoplankton, which have a large surface area to volume ratio.

---

terrestrial biome types present under the various climatic regimes. For example, rain-forests in the tropics could disappear if mean temperatures were to increase by a few degrees.

Terrestrial biomes are not static assemblages; succession of species is a common theme, reflecting the ability of plants to infiltrate newly created open spaces and compete against each other. The primary colonizers are generally outcompeted by later arrivals, during an increase in diversity, but eventually diversity decreases as a relatively few species dominate in the climax populations. At any time in a forest the entire succession is likely to be represented, as older trees die off and pioneer plants move in. Larger open areas can be created by climatic or other environmental changes, such as fires. When pasture is removed from grazing the most competitive grasses dominate, but gradually scrub takes over and eventually woodland. A dramatic demonstration of a tundra to grassland to boreal forest to temperate forest succession can be found in the pollen record for the end of the last glacial episode in the northern hemisphere. Grasses and herbs were succeeded by birch (pioneers among the trees), then by pines, and subsequently deciduous trees: hazel and elm at first and finally oak, lime and alder (Bennett 1983). It is assumed that the temperate forest species migrated northward from refuges during the rapid warming, at rates determined by factors such as the age at which reproductive maturity is reached and the range of seed dispersal.

3.3 Preservation and degradation of organic matter

3.3.1 Fate of primary production in the water column

Much of the organic material produced by phytoplankton is consumed within the euphotic zone by herbivorous zooplankton. The most important members of the

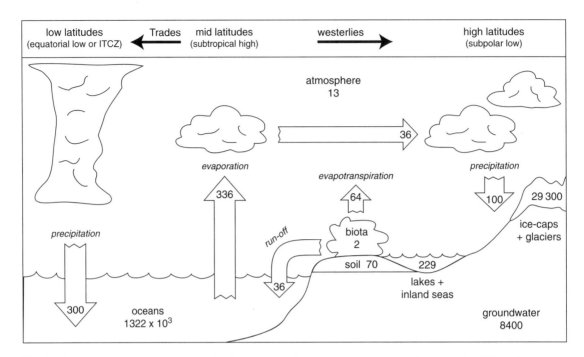

Fig. 3.12 Summary of the hydrological cycle, showing sizes of main reservoirs and annual fluxes in units of Tt, and relationship of major regions of evaporation and precipitation to prevailing winds and latitudinal atmospheric pressure zones (see Box 3.2 and Fig. 3.3a; after Colling et al. 1997). ITCZ = intertropical convergence zone.

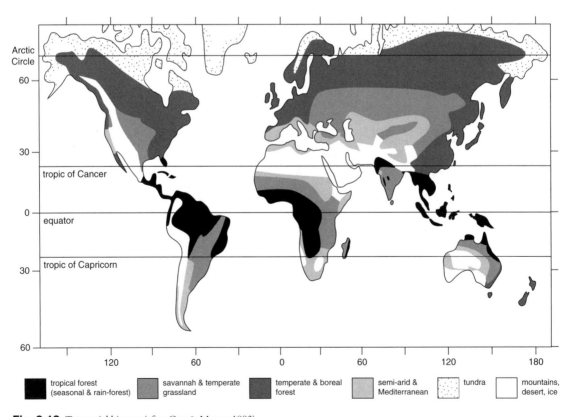

Fig. 3.13 Terrestrial biomes (after Cox & Moore 1993).

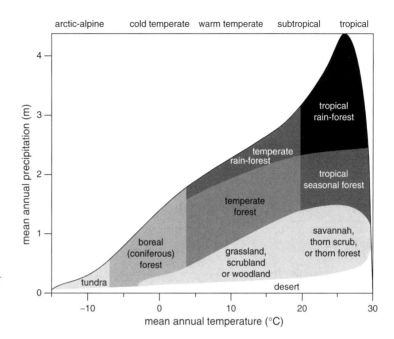

Fig. 3.14 Approximate correlation of terrestrial biomes with temperature and precipitation patterns (after Whitaker 1975).

marine zooplankton are the copepods, which belong to the macroplankton (Fig. 1.9). However, herbivorous protozoa also exist, which are members of the microplankton (Fig. 1.9), comprising chiefly ciliates in fresh water and rhizopods (mainly radiolarians and foraminiferans) in the oceans. These protozoa can be significant sources of sedimentary material, an example being the deep-sea calcareous oozes to which foraminiferans such as *Globigerina* may be major contributors. A simplified representation of carbon flow through microorganisms in the marine water column is shown in Fig. 3.15. It can be seen that viruses play a role in regulating bacterial populations.

Detritus, comprising the remains of various planktonic organisms and zooplanktonic faecal pellets, sinks down through the water column. Zooplanktonic faecal pellets, being larger, fall more quickly through the water than the phytoplanktonic remains ($c.160$ m day^{-1} for faecal pellets cf. 0.15 m day^{-1} for coccolithophores). However, the phytoplanktonic remains tend to form fluffy aggregates of 'marine snow', which settles more rapidly than the remains of individual organisms. Differences in exposure time to degradation in the water column are likely to influence the amount of organic matter reaching the sea floor, with large faecal pellets and 'snow flakes' (>200 μm diam.) being relatively more

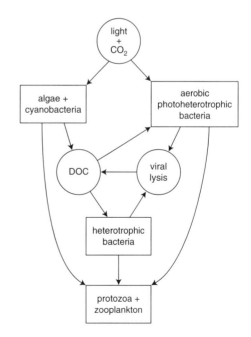

Fig. 3.15 Simplified carbon flow through microorganisms in the marine water column (after Fenchel 2001).

important in the formation of organic-rich sediments (Suess 1980). There is evidence in continental margin settings that the pelagic snow undergoes some disaggregation and recombination with terrigenous detrital material at or near the sea floor before its final incorporation into marine sediments, which may affect the degree of preservation of the organic matter (Ransom et al. 1998). Zooplanktonic grazing can be less efficient during phytoplanktonic blooms (particularly diatom blooms; see Section 3.2.7), and proportionally more organic matter may reach the sediment than in times of lower productivity. Particulate organic matter export from the base of the euphotic zone can reach $c.200\,\text{mg}\,\text{C}\,\text{m}^{-2}\,\text{day}^{-1}$ in upwelling regions, which is almost an order of magnitude greater than generally observed in open oceanic areas (Martin et al. 1987; Wefer & Fisher 1993).

The particulate material is colonized by bacteria, which further break down the organic matter, releasing dissolved organic compounds and nutrients that can be assimilated in the euphotic zone by phytoplankton. Some dissolved organic matter (DOM) is adsorbed on to clay and other mineral particles and is transported to the sediment. Most of the nutrients and organics are recycled within the water column (Wakeham et al. 2002), and details of the degradation of some lipid classes are considered in Section 5.3.2. The rate of decline in organic matter abundance and of change in lipid compositions with water depth appear to be similar for a wide range of primary production rates (e.g. there is a characteristic exponential decline in planktonic remains; Wakeham et al. 1997). Consequently, the amount of organic material that reaches the sediment is dependent on the depth of water through which it passes and the amount of primary production in the euphotic zone. The deeper the water, the longer the residence time and the longer the particles are exposed to degradation. Generally, the amount of organic material that reaches the bottom decreases by a factor of about ten for every tenfold increase in water depth.

The flux of particulate organic C seems to be linked to that of biominerals (silicate and carbonate) in the deep ocean, and suggests that the minerals present in various phytoplanktonic classes provide some protection to the organic matter during zooplanktonic herbivory and also aid transportation to the sea floor (Armstrong et al. 2002). It might be expected that the bulk composition of organic matter falling through the water column would change significantly, given that systematic changes are observed in lipids and other identifiable components, and that 98% of the organic matter is degraded (Wakeham et al. 2002). However, bulk compositional data from ^{13}C nuclear magnetic resonance (NMR) analyses reveal little change in the overall composition, and amino-acid-like material is a major component throughout the water column (Hedges et al. 2001). This seems to suggest that degradation of the organic matter is not selective of particular compound classes, and that the material that is not degraded undergoes only relatively minor chemical alteration, which could reflect a degree of protection within inorganic matrices.

A significant proportion ($c.80\%$) of the organic matter that accumulates in sediments resists molecular characterization by existing analytical techniques (Hedges et al. 2000). Identifiable molecules constitute $c.80\%$ of the organic C in phytoplankton and particulates sinking from the euphotic zone, but only $c.25\%$ in deep-water particulates and generally only $c.20\%$ in surface sediments in the equatorial Pacific (Wakeham et al. 1997). In feeding experiments it has been found that a chemically uncharacterized, but relatively stable, component of diatoms is concentrated in copepod faecal pellets (Cowie & Hedges 1996). The isotopic similarity of marine particulate organic carbon to extractable lipids suggests that the material is derived predominantly from lipid-like macromolecules (Hwang & Druffel 2003). An uncharacterized fraction of organic matter is ubiquitous in marine sediments, soils and natural waters. Another poorly characterized component of sedimentary organic matter that may be important in many environments is recalcitrant black carbon (which seems to be produced by varying degrees of charring of terrestrial organic matter: Masiello & Druffel 1998; Hedges et al. 2000).

(a) Dissolved organic matter

It is important to consider dissolved organic matter (DOM) in a little more detail because it is quantitatively very important and, as seen in Section 3.3.4b, may make a significant contribution to the organic matter preserved in sediments. The dissolved organic carbon (DOC; see Fig. 3.15) in the DOM is the single largest carbon pool in the oceans: at $c.700 \times 10^{15}\,\text{g}$ it represents $c.96\%$ of total marine organic C (Hedges 1992), although the accuracy of its measurement has been questioned (Suzuki 1993). DOM comprises truly soluble material and colloids ($\leq 0.45\,\mu\text{m}$ diam), and up to 50% may be high molecular weight (>1000) colloids (Lee & Wakeham 1992). However, most of it—apart from a minor proportion in the euphotic zone—seems to be refractory and is several thousand years old (Carlson & Ducklow 1995). It is poorly characterized and has been termed Gelbstoff ('yellow substances', synonymous with humic material; see Section 4.2). In common with soils and surface sediments, the organic nitrogen in marine DOM appears to be predominantly in the form of amides, which suggests a biotic origin, possibly with a major contribution from the peptidoglycans of eubacteria (McCarthy et al. 1998; Hedges et al. 2000; Ogawa 2001; see Section 2.2.2).

Photolysis in surface waters may transform a part of this recalcitrant DOM into low molecular weight compounds that are available for microbial uptake (Mopper et al. 1991; Hedges 1992). Manganese oxides have also been reported to oxidize humic substances spontaneously, forming some simple compounds such as acetaldehyde and pyruvate, which are readily assimilable by microorganisms (Sunda & Kieber 1994).

3.3.2 Sedimentary fate of organic material

(a) Aerobic decomposition

When detritus and colonized particles are deposited in sediment (whether aquatic sediments or soils), their fate depends on the rate of burial and amount of oxygen present at the sediment surface and within the interstitial spaces. Where there is sufficient oxygen, organic material is assimilated by the benthonic community of detritivores: deposit feeders and, additionally in subaquatic environments, suspension feeders. The detritus that escapes the deposit feeders passes to the decomposers: aerobic heterotrophic bacteria and fungi in the oxic zone. The organic matter excreted or partially disseminated by the detritivores has a high surface area to volume ratio and so is more readily available to the decomposers, which use extracellular digestive enzymes. Decomposers inevitably form part of the diet of deposit feeders, being ingested with the detritus upon which they are working. The action of burrowing detritivores (e.g. polychaete worms) mixes the sediment (**bioturbation**) and aids oxygenation (see also Section 3.3.4a).

After death, cells self-destruct (the process of **autolysis**) under the influence of hydrolytic enzymes, which, in life, aided the recycling of cellular components. This process makes proteins and other components more readily available to the decomposers. Bacteria and fungi preferentially remove the more labile components from detritus and the residue becomes increasingly refract-ory. Much of the soluble product of the microbial breakdown of organic matter diffuses upward within pore waters to the sediment–water interface and is returned to the water column. Bacteria are important in all environments, but fungi are relatively more important in the degradation of plant material on land than they are in the decomposition of planktonic remains. Single-celled forms (yeasts) dominate aquatic fungi (although the filamentous *Cladosporium* is common in fresh water, and various oomycetes and chytridiomycetes are present in fresh water and soils), whereas colonial types are most important on land, producing the familiar fruiting bodies found in woodlands (many ascomycetes and basidiomycetes, together with fungi imperfecti such as *Aspergillus* and *Geotrichum*).

Decomposition of organic material (**mineralization**) occurs rapidly under aerobic conditions. The rate of oxygen supply to the sediment is a critical factor and is influenced by sedimentary particle size. The restricted size of pores in fine-grained silts and clays results in rapid reduction of water circulation within the sediment with increasing depth and, ultimately, oxygen can only enter pore waters by diffusion (Box 3.8). The amount of organic matter in the sediment affects the balance between the rate of oxygen consumption during aerobic degradation and the rate of diffusion of oxygen into sedimentary pore waters. Pore waters in open marine (**pelagic**) sediments in which oxygen consumption is low are often oxygenated to depths of 0.5 m or more (Murray & Grundmanis 1980). In contrast, oxygenation is often limited to the top few millimetres of fine-grained shelf sediments, although oxygen penetration can extend to a few centimetres as a result of bioturbation and in the walls of well ventilated burrows (Revsbech et al. 1980).

As oxygen levels fall within the sediment and conditions become dysaerobic (see Box 1.5) bioturbation

---

Box 3.8 | Diffusion

Diffusion is a process caused by the random motion of molecules in a fluid, resulting in the net movement of the molecules of a particular compound from regions of higher to lower concentration of that compound. At constant temperature all the molecules share the same average kinetic energy ($\frac{1}{2}mv^2$), so the molecules of smallest mass have the highest mean velocities and will diffuse at the fastest rate. For ideal gases (see Box 1.11) the ratio of the rates of diffusion of two different types of molecules (a and b) is expressed simply by Graham's Law:

$$\frac{V_a}{V_b} = \frac{\sqrt{m_b}}{\sqrt{m_a}} \qquad [\text{Eqn 3.4}]$$

ceases. If oxygen demand outstrips supply the sediment and sometimes the overlying water column become completely anoxic. In eutrophic lakes anoxicity usually develops within the hypolimnion (Fig. 3.1). Under such conditions the activity of bacteria that are **obligate** (i.e. strict) aerobes (e.g. actinomycetes, which can form up to one-third of soil bacteria, and certain species of *Bacillus*, *Pseudomonas*, *Corynebacterium* and *Flavobacterium*) is severely restricted.

(b) Anaerobic decomposition

Mineralization of organic matter continues under anoxic conditions, due to the activity of various anaerobic bacteria (the main group being the Clostridia), but the overall rate is slower. As well as obligate anaerobes (e.g. *Clostridium*), these bacteria include aerotolerant anaerobes (e.g. *Lactobacillus*) and **facultative anaerobes** (i.e. bacteria that are usually aerobes but can function anaerobically, e.g. *Escherichia*).

Some bacteria (e.g. species of obligately anaerobic *Clostridium* and facultatively anaerobic *Bacillus*) break down the macromolecular components of detritus into simpler molecules by hydrolytic and fermentative processes. These products are the substrates used by other groups of anaerobic heterotrophic bacteria that complete the mineralization of organic matter. The most important of these bacteria are the denitrifiers (which include dissimilatory nitrate and nitrite reducers), sulphate reducers and methanogens (methane producers). While the term **fermentation** is sometimes applied to all anaerobic degradation processes, strictly it only describes reactions in which an internal source of electron acceptors is used, not an external source like nitrate or sulphate. An example is fermentation of glucose to carbon dioxide and ethanol, which can be considered to involve an internal redox reaction whereby part of the substrate is oxidized (to CO_2) and part reduced (to CH_3CH_2OH). A further example is shown in Eqn 1.17. Organic substrates cannot be fermented further than acetate (Fenchel et al. 1998).

Hydrolysis of macromolecular components releases sugars, amino acids and long-chain fatty acids, which undergo fermentation to yield a relatively limited range of compounds upon which subsequent bacterial metabolism is based, as summarized in Fig. 3.16. The main

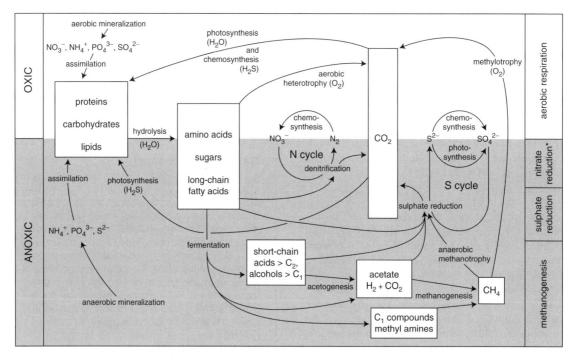

Fig. 3.16 Generalized scheme of the role of bacteria in the carbon cycle and its coupling to the nitrogen and sulphur cycles (after Fenchel & Jørgensen 1977; Jørgensen 1983a, b; Parkes 1987; Fenchel & Finlay 1995; Werne et al. 2002). For clarity, the forms of N, S and P liberated at each stage of mineralization are summarized on the left side of the diagram, where they contribute to the general mineral pools from which assimilation occurs. ★The nitrate reduction zone refers to the dissimilatory processes involved in denitrification.

products are short-chain volatile acids (e.g. formate, acetate, propionate, butyrate, lactate and pyruvate), together with alcohols (e.g. methanol and ethanol), methylated amines, carbon dioxide and water. Acetate (CH_3COO^-) is an important substrate and is produced, together with hydrogen and carbon dioxide, from other short-chain acids by **acetogens**. Some acetogenic bacteria can convert H_2 and CO_2 to acetate (e.g. some species of *Clostridium* and *Acetobacterium*):

$$2CO_2 + 4H_2 \rightarrow CH_3COOH + 2H_2O \quad [Eqn\ 3.5]$$

In the absence of molecular oxygen certain anaerobes oxidize the products of fermentation by using various inorganic oxidizing agents (which act as terminal electron acceptors): manganese(IV), nitrate, iron(III), sulphate and bicarbonate. These processes release less energy to decomposers than the complete aerobic degradation of organic matter to carbon dioxide and water. Oxidizing agents in anaerobic degradation tend to be utilized in order of decreasing energy return, as listed above (Nedwell 1984; Nealson & Saffarini 1994). The most important of these processes are dissimilatory nitrate reduction (i.e. denitrification), sulphate reduction and methanogenesis.

Denitrification generally follows rapidly upon depletion of oxygen and yields, ultimately, carbon dioxide, water and nitrogen (via nitrite, NO_2^-; Box 3.9). These

---

Box 3.9 | **Nitrogen cycle**

A summary of the nitrogen cycle is shown in Fig. 3.17. Nitrogen fixation, which converts N_2 into NH_4^+, is performed by a restricted number of eubacteria and archaebacteria (sometimes in symbiotic associations). Anaerobes such as the photosynthetic bacteria, together with some species of *Clostridium* and sulphate reducers, can perform the process. However, the most important nitrogen-fixing bacteria are aerobes: the cyanobacteria (e.g. *Oscillatoria*) in aquatic environments and *Rhizobium* in plant root nodules in soils (free-living soil bacteria such as *Azotobacter* are less prolific N-fixers). Because the nitrogenase enzyme is easily oxidized, many aerobic N-fixers actually perform better under low O_2 levels, deeper within soils and sediments or in O_2-depleted waters. The ammonia (which rapidly forms the ammonium ion in the presence of water) produced by N fixation in vascular plant roots is taken up by glutamate to yield glutamine, which acts as a carrier of the amino group and can pass it on to many other α-ketocarboxylic acids to produce various amino acids (Fig. 3.18).

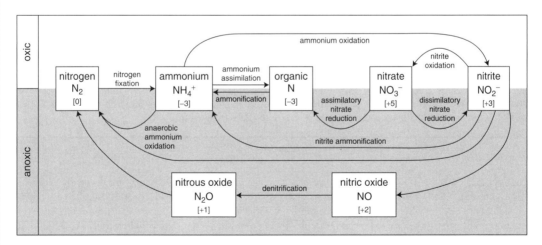

Fig. 3.17 Summary of the nitrogen cycle (oxidation states of nitrogen shown in parentheses). Ammonium assimilation and ammonification can occur in oxic and anoxic environments, as can nitrogen fixation (although the most prolific bacteria are aerobes).

Continued

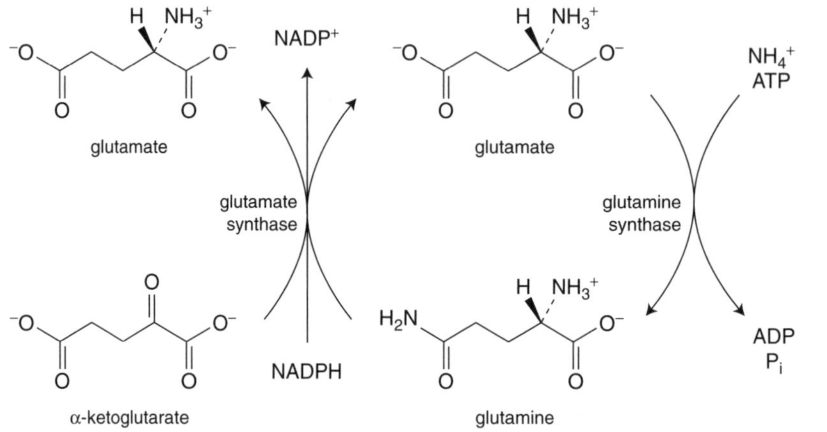

Fig. 3.18 Amino group transfer (amination of α-ketoglutarate; after Woodall et al. 1996). P_i = inorganic phosphorus.

Upon decay of organic matter the nitrogen is released first as ammonium (the process of ammonification) and then is either recycled by organisms or tends to become oxidized to various oxides of nitrogen. **Nitrification**, the conversion of ammonium to nitrate via nitrite, is carried out by the combined action of two groups of aerobic chemosynthesizers. In soils the first step (NH_4^+ to NO_2^-) is primarily performed by *Nitrosomonas* and the second step (NO_2^- to NO_3^-) by *Nitrobacter*. Some fungi can also carry out nitrification at low pH.

The nitrate produced by nitrification diffuses downwards into the upper layers of anoxic sediment, where it is used by denitrifiers (sometimes called nitrate reducers, but this term can lead to confusion between dissimilatory and assimilatory nitrate reduction) for the oxidation of organic matter, with nitrite as the by-product. Nitrate can also be converted into organic-N, by the process of assimilatory nitrate reduction, which is performed by many bacteria, fungi and algae. Under anaerobic conditions there are many genera of facultative anaerobes (e.g. *Escherichia, Bacillus, Pseudomonas, Vibrio* and *Nocardia*) that reduce NO_3^- to NO_2^-, and subsequently NO_2^- to NO or N_2O. Other denitrifying bacteria, such as *Thiobacillus denitrificans*, can take the reduction all the way to N_2, although often this final step is incomplete (Fenchel et al. 1998). In soils the main denitrifiers are *Pseudomonas* and *Alcaligenes*. Although denitrification occurs predominantly under strictly anaerobic conditions, it can also occur under mildly oxidizing conditions. The enzymes responsible for dissimilatory nitrate reduction (nitrate reductases) are bound to the bacterial membrane and their action is inhibited by oxygen, in contrast to the assimilatory nitrate reductases. (**Dissimilatory** processes do not result in the incorporation of substrate (e.g. N) into the cellular material, whereas **assimilatory** processes do.) Nitrogen can also be formed by anaerobic oxidation of ammonia by nitrite (the anammox reaction), probably by planktomycete bacteria in the water column and sediments. This process may account for 30–50% of oceanic N_2 production (Devol 2003).

denitrifiers (e.g. species of *Escherichia, Pseudomonas, Bacillus, Micrococcus* and *Thiobacillus*) are facultative anaerobes and use oxygen when it is in sufficient supply (production of the enzyme nitrate reductase is inhibited by O_2). Nitrate reduction provides greater energy returns than sulphate reduction from the same carbon-containing substrates, and the dentrifiers have lower thresholds (i.e. have a lower minimum abundance requirement) for the substrates, so they generally outcompete sulphate reducers. However, the vertical extent of the denitrification zone is usually very limited in marine sediments because the nitrate concentration in pore water is typically very low (only a few tens of μmol l^{-1}) and is rapidly depleted.

Sulphate reduction becomes important when nitrate is depleted; its products are carbon dioxide, water and hydrogen sulphide (Box 3.10). **Sulphate reducers** (e.g. *Desulfovibrio, Desulfotomaculum, Desulfobacter*,

Box 3.10 | Sulphur cycle

A summary of the sulphur cycle is shown in Fig. 3.19, incorporating important geological pathways as well as major biologically mediated processes. In organisms thiol groups are the dominant form of sulphur, as in the amino acids cysteine and methionine (see Fig. 2.11) and various co-enzymes. Consequently, S-incorporation into organic tissue generally requires assimilatory sulphate reduction to form sulphide (see e.g. Erlich 1995). The largest reservoir of sulphate is in the oceans. Direct uptake of sulphide is unfeasible for most organisms because H_2S is toxic. This problem is avoided during assimilatory sulphate reduction by the immediate reaction of the sulphide with an acceptor (e.g. serine, producing cysteine; see Fig. 2.11).

Decomposition of organic matter results in sulphide (and some thiol) generation through the action of desulphydrases. This process of desulphuration is analogous to ammonification in the N cycle. In the marine environment a major product is dimethylsulphide (DMS; see Section 6.3.1d). The sulphides are oxidized to sulphate, the dominant volatile forms (H_2S, DMS and thiols) mostly via photo-oxidation in the atmosphere. Hydrogen sulphide may be microbially oxidized under aerobic conditions by chemosynthesizers or under anaerobic conditions by photosynthesizers.

Aerobic and anaerobic communities are, to a large extent, interdependent, and their relative positions in the water column and sediment reflect their reliance on particular substrates as well as oxygen levels. For example, photosynthetic bacteria such as the green and purple sulphur bacteria (*Chlorobium* and *Chromatium*, respectively) live just within the anaerobic zone, where there is sufficient light penetration for them to make use of the products of anaerobic degradation processes (e.g. H_2S and CO_2). These phototrophic sulphur bacteria oxidize S^{2-} to elemental sulphur (during the process of anoxygenic photosynthesis), but have only limited ability to oxidize the S to

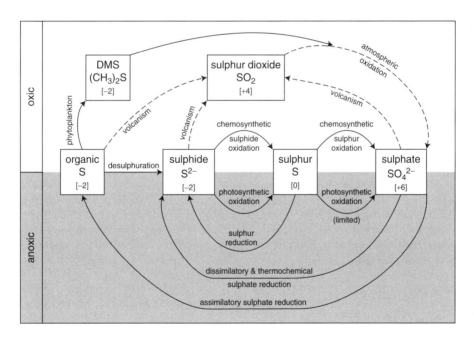

Fig. 3.19 Summary of the sulphur cycle (oxidation states of sulphur are shown in parentheses). Solid lines show major microbial pathways; broken lines show geological processes involving sedimentary organic-S, calcium sulphate (gypsum/anhydrite) and iron(II) sulphide (pyrite).

Continued

SO_4^{2-}, so S accumulates. This S can be oxidized to SO_4^{2-} by *Thiobacillus denitrificans*, a chemosynthetic bacterium that is unusual for *Thiobacillus* species in being a facultative anaerobe (a nitrate reducer). The S can also be reduced to S^{2-}; for example, by *Desulfuromonas acetoxidans*, which lives in syntrophic association with *Chlorobium* (Kuenen et al. 1985).

At the oxic–anoxic boundary a group of chemosynthesizers (e.g. species of *Thiobacillus*, *Beggiatoa*, *Thiothrix*), all obligate aerobes, converts the S^{2-} produced by dissimilatory sulphate reduction back into SO_4^{2-}. If the water immediately overlying the sediment is anoxic, conversion of S^{2-} to SO_4^{2-} often does not occur, and when it does it is minor and due entirely to the activity of anaerobic photosynthesizers. Under such conditions there is a build-up of S^{2-} and an increase in acidity, which results in the decalcification of calcareous tests and shells.

The fluids emanating from deep-sea hydrothermal vents are enriched in H_2S and S, and some hyperthermophilic archaebacteria can reduce S with H_2 to produce more H_2S (e.g. *Thermoproteus*, *Pyrobaculum* and *Pyrodictium*; Fischer et al. 1983). Where the vent fluids meet oxygenated seawater various eubacteria (e.g. *Beggiatoa*) harvest energy from oxidation of the reduced S forms (Jannasch & Mottl 1985).

If iron is present within sediments it can combine with S^{2-} to form insoluble pyrite, removing sulphur from the system. If iron is not available sulphur can be deposited, a process that can occur during the formation of carbonate sediments. Sulphate can be precipitated from saline waters as the calcium salts gypsum or anhydrite under suitable conditions (evaporite deposition). Subduction of S-containing sediments leads to emission of SO_2 during volcanism, which can be oxidized in the atmosphere to sulphate. The sedimentary reservoirs of gypsum and pyrite amount to *c*.2.47 Pt and 2.47 Pt of S, respectively, while some 1.28 Pt of S is present as oceanic sulphate (Holser et al. 1989a).

Desulfobulbus) are generally obligate anaerobes, mostly belonging to the proteobacteria, but there are also members among the Gram-positive eubacteria and the thermophilic archaebacteria. As a group, these bacteria can tolerate wide ranges of pH, pressure, temperature and salinity. Their growth is generally limited by the rate of sulphate supply (by diffusion of fresh sulphate into sediment pore waters and/or bacterial oxidation of sulphide; Box 3.10), so the sulphate reduction zone may be relatively shallow in organic-rich areas, where sulphate is rapidly depleted, but can occupy several metres in pelagic sediments with lower organic content. Acetate is the most important substrate (accounting for >60% of total sulphate reduction; Winfrey & Ward 1983), and is completely oxidized to CO_2 by the group of sulphate reducers that utilize the citric acid cycle (e.g. *Desulfobacter*; see Section 2.2.2):

$$2H^+ + CH_3COO^- + SO_4^{2-} \rightarrow 2CO_2 + 2H_2O + HS^- \quad [Eqn\ 3.6]$$

Some sulphate reducers do not have a citric acid cycle (e.g. *Desulfovibrio*) and use other substrates, such as lactate (CH_3-CH(OH)-COO$^-$), which yields acetate (Hines et al. 1994).

Methanogens are also obligate anaerobes and synthesize methane from the smallest fermentation products. Carbon dioxide and hydrogen are important substrates (e.g. for *Methanobacillus*, *Methanococcus*), with CO_2 being the electron donor in the reaction:

$$CO_2 + 4H_2 \rightarrow CH_4 + 2H_2O \quad [Eqn\ 3.7]$$

This equation is similar to Eqn 3.5, and suggests that there could be competition for substrates between certain methanogens and acetogens. The acetogenic reaction liberates less energy than the methanogenic reaction, but the acetogens compensate in general by using a broader range of substrates, and they are also more tolerant of low pH. Another group of methanogens can perform what is effectively a fermentation of acetate (see Eqn 1.17). Some methanogens can use simple compounds that are either C_1 compounds (e.g. methanol and formate) or readily yield C_1 units (methylated amines). In freshwater sediments *c*.70% of methane generation may result from utilization of acetate (produced by acetogens), with the remainder deriving from CO_2 and H_2. Methanogens have a lower utilization efficiency and higher thresholds for H_2 (i.e. require higher minimum amounts of H_2) than sulphate reducers, so generally do not compete well against sulphate reducers until sulphate is well depleted.

Compounds containing a single C atom, usually in the form of a methyl group, can be oxidized back to carbon dioxide by **methylotrophes** (e.g. *Methylomonas*, which are obligate aerobes). The **methanotrophes**, a subgroup of the methylotrophes, use only methane as their carbon and energy source (Hanson & Hanson 1996). All methylotrophes were once thought to be aerobes, but recently some methanotrophes have been implicated in anaerobic methane oxidation (Fig. 3.16). The anaerobic route appears to be important in the recycling of a significant proportion of methanogenic methane back into sedimentary organic matter (some 90% of methane is recycled within sediments), via the action of bacterivorous ciliates, in methane-rich, anoxic, marine environments, and is associated with sulphate depletion (Reeburgh 1980; Jørgensen 1983a). This can be likened to a reversal of methanogenesis, involving a consortium of methane-oxidizing archaebacteria and sulphate reducers (Hoehler et al. 1994; Hinrichs et al. 1999; Boetius et al. 2000; Orphan et al. 2001; Werne et al. 2002; see Section 5.8.5). The methane that escapes into the aerobic zone is efficiently oxidized by other methylotrophes. The carbon dioxide produced by the various methylotrophes may be reintroduced into the biological cycle by phototrophes and chemotrophes.

Reliance of one group of bacteria on the products of another group is a general feature of bacterial communities (**consortia**), involving photosynthetic and chemosynthetic bacteria as well as members of the decomposer system. This interdependence is vital, because any single group of bacteria would eventually be poisoned by a build-up of their own by-products. Perhaps the best example is the sulphate reducers, the main product of which, H_2S, is extremely toxic to most organisms (it poisons the metal-containing groups of cytochromes) but is oxidized by anaerobic photosynthesizers (Fig. 3.19). Figure 3.16 provides a general idea of the interrelationships within bacterial consortia, and readers are referred to specialized texts for a more detailed discussion (e.g. Atlas & Bartha 1998; Fenchel et al. 1998). From Fig. 3.16 it can be seen that the carbon, nitrogen and sulphur cycles are interlinked through bacterial activity. The aerobic and anaerobic respiration processes of bacteria release nitrogen and sulphur from detritus and convert them into forms assimilable by plants. Phosphorus is also released for uptake but remains in the form of phosphate throughout its cycle. The nitrogen and sulphur cycles are considered in more detail in Boxes 3.9 and 3.10, respectively.

The availability of substrates and the competition for them between various anaerobes is important. Sulphate is plentiful in marine environments ($c.28\,\mu mol\,l^{-1}$) and so sulphate reduction is the dominant anaerobic process (Jørgensen 1982) and accounts for $c.50\%$ of carbon oxidation in marine shelf sediments (compared with $c.3\%$ by denitrifiers). The H_2S liberated during sulphate reduction may partially inhibit denitrification in marine environments and sometimes a denitrification zone may not be present (Jørgensen 1983b). It also appears that methanogens do not compete effectively with denitrifiers or, more importantly, with sulphate reducers for the main substrates (H_2, CO_2 and CH_3COO^-). Consequently, methanogens are largely confined to the area below the sulphate reduction zone, but even in marine environments methanogenesis may be significant in the overall reworking of organic matter because it can extend to considerable depth. Sulphate levels are usually low in freshwater lakes and soils, resulting in methanogens, and also denitrifiers, being more important than sulphate reducers.

Anaerobic and aerobic degradation reactions probably proceed at about the same rate under identical conditions (e.g. fresh organic matter and unlimited supply of electron acceptors), but anaerobic degradation often appears slower because most of the labile organic components have already been removed, leaving mainly recalcitrant material, and because electron acceptors (e.g. sulphate for the sulphate reducers) become depleted. In addition, some organic matter (e.g. lignin; see Section 3.3.3c) may be more resistant towards biodegradation under anaerobic conditions (Canfield 1994). The remains of dead bacteria also contribute to the organic material present in the sediment and are degraded, or preserved, in the same manner as the rest of the detritus. Total microbial numbers decrease with increasing depth in the sediment and, eventually, all biologically mediated degradation must cease, an important factor being the decreasing energy returns from oxidation of the increasingly recalcitrant organic matter. Significant bacterial populations can be found in sediments 500 m below the sea floor (Parkes et al. 1990), but the depth limit has yet to be established. Thermal degradation of organic matter liberates acetate that may stimulate sulphate reducer activity even at depths associated with oil generation (at temperatures possibly as high as $150\,°C$; Wellsbury et al. 1997). Various communities of anaerobic heterotrophic thermophilic eubacteria and hyperthermophilic archaebacteria have been isolated from deep boreholes (L'Haridon et al. 1995). Many of these communities are supported by sedimentary organic matter but others, particularly those discovered in igneous rocks, are not (Krumholz 2000). The primary producers in the latter appear to be lithotrophes (mainly methanogens and acetogens) that harvest energy from the H_2 generated by the interaction of water with iron-bearing minerals in basalt, with mantle-derived CO_2 presumably serving as the carbon source.

The composition of bacterial communities depends on a number of factors such as acidity and salinity, in addition to the availability and type of organic substrates and terminal electron acceptors. Bacterial species can be active over a wide temperature range (e.g. c.0–45 °C for most methanogens and *Desulfovibrio*), although most microorganisms exhibit optimum growth rates within the range 20–35 °C. Bacterial activity is generally suppressed when pH falls below c.5 (the exception being archaebacterial acidophiles, which grow at pHs <2; Fenchel et al. 1998) and, as a result, fungi are more important than bacteria in acidic soils. The role and interactions of bacteria in sedimentary environments are complex and not fully understood at present (e.g. Kaeberlein et al. 2002). Some types can grow in both aerobic and anaerobic environments and are able to carry out more than one of the functions shown in Fig. 3.16, and different species of the same genus may carry out markedly different processes.

3.3.3 Degradation of biopolymers

The size of the largest biopolymers dictates that decomposition can only take place in the presence of extracellular enzymes, because the molecules are too large to enter microbial cells. Hydrolytic enzymes are common, and reverse the dehydration step used in many biopolymerization reactions, such as the condensation of glucose to give cellulose, and they are specific to a particular substrate. Fungi and bacteria secrete a range of enzymes to hydrolyse most biopolymers (e.g. cutinase, chitinase, proteases, cellulases, amylases, pectinases). It is the presence of microbial communities in their gut that permits terrestrial herbivores to digest higher plant material.

DNA degrades in tissue shortly after death unless protected by dehydration, and it is inevitably modified/degraded after deposition in sediments and certainly is not preserved in any significant amounts over millions of years. It is far more sensitive to hydrolytic and oxidative processes than most other biomacromolecules.

(a) Proteins and amino acids

Proteins are readily hydrolysed by microbial enzymes, and their preservation requires desiccation, freezing or pickling, to retard microbial action. The key degradation step is peptide-bond hydrolysis, which is temperature dependent. The more highly ordered and cross-linked proteins—such as keratin, fibrinogen (in silk and mollusc-shell matrix) and collagen (comprising triple α-chains)—are insoluble and so less accessible to extracellular enzymes. Proteins within a mineral matrix, such as bone and shell, are protected to an extent from contact with microbial proteases. The mineral matrix may also reduce the conformational flexibility of a protein (i.e. the degree to which it can bend and twist to permit close approach of enzymes to peptide linkages), lowering the rate of hydrolysis. Even this protection is unlikely to result in protein preservation for more than a few million years.

Over time water can leach the amino acids and small peptides released by hydrolysis, but it may also introduce amino acids, such as the major metabolic by-product serine (Mitterer 1993). The introduced amino acids are likely to be dominated by the L enantiomers characteristic of living organisms (Section 2.3.1), whereas the leached amino acids are likely to experience racemization (Section 2.1.3). The more functional groups there are in an amino acid the more reactive it is, so neutral amino acids are the most stable and tend to dominate in older fossils, but even their concentrations decrease over a few Myr and their degree of racemization increases. Fossils of greater than Neogene age from deep-sea sediments and high-latitude continental environments may contain some of their original amino acids, but those from low- and mid-latitude continents are unlikely to (Mitterer 1993).

(b) Polysaccharides

In the absence of enzymes, polysaccharide hydrolysis is relatively slow (Table 3.7), with the α-linked polymers (e.g. amylose) being slightly more stable than their β-linked counterparts (e.g. cellulose and chitin). It appears that glycosidases can accelerate hydrolysis by a factor of c.10^{17} (Wolfenden et al. 1998).

Cellulose is attacked by several types of fungi and bacteria, with basidiomycetes (fungi) being the major

Table 3.7 Comparison of stabilities of biopolymers towards hydrolysis in the absence of enzymes (after Wolfenden et al. 1998; Thompson et al. 1995). Protein values are based on glycine-glycine units; see Box 5.4 for explanation of half-life

	half-life (25 °C)
RNA	4 yr
proteins	460 yr
DNA	140 kyr
β-glucopyranosides	4.7 Myr
α-glucopyranosides	12 Myr

degraders in wood and litter at the soil's surface. The cellulose degrader assemblage is affected by pH: at pH <5.5 filamentous fungi dominate, whereas at pH ≥7 bacteria belonging to *Vibrio* species as well as fungi are important. Under aerobic conditions fungi and aerobic and facultatively anaerobic bacteria degrade cellulose, but under anaerobic conditions *Clostridium* (obligate anaerobes) dominate. Hemicelluloses (e.g. xylans, mannans and galactans) are degraded by fungi and bacteria (e.g. actinomycetes and *Bacillus*), and appear more susceptible than cellulose (Hedges et al. 1985).

The important bog moss *Sphagnum* contains the acidic polysaccharide sphagnan, which resembles the pectins of higher plants, but in addition to residues of galacturonic acid and neutral sugars it also has side chains with residues of D-lyxo-5-hexosulouronic acid (which can also be called 5-keto-D-mannuronic acid; see Section 2.2.1). Under mildly acidic conditions the cross-linking formed by the 5-keto-D-mannuronic acid residues spontaneously hydrolyse, slowly releasing sphagnan, which represents 55% of cell-wall polysaccharides (known as holocellulose). The released sphagnan can undergo polymerization reactions with amine groups on other fragments to form melanoidins (see Section 4.1.3) with powerful chelating capabilities (see Box 3.7). *Sphagnum* appears not to be degraded by bacteria for a combination of reasons: the low pH in the bog ($c.3–5$); the fact that the plant is very efficient at removing all available nitrogen; and because the various extracellular digestive enzymes are deactivated by irreversible binding to holocellulose.

Among other polysaccharides, chitin, the annual production of which is estimated at $c.10^{11}$ t – mostly from arthropods – is degraded by fungi and bacteria (again actinomycetes are important). It is protected to a degree in arthropod cuticles by a wax coating and a protein complex that is cross-linked by catechol and aspartic/histidyl moieties (Briggs et al. 1998), but nevertheless it is still mostly decomposed. Agar, which is produced by many marine algae, is decomposed by relatively few species of bacteria (which often occur as epiphytes).

Lipids are hydrolysed; for example, wax esters are converted into their constituent fatty acids and alcohols. The less soluble components tend to exhibit the greatest inhibition to ingress of hydrolytic enzymes, and so are the most stable towards biodegradation. The waxy components forming the cuticles of plant leaves and stems are among the best preserved biomacromolecules over geological periods (see Section 4.4.1a). The transformations of lipids are examined in more detail in Chapter 5.

(c) Lignin

Lignin, the major component of woody tissue (see Section 2.5.1), is degraded by a much smaller range of microbes than other major biopolymers and is one of the most resistant. Its degradation, which appears to involve non-specific oxidases, offers some insight into the preservation potential of sedimentary organic matter in general (see Section 3.3.4). Most lignin degradation is achieved by white-rot fungi, among which the basidiomycetes are the most numerous group as well as the most efficient of the lignin degraders. The white-rot fungi also comprise a few species of ascomycetes (e.g. *Xylaria*, *Libertella* and *Hypoxylon*), but they do not appear to degrade the vanillyl lignin of gymnospermous wood and preferentially attack the syringyl units of angiospermous lignin (see Fig. 2.30). As yet, there is little evidence from simple culture enrichment experiments to suggest that significant lignin biodegradation occurs under anaerobic conditions (e.g. Odier & Monties 1983). Laboratory studies suggest that aerobic bacteria (e.g. actinomycetes) are not particularly efficient lignin decomposers either, although bacterial consortia may be more effective than single species, particularly when presented with smaller lignin fragments (Kirk & Farrell 1987).

Decomposers do not seem to be able to use lignin as their only source of C or energy; they consume polysaccharides at the same time. For example, although lignin degradation occurs during soft-rot wood decay by some species of ascomycetes and fungi imperfecti, polysaccharides are preferentially degraded (Kirk 1984). The basidiomycetes that are responsible for brown-rot wood decay also mainly decompose cellulose and hemicelluloses, but they seem to require the presence of lignin. They do cause some relatively minor degradation of lignin, through limited hydroxylation of aromatic rings and some ring cleavage. Although the brown-rot basidiomycetes have a minimal effect on the framework of lignin they are capable of causing significant demethylation of the methoxy groups on aromatic rings (Kirk 1984), leading to an initial increase in OH group content (Filley et al. 2002).

Some lignin linkages are unstable, such as those involving the αC (Fig. 2.30), which degrade slowly without microbial intervention at elevated temperatures or in acidic or alkaline environments. In contrast to most polysaccharides and proteins, the structure of lignin is random and its formation does not involve simple dehydration–condensation reactions. This structural complexity dictates that the initial stages of lignin biodegradation must involve a variety of non-specific enzymes that are also not generally hydrolytic. The oxidative changes to lignin brought about by the white-rot basidiomycetes include aromatic-ring cleavage (Higuchi 1985) and progressive depolymerization, releasing a wide range of smaller fragments. The range of enzymes involved include ligninases, manganese peroxidases and phenol-oxidizing enzymes (De Jong et al. 1992). Oxygen availability is a key factor in the generation by hydrogen peroxide-producing enzymes of the H_2O_2 required by the ligninases (Tien & Kirk 1983, 1984), which alternatively can be called lignin peroxidases (and contain a haem-like unit). All the lignin-degrading enzymes cause a comparatively simple one-electron oxidation of susceptible aromatic nuclei to give unstable cationic radicals (see Box 2.10), which in turn undergo a variety of non-enzymatic reactions, yielding a diverse range of products (e.g. Hammel et al. 1986).

Phenol-oxidizing enzymes, such as laccase (which contains copper) and manganese peroxidase (which contains a haem-like protein), oxidize the phenolic residues in lignin to phenoxy radicals. The resulting one-electron oxidation causes: αC oxidation; limited demethoxylation; aryl-αC bond cleavage; and, in phenolic syringyl units, cleavage of the αC–βC bond (Higuchi 1985). The one-electron oxidation of phenolic compounds might be expected to result in extensive repolymerization, but such reactions appear to be limited during lignin degradation, as discussed in Section 4.3.2a.

Ligninases act on the aromatic nucleus of non-phenolic units; whether the aromatic nucleus is oxidized depends upon the effect of its substituents. Strong electron-withdrawing groups like an αC-carboxyl group tend to deactivate the aromatic nucleus, whereas an alkoxy group activates it towards ligninase attack (see Box 2.10). The type, position and number of groups are important. The position of alkoxy groups in lignin precursors is fixed, but the number varies from one to three; the more groups the faster the oxidation rate, which consequently increases in the order p-hydroxyphenyl < vanillyl < syringyl (see Fig. 2.30). The lower stability of syringyl units has been observed during natural decomposition of hardwoods (Hedges et al. 1985). The nature of the substituents on the aromatic nucleus also affects the subsequent reactions of the cations produced by the initial oxidation step, which include nucleophilic attack (see Box 2.10) by H_2O or internal OH, loss of an acidic proton at αC and αC–βC cleavage. The wide range of possible reactions generates a variety of intermediate products, as shown in Fig. 3.20 for the dominant β-O-4 linked units and in Fig. 3.21 for the less common β-1 linked units, demonstrating the non-specific nature of the oxidation. No ring-cleavage products have been observed from the decomposition of model compounds. One lignin degradation product from *Phanerochaete chrysosporium* is veratryl alcohol (Fig. 3.22), which stimulates the synthesis of lignin peroxidase and so exerts a positive feedback on the degradation process (Carlile et al. 2001). It is not yet clear whether ligninase oxidation alone accounts for the large reduction in methoxy group content after partial degradation of lignin by white-rot fungi. Although anaerobic decay has yet to be proven to be important, synthetic compounds that model aspects of the lignin structure have been found to undergo significant demethoxylation with ligninase under anaerobic conditions (Miki et al. 1986).

The overall effect of fungal attack of lignin is: decrease in methoxy, phenolic-OH and aliphatic-OH content; cleavage of aromatic nuclei forming aliphatic carboxyl-containing residues; creation of new αC-bonded carbonyl and carboxyl groups; formation of alkoxyacetic acid, phenoxyacetic acid and phenoxyethanol structures (Higuchi 1985). Many of these reactions can be mediated by ligninase, but some cannot, suggesting that other oxidases are involved.

3.3.4 Factors affecting sedimentary preservation of organic matter

(a) Organic-rich sediments

Our ideas about the factors responsible for the preservation of organic matter in sediments have largely focused on the conditions under which organic-rich petroleum source rocks are likely to have formed. In this context the relative importance of productivity (Pedersen & Calvert 1990) and anoxicity (Demaison & Moore 1980) has been hotly debated. Although most organic-rich sediments appear to be associated with anoxicity there are certainly examples of good preservation of organic matter in oxic sediments. As seen in Section 3.3.1, the flux of organic carbon to the sea floor is approximately proportional to primary production and inversely proportional to water depth (Suess 1980), so sufficiently high productivity is clearly important if organic material is to reach the sediment in any quantity and stand a chance of preservation (e.g. during algal blooms).

Organic-rich (and particularly lipid-rich and hence hydrogen-rich), fine-grained sediments are generally finely laminated (i.e. non-bioturbated), consistent with anoxicity in sediments and bottom waters. However, this anoxicity may be as much a consequence of the

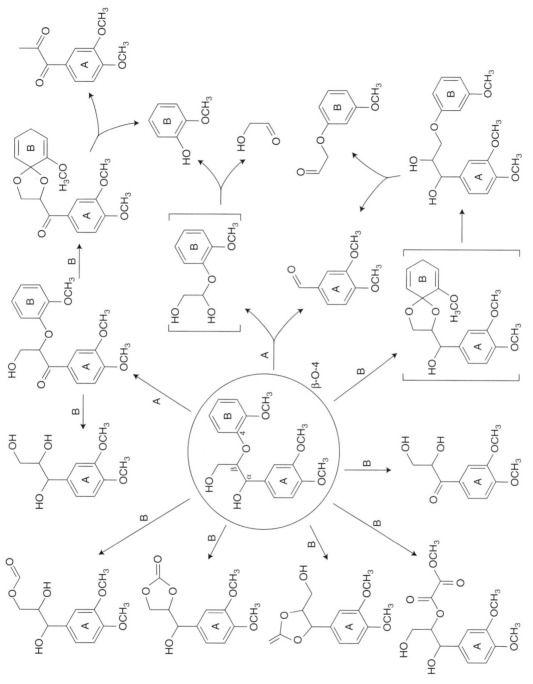

Fig. 3.20 Oxidation of a β-O-4 model lignin compound by ligninase/H_2O_2 (after Kirk & Farrell 1987). A and B on arrows show which ring is oxidized; [ ] = postulated intermediates.

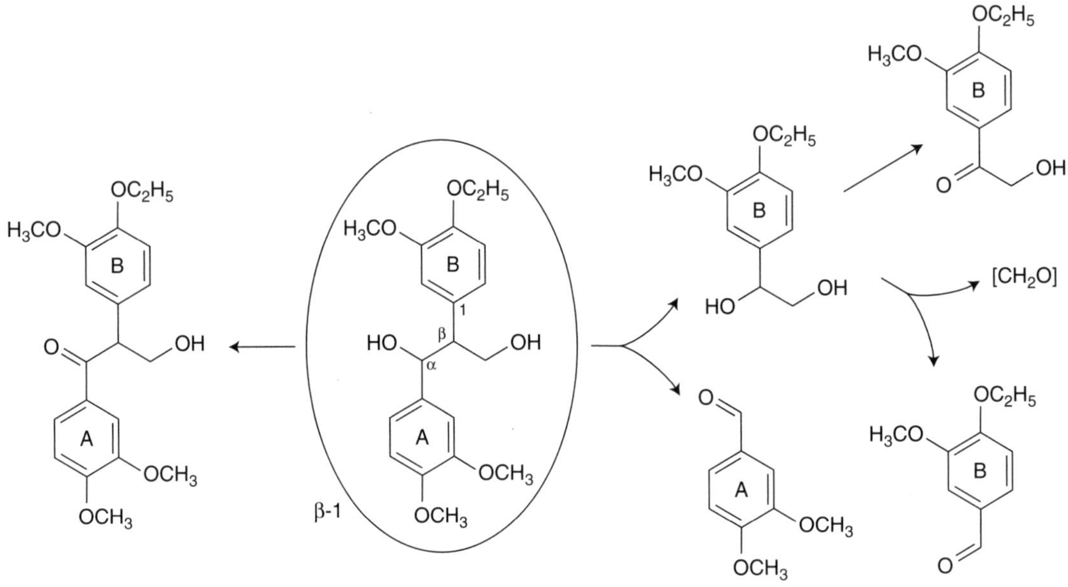

Fig. 3.21 Ligninase/H_2O_2 mediated oxidation of a β-1 model lignin compound (after Kirk & Farrell 1987).

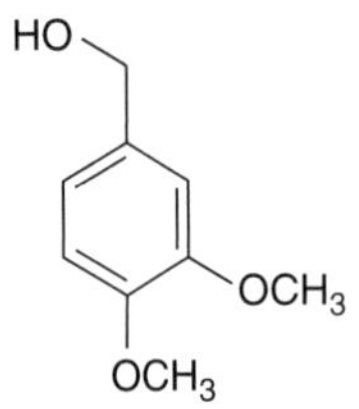

Fig. 3.22 Veratryl alcohol.

degradation of large amounts of organic matter, aided by stratification of the water column, as a prerequisite for organic preservation. The activity of detritivores is probably a key factor: first, they comminute organic detritus, which increases its surface area and facilitates microbial degradation; second, they bioturbate sediments, which tends to keep the organic matter in the oxidized surface layer (Hartnett et al. 1998). Anoxicity drastically reduces the degree of reworking of sedimentary organic matter by detritivores, even if the extent of bacterial degradation is not significantly affected. It is likely, therefore, that the formation of organic-rich sediments is influenced by both anoxicity and the level of primary productivity. In addition, pore waters in fine-grained sediments generally become isolated from exchange with the oxygenated water column more rapidly during burial and compaction than coarser-grained sediments. The clays in fine-grained sediments can bind and inactivate the extracellular digestive enzymes of bacteria, increasing the preservation potential of organic matter.

Most soils are oxygenated and the organic matter they contain is efficiently recycled by organisms of the detrital food chain, resulting in little long-term preservation of organic material in soil. However, waterlogging aids the development of anoxicity and also the build-up of acidity, and hence the preservation of organic matter as peat (Dean & Gorhan 1998).

In both marine and lacustrine environments oxygen consumption increases with depth under areas of high primary productivity, mainly due to microbial respiration during the degradation of organic matter. In oceanic environments an **oxygen-minimum layer** (OML) may be produced beneath the euphotic zone, where oxygen demand is greater than supply and there is a maximum in carbon dioxide and nutrient concentrations because of the absence of phototrophic assimilation. In the present-day oceans the OML generally forms within a depth range of $c.300-1500$ m. When upwelling occurs nutrients from this zone are taken up into the oxygenated euphotic zone, increasing phytoplanktonic production. The intensity of oxygen depletion within the OML is dependent on the residence time of this layer within the water column and the productivity of the overlying euphotic zone. Where the

OML intercepts the continental slope, the bottom waters and sometimes the sediment become anoxic (Fig. 3.23). An example is the accumulation of organic-rich sediments under the Peru upwelling. OMLs are not always directly related to high rates of *in situ* degradation of organic matter, and can be found at intermediate depths in parts of the open ocean well away from coasts, due to **advection** (the horizontal movement of water bodies) resulting from oceanic circulation patterns (Box 3.2). Oxygen will slowly diffuse into the OML from surrounding water bodies (Box 3.8), but it can be removed by relatively moderate levels of aerobic respiration.

At present, bottom waters and the upper layer of sediment throughout most of the oceans are oxygenated, as a result of the deep-water circulation pattern (see Box 3.2). The oxygen content of the deep oceans is controlled mainly by the balance between oxygen supply in downwelling areas (primarily the main zones of deep-water formation) and oxygen consumption during oxidative decomposition of organic matter (Sarmiento et al. 1988). Anoxia in deep water could occur if: (a) the amount of dissolved oxygen supplied by downwellings was reduced sufficiently (i.e. the rate of ventilation declined); (b) nutrient supply increased, fuelling primary production and so increasing the rate of organic decomposition; or (c) nutrient utilization increased (it is presently poor in high-latitude oceans; see Section 3.2.7). At the present oceanic ventilation rate, phosphate level and Redfield ratio (which control primary production and organic decomposition; see Section 3.2.7), the maintenance of deep-water oxygenation requires a minimum dissolved oxygen concentration in downwelling surface waters of $c.5\,mg\,l^{-1}$ (Sarmiento et al. 1988). It is presently about double this value.

Rapid burial can aid the development of anoxicity in sediments but can lead to the dilution of organic matter by large amounts of inorganic material. If the energy of the water system is too high, deposition of fine-grained material is limited and any organic matter present is susceptible to erosion and degradation. This can be the fate of much sedimentary organic matter in productive areas, such as estuaries.

Among the most prolific primary producers are mangrove swamp and salt marsh communities (see Table 3.2), and anoxicity can develop in the sediments trapped by the root systems of the macrophyte stands. However, the formation of organic-rich sediments in these environments is hindered by dilution with clastic material, a limited vertical extent of accumulation (if sea level remains constant) and subaerial exposure. Similar limitations can apply to the preservation of organic-rich sediments in freshwater swamps and marshes. Accumulation of significant thicknesses of organic-rich sediments in these areas requires a gradual rise in water level (see Section 3.4.2).

Coral reefs are also productive areas, but are confined to oligotrophic waters, where the depth of light penetration is at its maximum and coral polyps are not choked by detrital material. However, all nutrients are effectively recycled within the living reef system, and the high energy, well oxygenated environment again limits the potential for formation of organic-rich sediments.

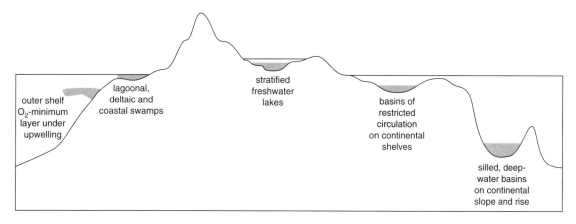

Fig. 3.23 Important oxygen-depleted environments (shaded areas) associated with deposition of organic-rich sediments (after Brooks et al. 1987).

To summarize, in the modern oceans it is likely that <1% of primary production reaches the sea floor to be preserved in sediments (Müller & Suess 1979). However, the preservation potential of this marine organic matter can be enhanced by high primary productivity, an intensified OML, accelerated sinking rates of detrital particles and rapid burial (Emerson & Hedges 1988). It appears that the accumulation of organic-rich sediments in general is mostly associated with relatively high productivity in low-energy, oxygen-depleted environments, some examples of which are shown in Fig. 3.23. Basins are important sites for the potential large-scale accumulation of sedimentary organic material because anoxicity can develop in bottom waters during the degradation of large amounts of detritus and thick sedimentary deposits can form (Box 3.11). Among such depositional settings are large eutrophic lakes and marginal marine basins in which the inflow of oxygenated bottom waters is obstructed by a sill or similar barrier (Fig. 3.23).

(b) General controls on the organic content of marine sediments

At present, only c.6% of all the organic carbon being deposited in marine sediments is accumulating under high productivity (and low oxygen) zones and <1% in anoxic basins (see Table 3.8). The bulk (c.90%) is accumulating at lower concentrations in coastal margin sediments, and understanding the conditions responsible for its preservation will provide an insight into the conditions under which the low levels of finely disseminated organic matter found in ancient sedimentary rocks probably accumulated. This finely disseminated material accounts for most of the organic carbon present in the Earth's crust (some 15 Pt).

The sedimentary organic matter that is preserved in coastal marine sediments is intimately associated with mineral surfaces (Mayer 1994), with c.83% being apparently irreversibly adsorbed (Mayer's comment on Hedges & Keil 1995). Away from deltas the abundance of this organic material appears to be directly related to the surface area of sedimentary minerals. It is equivalent to a monolayer coating of the mineral surfaces (c.0.86 mg C_{org} m^{-2}), although much of it may be located within small pores (<10 nm diameter), preventing access by hydrolytic enzymes (Mayer 1994). Strong adsorption is capable of protecting even labile components such as amino acids and sugars (Henrichs 1992), slowing remineralization rates by up to five orders of magnitude (Keil et al. 1994). Strong adsorption involving alkylammonium cations may account for the enrichment of the non-protein amino acids β-alanine and γ-aminobutyric acid in the oxidized surface layer of abyssal sediments (Whelan 1977). The presence of an adsorbed monolayer would suggest that the organic material had originally been in solution (as DOM; see Section 3.3.1a), but adsorption must have been rapid and irreversible for the most labile organic compounds, such as polyunsaturated fatty acids, to be preserved, because continual desorption and readsorption would result in complete degradation of such compounds in <1 kyr (Lee 1994). Sorption is likely to promote condensation reactions, producing more refractory macromolecules as diagenesis progresses, and the polymers would also be able to bind to more adsorption sites.

Deltaic settings are generally characterized by concentrations of organic matter significantly less than that required to form a complete monolayer; this probably simply reflects a deficiency in the supply of organic matter relative to the very large amounts of mineral matter (Mayer's comment on Hedges & Keil 1995). The proportion of terrestrial organic material within deltaic sediments decreases away from the shore, and appears to be replaced by marine organic matter. A possible explanation is that the concentration of terrestrial DOM in seawater is very low, which promotes desorption of the terrestrial organic matter from particulates and partitioning of the more plentiful marine DOM on to the available adsorption sites (Henrich's comment on Hedges & Keil 1995). Terrestrial organic matter seems slightly more recalcitrant than its marine counterpart in deltaic settings (e.g. Amazon delta), although both types are fairy efficiently degraded (≥70% and ≥90% respectively). The degradation efficiency can be attributed to the intense and large-scale physical reworking of the sediments by currents and waves, creating a massive, fluidized bed, which subjects the organic matter to repeated cycling from oxic to anoxic conditions (Aller 1998).

Table 3.8 Approximate proportions of sedimentary organic carbon accumulating in various present-day marine depositional environments (based on a total C_{org} burial rate of 160 Mt C yr^{-1}; after Hedges & Keil 1995, based on Berner 1989)

environment	%
delta	44
shelf and upper slope	42
shallow-water carbonate	4
high-productivity slope	4
high-productivity pelagic	2
low-productivity pelagic	3
anoxic basin	<1

Box 3.11 | Basins and accommodation space

A **sedimentary basin** is a depression in the Earth's crust, usually the result of subsidence over a considerable period of time, forming a depositional site where a large thickness of sediment can accumulate. Lithospheric movement is involved (Fig. 3.24), resulting from thermal and tectonic (crustal plate movement) processes. There are three main causes:
1 Thermal subsidence, caused by the contraction of cooling lithosphere (e.g. as newly formed oceanic lithosphere moves away from the constructive ridge).
2 Crustal thinning by extensional tectonics, causing fault-controlled subsidence (e.g. rift basins like the North Sea, and strike-slip basins as formed along the San Andreas fault zone).
3 Flexural loading adjacent to mountain ranges (where compressional forces thrust up a mountain range, the mass of which down-warps the lithosphere ahead of it, creating a foreland basin as in the Basin and Range Province of western North America) and adjacent to subduction zones (where the mass of the magmatic arc and the drag exerted by the descending slab result in a fore-arc basin).

Combinations of the above mechanisms can be involved in basin formation. For example, the crustal thinning caused by extension is generally associated with higher temperatures than in surrounding areas of normal crustal thickness due to the rise of hot asthenosphere nearer to the surface (see Box 1.2). As tectonic activity subsides, cooling (thermal relaxation) results in crustal sagging and basin formation. Another example is erosion and subsequent subsidence during cooling of a crustal bulge caused by a local hot spot in the mantle. On a more localized scale, sediment loading where a river delta progrades on to adjacent oceanic crust can cause the development of a depression. There is a variety of classification systems for the basin types that can be formed (e.g. Klemme 1980). The degree of lithospheric stretching in a basin is often represented by the **β factor**, which is the ratio of the length of a zone after extension to that before (or, conversely, the thickness of a unit before extension to that after).

Regional tectonic processes exert a major control on the room available for sediment to accumulate (**accommodation space**) by causing either subsidence, which permits continued deposition of sediments, or uplift, which halts deposition and potentially leads to erosion of exposed deposits. Accommodation space in marine basins is also affected by global sea-level (**eustatic**) changes, which result from changes in the volumes of both the ocean basins and the seawater they hold. The former are tectonically related and potentially can have the greatest effect.

When the total length of mid-ocean ridges and the spreading rate are high, a large proportion of the sea floor is elevated (thermal subsidence occurs further from the ridge axis than at less tectonically active times), and sea level can rise 600 m (as during the Cretaceous), leading to widespread marine transgressions. The volume of water is mainly controlled by variations in the size of continental ice sheets and thermal expansion of surface waters, which can cause a eustatic change of up to c.200 m.

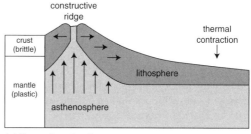

(a) thermal contraction

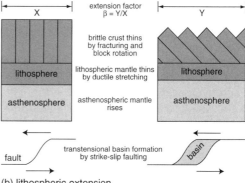

(b) lithospheric extension

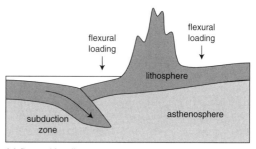

(c) flexural loading

Fig. 3.24 Major mechanisms of basin formation.

The sediments of the abyssal plains contain <10% of the amount of organic matter needed to form a monolayer on mineral grains, although there should be sufficient pelagic supply of organic matter to provide full monolayer coverage. It appears that oxidative degradation is responsible, either directly by the action of molecular oxygen or mediated by Fe^{3+} and Mn^{4+} (Mayer's comment on Hedges & Keil 1995). Evidence for such oxidation is found in shelf/slope sediments redeposited on to the abyssal plains as **turbidites** (accumulations of sediment, often thick, that have been transported downslope in the form of a sediment–water slurry, also called a density flow; Hedges & Keil 1995). The upper layers of the turbidites are highly oxidized, with depleted organic content, but below the sharp redox gradient there is little sign of oxidative degradation or of significant sulphate reduction, and organic matter content is equivalent to that of the parent shelf/slope sediments. The degradation appears to be generally non-selective, similar to the oxidation of lignin, and so may be the result of small oxidants such as H_2O_2, which would be able to access small pores. The preservation of a monolayer-equivalent of adsorbed organic matter appears to be related to oxygen-exposure time, which is least on the shelf and greatest on the abyssal plain (Hedges & Keil 1995).

Overall, primary production is important in determining the sedimentary preservation of organic matter; it must be high enough for sufficient organic matter to reach sediments. The length of time for which the organic matter is exposed to oxygen plays a secondary role, which is modified by the degree to which adsorption of the organic matter on to mineral surfaces protects it from microbial degradation. Whether the detrital organic matter on mineral grains forms a smooth monolayer or a more patchy distribution has yet to be established, and the potential contribution to it from bacterial biofilms should not be overlooked (Parkes et al. 1993). However, a general sorption model is consistent with the approximately constant C_{org}:S weight ratio of c.2.8 observed in marine sediments (Berner 1984). This coupling between carbon and sulphur could arise from the preferential sorption of both iron hydroxyoxides (see Box 3.7) and organic matter on fine-grained sedimentary minerals, and conversion of the iron to pyrite by the H_2S from sulphate reduction (Hedges & Keil 1995). The organic matter in organic-rich sediments may be present as discrete packages or as multiple layers adsorbed on to mineral surfaces. The latter seem to dominate in marine sediments, and it would be expected that the strength of adsorption would decrease as the number of layers increases. Under such circumstances the upper layers would be more prone to degradation, so the protection afforded by anoxia may be particularly important (Hedges & Keil 1995).

From the negative correlation between the organic C content of sediments and exposure time to oxygen in Fig. 3.25 it appears that the development of anoxicity aids the preservation of sedimentary organic matter in all marine environments. The degree of concentration of organic matter is significantly greater in coastal areas—where burial rate and hence the duration of exposure to oxygenated pore waters is restricted—than in deep-sea settings with very low sediment depositional rates. In addition, the concentration of non-protein polymethylenic material (e.g. algaenans; see Section 2.4.2b) seems to be favoured by low levels of exposure to oxygen (Gélinas et al. 2001). As this material is believed to be the main precursor of oils (see Section 4.4.3a), it appears that the long-held belief that anoxic diagenesis aids the development of oil potential is correct.

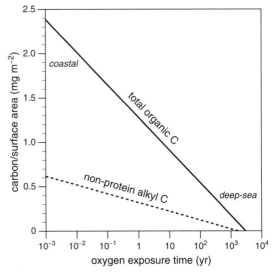

Fig. 3.25 Inverse relationship between carbon content of sediments (normalized to grain surface area) and exposure time to oxygen (after Gélinas et al. 2001). Oxygen exposure time (OET) = (depth of O_2 penetration in sediment pore waters)/(sediment accumulation rate). $C_{total\ organic}$ (mg m^{-2}) = $-0.16\log_e OET + 1.28$ and $C_{non-protein\ alkyl}$ (mg m^{-2}) = $-0.043\log_e OET + 0.32$ ($r_2 = 0.96$ for both; OET measured in years). Low OETs are typical of coastal sediments and high OETs of deep-sea sediments.

(c) Feedbacks to atmospheric oxygen

As seen in Section 1.4.2, the amount of oxygen in the atmosphere is linked to the amount of organic carbon buried in sediments. If organic monolayer coverage of mineral surfaces is the dominant process in preservation of sedimentary organic carbon, and the surface area of sedimentary minerals has remained fairly constant during the Phanerozoic, it could help to explain the apparent stability of atmospheric oxygen levels (Hedges & Keil 1995; and Berner's comment). However, this is an extremely conjectural proposition. Less controversial is the influence of oxygen exposure time on the degradation of organic matter in deep-water sediments and its potential negative feedback, and hence stabilizing influence, on atmospheric oxygen levels (Harnett et al. 1998), as shown in Fig. 3.26.

3.4 Depositional environments associated with accumulation of organic matter

In the earlier parts of this chapter we considered factors directly related to the production, degradation and preservation of organic matter. This section is concerned with a more detailed examination of some of the environments where organic-rich sediments accumulate.

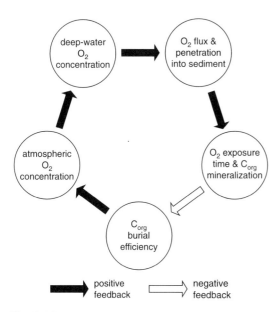

Fig. 3.26 Negative feedback between oxidation of sedimentary organic matter and atmospheric oxygen levels (after Harnett et al. 1998).

3.4.1 Lacustrine environments

Lakes contain only 0.02% of the water in the hydrosphere, yet their deposits of organic-rich sediments can be important sources of petroleum. There is a great diversity of lake types and sizes but an important feature of all lakes is that they are ephemeral, on geological timescales, and conditions in individual lakes may change over relatively short periods. This can lead to a variety of organic facies (see Box 1.1) in ancient lake basins, reflecting different communities of organisms (i.e. autochthonous sources), allochthonous contributions and depositional conditions. An example of such a complex lake system is the ancient Green River Formation of Utah, Wyoming and Colorado, which contains huge reserves of oil shales. The major source of organic matter is usually phytoplankton (see Section 3.2.6), but there can also be significant contributions from terrestrial higher plant material.

Small lakes can have a variety of origins: they can be formed by volcanic or glacial action, by river meandering (ox-bow lakes) or by coastal processes (lagoons). Lakes formed in areas adjacent to glaciated regions (pluvial lakes) are now found in present-day dry belts (e.g. Salt Lake, Utah, USA) and were formerly much larger than at present. Very large-scale lakes are tectonic in origin (see Box 3.11) and are formed either in active tectonic areas such as extensional rift valleys (e.g. East African Rift) or as intracratonic sag basins (i.e. within a stable continental plate, e.g. Lake Chad, Africa).

Lake size and morphology influence the thermally induced stratification of the water column and its stability, as there is a relationship between the depth of the seasonal thermocline and the maximum fetch of the water body: the longer the fetch (i.e. the distance over which wind acts on the water), the deeper the thermocline. The potential effects of stratification on productivity and anoxicity have been discussed in Sections 3.2.2, 3.2.3, 3.2.4 and 3.3.4.

Lake sedimentology is dependent on the size of the lake, its water chemistry and the amount of allochthonous, river-derived, clastic material (Kelts 1988). Organic material found in lakes ranges from microbial (i.e. algal and bacterial) remains through to degraded terrestrial material. Lakes can be classified according to whether there is a through-flow of water: they are hydrologically **open** if they have an outlet and are hydrologically **closed** if they lack one. However, some lakes have been both closed and open during periods of their history (e.g. Lake Kivu, Africa).

(a) Open lakes

Open lakes are characterized by having fairly stable shorelines, because inflow and precipitation are balanced by outflow and evaporation. Most of the allochthonous material in hydrologically open lakes is transported there by rivers. This input is, therefore, controlled by river drainage (or catchment) areas and the climatic influence on run-off. During the early summer the load can be heavy, but during the winter it may be only very minor, if much of the precipitation is held as snow and ice on higher ground. These conditions give rise to layered sediments, which are discussed more fully below. Examples of depositional elements in a hydrologically open, freshwater lake are shown in Fig. 3.27.

In large lakes siliciclastic sediments (see Box 3.11) are normally concentrated around river mouths as beaches, spits and barriers fashioned by wave and current action. Deltas may also be present in relatively deep freshwater lakes and resemble those found in marine environments. Nearshore deposition is characterized by accumulations of terrestrial plant material, deposited on delta tops, which can form peats (see Section 3.4.2). Decomposer communities may actively rework the higher plant material deposited within the delta, their remains augmenting the organic material (particularly the lipids). Delta fronts can become over-steep and collapse, depositing turbidites in deeper water. While this results in conditions suitable for preservation of organic matter, dilution by large volumes of clastic material tends to occur (see Box 3.11).

In offshore regions, clastic sediment is deposited by pelagic rain (i.e. fall-out of material from the water column) and by various forms of sediment gravity flow (e.g. slides and slumps as well as turbidites; Reading 1986). Organic material can be mostly autochthonous, of planktonic origin. It may be deposited in anoxic or poorly oxygenated bottom waters, due to water column stratification, in which bioturbation is inhibited. This often results in the formation of organic-rich shales and marls, which may be of great thickness and have a very high organic carbon content (Powell 1986). Large permanently stratified lakes with anoxic bottom waters are mainly associated with tropical climates (see Section 3.2.2b and Box 3.3).

The offshore deposition of many lakes is characterized by rhythmical sedimentation of alternating layers (or **varves**; Reading 1986). Varves arise because fine-grained sediment may be suspended for long periods in stratified lakes, depending on the degree of circulation and the availability of flocculating agents. The coarser-grained layers of sediment result from fluvial summer deposits and the fine-grained layers from fall-out during the winter. The thickness and nature of the layers depend on the proximity of the inflowing river.

Sedimentary lamination is also caused by the deposition of calcium carbonate and is important in most freshwater lakes where the clastic input from rivers is low. Carbonate deposits may be chemical or biogenic (see Box 1.1). As with phytoplanktonic production, the abiogenic (chemical) precipitation of calcium carbonate is seasonal, occurring during the warmer months. During this period carbon dioxide is removed from the water by planktonic photosynthesis, which causes water acidity to decrease and calcium carbonate to precipitate. From the equilibrium in Eqn 3.9c (Box 3.12) it can be seen that if the amount of $CO_2(aq)$ decreases, more $CO_2(aq)$ and carbonate are produced from bicarbonate, which increases the concentration of carbonate and promotes its precipitation. Subsequently, the water becomes more acidic and no longer supersaturated with respect to calcite (the main form of crystalline calcium

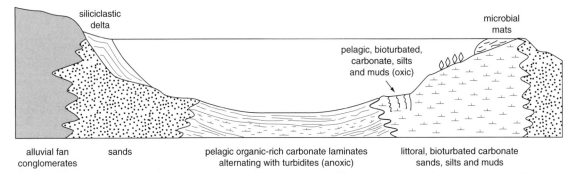

Fig. 3.27 Generalized depositional components in a hydrologically open, freshwater lake (after Eugster & Kelts 1983).

Box 3.12 | Alkalinity of seawater and dissolved carbon

When atmospheric CO_2 dissolves in water, a sequence of equilibria are involved, which can be represented by:

$$CO_2 + H_2O \rightleftharpoons \underset{\text{carbonic acid}}{H_2CO_3} \rightleftharpoons \underset{\text{bicarbonate}}{HCO_3^-} + H^+ \rightleftharpoons \underset{\text{carbonate}}{CO_3^{2-}} + 2H^+$$

[Eqn 3.8]

The distribution of the carbon-containing species (carbonic acid, bicarbonate and carbonate) depends upon pH, as shown in Fig. 3.28 (see Box 3.4). Seawater is slightly alkaline, and its pH rarely falls outside the range 7.5–8.5 owing to the buffering effect of the equilibrium between bicarbonate and carbonate ions. Consequently, seawater contains negligible carbonic acid, and OH^- ions are more abundant than H^+, so the main equilibria can be represented by Eqns 3.9a and 3.9b, which are combined to give Eqn 3.9c:

$$CO_2(aq) + OH^- \rightleftharpoons HCO_3^- \quad \text{[Eqn 3.9a]}$$
$$CO_3^{2-} + H_2O \rightleftharpoons HCO_3^- + OH^- \quad \text{[Eqn 3.9b]}$$
$$\overline{CO_2(aq) + H_2O + CO_3^{2-} \rightleftharpoons 2HCO_3^-} \quad \text{[Eqn 3.9c]}$$

The concentration of the positive charges for the cations Na^+, K^+, Mg^{2+} and Ca^{2+} present in seawater is slightly greater than the total concentration of the negative charges from the anions Cl^-, SO_4^{2-} and Br^-. This excess positive charge is balanced primarily by bicarbonate and carbonate anions, which can be considered to derive from the weak acid carbonic acid.

Alkalinity may be defined as the excess of anions of weak acids in seawater, so it is a measure of the ability to react with H^+ ions: the higher the alkalinity of a solution, the greater its capacity to react with H^+. There are also minor contributions to alkalinity from anions of other weak acids such as borate, silicate and phosphate. Water can be considered a weak acid; its contribution to alkalinity is the difference between the hydroxyl and proton concentrations. Total alkalinity can be defined as:

$$T_{alk} = [HCO_3^-] + 2[CO_3^{2-}] + [H_2BO_3^-] + [H_3SiO_4^-]$$
$$+ [H_2PO_4^-] + 2[HPO_4^{2-}] + 3[PO_4^{3-}] + [OH^-] - [H^+]$$

[Eqn 3.10]

where [ ] denotes concentration.

Because bicarbonate and carbonate are by far the most abundant of the anions in Eqn 3.10, alkalinity (A) can be simplified to:

$$A = [HCO_3^-] + 2[CO_3^{2-}] \quad \text{[Eqn 3.11]}$$

Total dissolved carbon is usually represented by the term ΣCO_2, and is dominated by bicarbonate and carbonate (dissolved CO_2 and organic C are minor components), so its concentration can be approximated by:

$$[\Sigma CO_2] = [HCO_3^-] + [CO_3^{2-}] \quad \text{[Eqn 3.12]}$$

Subtracting Eqn 3.11 from 3.12 gives:

$$[\Sigma CO_2] = A - [CO_3^{2-}] \quad \text{[Eqn 3.13]}$$

Removal of dissolved C does not affect the amount of cations in seawater, so alkalinity (A) remains constant and Eqn 3.13 indicates that the concentration of carbonate ions will increase. Dissolved CO_2 levels decrease as both salinity and temperature increase (Weiss 1974) and, ultimately, seawater can become supersaturated with respect to calcium carbonate. Removal of $CaCO_3$ affects both alkalinity (because of the decrease in Ca^{2+} ions) and dissolved C concentrations.

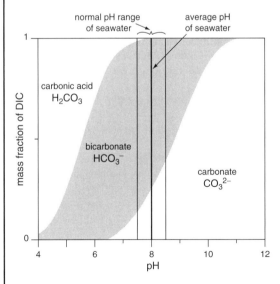

Fig. 3.28 Summary of the approximate equilibration of carbonic acid, bicarbonate and carbonate in water of varying pH. (DIC = dissolved inorganic carbon.)

carbonate), resulting in cessation of precipitation. In large, deep, seasonally stratified, anoxic lakes this seasonal variation gives rise to laminated organic-rich sediments showing annual couplets. The lighter laminae represent the spring and summer production of carbonate with low magnesium content following phytoplanktonic blooms, whereas the dark, carbonate-poor, organic-rich layers arise from the settling out of organic material (particularly diatom remains) during winter (Allen & Collinson 1986). Most organic material in these lakes is preserved as carbonate laminates in the offshore lacustrine depositional region. Moving shorewards into shallower water, biogenic carbonate may form, consisting of shells or algal carbonates.

(b) Closed lakes

Hydrologically closed lakes can be divided into perennial saline lake basins and ephemeral salt-pan basins. Changes in water level in lakes with closed drainage cause substantial fluctuations in shorelines and much reworking of sediment in the nearshore zone. However, pelagic, organic-rich, carbonate laminates, interbedded with evaporitic minerals, can form in the centre of perennial saline lake basins when anoxic conditions exist due to stratification at times of high water levels (Eugster & Kelts 1983). The stratification results from salinity differences, and the water below the halocline becomes depleted in oxygen.

In ephemeral continental **sabkhas**, evaporitic minerals (e.g. high magnesium calcite and gypsum) form on the edges of the water body, and during periodic rains they may be washed into the lake and deposited. The organic-rich facies of these lakes may be formed during more arid periods, when the growth of cyanobacteria is favoured. Mat-forming communities at the water line, dominated by cyanobacteria, can be highly productive in hypersaline environments and form considerable thicknesses of laminated stromatolitic deposits, because desiccation discourages colonization by grazing invertebrates. As well as colonial genera of cyanobacteria (e.g. *Lyngba*, *Microcoleus*, *Oscillatoria*), many other microbes are present. For example, diatoms are present in the upper layers, along with cyanobacteria, but lower down there are photosynthetic purple and then green sulphur bacteria, and towards the bottom anaerobic respiration by sulphate reducers and methanogens occurs (van Gemerden 1993). Heterotrophes are found in the uppermost layer and methane and sulphur oxidizers are also present, leading to efficient nutrient recycling by the microbial consortia (e.g. Canfield & Des Marais 1993), which form a complex syntrophic community.

The greater part of the energy flow occurs in a surface layer <1 cm thick, driven primarily by the rapid turnover of cyanobacterial biomass.

3.4.2 Coal-forming mires

Coal is a carbonaceous material formed by compaction and induration of plant remains that were most probably originally deposited as peat. Coal seams date back to the Middle Devonian, when extensive development of land plants occurred (see Section 1.5.4), but the first economic deposits originate from the Early Carboniferous. The number of peat-forming plant varieties was small until the Early Cretaceous, when a great increase in species diversity occurred. Although most peats are autochthonous, some are thought to have an allochthonous element and are associated with enhanced mineral content. The latter peats may be formed from reworked peats or from vegetation that has been transported, often by rivers and wave action, and the coal formed under these conditions is termed detrital. However, there is debate about whether detrital deposits have been correctly identified and whether such deposits are mostly autochthonous but reflect particular environmental conditions during peat formation.

(a) Peat formation

A variety of terms is used for the depositional setting of peats—bogs, swamps and mires—but **mire** is used in this section. Accumulation of peat in a mire requires the following conditions (Stach et al. 1982):
• a gradual and continuous rise in the groundwater table;
• protection of the mire against flooding by sea and rivers;
• a low-relief hinterland limiting supply of fluviatile sediments.

To create favourable peat-forming conditions the water balance must be right, with precipitation and inflow exceeding evaporation and outflow. Although compaction produces some accommodation space (see Box 3.11), there generally needs to be a continuous rise in the groundwater, keeping it at or above the sediment surface, if accumulation of peat is to continue. If the water table rises too rapidly the mire drowns and peat formation ceases. Marine transgressions are often associated with an increase in sulphur content towards the top of a coal bed and the presence of overlying sands. If the water table falls the peat already formed dries out and is subject to erosion and, eventually, subaerial oxidation. Peat deposition may be cyclical, related to fluctuations in relative sea level. Figure 3.29 summarizes conditions associated with coal-forming mires during the Holocene.

Clearly, the rate of production of plant material must exceed its rate of decay if peat is to form, but this is effectively a function of the water table. Waterlogging and stagnation allow anoxia to develop rapidly due to the activity of the decomposers and, as noted in Section 3.3.3c, woody tissues are far less readily degraded under anaerobic conditions.

Climatic factors play an important role in peat formation because plant production and decay both tend to increase with rising temperature and humidity. So although productivity is higher in tropical than temperate mires (Fig. 3.29), peat accumulation rates can be significantly lower because of elevated decay rates (Table 3.9). At present, large peat deposits are forming under cool conditions at latitudes >45° in high moors and raised bogs. This is because plant growth is less dependent on temperature than is bacterial decay. Bryophytes (mosses) are particularly important in such areas, especially *Sphagnum*. The acidity of bog pools creates an unfavourable environment for the decomposer bacteria (see Section 3.3.2b) and results from moss exudates as well as the by-products of the limited bacterial activity that does occur. In tropical regions, such as the Mahakam Delta (Borneo, Indonesia), peat preservation results from the extremely high rainfall, leading to waterlogging and anoxia.

Peat deposited in **paralic** (i.e. marginal marine) environments, such as river floodplains and deltas, needs protection by bars and spits from major marine inundations and by levees from river floodwaters. Peat formation can be halted by the deposition of excessive fluviatile sediments (Fig. 3.29): the siliciclastic material is responsible for the ash formed during coal combustion, and ash content of *c*.20% seems to be the upper limit for coal formation. Consequently, peat formation is favoured by a river hinterland of low relief (and hence low energy), which reduces the sediment load of rivers.

A variety of peat types can be identified according to the main plant genera present. These include reed/sedge types (e.g. *Cladium*, sawgrass), woody types (e.g.

Table 3.9 Comparison of peat-forming environments (after Bend 1992)

	temperate boreal mires	tropical mires
growth rate (kg dry wt m^{-2} yr^{-1})	1–2	<6
plant debris/peat accumulation rate (mm yr^{-1})	<5	0.5–2.0

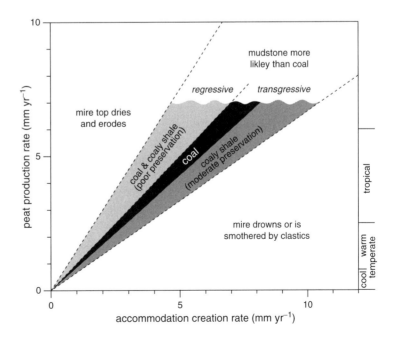

Fig. 3.29 Conditions associated with coal-forming mires in the Holocene (after Bohacs & Suter 1997).

Taxodium distichum, swamp cypress; *Rhizophora mangle*, mangrove) and moss types (e.g. *Sphagnum*). Not only does the flora contributing to peats vary with environmental factors (particularly climate) and over geological time, it also varies with time during the evolution of a mire. Succession of plant communities can occur in mires as a consequence of changing environmental factors that control the level of the water table relative to the height to which the substrate of dead plant material has built up. Initially, therefore, water plants such as reeds may dominate. As the ground builds up and dries out, herbs and then scrub, characteristic of drier ground, can develop and eventually trees become established. If this succession occurs over the entire area of a mire, the mire disappears and peat accumulation ceases. However, different patches of a mire can be at different stages in this cycle and, overall, peat accumulation continues. Horizons can then be recognized in the peat due to formation from different plant types. A rapid rise in the water table can cause drowning of the mire and again peat deposition stops.

(b) Mire types

Mires can be classified according to the hydrological regime as **ombrotrophic**, where the water supply is primarily rain, and **rheotrophic**, where groundwater is the main supply. Ombrotrophic mires are associated with surface doming, forming a raised mire (or bog), which is low in nutrients (oligotrophic) but is also removed from the influence of major influxes of inorganic minerals and so produces coals with very low ash contents. Persistent heavy rainfall is responsible for maintaining water saturation within raised mires, which lie above the regional water table. The surfaces of raised mires grow up to 10 m above the coastal plain in the Mahakam Delta. Rheotrophic mires are generally rich in nutrients and minerals, and tend to yield high-ash coals. This type of mire produces a more uniformly low-lying swamp, or planar mire, in which the surface is at or below the water table most of the time (e.g. Florida Everglades).

The major areas of peat formation today are raised mires, which are found in the tropics (e.g. Malaysia and Indonesia) and the temperate zone of the northern hemisphere. The boreal raised mires are dominated by moss (*Sphagnum*), whereas their tropical counterparts bear trees. As the raised mires develop, plant diversity decreases; for example, the dipterocarp *Shorea albida* is often the only tree on raised mires in north-west Borneo.

The planar mires (swamps) of modern tropical and subtropical deltas (e.g. Mississippi and Niger) do not seem to be good models for ancient coal-forming environments because they contain too much inorganic mineral matter and experience seasonal fluctuations in water level that either prevent much peat accumulating or result in its oxidation. Most Permian, Carboniferous and Tertiary coal deposits have a low ash content, and so appear more akin to raised mires. However, they occupied a much larger area than present-day peat-forming environments (modern boreal raised mires are mostly sited in depressions formed by Quaternary glaciation). The ancient coal-forming environments appear to be associated with high and rising sea level (see Section 4.6.1), which generally leads to relatively low rates of supply of terrigenous clastic material from smaller subaerial sources, so a conventional planar-mire setting cannot be discounted.

The Okefenokee swamp of south Georgia, USA, is believed to be similar to planar mires existing in the Tertiary. It covers $c.1500 \text{ km}^2$ and has peat deposits up to 5 m deep, which are accumulating at $c.0.5 \text{ mm yr}^{-1}$. Beneath the peats are Pleistocene sands and muds, which originally formed the bed of a shallow lagoon at a time when the Atlantic reached much further inland, but the swamp is now 30 m above sea level. There are two distinct environments: forests (covering most of the swamp) and open-water marshes (called prairies).

The prairies are rarely flooded to >30 cm and only dry out during major droughts (at $c.25$-yr intervals). The fibrous peat in the prairies is thickest where it overlies depressions in the sands and muds of the ancient lagoon, and is mainly formed from the roots, leaves and stems of the abundant water lilies. Marsh gas can cause large areas of peat (batteries) to rise to the surface. Water plants on these batteries die as the surface dries out and are replaced sequentially by grasses and sedges, then shrubs and finally large trees (mainly cypress), which form tree houses anchored by tree roots to the underlying peat. The fringing growth of maiden cane and sedges rots away quickly and contributes little to the peat.

Eventually tree houses may merge and form expanses of forest. Because these areas are relatively dry, oxygen is generally available for the degradation of litter, although a granular peat still accumulates ($c.1$ m thickness), formed mostly from cypress remains (needles and wood). Major droughts can cause the surface of the prairies to dry out and the peat may even ignite. The existence of prairies, therefore, is due mainly to fires that halt forest establishment, leaving charcoal behind as evidence. The relative abundance of prairies and forests can be seen in the vertical profile of repeating sequences, grading upward: cypress peat, charcoal zone, fibrous water lily peat.

3.4.3 Marine environments

Oceanic environments can be divided into areas according to water depth: **continental shelf**, 0–200 m; **continental slope** (which includes the continental rise), 200–4000 m; **abyssal plain**, 4000–5000 m (see Fig. 3.6). Deeper areas are present—the trenches associated with subduction zones (destructive plate margins)—but they are of relatively minor area and importance. Shelf, slope and continental rise are collectively termed the **continental margin** and, as we have seen, primary productivity is at its highest in this area of the oceans (Fig. 3.8). Basins on continental margins in which water exchange with the open sea is restricted by physical barriers are important settings for the formation of organic-rich sediments, together with upwelling areas (Fig. 3.23).

(a) Marine shelf deposits

Mud-dominated, but not necessarily organic-rich, offshore shelf deposits are abundant in the sedimentary record. Currently, thick shelf deposits are forming in areas that have low wave and current energies and where suspended sediment is at high concentrations. The type of sediments deposited is dependent upon climate, latitude, major wind belts and tidal energies.

Major periods for the widespread formation and preservation of organic-rich sediments in shelf deposits include the mid part of the Cambrian, the early Mid-Ordovician, the Devonian, Early and Late Jurassic and the Mid and Late Cretaceous (see Section 6.3.4a). These periods may have coincided with large global (eustatic) rises in sea level resulting in marine **transgressions** and the extensive deposition of organic-rich, shallow, marine mudstones as the oceanic mid-water oxygen-minimum layer impinged on large, epicontinental, shelf areas. Such transgressions can release large amounts of nutrients into epeiric (shallow inland) seas, enhancing primary productivity and generating large quantities of organic detritus. It is likely that areas of the flooded shelves had restricted water circulation, providing the potential for the development of anoxia and the preservation of sedimentary organic material. In the present-day marine environment, shelves are also the setting for upwelling currents, which off the Namibian coast result in an oxygen-deficient zone measuring 340 km long by 50 km wide. Sediments deposited in upwelling zones have distinctive features, including laminated non-bioturbated muds, phosphorites and uranium minerals (Allen & Allen 1990). Uranium is an unusual metal in that salts of its higher oxidation state are soluble, but those of its lower oxidation state are insoluble. Consequently, uranium enrichment in sediments is generally associated with anoxicity and often elevated organic content.

The coastal shelf area has an intimate association with the land at the deltas of major rivers (e.g. Mississippi and Amazon). Peat and lignite formation is often a feature of the delta plain, although freshwater phytoplankton and bacteria that rework detritus may be locally important in lakes, swamps and abandoned channels. Brackish conditions may also occur, favouring those bacteria that rework organic debris (chiefly from higher plants) and giving rise to organic-rich sediments with a high wax content. On the seaward slopes of the deltas a large proportion of organic material is of terrestrial origin and at least partially oxidized. Nutrients are abundant and productivity may be high, but preservation of organic material in high concentration is not favoured because it is often greatly diluted by organic-poor fluvial sediments derived from a hinterland that is not covered with lush vegetation. However, organic-rich sediments are presently forming in the tide-dominated portion of the Mahakam Delta plain. This is because there is a large amount of fine-grained organic-rich sediment in the delta, which is derived from the fluvially dominated part of the delta plain that supports rich equatorial forests of mangrove and palm (Elliott 1986).

Deposition of organic-rich sediments further down the shelf and on to the continental slope and rise often occurs as a result of turbidite flows, redistributing organic-rich sediments from delta fronts or from further up the shelf and slope (Summerhayes 1983). While there is a certain amount of pelagic sedimentation, primary production decreases away from the coastline as nutrient levels decline, and detritus is largely recycled before it settles to the sea floor. However, this may not always have been so in the past, when the thermohaline circulation (Box 3.2) did not operate and there may have been widespread anoxia in bottom waters, aiding preservation of sedimentary organic matter (e.g. Cretaceous oceanic anoxic events; Section 6.3.4).

In hot arid areas where the continental shelf meets the land, marine sabkha-type environments exist. For example, on the Trucial coast of Abu Dhabi shallow marine carbonate sediments are reworked and bound by microbial mats. In the upper intertidal zone these mats are virtually undisturbed by predators and can grow and be preserved, eventually forming extensive organic-rich deposits (stromatolites; see Section 3.4.1).

(b) Enclosed and silled basins

Apart from nearshore areas and deltas, continental shelves are not important as sites of organic-rich sediment accumulation except where basins are present. Whether or not such a basin will contain organic-rich sediments depends on factors such as climate, water depth (200–1000 m is ideal) and hydrology. The nature of water exchange with the open sea is important when establishing whether a restricted basin has the potential to accumulate organic-rich sediments. A negative water balance occurs in arid hot climates, where surface seawater flows in to make good losses due to both evaporation and the sinking of hypersaline water, which then flows out of the bottom of the basin. The bottom waters of the basin are, therefore, oxygenated and depleted in nutrients, providing unfavourable conditions for the formation of organic-rich sediments. Examples of such basins are the Red Sea, the Mediterranean Sea and the Persian Gulf.

When the water balance is positive the outflow of relatively fresh water exceeds the inflow of denser saline water. The latter tends to be relatively rich in nutrients but depleted in oxygen (due to the decay of organic matter sinking out of the overlying water column). A pronounced salinity gradient often occurs in the water column of the basin, allowing stratification to occur and the development of anoxia in the lower layers. Nutrients are often trapped in such basins, potentially resulting in high productivity and preservation of sedimentary organic matter in oxygen-poor conditions. Examples of such basins are the Black Sea, the Baltic Sea and Lake Maracaibo (Venezuela) (Demaison & Moore 1980).

(c) Production and preservation of organic matter in the Black Sea

The Black Sea is often used as a model for the preservation of organic-rich sediments of the type that can yield petroleum. It is the largest body of anoxic marine water in the world. Productivity is high in surface waters and degradation of the resulting detritus leads to anoxicity in the deep water of the isolated basin. Preservation of organic carbon is consequently high; the estimated production and preservation rates of organic matter are

Table 3.10 Sources and fate of organic carbon in the Black Sea over the past 2000 yr (data after Tissot & Welte 1984)

	annual flux (g C m^{-2})
inputs	
photosynthesis	100
chemosynthesis	~15
rivers	5
Azov and Marmara Seas	2
outputs	
aerobic respiration	~100
Azov and Marmara Seas	3
sulphate reduction	10
dissolution	5
sedimentary preservation	4

given in Table 3.10. Primary production is largely attributable to phytoplankton, but there are also small allochthonous inputs of detrital organic matter to the euphotic zone from rivers. Photosynthetic and chemosynthetic bacteria also make significant contributions to autochthonous organic matter. Most of the photosynthetic production and allochthonous inputs are recycled within the euphotic zone, while sulphate reduction accounts for most of the anaerobic degradation. However, $c.4\,g\,C\,m^{-2}$ is preserved in sediments annually.

Some interesting changes seem to have occurred in the water column of the Black Sea in recent years (Repeta et al. 1989). The hydrogen sulphide interface (which is also termed a **chemocline**, because the sulphide concentration increases rapidly across it) has moved up into the euphotic zone, at $c.80$–$100\,m$ depth. Concentrations of bacteriochlorophyll-e (Fig. 2.27) and bacterial aromatic carotenoids (e.g. isorenieratene; Fig. 2.23) reach a maximum at this depth, providing evidence for anaerobic bacterial photosynthesis by *Chlorobium* in this region of the water column. The biomass of photosynthetic bacteria (estimated at $c.0.5\,g\,m^{-2}$) may amount to half that of the phytoplankton, suggesting that the Black Sea has the largest and deepest community of photosynthetic bacteria presently in existence. There is evidence that shoaling of the chemocline, leading to euxinic conditions in the lower euphotic zone, has occurred periodically over at least the past 6 kyr (Sinninghe Damsté et al. 1993c). It seems likely that anaerobic bacterial photosynthesis has been important in similar ancient environments (see Section 6.3.4a).

4 | Long-term fate of organic matter in the geosphere

4.1 Diagenesis

4.1.1 Introduction

The preceding chapters have examined the main contributions to the organic matter reaching the upper layers of sediment, the types of organic compounds involved and the environments associated with the large-scale production and degradation/preservation of this material. In Chapter 3 we considered the biological processes involved in the recycling of organic matter, and now we review the transformations that the organic matter undergoes during its long-term incorporation into sediments and subsequent lithification (see Box 1.1). In organic geochemistry the term **diagenesis** is applied to the processes affecting the products of primary production that take place prior to deposition and during the early stages of burial under conditions of relatively low temperature and pressure (in contrast to the general geological usage, which extends to somewhat higher temperatures and pressures). Biological agents are mainly responsible for diagenetic transformations, although some chemical transformations are possible, with the possibility of catalysis by mineral surfaces.

During diagenesis burial depth increases and the sediment undergoes compaction and consolidation. There is an associated decrease in water content and an increase in temperature. The rising temperature increasingly restricts biological activity but the thermal energy becomes sufficient to break chemical bonds, and ultimately, as the phase of **catagenesis** is entered, thermally mediated alteration of organic matter dominates. The boundary between diagenesis and catagenesis is generally not sharp; it is often correlated with the onset of oil formation, which may correspond to temperatures as low as 60 °C but is usually closer to 100 °C (see Section 4.5.2). As we have seen in earlier chapters, thermophilic and hyperthermophilic bacteria are active at relatively high temperatures, and certainly within the early stages of catagenesis. The evolution of sedimentary organic material with increasing burial and temperature is generally termed **maturation**.

4.1.2 Microbial degradation of organic matter during diagenesis

The composition of the organic matter ultimately preserved in sediments is controlled by the diagenetic transformations of primary production, particularly by heterotrophic microorganisms. Amino acids together with low molecular weight (LMW) peptides and carbohydrates (sugars) are soluble and can be assimilated directly by the decomposer community, so they are usually rapidly recycled. They are, consequently, generally minor constituents in ancient sediments. Insoluble proteins and polysaccharides can be hydrolysed by the extracellular enzymes of fungi and bacteria into their water-soluble constituent amino acids and monosaccharides, respectively (Section 3.3.3). In this way, high molecular weight (HMW) compounds forming storage (e.g. lipids and starch) or structural (e.g. exoskeleton and cell wall) materials can be transformed into assimilable components. As a result, the initial input of biological macromolecules (e.g. proteins, polysaccharides, lipids and lignin) often comprises no more than 20% of the total organics in the upper layers of sediments.

The rates at which the various compound classes are degraded differ. Proteins generally appear to degrade faster than the other major sedimentary organic compound classes, and the highest concentrations of amino acids are found at the water–sediment interface. Within this general lability there is variation in the relative stabilities of individual amino acids (e.g. Bada & Mann 1980). Highly cross-linked fibrous proteins such as keratin resist microbial attack but can be degraded by some fungi and most actinomycetes. Carbohydrates seem only slightly less readily degraded than proteins, and the predominant sugars formed by their hydrolysis are hexoses and pentoses. Carbohydrates and proteins are not necessarily completely degraded; traces have been found in Precambrian sediments. Lipids can also be recycled (e.g. fatty acids) but some appear more resistant towards degradation than proteins and carbohydrates, and can survive diagenesis, undergoing only minor alteration. Lignin can be microbially degraded, even though it is one of the more resistant biopolymers.

The general effect of microbial attack is, therefore, to reduce the concentrations of all compound classes with increasing sediment depth. However, the preservation potential of even the most readily degradable substances can be increased if they are intimately associated with resistant structures. For example, the proteins that form the matrix for mineralization in invertebrate shells are relatively protected against degradation by their mineral covering. Similarly, cellulose and hemicellulose from higher plants can be protected by lignified layers. Lipids can also be protected where they are incorporated in resistant coatings, such as in spores, pollen and leaf cuticles (e.g. Killops & Frewin 1994).

4.1.3 Geopolymer formation

The chemical residues from microbial degradation, if not rapidly assimilated, undergo condensation reactions to form new polymeric material. For example, irreversible condensation reactions between sugars and amino acids can occur readily, but via an exceedingly complex and incompletely understood series of reactions (often termed Maillard reactions). The types and relative abundances of the precursor amino acids and sugars and the environmental conditions all influence the condensation reactions, the products of which are brown polymeric substances called **melanoidins** (Hedges 1978; Rubinsztain et al. 1984). From laboratory syntheses it appears that much of the 'backbone' of melanoidins is formed from furanyl units, derived from modified (partially dehydrated) sugars, and there is cross-linking of chains via condensation with amino acids (Fig. 4.1). Simple amide formation between amino acid units (see Eqn 2.5) does not seem an important reaction. The extent to which melanoidins actually contribute to sedimentary organic matter has been questioned, given the rapid microbial recycling of sugars and amino acids (Collins et al. 1992).

Lipid moieties and lignin degradation products (see Section 3.3.3c) can also be incorporated in the formation of new macromolecular material by condensation reactions. The extent of condensation increases during diagenesis, finally yielding large amounts of brown organic material, which becomes increasingly insoluble as hydrolysable components, such as carbohydrates and proteins, disappear. At the end of diagenesis the polycondensed organic residue, or geopolymer, is called humin in soils, brown coal in coal mires and kerogen in marine and lacustrine sediments. These diagenetic products probably also contain varying amounts of largely unaltered refractory organic material. Sometimes the term kerogen is applied to all forms of insoluble organic geopolymers produced during diagenesis.

As we have just seen, the classical model for the formation of kerogen involves the condensation of the various products of microbial degradation that escape

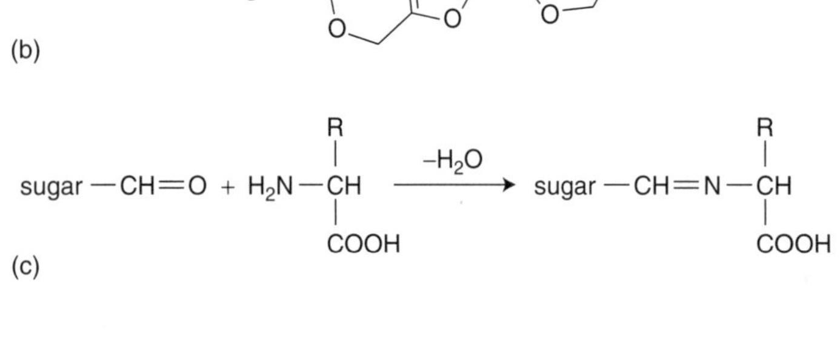

Fig. 4.1 Reactions and structures involved in melanoidin formation: (a) example of sugar rearrangement; (b) typical units in the carbohydrate backbone of melanoidin; (c) common initial condensation reaction between a sugar residue and an amino acid (after Rubinsztain et al. 1984).

further degradation by the decomposer community. It has been modified in the light of recent evidence suggesting that kerogen is formed at a relatively early stage of diagenesis (earlier than previously suspected) and that a significant fraction of it may derive from mixtures of selectively preserved, sometimes partly altered, resistant biomacromolecules, rather than from random recondensation of the constituents of depolymerized biomacromolecules (Tegelaar et al. 1989). It is likely that kerogen is formed by a combination of geopolymer formation and the selective preservation of biopolymers that are particularly resistant towards microbial degradation, as discussed in more detail in Section 4.4.1a.

Structural determination of geopolymers is extremely difficult, owing to their low solubility and the random assemblage of many different units. Bulk properties, such as molecular weight range, functional group distributions and proportions of aliphatic and aromatic C atoms, provide some information on the types of units involved. Various degradative techniques (e.g. pyrolysis or chemical oxidation) can be used to break up the macromolecules, providing more detail of the structural units, but determining how these units were originally arranged is problematical (Rullkötter & Michaelis 1990; Kögel-Knabner 2000). Modelling of geopolymer formation in the laboratory provides only a limited insight into the behaviour of the complex substrate mixtures under natural conditions.

4.2 Humic material

4.2.1 Occurrence and classification

Humic substances are found in soils, brown coals, fresh water, seawater and both marine and lacustrine sediments. The humics in soils are originally derived largely from plant matter. They can be leached, particularly from acidic soils in cold wet upland regions of the UK (and similar environments elsewhere), and account for almost all the organic carbon in fresh waters, imparting the characteristic brown coloration to upland waters. In addition, most of the dissolved organic matter (DOM) in seawater is humic or humic-like material (the Gelbstoff mentioned in Section 3.3.1a), as is the greater part of sedimentary organic matter (Ishiwatari 1985).

The term humic substances describes three groups of material: **fulvic acid**, **humic acid** and **humin**. This distinction is based on the traditional fractionation of soil humic material. Treatment of bulk humic material with dilute alkali dissolves the fulvic and humic acids, leaving a residue of insoluble humin. Acidification of the alkaline extract precipitates humic acid, leaving fulvic acid in solution. Fulvic acid comprises smaller and more oxygenated units than humic acid, and so has been proposed as possibly the oxidized degradation product of humic acid. Both humic and fulvic acids are thought to be intermediates in the diagenetic formation of humin. Humic substances can occur in different physical states. For example, humic acids are found in solution in fresh water, as solids or gels (colloids) in soil and as dry solids in coal.

4.2.2 Composition and structure

The complex polymeric nature and interaction between component chains of humic material make structural analysis difficult; however, compositional information can be obtained from elemental and functional group analysis. Elemental analysis of humic and fulvic acids from a range of soils is presented in Table 4.1. The atomic H/C ratio is quite low, and is lower for humic acid ($c.0.8$) than fulvic acid ($c.1.3$), which is consistent with a higher aromatic content for humic acid. The atomic O/C ratio is also lower for humic acid ($c.0.5$) than fulvic acid ($c.0.8$), reflecting the higher content of polar groups in fulvic acid. There are comparable differences between marine humic and fulvic acids. Although nitrogen and sulphur levels are relatively low in humic substances in general, they can be higher in marine than terrestrial humics.

Oxygen is the major heteroatom in humic substances and occurs predominantly in the following functional groups: COOH, phenolic and alcoholic OH, ketonic and quinoid C=O, and OCH_3 (ether and ester). The estimated abundances of these groups in soil humic and fulvic acids are given in Table 4.2. The greater water solubility of fulvic acids compared to humic acids can be attributed to the higher content of polar groups, particularly carboxyl groups. Among the other functional groups present in smaller quantities are ether, aldehyde and amine.

Table 4.1 Typical elemental compositional ranges for soil humic and fulvic acids (after Schnitzer 1978)

element	humic acid (wt %)	fulvic acid (wt %)
C	53.6–58.7	40.7–50.6
H	3.2–6.2	3.8–7.0
N	0.8–5.5	0.9–3.3
O	32.8–38.3	39.7–49.8
S	0.1–1.5	0.1–3.6

The macromolecular structure of fulvic and humic acids probably consists of a flexible, extended chain with only limited branching and cross-linking (Hayes et al. 1989). Attached to this backbone are smaller compounds, particularly sugar and amino acid residues, which can be released by hydrolysis. Concentrations of these sugars and amino acids are greater in humic than fulvic acids and are also greater in soil humic acids than freshwater humic acids. Sugars are usually more abundant than amino acids but their combined abundance does not normally exceed 20% of humic substance weight.

There are significant differences between terrestrial and marine humics; for example, in contrast to fluvial humics, only trace amounts of identifiable lignin derivatives are present in marine humics, so the terrestrial contribution to marine DOM either is minor or has been altered beyond recognition (Lee & Henrichs 1993). Units derived from lignin appear to be important in the backbone structure of terrestrial humic substances, some being largely unaltered lignin components, while others show signs of microbial alteration. Some likely structural features of terrestrial fulvic acid, based on predominantly lignin-derived polycarboxyl-phenol components, are shown in Fig. 4.2. Aromatic systems are dominated by single aromatic rings in both fulvic and humic acids, although there may also be small amounts of condensed (i.e. polycyclic) aromatic rings and furan derivatives. Aromatic units can be joined by ether (-O-) links and also sometimes by short-chain alkyl groups (particularly the C_3 group of lignin components). There is evidence for the involvement of polysaccharide-like material in the macromolecular backbone, suggesting that carbohydrates are incorporated during diagenesis. A minor degree of amino acid incorporation into the backbone may also occur.

Analysis of the degradation products obtained from pyrolysis and oxidative degradation studies suggests that terrestrial humic substances are highly aromatic (c.70% of total C). However, these techniques tend to overemphasize the importance of aromatic and phenolic units. Non-degradative techniques, such as ^1H and ^{13}C nuclear magnetic resonance, probably give more realistic aromatic carbon contents of 20–50% for soil humics,

Table 4.2 Estimated abundance of functional groups (mequiv g^{-1})* in soil humic and fulvic acids (after Schnitzer 1978)

functional group	humic acid	fulvic acid
total acidic groups	5.6–8.9	6.4–14.2
carboxyl COOH	1.5–5.7	5.2–11.2
phenolic OH	2.1–5.7	0.3–5.7
alcoholic OH	0.2–4.9	2.6–9.5
quinoid/keto C=O	0.1–5.6	0.3–3.1
methoxy OCH_3	0.3–0.8	0.3–1.2

*mequiv g^{-1} is equivalent to mmol of each group per g humic substance.

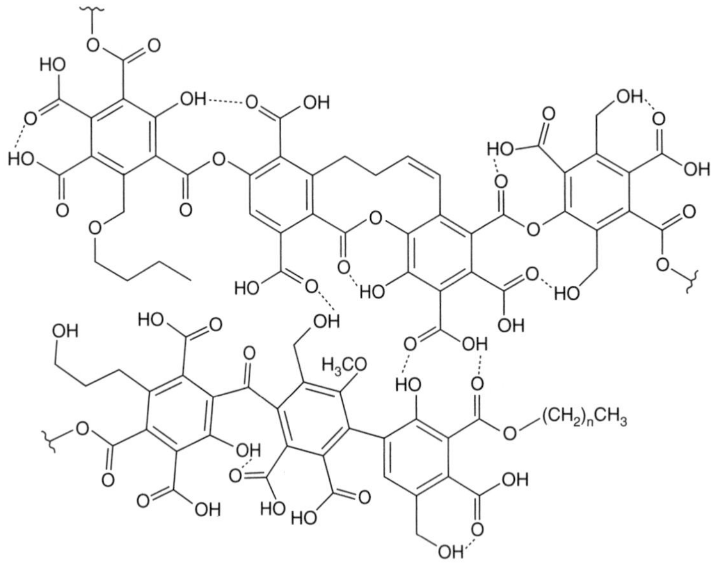

Fig. 4.2 Some examples of structural units and bonding modes in fulvic acid (hydrogen bonding is denoted by dotted lines; after Hayes et al. 1989).

20–35% for peat humics and <15% for marine humics (Hatcher et al. 1980). These data suggest that aromaticity can offer a means of discriminating terrestrial and marine sources of humic material (Dereppe et al. 1980), and that although aliphatic structures are important in all humic material, they are more abundant in marine than terrestrial humics (Hatcher et al. 1981). In addition, humic acids contain a greater proportion of alkyl C atoms than their fulvic acid counterparts (Rasyid et al. 1992). Important contributions to aliphatic structures are made by fatty acids, derived from microbial or higher plant sources, which are bonded by ester linkages to the macromolecular backbone. As a result of their greater ratio of aliphatic to aromatic structures, humic acids in marine sediments have higher H/C atomic ratios (1.0–1.5) than their soil counterparts (0.5–1.0) and also contain fewer phenolic constituents but often more sulphur.

Soil fulvic and humic acids are larger than their aquatic counterparts. Freshwater humic acids tend to have molecular weights of $\leq 10^3$, whereas soil humic acids generally have larger molecular weights, sometimes in excess of 10^6 (Hayes et al. 1989). Hydrogen bonding appears to play a role in molecular aggregation (see Fig. 4.2). An important factor influencing the shape of humic and fulvic acid molecules in solution is pH (see Box 3.4). At low pH fulvic acids are fibrous, but as the pH rises the fibres tend to mesh together to form a sponge-like structure by pH 7, whereas at higher pH the structure becomes plate-like (Schnitzer 1978). Humic acids behave similarly but precipitate when the pH falls below c.6.5.

The mesh structure formed by humic substances is capable of trapping smaller chemical species. For example, minor amounts of acyclic alkanes are found in most samples of humic and fulvic acids, and some of the fatty acids associated with humics may be similarly trapped components rather than bonded to the macromolecular backbone. Humic substances also usually contain a variety of metals, which are incorporated into the macromolecular structure. Metal ions can be surrounded by and bonded to suitable chelating groups, chiefly carboxylic acids, on humic molecules that stabilize the ions and allow them to be transported with the organic material. This important property of humic substances is examined again later (Section 7.6.5) in relation to the environmental fate of heavy metals.

4.2.3 Formation of humic substances

The availability of different types of potential precursor substances in particular environments of humic formation appears to affect the composition of humic substances. The abundance of polycarboxyl-substituted phenols shows a positive correlation with terrestrial inputs, indicating that lignins and probably also tannins are important in humic formation in terrestrial environments (Hayes et al. 1989). These components are resistant towards biodegradation and so their dominance in aerobic environments is not surprising. Refractory lignified plant tissue, altered to varying degrees by microbial degradation, is the major substrate of soil humic material and contributes the bulk of the aromatic units. The by-products of the decomposer organisms and their remains are also likely to contribute to humic substance formation and may account for the major part of the aliphatic components (e.g. fatty acids and alkanes), supplementing the aliphatic contribution from higher plants (e.g. waxes). The higher aliphatic content of marine humics may be the result of an increased contribution from microbial lipids, particularly algal.

As diagenesis proceeds in naturally occurring humic material the amount of hydrolysable components, sugars and amino acids, decreases, suggesting either that they are lost from the humic structure or that they are being permanently incorporated into the macromolecular backbone. While carbohydrates seem to be incorporated into the humic backbone, evidence for amino acid incorporation is less conclusive, because the nitrogen content of humic material is low and mostly in the form of hydrolysable amino acids superficial to the macromolecular backbone (Hayes et al. 1989). Synthetic melanoidins subjected to artificial diagenesis develop some of the properties of naturally occurring humic substances (Rubinsztain et al. 1984). However, there is little evidence to suggest that melanoidins make a significant contribution to natural humic material.

Condensation of the various units present in humic material progresses during diagenesis, yielding the insoluble humin residue in soils. Humic or humic-like substances are likely to make some contribution to kerogen, just as they do to brown coals. Sulphur levels generally appear to increase as diagenesis progresses, possibly involving incorporation of sulphide (from bacterial sulphate reduction) into the macromolecules (Nissenbaum & Kaplan 1972). For example, in marine humic substances sulphur levels increase with depth (Francois 1987) and in peat the humin fraction, which increases in relative abundance with depth, contains the greatest amounts of sulphur.

4.3 Coal

4.3.1 Classification and composition

(a) Classification

Coals are usually classified as either humic or sapropelic. **Humic coals** are formed mainly from vascular plant remains (e.g. Westphalian coals of northern Europe). They tend to be bright, exhibit stratification and go through a peat stage involving humification (i.e. formation of humic material). The major organic components derive from the humification of woody tissue and have a lustrous, black/dark brown appearance, although resin bodies are also sometimes present. In contrast, the less common **sapropelic coals** are not stratified macroscopically and are dull. They are formed from fairly fine-grained organic muds in quiet, oxygen-deficient, shallow waters. Normally they do not go through a peat stage but follow the diagenetic path of hydrogen-rich kerogens (see Section 4.4.3a).

Sapropelic coals contain varying amounts of allochthonous organic and mineral matter, as do H-rich kerogens. However, the organic fraction is dominated by autochthonous algal remains and varying amounts of the degradation products of peat swamp plants. Sapropelic coals are subdivided into cannel and boghead coals. **Boghead coals** (or **torbanites**) contain larger amounts of algal remains together with some fungal material (e.g. the oil shale in the Midland Valley of Scotland). **Cannel coals** have a higher concentration of spores.

(b) Petrology

The individual morphological constituents of coal, its petrological components, represent the preserved remains of, primarily, plant material and are known as **macerals**. They are observed by transmitted or reflected light microscopy and can be recognized by their differing optical properties (a study known as **petrography**). There are three main groups of macerals, which are listed for the hard (or bituminous) coals in Table 4.3 together with their likely origins.

The major maceral group in humic coals is **vitrinite**, the lustrous, black/dark brown constituents from the humification of woody tissue. Where the morphological structure of the woody tissue is preserved the vitrinite is termed telinite, which is transparent and orange-red with a vitreous lustre, and any voids are generally filled with colloidal humic material or resins. Where the structure of the vitrinite is lost but it is still a translucent brown it is termed collinite. The term vitrinite is usually applied to the higher maturation stages, whereas **huminite** is the commonly used name for this maceral group during diagenesis (i.e. peat, lignite and sub-bituminous coal stages in Table 4.6).

Among the **inertinite** macerals are opaque modifications of vitrinite, termed fusinite and semifusinite,

Table 4.3 Hard (bituminous) coal maceral groups (modified Stopes–Heerlen system; after Bend 1992)

maceral group	maceral	origin
vitrinite	telinite	cellular structure of wood, leaf and root tissue
	collinite	structureless, infilling gel
	vitrodetrinite	unidentified cell fragments
liptinite	sporinite	spore and pollen cases
	cutinite	waxy coating of leaves and stems
	suberinite	cork tissues, e.g. bark and root walls
	resinite	resin bodies
	alginite	algal tests
	liptodetrinite	unidentified liptinite fragments
	fluorinite	lenses/layers, possibly plant essential oils
	bituminite	wisps or groundmass, from lipids
	exudatinite	veins of expelled bitumen-like material
inertinite	fusinite	charred wood and leaf tissue
	semifusinite	partially charred wood and leaf tissue
	macrinite	charred gel material
	micrinite	charred liptinitic material
	sclerotinite	fungal remains (e.g. sclerotia)
	inertodetrinite	unidentified intertinite fragments

which retain the morphological structure of woody tissue and are dull brown/black and friable. They seem to represent highly carbonized, non-fusible, vitrinitic material, of the sort that can be formed by charring. Semifusinite appears to be intermediate in degree of charring between fusinite and vitrinite. Equivalent 'charred' forms of what appear to be exinitic material can also be recognized as micrinite and macrinite. Fungal remains of moderate reflectance are also included among the inertinites, as sclerotinite.

Liptinite is composed of lipid-rich, translucent, yellow/red macerals. This group of macerals was originally termed exinites, because it contained only material derived from the casings of spores and pollen (i.e. exines). However, the group has subsequently been enlarged to include leaf cuticular material (cutinite), resin bodies (resinite) and algal remains (alginite), and so the collective term liptinite is more appropriate. More recently, new macerals within the liptinite group have been described, comprising three amorphous bitumen-like bodies—fluorinite, bituminite and exudatinite—that can be distinguished by their fluorescence properties (Teichmüller 1974).

The three maceral groups exhibit different optical properties under transmitted and reflected light, and when subjected to UV-induced fluorescence (Table 4.4), which provides a basis for identification of the various macerals. The maceral groups also behave differently when heated. Normally, vitrinites evolve some gas and leave a fused carbon residue. Liptinites are generally transformed into gas and tar. Inertinites are, as the name implies, inert; they do not generate any hydrocarbons or appear to change in form (they are not fusible). With increasing maturity the recognition of macerals becomes more difficult as structural features become less distinct. For example, the reflectances of all macerals increase and converge, while fluorescence decreases.

Various combinations of maceral groups occur in humic coals but four main lithotypes are recognized on the macroscopic scale: vitrain (comprising mainly vitrinite), durain (comprising liptinite and inertinite), clarain (comprising vitrinite and liptinite) and fusain (mainly fusinite, with some semifusinite). On the microscopic scale, macerals often occur in certain associations, recognized as microlithotypes (e.g. vitrite = mainly vitrinite, clarite = vitrinite + liptinite, durite = liptinite + inertinite). Fusain bodies have the appearance of charcoal and there is debate about whether fusain represents fossil charcoal from ancient vegetation fires (Chaloner 1989; Scott 1989).

(c) Chemical composition

Various standardized analyses have been developed to determine the chemical composition of coals. Among them are the proximate analyses, which quantify the volatile and non-volatile components, and the ultimate analyses, which determine the elemental composition. These, and examples of other types of analyses, are listed in Table 4.5. Data are often recorded on a dry and ash-free (daf) basis, because of the variable amount of unbound water (particularly in brown coals) and inorganic minerals that may be present. A mineral-matter-free (mmf) rather than simple ash-free basis is often used for elemental composition in order to take account of the oxides, sulphides etc., and also the water of crystallization in inorganic minerals, when calculating the composition of the organic matter.

The main elements in coal are carbon, hydrogen, oxygen, nitrogen and sulphur. Oxygen is present mainly in carboxyl, ketone, hydroxyl (phenolic and alcoholic) and methoxy groups, but the distribution of these functional groups varies with increasing burial, as described below. Nitrogen is found in amines and in aromatic rings (e.g. pyridyl units). Sulphur is found in thiols, sulphides and aromatic rings (e.g. thiophenic units). Sulphur is also a common constituent of coal in inorganic form, usually as pyrite. A variety of metals can be present in coal; they were either present in the original biogenic material or incorporated into the matrix of humic material during diagenesis.

Van Krevelen diagrams, of atomic O/C versus H/C ratios, are frequently used to compare the compositions of coals and their components. Typical initial (immature) positions of some of the major macerals and plant tissues in such a plot are shown in Fig. 4.3. As can be seen

Table 4.4 General optical properties of maceral groups

maceral group	transmittance	reflectance	fluorescence
liptinite	high	low	intense (at low maturity)
vitrinite	moderate	intermediate	usually absent
inertinite	low (opaque)	high	absent

Table 4.5 Coal analysis (ASTM procedures; after Ward 1984)

proximate analyses	
moisture	drying at 110 °C (avoiding oxidation and decomposition)
volatile matter	volatiles liberated at 950 °C in absence of air, excluding moisture
ash	inorganic residue from combustion
fixed carbon	C remaining after volatiles determination (i.e. coking potential)
ultimate analyses	
elemental analysis	C, H, N and S content
oxygen	determined by difference (i.e. total minus C, H, N, S content, and correcting for inorganic mineral content); more rarely determined directly
examples of other analyses	
forms of sulphur	organic, sulphide, native S, sulphate
other elements	e.g. trace elements, phosphorus, chlorine, carbonate CO_2
relative density	depends on ash content and maturity
specific energy	energy liberated upon combustion

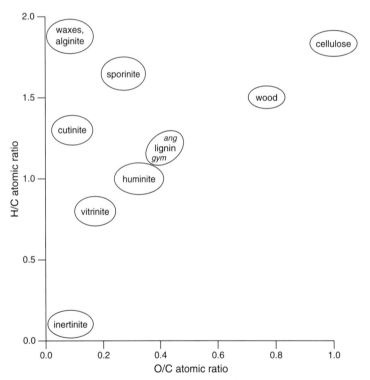

Fig. 4.3 Chemical composition of major coal macerals and plant tissues plotted on a van Krevelen diagram (after Tissot & Welte 1984; Hedges et al. 1985). Lignin incorporates differences between angiosperms (*ang*) and gymnosperms (*gym*) (after Hatcher 1990).

in this figure, the tissues that comprise the liptinite group of macerals have diverse compositions. The vitrinite group appears more uniform in composition, although the amorphous material that constitutes some forms of collinite (e.g. desmocollinite) can have varying atomic H/C ratio, the more hydrogen-rich material possibly reflecting a contribution from cuticular waxes. If sufficiently abundant, this material can cause suppression of vitrinite reflectance and will fluoresce under UV/blue light when immature (Price & Barker 1985; Wilkins et al. 1992). With increasing maturity all the components in Fig. 4.3 become increasingly similar in chemical composition and follow evolution paths leading towards the origin. The chemical changes lead to increasing vitrinite reflectance and a rapid decrease in fluorescence of liptinites (Box 4.1).

4.3.2 Formation

There are two main phases in the formation of humic coals: **peatification**, followed by **coalification**. Coalification can be subdivided into a **biochemical stage** and a **geochemical stage**. The main agents during peatification and early coalification (biochemical stage) are biological, and these stages are equated with diagenesis. Late coalification (geochemical stage) is primarily the result of increasing temperature and, to a lesser extent, pressure, and may be equated with catagenesis. As previously noted, there is usually some overlap between the action of biological and physicochemical agents, so the boundary between the two stages of coalification may be indistinct. The biochemical stage generally ends at depths of the order of several hundred metres, at the limit of penetration of percolating waters supplying *in situ* microorganisms with nutrients.

Although the term maturity can be used to describe how far a coal has progressed through peatification/coalification, another term, **rank**, is routinely applied. In order of increasing rank, the main stages of humic coal formation can be termed: **peat, lignite, sub-bituminous coal, bituminous coal** and **anthracite** (Table 4.6). Variations do, however, exist in the use of rank terminology, with **brown coals** being applied to lignites and sub-bituminous coals, and **hard coals** being substituted for bituminous coals (German DIN classification). For our purposes, the boundary between brown and hard coals is a convenient approximation to the diagenetic/catagenetic boundary. Various subclassifications of the main groups are also recognized but are not considered here (e.g. Stach et al. 1982; Ward 1984).

The extent of peatification and coalification can be examined using a van Krevelen diagram, as shown in Fig. 4.5 for both sapropelic and humic coals. Loss of oxygen-containing functional groups during diagenesis can clearly be seen in the decreasing O/C ratios of the humic coal band, which is primarily attributable to the evolution of CO_2 and H_2O (Boudou et al. 1984), while the H/C changes very little. In contrast, the sapropelic coals initially contain little oxygen but relatively large amounts of hydrogen in the dominant lipid-rich macerals, and their evolution at all stages involves hydrogen

Table 4.6 ASTM rank classification of coals (approximate boundary values after Stach et al. 1982; Ward 1984)

coal rank (ASTM)*	volatiles (% dmmf)†	fixed C (% dmmf)†	vitrinite reflectance (% R_o)‡	moisture (%)	calorific value (kcal kg^{-1} mmf)§
peat					
	~63	~37	~0.25	~75	3500
lignite					
	~53	~47	~0.4	~30	4600
sub-bituminous coal					
	42–47	53–58	0.5–0.65	~10	6500
high-volatile bituminous coal					
	31	69	1.1		8600
medium-volatile bituminous coal					
	22	78	1.5		
low-volatile bituminous coal					
	14	86	1.9		
semi-anthracite					
	8	92	2.5		
anthracite					
	2	98			
meta-anthracite					

*American Society for Testing and Materials; †dry mineral-matter-free basis; ‡see Box 4.1; §mineral-matter-free basis.

Box 4.1 | Vitrinite reflectance and liptinite fluorescence

Vitrinite reflectance

Reflectance measurements are usually made on isolated macerals immersed in oil, to prevent stray reflections, and illuminated with monochromatic light (546 nm, in the green region of the visible spectrum). **Vitrinite reflectance** values are then expressed as a percentage by the term R_o. The method was initially developed for measuring the rank of coals (McCartney & Teichmüller 1972), but can be generally applied in assessing thermal maturity in types II and III kerogen (see Section 4.4.3b, c).

Vitrinite reflectance increases with maturity, reflecting the increase of planar aromatic sheets in the kerogen/coal structure, and so it provides a measure of the thermal stress experienced by the organic matter. As a general guide, the main phase of oil generation usually occurs in the range 0.65–1.30% R_o, with peak generation occurring around 1.0% R_o. However, oil generation can commence around 0.5% R_o, which is usually considered to be the diagenesis–catagenesis and brown coal–bituminous coal boundaries. Wet gas generation from kerogen occurs in the range 1.3–2.0% R_o, giving way to dry gas at >2.0%. The latter value is close to the boundary between bituminous coal and anthracite. Under advanced metagenesis, vitrinite reflectance values of $c.11\%$ can be associated with graphite formation (see Table 4.6).

A range of reflectance values is usually observed for vitrinite in a kerogen sample, and it is important to determine statistically the accuracy of the mean values used in maturity assessment and the significance of depth trends based on such values. Data can be recorded in the form of a plot of frequency (i.e. number of macerals exhibiting a given reflectance value) versus reflectance (Fig. 4.4). In such plots two reflectance maxima can sometimes be observed for vitrinite, as in Fig. 4.4. That at lower reflectance represents the real maturity of the kerogen, while that at higher reflectance probably represents an input, during deposition of the sediment, of reworked vitrinite from an eroded, older and more mature source. It can be difficult to identify all the finely disseminated vitrinite macerals in a kerogen sample, particularly if there is a paucity of other maceral groups for comparison of reflectance values. In addition, the reflectance values and morphology of vitrinite and inertinite macerals form a continuum, making distinction between these two groups difficult on occasion. It is, therefore, best to apply statistical evaluation to a large data set obtained from isolated and concentrated kerogen macerals.

Liptinite fluorescence

Vitrinite reflectance cannot be used for type I kerogen (see Section 4.4.3a) because vitrinite is absent. However, liptinite macerals fluoresce under blue/UV light and the fluorescence is characterized by its intensity and wavelength. Fluorescence is intense in immature samples but decreases during diagenesis and catagenesis, and by the end of the oil window it has usually disappeared. The intensity of fluorescence can, therefore, be used as a maturity indicator. In addition, the wavelength of fluorescence progressively increases (i.e. moves to the red end of the visible spectrum) with increasing degree of catagenesis. Fluorescence measurements are most accurate when vitrinite reflectance measurements are least accurate—i.e. the decrease in fluorescence is greatest where vitrinite reflectance is lowest and changes slowly—at the beginning of the oil window, so the two sets of measurements are complementary.

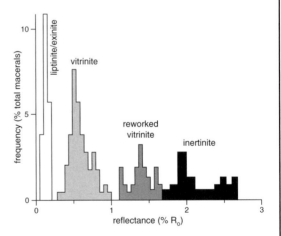

Fig. 4.4 Example of reflectance distributions for various maceral groups in a kerogen sample.

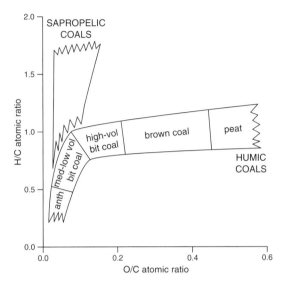

Fig. 4.5 Van Krevelen diagram showing the main evolutionary trends of sapropelic (cannel and boghead) and humic coals (after Durand et al. 1983). Rank increases towards the origin; brown coal = lignite + sub-bituminous coal).

loss and declining H/C ratios. For the main sequence of humic coal transformation, the end of diagenesis corresponds to an O/C atomic ratio of $c.0.1$. At this point the atomic ratios of different maceral groups begin to become similar, and subsequently the main change is a decline in the H/C ratio. Comparison of Figs 4.3 and 4.5 shows that alginite and wood lie at the low maturity ends of the sapropelic and humic coal evolution paths, respectively, and that the evolution path of humic coal passes through huminite (equivalent to the lignite rank) and vitrinite (equivalent to sub-bituminous coal rank) towards inertinite.

(a) Peatification

Peat formation is an inefficient process: less than 10% of plant production in a typical peat-forming environment accumulates as peat, the rest being recycled by the associated microbial community or lost from the system (by mineralization and leaching). The products of peatification are humic substances: brown hydrated gels with no internal structure. Oxidizing conditions generally prevail at the surface of peat bogs but reducing conditions develop with increasing compaction and depth, and are favoured by stagnant surface waters. At the surface conditions are usually neutral or mildly acidic but with increasing burial depth acidity increases, bacterial communities change and their activities decrease and eventually cease. In this way, conditions favourable for large-scale preservation can develop just beneath the sediment surface.

Peatification begins in the surface litter with the mechanical disintegration of plant material by invertebrates. This comminution greatly aids microbial degradation, mainly by increasing the surface area that can be attacked. Depolymerization of polysaccharides by decomposers occurs during early peatification: hemicellulose constituents are rapidly removed, followed by conversion of cellulose into glucose units (Stout et al. 1988). Lignin is more resistant, but a proportion is degraded, mostly under aerobic conditions, yielding large amounts of aromatic, phenolic and carboxylic acid (COOH) units. In addition to these residues, microbial metabolites and the decaying remains of fungi and bacteria invariably contribute to the organic matter (e.g. fungal carbohydrates). A degree of condensation of the products of the various biodegradation reactions occurs, with loss/modification of functional groups, resulting in peat formation. During these biochemical changes gases are the main expelled products, such as CH_4, NH_3, N_2O, N_2, H_2S and CO_2, together with H_2O. A large proportion of the CO_2 derives from the breakdown of carbohydrates, and demethylation of the methoxy groups of lignin is probably a significant contributor of methane. Much of the lignin in peatified wood may be only slightly altered and becomes the primary precursor of vitrinite (Stout et al. 1988). However, the groundmass (matrix material) in low-rank coals generally contains some macerated wood, which exhibits greater alteration, involving depolymerization and defunctionalization (Stout & Boon 1994).

With increasing diagenesis the lignin and polysaccharide content of peat decreases but the content of humic substances increases. Cellulose is still found in peat but is absent from brown coal, which is formed at a more advanced phase of diagenesis, during the biochemical stage of coalification. Recognizable lignin can constitute up to 35% of peat but decreases to <10% in brown coal. Lipid material is confined largely to leaves, spores, pollen, fruits and resinous tissues (resins are particularly abundant in conifers). Although these components are very minor in higher plants they are concentrated during peatification because they are fairly resistant towards degradation. Peats have a high water content ($c.95\%$) and usually contain 5–15% bitumen (see Box 4.2), which derives from lipid components and comprises a mixture of waxes, paraffins and resins.

Box 4.2 | Bitumen

The term bitumen is usually applied to naturally occurring solid or liquid hydrocarbon deposits that are soluble in an organic solvent such as carbon disulphide, and the solid types are fusible. The term petroleum is used to describe naturally occurring liquid (i.e. oil) and gaseous hydrocarbon deposits. Oil can, therefore, be classified as a bitumen. In contrast, pyrobitumen is an insoluble and infusible solid that gives rise to petroleum-like products upon heating. Therefore, kerogen that is not totally inert can be classified as a pyrobitumen, as can the altered residue of bitumen remaining in a petroleum source rock that has passed beyond the maturity range associated with oil generation.

Bitumen is composed of three main fractions (see Fig. 4.12): asphaltenes, resins and hydrocarbons. Asphaltenes and resins are heavy N,S,O-containing molecules (molecular weight >500), whereas the hydrocarbons are usually of lower molecular weights.

The resins fraction of bitumen should not be confused with plant resins. When bitumen is mixed with a light hydrocarbon (e.g. pentane) it separates into a soluble fraction, maltenes, and an insoluble fraction, asphaltenes. The maltenes can then be separated chromatographically into hydrocarbons and resins on the basis of the greater polarity of the resins.

The hydrocarbons fraction can be separated into aliphatic and aromatic hydrocarbon subfractions, again on the basis of the greater polarity of the aromatics. Traditionally, the aliphatic hydrocarbons (or saturates) are described in terms of their paraffinic (acyclic alkane) and naphthenic (cycloalkane) content. Strictly, the term hydrocarbon should only be applied to compounds containing H and C atoms, but there are usually other compounds present in the hydrocarbons fraction isolated by simple chromatographic procedures that contain an atom of S, O or N (generally in the aromatics subfraction).

(b) Biochemical stage of coalification

Alteration and condensation of organic residues progress during the early stage of coalification with further loss of oxygen-containing functional groups. Bacterial action continues to be an important factor. With the loss of functional groups the residue becomes relatively concentrated in carbon and hydrogen. The final preserved organic-rich product of diagenesis, brown coal, contains $c.50-70\%$ C and $c.5-7\%$ H. Aromatic units are an important part of the coal structure, and initially they are mainly inherited from lignin, but with increasing diagenesis the aromatic content of the macromolecular organic material increases generally. Modified lignin is a major constituent of huminite macerals. Aliphatic chains are restricted in abundance and in length, reflecting the initial paucity of lipid-rich components. A generalized picture of changes in the content of oxygen-containing functional groups during coalification as a whole is shown in Fig. 4.6.

Because polysaccharides have largely been degraded and so do not make a major contribution to coalified woods (Hatcher et al. 1989a), the chemical changes during lignite formation can be viewed in terms of modification of the lignin structure by microbial enzymatic attack. The sequence of reactions is quite well understood (Hatcher 1990), and is summarized in Fig. 4.7.

The major early reactions involve the cleavage of aryl ether bonds: the hydrolysis of methoxy groups and cleavage of β-O-4 linkages, both yielding phenolic OH groups. For vanillyl units this results in the formation of catechol-like (i.e. dihydroxybenzene) structures. The β-O-4 cleavage leaves an OH group at C-4 in both angiospermous and gymnospermous lignin (Bates & Hatcher 1989; Hatcher et al. 1989b), while generating a carbocation on the βC atom of the propyl chain. The

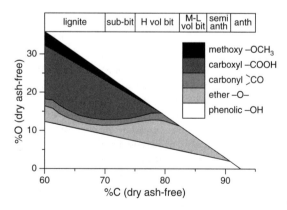

Fig. 4.6 Variation in oxygen-containing functional groups in coal with carbon content (after Whitehurst et al. 1980).

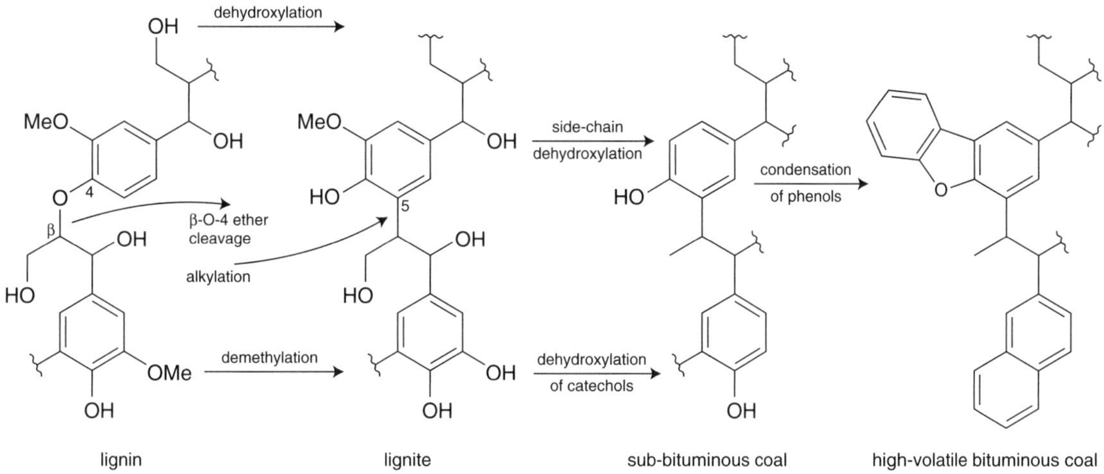

Fig. 4.7 Major reactions affecting the vanillyl units of lignin during coalification (after Hatcher & Clifford 1997).

carbocation is a strong electrophile (see Box 2.10) and rapidly attacks the adjacent aromatic ring of vanillyl units, predominantly at C-5 (Botto 1987). This alkylation reaction preserves the general macromolecular structure (i.e. the order in which the units are joined remains the same, only the site of linkage changes slightly), in contrast to a purely random recondensation process. However, the electrophilic attack is blocked by a methoxy group in syringyl units (Hatcher et al. 1989b), so the water-soluble phenols produced by the β-O-4 cleavage of angiospermous lignin can potentially be lost. This may explain why syringyl units decrease in relative abundance as lignin is progressively degraded, and why angiosperm-derived wood is generally less well preserved than its gymnospermous counterpart. The differences between lignins from angiosperms and gymnosperms (Fig. 4.3) are virtually eliminated by the lignite rank stage (Killops et al. 1998). The hydrolytic loss of methyl from methoxy groups can potentially convert syringyl to vanillyl units, but the process may not be particularly important until after the β-O-4 cleavage and C-5 alkylation reactions are virtually complete (Hatcher & Clifford 1997).

By the early part of the lignite stage (sometimes referred to as lignite B) about half the methoxy groups have been altered to OH groups. In addition, the β-O-4 cleavage and subsequent C-5 alkylation reaction have caused an increase in the degree of aromatic ring substitution, from three C atoms substituted on average in gymnospermous lignin to four in the associated lignite B. Unlike the parent lignin, lignite B contains small amounts of COOH and CO groups ($c.2$–4% of total C), probably resulting from oxidation of the αC in the propyl chain. The gain in COOH and CO groups is probably counterbalanced by loss of OH from the γC of the propyl chain, because overall oxygen content is fairly constant (Hatcher 1990). The carboxyl content subsequently decreases with increasing rank (McKinney & Hatcher 1996). By the end of the lignite stage (sometimes termed lignite A) removal of methoxy groups has progressed ($c.75$% removed) and catechol units dominate in the modified lignin residue, reaching their maximum abundance at this stage (Hatcher 1990). The number of OH groups on the propyl chains seems to decrease, apparently via reduction, because the increase in aromaticity that would be expected from elimination reactions is not observed (Hatcher 1990).

Formation of sub-bituminous coal seems to involve O loss through conversion of dihydroxy phenolic units (catechols) to monohydroxy units (phenols and alkylphenols), as shown in Fig. 4.7, based on the simple distribution of pyrolysis products, which are dominated by phenol, *ortho*-cresol (2-methylphenol) and 2,4-dimethylphenol (Hatcher 1990). Oxygenated aliphatic structures (alkyl hydroxyls and ethers) seem to be absent. Figure 4.8 shows the types of units present at various stages of biochemical coalification, based on a random lignin polymer.

At the end of diagenesis the resulting brown coal contains no carbohydrates, and the amount of relatively unaltered lignin has decreased to <10%. As well as the macromolecular material, brown coals also contain

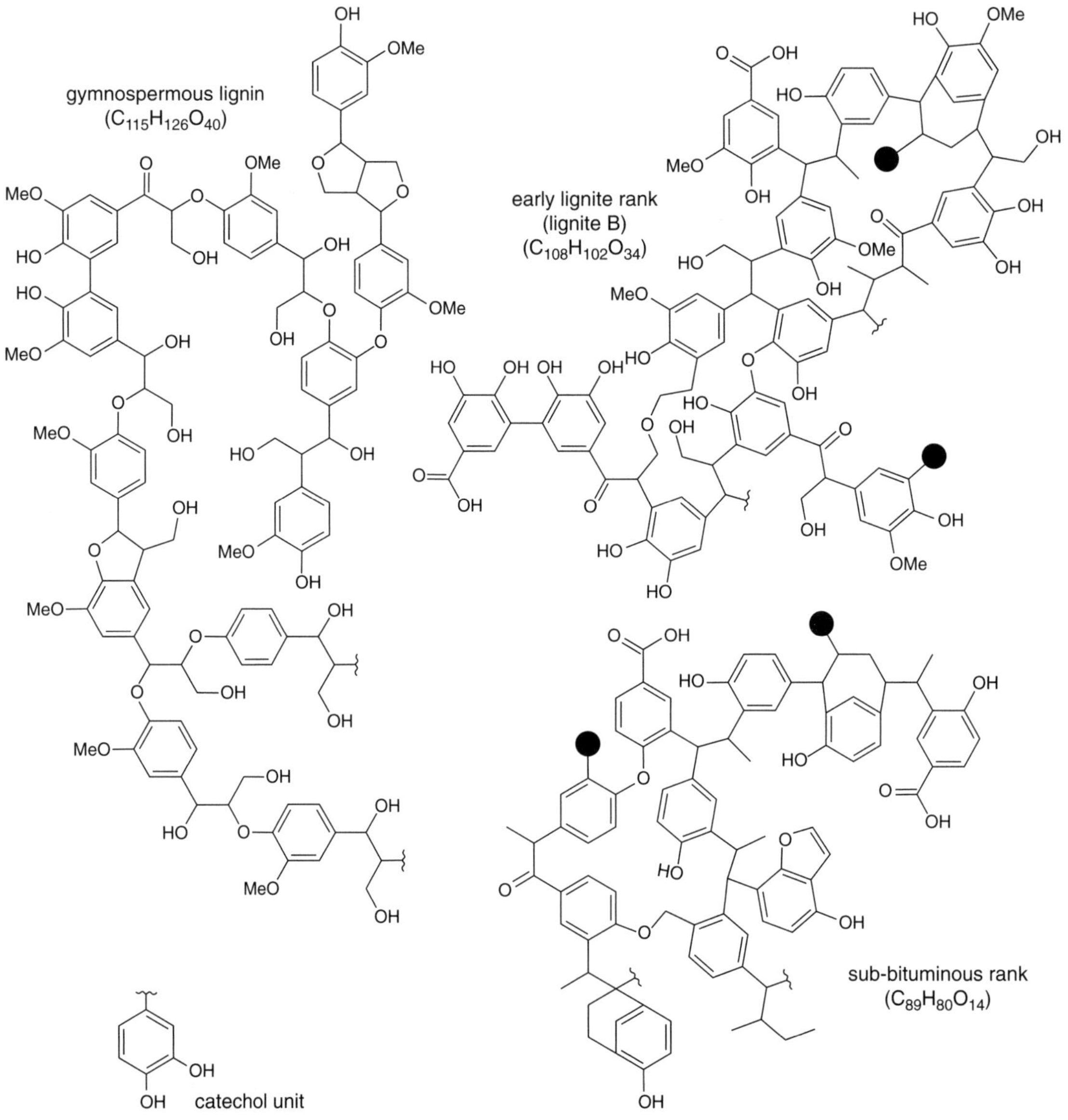

Fig. 4.8 Partial model structures showing the evolution of gymnospermous lignin up to the sub-bitminous rank stage, and the catcheol units that dominate during early diagenesis (the filled circles represent an intramolecular linkage; after Adler 1977; Hatcher 1990).

small amounts of LMW volatile substances (e.g. resin and wax constituents), encapsulation within resistant structures aiding their preservation during diagenesis. Metal ions inherited from humic substances seem to be largely retained in the macromolecular structure during peatification and coalification. The original morphology of wood is largely retained at the lignite stage, and is still recognizable at the sub-bituminous coal rank, suggesting that the chemical structure of lignin, although modified, has not been greatly altered (Faulon et al. 1994; Hatcher & Clifford 1997). This observation is consistent with the proposed proximal C-5 alkylation following β-O-4 cleavage, which tends to retain the overall macromolecular shape. Similarly, the demethylation reaction and subsequent reduction of hydroxy groups would not be anticipated to affect the gross macromolecular structure. A more ordered polymeric structure for lignin than shown in Fig. 4.8, as suggested in Section 2.5.1, would seem to be required to permit the retention of morphology, which must be based on preservation of gross chemical structure (Hatcher & Clifford 1997).

(c) Geochemical stage of coalification

With the termination of biologically mediated transformations the diagenetic stage of coal formation ends. As temperature and pressure become the dominant agents in the transformation of coal the geochemical stage of coalification begins. The transition from sub-bituminous to high-volatile bituminous coal rank is marked by a further decline in O content, but the aryl-O content remains approximately constant or may even increase slightly (Fig. 4.6), suggesting the condensation of phenols to aryl ethers or dibenzofuran-like structures (Fig. 4.7; Hatcher et al. 1992). Aromaticity also increases, which would be consistent with the cyclization and aromatization of alkyl side chains (Hatcher 1988). Aromatization, together with the influence of rising temperature and pressure, increasingly disrupts the chemical and morphological structure of wood. Carboxylic acid and hydroxy groups are important in brown coals, but during the sub-bituminous and high-volatile bituminous stages significant decarboxylation occurs, which is probably mainly thermally mediated owing to the relatively low energy demands of this reaction (Alexander et al. 1992). By the medium–low-volatile bituminous coal stage only ether and phenolic groups remain, and when the anthracite stage is entered generally only a few phenolic OH groups are left (Fig. 4.6).

Throughout the geochemical stage carbon content increases and there is a concomitant decrease in oxygen content. Hydrogen levels remain fairly constant for most of the geochemical stage but begin to decrease at an increasing rate at the highest levels of maturity. Nitrogen content remains low and fairly constant at $c.1–2\%$ throughout, while levels of sulphur are variable but generally <1%. At the beginning of this stage the aromatic nuclei appear to bear peripheral functional groups (e.g. methyl, hydroxyl, carboxyl, carbonyl, amino) and to be partly linked together by aliphatic rings and CH_2 (methylene) units. The changes in elemental composition during the geochemical stage reflect both the continued elimination of functional groups and the growth of aromatic nuclei that results from increasing aromatization of cycloalkyl structures and condensation between individual aromatic nuclei. The aromatic carbon content of vitrinite increases from $c.70\%$ in bituminous coals to $c.90\%$ in anthracite.

Although liptinite macerals initially have a much lower aromatic content than vitrinite, by the later stages of coalification their aromatic content parallels that of vitrinite. In inertinites the amount of carbon involved in aromatic structures is >90% at all ranks. Methane, carbon dioxide and water are regarded as the main volatile products of coalification, and bitumen is also generated during the bituminous coal stage. Compaction continues with increasing burial depth, resulting in decreased porosity and increased density. Water content decreases with increasing rank; from up to 95% in peat to as little as 1% in anthracites. The production of methane and carbon dioxide increases but that of other volatile components decreases during coalification. Some of these bulk changes with increasing rank are shown in Table 4.6.

Although bitumen is evolved during the high-volatile bituminous stage, there is only a slight decrease in the atomic H/C ratio (Fig. 4.5), suggesting that the bitumen is mostly retained within the coal matrix. The H/C ratio decreases at an increasingly rapid rate from the medium-volatile bituminous stage onward, reflecting the loss of hydrogen-rich material, mainly as gas dominated by methane (from the methyl and propyl units on lignin residues), together with some bitumen (from the more hydrogen-rich coals). Much of the hydrogen needed to form the saturated hydrocarbons that are released (e.g. conversion of methyl groups to methane) is provided by the aromatization of cycloalkyl structures and progressive condensation of aromatic clusters. Dehydration, decarboxylation and demethylation are, therefore, the principal reactions during early coalification, associated with some aromatization, while during late coalification aromatization becomes the dominant reaction.

The complex polymeric nature of vitrinite, the major maceral of humic coals, imposes difficulties in the determination of its chemical structure. However, Fig. 4.9 presents a possible partial structure for vitrinite containing 83–84% C, corresponding to the start of the medium-volatile bituminous coal stage, based on general elemental composition and functional group analysis and bearing in mind the likely precursor molecules. At the macromolecular level, the change in structure of vitrinite during the geochemical stage of coalification begins with the aggregation of aromatic nuclei into

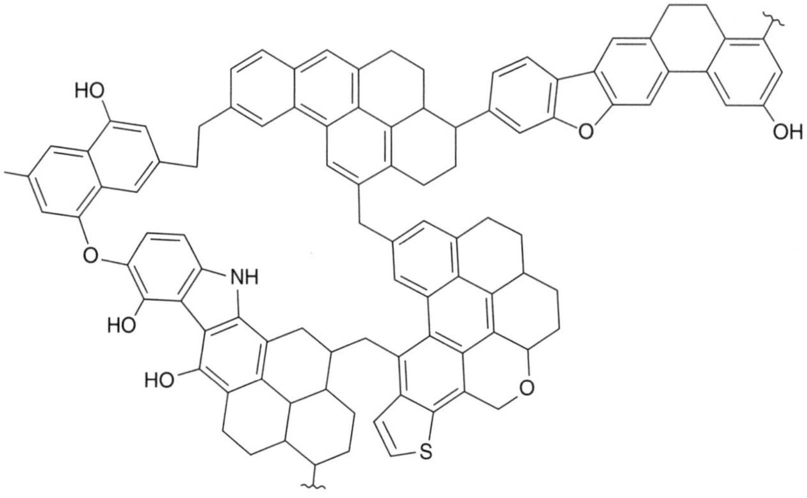

Fig. 4.9 A partial structure for vitrinite (83–84% C; after Heredy & Wender 1980).

clusters by the processes of aromatization and condensation. These clusters form flat lenses (due to the inherently flat benzenoid ring systems involved), which are initially randomly orientated but gradually become more ordered, tending towards a parallel arrangement as pressure increases. Further condensation occurs between laterally adjacent lenses so that, by the anthracite stage, a generally parallel sheet arrangement has been attained. Ultimately, amorphous carbon can be produced that is part way towards the idealized structure of graphite (the most stable form of carbon under surface conditions, although it is generally associated with metamorphic shear zones).

4.4 Kerogen

4.4.1 Formation

(a) Geopolymer formation during diagenesis

Kerogen is the polymeric organic material from which hydrocarbons are produced with increasing burial and heating. It occurs in sedimentary rocks mostly in the form of finely disseminated organic macerals and it is by far the most abundant form of organic carbon in the crust. As mentioned in Section 4.1.3, it is likely that kerogen is formed by a combination of the selective preservation of biopolymers and the formation of new geopolymers.

In the classical theory of geopolymer formation, kerogen was believed to derive from humic substances similar to the humic and fulvic acids found in soils (Durand 1980; Tissot & Welte 1984). During early diagenesis in aquatic environments the organic material from primary production is broken down by microbial action into smaller constituents, which then undergo condensation reactions, giving rise to humic substances. Microbial degradation and condensation follow in immediate succession, leading to a zone in the top few metres of sediment (and possibly also in the water immediately overlying it) where both processes are active at the same time. With increasing time and burial depth most of this humic material becomes progressively insoluble due to increasing polycondensation, which is associated with the loss of superficial hydrophilic functional groups (e.g. OH and COOH). Insolubilization can begin quite early and apparently continues to significant depths, on the basis that humic acids can still be found at several hundred metres depth in sediments that contain abundant terrestrial detrital material. The humin-like material that is formed continues to undergo condensation and defunctionalization, resulting in kerogen. Lipid components are also incorporated into the kerogen structure. In this model of kerogen formation, fulvic acids are viewed as the precursors of humic acids, in contrast to some theories of the evolution of soil humic substances (see Section 4.2.1).

The selective preservation of resistant biomacromolecules (with the possibility of minor microbial alteration) is likely to be an important process in the formation of most kerogens, and particularly those formed at very early stages of diagenesis (Tegelaar et al.

1989). Some possible resistant biomacromolecular precursors of kerogen and their related macerals are given in Table 4.7. Highly aliphatic, insoluble and non-hydrolysable biopolymers that resist biodegradation have been detected in the protective outer layers of some extant higher plants and algae as well as in the corresponding fossil remains (see Section 2.4.2b). These materials are termed cutan (Nip et al. 1986a, b; Fig. 4.10a) and suberan (Tegelaar et al. 1995) in terrestrial plant cuticles, and are believed to make significant contributions to kerogen (Hatcher et al. 1983; Tegelaar et al. 1991). Equivalent materials, termed algaenans, are found in the cell walls of eustigmatophytes (an order of the Xanthophyceae; Table 3.6) and many chlorophytes, but have yet to be detected in diatoms and haptophytes, and may be rare in dinoflagellates (Gelin et al. 1999). Comparable material derived from heterotrophic bacterial cell walls has proven elusive so far (Allard et al. 1997).

The algaenans from most algae contain long n-alkyl chains, such as that from the chlorophyte *Tetraedron*, which appears to make a major contribution to the Messel oil shale (Goth et al. 1988). Initial studies suggested that these algaenans were polyesters with n-alkyl chains of up to C_{33} (Fig. 4.10b; Blokker et al. 1998), but more recently it has been suggested that the alkyl chains may be extremely long, up to C_{120} (Allard et al. 2002). In contrast, the characteristic algaenans from *Botryococcus braunii* contain long isoprenoidal alkane chains, as found in several torbanites (Largeau et al. 1986; Derenne et al. 1994), and seem to involve aldehyde linkages (Berthéas et al. 1999). Highly aliphatic macromolecules have also been identified in the cell walls of cyanobacteria and may make an important contribution to amorphous kerogen (Chalansonnet et al. 1988). Algae can also contribute aromatic material to kerogen, such as polyalkylphenolic macromolecules, which may be related to the phlorotannins (polymers of phloroglucinol, which can also be called 1,3,5-trihydroxybenzene; van Heemst et al. 1996).

Although they are relatively minor components in organisms, the resistant biomacromolecular structures discussed above are likely to become concentrated as more abundant and readily hydrolysable biopolymers, such as proteins and carbohydrates, are degraded. There is also the possibility for less resistant material to be protected against biodegradation within coatings of resistant material (e.g. polysaccharides within higher plant cuticular membranes and lipids within microbial cell walls). Selective preservation seems a reasonable proposition because preserved parts of organisms—such as cuticles, spores and pollen—can be recognized in both coal and kerogen. However, cutan is difficult to find in most extant terrestrial plants, apart from *Agave americana* (which is an unlikely candidate for coalification). Although the cuticles of various Cretaceous conifers and

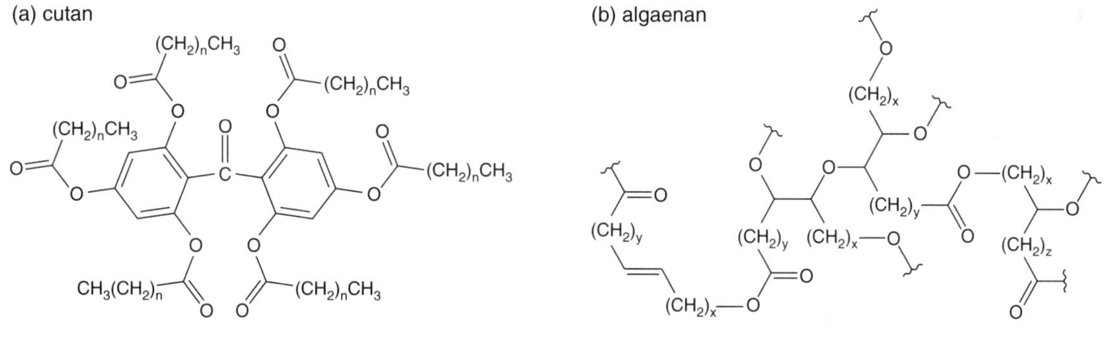

Table 4.7 Some resistant biomacromolecules and their related kerogen macerals (after Tegelaar et al. 1989)

resistant biomacromolecule	maceral
algaenan	alginite
cutan	cutinite
lignin	vitrinite/semifusinite
polyterpenoids*	resinite
sporopollenin	sporinite
suberan	suberinite
tannin	vitrinite

*Mainly from sesqui- and diterpenoids.

Fig. 4.10 Proposed structures for (a) cutan (from *Agave americana*; n is large but unknown, and the central carbonyl link has yet to be confirmed; after McKinney et al. 1996); and (b) algaenan (from freshwater species *Tetraedron minimum*, *Scenedesmus communis* and *Pediastrum boryanum*; x + y = 27, 29, 31 and x + z = 28, 30, 32; after Blokker et al. 1998).

Ginkgo contain cutan, their counterparts in Recent sediments do not. There appears to be no evidence for the amorphous cutan originating elsewhere and being forced into vacant cuticular structures, which has led to the suggestion that it is probably a diagenetic product of cutin and waxes (Mösle et al. 1998). This proposition is supported by the fact that arthropod cuticles contain a chitin–protein complex and no aliphatic macromolecules, but diagenetic transformation leads to the development of the latter, presumably from *in situ* polymerization of cuticular lipids (Stankiewicz et al. 2000). A similar diagenetic origin is possible for sporopollenin (Briggs 1999).

The potential pathways to kerogen formation are combined in the scheme shown in Fig. 4.11, in which kerogen is seen to be the combination of resistant biomacromolecules, geomacromolecules, sulphur-rich macromolecules and incorporated LMW biomolecules. The extent to which humic material contributes to kerogen is uncertain. Partial alteration of biomacromolecules (e.g. the possible cross-linking of soluble components) is represented in Fig. 4.11 by the arrows via humic substances or directly to geomacromolecules. The relatively minor amount of bitumen present at the end of diagenesis derives predominantly from preserved lipid components, shown on the lower right-hand side of Fig. 4.11. The classical and selective preservation models can be considered extremes (i.e. end members) of the scale of alteration that may be undergone by the biomacromolecular precursors of kerogen. Quantitatively, the major pathway of the classical model is the route to geomacromolecules (via LMW biomolecules and humic substances) and for selective preservation it is the route to resistant biomacromolecules. Both these models share the quantitatively less significant pathways to incorporated, LMW biomolecules and preserved lipid components. The importance of the vulcanization pathway to sulphur-rich macromolecules in kerogen varies, depending on diagenetic conditions, and is discussed below.

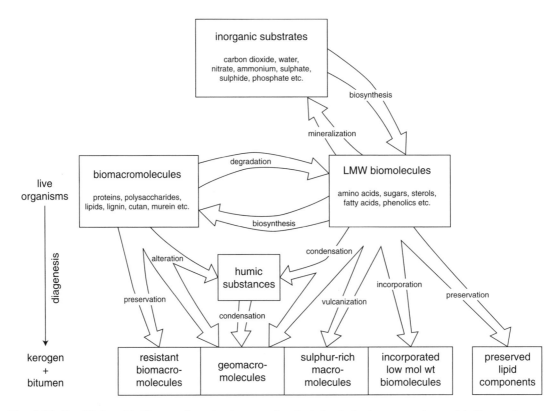

Fig. 4.11 Simplified model of kerogen formation incorporating classical and selective preservation models. Kerogen is represented by the combined four boxes at the bottom, whereas the preserved lipid components box represents a constituent of bitumen.

At the end of diagenesis, sedimentary organic matter is mainly composed of kerogen (Fig. 4.12), which is insoluble in organic solvents. There are also smaller amounts of organic material that is soluble in organic solvents and is termed bitumen (see Box 4.2). This bitumen comprises some fragments of polymeric material (polar N, S, O-containing compounds termed asphaltenes and resins) that may have been thermally cleaved from the main kerogen structure (or may have failed to be incorporated into the original kerogen macromolecule), together with some free (i.e. not chemically bound to kerogen), relatively small molecules (molecular weights $c.<600$), that are mainly hydrocarbons of lipid origin (Fig. 4.12). Kerogen formation involves mild temperature and pressure conditions, and so the composition of the original organic input has an important influence on the chemical nature of the resulting kerogen. As in coal formation, microbial metabolism and the elimination of functional groups during condensation are accompanied by the evolution of volatiles such as methane, carbon dioxide and water.

(b) Biomarkers

The preserved lipid components in Fig. 4.11, which are found within the bitumen in coals as well as kerogens, have also been called **biomarkers** (other pseudonyms being **geochemical fossils** and biological marker compounds) because they can be unambiguously linked with biological precursor compounds, owing to the preservation of their basic skeletons in recognizable form throughout diagenesis and much of catagenesis. Many biomarkers originally possess oxygen-containing functional groups and undergo the same defunctionalization processes as the bulk of the organic matter. Diagenetic products are, therefore, generally hydrocarbons, although lesser amounts of functionalized components (e.g. fatty acids) can survive diagenesis. In addition, unsaturated compounds (i.e. containing C=C bonds) tend either to become reduced (hydrogenated), resulting in the formation of aliphatic hydrocarbons (e.g. steranes and hopanes), or to undergo aromatization (if they are part of a ring system). A proportion of the biomarkers may become a chemically bound part of the kerogen structure (incorporated LMW biomolecules in Fig. 4.11), but otherwise these compounds remain relatively unaltered, encapsulation within resistant macromolecular structures being an important factor in their preservation. The use of biomarkers in molecular palaeontology is discussed in Chapter 5.

(c) Sulphur incorporation

Incorporation of sulphur into macromolecules that are an integral part of kerogen (Fig. 4.11) can be important, depending on the sedimentary environment. Conditions favouring the activity of sulphate-reducing bacteria result in the production of sulphide, which is usually taken up rapidly by iron(II) ions (Hartgers et al. 1997), especially in some types of clastic and argillaceous sediments. Where there is a limited supply of iron(II) ions, free hydrogen sulphide and polysulphides are produced (HS_4^-, HS_5^-, S_4^{2-} and S_5^{2-} are the most common naturally occurring types of polysulphide ions in aqueous environments), which can react with organic matter during diagenesis (Kohnen et al. 1989, 1991b). The formation of sulphur-rich kerogens is, therefore, more likely in non-clastic sediments (e.g. carbonates, siliceous oozes and evaporites) where only very small amounts of iron are available. Inclusion of sulphide (or perhaps even native sulphur) into LMW unsaturated compounds

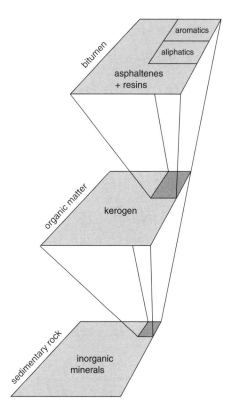

Fig. 4.12 Composition of disseminated organic matter in sedimentary rocks (after Tissot & Welte 1984). The aliphatic and aromatic fractions of bitumen comprise mainly hydrocarbons ($c.C_5–C_{40}$; see Box 4.2).

appears to occur during early diagenesis and these compounds can then go on to become incorporated into the macromolecular structure of kerogen (Sinninghe Damsté et al. 1989).

A summary of the possible mechanisms of sulphur incorporation is given in Fig. 4.13. There is a wide range of unsaturated precursor compounds that are potentially capable of incorporating sulphur, including isoprenoidal alkenes from archaebacteria, bacterial hopanoids, long-chain (C_{37} and C_{38}) unsaturated ketones from coccolithophores (e.g. *Emiliania huxleyi*) and phytadienes from chlorophyll diagenesis (Sinninghe Damsté & de Leeuw 1990). Addition of hydrogen sulphide across a C=C bond yields a thiol. Reaction of this compound with another unsaturated compound (i.e. intermolecular addition) can lead to the formation of sulphur-rich, HMW substances (left-hand side of Fig. 4.13). Where there is more than one double bond present in the precursor molecule, addition and then loss of H_2S can lead to movement of double bonds along the chain (isomerization). In addition, a thiol group formed from one C=C bond can react with an adjacent double bond in the molecule (intramolecular addition) to yield cyclic structures containing sulphur (central column of Fig. 4.13). The size of the ring will depend on the relative positions of the double bonds. Initially, the cyclic centres are aliphatic, but undergo aromatization with increasing maturity. Polysulphide addition leads to similar products, upon loss of surplus sulphur. Incorporation of sulphur into individual biomarkers can also occur.

4.4.2 Kerogen composition

Because kerogen is finely dispersed in sedimentary rock it is often isolated prior to microscopic examinations of the kind used in coal petrography. However, it is quite difficult to isolate kerogen quantitatively without alteration. The inorganic matrix is usually removed by successive treatment with hydrochloric and hydrofluoric acids. While microscopic examination reveals the presence of defined organic remains, such as algae, spores, pollen and vegetative tissue, these are usually only a minor part of kerogen and are fairly well dispersed. A large proportion of the kerogen is often amorphous. The same three maceral groups recognized in coals are found in kerogen: liptinite, vitrinite and inertinite.

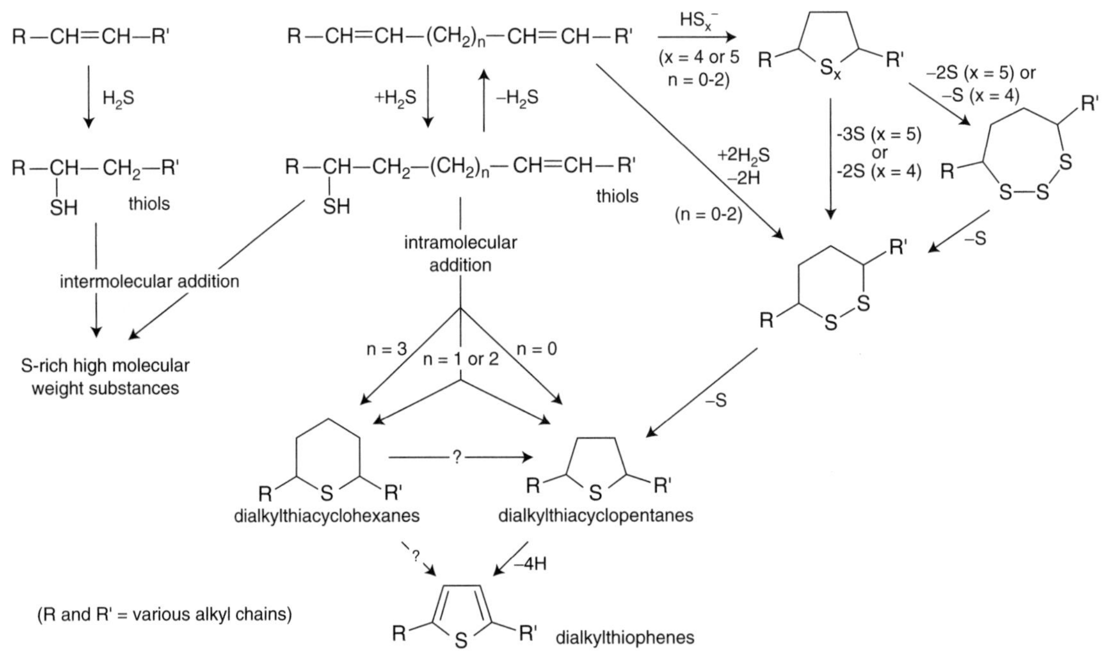

Fig. 4.13 Possible mechanisms for sulphide incorporation into unsaturated substrates during diagenesis (after Sinninghe Damsté et al. 1989; Kohnen et al. 1991c).

Carbon and hydrogen are the main elements of kerogen, but oxygen is also important in the structure of kerogen and comprises 10–25% by weight in shallow immature sediments. For every 1000 C atoms there are c.500–1800 H, c.25–300 O, c.5–30 S and c.10–35 N atoms (Table 4.8). The oxygen-containing functional groups include carboxylic acid, alcohol, carbonyl, ester, ether and amide. The aliphatic content of kerogen is generally higher than that of coal, reflecting the input of highly aliphatic planktonic and microbial lipids to the original sedimentary organic material.

The potential precursors of kerogen are more numerous than for coal, so not surprisingly the structural analysis of kerogen is difficult and various destructive (e.g. oxidative and pyrolytic degradation) and non-destructive (e.g. solid-state infrared spectroscopy and ^{13}C nuclear magnetic resonance) techniques have been employed to identify the main units present. Kerogen is a three-dimensional macromolecule formed from nuclei, in which clusters of aromatic sheets are an important part, that are cross-linked by chain-like bridges, which comprise aliphatic chains as well as various functional groups containing O or S. The aromatic rings can contain nitrogen, sulphur and oxygen. Lipids can be trapped within the kerogen matrix, a property that is shared with coal.

The composition of shallow, immature kerogen depends upon the nature of the original organic matter incorporated into the sediments from which it is formed and also upon the extent of microbial degradation. For example, humic kerogen is formed from organic matter with a large allochthonous higher plant contribution and has a lower aliphatic content than sapropelic kerogen formed largely from phytoplanktonic remains.

4.4.3 Kerogen classification

Traditionally, three general types of kerogen are distinguished, types I, II and III (Tissot et al. 1974; Tissot & Welte 1984). **Type I kerogen** comprises mostly liptinite and so follows an evolution path similar to the sapropelic coals in Fig. 4.5. **Type III kerogen** is the equivalent of humic coal in Fig. 4.5, although the terrestrial plant material is finely dispersed within an inorganic mineral matrix. **Type II kerogen** has intermediate properties between types I and III kerogens, and plots between the sapropelic and humic coals in Fig. 4.5 (see also Fig. 4.14). Occasionally a **type IV kerogen** is recognized, with extremely low H content and low O content. Types II and III kerogen have higher oxygen content than type I, and the distribution of oxygen within the various functional groups differs for each type of kerogen (Fig. 4.14). There are also S-rich subtypes of the three main kerogens types.

(a) Type I kerogen

Type I kerogen is relatively rare, and initially has a high H/C ratio (≥1.5) and a low O/C ratio (<0.1). It contains a significant contribution from lipid material, especially long-chain aliphatics. These lipids are predominantly derived from alginites, and particularly algaenan, although amorphous bacterial matter may contribute to a degree. Compared with the other kerogen types it contains low amounts of aromatic units and

Table 4.8 Bulk compositional changes in the three main kerogen types with increasing maturity (after Béhar & Vandenbroucke 1987). Maturity levels are shown by positions of samples in Fig. 4.15. Type I kerogen is based on the Eocene, upper Green River Formation shale from the Uinta Basin; type II is based on the Toarcian shales of the Paris Basin (France) and Liassic α shales from Germany; and type III on the deltaic sequences from the Upper Cretaceous of the Douala Basin (Cameroon) and from the Tertiary of the Mahakam Delta (Borneo)

	type I			type II			type III		
maturity	low	medium	high	low	medium	high	low	medium	high
C (% wt)	81.5	85.3	91.0	70.0	77.6	85.9	66.2	76.9	86.0
H (% wt)	11.1	11.1	6.3	7.8	8.1	5.2	5.8	6.3	4.8
O (% wt)	6.5	2.8	1.6	18.3	9.2	3.0	24.8	14.2	6.8
N (% wt)	0.3	0.3	0.3	3.3	1.1	1.4	1.2	1.4	0.9
S (% wt)	0.2	0.5	0.7	0.6	3.7	4.4	0.2	0.8	1.4
% initial C lost	0	19	79	0	15	62	0	16	34
% paraffinic C	74	–	36	51	45	28	38	34	19
% naphthenic C	12	–	0	19	14	0	13	6	0
% aromatic C	14	–	64	30	41	72	49	60	81

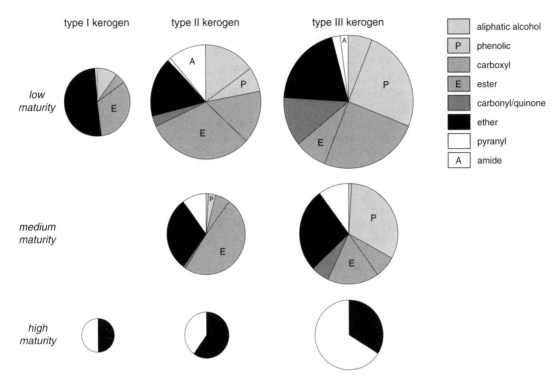

Fig. 4.14 Distribution of oxygen-containing functional groups in the three main types of kerogen with increasing maturity (after Béhar & Vandenbroucke 1987). See Fig. 4.15 for positions of samples on a van Krevelen diagram and Table 4.8 for additional compositional data.

heteroatoms (see Box 2.1). The limited amount of oxygen present is mainly in ester and ether groups (Fig. 4.14).

Type I kerogen is dark, dull and either finely laminated or structureless. It is usually formed in relatively fine-grained, organic-rich, anoxic muds deposited in quiet, oxygen-deficient, shallow-water environments (e.g. lagoons and lakes). For example, tasmanite oil shales were formed from predominantly algal (*Tasmanites*) remains in low-energy marine/brackish water environments (Revil et al. 1994). The freshwater alga *Botryococcus braunii* appears to be a major contributor to some type I kerogens, and its Carboniferous equivalent formed the torbanites of Scottish oil shales. The kerogen in the Eocene Green River oil shale of Colorado, Utah and Wyoming is a further example of type I kerogen, although it is not homogeneous, due to varying environmental conditions, and its organic content varies from <1% up to 40%. Although type I kerogens are mostly associated with lacustrine deposition (Talbot 1988), the type of kerogen deposited in a lake can vary with the evolution of the lake and its immediate environs.

Simulation of natural maturation by rapid pyrolysis of type I kerogen yields more LMW material than other types (up to 80% by weight from shallow, immature samples), indicating that type I kerogen has the highest oil potential. The main pyrolysis products are straight and branched acyclic alkanes (isoprenoids from *Botryococcus* algaenan and probably also from the phytol unit of chlorophylls). Sulphur-rich type I kerogens (type I-S) exist, examples being the Tertiary Ribesalbes and Campins kerogens (from Catalonia, Spain; Sinninghe Damsté et al. 1993a). The Ribesalbes kerogen appears to be derived from *Botryococcus*, which required a freshwater environment, so the incorporation of significant amounts of S during kerogen formation would have necessitated a much greater supply of sulphate than found in the ambient growth environment of the alga (possibly by later infiltration of saline waters) to fuel the necessarily high levels of sulphate reduction.

(b) Type II kerogen

Type II kerogen is more common than type I, and has relatively high H/C and low O/C ratios. Polyaromatic nuclei and ketone and carboxylic acid groups are more important than in type I but less so than in type III, while ester bonds are abundant (Fig. 4.14). Aliphatic structures are important and comprise abundant chains, mostly of moderate length (up to $c.C_{25}$), and ring systems (naphthenes). Sulphur is often found in substantial amounts; in cyclic systems and probably also in sulphide bonds. Associated bitumens contain abundant cyclic structures—aliphatic and aromatic hydrocarbons, and thiophenes—and have a higher sulphur content than other types. A further classification of **type II-S** can be used for sulphur-rich type II kerogens (8–14% organic S by weight) in which the atomic S/C ratio is >0.04 (Orr 1986).

Type II kerogen can potentially be formed in any environment, but in marine settings a major source is the mixture of autochthonous organic matter from phytoplankton (and possibly also zooplankton and bacteria) together with an allochthonous contribution of higher plant material. The pyrolysis products from this blend of organic matter reflect the combined characteristics of types I and III kerogens. Type II kerogen has a lower yield of hydrocarbons upon pyrolysis than type I, but is nevertheless the source of hydrocarbons in many oil and gas fields.

(c) Type III kerogen

Type III kerogen is common, and has a low H/C (<1.0) and a high O/C (up to 0.3) ratio initially. Polyaromatic nuclei and ketone and carboxylic acid groups are important but there are few, if any, ester groups. A significant proportion of oxygen is in non-carbonyl groups (possibly heterocycles, quinones, ethers and methoxy groups; Fig. 4.14). Only minor amounts of aliphatic groups are present, dominated by methyl and other short chains, and often bound to oxygen-containing groups. A few long alkyl chains are present, originating from cuticular coatings (cutan and suberan).

Type III kerogen is essentially formed from vascular plants and contains much identifiable plant debris, so vitrinite macerals predominate. Sulphur-rich brown coals can be formed, which have been described as type III-S kerogen (Sinninghe Damsté et al. 1992).

Type III kerogen is less productive upon pyrolysis and is much less likely to generate oil than types I and II, but it may be a source of gas (primarily methane) if buried deeply enough (i.e. sufficiently high maturity is reached). Pyrolysis products are characterized by phenols and simple aromatic hydrocarbons of lignin origin. Straight-chain aliphatic hydrocarbons (n-alkanes) are also present, dominated by the carbon number range characteristic of higher plant leaf waxes ($c.C_{23}$ to C_{35}).

(d) Type IV kerogen

Type IV kerogen comprises primarily black opaque debris, largely composed of inertinite, with minor amounts of vitrinite. As this has no hydrocarbon-generating potential it is sometimes not considered as a true kerogen. It is probably formed from higher plant matter that has been severely oxidized on land and then transported to its deposition site.

(e) Improved kerogen typing

The main kerogen types and their evolution paths are shown in the van Krevelen diagram in Fig. 4.15. Coals are also shown, which occupy a large area of the plot and can be divided into high- and low-H content groups. The high-H coals overlap with type III kerogen, but also extend towards the higher H/C values of type II kerogen, whereas the low-H coals lie mostly below type III kerogen. European Carboniferous coals tend to plot within the low-H band and Late Cretaceous to Tertiary New Zealand coals in the high-H band (Killops et al. 1996). The classical petroleum source rock is formed from marine organic matter deposited in a reducing environment and broadly corresponds to type II kerogen, with a high petroleum-generation potential. However, most kerogens fall between types II and III in the van Krevelen diagram.

It is apparent that this kerogen classification system is an oversimplification of the variety of sedimentary organic material and its degree of diagenetic alteration leading to kerogen formation. In terms of understanding the types of compounds likely to be evolved as maturation progresses, it is better to consider the types of submacerals or, better still, the macromolecules that contribute to the kerogen matrix. Detailed analysis of pyrolysis products by gas chromatography (see Box 4.3) enables the major petroleum-generating components in a kerogen to be determined—such as long-chain n-alkanes from algaenan and cutan, and phenols and simple aromatics from lignin residues—and permits more accurate estimation of the types of hydrocarbons likely to be generated under natural conditions (Horsfield 1989). A wider range of kerogens than the basic three on van Krevelen diagrams can be identified on the basis

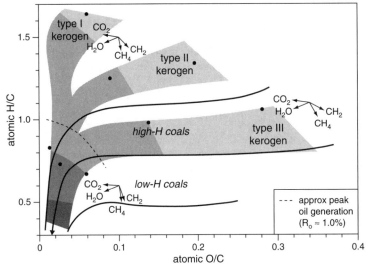

Fig. 4.15 A van Krevelen plot showing the evolutionary trends of the main kerogen types (shaded bands; after Béhar & Vandenbroucke 1987) and most coals (between solid lines; after Killops et al. 1998). With increasing maturity kerogens follow a path towards the origin, as shown for the low/high-H coals boundary. The main hydrocarbon-generation zones are shown, with their approximate rank boundaries (as vitrinite reflectance values, %R_o), together with the effect of loss of various volatiles from different kerogen compositions (CH_2 = oil). Points represent the composition of kerogens in Fig. 4.14.

Box 4.3 | Chromatography

Chromatography is concerned with separating a mixture of components during passage over the surface of an immobile material (a solid or liquid stationary phase) that has varying affinity for the different compound types in the mixture. The mixture is moved over the surface of the stationary phase (the process of **elution**) in a suitable fluid (a gas or liquid), which is termed the mobile phase, and separation of components results from their differing degrees of retention on the stationary phase.

The fractionation of bitumen (see Box 4.2) can be achieved by liquid chromatography, using a simple column containing suitably activated alumina (the stationary phase) and various solvents as mobile phase, moving under gravity. By increasing the polarity of the solvent it is possible to elute sequentially from the column the saturates (with hexane), aromatics (with toluene) and resins (with methanol).

Gas chromatography

In gas chromatography, the stationary phase is in the form of a thin film lining the interior wall of a long, open tubular, capillary column. Coiled columns are typically made of vitreous silica, c.25 m long with an internal diameter of c.0.25 mm. The stationary phase is often a methylsilicone liquid film of c.0.25 μm thickness, which can be immobilized by direct chemical bonding to the column wall. The mobile phase is an inert gas (e.g. helium), passed through the column under pressure. This separation technique is strictly called gas–liquid chromatography because it involves a gaseous mobile phase (also termed the carrier gas) and a liquid stationary phase (McNair & Miller 1997).

Ideally, the individual compounds present in a mixture emerge from the end of the column at varying intervals, with no two compounds eluting at the same time. Because a compound can move along the

Continued

column only when in the gaseous phase, the time it takes to elute (its retention time) depends on its vapour pressure (i.e. boiling point) and its chemical affinity for the stationary phase (i.e. partition coefficient). The lower a component's vapour pressure (i.e. the higher the boiling point) and/or the higher its affinity for the stationary phase, the longer it will take to elute. For a homologous series like the n-alkanes, each member has a similar affinity for the stationary phase, so the elution order is effectively governed by volatility, which decreases with increasing carbon number. The structure of a hydrocarbon can influence its interaction with the stationary phase and so diastereomers can potentially be resolved. Increasing the column temperature uniformly throughout the analysis enables the compounds of lower volatility to elute within a reasonable time, limiting the extent to which diffusion (see Box 3.8) can spread them out on the column, causing peak broadening and low signal:noise ratios in the detection system.

For the more abundant and readily identifiable hydrocarbon components in oil, a flame ionization detector (FID) is suitable (which is sensitive to c.100 pg). Using this detector, components such as n-alkanes are identified from their recognizable elution patterns. Terpenoidal and steroidal hydrocarbons are generally present only in trace quantities, at concentrations of around two orders of magnitude lower than the more abundant components, and require a more specific and sensitive detector, a mass spectrometer. The technique is then referred to as gas chromatography–mass spectrometry (GCMS). In the simplest form of GCMS, compounds emerging from the column are bombarded by high-energy electrons, which expel an electron from each molecule, producing positively charged molecular ions, which tend to fragment into smaller, more stable ions, characteristic of the particular structural units present in the parent molecule. All steranes yield an abundant fragment ion with a mass:charge (m/z) ratio of 217, monitoring of which provides the necessary sensitivity and specificity to examine sterane distributions. Other biomarker families have different characteristic fragment ions. The plot of time versus intensity obtained from whatever detector is used is termed a **chromatogram**. An example of a whole-oil FID chromatogram is shown in Fig. 4.16.

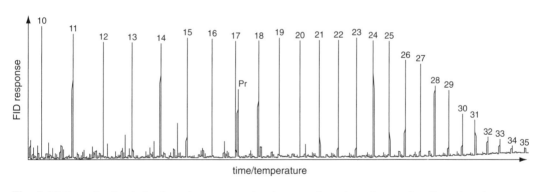

Fig. 4.16 Example of a whole-oil gas chromatogram (numbers = n-alkanes by carbon number, Pr = pristane).

of dominant maceral and major pyrolysis products (Table 4.9).

4.4.4 Structural and compositional changes

(a) Structural changes

As in coal formation, with increasing compaction and burial depth bacterial activity declines and temperature increases, so that the low-temperature, biologically mediated processes characteristic of diagenesis are replaced by mainly thermally mediated changes to the kerogen structure during catagenesis. However, again as in coal formation, there is some overlap between the zones of diagenesis and catagenesis. As temperature and pressure rise with increasing burial the structure of immature kerogen rearranges to reduce the increasing molecular strain by the formation of a more ordered and compact structure, again very much along the lines described for coal. This is brought about by the elimination of units that prevent the close packing of nuclei. Peripheral bulky groups on nuclei (such as non-planar cycloalkyl

Table 4.9 Kerogen classification according to pyrolysis products (after Larter & Senftle 1985)

kerogen type	major pyrolysis products	major macerals
I	paraffins	alginite (*Botryococcus* torbanite)
I	paraffins/naphthenes	alginite (tasmanite)
I	naphthenes	resinite
II	paraffins/naphthenes	amorphinite (amorphous matter, marine origin)
II/III	paraffins	vitrinite/liptinite
II/III	phenols/paraffins	sporinite
III	phenols/paraffins	vitrinite (marine/deltaic shales)
III	aromatics/phenols	vitrinite (coal swamps)
IV	aromatics	fusinite (inertinite)

systems) and bridging units between nuclei (which include aliphatic chains) are ejected. Functional groups are chiefly associated with peripheral units and so they are progressively eliminated from the kerogen structure with increasing maturity, and the nuclei become increasingly cross-linked (Larsen et al. 2002).

It is difficult to represent the complex structure of kerogen, but Fig. 4.17 attempts to show the various units present in type II kerogen at the start and towards the end of catagenesis. The elimination of bulky peripheral substituents and the increasingly compact and aromatic nature of the residual kerogen with increasing maturity are apparent. The increase in aromaticity can be seen in Table 4.8.

Ultimately, it is possible for a graphite stage to be reached if sufficiently extreme conditions associated with high-grade metamorphism are attained. The changes associated with this stage are termed **metagenesis**. However, the initial kerogen structure will remain relatively unchanged, even in ancient sediments, unless the temperatures associated with catagenesis and metagenesis, usually related to deep burial, are attained. Although changes do occur at lower temperatures, they are much slower (see Chapter 7).

(b) Changes in chemical composition

A wide range of compounds is eliminated from the kerogen structure, although the main elements are C, H and O, so the evolution of the different kerogen types with increasing maturity can be represented by the changes in atomic H/C and O/C ratios in a van Krevelen diagram (Durand 1985). Figure 4.15 shows examples of the trends resulting from the expulsion of the main products of diagenesis (CO_2 and H_2O), catagenesis (oil and some gaseous hydrocarbons, which can be represented by CH_2) and metagenesis (CH_4) from the macromolecular structure of kerogen. Although type III kerogen is shown yielding oil during the first part of catagenesis, the amounts may be very low, because the lignin-derived material on its own has only gas potential, which is attributable to the methyl and propyl units. High-H coals appear to have some limited oil potential, but low-H coals are generally only gas prone. The position of a kerogen on a van Krevelen diagram can sometimes be misleading. For example, a high contribution of inertinite will make an otherwise liptinite-rich kerogen plot at a relatively low H/C value. This will suggest a lower potential for oil generation than actually exists, and the kerogen will also appear to be more mature than it really is (given the low H/C and O/C ratios of inertinite and the slope of the isorank lines in Fig. 4.15).

As for humic coals, type III kerogens follow a pathway virtually parallel to the atomic O/C axis during diagenesis, which is consistent with the trend predicted by combined evolution of CO_2 and H_2O (Fig. 4.15). Infrared spectroscopy suggests that there is a major reduction in the carbonyl and carboxyl content (Fig. 4.14; Robin et al. 1977). Type I kerogens follow a diagenetic path like that of sapropelic coals, in which there are small declines in O/C and H/C ratios, while type II kerogens follow an intermediate trend. Some thermally induced bond-breaking occurs in all the main kerogen types towards the end of diagenesis, such as decarboxylation reactions like those noted for brown coals (Section 4.3.2c). Evolution of significant amounts of CO_2 also occurs at high maturities (Seewald et al. 1998).

During catagenesis the compositional changes of types I and II kerogen are chiefly related to the decrease in hydrogen content due to the generation and release of hydrocarbons (acyclic and cyclic), involving the breaking of C–C bonds and most of the remaining C–O bonds. Consequently, H/C ratios decrease significantly (from $c.1.25$–1.5 to 0.5). The aromaticity of

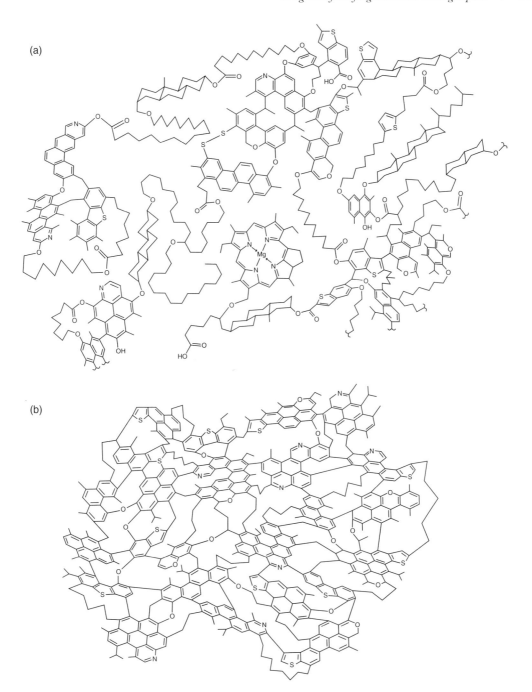

Fig. 4.17 Structural characteristics of type II kerogen: (a) at the end of diagenesis (atomic H/C 1.25, O/C 0.09); and (b) at the end of catagenesis (H/C 0.73, O/C 0.03) (after Béhar & Vandenbroucke 1987).

kerogen increases as a result of both the expulsion of aliphatic hydrocarbons and the increasing aromatization of naphthenic rings (i.e. aliphatic cyclohexyl groups are converted into benzene ring systems by dehydrogenation reactions). The expulsion of alkanes probably accounts for the abrupt increase in apparent rate of aromatization generally observed in kerogens and most coals during catagenesis (Fig. 4.18; Patience et al. 1992). There is a relative increase in the amount of methyl (CH_3) groups as long-chain aliphatic components are preferentially removed. The abundance of residual oxygen-containing groups decreases at a higher rate during catagenesis (Fig. 4.18). The O/C ratio eventually reaches a low level (c.0.05) and remains constant, indicating that the remaining oxygen is difficult to eliminate, and is probably mostly confined to ethers and aromatic units, such as pyrans (Fig. 4.14).

The fall in the H/C ratio during catagenesis is lower for type III kerogens, reflecting their smaller potential for oil generation, although quite large amounts of gaseous hydrocarbons (mainly methane) can be generated. However, oxygen-containing groups continue to be eliminated from type III kerogen, resulting in a further decrease in its O/C ratio. As catagenesis becomes more advanced, the structures of all three kerogen types become increasingly alike, as shown by the simple distribution of oxygen-containing groups in Fig. 4.14 and the van Krevelen diagram in Fig. 4.15.

Metagenesis is experienced by very deep samples or those near high geothermal gradients (see Box 4.4). By this stage hydrogen elimination has slowed (H/C ≤ 0.5) in all types of kerogen and 30–80% of the original carbon content has been lost; type I kerogen losing the most and type III the least (Table 4.8). Methane is practically the only remaining hydrocarbon to be generated and during its evolution virtually all the remaining sulphur in the kerogen structure is lost, mainly as H_2S. In extreme conditions the carbon content exceeds 90% (by weight), the H/C ratio is ≤0.4 and there are effectively no other elements in the kerogen structure. The remaining carbon atoms are confined to aromatic systems, and a major rearrangement of the kerogen occurs in which formerly random layers of aromatic nuclei cluster together and assume an orientation permitting maximum compactness.

4.5 Catagenesis and metagenesis

4.5.1 Petroleum generation

If a body of rock experiences uplift, exposure and erosion, the kerogen it contains is weathered, ultimately to

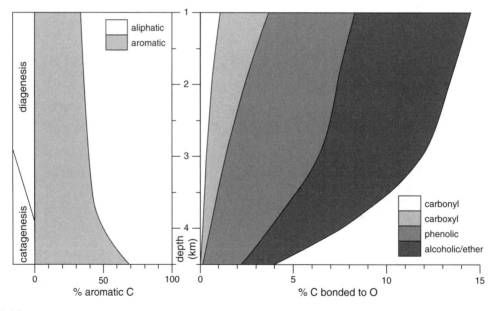

Fig. 4.18 Typical changes in oxygen-containing functional group content with increasing maturity (represented by burial depth) of type II kerogen from the Kimmeridge Clay Formation in the UK sector of the North Sea (after Patience et al. 1992). Data obtained from ^{13}C NMR analysis, so the functional groups are represented by the signal of the C atom(s) directly bonded to each O atom (n.b. two O atoms are associated with each carboxyl C, but two C atoms with each ester O).

Box 4.4 | Subsurface temperature and pressure gradients

Geothermal gradient, heat flow and thermal conductivity

As a general rule temperature increases with increasing burial depth because heat flows from the Earth's interior towards the surface. The temperature profile, or **geothermal gradient**, is related to the thermal conductivity of a body of rock and the heat flow through it by:

$$\text{heat flow} = (\text{geothermal gradient}) \times (\text{thermal conductivity of lithology}) \quad [\text{Eqn 4.1}]$$

Surface heat flow can be considered as the sum of two components: heat flowing up from the mantle and lower crust; and heat generated within the upper crust by radioactive isotopes (^{40}K, ^{232}Th, ^{235}U, ^{238}U). Regional tectonism usually causes a significant increase in heat flows and associated geothermal gradients. Geothermal gradients are not necessarily uniform because of the varying thermal conductivities of different rock types. An average geothermal gradient is $c.30\,°C\,km^{-1}$.

Thermal conductivity is dependent upon the mineralogical composition of the rock, its porosity and the presence of water or gas. Minerals have higher thermal conductivities than water, so non-porous rocks with a low water content are more conductive than porous rocks with a high water content. If the pores within the rock contain gas the conductivity is further reduced. Many evaporitic deposits have high thermal conductivities (halite and anhydrite $c.5.5\,W\,m^{-1}\,°C^{-1}$), whereas the values for shales with a high water content (and therefore high porosity) may be very low ($c.1.5\,W\,m^{-1}\,°C^{-1}$).

Differences in conductivity between adjacent lithologies can result in locally high temperatures and more mature kerogens. For example, where a salt dome is capped by organic-rich shale, the salt transports heat rapidly upwards through its structure until the cap rock is reached. The shale, being of lower thermal conductivity, is unable to dissipate this heat as quickly as the salt and its temperature rises, resulting in the kerogen in the shale reaching a higher level of maturity than if no salt dome were present. The difference in temperature between the top and bottom of a shale of low conductivity is relatively great and so the temperature gradient is large. For rocks with high conductivities the temperature range is reduced and the thermal gradient is low. Generally, conductivity increases with increasing depth as rocks become more compacted, resulting in a corresponding decrease in the geothermal gradient.

Hydrostatic and lithostatic pressure gradients

With increasing depth, pressure increases owing to the burden of the overlying sediments and rocks. Where permeability is sufficient for pore water to move freely, the water will experience a pressure equivalent to the weight of water above it. The resulting pressure gradient is the hydrostatic pressure gradient, which is $c.10\,MPa\,km^{-1}$ (although it rises slightly as salinity increases). When permeability reduces and pore waters can no longer move, pressure can increase above the hydrostatic pressure, resulting in over-pressure. The maximum over-pressure that can be reached is the lithostatic pressure, which corresponds to the weight of the rock burden. Assuming an average rock density of $2.65\,t\,m^{-3}$, the lithostatic gradient is $c.26.5\,MPa\,km^{-1}$.

---

carbon dioxide and water. However, if the rock experiences continued burial, eventually thermal alteration of the kerogen occurs. The hydrocarbon-rich fluids (liquids and gases) evolved from kerogen during catagenesis and metagenesis are collectively termed petroleum (see Box 4.2). Petroleum generation is, therefore, a consequence of the kerogen structure attempting to attain thermodynamic equilibrium as temperature and pressure increase during burial. As might be expected, the amount and composition of the petroleum generated depend on the amount and composition of the organic matter incorporated into the kerogen. In Section 4.4.3a it was seen that type I kerogens are rich in acyclic, medium- to long-chain, aliphatic structures of the type generally abundant in oil. Type II kerogens contain relatively fewer acyclic aliphatic units than type I, but they are still capable of generating large quantities of oil.

The predominant aliphatic units in type III kerogen are small chains (mostly methyl and propyl), primarily derived from lignin residues in vitrinite macerals, and so hydrocarbon gases, particularly methane, are the main petroleum products. There may be some potential for oil production if cutan and/or suberan has been incorporated into the kerogen. Similarly, there is now consensus that some coals can have oil potential (Law & Rice 1993; Scott & Fleet 1994), the highly paraffinic and waxy nature of such oils (examples of which are found in South-east Asia, Australia and New Zealand) again suggesting a cutan/suberan origin (e.g. Killops et al. 1998; Powell et al. 1991).

The influence of the distribution of acyclic aliphatic, cyclic aliphatic and aromatic units in the parent kerogens on the hydrocarbons generated at the peak of oil formation can be seen for examples of types I–III kerogen in Fig. 4.19. However, the distribution and amount of hydrocarbons generated vary throughout catagenesis (see Section 4.5.2).

Although it is widely considered that petroleum is generated from kerogen, this view is not universally held. There are proponents of a deep (mantle) source of hydrocarbons, based on thermodynamic considerations (e.g. Kenney et al. 2002). Under the temperature and pressure conditions that pertain in the upper regions of the crust, the hydrocarbons of petroleum are thermodynamically unstable, and so generation should be driven towards formation of the two most thermodynamically stable end members of the C–H system, methane and carbon (as graphite). Pressures in excess of $c.30$ kbar are theoretically required for n-alkanes to be stable, which corresponds to a depth of at least 100 km (and a temperature of $c.1200\,°C$). Under such conditions the C–H system can spontaneously evolve hydrocarbons (Kenney et al. 2002). Although thermodynamics dictate whether a hydrocarbon is stable, how long it may exist in a shallow crustal environment is determined by the rate at which a suitable reaction mechanism proceeds to bring about rearrangement to stable products (e.g. diamonds are not forever, but their transformation at the crust's surface into graphite is gratifyingly slow). There is extensive evidence for a relatively shallow crustal origin of petroleum, such as compositional similarities of oils and specific kerogenous sources (including carbon isotopic ratios, presence of biomarkers and distributions of stereoisomers). The uniform isotopic composition among, for example, light n-alkanes that would be expected for a deep origin under thermodynamic control is not observed (e.g. Whiticar & Snowdon 1999). It is likely that the bulk of petroleum originates from kerogen, but relatively minor contributions from deep sources cannot be discounted.

A summary of the evolution of carbon-rich fluids with increasing depth/temperature within the upper crust is shown in Fig. 4.20. Temperature appears to have the greatest effect on the breakdown of kerogen (which is considered further in Section 5.7.3a), with pressure playing a subordinate role. There is laboratory and field evidence that elevated pressure slows down fluid evolution from kerogen (Price & Wenger 1992; Fang et al. 1995), which is to be expected from the application of Le Chatelier's Principle (Box 4.5). The temperature scale in Fig. 4.20 has been equated with depth using a typical value of geothermal gradient of $30\,°C\,km^{-1}$. At temperatures $>160\,°C$ the distribution of carbon between CO_2 and CH_4 begins to approach equilibrium under the control of interactions with iron minerals in rocks, and at $>180\,°C$ CO_2 becomes the dominant component (Giggenbach 1997).

4.5.2 Kerogen maturity and hydrocarbon composition

The amount and composition of hydrocarbons generated from a particular kerogen vary progressively with increasing maturity. These changes are summarized for a typical type II kerogen in the right-hand panel of Fig. 4.20. During diagenesis the only hydrocarbons present are those inherited directly from living organisms, together with the methane produced by methanogens. With increasing burial depth during diagenesis the gaseous end products of the anaerobic degradation of organic detritus, CH_4 and CO_2 (see Section 3.3.2b), become less able to escape as water is expelled from the sediment and permeability is reduced with increasing

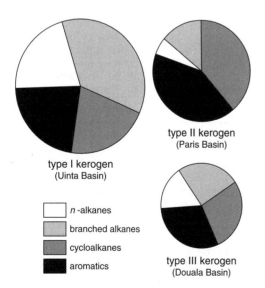

Fig. 4.19 Distribution of hydrocarbons generated from different kerogen types at the peak of oil formation. Areas are proportional to amount of hydrocarbons per unit mass of organic carbon (after Tissot & Welte 1984).

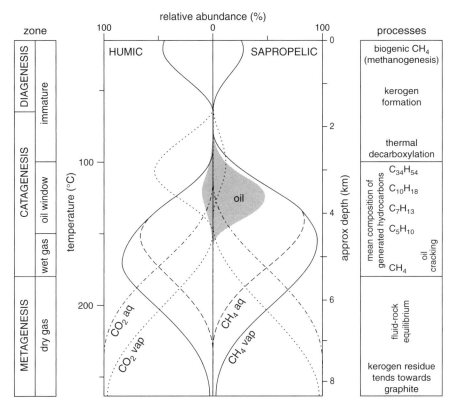

Fig. 4.20 Changes in the composition of C-containing fluids evolved from humic and sapropelic sources with increasing temperature. For temperatures >150 °C the methane and carbon dioxide distributions are shown for separate vapour and aqueous phases in equilibrium with crustal rocks (after Tissot et al. 1974; Giggenbach 1997). The diagenesis–catagenesis boundary is variable, and can be placed between c.50 °C and the onset of oil generation at c.100 °C. An average surface temperature of 15 °C and a geothermal gradient of 30 °C km^{-1} have been assumed.

Box 4.5 | Le Chatelier's Principle

This principle states that if a system at equilibrium is subjected to a change in conditions, it will respond in a way that counteracts the change. It can be demonstrated by changing the pressure applied to a fixed volume of the following system of gases at equilibrium:

$$2NO_2 \rightleftharpoons N_2O_4 \quad \text{[Eqn 4.2]}$$
$$\text{brown} \quad\quad \text{colourless}$$

Increasing the pressure will drive the equilibrium to the right, in favour of N_2O_4, because for every two molecules of NO_2 that combine, only one of N_2O_4 is formed, which reduces the total number of molecules of gas present and hence the pressure (see Box 1.11). As a result, the mixture becomes paler, as the amount of brown NO_2 decreases. Decreasing the pressure drives the equilibrium reaction in the opposite direction, generating more NO_2.

There is generally at least a small increase in volume associated with the generation of petroleum, and particularly methane, from kerogen. So pressures significantly higher than normally associated with the depth at which petroleum generation occurs in the subsurface will have the effect of suppressing the formation of hydrocarbon fluids.

overburden. Methanogenesis is the last truly biological process to operate at depth (at temperatures up to $c.75\,°C$).

Temperature steadily increases with burial depth until, during catagenesis, thermal energy is sufficient to cause certain hydrocarbon fragments to break off and be expelled from the kerogen (and possibly also from asphaltenes, given the similar distributions of compounds that can be produced during pyrolysis; e.g. Cassani & Eglinton 1986). As can be seen from Fig. 4.20, the size of hydrocarbons evolved decreases with increasing maturity, and on this basis hydrocarbon generation during catagenesis can be divided into two zones: the main zone of oil formation and the wet gas zone (see also Fig. 4.15).

During the main zone of oil generation (or '**oil window**') there is significant production of liquid hydrocarbons of low to medium molecular weight. The bitumen inherited during diagenesis and previously trapped within the kerogen matrix (see Section 4.4.1a) is progressively diluted by the thermally generated hydrocarbons, the bulk of which exhibit random structural and C-number distribution patterns. The first liquid hydrocarbons to be evolved have relatively high molecular weights (mean composition $c.C_{34}H_{54}$), but as temperature increases hydrocarbons of successively lower molecular weight are produced, as indicated by the mean compositions in Fig. 4.20. The major phase of oil generation is usually limited to a window of $c.100–150\,°C$ ($c.2.5–4.5\,km$ depth in Fig. 4.20; Mackenzie & Quigley 1988). Lower temperatures have been proposed for the onset of oil generation ($c.60\,°C$), but such temperatures do not appear to be typical and may, in part, result from underestimation of maximum burial depths. Most extensional basins (see Box 3.11) experience compression at some stage of their evolution, sometimes resulting in uplifted regions that can suffer erosion. Under such circumstances it is possible to underestimate maximum burial depths and related temperatures (e.g. Paris Basin; Mackenzie & McKenzie 1983).

Gas is produced at all depths down to the limit of hydrocarbon generation at $c.230\,°C$. Near the surface it comprises methane from methanogenesis, while at greater depths methane and other hydrocarbon gases are produced from the thermal alteration of kerogen. Gaseous hydrocarbons are considered to comprise C_1 to C_4 compounds, and liquid hydrocarbons $\geq C_6$. The C_5 members can be either liquids or gases at the surface (the boiling point of *iso*pentane is $28\,°C$ and of *n*-pentane is $36\,°C$). Oils contain varying amounts of dissolved gases and, similarly, gases can contain varying amounts of dissolved hydrocarbons that would normally be liquids. The term **condensate** (or gas condensate) is applied to the liquid hydrocarbons that condense from the gaseous phase at the surface during commercial recovery. The proportion of gaseous hydrocarbons generated increases in the later part of catagenesis, the **wet gas** zone, because the remaining alkyl chains in the kerogen structure are relatively short and yield light hydrocarbons that are mostly gases under surface conditions.

Towards the end of catagenesis the proportion of methane in the gaseous products rises rapidly with increasing temperature and kerogen evolution. During metagenesis, methane is the only hydrocarbon released, and is often referred to as the **dry gas** zone (because no condensate is produced). The whole zone of gas generation (wet and dry) is sometimes called the cracking zone, due to the presumed thermal cracking of previously evolved hydrocarbons into smaller, gaseous products.

A simple model of petroleum formation has been proposed that treats kerogen as being formed from labile, refractory and inert components (Mackenzie & Quigley 1988). **Labile kerogen** (i.e. the polymethylenic part of liptinite) yields mainly oil, **refractory kerogen** (the main lignin-derived component of vitrinite) yields gas and **inert kerogen** (inertinite) produces no hydrocarbons. Labile and refractory kerogen, therefore, generate petroleum and may be termed **reactive kerogens**, while only H, O, S and N are eliminated from inert kerogen (e.g. as H_2O, H_2S, SO_2, N_2), yielding a residue of carbon. Based on this model, it appears that:
- most oil is formed in the range $100–150\,°C$;
- most gas is formed in the range $150–230\,°C$;
- any oil left in the source rock undergoes cracking to gas at elevated temperatures.

These guidelines are summarized in Fig. 4.21. It was thought that cracking of oil to gas occurred over the range $150–180\,°C$ (e.g. Dieckmann et al. 1998), suggesting that oil does not survive for geological time periods at $>160\,°C$ (e.g. Mackenzie & Quigley 1988). However, in recent years several examples have been found of light oils dominated by *n*-alkanes existing in reservoirs at temperatures up to $c.200\,°C$ (McNeil & BeMent 1996; Vandenbroucke et al. 1999), and it has been argued from a consideration of free-radical mechanisms that oil could be stable up to $240–260\,°C$ (Dominé et al. 2002). The stability of *n*-alkanes towards cracking is significantly influenced by the composition of the oil (Burnham et al. 1997; McKinney et al. 1998). Although *n*-alkanes in oil that has migrated away from the source rock may be more stable than previously

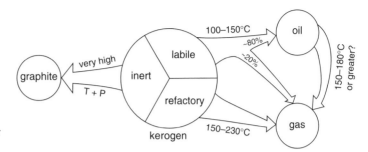

Fig. 4.21 Summary of temperature ranges for petroleum generation from kerogen (after Mackenzie & Quigley 1988).

Table 4.10 Elemental composition range typical of crude oils (after Levorsen 1967)

element	abundance (wt %)
C	82.2–87.1
H	11.8–14.7
S	0.1–5.5
O	0.1–4.5
N	0.1–1.5
others	≤0.1

supposed towards thermal cracking, it has yet to be established that liquid hydrocarbons retained within the kerogen do not undergo thermally induced cracking of C–C bonds in the wet gas zone (over the temperature range of c.150–180 °C in Fig. 4.21) to yield light, dominantly gaseous hydrocarbons.

4.5.3 Petroleum composition

Crude oils, like their associated bitumens in source rocks, contain hydrocarbons (aliphatic and aromatic), resins and asphaltenes (Box 4.2). Table 4.10 presents the typical range of elemental composition for crude oils. In terms of atomic ratios, for every 1000 C atoms there are around 1600–2200 H atoms and up to c.25 S atoms, 40 O atoms and 15 N atoms. Much of the nitrogen, oxygen and sulphur is associated with the resins and asphaltenes, and hence they are often collectively termed polar NSO compounds. However, a significant amount of sulphur can be present in compounds of medium molecular weight that are isolated in the hydrocarbons fraction when using simple chromatographic fractionation procedures. Important members of this group of compounds are aromatic thiophenes, deriving from sulphide incorporation into unsaturated compounds during diagenesis (see Fig. 4.13). Other elements in oil include various metals, particularly nickel and vanadium, which are generally present in trace amounts and are mostly associated with the polar NSO compounds. The average oil contains c.57% aliphatic hydrocarbons, 29% aromatic hydrocarbons and 14% resins and asphaltenes; while sulphur, incorporated into thiophenic compounds, accounts for c.2% (by weight) of the aromatic hydrocarbon fraction. There is, of course, a wide range of oil compositions, and hence properties and commercial uses (Box 4.6).

Although the composition of natural gas accumulations can vary significantly, methane is generally by far the most abundant component (e.g. see Tissot & Welte 1984). Lesser amounts of other hydrocarbon gases are usually present, chiefly ethane (CH_3CH_3), propane ($CH_3CH_2CH_3$), n-butane ($CH_3CH_2CH_2CH_3$), isobutane ($CH_3CH(CH_3)CH_3$), n-pentane ($CH_3CH_2CH_2CH_2CH_3$) and isopentane ($CH_3CH(CH_3)CH_2CH_3$). Natural gas can also contain carbon dioxide, hydrogen sulphide, nitrogen and traces of helium.

(a) Major hydrocarbons in oils

The aliphatic hydrocarbons (also known as **saturates** or **paraffins**) are divided into acyclic alkanes (normal and branched) and cycloalkanes (or **naphthenes**). Normal (i.e. straight-chain) alkanes (Fig. 4.23a) usually predominate in oils, with abundance often peaking around C_6–C_8. Usually found among the branched alkanes are *iso*alkanes (2-methylalkanes) and possibly lesser amounts of *anteiso*alkanes (3-methylalkanes); both groups exhibit similar carbon number ranges to the n-alkanes (Fig. 4.23a). Waxy oils have been found to contain alkanes up to about C_{120} (Carlson et al. 1993; Philp 1994), with the C_{40+} range dominated by n-alkanes in oils from coaly sources and by branched or cyclic alkanes in typical marine- and lacustrine-sourced oils (Killops et al. 2000a). Such long-chain components are

Box 4.6 | Oil composition and uses

Crude oils can be classified according to the relative amounts of acyclic alkanes, cycloalkanes and combined aromatic hydrocarbons plus NSO compounds present. This classification is represented by the ternary (triangular) plot in Fig. 4.22, and can be seen to distinguish between the main fields of marine and terrestrially sourced oils. The main classes of normal crudes resulting from this classification are:
1 Paraffinic oils, containing mainly acyclic alkanes and with <1% S.
2 Paraffinic–naphthenic oils, containing mainly acyclic alkanes and cycloalkanes, and with <1% S.
3 Aromatic–intermediate oils, containing >50% aromatic hydrocarbons and usually >1% S.

The density of an oil is one of its most frequently specified properties, and generally falls within the range 0.7–0.9 g cm^{-3}. Another commonly used measurement is specific gravity (more commonly termed relative density), which compares the mass of a substance with that of an equal volume of pure water at 4 °C and 1 atm pressure. Most oils are lighter than water and so have specific gravities <1. An alternative measurement of oil density is often used, which is expressed as degrees of API (American Petroleum Institute) gravity:

$$°\text{API} = \left(\frac{141.5}{\text{s.g. } 60/60\text{F}}\right) - 131.5 \quad [\text{Eqn 4.3}]$$

where s.g. 60/60F = specific gravity of oil at 60 °F relative to water at the same temperature. By definition, the API scale is inversely proportional to density: API gravities for conventional (or normal) oils are ≥20°, for light oils (and condensates) are >40° and for heavy oils are <20°.

The gross composition of any fossil fuel is important in terms of its uses and the resulting environmental impact. The most efficient fuels are capable of liberating the most energy from a given mass of fuel, i.e. they have the highest calorific values. Sulphur and nitrogen contents are also important considerations because the SO_2 and nitrogen oxides (NO_x) produced during combustion can lead to acid rain. In other applications of oil, e.g. as lubricants, pour point and specific gravity are important. Some of the basic uses of oil can be seen from the names given to the various fractions obtained upon distillation: gasoline (C_4–C_{10}), kerosine (C_{11}–C_{13}), diesel fuel (C_{14}–C_{18}), heavy gas oil (C_{19}–C_{25}) and lubricating oil (C_{26}–C_{40}). Although coal, oil and gas are all primary energy sources, oil is more valuable in terms of the uses to which many of its constituents can be put by the petrochemicals industry. There is increasing interest in coal as a source of similar compounds.

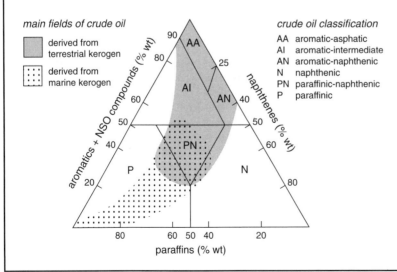

Fig. 4.22 Classification of crude oils based on distribution of paraffins, naphthenes and aromatics plus NSO compounds (after Tissot & Welte 1984).

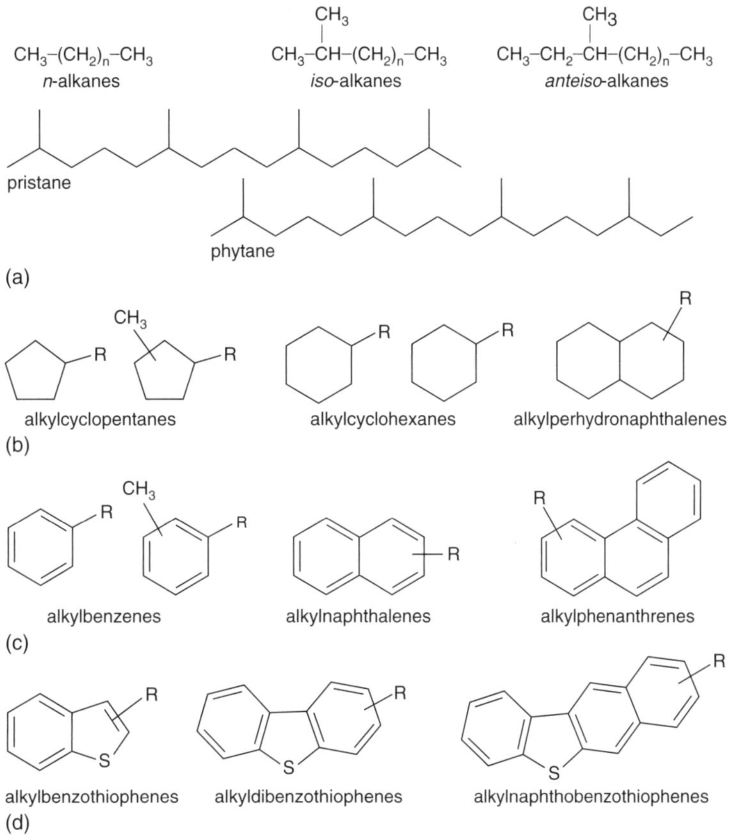

Fig. 4.23 Major hydrocarbons and simple heteroatomic compounds in crude oils: (a) acyclic alkanes (paraffins); (b) cycloalkanes (naphthenes); (c) aromatic hydrocarbons; (d) sulphur-containing aromatics. (R = alkyl group.)

consistent with an origin from cutan, suberan or algaenan.

Acyclic isoprenoidal alkanes may be important constituents, particularly pristane (2,6,10,14-tetramethylpentadecane) and phytane (2,6,10,14-tetramethylhexadecane) (Fig. 4.23a). Unfortunately, the term *iso*alkane is sometimes used to describe all the branched alkanes and not just the 2-methylalkanes. Acyclic alkanes <C_5 are gases under normal surface conditions, while those up to C_{15} are liquids. The *n*-alkanes with >15 C atoms tend to be viscous liquids grading into solid waxes (the transition depending upon the ambient temperature).

Major cycloalkanes (Fig. 4.23b) include cyclohexane and cyclopentane series with alkyl chains of similar carbon number range to the *n*-alkanes. Further series possessing additional ring-methyl groups are sometimes found. All these cycloalkanes are liquids, grading into solids with increasing alkyl chain length. Cycloalkanes with more than one ring system, such as the alkylperhydronaphthalenes in Fig. 4.23b, are generally present but their abundance tends to decrease as the number of rings increases.

LMW alkylbenzenes are generally the most abundant aromatic hydrocarbons and, again, alkyl chains often have similar carbon number ranges to the *n*-alkanes. As for the cycloalkanes, series of alkylbenzenes are sometimes observed with additional ring-methyl groups. Alkylnaphthalenes and alkylphenanthrenes (Fig. 4.23c) are also usually present but larger polycyclic aromatic hydrocarbons, such as chrysene (see Fig. 7.3 for structure), are less abundant. Alkyl substituents on these polycyclic aromatic hydrocarbons take the form of methyl and ethyl groups, in contrast to the long alkyl chains present in alkylbenzenes. Aromatic compounds with fused cycloalkyl rings (naphthenoaromatics) may also be present. The short-chain alkylbenzenes are liquids, but the longer-chain members and also polycyclic aromatic hydrocarbons are generally solids under normal surface conditions.

Nitrogen, oxygen and sulphur compounds can be found in varying amounts in the aromatic hydrocarbon fractions of oils isolated using simple chromatographic procedures, but they are usually less abundant than the major true hydrocarbons (benzene, naphthalene, phenanthrene and their alkylated derivatives). Sulphur-containing compounds, such as benzothiophene, dibenzothiophene, naphthobenzothiophene and their alkyl derivatives (Fig. 4.23d), are usually present in oils. Sulphur-rich kerogens can give rise to significant quantities of thiophenic compounds (see Fig. 4.13). Compounds containing oxygen and nitrogen are usually less significant components. Oxygen forms analogous compounds to sulphur, in which the furan unit (Table 2.2) replaces the thiophene unit (e.g. dibenzofuran). Nitrogen-containing compounds are usually the least abundant and, when present, include pyrrole and pyridine derivatives (Table 2.2) with additional benzene rings analogous to the sulphur aromatics.

The average hydrocarbon composition of crude oils is: acyclic (normal and branched) alkanes 33%; cycloalkanes 32%; aromatic hydrocarbons 35%; although, as noted earlier, the distribution of hydrocarbons expelled during catagenesis depends on the chemical composition of the source rock kerogen and its thermal maturity. The range of hydrocarbons generated during catagenesis is truly immense: the 17 C_7 structures identified in oils are shown in Fig. 4.24, and the number of possible compounds increases dramatically with C-number. In addition to the identified hydrocarbons (Fig. 4.23), oil also contains a complex mixture of hydrocarbons that is not resolved during gas chromatographic analysis and appears as a hump in the chromatogram (see Fig. 7.8). This unresolved complex mixture (UCM) becomes more apparent when oils are biodegraded and the n-alkanes are depleted (see Sections 4.5.6c and 7.3.2b), but the UCM is a component of the original oil (i.e. it is generated from kerogen during catagenesis), and appears to comprise mostly aliphatic and monoaromatic compounds (Killops & Al-Juboori 1990). It can be more abundant than the total n-alkanes and so accounts for a significant proportion of the hydrocarbons fraction of non-biodegraded oils.

The relatively minor hydrocarbon liquids that are generated by most humic coals are characteristic of higher plant material. Coal-sourced oils are generally dominated by cuticular-wax derived n-alkanes, although benzene, naphthalene and phenanthrene and their alkyl derivatives are usually important. Small amounts of larger aromatic hydrocarbons formed from

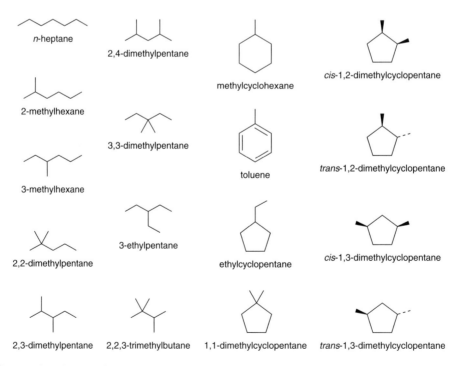

Fig. 4.24 C_7 Hydrocarbons in oil.

extensively fused benzenoid systems may also be present (e.g. White & Lee 1980). Such polycyclic aromatic hydrocarbons are also produced during combustion of oil and coal (see Section 7.3.1a).

(b) Biomarkers

Also present among the hydrocarbons of oils are relatively small amounts (usually <1% by weight) of biomarkers, generally of lipid origin (see Section 4.4.1b). By the end of diagenesis their functionalized precursor compounds have been transformed into hydrocarbons (e.g. hopanes, steranes and aromatic steroids), as shown in Fig. 4.25. While the bulk of hydrocarbons generated during catagenesis cannot readily be linked to specific precursors, biomarkers can. Pristane and phytane (Fig. 4.23a) are, therefore, included among the biomarkers and are often the most abundant of these compounds. At the end of diagenesis a significant proportion of the biomarkers is present as discrete molecules trapped within the kerogen matrix. However, some biomarkers may also be incorporated into the kerogen structure and are released during catagenesis. Trapped compounds may provide a diagenetic source of hydrocarbons, but during catagenesis they are progressively diluted by new hydrocarbons. The transformations undergone by biomarkers and the information that can be obtained from them are considered in Chapter 5.

4.5.4 Reactions involved in hydrocarbon generation

Free-radical reactions probably account for most of the thermally induced fragmentation of kerogen (Ungerer 1990). The formation of *n*-alkanes with a random carbon number distribution can be accounted for by such a mechanism, although *iso*alkanes are more likely to be produced by ionic reactions involving acid-catalysis, possibly involving clay-mineral rearrangement of carbocations (Box 4.7). An essential feature of hydrocarbon generation is the transfer of hydrogen from the residual kerogen, which becomes richer in condensed aromatic structures, to the expelled hydrocarbon fragments, thereby forming alkanes. Laboratory pyrolysis under dry conditions results in a deficiency of hydrogen and the production of alkenes as well as alkanes (Ishiwatari & Fukushima 1979), whereas hydrous pyrolysis yields predominantly alkanes, the water providing a supply of hydrogen (Leif & Simoneit 2000).

Examples of the types of reaction that can lead to the redistribution of hydrogen between the parent kerogen and thermally cleaved fragments (disproportionation reactions) under geological maturation conditions are shown in Fig. 4.27. Cyclization and aromatization of residual kerogen structures liberate hydrogen for the reduction of the alkyl chains cleaved from kerogen at C–O and C–C bonds. Ring-opening reactions can also

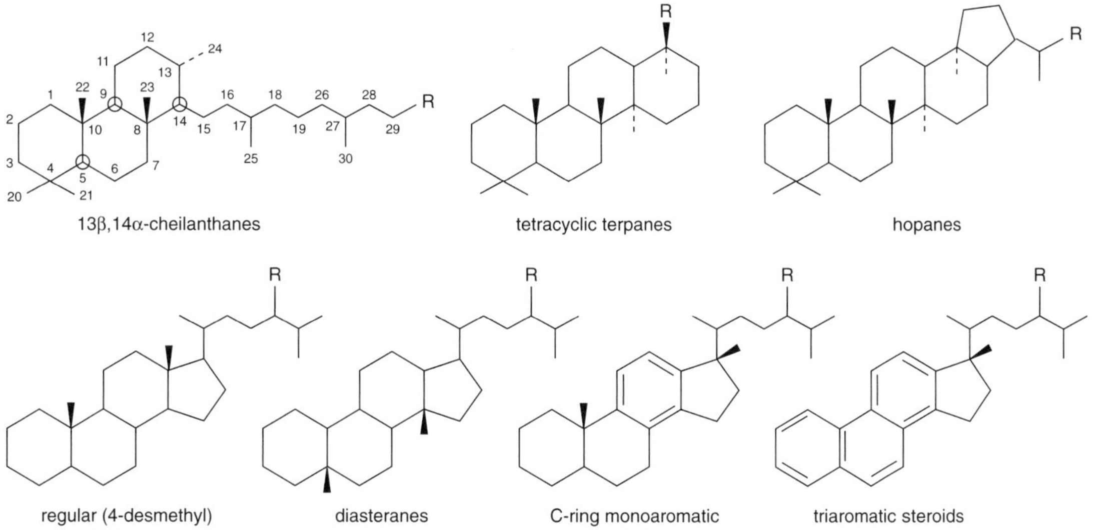

Fig. 4.25 Some important biomarker hydrocarbons in crude oils (n.b. numbering system for cheilanthanes is based on the C_{25} parent sesterterpane and so is different from the hopanoidal system).

Box 4.7 | Hydrocarbon-generating reactions during catagenesis

Two types of reactions have been postulated for the formation of the major groups of paraffins during catagenesis, based on their random distribution of chain lengths and methyl-substitution positions. The distribution pattern for *n*-alkanes suggests a free-radical mechanism (see Box 2.10). Free-radical reactions are chain reactions, which comprise chain initiation, propagation and termination processes. Chain initiation during hydrocarbon generation produces an unpaired electron, which is extremely reactive. Thereafter, a range of chain-propagation reactions is possible until the free radical is quenched (chain termination) by interaction with another radical, resulting in pairing of the electrons (Fig. 4.26a).

Oils contain series of alkanes with a randomly positioned single methyl branch. The origin of these compounds is best explained by rearrangement of carbocations formed from alkenes. The hydrogen transfer that results from cleavage of an alkyl chain produces an alkane and an α-alkene (i.e. the double bond is at the end of the alkyl chain). The α-alkenes are known to produce a range of methyl-substituted alkanes by acid catalysis, upon addition of a proton to yield a carbocation. The rearrangements involved are shown in Fig. 4.26b.

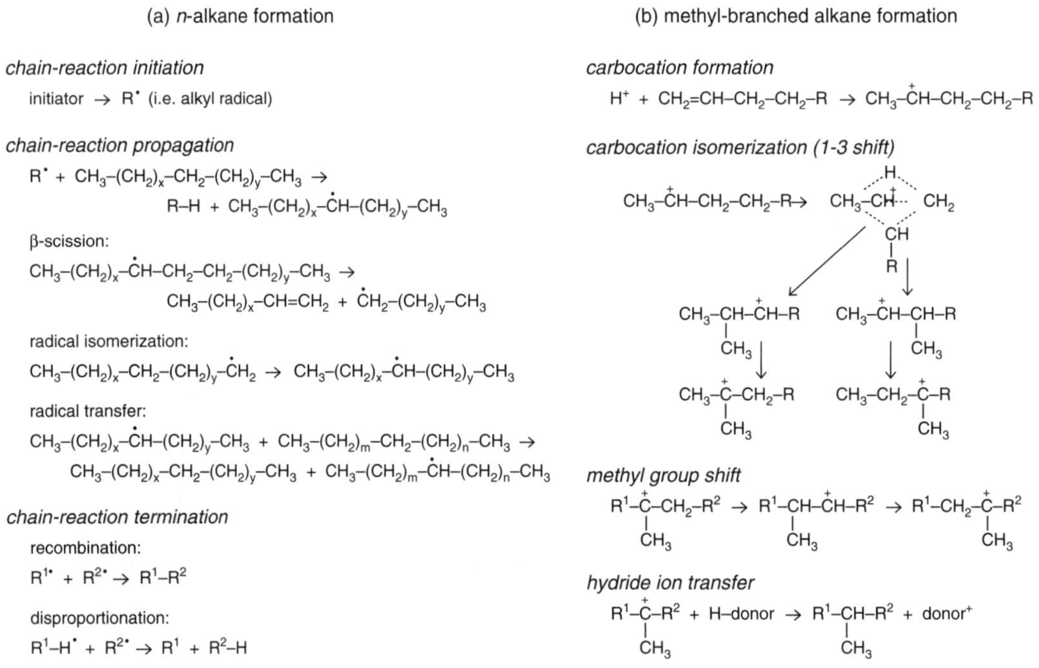

(a) *n*-alkane formation

chain-reaction initiation

initiator → R• (i.e. alkyl radical)

chain-reaction propagation

R• + CH$_3$–(CH$_2$)$_x$–CH$_2$–(CH$_2$)$_y$–CH$_3$ →
  R–H + CH$_3$–(CH$_2$)$_x$–ĊH–(CH$_2$)$_y$–CH$_3$

β-scission:

CH$_3$–(CH$_2$)$_x$–ĊH–CH$_2$–CH$_2$–(CH$_2$)$_y$–CH$_3$ →
  CH$_3$–(CH$_2$)$_x$–CH=CH$_2$ + ĊH$_2$–(CH$_2$)$_y$–CH$_3$

radical isomerization:

CH$_3$–(CH$_2$)$_x$–CH$_2$–(CH$_2$)$_y$–ĊH$_2$ → CH$_3$–(CH$_2$)$_x$–ĊH–(CH$_2$)$_y$–CH$_3$

radical transfer:

CH$_3$–(CH$_2$)$_x$–ĊH–(CH$_2$)$_y$–CH$_3$ + CH$_3$–(CH$_2$)$_m$–CH$_2$–(CH$_2$)$_n$–CH$_3$ →
  CH$_3$–(CH$_2$)$_x$–CH$_2$–(CH$_2$)$_y$–CH$_3$ + CH$_3$–(CH$_2$)$_m$–ĊH–(CH$_2$)$_n$–CH$_3$

chain-reaction termination

recombination:

R$^{1\bullet}$ + R$^{2\bullet}$ → R^1–R^2

disproportionation:

R^1–H• + R$^{2\bullet}$ → R^1 + R^2–H

(b) methyl-branched alkane formation

carbocation formation

H$^+$ + CH$_2$=CH–CH$_2$–CH$_2$–R → CH$_3$–CH$^+$–CH$_2$–CH$_2$–R

carbocation isomerization (1-3 shift)

CH$_3$–CH$^+$–CH$_2$–CH$_2$–R → CH$_3$–CH⋯CH$_2$
                                     ⋮
                                     CH
                                     |
                                     R

↓   ↓

CH$_3$–CH–CH$^+$–R   CH$_3$–CH$^+$–CH–R
       | |
       CH$_3$ CH$_3$

↓   ↓

CH$_3$–C$^+$–CH$_2$–R   CH$_3$–CH$_2$–C$^+$–R
       | |
       CH$_3$ CH$_3$

methyl group shift

R^1–C$^+$–CH$_2$–R^2 → R^1–CH–CH$^+$–R^2 → R^1–CH$_2$–C$^+$–R^2
       | | |
       CH$_3$ CH$_3$ CH$_3$

hydride ion transfer

R^1–C$^+$–R^2 + H–donor → R^1–CH–R^2 + donor$^+$
       | |
       CH$_3$ CH$_3$

Fig. 4.26 Possible reaction schemes involved in alkane formation from thermal degradation of kerogen: (a) radical reactions in *n*-alkane generation; (b) carbocation reactions in methyl-branched alkane generation (after Kissin 1987).

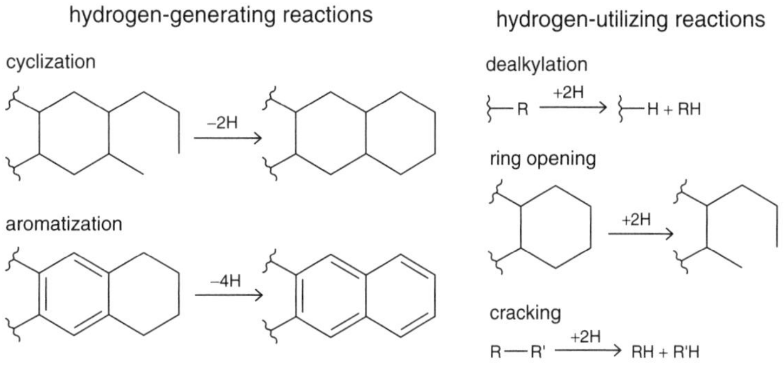

Fig. 4.27 Examples of hydrogen transfer reactions during catagenesis.

occur and require the addition of hydrogen. The evolution of CO_2 at high maturity may be a result of water providing hydrogen for hydrocarbon formation and oxygen for C elimination (Seewald et al. 1998).

Because the composition of kerogen varies according to the differing original inputs of organic matter, there is variation in the types of bonds that have to be broken during hydrocarbon evolution from types I, II and III kerogens. The predominant C–C bonds of type I kerogen are the strongest and so peak oil generation from this type of kerogen occurs at a higher relative maturity level than for types II and III, which contain larger amounts of the weaker C–O, C–N and C–S bonds (Tissot & Espitalié 1975; Waples 1984; see Section 5.7.3b and Fig. 5.51). In particular, sulphur-rich kerogens (type II-S) contain a large proportion of alkyl chains attached to the kerogen nuclei by C–S and S–S bonds. These bonds are significantly weaker than the C–C and C–O bonds that are abundant in other kerogen types. As a result, oil can be generated from type II-S (and other S-rich) kerogens at considerably lower temperatures (c. 80 °C) than the normal type II (Orr 1986). In addition to abundant sulphur-containing aromatics, the sulphur-rich oils produced contain large quantities of asphaltenes and resins, because breaking of the weaker S-containing bonds also results in relatively larger fragments than usual. The composition of the asphaltenes parallels that of the remaining kerogen with increasing maturity.

The cracking of aliphatic chains in oil retained within kerogen towards the end of catagenesis requires hydrogen (as can be seen from Fig. 4.27), which may again be supplied by increasing cyclization and aromatization of the residual kerogen structure. Water may also make a contribution, reducing the degree of aromatization of kerogen (Seewald 1994; Michels et al. 1995). The production of small aliphatic chains depends on both the breaking of particular C–C bonds and the statistical likelihood of a particular fragment being formed. For example, C–C bonds towards the centre of a relatively long chain are generally slightly weaker than those towards the end. However, as the chain gets shorter, proportionally more energy is required to break the central C–C bond (although this energy difference does not change greatly for components $>C_8$). Moreover, the more ways there are of forming a fragment of a particular size, the more likely it is to be produced. Consequently, with increasing temperature, progressively smaller molecules are produced by cracking. The rate of cracking of n-alkanes is affected by the overall composition of an oil, as a result of influences on the radical initiation and propagation reactions similar to those involved in the primary cracking of kerogen (Fig. 4.26; Burnham et al. 1997).

(a) Isotopic fractionation

Because the organic matter from which kerogen is formed is relatively enriched in ^{12}C and depleted in ^{13}C (Box 1.3), so too is the kerogen in sedimentary rocks. As might be expected, the hydrocarbons evolved from kerogen reflect this source-related isotopic signature. Methane generally exhibits an even lighter signature than the kerogen from which it is generated. This is because methane is generated by cleavage of C–C bonds in kerogen or during hydrocarbon cracking, and it requires less energy to break a ^{12}C–^{12}C bond than a ^{13}C–^{12}C bond. Hence, isotopically light methane is preferentially formed. Other hydrocarbons generated from kerogen are similarly relatively light, and so with increasing maturity the residual kerogen becomes isotopically heavier (Galimov 1980; see Section 5.8.6).

4.5.5 Movement of hydrocarbons from kerogen and coal

Petroleum is generally found in reservoirs at some distance from its source rock. The whole of the journey from source to reservoir is termed migration, but it is usually divided into primary and secondary stages. The expulsion of petroleum from the source rock into suitable adjacent carrier strata is often referred to as **primary migration**, and the subsequent transport through the carrier rock to the reservoir as **secondary migration**. Important rock properties are **porosity** (a measure of the total volume of pores able to accommodate generated hydrocarbons) within potential reservoirs and **permeability** (the degree to which pores are interconnected, allowing flow) within the carrier rocks.

(a) Mechanisms of expulsion

As burial increases so does compaction of the generally fine-grained source deposits and much of the water originally associated with the sediments during deposition is expelled. By the time of the main phase of petroleum generation the source rock is relatively dense, with low porosity and permeability. The remaining pores may even be smaller than some of the petroleum molecules. Some pore water is still present, largely bound to mineral surfaces by hydrogen bonding, which further restricts the size of pore throats and severely hinders movement of hydrocarbon fluids out of the source rock. Oil globules or gas bubbles must undergo distortion to pass through narrow pore throats, and this distortion is resisted by the interfacial tension between the hydrocarbon phase and the water lining the pore. This surface tension effect results in a **capillary pressure** that retards the flow of petroleum.

The precise mechanisms by which expulsion of petroleum occurs are not fully understood, although pressure and to some extent temperature are of importance (England et al. 1987). Different mechanisms may operate in different types of source rock (Stainforth & Reinders 1990). One possibility is that hydrocarbons move through microfractures in the source rock under the influence of over-pressure (Tissot & Welte 1984), and compaction plays a part (Braun & Burnham 1992). An increase in volume during the liberation of hydrocarbon fluids from the solid kerogen matrix would contribute to over-pressure development, but evidence for it is equivocal (Osborne & Swarbrick 1997). Microfracturing will reduce capillary pressure and so relieve over-pressure by allowing the escape of hydrocarbons.

Pressuring and fracturing processes may recur in a cyclical manner. Upon entering larger pores, oil particles tend to coalesce as globules, forming stringers (elongated globules) upon expulsion through narrow pore throats. The general lack of solubility of petroleum components in water and the small amounts of unbound water remaining in the source rock suggest that a pressure-driven expulsion of this kind would occur as a discrete hydrocarbon phase (England & Fleet 1991). Methane and carbon dioxide are supercritical fluids rather than gases at depths $c.>1$ km (Box 4.8) and so may significantly influence oil migration (Killops et al. 1996).

Expelled oil is generally enriched in paraffins, and to a lesser extent aromatic hydrocarbons, compared to the bitumen remaining in the source rock. It would appear that adsorption is important, and can occur within both the kerogen network and the water-coated inorganic mineral matrix (Sandvik et al. 1992; Killops et al. 1998; Ritter 2003). In the laboratory, solvent extraction does not recover all the bitumen present in a coal or kerogen; subsequent thermally induced desorption can liberate significant additional quantities (Killops et al. 2001). Expulsion of oil may not occur until adsorption sites have been effectively saturated, and a general guideline of 100 mg of hydrocarbons per gram of organic matter has been suggested (Pepper & Corvi 1995b). In effect, the adsorption by kerogen can be considered as dissolution of individual oil components in the kerogen. The predicted solubility of various compound classes in solid kerogen decreases with decreasing polarity in the order asphaltenes < resins < aromatic hydrocarbons < cyclic and branched alkanes < n-alkanes, which reflects the observed compositional differences between expelled and retained petroleum (Ritter 2003).

Another possible mechanism for primary migration is the thermally activated diffusion of hydrocarbons through the kerogen network of the source rock (Stainforth & Reinders 1990). The rate at which the hydrocarbons diffuse away from their site of origin within the source rock (i.e. the region in which they occur at highest concentration; Box 3.8) has been estimated to be of the same order of magnitude as their rate of thermal generation. Such a mechanism would, at least during the early phases of expulsion, result in the preferential loss of smaller hydrocarbons. Evidence for diffusion and pressure-driven expulsion effects has been reported for the Kimmeridge Clay (Upper Jurassic) shales in the North Sea, in the form of variations in C-number distributions in hydrocarbons across the interface between the shale units and adjacent sandstone

Box 4.8 | Gas density and supercritical fluids

The density of a gas is profoundly influenced by temperature and pressure (see Box 1.11). Sometimes gas densities are expressed in terms of gravity relative to air at STP (standard temperature and pressure, which corresponds to 0 °C and 1 atmosphere), and because the density of a gas is proportional to the average molecular weight, natural gases have gravities of mostly <1, and that for pure methane is 0.55 (air density at STP = 1.161 kg m^{-3}).

In the subsurface the density of a gas increases with depth, despite increasing temperature, because of the pressure-induced compression. When a fluid's critical temperature (T_c) and pressure (p_c) are exceeded there are no longer separate gas and liquid phases; only a single **supercritical fluid** can exist. For methane T_c = −82.6 °C and p_c = 4.6 MPa, whereas for carbon dioxide the corresponding values are −31.0 °C and 7.4 MPa (a typical phase diagram is shown in Fig. 4.28). A supercritical fluid has a much higher density than a gas and many of its properties are intermediate between those of a gas and a liquid. Consequently, supercritical methane and carbon dioxide are potentially excellent solvents for oil.

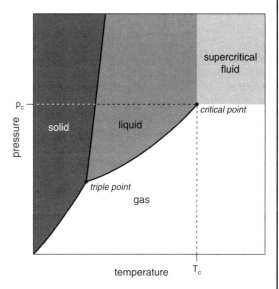

Fig. 4.28 Generalized phase diagram. Only a supercritical fluid exists above the critical temperature (T_c) and critical pressure (p_c), which has properties intermediate between liquid and gas. (Note: water is unusual in having a negative slope for the solid–liquid phase equilibrium line.)

---

carrier beds (Mackenzie et al. 1988). Both diffusion and pressure-driven expulsion mechanisms enable hydrocarbons to be expelled from the bottom as well as the top of a source unit.

Only if sufficient hydrocarbons are generated in the source rock will expulsion of hydrocarbons occur, and there is a lag between the onset of hydrocarbon generation and expulsion, whether diffusion- or pressure-driven mechanisms operate. The proportion of generated hydrocarbons that is expelled from the source rock during the main stage of oil formation is strongly dependent upon the type and initial amount of kerogen present. Nearly all of the oil generated by rich source rocks (containing >10 kg of labile kerogen per tonne of rock; e.g. Kimmeridge Clay, North Sea, and Bakken Shale, Williston Basin, North Dakota, USA) may be expelled (60–90% between 120 and 150 °C). The oil contains minor amounts of dissolved gas. Below a minimum initial concentration of labile kerogen ($c.$<5 kg t^{-1}; Mackenzie & Quigley 1988) most or all of the oil generated may remain in the source rock and be cracked to gas at higher temperatures ($c.$150–180 °C). Expulsion of gas may be aided by microfracturing due to the increasing fluid pressure within the pore spaces as cracking proceeds. Cracking may be the fate of most of the oil generated by humic coals. A summary of generation and expulsion of hydrocarbons for different concentrations of labile kerogen is shown in Fig. 4.29.

(b) Secondary migration

If suitably porous and permeable strata lie adjacent to the source rock, either above or below, the expelled oil may coalesce into larger stringers or globules within the carrier rock and may travel large distances, until it escapes to the surface or is trapped by a suitably impermeable barrier (Tissot & Welte 1984). The three important factors controlling secondary migration are buoyancy, capillary pressure and hydrodynamic flow (England & Fleet 1991). Oil and gas have specific gravities of

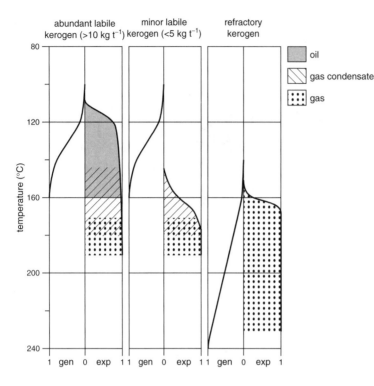

Fig. 4.29 Trends in petroleum generation (gen) and expulsion (exp) for source rocks of varying kerogen composition (at a mean heating rate of 5°C Myr^{-1}; after Mackenzie & Quigley 1988). Concentrations refer to kg of kerogen per tonne of rock.

0.7–1.0 and <0.0001, respectively, compared with 1.0–1.2 for the aqueous pore fluids (Tissot & Welte 1984). Petroleum compounds, therefore, undergo buoyant rise in water-saturated porous rocks. Upward migration is usually retarded by layers of less permeable rock, so secondary migration generally occurs along permeable strata in the direction of decreasing pressure. Extensive vertical migration generally requires suitable pathways produced by large-scale faulting.

Petroleum flow stops when capillary pressure exceeds buoyancy, due to a decrease in porosity and permeability. There is often hydrodynamic flow in sedimentary rocks, which can modify the buoyant flow of petroleum through carrier strata, depending on the relative directions of the two flows. Secondary migration can involve distances of 10–100 km and occasionally more. The size of an oil accumulation is related to the area of source rock from which it was generated (the drainage area) and hence to migration distance. Secondary migration distances are generally short in lacustrine sequences due to the limited scale of stratigraphic relationships between source rocks and reservoirs.

With decreasing depth during vertical migration, temperature and pressure fall, and at some stage the single supercritical hydrocarbon phase generally associated with expulsion may separate into liquid and gaseous phases, after the critical point has been passed (Box 4.8). Such changes can significantly affect the ability of the main hydrocarbon phase to carry the least soluble components in solution, and asphaltene precipitation may occur. Understanding the phase behaviour of petroleum in the subsurface is important if, for example, gas:oil ratios in reservoirs are to be predicted with accuracy. Such predictions can be obtained from approximations of fluid composition and the application of equations of state, assuming ideal behaviour of PVT (pressure, volume, temperature) characteristics (Meulbroek 2002; see Box 1.11).

Petroleum migration ends with either escape to the surface or entrapment owing to the inability of buoyancy to overcome capillary pressure (Box 4.9). Temperature and pressure conditions within a reservoir can be important in determining the hydrocarbon phases present (see Box 4.8). The solubility of gas in oil increases with increasing pressure, but decreases with increasing temperature. If the amount of gas generated exceeds the capacity of the oil to dissolve it, a separate gas phase will result that occupies the top part of the reservoir, due to its substantially lower density. A light oil can dissolve more gas than a heavy oil.

Box 4.9 | Reservoirs and traps

The rocks in which large volumes of petroleum are able to accumulate are termed **reservoir rocks**. They require suitable porosity (typically 10–25%) and permeability (typically 1–1000 mD; 1 mD or milliDarcy = $c.10^{-9}$ m^2), with reasonably sized pores and an impermeable cap rock or seal to prevent escape of petroleum over geological time periods. They must also be in place before the onset of oil generation. Sandstones often provide suitable reservoir characteristics. More than 60% of all oil occurrences are in clastic rocks, while carbonate reservoirs account for $c.$30%. The smaller molecules present in gases can escape through narrower pores than oil components, and seals are often slightly leaky with respect to gas.

A trap is a three-dimensional geological feature that obstructs the flow of petroleum, forming an accumulation. Traps can be of a variety of types, and a full discussion is beyond the scope of this book. The reader is referred to any general text on petroleum geology (e.g. Selley 1997; Gluyas & Swarbrick 2004). By way of a simple summary, traps are usually divided into two classes: structural and stratigraphic. **Structural traps** are the commonest and are caused directly by tectonism (e.g. anticlines, faults and folds). Petroleum is trapped below impervious strata because outward flow would require movement opposed to the direction of buoyant forces, although if sufficient petroleum is generated it can fill an anticlinal or similar trap and spill out. In fault traps, flow is opposed by large capillary forces in structures of very low porosity and permeability that cut across the carrier rock at the fault face. A variety of structural traps can be associated with salt domes (structures formed by plastic deformation of salt deposits under high pressure resulting in the formation of a dome that pierces the immediately overlying sediments). **Stratigraphic traps** are depositional features in which a reservoir unit is surrounded by less porous and permeable rocks, such as dense shales or limestones, preventing the outflow of petroleum. Typical reservoir units include barrier sand bars and islands (formed by wave and tidal action), channel sands (deposited in deltaic distributary systems), submarine fans (deep-water deposits from sediment density flows) and carbonate reefs.

4.5.6 Post-generation alteration of petroleum

A number of factors can affect the composition of oil encountered in a reservoir compared with the bitumen generated within the source rock, and those of most importance are considered in this section. Ratios of pairs of hydrocarbons can be used to investigate the extent of the various processes, although it is difficult to select compounds that are influenced by just one process. Among the most useful group of compounds are the gasoline-range hydrocarbons (e.g. Halpern 1995; see Box 4.6).

(a) Migration

The minerals of the pore walls and their associated bound pore waters present a polar surface to the compounds generated from kerogen. The more polar oil constituents, the asphaltenes and resins and to a lesser extent the aromatic hydrocarbons, are attracted to this polar surface, become relatively concentrated in the interfacial layer and are less readily expelled from the source rock. Adsorption of the more polar constituents can continue throughout secondary migration, leading to a relative increase in concentrations of apolar hydrocarbons, a process often termed geochromatography (e.g. Mackenzie 1984). Compared with the source rock bitumen, the oil that reaches the reservoir contains slightly less aromatics and significantly less resins and asphaltenes relative to the aliphatic hydrocarbons.

(b) De-asphalting

Because asphaltenes are insoluble in light hydrocarbons (C_1–C_8), the infiltration of gas or light oil into a fairly heavy oil, either in the reservoir or during migration, can result in precipitation of asphaltenes.

(c) Biodegradation

Aerobic bacteria may degrade petroleum if the temperature is not too high ($c.$<80 °C). There is evidence of anaerobic biodegradation in anoxic oil reservoirs (Connan et al. 1996) and sulphate reducers have been implicated (Wilkes et al. 2000). It is possible that the anaerobes utilize metabolites from prior aerobic degradation (Jobson et al. 1979). Thermophilic sulphate reducers preferentially remove C_8–C_{11} from crude oil (Rueter et al. 1994), and hyperthermophilic sulphate reducers recovered from oil reservoirs have been found to grow using oil as the sole source of carbon and energy (Stetter et al. 1993), so it is possible that biodegradation can occur at higher temperatures than previously supposed.

The various components of oil are not utilized at the same rate and there appears to be a general order of preference of removal. First to be degraded appear to be the *n*-alkanes, but long-chain alkylbenzenes are also affected at an early stage (Jones et al. 1983). However, the C_{45+} *n*-alkanes are relatively resistant (Heath et al. 1997), probably reflecting their highly hydrophobic nature, which limits accessibility of bacterial extracellular enzymes. The effects of biodegradation on the appearance of the gas chromatograms of saturates is most obvious when *n*-alkanes have been attacked significantly. As biodegradation progresses and the major resolved components are removed, the characteristic broad hump of the UCM becomes apparent in chromatograms (see Sections 4.5.3a and 7.3.2b, and Fig. 7.8). The isotopically light components are preferentially biodegraded, resulting in the $\delta^{13}C$ value of the residual petroleum becoming heavier (Connan 1984; Wilkes et al. 2000; George et al. 2002).

The susceptibility of petroleum hydrocarbons towards biodegradation generally decreases in the order: *n*-alkanes; alkylcyclohexanes; alkylbenzenes; *iso-* and *anteiso-*alkanes; acyclic isoprenoids; alkylnaphthalenes; bicyclic alkanes; C_{27}–C_{29} regular steranes; C_{30}–C_{35} hopanes; diasteranes; C_{27}–C_{29} hopanes; C_{21}–C_{22} regular steranes; tricyclic terpanes (Connan 1984; Volkman et al. 1984; Fisher et al. 1998; George et al. 2002). One of the several suggested scales for assessing the degree of biodegradation of oil is shown in Table 4.11.

It is possible for complete removal of hydrocarbon classes to occur. Although steranes and terpanes are less affected than other alkanes, they can also be totally removed under severe biodegradation. In general, aromatic components tend to be more resistant towards biodegradation than alkanes, their resistance increasing with number of aromatic rings and degree of alkyl substitution. Aromatic steroids appear only to be affected under the most severe conditions (Wardroper et al. 1984). Clearly, the properties of an oil can change significantly upon biodegradation, generally to the detriment of its commercial worth.

Diamondoids are hydrocarbons with a carbon skeleton approaching that of diamond, and include adamantane and diamantane (Fig. 4.30). They are particularly resistant towards biodegradation and may be used to monitor more extreme levels of biodegradation (Grice et al. 2000). The ratio of methyladamantanes (1- and 2-methyl isomers) to adamantane appears particularly useful as an indicator of biodegradation, increasing by a significant amount over biodegradation stages 4–8 (Table 4.11). The ratio of methyldiamantanes (1-, 3- and 4-methyl isomers) to diamantane does not exhibit such a range of values over the same biodegradation stages, but may be more useful at higher levels of biodegradation.

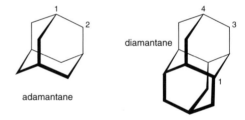

Fig. 4.30 Diamondoid hydrocarbons adamantane and diamantane (n.b. chair conformation of rings; variation in line thickness to emphasize three-dimensional structure).

Table 4.11 Oil biodegradation scale (after Peters & Moldowan 1993)

biodegradation rank	biodegradation stage	biomarker changes
light	1	short-chain *n*-alkanes depleted
	2	general depletion of *n*-alkanes
	3	only traces of *n*-alkanes remain
moderate	4	no *n*-alkanes remain, acyclic isoprenoids intact
	5	acyclic isoprenoids absent
heavy	6	steranes partly degraded
	7	steranes degraded, diasteranes intact
very heavy	8	hopanes partly degraded
	9	hopanes absent, diasteranes attacked
severe	10	long-chain aromatic steroids attacked

(d) Water washing

Water washing can occur in the reservoir when there is infiltration of meteoric waters. It can also occur during secondary migration, particularly where countercurrent hydrodynamic flow is encountered. The effect is the removal of the more water-soluble components from oil. The most susceptible components include the polar compounds preferentially adsorbed by pore water and mineral interactions during migration, but small hydrocarbons are also relatively soluble in water. Water washing and biodegradation can occur together in reservoirs, and have opposing effects, with washing tending to remove aromatics and biodegradation the saturates.

(e) Thermal alteration

Oil accumulations can continue to undergo thermal evolution, depending on the depth of the reservoir and subsequent geothermal history. As noted in Section 4.5.2, oil becomes susceptible to thermal cracking at temperatures above 160–200 °C. For example, paraffinic-naphthenic oils (Fig. 4.22) are degraded to aromatic-naphthenic oils (with moderate S content, <1%), and aromatic-intermediate oils degrade to aromatic-asphaltic oils (with high S content, >1%).

(f) Thermochemical sulphate reduction

The main source of H_2S below c. 100 °C is bacterial sulphate reduction (BSR; Section 3.3.2b), which is most efficient below 60 °C and virtually ceases by 100 °C or where H_2S levels reach 5–10%. Some H_2S is produced by the thermal breakdown of kerogen, particularly S-rich varieties, but it usually does not exceed 3% (by volume) of gas. Most natural gases contain negligible H_2S, and H_2S-rich (or sour) gas fields are mainly associated with carbonates, because clastic reservoirs generally contain sufficient iron(II) to remove the highly reactive H_2S as pyrite. Where H_2S levels exceed 10% and reservoir temperatures are >100 °C, the source is invariably thermochemical sulphate reduction (TSR). TSR involves the reduction of sulphate in aqueous solution by petroleum hydrocarbons (Machel et al. 1995). This results in the oxidation of the hydrocarbons to organic acids and organic-S compounds, and the formation of solid bitumens—except where methane is the only substrate—together with H_2S and bicarbonate/CO_2.

TSR appears to be associated with temperatures in the range 100–140 °C, and the saturated hydrocarbons seem the most reactive fraction. Anhydrite is the major source of the required sulphate, and during its consumption by TSR it is replaced by calcite with a characteristic $\delta^{13}C$ signature. Among the liquid hydrocarbons, TSR causes a decrease in the saturated: aromatic hydrocarbon ratio, an increase in abundance of organic-S compounds and a slight decrease in API gravity; all trends opposite to those usually associated with increasing thermal maturity (Manzano et al. 1997).

Kinetic isotope effects are involved in TSR (see Box 1.3), and the $\delta^{13}C$ values of methane, ethane and propane become a few permil heavier (Krouse et al. 1988). The conversion of sulphate to sulphide might also be expected to exhibit an isotopic fractionation, lowering $\delta^{34}S$ values by up to −20‰, but often zero fractionation is observed, and the $\delta^{34}S$ values for H_2S (and organic-S) approach those of the pore-water sulphate (and anhydrite). This is because the slowest (rate-determining) step for the overall reaction is dissolution of sulphate, and once in solution the sulphate is effectively completely reduced, so the sulphide has the same isotopic signature as the sulphate from which it originated (Manzano et al. 1997). The net TSR reaction can be described by the following sequential steps, involving native S as an intermediate and the recycling of CO_2:

$$CaSO_4 + 3H_2S + CO_2 \rightleftharpoons CaCO_3 + 4S + 3H_2O \quad \text{[Eqn 4.4a]}$$
$$4S + 2H_2O + CH_4 \rightleftharpoons CH_2S + CO_2 \quad \text{[Eqn 4.4b]}$$
$$\underset{\text{anhydrite}}{CaSO_4} + CH_4 \rightleftharpoons \underset{\text{calcite}}{CaCO_3} + H_2S + H_2O \quad \text{[Eqn 4.4c]}$$

(g) Gas diffusion

Light hydrocarbons, particularly methane, can diffuse relatively rapidly through imperfect seals. Because reservoir seals are usually not perfect the gases are likely to become depleted over geological time periods. Because diffusion is related to molecular mass, methane containing the ^{12}C isotope diffuses more rapidly than that containing ^{13}C, so there is an isotopic fractionation effect, with the residual gas becoming isotopically heavier (see Section 5.8.6b).

(h) Evaporative fractionation

Where oil and gas phases occur in a reservoir, each compound will distribute itself between the two phases according to its vapour–liquid equilibrium constant. For a particular hydrocarbon, this constant is affected by all of the components present in each phase, as well as the hydrocarbon's vapour pressure. If the oil and gas phases become separated through leakage of the gas out of the cap or spillage of the oil out of the bottom of the reservoir, the compounds remaining will attempt to reach a new partitioning equilibrium between the two phases, resulting in evaporative fractionation. Such a process has been suggested to occur in certain Tertiary reservoirs in the US Gulf Coast (Thompson 1987), and can lead to significantly different distributions from that present in the original oil. One major effect is the enrichment in aromatic hydrocarbons relative to *n*-alkanes containing the same number of C atoms.

(i) Tertiary migration

Petroleum may not remain in the first trap it encounters and spillage can occur: the most dense fluids can be displaced past the spill point if less dense fluids, especially gas, continue to reach the trap. This process of displacement—tertiary migration—can significantly affect the composition of hydrocarbons encountered in a sequence of stacked reservoirs, possibly leading to the presence of oil in the most shallow units and gas in the deeper units.

(j) Carbon dioxide

Although it is not usually considered a component of petroleum, it is worth considering the fate of the carbon dioxide generated during the diagenetic and earliest catagenetic phases of kerogen and coal maturation. Because of its high aqueous solubility, CO_2 is quickly transported away from kerogen in the presence of hydrodynamic flow. As a CO_2-rich aqueous solution migrates upward and both temperature and pressure fall, carbonate is likely to precipitate, and degassing may also occur, resulting in a rise in pH (Irwin & Hurst 1983). These processes are likely to cause a reduction in the porosity of secondary migration carrier beds and reservoirs, although the interactions are complex, and organic acids can create secondary porosity by dissolving feldspars and carbonates (Gautier 1986; Bjørlykke 1994; Killops et al. 1996). Acetate is by far the most important acid, and does not appear to be generated in sufficient quantity by kerogen to account for the observed concentrations in reservoir fluids (Barth et al. 1988), so it may originate from redox processes involving mineral surfaces within the reservoir initiated when the first charge of oil enters (Borgund & Barth 1994), or possibly even from anaerobic bacterial degradation of oil (Killops et al. 1996).

4.6 Temporal and geographical distribution of fossil organic carbon

4.6.1 Coal and kerogen

Throughout the Earth's history there have been periods when conditions have been particularly suitable for the deposition of coal and kerogen-rich deposits (suitable for petroleum generation), which are shown in Fig. 4.31. As discussed in Sections 3.4.2 and 3.4.3, these episodes correlate with rising sea level. The long-period (*c*.400 Myr) trend in eustatic sea level is related to the dispersal of continents and the relative volume of oceanic basins occupied by constructive ridges (continental glaciation has a subordinate effect; Box 3.11). Superimposed on this general sea-level trend are shorter period cycles of rises and falls, which appear to govern the deposition of major coal deposits and oil source rocks. These second-order sea-level rises, with periods of tens of Myr, are also caused by tectonic processes.

Major coal deposits coincided with periods following regression, which is consistent with the most suitable conditions for deposition—a steady increase in accommodation space—being predominantly found in lowland coastal swamps (Section 3.4.2a). Two main episodes of coal formation can be distinguished: the first during the Carboniferous–Permian and a second smaller episode spanning the Jurassic to early Tertiary. The majority of coals formed in the earlier episode are now bituminous coals or anthracites, whereas those from the Tertiary are mainly brown coals.

The major prolific oil source rocks were deposited during global marine transgressions. As for coals, there appear to have been two main episodes of oil source-rock deposition: one during the Palaeozoic, peaking around the Devonian, and another during the Mesozoic, peaking around the Cretaceous. Such periods were characterized by marine transgressions on to continental margins, creating suitable conditions for the production (e.g. high nutrient supply) and preservation (e.g. anoxic basins) of sedimentary organic matter (Klemme & Ulmishek 1991; Section 3.4.3a). This is reflected in the correlation of phytoplanktonic abundance with organic-rich sediments in Fig. 4.31. The Palaeozoic peak in phytoplanktonic production can be attributed to the dominant, organic-walled organisms of the era (acritarchs, green algae and cyanobacteria). The latter productivity peak was characterized by major contributions from, initially, calcareous (coccolithospores) and, subsequently, siliceous phytoplankton (silicoflagellates and diatoms; see also Fig. 1.12).

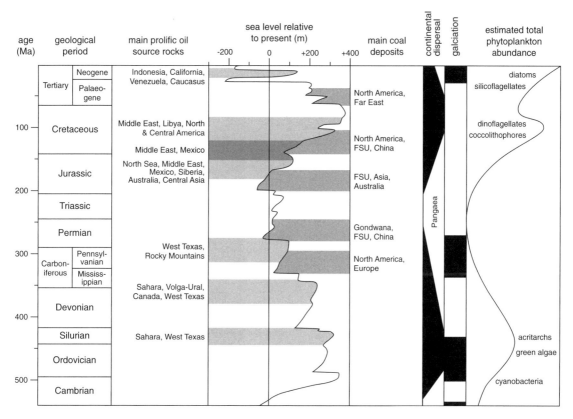

Fig. 4.31 Relationship between eustatic sea level and major depositional periods of coals and petroleum-source rocks (after Tissot 1979; Vail et al. 1978; FSU = former Soviet Union).

4.6.2 Oil and gas

The estimated total reserves of coal, conventional oil (produced as a liquid; see Box 4.6) and gas at the end of 1999 are shown in Table 4.12 (World Energy Council 2002). These values clearly change over time as consumption and exploration continue (oil generally diminishing, but coal sometimes increasing), and the potentially vast reserves of coal in Antarctica are omitted from Table 4.12. More useful figures in terms of understanding the amounts of carbon residing in the crust under steady-state conditions are the sizes of the various reservoirs of fossil carbon that existed prior to Mankind's large-scale exploitation of them, which are estimated in Table 4.13. There is significantly less oil than coal and it is less evenly distributed on a global basis. About 40% of the total estimated coal reserves are likely to be recoverable.

As well as conventional oil, Table 4.12 includes heavy oils from tar-sand deposits (see Box 4.6) and the oil that

Table 4.12 Estimated total reserves of fossil fuels (after World Energy Council 2002). Approximate conversion to tonnes of oil equivalent based on energy equivalence factors of 1.07 for natural gas liquids (condensate), 0.9 for heavy oils, 0.7 for hard coals, 0.47 for brown coals, 0.23 for peat and 0.86 t per $10^3 \, m^3$ for gas (1 tonne crude oil = c.7.3 barrels = c.1.16 m^3)

coal	Gt	Gt oil equiv.
bituminous + anthracite	1441	1010
sub-bituminous	699	330
lignite	839	390
peat	461	110
oil	Gt	Gt oil equiv.
conventional	287	287
gas liquids	16	17
heavy oils/bitumens	99	90
shale oil	3604	3604
gas	$10^{12} \, m^3$	Gt oil equiv.
natural gas	152	130

Table 4.13 Estimated amounts of fossil carbon in the crust (after Kempe 1979; Tissot & Welte 1984; Ward 1984; Kvenvolden 1998; Falkowski et al. 2000). Approximate conversions to carbon basis assume: densities of $0.8\,t\,m^{-3}$ for conventional oils, $0.9\,t\,m^{-3}$ for heavy oils, $0.65\,kg\,m^{-3}$ for gas, and corresponding C contents of 85, 80 and 75%; coal content of 80% C (dmmf) 10% ash and 5% moisture

mineral carbon	Gt	Gt C
carbonate	–	60×10^6
kerogen	–	15×10^6
coal	*Gt*	*Gt C*
coal	10 000	7 000
peat	–	250
oil	*$10^9\,m^3$*	*Gt C*
conventional	500	350
heavy	600	400
gas	*$10^{12}\,m^3$*	*Gt C*
coal + kerogen related	250	120
methane hydrates	21 000	10 000

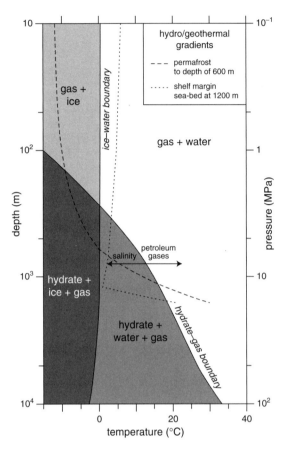

Fig. 4.32 Phase diagram showing stability fields of methane as hydrate or free gas, together with associated water phases. Depth scale assumes hydrostatic pressure gradient of $10\,MPa\,km^{-1}$. Increase in salinity shifts the hydrate–gas boundary to lower temperature, and addition of CO_2, H_2S, C_2H_6 or C_3H_8 shifts it to higher temperature. Representative hydrothermal and geothermal gradients are shown for continental permafrost and shelf margin settings (after Kvenvolden 1988, 1998).

can be recovered by pyrolysing oil shales (the majority of which are located in North and South America and the former Soviet Union). Oil shales are often not true shales, but they are kerogen-rich rocks, and so are classified under kerogen in Table 4.13. As a proportion of proven conventional oil reserves ($c.105 \times 10^9\,m^3$ or $c.660 \times 10^9$ barrels), $c.50\%$ is accounted for by just 33 supergiant fields, >60% is located in the Arabian–Iranian province (Middle East) and $c.25\%$ is recoverable using currently available technology. In contrast, much of the heavy oil reserves (mostly in Canada and Venezuela) may not be recoverable, although substantial amounts of bitumen have been recovered from the Athabasca oil sands. Heavy oils derive from degradation of conventional oils, and so are usually aromatic-naphthenic or aromatic-asphaltic (Fig. 4.22).

Of the two main episodes of oil source-rock deposition, the Palaeozoic accounts for only 10–15% of the total reserves of conventional oils (Bois et al. 1982). The smaller size of the Palaeozoic oil reserves compared with those of the Mesozoic is probably mainly attributable to escape of the older oils from reservoirs due to later tectonic events and to various degradation processes (as described in Section 4.5.6). The surviving Palaeozoic oils are found in stable continental platforms that have escaped significant tectonic activity. The relationship of relative sea level to the occurrence of oil source rocks is reflected in the fact that most oil is derived from marine sources. In comparison, lacustrine source rocks are usually of much more restricted extent; the largest lacustrine-sourced province, Songliao (China), accounts for <1% of global recoverable resources ($c.1.36 \times 10^9\,m^3$).

Gas reserves are less easy to associate with their source rocks because gas is able to migrate more easily and over greater distances than oil. Consequently, conventional reserves (existing as gas in reservoirs) are dominated by contributions from coal and kerogen sources of Cretaceous–Tertiary age (Bois et al. 1982). The largest proven gas reserves are in the Middle East and the former Soviet Union, each of which accounts for $c.35\%$ of the total. There are also large amounts of methane in coal beds, some of which may be exploitable, as in the USA (Kaiser et al. 1994).

(a) Methane hydrates

There is another, very important and large repository of methane: **methane hydrates** (also known as gas hydrates or clathrates; Kvenvolden 1988). They comprise ice in which the interstices of the lattice house small molecules, such as methane, ethane, carbon dioxide and hydrogen sulphide. In fact, enough gas needs to be present to fill 90% of the interstices in order for the hydrate to form, and it has a different crystal structure from normal ice (Sloan 1990). If fully saturated, the most common crystalline structure can hold one molecule of methane for every 5.75 molecules of water, so 1 m^3 of hydrate can contain 164 m^3 of methane at STP (see Box 4.8). The solubility of methane in water is insufficient to account for hydrate formation, and a major nearby source is required, typically methanogenesis, based on the dominance of methane (99%) and its very light isotopic composition (δ^{13}C generally <−60‰; see Section 5.8.2).

Methane hydrates are mostly found in permafrost (onshore and offshore shelf sediments) in polar regions and in deep oceanic settings (sediments on continental slopes and rises), where cold bottom water is present (e.g. Gulf of Mexico). The stability field for methane hydrate formation in pure water is shown in Fig. 4.32; addition of NaCl or N_2 shifts the hydrate–gas boundary to lower temperature, decreasing the stability range, whereas CO_2, H_2S, C_2H_6 or C_3H_8 shifts it towards higher temperature, increasing the stability range. Under typical hydrostatic pressure gradients (Box 4.4), methane hydrate formation in polar regions requires a surface temperature <0 °C and a minimum burial depth of *c.*150 m (see representative permafrost geotherm in Fig. 4.32), whereas in oceanic sediments with a bottom water temperature approaching 0 °C the water depth needs to be at least 300 m. The base of the stability field is controlled by the geothermal gradient (Box 4.4); it is unlikely to be >2000 m below the sediment surface, and is typically much shallower (as shown for the representative shelf margin hydro/geotherm in Fig. 4.32). The base of the stability zone can be recognized by seismic stratigraphy, and such surveys suggest that methane hydrates are widely distributed around the globe in continental margin sediments (Kvenvolden 1998). Current estimates of the amount of carbon present in methane hydrates exceed that in all other fossil fuels, at *c.*10^4 Gt (Table 4.13), although they are poorly constrained.

5 Chemical stratigraphic concepts and tools

5.1 Biologically mediated transformations

5.1.1 Introduction

It is possible to assess sedimentary contributions in contemporary environments from the microscopically identifiable remains of organisms, particularly inorganic skeletal material (e.g. calcareous and siliceous tests). Unfortunately, these components do not always survive in older sediments, but molecular evidence may, particularly in the form of biomarkers: lipid-derived compounds that can be traced to particular biological precursor molecules (e.g. Cranwell 1982). The application of molecular palaeontology (or chemotaxonomy) to ancient sediments requires an understanding of how individual compounds and groups of compounds can be used to identify contributions from extant organisms in **Recent** (i.e. Holocene) sediments and of what changes these compounds undergo during diagenesis and subsequent catagenesis.

Early on in the sedimentary process—in water-column particulates and in the top of the sediment pile—some organic compounds exist as largely unaltered constituents of the source organisms. Depending on the specificity of these compounds, it may be possible to gain an idea of the types of organisms contributing to the sedimentary organic matter and even to estimate their relative contributions. During the discussion of the composition of organic material in organisms in Chapter 2 some examples of the specificity of compounds to groups of organisms were considered. This specificity is not restricted to lipids, although certain lipid classes have received most attention because of their general ease of analysis (see Box 5.1).

In young sediments, biomarker distributions can usually be used to distinguish between contributions from major groups of organisms (e.g. eubacteria, archaebacteria, algae, higher plants), broadly corresponding to the division (phylum) level (e.g. algae,

---

Box 5.1 | Free and bound lipids

Analysis of sedimentary biomarkers involves extracting the sediment with a suitable combination of organic solvents, usually followed by some separation procedure(s) in order to aid component identification by providing less complex mixtures. Solvent extraction on its own removes the **free lipids**, but a proportion, the **bound lipids** fraction, remains bonded to insoluble polymeric material in the sediment. Bound lipids are subsequently extracted following hydrolysis of the sediment residue, which breaks the bonds between the remaining lipids and insoluble matrix (e.g. Goossens et al. 1986). The free lipids usually contain a substantial proportion of compounds in which two classes of components are chemically bonded together, such as fatty acids and fatty alcohols in wax esters. In contrast, the hydrolysis step during extraction ensures that the bound lipid fraction does not contain such combined components. The amount of combined components in the free lipids fraction depends on the degree of hydrolysis that has occurred during diagenesis, a process that releases individual components (e.g. fatty acids and sterols from steryl esters). Because of this, hydrolytic cleavage of the remaining combined components in the free lipids is often undertaken in the laboratory, enabling distributions within a compound class (e.g. total fatty acids) to be examined, although techniques are available for analysing combined components. With increasing burial, the amount of free lipids usually decreases during the formation of insoluble kerogen, while changes in bound lipids reflect changes in kerogen composition. It can be informative, therefore, to analyse free and bound components separately.

spermatophytes). Distinction is also usually possible at the next taxonomic level (e.g. algal classes of Chlorophyceae, Phaeophyceae and Dinophyceae; spermatophyte subdivisions of angiosperms and gymnosperms; see Box 1.8), but becomes increasingly difficult as the species level is approached. This is because there are relatively few examples of compounds exclusive to a family of organisms, which demonstrates the evolutionary relationship between the extents of biochemical and physiological diversification. However, a distinctive combination of biomarkers may be sufficient to identify a contribution from a family (or even species), providing it can be reasonably ensured that such a pattern may not arise from the combined contributions of other organisms (C isotopic composition can play a vital role here; see Section 5.8). Absolute quantification of inputs of sedimentary organic matter using biomarkers alone is difficult, because it requires information on their abundance relative to total organic matter in a given organism (which can vary, depending on environmental conditions; see Section 2.7.1) and the relative stabilities of biomarkers in the sedimentary environment.

5.1.2 General differences between major groups of organisms

The more obvious compositional differences occur between the major groups of organisms; for example, lignin, composed of polyhydroxyphenol units (Section 2.5.1), is present only in the higher plants, and particularly the woody varieties. Another characteristic of higher plants is the presence of waxes as protective coatings on leaves. In other organisms such coatings are usually biopolymers of carbohydrates or amino acids. The major wax components are saturated, straight-chain fatty acids with >22 C atoms together with components of similar structure that are biosynthetically derived from these acids, including straight-chain alcohols (n-alkanols) and alkanes (n-alkanes). This biosynthetic relationship is reflected in an even-over-odd predominance (**EOP**) in carbon numbers for the fatty acids and alcohols, whereas the alkanes, which result from decarboxylation of fatty acids (i.e. loss of CO_2), exhibit a corresponding odd-over-even predominance (**OEP**; see Eqns 2.7 and 2.8, Section 2.4.2a). In contrast, in microorganisms and multicellular algae the major saturated, straight-chain fatty acids and their biosynthetically related n-alkanes and n-alkanols generally have <22 C atoms. A higher plant contribution to Recent sediments is, therefore, usually readily identified, although there are exceptions to this general rule. For example, the fresh/brackish-water alga *Botryococcus braunii* can potentially produce an OEP in the nC_{25}–nC_{31} range (Metzger et al. 1985).

Contributions from photosynthesizing organisms are characterized by various chlorophyll and related tetrapyrrole pigments. Chlorophyll-a is present in all plants and the cyanobacteria, while various bacteriochlorophylls are dominant in photosynthetic bacteria. Higher plants and green algae also contain chlorophyll-b, whereas brown algae contain chlorophyll-c. As demonstrated in Section 2.4.3g, carotenoid accessory pigments can be useful markers for groups of photosynthetic organisms. Some non-photosynthetic microorganisms contain characteristic carotenoids, such as the C_{50} bacterioruberin (Fig. 5.1) that confers the red colour on colonies of halophilic bacteria.

Various terpenoids can be specific to certain groups of organisms. Higher plants contain many types of terpenoids, some of which may be present in relatively large amounts in the more resinous woody plants. Bacteria do not appear to contain sterols, unlike eukaryotes, and make use of other compounds in the role of cell membrane rigidifiers. Among these compounds are hopanoids, which are found in eubacteria – mainly aerobes, but also some anaerobes (purple non-sulphur bacteria and planctomycetes; Pancost & Sinninghe Damsté 2003; Sinninghe Damsté et al. 2004) – but are rare in other organisms. Archaebacteria contain phospholipid ethers, and the biphytanyl ethers present in thermoacidophilic and methanogenic species are diagnostic.

In Section 5.2 the chemotaxonomic application of some compound groups is examined in a little more detail, but first it is useful to consider some of the variations that can occur in the lipid constituents of individual organisms.

5.1.3 Factors affecting the lipid composition of organisms

There are various factors that affect the biomarker distributions in individual organisms, some of which were mentioned in Section 2.7.1, such as the decline in storage lipids in zooplankton during times of poor nutrient availability. Light levels are important for plants because they affect the production of chloroplast pigments in green tissue, while the lipid content of roots increases during periods of salt stress and drought due to a proliferation of internal membranes.

Lipid content varies with tissue type within organisms and so the degree to which different tissues are preserved can be important. For example, 18:1 and 18:2 account for c.75% of the fatty acids in the beans of soya plants, whereas its leaves contain only c.20% of these

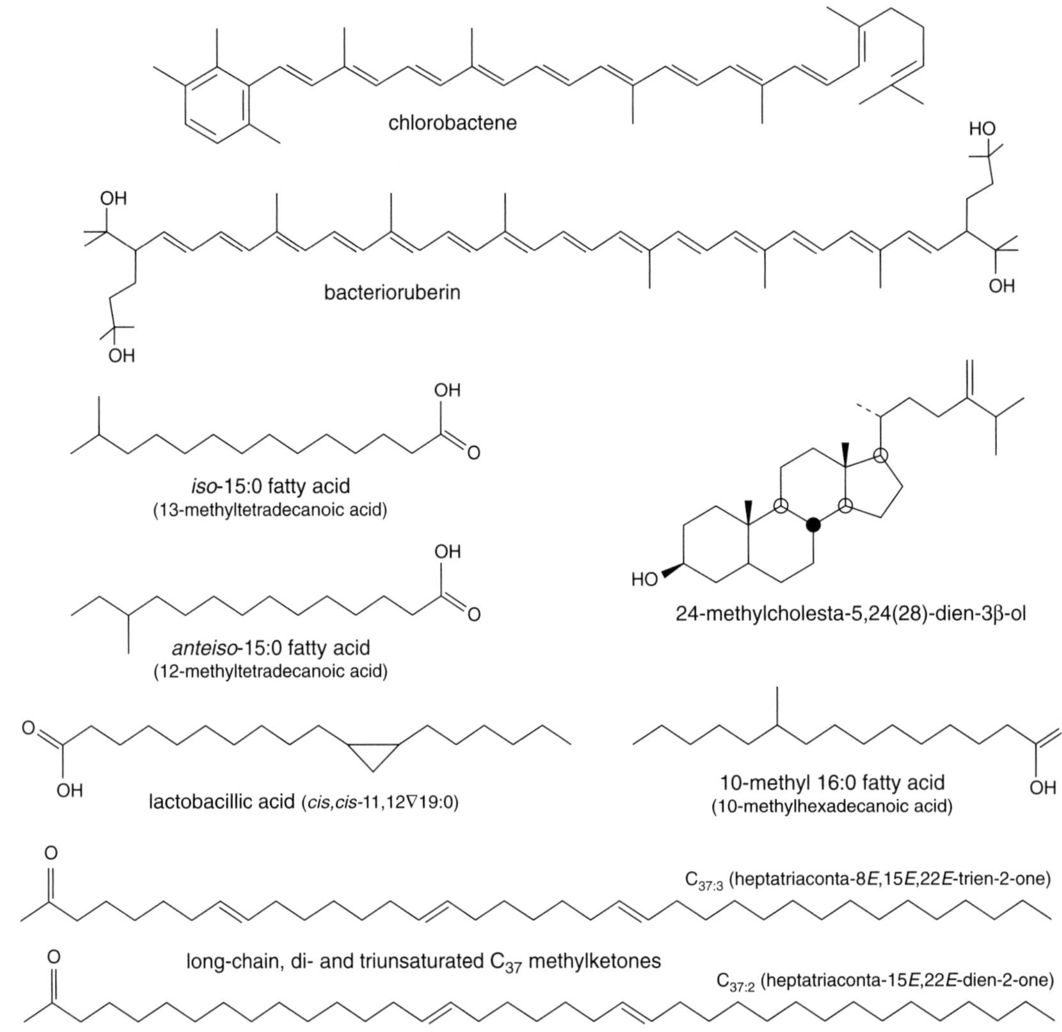

Fig. 5.1 Some biomarkers of microbial origin.

acids but c.55% of 18:3. There are also morphological changes in the life-cycles of plants, fungi and bacteria during which membranes, organelles and storage tissues appear and disappear. Lipids are important constituents of such tissues and so they also vary. For example, specialized pigments (e.g. carotenoids) are important during flowering but triglycerides are important as energy reserves when seeds are set. Most plant lipids (c.80%) are found in the chloroplasts (particularly as chlorophylls). In bacteria, lipid synthesis increases rapidly just prior to cell division, while the lipid content of mycelium generally increases rapidly during the vegetative growth of fungi, and fungal spores also contain lipid-rich bodies, which serve as energy reserves.

As temperature decreases higher plants and microorganisms maintain the fluidity of their cell membranes by lowering the melting point of the constituent lipids. This can be achieved by increasing the proportion of unsaturated fatty acids and, in bacteria, by producing more *anteiso* (relative to *iso*) fatty acids of shorter average chain length. The nature of the substrate can also significantly affect the lipids synthesized by bacteria. High L-valine or L-leucine content leads to an increase in *iso*-branched acids, whereas *anteiso* acid production is favoured by high concentrations of L-isoleucine. Among the sulphate-reducing bacteria, *Desulfobulbus* produces mainly even-numbered acids from combined H_2 and CO_2, mainly odd-numbered acids from propionate ($CH_3CH_2COO^-$), but a mixture of odd and even acids when grown on lactate ($CH_3CH(OH)COO^-$; Taylor & Parkes 1983).

5.2 Examples of source indicators in Recent sediments

In this section two of the most intensively studied classes of biomarkers are examined: fatty acids and sterols. The units from which polysaccharides and lignins are formed (i.e. monosaccharides and phenolic compounds, respectively) can be analysed as readily as biomarkers and the source-related information that their distributions can convey is also considered.

5.2.1 Fatty acids

Fatty acids are widely occurring compounds that fulfil a variety of roles, such as cellular membrane components (e.g. phospholipids), energy stores (e.g. triglycerides) and protective coatings (e.g. wax esters). They occur in either free or bound forms, although mostly the latter within organisms (bonded through the ester linkage to other compounds). More than 500 fatty acids are known in plants and microorganisms, but the most abundant are relatively few in number, palmitic acid (16:0) being the most common (see Table 2.4 for examples and Box 2.7 for notation scheme). In higher plants seven fatty acids account for c.95% of the total in combined leaf tissues and seed oil: 12:0, 14:0, 16:0, 18:0, cis-18:1ω9, cis,cis-18:2ω6 and cis,cis,cis-18:3ω3 (Harwood & Russell 1984). The fatty acids of multicellular algae are generally similar to those of the higher plants. The three C_{18} unsaturated acids are also often abundant in unicellular green algae, together with 16:4ω3 (Johns et al. 1979). All fungi appear to contain 16:0, 18:0 and C_{18} unsaturated fatty acids (Harwood & Russell 1984).

As noted in Section 5.1.2, among the saturated straight-chain fatty acids, long-chain (generally C_{24}–C_{36}) components with an EOP are characteristic of contributions of higher plant detritus to sediments. Chain length is, therefore, a useful broad—although not infallible—indicator of source type, as will become apparent. More specific information can, however, be obtained from unsaturated, branched and hydroxy acids.

(a) Monounsaturated fatty acids

There is biological preference for the cis configuration at C=C bonds (although some clay-catalysed isomerization to the trans form can occur in sediments). The fatty acid 18:1 is generally either ω9 (oleic acid), which is common in animals, higher plants and algae, or ω7 (cis-vaccenic acid), which is particularly abundant in, although not exclusive to, bacteria (Harwood & Russell 1984). Like 18:1, 16:1 is also often an abundant acid in Recent sediments and exhibits a similar differentiation between predominantly bacterially derived ω7 and algally derived ω9 isomers (although 16:1ω7 is a major fatty acid of diatoms). This differentiation arises because of different pathways of fatty acid biosynthesis. Along the aerobic pathway (described in Section 2.4.1a, and Fig. 2.13), unsaturated acids can be formed by the action of desaturase enzymes, resulting in ω9 compounds. In contrast, the anaerobic pathway operates in all anaerobic and many aerobic and facultatively aerobic eubacteria, again resulting in fatty acids with even numbers of C atoms. However, in this pathway enzymatic dehydration of an intermediate gives rise to a C=C bond with either cis-ω8 or trans-ω7 configuration. The trans isomer can undergo the usual enzymatic reduction to saturated fatty acids but the cis isomer cannot, leading to ω7 unsaturated products (Harwood & Russell 1984). This biosynthetic route also appears to produce ω5 C=C bonds.

Not all monounsaturated fatty acids are biosynthesized as cis isomers with even numbers of C atoms. For example, trans-16:1ω7 and trans-18:1ω7 may be bacterial markers, whereas trans-16:1ω13 is produced by photosynthetic bacteria and some phytoplankton but not cyanobacteria (Johns et al. 1979). The fatty acid distributions of cyanobacteria can be quite variable, with some exhibiting major 16:0 and 16:1ω7, while others have abundant 18:1ω9. The odd-numbered acids 15:1 and 17:1 with ω6 or ω8 C=C bonds are bacterial markers, produced by the anaerobic biosynthetic route. Among the sulphate-reducing bacteria, 17:1ω8 appears to be characteristic of Desulfobulbus (Taylor & Parkes 1983).

(b) Polyunsaturated fatty acids

Polyunsaturated fatty acids (PUFAs) are biosynthesized from saturated fatty acids by the action of desaturase enzymes (Section 2.4.1a), and many are probably formed by chain-shortening of 24:6ω3 (Volkman et al. 1998). Most bacteria do not contain fatty acids with more than one C=C bond, but C_{20} and C_{22} PUFAs are common among phytoplankton (Harwood & Russell 1984). For example, many planktonic algae (e.g. eustigmatophytes, haptophytes and dinoflagellates) have high concentrations of 20:5ω3 and/or 22:6ω3, although marine invertebrates can also be important contributors of 22:6ω3. In contrast, chlorophytes contain mainly 18:2ω6 and 18:3ω3, the latter also being common in higher plants (Harwood & Russell 1984; Volkman et al. 1998). Dinoflagellates often contain abundant 18:5ω3, but it is also present in many other phytoplankton (e.g. some prasinophytes, raphidophytes and haptophytes).

Of the three commonly occurring 16:2 isomers in marine phytoplankton (ω4, ω6 and ω7) only the ω4 and ω6 isomers appear to be present in the macroscopic green, red and brown algae, while ω4 and ω7 are the common isomers in diatoms (Volkman & Johns 1977). Abundant 16:3ω3 suggests an input from green algae. Diatoms are different from other algae in exhibiting a characteristic distribution of 16:0, 16:1ω7, 16:3ω4, 20:4ω6 and 20:5ω3, together with low amounts of C_{18} components. The presence of 16:3ω4 may be diagnostic of diatoms.

(c) Iso and anteiso methyl-branched fatty acids

These branched acids can be source-specific and they are rarely unsaturated. They are formed by the incorporation of branched amino acids into the biosynthetic pathway, yielding *iso* and *anteiso* acids, as noted in Section 5.1.3. *Iso* and *anteiso* saturated fatty acids are found in fungi, molluscs and phytoplankton, but they are generally in higher levels in bacteria and are often observed in the C_{13}–C_{17} range (Harwood & Russell 1984). The C_{15} isomers (Fig. 5.1) are usually particularly abundant in bacteria and the ratio (*iso* + *anteiso*)/normal derived from C_{15} components can be used as an indication of relative bacterial contributions (Parkes & Taylor 1983). Similarly, the ω8 isomers of *iso*-15:1 and *iso*-17:1 are bacterial markers (Perry et al. 1979), and *iso*-17:1ω7 is characteristically a major fatty acid in the sulphate-reducing bacteria *Desulfovibrio desulfuricans* (Taylor & Parkes 1983).

(d) Internally branched and cycloalkyl fatty acids

Internally branched acids occur naturally, such as the 10-methyl (numbering from the acid-group end) isomers of 16:0 and 18:0 found in fungi and bacteria. For example, 10-methyl 16:0 (Fig. 5.1) is characteristic of *Desulfobacter* (Taylor & Parkes 1983). There are also cycloalkyl acids, such as 17:0 and 19:0 cyclopropyl moieties. Cyclopropyl acids are more common in eubacteria than other organisms, but the position of the cyclopropyl group is important. Out of the more commonly occurring *cis*-11,12 and *cis*-9,10 isomers of 19:0 and the *cis*-9,10 isomer of 17:0, only the *cis*-11,12 isomer (lactobacillic acid; Fig. 5.1) appears to be a specific (probably aerobic) bacterial marker (Perry et al. 1979); *cis*-9,10 isomers have also been found in some terrestrial plants (Harwood & Russell 1984).

(e) Hydroxy fatty acids

α-Hydroxy acids (i.e. 2-hydroxy, the OH group is on the C atom next to the COOH group) and β-hydroxy acids (i.e. 3-hydroxy, the OH group is situated on the next but one C atom to the COOH group) can be formed as intermediates in the oxidation of monocarboxylic acids, although β-oxidation is more widespread. In contrast to the fatty acids of most Gram-positive bacteria, which comprise mainly *iso* and *anteiso* fatty acids <C_{22} with odd C numbers, those of most Gram-negative bacteria (which include most planktonic bacteria) are generally dominated by the even C numbered, <C_{22}, β-hydroxy acids present in lipopolysaccharides (14:0 often predominates; Harwood & Russell 1984; Volkman et al. 1998). However, β-hydroxy acids in the range C_8–C_{26} have also been found in cyanobacteria (predominantly C_{14} and C_{16} in lipopolysaccharides) and some microalgae (Matsumoto & Nagashima 1984). *Iso* and *anteiso* isomers of both C_{15} and C_{17} β-hydroxy acids are probably bacterial markers. Some freshwater eustigmatophytes produce 26:0 to 30:0 β-hydroxy acids (Volkman et al. 1998).

Hydroxy acids are relatively minor components in fungi but they are the most common substituted fatty acids in higher plants, particularly α-hydroxy acids (Harwood & Russell 1984). Saturated and monounsaturated, C_{22}–C_{26}, α-hydroxy acids are major components in the cell walls of many marine chlorophytes (Volkman et al. 1998). The 9,16- and 10,16-dihydroxy C_{16} and 9,10,18-trihydroxy C_{18} acids that are abundant in cutin can be diagnostic of higher plant contributions to sediments (Cardoso & Eglinton 1983). Hydroxy acids >C_{22} with an ω-OH group (i.e. at the opposite end of the molecule to the COOH group) are also vascular plant markers, derived from suberin (see Section 2.4.2b). ω-Hydroxy acids <C_{22} are not specific and may be derived from higher plants or bacteria. Even C-numbered C_{26}–C_{30} (ω-1)-hydroxy fatty acids have been found in some methylotrophes (Skerrat et al. 1992) and the C_{26} member is produced by some cyanobacteria (Volkman et al. 1998). Higher plants also produce α,ω-diacids, although these diacids can arise from bacterial oxidation of other compounds (see Section 5.3.3).

Mid-chain hydroxy acids are also found. Mosses (Bryopsida, a class of the Bryophyta) contain monohydroxy acids, such as the 7- and 8-hydroxy C_{16} diacids in *Sphagnum* (Cardoso & Eglinton 1983). Some marine eustigmatophytes contain C_{30}–C_{34} ω18-hydroxy acids (Volkman et al. 1998).

5.2.2 Sterols

Sterols are found in both free and bound form (e.g. steryl esters and glycosides) in organisms. The carbon-number distribution of regular (i.e. 4-desmethyl) sterols in young sediments appears to permit a degree of distinction between the contributions of some groups of organisms. Phytoplankton usually contain abundant C_{28} sterols (although diatoms can contain approximately equal amounts of C_{27}, C_{28} and C_{29} sterols), while zooplankton, of which the crustaceans are the dominant class (copepods account for *c.*90% by weight of the zooplankton), often contain abundant C_{27} sterols, particularly cholesterol (cholest-5-en-3β-ol; Fig. 2.22). In contrast, the major sterols in higher plants are C_{29} compounds, β-sitosterol (24β-ethylcholest-5-en-3β-ol) and stigmasterol (24β-methylcholesta-5,22*E*-dien-3β-ol), although campesterol (24β-methylcholest-5-en-3β-ol), a C_{28} sterol, is also often abundant (see Fig. 2.22 for structures). Fungi contain C_{27}–C_{29} sterols but ergosterol (Fig. 2.22), a C_{28} compound, often predominates. Terrestrial and marine invertebrates probably contribute a range of C_{27}–C_{29} sterols.

A triangular plot of C_{27}, C_{28} and C_{29} sterols can be an aid to differentiating marine, estuarine, terrestrial and lacustrine environments (Fig. 5.2), based on the characteristic associations of contributing organisms (Huang & Meinshein 1979). However, this approach to sterol distributions is very simplistic and may not always provide accurate indications of contributing organism groups. For example, the C_{29} compound 24-ethylcholest-5-en-3β-ol is found in higher plants and many unicellular algae (Table 5.1). Nevertheless, the configuration at C-24 can provide a degree of source specificity, because it appears that higher plant-derived C_{29} sterols usually possess a 24β-alkyl configuration (e.g. 24β-ethyl in β-sitosterol and stigmasterol; Fig. 2.22),

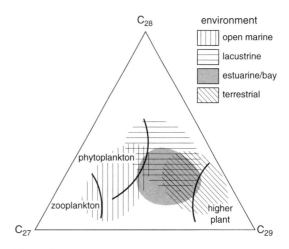

Fig. 5.2 Sterol distributions (expressed as relative amounts of C_{27}, C_{28} and C_{29} components) in relation to source organisms and environments (after Huang & Meinshein 1979).

Table 5.1 Characteristic major sterols in phytoplankton classes (after Volkman et al. 1998). Positions of unsaturation are given by Δ nomenclature (see Box 2.8), 4Me = 4-methylsterols, dinos = dinosterol. The Prochlorophyceae and Cyanophyceae do not appear to synthesize sterols

phytoplankton class	C_{27} Δ^5	$\Delta^{5,22}$	C_{28} Δ^5	$\Delta^{5,22}$	$\Delta^{5,24(28)}$	Δ^7	$\Delta^{7,22}$	$\Delta^{5,7,22}$	C_{29} Δ^5	$\Delta^{5,22}$	$\Delta^{5,7}$	C_{30} $\Delta^{5,24(28)}$	4Me	dinos
Bacillariophyceae	•	•		•	•				•					
Bangiophyceae	•	•					•							
Chlorophyceae				•					•	•				
Chrysophyceae					•				•	•		•		
Cryptophyceae				•										
Dinophyceae	•				•								•	•
Euglenophyceae						•	•	•	•		•			
Eustigmatophyceae	•								•					
Haptophyceae	•			•					•	•				
Prasinophyceae			•	•	•									
Raphidophyceae					•				•					
Xanthophyceae	•								•					

whereas many algae (including the chlorophytes and dinoflagellates) biosynthesize sterols with the 24α-alkyl configuration (Goad et al. 1974; Volkman 1986). Unfortunately, C-24 epimeric pairs are not readily separated by the gas chromatographic techniques routinely employed in sterol analysis (Maxwell et al. 1980; see Box 4.3). In the absence of information on C-24 configuration it can be difficult to determine the precise origin of some sterols in Recent sediments.

Sterol distributions in phytoplankton vary considerably, from dominance of a single component, as in marine eustigmatophytes and some diatoms and haptophytes, to ten or more major sterols, as in some dinoflagellates. Many of the Chlorophyceae contain Δ^7, $\Delta^{5,7}$ and $\Delta^{7,22}$ sterols (see Box 2.8 and Fig. 2.20a for numbering convention), but a few contain mainly 24β-ethylcholesta-5,22E-dien-3β-ol and 24β-methylcholest-5-en-3β-ol (i.e. the 24α(H) epimers of stigmasterol and campesterol, respectively). Cholesterol occurs in variable amounts in most phytoplankton and is sometimes a major component. Epibrassicasterol (24β-methylcholesta-5,22E-dien-3β-ol; Fig. 2.22), which is sometimes called diatomsterol, is generally only a minor sterol in diatoms, and it is also found in other planktonic algae, but either 24-methylcholesta-5,24(28)-dien-3β-ol (Fig. 5.1) or cholesterol is often a major component in diatoms. Either cholesterol or epibrassicasterol is often abundant in haptophycean algae (including coccolithophores). Dinoflagellates appear to be the most important source of C_{30} 4α-methylsterols in sediments, although some lack or contain only minor amounts of dinosterol (4α,23,24-trimethyl-5α-cholest-22-en-3β-ol; Fig. 2.22), and the freshwater species *Ceratium furcoides* has no 4-methylsterols (Robinson et al. 1987). Saturated (5α) sterols are common in dinoflagellates but not in other phytoplankton, apart from some diatoms and raphidophytes (Volkman et al. 1998).

Sterols have been reported in cyanobacteria, with Δ^5 C_{27} and C_{29} usually dominating (Volkman et al. 1998). However, the biosynthesis of sterols by cyanobacteria is debatable (Ourisson et al. 1987; see Section 2.4.3f), and recent studies have shown that the absolute abundance of sterols is usually low and probably attributable to contamination (Summons et al. 2002).

5.2.3 Carbohydrates

The most abundant aldopentoses and aldohexoses resulting from hydrolysis of polysaccharides in sedimentary organic matter are usually lyxose, arabinose, rhamnose, ribose, xylose, fucose, mannose, galactose and glucose (see Fig. 2.6). The distributions of these monosaccharides can be used to differentiate between various higher plant sources and to distinguish marine from terrestrial sources (Cowie & Hedges 1984). However, the variability in carbohydrate composition that can occur in algae, zooplankton and bacteria can make it difficult to differentiate these groups of the plankton.

Marine markers among the above pentoses are ribose and fucose. Ribose is present in many nucleotides and RNA, which are relatively abundant in the metabolically active organisms of the plankton. Fucose is used as a storage sugar by some plankton and bacteria, but it is rarely present in higher plants. In vascular plants structural polysaccharides predominate and the total level of carbohydrates is significantly higher than in marine organisms. Glucose is an important constituent of structural polysaccharides in vascular plants, leading to high glucose:ribose ratios (>50 by weight). This ratio is usually lower in phytoplankton but not always so, because of the extremely variable glucose levels in phytoplankton. However, the amount of ribose or (ribose + fucose) relative to total monosaccharides excluding glucose is usually reliably greater for planktonic than higher plant sources (Cowie & Hedges 1984).

Angiosperms (e.g. grasses and many deciduous trees) and gymnosperms (e.g. conifers) have characteristically different hemicellulose composition. Xylose-containing components are in greater relative abundance in angiosperms, whereas mannose-containing components are more abundant in gymnosperms. A plot of xylose versus ribose, expressed as levels relative to total monosaccharides excluding glucose, can allow distinction of marine, angiospermous and gymnospermous sources (Fig. 5.3a).

Non-woody vascular plant tissue (e.g. leaves and grasses) contains more pectin than woody tissue, resulting in higher relative abundances of arabinose and galactose. A plot of (arabinose + galactose) versus mannose, again evaluated as levels relative to total monosaccharides excluding glucose, allows distinction of woody and non-woody angiospermous tissues and of the corresponding gymnospermous tissues (Fig. 5.3b).

Minor monosaccharides released upon hydrolysis of sediments can also provide source indications. For example, the methoxy (i.e. O-methyl) monosaccharides 3-O-methylxylose and 6-O-methylmannose are abundant in the coccolithophore species *Emiliania huxleyi*. Bacteria synthesize a wider range of these components together with deoxy monosaccharides (Klok et al. 1984a).

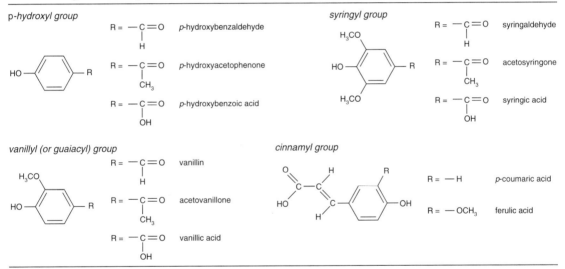

Fig. 5.3 Plots of some monosaccharides (expressed as wt % of total amount of combined lyxose, arabinose, ribose, xylose, rhamnose, fucose, mannose and galactose), showing (a) differentiation of angiospermous and gymnospermous vascular plant sources from total marine plankton, and (b) distinction between different types of higher plant sources (after Cowie & Hedges 1984).

Table 5.2 Lignin oxidation products

5.2.4 Lignins

Lignin is an abundant component of vascular plants, and the phenolic units from which it is synthesized are relatively stable towards chemical alteration but exhibit some compositional variation with plant type. Lignin constituents are, therefore, potentially useful indicators of different types of vascular plant sources. Alkaline oxidation of lignins from vascular plant tissue using copper(II) oxide (e.g. Hedges & Mann 1979; Hedges & Ertel 1982) yields four groups of structurally related products: p-hydroxyphenyl, vanillyl (also called guaiacyl), syringyl and cinnamyl groups (Table 5.2). There are sources of the p-hydroxyphenyl group other than lignins, but the other three groups are usually diagnostic of lignin. Other than some rare exceptions for the cinnamyl group, all members of a group are either present or absent in the oxidation products of a particular vascular plant tissue. The levels of these groups in the woody and non-woody tissues of gymnosperms and angiosperms are presented in Table 5.3.

Table 5.3 Lignin oxidation product distributions in different vascular plant tissues (amounts are expressed as wt % of vascular plant organic carbon; after Hedges & Mann 1979)

plant tissue	syringyl	cinnamyl	vanillyl
non-vascular plants	0	0	0
non-woody angiosperms	1–3	0.4–3.1	0.6–3.0
woody angiosperms	7–18	0	2.7–8.0
non-woody gymnosperms	0	0.8–1.2	1.9–2.1
woody gymnosperms	0	0	4–13

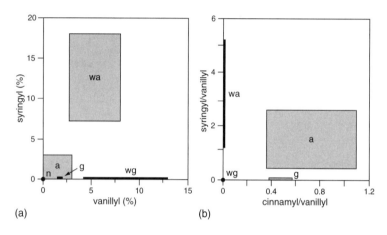

Fig. 5.4 Plots of lignin oxidation product parameters permitting distinction of different higher plant tissues (a = non-woody angiosperm, wa = woody angiosperm, g = non-woody gymnosperm, wg = woody gymnosperm, n = non-vascular plant tissue). Amounts of syringyl and vanillyl components in (a) are expressed as wt % of vascular plant organic carbon (after Hedges & Mann 1979).

Plots of the concentrations of any pair of the three oxidation product groups in Table 5.3 can be made. That of syringyl versus vanillyl generally allows differentiation of the four types of vascular plant tissue and non-vascular plant material (Fig. 5.4a). It is not always simple to determine concentrations relative to the total contribution of organic carbon from vascular plants in a sediment. It is simpler to calculate ratios of the amounts of pairs of oxidation product groups, and the most useful plot of this type in distinguishing the four main types of vascular plant tissues is cinnamyl:vanillyl versus syringyl:vanillyl (Fig. 5.4b).

Similar distinction between lignin types is possible from the monomeric units obtained by pyrolysis (Saiz-Jimenez & de Leeuw 1986). For example, softwood (i.e. woody gymnosperm) lignins yield vanillyl derivatives (mainly coniferaldehyde and coniferyl alcohol), whereas hardwood (i.e. woody angiosperm) lignins are characterized by vanillyl and syringyl units (mainly syringaldehyde, coniferyl alcohol and sinapyl alcohol). The major pyrolytic products from bamboo lignin, which is representative of grass lignins, is p-vinylphenol (vinyl = —CH=CH$_2$).

5.3 Diagenesis at the molecular level

5.3.1 General diagenetic processes

(a) Carbohydrates and lignins

As noted in Section 4.1.2, carbohydrates are relatively labile sedimentary components and readily undergo diagenetic changes; this can affect the interpretation of source indications based on monosaccharide distributions. For example, glucose in its role as an energy storage carbohydrate is readily utilized, whereas bacterial cell wall components are less readily degraded. The lignified material of higher plants is more resistant towards degradation and can protect incorporated carbohydrates. The following order of increasing stability has been noted for hardwoods: hemicellulose < cellulose < pectin < syringyl lignin units < vanillyl and p-hydroxyphenyl lignin units (Hedges et al. 1985). Lignin and polysaccharide composition changes from the inside to the outside of woody cell walls and the middle lamellar region appears to be preferentially preserved, affecting the distribution of preserved components during diagenesis. The different relative stabilities of the polysaccharides are reflected in the relative stabilities of their major constituent monosaccharide units: xylose and mannose in hemicellulose; glucose in cellulose (often referred to as α-cellulose); and arabinose, galactose and galacturonic acid in pectin.

(b) Biomarkers

During diagenesis biomarkers undergo the same main types of reactions as other biogenic organic compounds: defunctionalization, aromatization and isomerization. Oxygen-containing functional groups predominate among the lipid components at the start of diagenesis and their loss involves reactions such as dehydration and decarboxylation. By the end of diagenesis these defunctionalization processes lead to the formation of hydrocarbons, either saturated or aromatic. Non-aromatic, unsaturated hydrocarbons (alkenes) can be formed initially, such as sterenes from stenols, but they are reactive and do not survive in the longer term. They may undergo rearrangement reactions, involving migration of C=C bonds before being either hydrogenated to yield saturated hydrocarbons (alkanes) or, if the C=C bond is in a six-membered cyclic system, dehydrogenated to yield an aromatic system. For example, the diagenetic products of a sterol with a ring C=C bond (e.g. cholesterol) include the steranes, C-ring monoaromatic steroidal hydrocarbons and their rearranged counterparts the diasteroids (see Fig. 4.25).

Biomarkers can undergo **isomerization** (the interconversion of isomers) during diagenesis and catagenesis. A major isomerization process during diagenesis is the migration of double bonds in unsaturated biomarkers, such as sterenes. Configurational isomerization at stereogenic centres (Section 2.1.3) can also occur during diagenesis but is mainly experienced at the higher temperatures associated with catagenesis and the geochemical stage of coalification. Both isomerization processes involve the movement of hydrogen, in the form of either a hydrogen radical (H·, i.e. a hydrogen atom bearing a single electron) or a hydride ion (H⁻, i.e. a hydrogen atom with an additional electron) depending upon whether the hydrogen nucleus departs, respectively, with one or both of the electrons involved in its bonding to a C atom. Mineral surfaces (e.g. clay particles) are believed to play an important role in these processes, which are therefore geochemical rather than microbial.

Epimerization (see Section 2.1.3) converts the single, biologically conferred configuration at a stereogenic centre into an equilibrium mixture of the two possible stereoisomers. The mechanism for the loss and subsequent readdition of hydrogen at an acyclic stereogenic C atom is shown in Fig. 5.5. The intermediate involves a C atom bonded to only three groups and it has a trigonal planar geometry. The hydrogen species can re-enter the structure from either side and so there is an approximately equal probability that the original or the opposite configuration is formed. The same process operates at cyclic stereogenic centres, but the orientation of atoms on the groups surrounding the stereogenic C in the trigonal intermediate may partly hinder addition of H from one side (**steric hindrance**). This leads to a higher probability for the formation of one configuration, which will be present in greater abundance in the final equilibrium mixture. Configurational isomerization is a reversible reaction (see Box 5.2) and so, ultimately, a dynamic equilibrium is achieved between the two possible configurations at a stereogenic centre. The relative proportion of each configuration in the equilibrium mixture depends on the effect of steric hindrance; it is approximately 50 : 50 for acyclic stereogenic centres because steric effects are about the same for both isomers. Isomerization resulting in a 50 : 50 mixture of enantiomers (a racemic mixture) is termed **racemization**.

5.3.2 Lipid diagenesis in the water column

It was noted in Section 3.3.1 that phytoplankton and their remains upon death sink only very slowly, allowing ample opportunity for predation by zooplankton

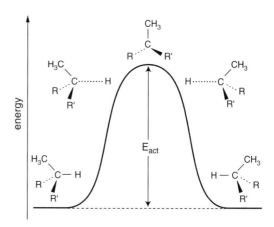

Fig. 5.5 Energy profile for epimerization at an acyclic stereogenic centre, showing the transition from a tetrahedrally bonded C atom to a trigonal planar geometry in the intermediate (E_{act} = activation energy).

Box 5.2 | Reversible reactions and chemical equilibria

As the term implies, reversible reactions can proceed in either direction. An important consideration is the activation energy: for the simple reaction in Fig. 2.12 the activation energy for the reverse reaction is the sum of ΔH and E_{act} (i.e. overall energy change plus activation energy for the forward reaction). If this is large the chance that a product molecule will possess the necessary energy for the reverse reaction to occur is small. However, as the number of product molecules increases and reactant molecules decreases, the probability of the reverse reaction occurring becomes greater, while the probability of the forward reaction occurring diminishes. Eventually, the probability of each reaction occurring is equal, the forward and reverse reactions occur at the same rate and the proportions of reactants and products remain constant. This is a dynamic rather than a static equilibrium, with reactants and products being continuously interconverted.

Isomerization at a stereogenic centre in a biomarker is an example of a reversible reaction in which reactant and product have similar energies. Initially, the concentration of the reactant is high and that of the product is low, so there is greater probability that a reactant rather than a product molecule will undergo hydrogen loss to form the trigonal intermediate. The ratio of product:reactant when dynamic equilibrium has become established reflects the balance between the probability of formation of the intermediate by each isomer and the probability of which configuration is most likely to be formed from the intermediate. Greater steric hindrance between the H atom and other atoms on surrounding groups in one configuration results in a higher energy for that isomer than the other and consequently a lower activation energy for intermediate formation. As a result, the equilibrium favours formation of a greater proportion of the less hindered configuration. For a single acyclic stereogenic centre there is little or no difference in steric hindrance between the two possible configurations, and the hydrogen can attack the intermediate from either side with equal probability, so the forward and reverse reactions have equal activation energies (Fig. 5.5) and a 50:50 mixture results.

and degradation by microbial communities. Sediment traps enable falling debris to be collected (which can be graded using filters of varying mesh size) at different depths in the water column, providing information on the different communities present at various depths and on the transportation of organic matter from the euphotic zone to the underlying sediments, although advection (horizontal transport) can complicate the analysis. Sediment trap experiments have confirmed that only a small fraction of the primary production of the oceans reaches the sediment, with most being recycled in the euphotic zone. Only the larger particulates reach the sediment (generally >50 µm diameter), among which zooplanktonic faecal pellets can make a significant contribution. After removing certain components from their diet of phytoplankton, the zooplankton excrete by-products and unaltered phytoplanktonic remains (see Section 3.3.1). The distribution of lipids reaching marine sediments can, therefore, be significantly modified by zooplanktonic grazing. In experimental feeding of the copepod *Calanus pacificus* with the diatom *Thalassiosora weissflogii* the digestion efficiency increased in the order chlorophyll pigments < total organic C < total N, reflecting the importance of polysaccharides and amino acids as nutrients for the copepod (Cowie & Hedges 1996). The intracellular components of diatoms were also more efficiently digested than cell-wall polymers by copepods.

The role that zooplankton play in the transport of organic matter to sediments is particularly important in the open oceans, where water depth is around 4000 m. The faecal pellets of zooplankton are considerably bigger than phytoplanktonic remains and sink at a substantially faster rate, providing a more efficient system of transport for both labile and refractory organic components to the sediment (Suess 1980). Although living phytoplankton stay at a fairly constant depth in the water column (within the euphotic zone), zooplankton migrate over considerable vertical extents, often feeding near the surface at night and migrating to depth during the day. The migration patterns are different for different species and can also vary with sex and growth stage for the same species (e.g. Barnes & Hughes 1988). Concentrations of cholesterol (the dominant zooplanktonic sterol) in particulates collected by sediment traps may sometimes reflect this behaviour, often reaching a maximum at a depth below the euphotic zone (Gagosian et al. 1982).

(a) Sterols

Crustaceans, such as copepods, cannot biosynthesize all the sterols they need from basic components and so they ingest and convert various algal sterols into components such as cholesterol. The modification of phytoplanktonic lipids in general by zooplankton during early diagenesis can be important. For example, the copepod *Calanus helgolandicus*, when fed on a variety of phytoplankton in laboratory studies, was found to be able to remove all algal hydrocarbons and a significant amount of fatty acids, particularly the polyunsaturated components (Prahl et al. 1984a; Neal et al. 1986). There appears to be some selectivity towards the components that are digested and some phytoplanktonic lipids can reach the sediment unaltered after passing through zooplankton. For example, while *Calanus* seems able to digest most algal stenols, it appears to have a preference for certain positions of unsaturation, such as $\Delta^{5,7}$. In contrast, stanols and Δ^7 stenols appear not to be digested. Stanols are, therefore, likely to be more accurate indicators of phytoplanktonic inputs to sediments than most stenols (Harvey et al. 1987). The fate of Δ^7 stenols is considered in Section 5.5.1.

Coprophagous feeding (ingestion of faecal pellets) is not uncommon among zooplankton and appears to aid the uptake of algal hydrocarbons and polyunsaturated fatty acids by zooplankton. In addition, coprophagy seems to result in the preferential removal of C_{26} and C_{27} zooplanktonic sterols from faecal material relative to C_{28} and C_{29} algal sterols (Neal et al. 1986). Sterenes can be produced in particulate organic matter in the water column (Wakeham et al. 1984a; Wakeham 1987), and they are also found in surface sediments (Gagosian & Farrington 1978).

(b) Chlorophylls

Among the early diagenetic transformations of chlorophylls in the water column immediately below the euphotic zone is the loss of magnesium to give phaeophytins, or the loss of the phytyl ester group to give chlorophyllides. Both processes occur during **senescence**, the period of intercellular changes accompanying cessation of photosynthetic activity prior to death, as phytoplankton sink out of the euphotic zone. However, enzymatic hydrolysis also occurs during zooplanktonic metabolism, and it is difficult to be sure of the extent to which herbivory modifies the imprint of senescence (Spooner et al. 1994). As a rule of thumb, chlorophyllides are the major enzymatic transformation products when senescence dominates, whereas phaeophorbides are the main products of zooplanktonic predation. This general picture is complicated by the different activity levels exhibited by chlorophyllase (the enzyme responsible for chlorophyllide formation), which can vary significantly for species within the same phytoplanktonic genus. High chlorophyllase activity tends to result in rapid formation of chlorophyllides, and their further transformation into phaeophorbides (by Mg and phytyl loss) and/or pyrophaeophorbides (by methyl formate loss from the C_5 ring), whereas low activity usually yields phaeophytins as the major products (Louda et al. 1998). Information on the physiological status of phytoplanktonic populations and the importance of grazing processes can be obtained from these early transformational processes. For example, chlorophyll-*a* levels provide a record of phytoplanktonic biomass and productivity, while phaeophorbide-*a* concentrations in zooplanktonic faecal pellets captured by sediment traps provide information on grazing efficiency.

The above transformations can be considered as catabolic, and they are summarized in Fig. 5.6 for chlorophyll-*a*, but appear equally applicable to chlorophyll-*b* and the corresponding bacteriochlorophylls. The later products, the pyro-pigments, may continue to be formed during early sedimentary diagenesis, as discussed in Section 5.3.3. It appears that steryl esters of pyrophaeophorbides are commonly formed during zooplanktonic herbivory and provide an important pool of sterols and tetrapyrrole pigments in sediments (Harradine et al. 1996; Talbot et al. 1999). The distribution of the esterified sterols probably reflects that of the original phytoplanktonic community better than the free sedimentary sterols. Steryl esters of phaeophorbides have also been identified, but it is not yet known how widespread they are (Riffé-Chalard et al. 2000).

The scheme in Fig. 5.6 is based on reactions in which the chlorophyll-*a* molecule is modified, but it should be remembered that the fate of by far the greater part of these pigments involves oxidative cleavage of the macrocyclic ring, with the tetrapyrroles ultimately being broken down into non-fluorescing, colourless compounds. Catabolic rupturing of the carbon bridge between C-4 and C-5 (see Fig. 2.27 for numbering scheme), yielding 19-formyl-1-oxobilanes (Fig. 5.7), has been observed during senescence, but oxidative ring-cleavage may also occur between C-1 and C-20 (Matile et al. 1996). Most chlorophyll-*a* is destroyed by photo-oxidation in surface waters (Carpenter et al. 1986). The degradation of chlorophylls, and pigments in general, is retarded in the absence of oxygen (Louda et al. 2002).

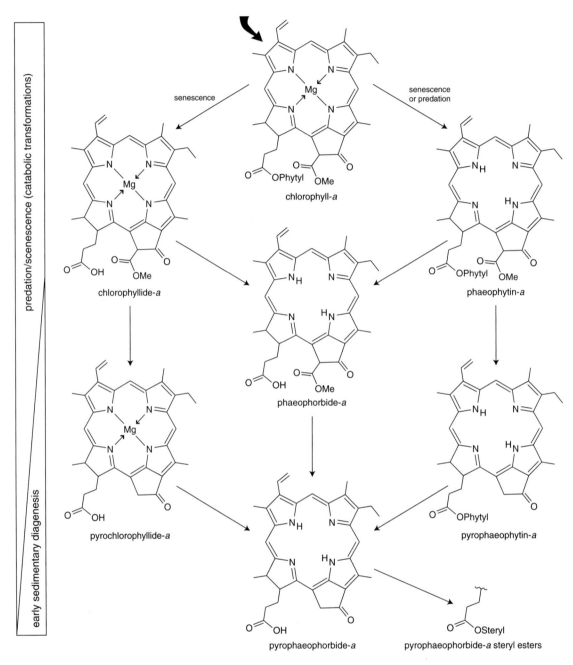

Fig. 5.6 Major early diagenetic (water column and very early sedimentary) pathways for chlorophyll-*a* transformation (after Baker & Louda 1986; Keely et al. 1990; Harradine et al. 1996; Louda et al. 1998; Louda et al. 2000).

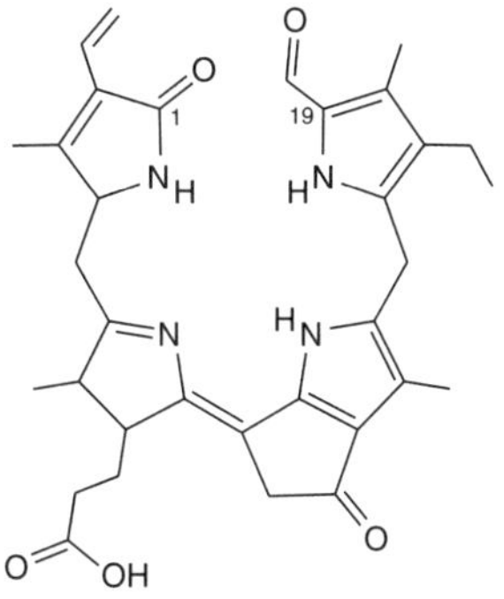

Fig. 5.7 Possible 19-formyl-1-oxobilane structure produced by oxidative cleavage at C-5 of the phaeophorbide macrocycle (after Matile et al. 1996).

Not surprisingly, the phytyl side-chains of chlorophylls suffer a similar fate to the tetrapyrrole units. The photosensitizing properties of the tetrapyrrole units promote the photodegradation of chlorophylls in senescent or dead phytoplankton in surface waters (Rontani 2001), but photodegradation ceases once the tetrapyrroles have been destroyed or the detritus has settled out of the euphotic zone. Autoxidation processes are then important, generally involving addition of ·OH or ·OOH radicals to the phytyl double bond (Fossey et al. 1995). Zooplankton assimilate the phytol side chain of chlorophylls (e.g. Prahl et al. 1984a, b), and so can various herbivorous and coprophagous feeders among the benthonic invertebrates (Bradshaw et al. 1990, 1991). Among the major products are the C_{19} compounds pristane, pristenes and pristanic acid, together with various oxidized C_{20} compounds such as dihydrophytol, phytanic acid and phytenic acids.

(c) Carotenoids

Like chlorophylls, carotenoids can provide information on primary productivity and grazing rates, although the basic carotenoid structure is less stable towards enzymatic degradation during herbivorous grazing than the porphyrin ring system of chlorophylls. For example, fucoxanthin is converted by enzymatic hydrolysis during zooplanktonic metabolism (e.g. by copepods and euphausiids) into fucoxanthinol (see Fig. 5.8), so the relative amounts of these compounds have been used to provide similar information to the ratio of chlorophyll-*a* to phaeophorbide-*a* (Repeta & Gagosian 1982, 1984), although the carotenoid transformation also occurs during senescence (Louda et al. 2002). Diadinoxanthin, which is generally the second most abundant carotenoid in diatoms and fucoxanthin-containing dinoflagellates, disappears rapidly during senescence, whereas diatoxanthin increases in abundance (see Fig. 2.25; Louda et al. 2002).

(d) Bacterial action

In addition to the modification of phytoplanktonic biomarkers by zooplanktonic metabolism, the lipids and other organic compounds in particulate matter are subject to alteration and supplementation by bacterial action (enzymatic hydrolysis being particularly important), although the extent of this action is generally lowest in the more rapidly sinking, larger particulates. The effects of bacterial action in the water column can be considered an extension of those in the sediment, which is considered in the following section. For example, the microbial hydrogenation of stenols to stanols can occur in anoxic sediments and beneath the oxic–anoxic boundary in the water column (where anoxicity develops in bottom waters).

5.3.3 Sedimentary diagenesis of lipids

(a) Fatty acids

As noted in Section 4.1.2, when labile compounds are associated with refractory material they are protected to a degree from the effects of diagenesis. This may account for the apparently greater stability of long-chain fatty acids ($>C_{22}$) from higher plant material, which are abundant in waxy leaf cuticles, compared with the shorter-chain fatty acids ($<C_{22}$) from microorganisms. Unsaturated acyclic components, such as polyunsaturated fatty acids, are degraded relatively quickly, both microbially and chemically, during diagenesis (Wakeham et al. 1984b). The more resistant saturated fatty acids, ω-hydroxy acids and α,ω-diacids, can survive in **ancient sediments** (i.e. sediments of greater than Holocene age).

Fatty acid distributions can be affected by microbial oxidation in sediments, so care must be exercised when certain components are used as source indicators. For example, β-hydroxy fatty acids are produced from β-oxidation of saturated fatty acids. In addition, ω-oxidation of saturated fatty acids to ω-hydroxy acids and of ω-hydroxy acids to α,ω-diacids is performed by yeasts (unicellular fungi, mainly belonging to the ascomycetes) and bacteria. Consequently, while the long-chain ($>C_{22}$) ω-hydroxy acids are reliable indicators of higher plant sources, the short-chain components

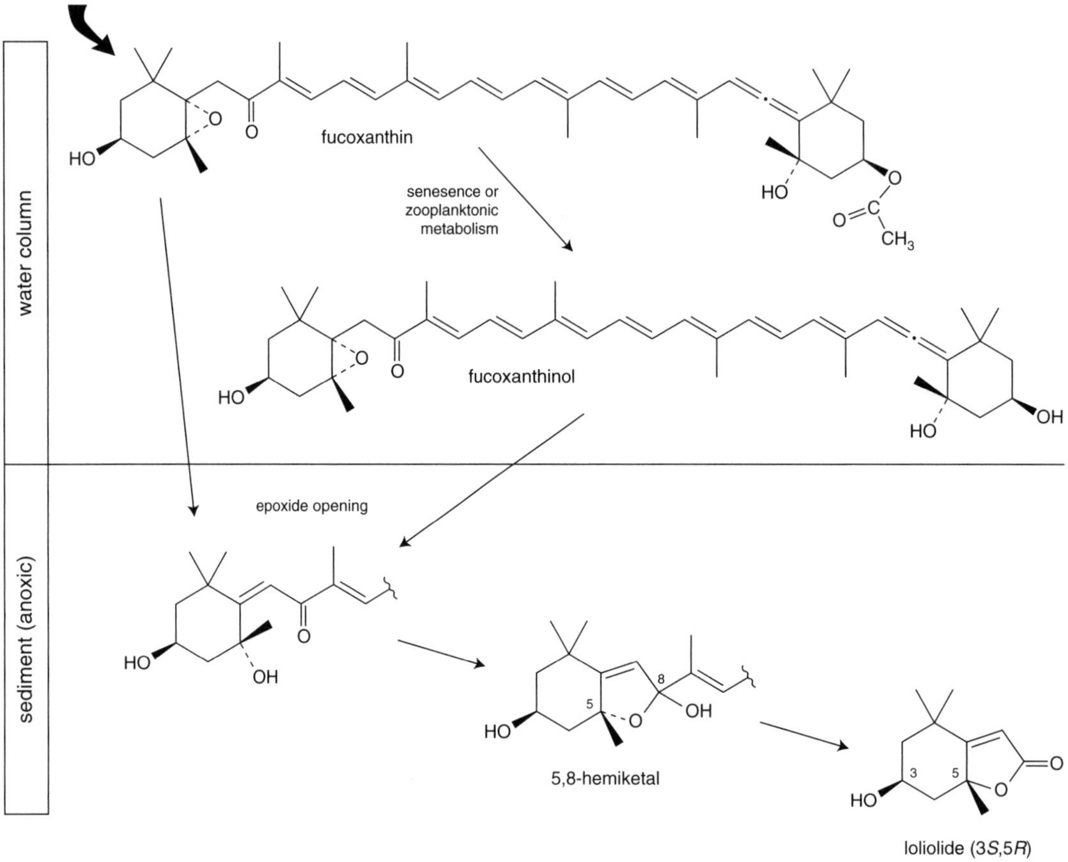

Fig. 5.8 Possible early diagenetic pathways for fucoxanthin under anoxic sedimentary conditions. Only the left side of the molecule is shown for intermediates in loliolide formation by epoxide opening; large arrow indicates biogenic input (after Repeta 1989).

($<C_{22}$) may be either microbial in origin or derived from higher plant suberin and cutin. Similarly, long-chain α,ω-diacids may be derived from vascular plant suberin and cutin or from microbial oxidation of long-chain ω-hydroxy acids, although in either event these diacids derive directly or indirectly from higher plant components. Confirmation of possible higher plant contributions of these components requires comparison with distributions of long-chain n-alkanes (exhibiting an OEP), an unambiguous source indicator.

(b) Tetrapyrrole pigments

The main inputs of chlorophyll-derived compounds to sediments under oxidizing conditions are usually pyrophaeophorbides, due to the activity of zooplanktonic herbivores in the water column and metazoans in the surface sediment. Where surface sediments are anoxic, phaeophytins can dominate, but are subsequently converted to pyrophaeophorbides (Fig. 5.6). Photosynthetic pigments generally do not survive in recognizable form in oxic sedimentary environments, but in anoxic settings the sequence of reactions shown in Fig. 5.9 has been recognized for tetrapyrroles. The pyrophaeophorbides can undergo reduction of the vinyl group at C-3, yielding mesopyrophaeophorbides. Alternatively, in anoxic carbonates, dehydration can result in cyclization, which may ultimately give rise to bicycloalkylporphyrins (Fig. 5.10; Louda et al. 2000).

Reduction (hydrogenation) and loss of oxygen (e.g. by decarboxylation) continue throughout early sedimentary diagenesis, at *in situ* temperatures up to

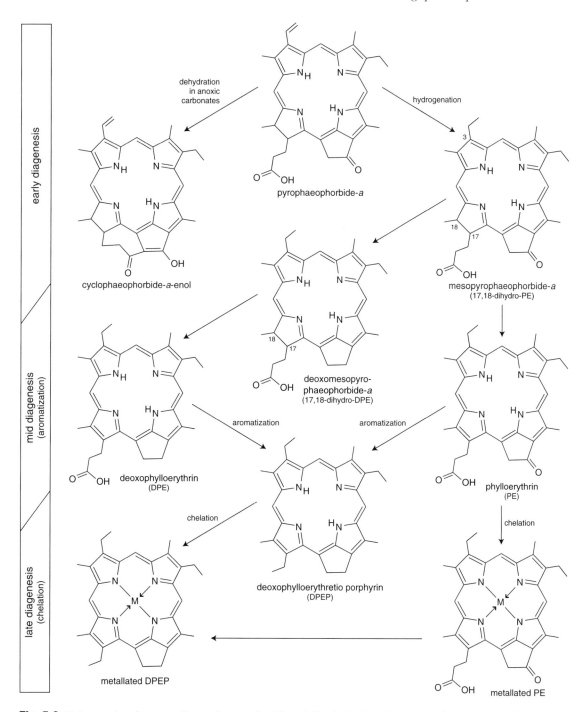

Fig. 5.9 Major anoxic sedimentary diagenetic routes for chlorophyll-*a* derivatives subsequent to those in Fig. 5.6 (after Barwise & Roberts 1984; Baker & Louda 1986; Louda et al. 1998, 2000). M = cations (e.g. Ni^{2+}, VO^{2+}, Cu^{2+}, $GaOH^{2+}$).

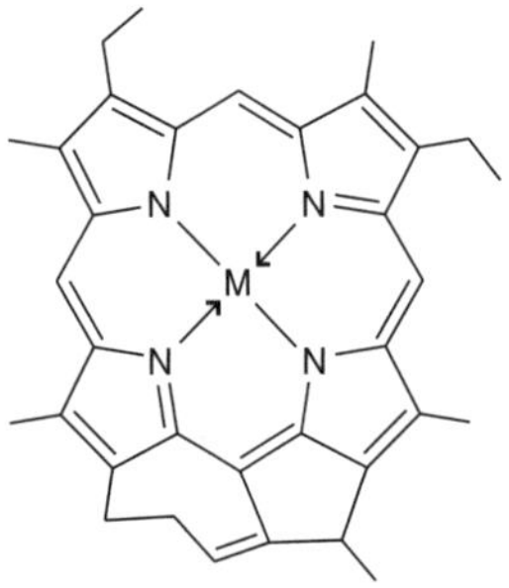

Fig. 5.10 Bicycloalkylporphyrin (BiCAP) possibly formed via a cyclophaeophorbide-enol intermediate (as in Fig. 5.9). M = VO or Zn (after Chicarelli & Maxwell 1986).

$c.25–30\,°C$. As temperature rises during mid-diagenesis the tetrapyrrole system of phorbides becomes fully aromatized, yielding free-base porphyrins, with deoxophylloerythroetio (DPEP) porphyrins usually predominating (Fig. 5.9). The long-term survival of porphyrins is enhanced by metal chelation during late diagenesis, involving cations such as Ni^{2+}, VO^{2+}, Cu^{2+} and $GaOH^{2+}$. Nickel and vanadyl (VO) porphyrins are usually the most common and seem to be formed by different routes. Nickel porphyrins appear to be readily solvent-extractable from sediments, i.e. they are present as chemically free components, whereas the vanadyl porphyrins are generally not extractable, i.e. they are bound to a polymeric matrix (e.g. kerogen; Baker & Louda 1986). Nickel chelation, as measured by the ratio of nickel-complexed porphyrins to total porphyrins, increases with sediment depth during diagenesis, starting at $c.45–55\,°C$ and finishing at $c.60\,°C$. Vanadyl complexation does not appear to follow a similar simple trend, although in immature sediments the distributions of vanadyl DPEP porphyrins more closely resemble the pattern expected from free-base porphyrin distributions than do those of the nickel DPEP porphyrins. The complexity of vanadyl chelation is probably due to both the fact that it occurs later than for other metals (laboratory modelling suggests that complexation occurs in the order Cu, Ni, VO) and the influence of the organic matrix to which the porphyrin is bound. Levels of vanadyl porphyrins increase with rising temperature, continuing into catagenesis. Late diagenesis, therefore, corresponds to the complete chelation of all free porphyrins and the onset of the evolution of vanadyl porphyrins ($c.65\,°C$).

Porphyrins of the DPEP type, with carbon number ranges of C_{26}–C_{38}, predominate during diagenesis in anoxic aquatic sediments and in sapropelic coals (Baker & Louda 1986). However, in humic coals the main porphyrins are the etio type (see Fig. 5.11), $\leq C_{32}$ and complexed with Fe(III), Ga(III) or Mn(III) (Bonnett et al. 1984). The reason for this difference is believed to be that plant porphyrins are largely degraded during the initially aerobic depositional conditions in peat bogs, and those that survive may undergo oxidative cleavage of the five-membered ring that does not contain nitrogen (isocyclic ring in main reaction scheme; Fig. 5.11). However, such cleavage cannot lead to members $>C_{30}$, which are probably generated under anoxic conditions from microbial cytochromes rather than chlorophylls (e.g. haemin in Fig. 5.11; Bonnett et al. 1984). Bacteriochlorophylls may also make a contribution to porphyrin formation. Aquatic sediments and kerogens often contain some etio porphyrins in addition to the DPEP type.

Only free phytol seems to be amenable to biodegradation, so the chlorophyllase-induced release of phytol from chlorophyll is an important process (Rontani et al. 2000). Aerobic bacteria can completely mineralize the phytol, via phytenal, mostly by a combination of β-oxidation and, when a β-methyl group blocks β-oxidation, α-oxidation plus decarboxylation (i.e. β-decarboxymethylation; Fig. 5.12; Rontani & Volkman 2003). Anaerobic bacteria can likewise mineralize phytol, with phytadienes, phytenes and sometimes dihydrophytol (see Fig. 5.33) dominating among the initial degradation products of sulphate reduction (Grossi et al. 1998; Schulze et al. 2001). Anaerobic degradation by denitrifying bacteria, like aerobic degradation, involves β-oxidation and, where there is a β-methyl block, β-decarboxymethylation (Rontani et al. 1999). Particulate associations can hinder certain early biodegradation routes (e.g. involving addition of water; Rontani & Bonin 2000).

(c) Carotenoids

Carotenoids do not generally survive the whole of diagenesis in recognizable form, even under anaerobic conditions. Defunctionalization occurs, as for other lipid groups, but is often accompanied by chain fragmentation and ring opening, which destroy any recognizable link between a sedimentary component and its biological precursor. However, there are some exceptions in which the complete carotenoid structure is retained in the form of a saturated (e.g. lycopane and β-carotane) or aromatic (e.g. isorenieratane) hydrocarbon (Fig. 5.13). Methanogenesis appears to be the dominant oxidation process rather than sulphate

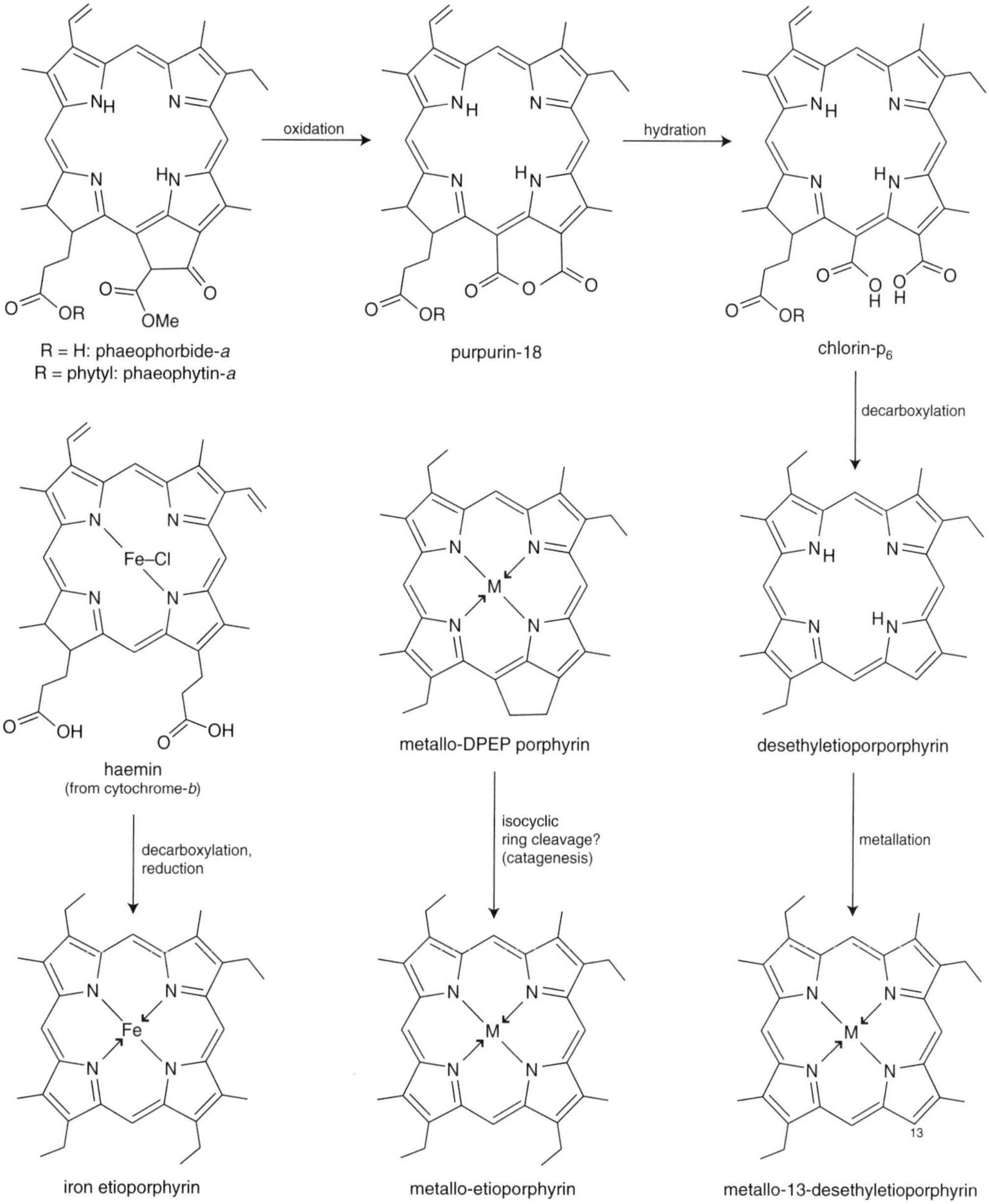

Fig. 5.11 Possible origins of etio porphyrins in sediments and humic coals (after Barwise & Roberts 1984; Bonnett et al. 1984; Baker & Louda 1986; Louda et al. 1998).

reduction in marine and lacustrine sediments containing saturated or partially saturated carotenoids (Repeta 1989).

Occasionally the carotenoid structure is partially preserved, in the form of one cyclic system and part of the isoprenoidal chain. Examples shown in Fig. 5.13 include C_{11}–C_{31} 1,1,3-trimethyl-2-alkylcyclohexanes, which may arise from degradation of β-carotane (Jiang & Fowler 1986), and C_{13}–C_{31} 1-alkyl-2,3,6-trimethylbenzenes, which are probably derived from degradation

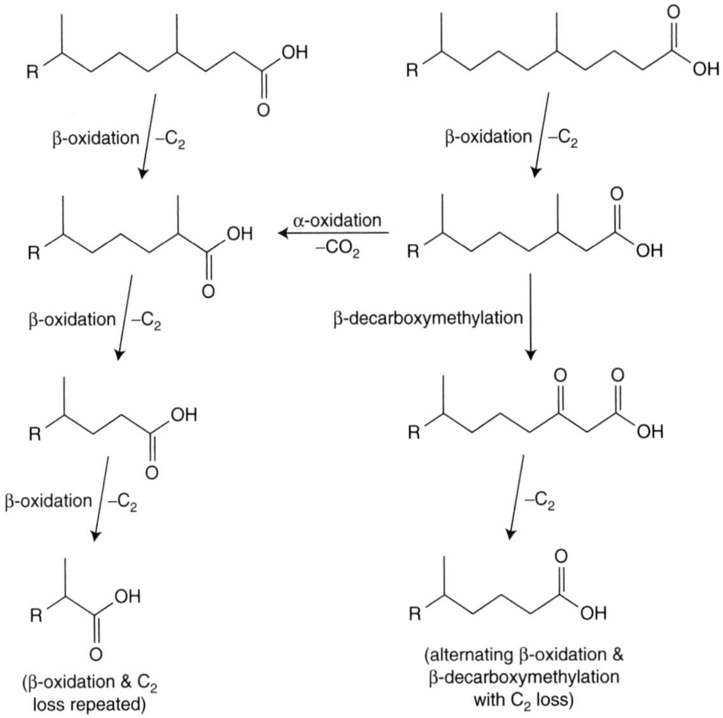

Fig. 5.12 Biological oxidation pathways involved in the degradation of phytyl chains (after Rontani et al. 1999).

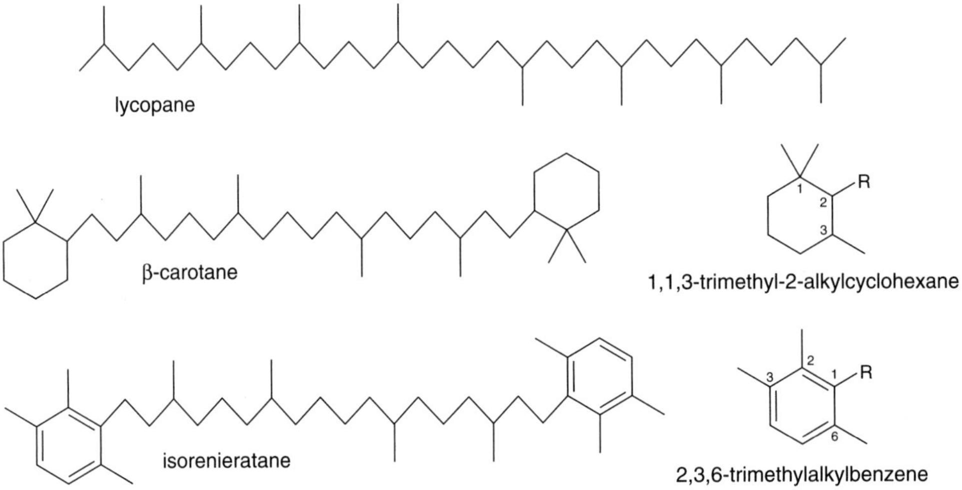

Fig. 5.13 Some diagenetic products of carotenoids (R = isoprenoid chain).

of isorenieratene (Summons & Powell 1987). The position of methyl groups on the ring of aromatic carotenoids can provide information about source organisms. A 2,3,6-trimethyl substitution pattern, as present in chlorobactene (Fig. 5.1) and isorenieratene (Fig. 2.23), is characteristic of the photosynthetic green sulphur bacteria (*Chlorobium*), whereas 2,3,4-trimethyl substitution, as present in okenone (Fig. 2.23), is characteristic of the purple genus (*Chromatium*). Isorenieratene can also undergo a series of internal ring-formation reactions leading to a variety of polycyclic hydrocarbons, which can represent a significantly larger proportion of the early diagenetic products than isorenieratane and the alkyl-trimethylbenzenes (Sinninghe Damsté et al. 2001).

Individual carotenoids exhibit varying stabilities during early sedimentary diagenesis. Degradation rates increase in the order: β-carotene and diatoxanthin < peridinin and diadinoxanthin < fucoxanthin (Repeta 1989). Carotenoids containing 5,6-epoxide groups (i.e. a three-membered ring containing an oxygen atom), such as fucoxanthin and diadinoxanthin, rapidly disappear in anoxic sediments. The 5,6-epoxide group is highly reactive, due to the presence of an oxo group at C-8, and undergoes the chemical epoxide opening shown in Fig. 5.8. This is rapidly followed by fragmentation of a 5,8-hemiketal intermediate to yield loliolide (Repeta 1989). Loliolide has a 3S,5R configuration, as do all epoxide-containing carotenoids. In contrast, carotenoids containing a C=C bond in the 5,6 position (e.g. diatoxanthin, β-carotene, lutein and zeaxanthin) are degraded more slowly in anoxic Recent sediments, via microbially mediated oxidation (Fig. 5.14; Repeta 1989). The initial epoxide-forming step of this microbial oxidation can involve attack by an oxygen atom from either side of the ring system, ultimately yielding equal amounts of the epimers loliolide and isololiolide, where a 3-hydroxy group is present in the original carotenoid (Fig. 5.14). The corresponding oxidation of β-carotene, which lacks an OH group at C-3, yields the 5R and 5S epimers of dihydroactinidiolide in equal amounts (Fig. 5.14; Klok et al. 1984b). Other carotenoids (e.g. acyclics like lycopene, aromatics like okenone, and those containing a 4-oxo group like astaxanthin; see Fig. 2.23) do not undergo such 5,6-epoxidation reactions.

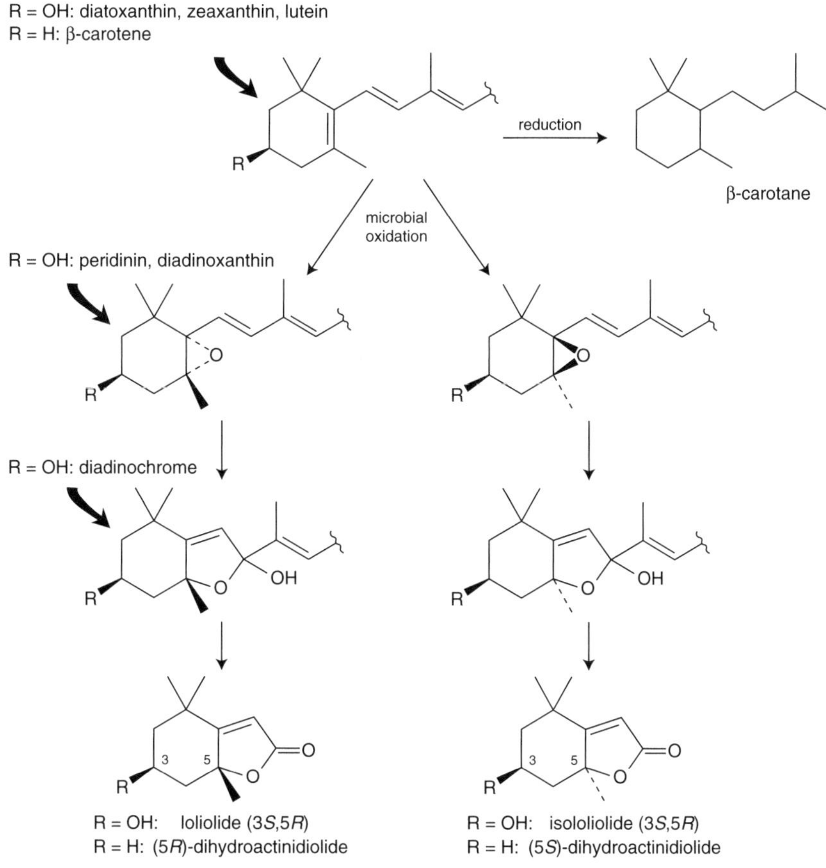

Fig. 5.14 Diagenetic pathways of carotenes and β-cyclic xanthophylls in anoxic sediments (large arrows indicate biogenic inputs; after Repeta 1989).

(d) Reduction of regular steroids

The diagenetic fate of steroids has received much attention (e.g. Mackenzie et al. 1982). As for lipid transformations in general, our knowledge is largely founded on apparent steroid product–precursor relationships in natural systems, such as the similarity of carbon-number distributions in the supposed product and precursor compound classes, and inverse abundance trends in product–precursor compound pairs with increasing burial depth. Where possible, transformations have been confirmed by simulated diagenesis in the laboratory using isolated or synthesized individual compounds (sometimes isotopically labelled). In this way, many diagenetic transformations have been recognized (e.g. de Leeuw & Baas 1986; Peakman & Maxwell 1988), but we restrict our attention to the transformations of some of the more widely occurring steroids in sediments. Δ^5-Stenols are the predominant biogenic sterols and, along with stanols, constitute the major steroidal input to sediments. Free sterols appear to undergo diagenetic alteration much more readily than esterified sterols and so hydrolysis of the latter may be considered the initial step in early diagenetic transformations. The subsequent transformations of free sterols are summarized in Fig. 5.15 for the major products of diagenesis.

Hydrogenation (reduction) of free unsaturated sterols (stenols) to their saturated counterparts (stanols) occurs at an early stage of diagenesis. It is a microbially mediated process, occurring under anaerobic conditions, both in sediments and in particulate matter in the water column (Wakeham 1989). The reduction may proceed via ketone (stanone) intermediates or by direct reduction of the C=C bond and yields predominantly 5α-stanols (Fig. 5.15; Mackenzie et al. 1982), with much smaller amounts of 5β-stanols. The ratio of stanols:stenols, therefore, increases during early diagenesis. However, a contribution of 5α-stanols (e.g. cholestanol) can reach the sediment, resulting from partial heterotrophic transformation of algal sterols and from direct algal inputs. The sterols of phytoplankton contain *c.*5–20% stanols, although the exact amount can vary significantly with growth conditions. For example, the chrysophyte alga *Monochrysis lutheri* does not appear to produce stanols in the exponential growth stage but its stanol content reaches *c.*50% in the stationary phase (Ballantine et al. 1979; see Box 5.8).

The next diagenetic transformation, again under anoxic conditions in the sediment or water column, is dehydration of stanols and stenols to **sterenes**, mainly the Δ^2 (via stera-3,5-dienes), Δ^4 and Δ^5 isomers (Fig. 5.15; van Kaam-Peters et al. 1998). The latter two isomers readily interconvert. There are several divergent reaction pathways leading from sterenes to significantly different products, the major pathway usually being reduction (hydrogenation) to the fully saturated **steranes**, with predominantly the 5α configuration (Fig. 5.15). The configuration at most of the stereogenic carbon centres in the newly formed steranes is unaffected by diagenetic reactions and so remains that inherited from the original biogenic steroid. At the end of diagenesis, therefore, steranes have predominantly a $5\alpha(H), 8\beta(H), 9\alpha(H), 10\beta(CH_3), 13\beta(CH_3), 14\alpha(H), 17\alpha(H), 20R$ configuration. Isomerization reactions involving stereogenic carbon centres bearing a hydrogen atom (Section 5.3.1 and Box 5.2) occur later for steranes.

(e) Rearrangement and aromatization of regular steroids

Rather than reduction, the intermediate Δ^4 and Δ^5 sterenes may undergo various rearrangement reactions, which are apparently catalysed by acidic sites on certain clay minerals (e.g. kaolinite and montmorillonite). The most important rearrangement products are **diasterenes** (Fig. 5.15; Rubinstein et al. 1975; Sieskind et al. 1979), while minor products include spirosterenes (de Leeuw et al. 1989) and steroids that have lost part or all of the A ring (van Graas et al. 1982). These minor products are formed by rearrangement of the cyclic backbone, whereas diasterene formation involves methyl group and C=C bond migrations, but the tetracyclic backbone remains unchanged. Diasterenes are more resistant towards reduction than sterenes and so they survive for longer during diagenesis. They are affected by isomerization at the C-20 position, the initial $20R$ isomer being converted into an equilibrium mixture of $20S$ and $20R$ epimers in approximately equal amounts before reduction to **diasteranes** occurs; which is the converse of the observed sequence for sterenes–steranes. Reduction yields mainly $13\beta,17\alpha$-diasteranes (Fig. 5.15), with smaller amounts of the $13\alpha,17\beta$ isomers.

As an alternative to reduction, biogenic stenols may undergo dehydration to yield diunsaturated hydrocarbons, **steradienes**. Double-bond migration can then occur followed by aromatization, yielding various monoaromatic steroidal hydrocarbons. The major products are C-ring monoaromatics, and by the time they are formed isomerization at the C-20 position has already occurred (Fig. 5.15). The intermediate steradienes in the formation of C-ring monoaromatics have yet to be conclusively identified, but may include the $\Delta^{3,5}$ isomers. Minor aromatic products include A-ring monoaromatic steroids (Hussler et al. 1981) and B-ring

Chemical stratigraphic concepts and tools | 187

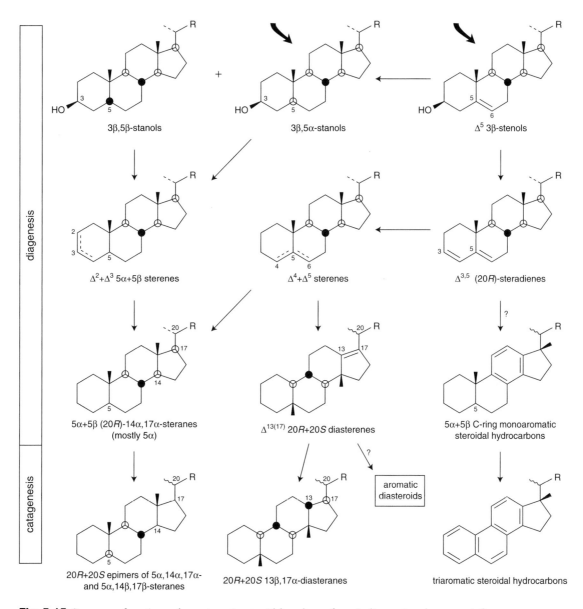

Fig. 5.15 Summary of reaction pathways to major steroidal products of anoxic diagenesis and catagenesis (large arrows indicate biogenic input; after Brassell et al. 1984; Mackenzie 1984; Peakman & Maxwell 1988; de Leeuw et al. 1989; van Kaam-Peters et al. 1998).

monoaromatic anthrasteroids (Hussler & Albrecht 1983), both of which may derive from stera-3,5-dienes by competitive reactions. These steradienes may also undergo rearrangement and/or reduction to yield saturated products (Fig. 5.15).

An important attribute of the diagenetic transformations of steroids is their sensitivity to temperature. For example, diasterene isomerization at C-20 exhibits changes over small depth intervals of around 20 m (Brassell 1985). Therefore, the sequence of diagenetic reactions can be used to determine the extent of diagenesis in an immature sediment. The balance established between the products of the competing reactions occurring during diagenesis is affected by factors such as the geothermal gradient and the availability of mineral surfaces (particularly clays) for involvement in catalysed rearrangements. The products of subsequent catagenetic transformations will depend to a large extent on the hydrocarbon distributions present at the end of diagenesis.

(f) 4-Methylsteroids

In addition to 4-desmethylsterols, 4-methylsterols are abundant in nature and they follow broadly similar diagenetic trends to their demethylated (regular) counterparts, as summarized in Fig. 5.16. The main precursors are 4α-methylsterols, such as those in dinoflagellates (Robinson et al. 1984), which upon dehydration yield various 4-methylsterenes. Reduction of the Δ^2 4α-methyl-5α-sterenes (derived from 4α-methyl-5α-stan-3β-ols) and Δ^4 4β-methylsterenes (derived from Δ^5 4α-methylsten-3β-ols) can then occur, yielding the 20R isomer of predominantly the 5α isomer of the 4-methylsteranes (Rechka et al. 1992). With increasing maturity the relative abundance of the thermally more stable 4α epimer increases, presumably via a hydrogen-exchange isomerization process.

As an alternative to reduction the Δ^4 4β-methylsterenes, via equilibration with their Δ^5 counterparts, may undergo rearrangement, one pathway producing 4-methyl-10α-diaster-13(17)-enes with predominantly a 4β-methyl configuration (Peakman et al. 1992). Like their demethylated counterparts, these 4-methyldiasterenes isomerize at the C-20 position before undergoing reduction to 4-methyldiasteranes, which possess mainly the 13β,17α configuration. Small amounts of 4-methylspirosterenes and their rearranged counterparts are also produced, which can undergo further rearrangements analogous to their 4-desmethyl counterparts. The rearrangement reactions are in competition with aromatization of $\Delta^{3,5}$ 4-methylsteradienes, intermediates in the formation of the Δ^4 and Δ^5 4-methylsterenes (Fig. 5.16).

There is, however, a difference between the behaviours of 4-methyl and 4-desmethyl steroids during diagenesis. Whereas sterenes are produced at an early stage of diagenesis, 4-methylsterenes are formed later. One possible explanation lies in the observation that dehydration occurs more readily for 3α than 3β hydroxy groups, and that although both configurations are found among stanols, to date only the 3β-hydroxy isomers of 4-methylstanols have been detected in immature sediments (Wolff et al. 1986). Consequently, 3α-stanols may provide an early diagenetic source of sterenes, whereas 3β-stanols undergo dehydration at a later stage; which is consistent with the observed more rapid disappearance of 3α-stanols than 3β-stanols from sediments. In contrast, only a later stage formation of 4-methylsterenes is possible from 4-methylstan-3β-ols.

(g) Mono-, sesqui- and diterpenoids

Of the major terpenoidal classes the monoterpenoids are quite volatile and labile, and do not generally survive diagenesis in recognizable form. The sesquiterpenoids are less volatile and can survive diagenesis, undergoing defunctionalization and either reduction (hydrogenation) or, in the case of the cyclic sesquiterpenoids, aromatization, to yield hydrocarbons such as eudesmane and cadalene (Fig. 5.17). The precursors of the cyclic sesquiterpenoidal hydrocarbons are often abundant in higher plant resins, but other sources are possible (Simoneit et al. 1986a).

Diterpenoids can be an important source of the saturated and aromatic hydrocarbons remaining at the end of diagenesis, following the usual defunctionalization and reduction/aromatization reactions. Examples of the diagenetic transformations involving aromatization of some higher plant diterpenoids are shown in Fig. 5.18. The precursor diterpenoids often exist as several isomers, resulting from variations in configuration or C=C bond position. For example, abietic acid has a 10β-methyl and $\Delta^{7,13}$ unsaturation, whereas epiabietic acid has a 10α-methyl group, neoabietic acid has $\Delta^{8(14),13(15)}$ unsaturation and levopimaric acid has $\Delta^{12,8(14)}$ unsaturation. However, upon aromatization these differences vanish, leading to a simplified range of products; for example, retene is formed by all the above abietic acid isomers. Similarly, pimaric acid and its 13α-methyl isomer, sandaracopimaric acid, both yield 1,7-dimethylphenanthrene.

In contrast to aromatization, hydrogenation initially preserves the biologically conferred stereochemistry, as previously observed for steranes. New stereogenic centres can also be generated by the addition of hydrogen across a C=C bond, such as at the C-16 position in phyllocladane (see Fig. 5.18). Reduction of phyllocladane initially leads to a 16α configuration but, with increasing maturity, this is largely converted to the more stable 16β epimer (Alexander et al. 1987).

(h) Triterpenoid reduction and rearrangement

As for steroids, defunctionalization of pentacyclic triterpenoids occurs during early diagenesis and reduction or aromatization later. Some bacterial degradation of triterpenols has even been observed in the leaves of

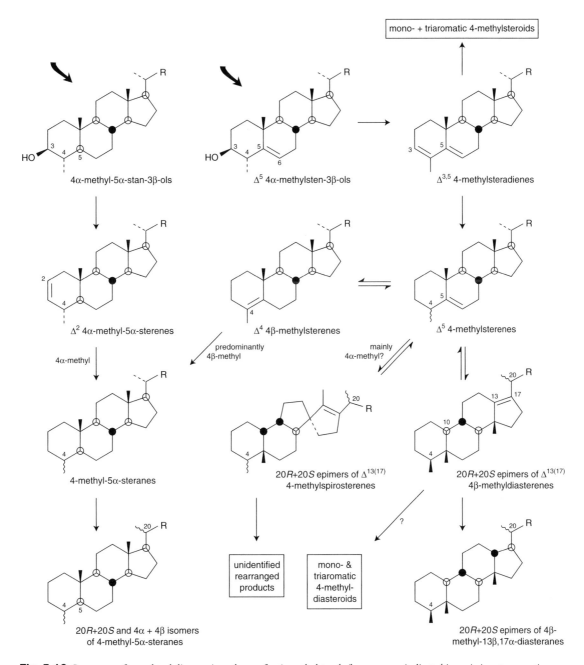

Fig. 5.16 Summary of postulated diagenetic pathways for 4-methylsterols (large arrows indicate biogenic input; aromatic steroid formation as for regular steroids in Fig. 5.14; after de Leeuw et al. 1989; Peakman et al. 1992; Rechka et al. 1992).

living plants (Killops & Frewin 1994). Higher plant triterpenoids often contain a ring C=C bond, which can migrate around the pentacyclic skeleton. For example, the triterpenes formed from taraxerol and β-amyrin rapidly isomerize to 18β-olean-12-ene and then to a mixture of predominantly the $\Delta^{13(18)}$ and 18α Δ^{12} isomers, with smaller amounts of Δ^{18}, as shown in Fig. 5.19 (Rullkötter et al. 1994). Subsequent reduction

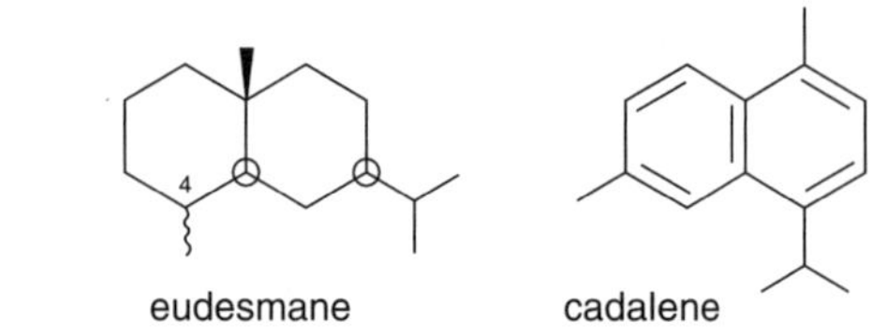

Fig. 5.17 Examples of bicyclic sesquiterpenoidal hydrocarbons.

yields 18α-oleanane together with smaller amounts of its 18β counterpart. Little taraxerane, the direct reduction product of taraxerol, is found in sediments. Instead, the intermediate taraxer-14-ene appears predominantly to undergo isomerization to olean-12-ene, finally yielding oleanane (Fig. 5.19; Rullkötter et al. 1994).

In laboratory dehydration experiments, 3α-triterpenols are found to yield Δ^2-triterpenes, whereas 3β-triterpenols form A-ring contracted triterpenes.

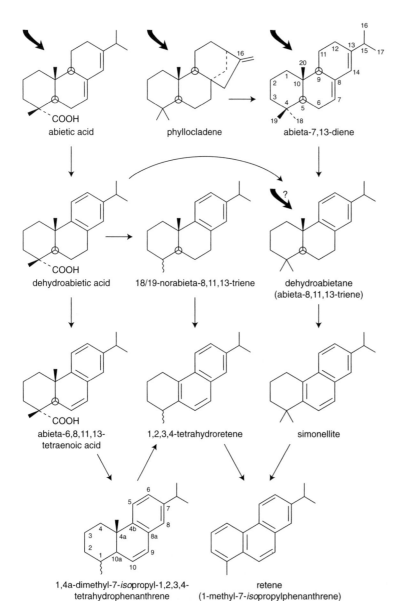

Fig. 5.18 Possible diagenetic pathways for the formation of retene from diterpenoidal resin components (large arrows indicate biogenic inputs; after Simoneit 1977, 1986; Simoneit et al. 1986a; Alexander et al. 1987). Note different number schemes for abietoids and phenanthrenes. (18-norabieta-8,11,13-triene is also known as dehydroabietin.)

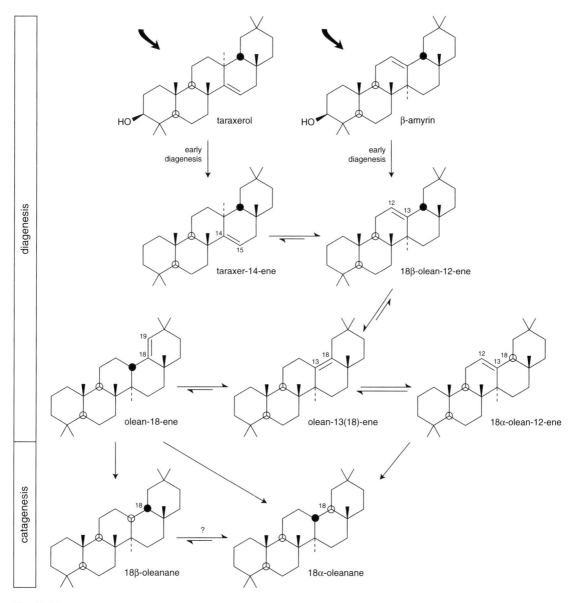

Fig. 5.19 Formation of oleanane from higher-plant triterpenoids (large arrows indicate biogenic inputs; after ten Haven & Rullkötter 1988; Ekweozor & Telnaes 1990; Rullkötter et al. 1994).

However, the dominant natural triterpenols have the 3β configuration, but Δ^2-triterpenes dominate in sediments, suggesting that another mechanism operates, possibly bacterially mediated, favouring formation of Δ^2-triterpenes (ten Haven et al. 1992). A possible reaction scheme is shown in Fig. 5.20.

Hopanoids (alkenes, acids, ketones and alcohols) are ubiquitous components of sediments and soils, and are more widespread than higher plant derived triterpenoids. Polyhydroxy-bacteriohopanes are major precursors, yielding ketones and carboxylic acids during early diagenesis. The latter are the chief products, with

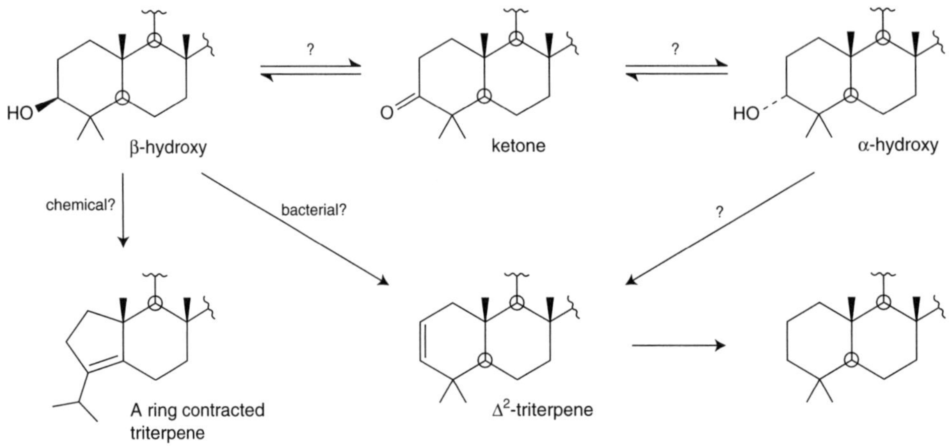

Fig. 5.20 Possible diagenetic pathways for the transformation of 3α- and 3β-triterpenols (after ten Haven et al. 1992).

C_{31}–C_{33} components occurring widely in Recent sediments. These acids are more resistant towards alteration than other oxygenated hopanoids during diagenesis. Often the C_{32} acid is dominant (31,32-dihomohopanoic acid; n = 2, Fig. 5.21), as in *Sphagnum* peats (Quirk et al. 1984) and soils (Ries-Kautt & Albrecht 1989). Diploptene is also often a major hopanoidal component in soils. As shown in Fig. 5.21, reduction of the hopanoidal acids and polyols yields **hopanes** with a 17β,21β configuration. Where the R group is C_2H_5 or larger (i.e. C_{31}–C_{35} components) in Fig. 5.21 there is an additional stereogenic centre at C-22, which is biosynthesized with the *R* configuration. Commonly, C_{29}–C_{35} hopanes are produced, arising from fragmentation of the *n*-alkyl chain attached at C-22 on the E ring of the precursor polyols (Ensminger et al. 1973). A C_{27} component is also produced by complete loss of the side chain, but a C_{28} product is only rarely formed.

The 8(14) bond in the C-ring of pentacyclic triterpenoids generally appears to be susceptible to acid-catalysed rupture. The **8,14-secohopanes** produced from hopanoids by this bond cleavage are shown in Fig. 5.21. Similar **8,14-secotriterpanes** can be formed from higher plant and other triterpenoids, the reaction appearing to be favoured by the presence of an oxygenated functional group at C-3 in the precursors. Another group of triterpanes, the onoceranes, lacks the 8(14) bond, but this is not due to cleavage during diagenesis. The plant-derived precursor compounds, onocerins (see Fig. 2.19), are biosynthesized without an 8(14) C–C bond, so simple reduction yields the onoceranes.

The 11(12) bond in all the above C-ring-opened compounds appears to be relatively weak and can undergo cleavage (Schmitter et al. 1982). Drimane, a sesquiterpane, can be produced from the A,B-rings fragment of secohopanes and secotriterpanes upon breaking of the 11(12) bond (Alexander et al. 1984a). However, the presence of drimane in sediments is not conclusive of such fragmentation reactions because it may also be produced from reduction of drimenol (Fig. 5.21). Complete C ring cleavage of higher plant triterpanes can also generate a range of dicyclic sesquiterpanes, similar to drimane, from the D,E-rings fragment.

(i) Triterpenoid aromatization

As with the steroids, triterpenoids may undergo aromatization rather than reduction during diagenesis. Higher plant triterpenoids, such as α-amyrin (Fig. 5.22), can ultimately form fully aromatized pentacyclic or tetracyclic hydrocarbons (polymethylpicenes or polymethylchrysenes, respectively). The latter are formed by degradation of the A ring, which appears to be a common process and, as mentioned in the previous section, is expected during chemical dehydration of 3β-triterpenols (ten Haven et al. 1992). Regardless of the initial position of any C=C bond, aromatization of higher plant triterpenoids appears to begin in the A ring upon loss of the oxygenated functional group at C-3, and progresses sequentially through the B, C, D and E rings (Laflamme & Hites 1979).

In contrast, hopanoidal precursors have no removable functional group at C-3 and aromatization begins

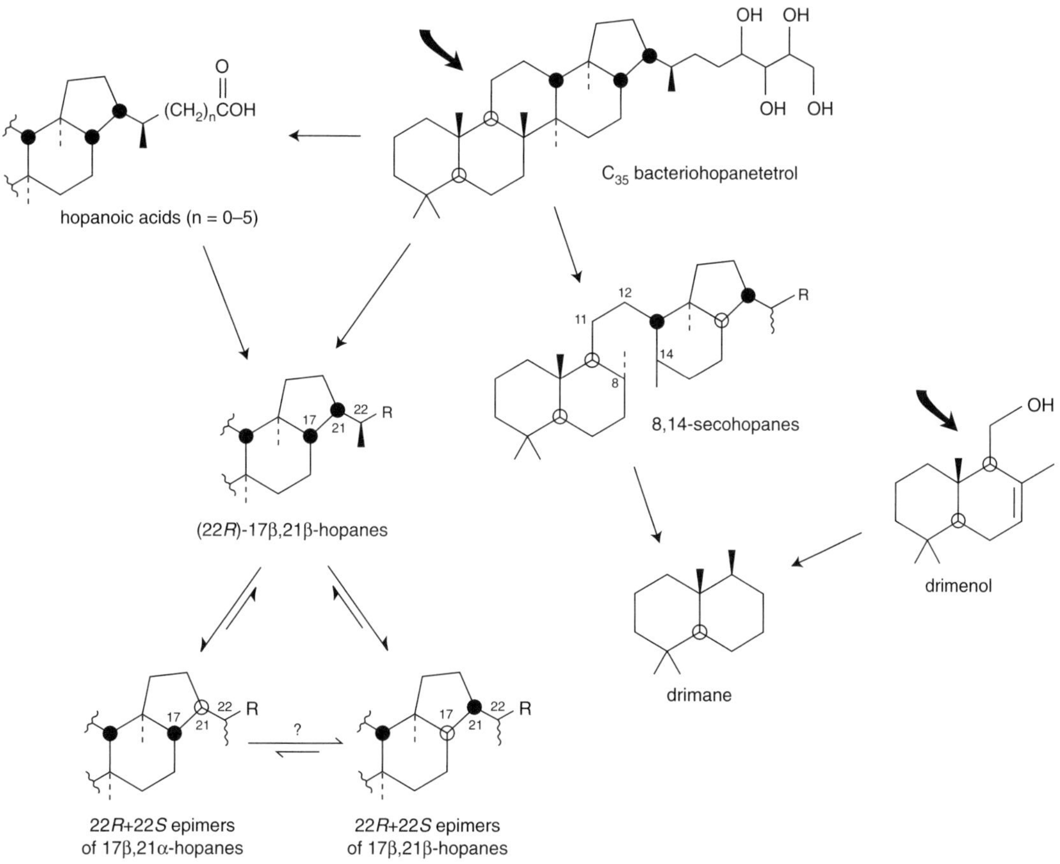

Fig. 5.21 Examples of pathways to saturated hydrocarbons from bacteriohopanetetrol (large arrows indicate biogenic input). The A, B and C rings of hopanes and hopanoic acids are omitted but are identical to those of the C_{35} tetrol precursor (R = H to C_6H_{13}; after Schmitter et al. 1982; Mackenzie 1984; Brassell 1985).

in the D ring and progresses through the C, B and A rings (Greiner et al. 1976). This sequence seems consistent with the C=C bond isomerization that has been observed for diploptene (Fig. 5.22). Unsaturated hopanoidal hydrocarbons, or **hopenes**, are common in Recent sediments, particularly the $\Delta^{17(21)}$ and $\Delta^{13(18)}$ isomers. The latter is the most stable, and would be anticipated to yield a D ring monoaromatic species, which may then undergo further aromatization from the C ring through to the A ring. The major aromatic hopanoidal products have an ethyl (C_2H_5) group attached to the E ring, rather than the *iso*propyl (i-C_3H_7 or $CH(CH_3)_2$) that would be expected from diploptene, and it remains to be established whether these compounds are derived from diploptene by loss of a methyl group from the side chain, as suggested in Fig. 5.22. Aromatic hopanoids with the same carbon-number distributions as the hopanes are sometimes observed, although they are usually minor components. They must arise from the same precursors as the hopanes.

Aromatization can also occur in 8,14-secotriterpenoids during diagenesis (de las Heras et al. 1991) and, as for the saturated counterparts, complete cleavage of the C ring can occur (Killops 1991). Examples of these processes are shown in Fig. 5.23 for β-amyrin and hopanoids. D ring monoaromatic 8,14-secohopanoids are common in sediments and oils, although usually only as trace components (Nytoft et al. 2000). However, they are often abundant in carbonates, along with a series of C_{32}–C_{35} benzohopanes (Fig. 5.23; Hussler et al. 1984). Both series of compounds probably originate

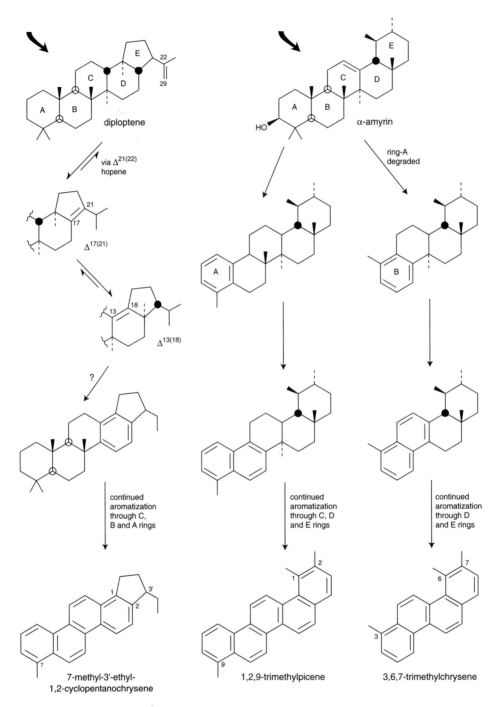

Fig. 5.22 Examples of triterpenoid aromatization (large arrows indicate biogenic inputs). For $\Delta^{17(21)}$ and $\Delta^{13(18)}$ hopenes only D and E rings are shown; A, B and C rings are identical to those in diploptene (after Greiner et al. 1976; Tan & Heit 1981; Chaffee & Johns 1983; Simoneit 1986; Hayatsu et al. 1987).

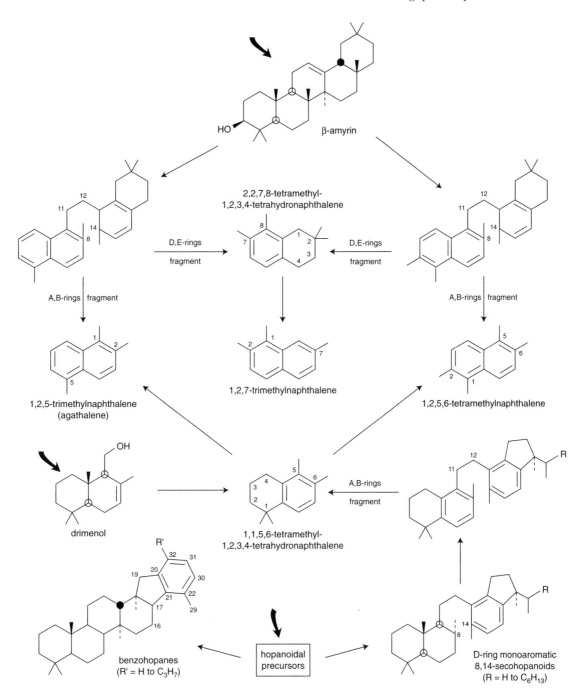

Fig. 5.23 Examples of the formation of polymethylnaphthalenes from terpenoidal precursors (large arrows indicate biogenic inputs; after Chaffee et al. 1984; Hayatsu et al. 1987; Püttmann & Villar 1987; Strachan et al. 1988).

from the same C_{35} bacteriohopanoidal precursors, but the benzohopanes seem to be formed at an earlier stage of diagenesis by reactions involving cyclization of the side chain on the E ring (Hussler et al. 1984).

The initial product from the A,B-rings fragment upon complete C-ring cleavage in aromatic secohopanoids is 1,1,5,6-tetramethyl-1,2,3,4-tetrahydronaphthalene, which can also be formed from the sesquiterpenoid drimenol (Püttmann & Villar 1987). This hydrocarbon becomes fully aromatized with increasing maturity, yielding 1,2,5-trimethylnaphthalene (agathalene) and 1,2,5,6-tetramethylnaphthalene (Fig. 5.23). Formation of the latter requires a methyl group to move between adjacent carbon atoms (termed a **1,2-methyl shift**), which is not an uncommon geochemical process (Püttmann & Villar 1987). During aromatization a methyl group at C-1 must either be lost or migrate to an adjacent C atom (where it replaces an H atom), because the C-1 atom can only be bonded to three other atoms or groups upon aromatization. Similarly, methyl groups cannot remain at ring junctions during aromatization and in Fig. 5.23 it can be seen that they are generally lost (rather than undergoing migration), as in the formation of 1,1,5,6-tetramethyl-1,2,3,4-tetrahydronaphthalene from drimenol.

Aromatic secotriterpenoids of higher plant origin give the same two polymethylnaphthalene products from the A,B-rings fragment as their hopanoidal counterparts (Fig. 5.23). There is a slight difference in the mechanism of formation in that the 1,2-methyl shift is believed to occur at the same time as loss of the OH group at C-3 and while the 11(12) bond of the higher plant secotriterpenoid is still intact (Püttmann & Villar 1987). A further polymethylnaphthalene can arise from the D,E-rings fragment of aromatic triterpenoids with a six-membered E ring, its methyl substitution pattern being characteristic of the structural type of triterpenoid from which it is derived. In Fig. 5.23 this compound is 1,2,7-trimethylnaphthalene, deriving from β-amyrin (Strachan et al. 1988). Angiosperms are often rich in β-amyrin and so large amounts of 1,2,7-trimethylnaphthalene relative to other polymethylnaphthalenes can be a good indicator of an angiospermous input in immature sediments. At higher maturity levels, typical of catagenesis, this source indicator tends to be overwhelmed by larger amounts of non-specific polymethylnaphthalenes expelled from kerogen, so the input signature is masked.

The greatest abundances of higher plant derived aromatic triterpenoids and their degradation products are, not surprisingly, found in brown coals. However, aromatic hopanoids and related compounds can also be abundant in coaly sediments, demonstrating the importance of bacterial activity during diagenesis.

The diagenetic fate of hopanoids is less well understood than that of steroids. However, the formation of hopanes, aliphatic and aromatic secohopanoids and dicyclic sesquiterpanes from the complete cleavage of the C ring may represent an array of competing reactions that parallel the transformations of steroids. There is even an apparent rearrangement of hopenes yielding diahopanes (15α-methyl-27-nor-17α-hopanes), analogous to diasteranes (Fig. 5.24; Killops & Howell 1991; Moldowan et al. 1991).

5.4 Source and environmental indicators in ancient sediments and oil

5.4.1 Introduction

By the end of diagenesis the biogenic organic material in sediments either has been degraded and recycled by microorganisms or has been largely converted into insoluble polymeric material (kerogen or brown coal). The most useful molecular source indicators at this stage of maturity in ancient sediments are biomarkers, which are mostly in the form of hydrocarbons. Some source-related information is lost with the disappearance of functional groups (including C=C bonds) and their associated stereochemistry. As seen in the previous section, it is possible for a number of biogenic precursor molecules to yield the same hydrocarbon at the end of diagenesis (e.g. drimane can be formed from 8,14-secotriterpanes and drimenol; Fig. 5.21). However, in cyclic hydrocarbons, both aliphatic and aromatic, the number of carbon atoms and the position of substituents on the basic carbon skeleton can still convey some source-related information. In addition, the biologically conferred configuration at most stereogenic centres is conserved during diagenesis, although isomerization does occur with increasing thermal maturity.

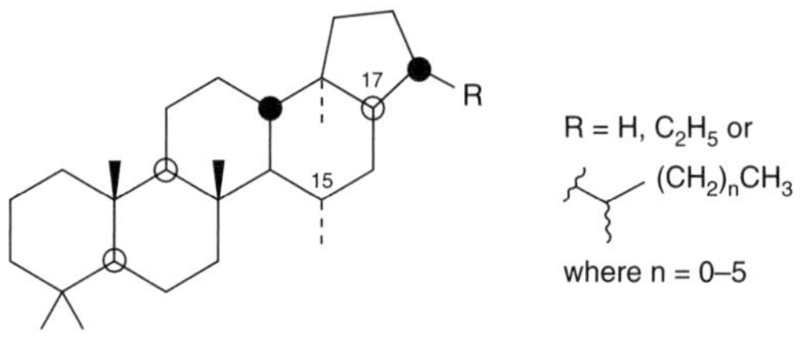

Fig. 5.24 C_{27}–C_{35} 17α-diahopanes (after Moldowan et al. 1991).

Caution must be applied when using biomarker distributions in bitumen extracted from ancient sediments to determine the contributions from different groups of organisms and the environments in which they are likely to have lived. Biomarkers represent only a small and variable fraction of the original biomass of the contributing organisms, so only indications of relative contributions is possible. In addition, the distribution of biomarker hydrocarbons can be distorted by preferential sulphurization (with H_2S or polysulphides; see Section 4.4.1c) of certain of the precursor functionalized lipids during early diagenesis, and their consequential removal from the free hydrocarbon pool (Kohnen et al. 1991a). The resulting organic sulphur compounds (OSCs), which may be incorporated into the kerogen matrix or may exist as free compounds within the bitumen, can provide much valuable information about ancient ecosystems (Kohnen et al. 1992; de Leeuw et al. 1995).

All the source-related biomarkers, and indeed biomarkers related to depositional environments, can provide a useful means of correlating oils with their likely source rocks, provided the effects of differences in maturity are allowed for. Because secondary migration can occur over significant distances (see Section 4.5.5b) it is possible, in a given exploration area, to have a number of reservoirs containing different types of oils and several potential source rocks. It is important to determine which source rocks have given rise to which oils, in order to obtain information about migration routes and the timing of oil generation in relation to trap formation. Biomarkers are the best tools for such correlation studies.

5.4.2 Source-related biomarkers

(a) Normal and methyl-branched alkanes

In Section 5.1.2 it was noted that some acyclic alkanes are produced directly by organisms (although usually at lower levels than fatty acids) and so the usefulness they may have as source indicators in Recent sediments can be retained in older sediments that have not reached the stage of catagenesis when cracking of alkyl chains occurs. For example, the mid-chain methyl-branched C_{17} alkanes 7- and 8-methylheptadecane are characteristic of cyanobacteria (Han & Calvin 1969; Shiea et al. 1990). Although normal (i.e. straight-chain), *iso* and *anteiso* alkanes are widely distributed in organisms, the presence of an odd-over-even predominance (OEP) in carbon number can provide source information. In particular, an OEP for *n*-alkanes >C_{22} is characteristic of higher plant waxes (Eglinton & Hamilton 1967), whereas phytoplankton are generally characterized by odd-numbered *n*-alkanes <C_{22} (Blumer et al. 1971). During catagenesis, however, such *n*-alkane source indicators become less distinct with the generation of large amounts of non-specific *n*-alkanes from kerogen.

The acyclic alkanes produced by defunctionalization and reduction of fatty acids retain some source-related information by way of their carbon numbers and positions of any branch points in the alkyl chains. For example, the OEP observed in higher plant wax *n*-alkanes (>C_{22}) is augmented by the decarboxylation (i.e. loss of one C atom) of wax-derived fatty acids with an EOP.

Most Ordovician petroleum source rocks are characterized by accumulations of the marine colonial microorganism *Gloeocapsomorpha prisca* (Fowler 1992), which appears to be an alga (Derenne et al. 1992). The fossilized cell walls contain a highly aliphatic macromolecular material based on *n*-alkylresorcinol units, although it is not clear whether this material was present as algaenan in the living organism or was formed by diagenetic polymerization of C_{21} and C_{23} *n*-alkenylresorcinols (Fig. 5.25; Blokker et al. 2001). Whatever the origin, bitumen and oils derived from the resulting kerogen are characterized by abundant *n*-alkanes and *n*-alkylcyclohexanes up to C_{19} with an OEP (Hoffmann et al. 1987; Douglas et al. 1991; Fowler 1992).

(b) Acyclic isoprenoids

Acyclic isoprenoidal alkanes occur widely in organisms but there are some source-specific examples. The biphytanyl ethers characteristic of some members of the archaebacteria (e.g. methanogens and thermoacidophiles) can be preserved as a series of long-chain (>C_{20}) isoprenoidal alkanes (e.g. C_{38} alkane, Fig. 5.26), which retain the head-to-head linkage where the two phytanyl units join together (Moldowan & Seifert 1979; Albaigés 1980). The C_{25} component 2,6,10,15,19-pentamethyleicosane appears to be diagnostic of methanogens (Schouten et al. 2001). Lycopane (Fig. 5.13), a saturated C_{40} isoprenoidal alkane, is also likely to be a bacterial marker, deriving from reduction of lycopene (Fig. 2.23). However, squalane (Fig. 5.26), a saturated C_{30} isoprenoidal alkane, may represent a direct archaebacterial input (Brassell et al. 1981), or it may derive from diagenetic reduction of squalene (Fig. 2.21), which occurs in a variety of organisms.

Botryococcane, a saturated C_{34} isoprenoidal alkane, is a particularly useful source- and environment-specific indicator. It appears to be derived only from the alga *Botryococcus braunii*, which is widely distributed in

Fig. 5.25 *N*-alkylresorcinol unit and partial structure for the algaenan of *Gloeocapsomorpha prisca* in kukersites (after Blokker et al. 2001).

freshwater lakes and often forms large mats. Three races of this organism are known, in each of which it produces different hydrocarbons (Grice et al. 1998). Race B produces the characteristic C_{34}, C_{36} and C_{37} alkenes (botryococcenes) containing six double bonds, of which the C_{34} component in Fig. 5.26 is the most abundant and undergoes complete hydrogenation during diagenesis to yield botryococcane (Maxwell et al. 1968; Metzger et al. 1985). In contrast, in race A odd numbered C_{25}–C_{31} *n*-alkadienes and *n*-alkatrienes domi-

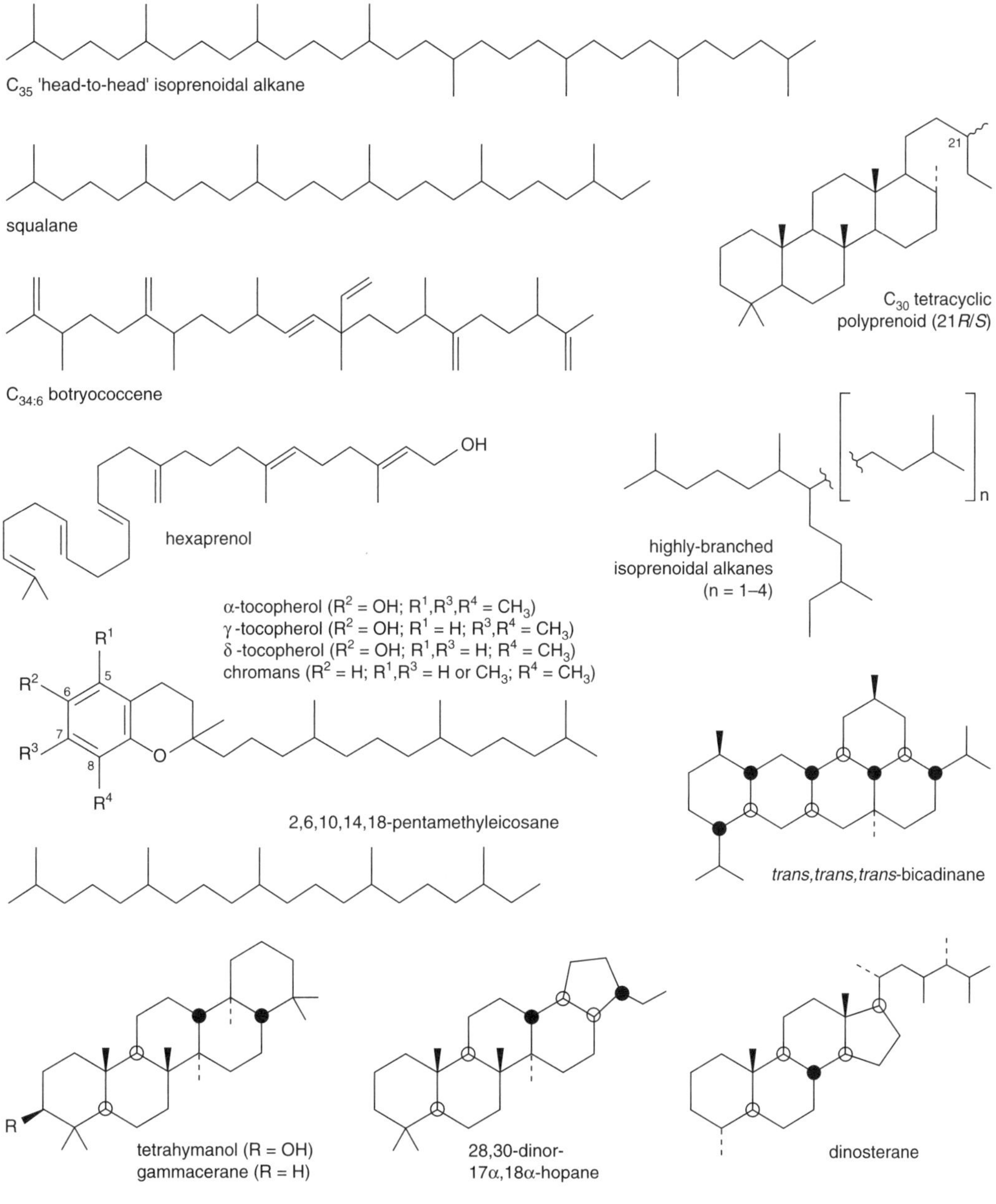

Fig. 5.26 Some sedimentary biomarker hydrocarbons and precursors.

nate (Knights et al. 1970), whereas race L produces lycopa-14E,18E-diene (Metzger & Casadevall 1987). Botryococcane is the dominant aliphatic component in some oils from Sumatra (Moldowan & Seifert 1980) and South Australia (McKirdy et al. 1986), which suggests that the oils are sourced largely by the remains of *Botryococcus* deposited in freshwater lakes (see Section 4.4.3a). The ancestors of race A may have been major contributors to some torbanites (Derenne et al. 1988).

Highly branched C_{20}, C_{25}, C_{30} and C_{35} isoprenoidal alkanes and their alkene counterparts with 1–4 double bonds (Fig. 2.18) occur widely in Recent marine and lacustrine sediments, and they can be abundant (Robson & Rowland 1986; Rowland & Robson 1990; Hoefs et al. 1995). Their main source is likely to be phytoplankton, probably diatoms (see Section 2.4.3d), which produce the corresponding polyunsaturated precursors such as the C_{25} components shown in Fig. 2.18. The di-, tri- and tetraunsaturated alkenes disappear rapidly from older sediments, possibly as a result of sulphurization (Sinninghe Damsté & de Leeuw 1990).

(c) Microbial tricyclic and tetracyclic alkanes

Extended tricyclic alkanes with isoprenoidal side chains (C_{19}–C_{30}; Fig. 4.25), which can be named after the C_{25} member 13β,14α-cheilanthane (a sesterterpane), are common in oils and ancient sediments and they may be microbial in origin (Aquino Neto et al. 1983). The hexaprenol in Fig. 5.26, a common eubacterial and archaebacterial component, seems a likely precursor (Ourisson et al. 1982). Occasionally, the cheilanthane series has been found to extend to C_{54} (de Grande et al. 1993), suggesting that hexaprenol either undergoes further polymerization reactions or is not the sole precursor.

Tetracyclic alkanes, generally in the range C_{24}–C_{27} (Fig. 4.25), are also frequent constituents of oils and bitumens. They are structurally related to the hopanes, from which they may be formed by thermal or bacterial cleavage of the 17(21) C–C bond in the E ring, although a direct bacterial source cannot be discounted. They can be called de-E-hopanes, or 17,21-secohopanes, and the C_{24} component (18β-de-E-hopane) is usually the most abundant of the series. The C_{25}–C_{27} members differ from the C_{24} in possessing an 18α-methyl group in addition to an 18β-alkyl group (C_1–C_3).

These tricyclic and tetracyclic series are absent from Recent sediments, suggesting that their formation from kerogen requires higher temperatures than those associated with diagenesis (Aquino Neto et al. 1983), assuming that the organisms that produce them still exist.

(d) Steranes

The carbon number of biogenic precursor sterols is largely preserved in steranes, but some source specificity is lost; for example, the same C_{29} sterane is produced by stigmasterol, sitosterol and fucosterol (Fig. 2.22). Consequently, a ternary plot of C_{27}:C_{28}:C_{29} steranes requires even more caution than the corresponding plot from sterols (Fig. 5.2) when evaluating sources of organic matter. However, such plots are useful for correlating oils with their source rocks.

Some systematic changes appear to have occurred in the sterane C-number distributions in oils derived from marine source rocks over geological time, with C_{29} steranes being relatively abundant in oils from Palaeozoic sources and C_{28} steranes becoming more important thereafter (Fig. 5.27; Grantham & Wakefield 1988). It is tempting to correlate the increasing relative importance of C_{28} steranes with the radiation of dinoflagellates and coccolithophores in the early Mesozoic, and the silicoflagellates and diatoms in the late Mesozoic/Tertiary (see Fig. 1.12), but the sterol distributions in extant phytoplanktonic classes (Table 5.1) do not provide unequivocal support. The effects of herbivory and the likely dominance of phytoplanktonic blooms by just a few species, possibly with atypical sterol distributions, may have a major influence on the ultimate sterane distributions (Volkman 1986). The two groups of Omani oils from Proterozoic sources (c.650 Myr old) are notable for exhibiting sterane distributions that lie off the main trend in Fig. 5.27.

Dinoflagellate contributions can usually be recognized from dinosterane (Fig. 5.26) and related C_{30} 4-methylsteranes (Summons et al. 1987), and their presence usually implies a Mesozoic or younger age (Summons et al. 1992; Moldowan et al. 1996). The ratio of 24-norcholestanes to 27-norcholestanes can provide age information too, because the ratio increases after the Triassic and particularly during the Cretaceous and Tertiary, and appears to be attributable to diatom contributions (Holba et al. 1998).

(e) Hopanes

Hopanes are ubiquitous components of sediments and sedimentary rocks, demonstrating the importance of bacterial activity during diagenesis. Other series of hopanoidal alkanes have been detected in which the number and/or position of ring-methyl groups differ from those in the regular hopanes. As can be seen in Section 5.4.3, some series of norhopanes (i.e. bearing one fewer ring-methyl group than the regular hopanes) are associated with particular depositional environments. Among the rearranged hopanes a C_{29} homologue of Ts (Fig. 2.20c), i.e. 30-nor-18α,21β-neohopane, has been identified (Moldowan et al. 1991) but an extended series equivalent to the regular hopanes has not been detected, so the neohopanes do not appear to be rearrangement products of the regular hopanoids, unlike the 17α-diahopanes (Moldowan et al. 1991).

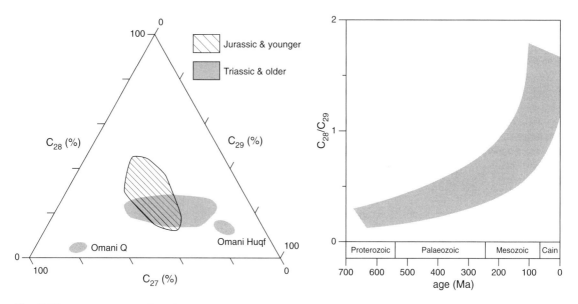

Fig. 5.27 Variations over geological time in sterane C-number distributions in oils from marine source rocks (age refers to inferred source; after Grantham & Wakefield 1988).

The most prominent of the methylhopanes (i.e. hopanes bearing an additional ring-methyl group) in ancient sediments and oils belong to the 2α-methyl-17α-hopane series, which derive from isomerization of less thermodynamically stable 2β precursors in certain types of bacteria. Some methylotrophic and nitrogen-fixing bacteria biosynthesize 2-methylhopanoids, the dominant members of which are the C_{31} analogues of diplopterol and diplotene, whereas 2-methylbacteriohopanepolyols are dominant in cyanobacteria and probably also in prochlorophytes (Summons et al. 1999). The polyfunctional nature of the cyanobacterial 2-methylhopanoids confers greater potential for incorporation into kerogen and preservation than is possible for the diplopteroids, and so cyanobacteria are likely to be the major source of 2-methylhopanes in ancient sediments and oils. Further circumstantial evidence is provided by the types and age distribution of sediments in which they have been detected. High abundances of 2-methylhopanes relative to regular hopanes are found in carbonates deposited in marginal marine settings and saline lakes, in which cyanobacterial mats have flourished since the Precambrian. Examples from more open marine settings, characterized by shale deposition, are more abundant for the Proterozoic than the Phanerozoic, probably owing to the greater importance of cyanobacteria and prochlorophytes among the phytoplankton during the Proterozoic (Summons et al. 1999). The earliest documented occurrence of 2-methylhopanes is in 2.7 Ga shales from the Pilbara Craton, Australia (Brocks et al. 1999).

(f) Higher plant diterpanes and triterpanes

Cyclic diterpanes with two (labdane), three (abietane, isopimarane and rimuane) or four rings (*ent*-beyerane, *ent*-kaurane and phyllocladane) can be particularly abundant where there has been a significant contribution from gymnospermous resins, such as in coals (Fig. 5.28; Aplin et al. 1963; Noble et al. 1985, 1986; Weston et al. 1989; Otto & Simoneit 2001).

As seen in Fig. 5.19, contributions from woody angiosperms are likely to lead to the presence of oleanane in ancient sediments and oils, often accompanied by its ring-A degraded counterpart, 10β-de-A-oleanane. The ring-A degradation products of lupeol (10β-de-A-lupane) and α-amyrin (10β-de-A-ursane) may also be present (Fig. 5.29; Woolhouse et al. 1992).

Angiosperms became widespread during the Late Cretaceous–early Tertiary (Crane & Lidgard 1989), but prior to that gymnosperms dominated, and their distinctive diterpane distributions have been traced back to Carboniferous rocks (Disnar & Harouna 1994). With appropriate regional palynological calibration, the ratio

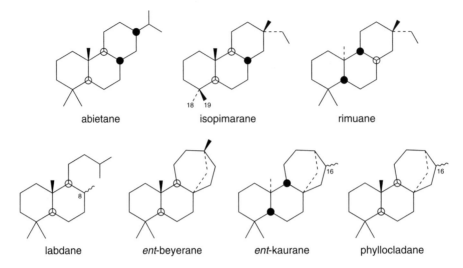

Fig. 5.28 Examples of cyclic diterpanes generally abundant in mature woody-gymnospermous kerogen. 18-Nor- and 19-nor-isopimaranes can also be abundant, deriving from norisopimaroid precursors.

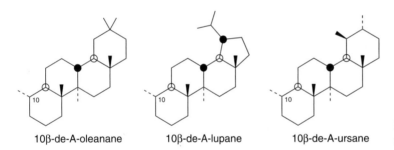

Fig. 5.29 Ring-A degraded, angiosperm-derived triterpanes.

of angiospermous to gymnospermous terpanes in oils with major terrestrial contributions, together with variations in the distributions within each of the two terpane groups, can help to constrain the likely age of its source rocks (Killops et al. 1995, 2003).

Often the higher plant derived terpanes are found in association with their aromatic counterparts. For example, among the aromatic diterpenoids that can dominate the methylated phenanthrenes of bitumens from humic coals are retene, formed from abietoids and phyllocladoids (Fig. 5.18), and pimanthrene from pimaroids (Fig. 5.30). Similarly, diagnostic pentacyclic aromatic hydrocarbons may be present in ancient sediments and oils, such as 1,2,9-trimethylpicene (derived from α-amyrin; Fig. 5.22) and 1,8-dimethylpicene (derived from β-amyrin; Fig. 5.30). Abundant 1,2,7-trimethylnaphthalene relative to other trimethylnaphthalenes

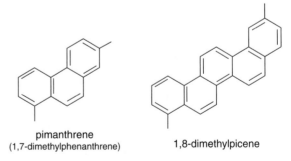

Fig. 5.30 Two higher-plant derived aromatic hydrocarbons that can be found in bitumens from humic coals.

probably also indicates an angiospermous contribution, resulting from fragmentation of the C ring of β-amyrin during diagenesis (Fig. 5.23). However, with increasing thermal maturity during catagenesis (or coalification) the relative abundance of specific polymethylnaphthalenes decreases, until they are no longer conclusive source indicators. This occurs because of the generation from kerogen (or coal) of polymethylnaphthalenes with random substitution patterns, and possibly also because of the conversion of the source indicators into other isomers by methyl group migration (see Section 5.5.4).

Other higher plant derived triterpanes include a series of bicadinanes, which can sometimes be observed in Tertiary oils and source-rock extracts from South-east Asia. They are believed to derive from thermal breakdown of polycadinanes, a class of biopolymer that has been identified in dammar resins in some South-east Asian angiosperms (particularly dipterocarps; van Aarssen et al. 1992). Other cadinane oligomers may also be present, but the bicadinanes are usually dominant and the major component in oils is generally the most thermally stable *trans,trans,trans* isomer (Fig. 5.26).

5.4.3 Indicators of depositional environment

To some extent molecular source indicators provide information on depositional environments, because the composition of organism assemblages varies with environmental conditions. These conditions will also influence the transformations that the molecular source indicators undergo during diagenesis (e.g. E_h and pH conditions (Box 3.4), and rearrangements catalysed by inorganic minerals (e.g. Section 5.3.3e)). As a general example, the ternary plot of branched and cyclic C_7 hydrocarbons in Fig. 5.31 can help to distinguish between marine, terrigenous and lacustrine environments (ten Haven 1996; Mango 1997). Terrigenous samples plot closest to the cyclohexyl apex (100% methylcyclohexane + toluene) because higher plant derived material tends to yield greater quantities of benzene and toluene than marine samples (Leythaeuser et al. 1979; Odden et al. 1998).

Environmental conditions can change as a basin evolves, examples of which are Brazilian marginal offshore basins formed during the opening of the South Atlantic by the rifting of the African and South American continental plates. The basins resulted from the progressive subsidence of the crust during the onset of rifting in the Early Cretaceous (see Box 3.11). Initially, freshwater and then saline lacustrine deposits were formed, followed in the mid-Cretaceous by marine evaporite deposition as rifting proceeded, resulting

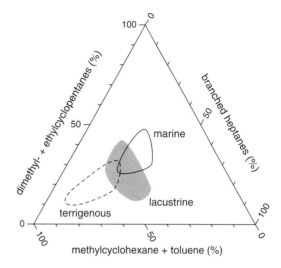

Fig. 5.31 Ternary plot of Mango's ring-preference parameters, showing depositional environment fields (after ten Haven 1996).

from the combined effects of intermittent marine transgressions and evaporation of the waters trapped by topographical barriers in a hot dry climate. Biomarkers have been successfully used to differentiate these depositional environments by applying a number of the concepts outlined in the rest of this section (Mello et al. 1988a, b).

(a) Freshwater and marine environments

Unfortunately, there are few types of depositional environments that are exclusively characterized by the presence of just one or two specific biomarker hydrocarbons. One example is botryococcane, which is diagnostic of contributions from *Botryococcus* in freshwater lakes, although it is not present in all such lacustrine sediments. Freshwater lakes often exhibit characteristics of vascular plant and phytoplanktonic inputs, and significant contributions of long-chain acyclic isoprenoids from methanogenic bacteria may occur where sulphate levels are very low. In contrast, bacterial activity in marine and saline lake sediments is dominated by sulphate reducers, and the S content (including S-containing compounds) is greater than in freshwater settings ($c. >0.3\%$).

Sometimes the presence (or even the absence) of a range of biomarkers can be diagnostic. For example, despite the reservations about sterane C-number distributions noted in Section 5.4.2, dominant C_{29} steranes (c.>60%) in post-Palaeozoic oils usually indicate a major contribution from terrestrial higher plants. Corroboration is provided by a high hopane:sterane ratio (reflecting bacterial activity), n-alkanes dominated by >C_{22} members with an OEP, high pristane:phytane ratios (discussed below) and abundant higher plant derived terpanes. Such a combination of biomarker distributions is characteristic of deposition in coastal floodplains or nearshore environments (e.g. marine or lacustrine deltas), as in oils from the Niger delta, Africa (Ekweozor et al. 1979) and Mahakam delta, Indonesia (Grantham et al. 1983). The absence of terrestrial indicators in marine sediments does not necessarily imply distal (offshore) deposition. Nearshore (proximal) areas virtually devoid of vegetation (e.g. deserts) or with limited river drainage (e.g. west side of the Andes) will not exhibit a significant vascular plant fingerprint.

The extended cheilanthanes (>C_{30}) appear to derive from bacteria that thrived in conditions of moderate salinity (De Grande et al. 1993; Dahl et al. 1993). The shorter chain (≤C_{30}) cheilanthanes seem to share a similar source, because they are generally in low abundance or absent in bitumen and oils from terrigenous source rocks (Aquino Neto et al. 1983; Philp & Gilbert 1986). However, even relatively minor marine incursions during early diagenesis can result in low cheilanthane abundances in coal-sourced oils (Killops et al. 1994).

Based on the apparent occurrence of 24-n-prop-24(28)-enylcholesterol (a C_{30} sterol) only in certain extant marine members of the chrysophytes (order Sarcinochrysidales; Billard et al. 1990), the presence of the corresponding 24-n-propylcholestanes in oils and ancient sediments is considered to indicate a contribution from marine algae (Moldowan et al. 1990). Low levels of these C_{30} steranes have been detected in some coals, in which they can be interpreted as resulting from a marine transgression, and are often accompanied by other signs of marine contributions (e.g. lower relative abundance of C_{29} steranes and elevated cheilanthane content; Killops et al. 1994).

A broad distinction between freshwater and marine environments can be obtained from the relative abundance of S-containing compounds; the greater abundance of these compounds in marine sediments being related to the sulphate content of seawater and the activity of sulphate reducers. A particularly useful plot is shown in Fig. 5.32, combining an S-containing parameter with the pristane:phytane ratio, which can help to distinguish between some basic marine and freshwater sub-environments (Hughes et al. 1995). Lacustrine oils in general have been found to contain elevated levels of a C_{30} tetracyclic polyprenoidal alkane of algal

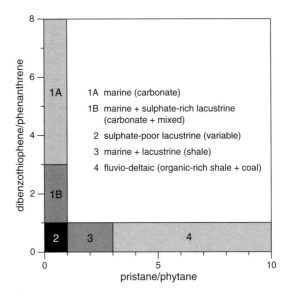

Fig. 5.32 Plot showing grouping of oils according to source-rock depositional environment and lithology (after Hughes et al. 1995).

origin (Fig. 5.26) relative to total 27-norcholestanes, which, together with the absence of 24-n-propylcholestanes, distinguishes lacustrine (fresh or brackish water) from all other depositional environments (Holba et al. 2000).

(b) Argillaceous and carbonate sediments

High diasterane:sterane ratios are usually attributed to abundant clay minerals (Section 5.3.3), so it is not surprising that most carbonates, which have low argillaceous contents, have low ratios. However, high ratios can also be found in some carbonate rocks with low clay content, and it appears that high diasterane content is better explained in terms of a high clay:organics ratio (van Kaam-Peters et al. 1998). The relatively low sterane content but high diasterane:sterane ratio of low-ash coals may primarily reflect the greater resistance of diasteroids during biodegradation, because most of the phytosterols are associated with mesophyllic tissues, which are readily biodegraded (Killops et al. 1994).

Carbonates can exhibit a number of biomarker characteristics. Abundant de-E-hopanes may be present (as they are in evaporites; Connan et al. 1986), although the C_{24} component (18β-de-E-hopane) is also often relatively important in coaly sediments (Philp & Gilbert 1986; Killops et al. 1994). Abundant 30-nor-17α-hopanes (C_{28}–C_{34}) are associated with carbonate sources (Subroto et al. 1991). Because of the low levels of iron in carbonates, high-sulphur levels generally result in abundant OSCs and hence high dibenzothiophene:phenanthrene ratios (Fig. 5.32; Hughes et al. 1995).

(c) Hypersalinity

Hypersaline environments are interesting in that very few organisms are tolerant of such extreme conditions and so relatively simple communities are usually found. Cyanobacteria can flourish in these environments because the mats they form are not subject to extensive grazing by herbivores, although the associated communities usually include some invertebrates together with algae. The halophilic archaebacteria also thrive under these conditions (they actually require salinities >15% in order to synthesize ATP). Bacterial inputs can, therefore, be significant in organic-rich sediments deposited in hypersaline lakes (see Section 3.4.1b). This organic material is less subject to degradation than in other sedimentary environments because most microbial activity, including sulphate reduction, is suppressed by hypersalinity (hence the use of salt in food preservation). Oil source rocks deposited in salt lakes occur widely in China (Jiamo et al. 1986; Jiamo & Guoying 1989). Biomarker distributions are characterized by relatively large amounts of phytane, β-carotane, C_{35} hopanes and gammacerane (Fig. 5.26). In addition, C_{22} n-alkane is often relatively abundant (ten Haven et al. 1985). Type II kerogen in marine evaporitic deposits can contain very high concentrations of squalane and a related C_{25} isoprenoidal alkane (2,6,10,14,18-pentamethyleicosane; Fig. 5.26), in addition to abundant phytane, gammacerane and β-carotane, and sometimes an EOP among the n-alkanes (Mello et al. 1988a, b).

Gammacerane (Fig. 5.26) has been found in sediments dating back to the Proterozoic, and most oils appear to contain trace amounts. Particularly high concentrations seem to be associated with highly reducing, hypersaline depositional conditions, although not all such environments result in abundant gammacerane (Moldowan et al. 1985; Mello et al. 1988a, b). Gammacerane is believed to be the diagenetic product of tetrahymanol (Fig. 5.26; ten Haven et al. 1989; Venkatesan 1989), a triterpenoidal alcohol that occurs widely in Recent marine sediments. Analogous to steroids and hopanoids, tetrahymanol functions as a rigidifier in the membranes of certain protozoa (Ourisson et al. 1987), phototrophic bacteria (Kleeman et al. 1990) and possibly other organisms. Several commonly occurring species of bacteria-eating marine ciliates have been found to contain abundant tetrahymanol, which may explain the ubiquity of gammacerane in marine and lacustrine settings (Harvey & McManus 1991). Water-column stratification promotes the development of hypersalinity and results in the ciliates that live in anaerobic conditions beneath the thermocline synthesizing tetrahymanol in the absence of a supply of dietary sterols. Hence gammacerane can be considered as an indicator of stratification (Sinninghe Damsté et al. 1995).

β-Carotane (Fig. 5.13) is found only in ancient lacustrine sediments and it has been suggested that it may be a by-product of methanogenesis (Repeta 1989). Reduction of carotenoids to form saturated hydrocarbons does not appear to occur in Recent sediments and, as sulphate reduction causes degradation of carotenoids, the survival of potential precursors of β-carotane until reduction occurs requires the absence or limitation of sulphate reduction. The most likely precursors are those that do not contain a 5,6-epoxide group, such as β-carotene, and so are degraded less rapidly by anaerobic oxidation (see Section 5.3.3c). The major phytoplanktonic carotenoids that contain a 5,6-epoxide grouping, such as fucoxanthin (Fig. 2.23), are rapidly degraded at the beginning of diagenesis even in anoxic sediments. The exceptionally high levels of β-carotane in salt-lake sediments reflect the abundance of suitable precursors in contributing organisms and the conditions necessary for their preservation during diagenesis (e.g. presence of methanogens but virtual absence of sulphate reducers).

Hypersaline conditions can also be distinguished by the degree of methylation of chromans (oxygen-containing aromatic isoprenoids; Fig. 5.26), which can be expressed by the chroman ratio:

$$\text{chroman ratio (CR)} = \frac{5,7,8\text{-trimethylchroman}}{(5,7,8\text{-trimethyl} + 7,8\text{-dimethyl} + 8\text{-methylchroman})} \quad [\text{Eqn 5.1}]$$

Values of CR <0.6 generally indicate hypersaline conditions, whereas greater values are found in marine to freshwater environments (Sinninghe Damsté et al. 1987, 1993b). The origins of the methylated chromans have yet to be identified, but are probably algal or cyanobacterial.

(d) Redox conditions and phytol diagenesis

While anoxic conditions generally favour the formation

of organic-rich sediments, oxygen is unlikely to be absent at all stages of diagenesis. Levels of oxygen in the water column can fluctuate over time and so can the position of the oxic–anoxic boundary. It is possible to gain some information on redox conditions (see Box 3.4) during diagenesis from molecular indicators. For example, the 25-nor-17α-hopanes, which lack the methyl group at the A/B-ring junction (C-10) that is present in the regular hopanes, appear not to be alteration products of regular hopanoids. They are generally associated with marine and lacustrine source rocks that were deposited under dysoxic (see Box 1.5), but not significantly hypersaline, conditions (Blanc & Connan 1992).

High concentrations of 28,30-dinorhopane (Fig. 5.26) are common in petroleums generated from source rocks deposited in highly reducing to anoxic environments (Grantham et al. 1980; Mello et al. 1988a, b). There is evidence that the compound is synthesized by bacteria utilizing CO_2 (Schoell et al. 1992; Schouten et al. 1997), a substrate that is likely to originate from sulphate reduction in marine environments, which suggests that 28,30-dinorhopane could originate within oxygenated sediments immediately overlying the anoxic zone.

Phytol is an important part of phytoplanktonic chlorophylls, from which it appears to be released early on during diagenesis in the water column (see Section 5.3.2b). Early studies of the diagenetic fate of phytol (Didyk et al. 1978) suggested that different products are formed under different redox conditions (Fig. 5.33).

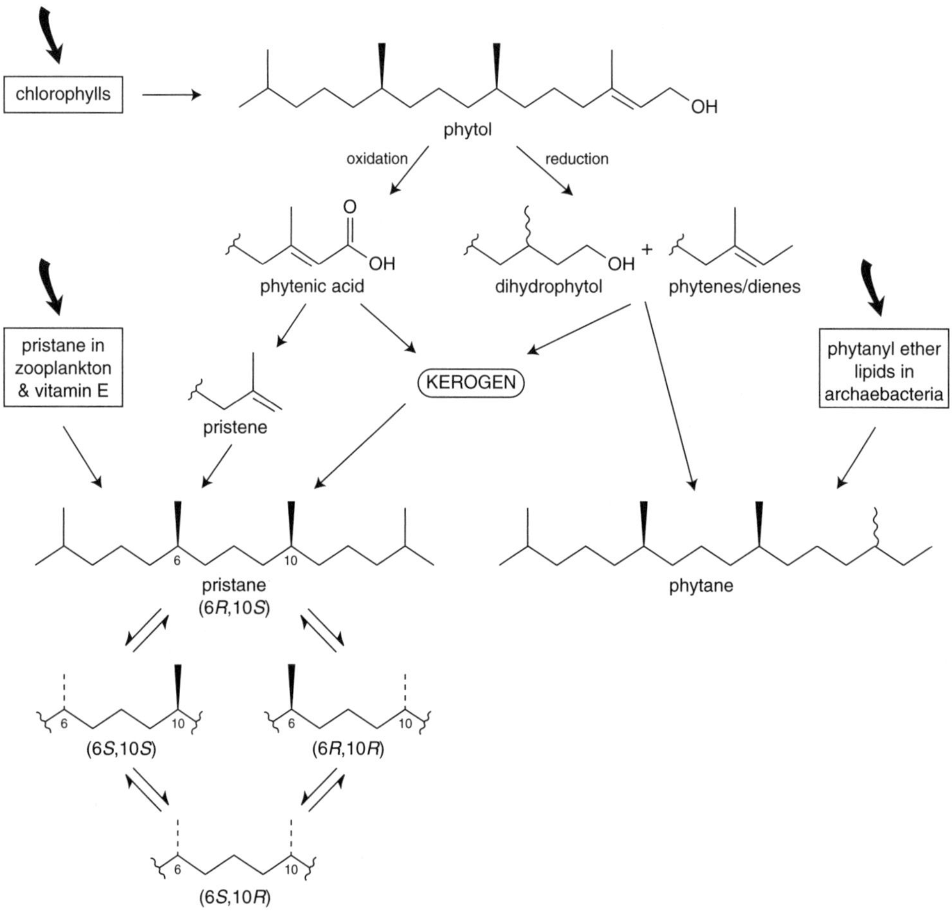

Fig. 5.33 Summary of phytol sedimentary diagenesis, showing pristane isomerization (phytane undergoes similar isomerization; large arrows indicate biogenic inputs).

Under relatively oxidizing conditions a significant proportion of phytol, a C_{20} compound, can be oxidized to phytenic acid, which may then undergo decarboxylation to pristene, a C_{19} compound, before finally being reduced to pristane. In contrast, under relatively anoxic conditions phytol is more likely to undergo reduction and dehydration to phytane, via dihydrophytol (phytanol) or phytene, with the preservation of all 20 C atoms in the product. On this basis it was suggested that the ratio pristane:phytane may provide a measure of redox conditions during diagenesis, with values <1 being typical of anoxic conditions and values >1 suggesting oxicity (Didyk et al. 1978). Isoprenoidal hydrocarbons with <20 C atoms (e.g. C_{14}–C_{16} and C_{18}) are common in sediments and may also derive from phytol.

As seen in Sections 5.3.2b and 5.3.3b, phytol geochemistry is complex (Rontani & Volkman 2003) and the scheme described above is an oversimplification. For example, much of the phytol can reach the sediment in esterified form (Johns et al. 1980) and so when hydrolysis liberates phytol in the anoxic zone phytane may be produced but not pristane (ten Haven et al. 1987). In addition, the effects of geological constraints on redox conditions have been overlooked. These include the effect of variations in sedimentation rate on the thickness of the oxic sediment layer, the interbedding of turbidites and pelagic sequences, and the effect of micro-environments (the conditions within two adjacent micro-environments can vary signifi-cantly). Clay minerals can catalyse the hydrolytic liberation of phytol from chlorophylls and also its subsequent dehydration to phytadienes (Rontani & Volkman 2003), which may then react with sulphides to produce thiophenes and other OSCs (Sinninghe Damsté & de Leeuw 1990; Schouten et al. 1994). Furthermore, there can be sources of pristane and phytane other than chlorophylls (ten Haven et al. 1987). These include methanogenic or halophilic phytanyl lipid sources for phytane and zooplanktonic or tocopherol (i.e. vitamin E; Fig. 5.33) sources for pristane (Goossens et al. 1984). In hypersaline environments halophilic bacteria are very important and they contain more phytanyl lipids than methanogens. Consequently, a low pristane:phytane ratio is usually recorded in these environments, apparently reflecting the salinity-dependent growth rate of bacteria and not the degree of anoxicity. Despite all these possible restrictions on the use of pristane:phytane as an indicator of redox conditions, this ratio appears to work surprisingly well for a large number of oil source rocks, as shown by Fig. 5.32. During catagenesis, however, the pristane:phytane ratio tends to increase as additional pristane appears to be generated from kerogen (Goossens et al. 1988a, b).

There are inorganic indicators of redox conditions, an example of which is considered in Box 5.3.

5.5 Thermal maturity and molecular transformations

5.5.1 Configurational isomerization

Many of the stereogenic centres remaining in hydrocarbons during diagenesis preserve the biologically conferred configuration. However, as temperature increases isomerization occurs, ultimately resulting in equilibrium mixtures of isomers, the ratios of which reflect their relative thermodynamic stabilities. Isomerization mostly involves hydrogen exchange, as described in Box 5.2.

(a) Acyclic isoprenoidal alkanes

Configurational isomerization begins during diagenesis for some components. For example, in pristane the biologically conferred $6R,10S$ isomer is converted into equal amounts of the four possible isomers ($6R,10S$; $6R,10R$; $6S,10S$; $6S,10R$) prior to the onset of catagenesis (Fig. 5.33). The $6S,10S$ and $6R,10R$ forms of pristane are, in fact, identical and can be interconverted simply by rotating the molecule about a vertical axis through the central C atom (and so are not chiral). This isomer is known as meso pristane (rather than $6S,10S$ or $6R,10R$).

(b) Steranes

At the end of diagenesis the configuration at the stereogenic carbon centres in the sterane skeleton is largely that inherited from the biogenic precursor, which provides the degree of flatness required for the function of a cell membrane rigidifier. At certain of the stereogenic centres the stereochemistry already corresponds to the thermodynamically most stable configuration (e.g. at C-8 and C-9; see Fig. 2.20a for carbon numbering scheme). At C-10 and C-13 hydrogen exchange cannot alter the configuration because there is no hydrogen directly bonded to the stereogenic carbon atom to allow the necessary process to occur (see Section 5.3.1 and Box 5.2). Isomerization is, therefore, limited to the C-14, C-17 and C-20 positions. A mixture of isomers is usually present at C-5 and C-24, although the 5α configuration is normally more abundant.

Box 5.3 | Nickel and vanadium distributions and sedimentary redox conditions

The interaction of metal ions with organic moieties is an important aspect of diagenesis and is affected by redox potential (E_h) and acidity (pH) (see Box 3.4). These factors control the oxidation state of a metal ion and the availability of anions that may compete with organic moieties for the metal. The distribution of metals in oils can, therefore, provide information on the depositional environment of its source rock. Two metals that are particularly useful are nickel (Ni) and vanadium (V). The bulk of these two metals in oils appears to be strongly associated with high molecular weight organic components, especially asphaltenes, and the ratio of Ni to V appears not to be affected by post-diagenetic processes. Although Ni and V are also associated with porphyrins, which do undergo some changes with increasing maturity, the amounts in porphyrins are generally relatively minor and do not affect the overall interpretation of Ni : V ratios (but there are exceptions, such as Venezuelan oils). Figure 5.34 shows the main field of E_h and pH conditions for marine organic sediments, which is divided into three areas (Lewan 1984). Whereas the average pH of seawater is 8.2, the pH of sedimentary pore water can vary significantly, as shown in Fig. 5.34.

The chemical form of sulphur is important in determining the availability of metal ions in marine environments. In area I of Fig. 5.34 sulphur is generally in the form of SO_4^{2-} and nickel is available for bonding (as Ni^{2+}), but vanadium is unavailable. Oils from source rocks deposited under these conditions, such as those from Mesozoic and Tertiary reservoirs in the Uinta Basin (USA), have V/(V + Ni) values <0.1 and S content <1%. In area II sulphate is again the major form of sulphur, while Ni^{2+} and VO^{2+} (vanadyl) are available for bonding, but the vanadyl species may

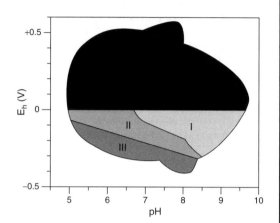

Fig. 5.34 E_h and pH stability fields in sediments. Total shaded area is the natural stability field for marginal and open marine sediments. Areas I, II and III are discussed in the text, and their combined area encompasses the field for organic sediments (after Lewan 1984).

be partially hindered by sulphide complexation. Crude oils associated with these conditions include those from Tertiary reservoirs in eastern Nigeria, which contain low amounts of S (c.0.2–0.5%) and have V/(V + Ni) values in the range 0.1–0.9. In area III sulphur exists mainly as H_2S at pH <7 and as HS^- at higher pH. Vanadium is available in the form of VO^{2+} and V^{3+}, but Ni^{2+} may be partially hindered by sulphide complexation. Associated oils, such as those from Jurassic reservoirs in Saudi Arabia, have high S content (>1%) and V/(V + Ni) values >0.5.

Only the 20R isomers of the regular steranes and 4-methylsteranes exist initially, but they undergo isomerization to form an equilibrium mixture containing approximately equal amounts of 20R and 20S isomers (Figs. 5.15 and 5.16). This process appears to require higher temperatures than the isomerization of pristane. Similar temperatures to those at which the 20S/R isomerization occurs are required for another isomerization process of the regular steranes that affects the C-14 and C-17 cyclic positions in concert. The 14α,17α isomer is converted into an equilibrium mixture of 14α,17α and 14β,17β isomers (Fig. 5.15) that contains rather more of the latter.

A variable amount of 5α,14β,17β-steranes may be present at the end of diagenesis. A possible source of these components at a relatively early stage of maturity is the reduction of Δ^{14} sterenes produced from Δ^7 stenols (synthesized by several common freshwater green microalgae; see Table 5.1), a much simplified scheme for which is shown in Fig. 5.35 (Peakman & Maxwell 1988; Peakman et al. 1989). Spirosterenes are involved as intermediates, and they may be important intermediates in many sterene rearrangements, even though they are not themselves preserved in substantial amounts at the end of diagenesis (e.g. as spirosteranes).

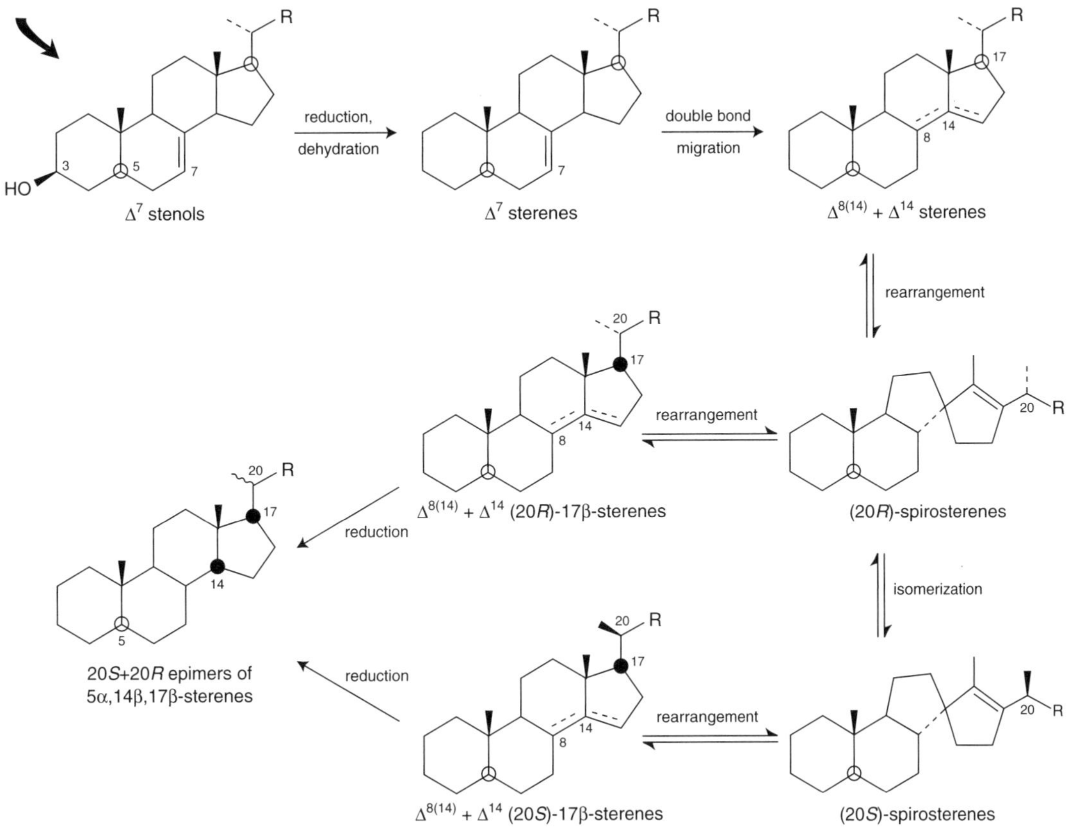

Fig. 5.35 Possible origin of 20S and 20R epimers of 14β,17β-steranes in immature sediments from Δ^7 stenols. Note: the Δ^{14} (20R)-17α-sterene intermediates can give rise to (20R)-14α,17α-steranes upon reduction, as an alternative to undergoing rearrangement to spirosterenes (large arrow indicates biogenic input; after Peakman & Maxwell 1988; Peakman et al. 1989).

(c) Triterpanes

Isomerization via hydrogen exchange also affects terpanes. For example, S/R interconversion can occur at C-16 in phyllocladane and *ent*-kaurane (Fig. 5.28) and at C-18 in oleanane (Fig. 5.19). In C_{31}–C_{35} hopanes the biologically conferred 22R configuration is preserved during the initial stages of diagenesis. Subsequent isomerization results in a final equilibrium mixture containing approximately equal amounts of 22R and 22S isomers. At the C-17 and C-21 positions (see Fig. 2.20a for carbon numbering scheme) the configurations are initially mainly 17β,21β. The 17β,21β isomer is much less thermally stable than either the 17β,21α or 17α,21β isomer and is rapidly converted into a mixture of the latter two isomers. The other possible isomeric configuration, 17α,21α—the theoretical stability of which lies between that of 17β,21β and 17β,21α—makes only a minor contribution at all maturity levels (Nytoft & Bojesen-Koefoed 2001). With increasing temperature the 17β,21α isomer is converted almost completely into the more stable 17α,21β isomer, so that the final equilibrium mixture is dominated by the latter (Fig. 5.21). It should be remembered that these compounds are usually collectively referred to as hopanes, regardless of configuration at the various stereogenic centres. Strictly speaking, however, only those with a 21β configuration are hopanes, whereas those with a 21α configuration are called **moretanes**.

(d) Alternative explanations of apparent isomerization

Studies of the changes in absolute abundance of sterane and hopane isomers during oil generation under different heating regimes (i.e. average maturation conditions, more rapid heating by igneous intrusions and extremely rapid heating during hydrous pyrolysis in the laboratory) have provided evidence that the variation in isomer ratios may not be entirely the result of isomerization in the bitumen. Differences in generation rates from kerogen and in thermal stabilities (i.e. tendency to disintegrate) of the various isomers may also be important (Abbott et al. 1990; Bishop & Abbott 1993; Requejo 1994; Farrimond et al. 1998). It is likely that isomerization occurs in biomarkers incorporated into kerogen (e.g. at C-17, C-21 and C-22 in bound hopanoic acids; Abbott et al. 2001) as well as in those in the free bitumen, although there is some evidence for preservation of an elevated $R:S$ ratio at C-22 in hopanes and at C-20 in steranes among the last evolved members of these hydrocarbon classes in some shales and coals (Peters et al. 1990; Farrimond et al. 1996; Killops et al. 1998). Such an immature signature could result from steric hindrance restricting the hydrogen exchange within certain parts of the kerogen matrix, but would require the affected biomarkers to be among the last released from the kerogen into the bitumen, after most of the earlier hopanes and steranes have been expelled from the host rock.

The apparent isomerization at C-14 and C-17 in steranes occurs over a higher maturity range than the isomerization at C-20 in steranes and C-22 in hopanes. It may be at least partially controlled by differences in thermal stability towards the higher maturity limit, similar to that suggested for the increasing dominance of the 18α over the 18β epimer of oleanane (Rullkötter et al. 1994).

5.5.2 Aromatization

Another important, thermally mediated, transformation is aromatization, which we have already examined for diterpenoids (Fig. 5.18) and triterpenoids (Fig. 5.22). One of the major classes of diagenetic products of steroids is the C-ring monoaromatics (Fig. 5.15), which include diasteroids and 4-methylsteroids as well as regular steroids (Riolo et al. 1986). As temperature rises with increasing burial, complete aromatization of the three six-membered rings occurs, generally with the loss of the C-19 methyl group at the A,B-ring junction (Fig. 5.36). Only small amounts of B,C-ring diaromatics have been detected, suggesting that these intermediates in the aromatization process are short-lived (Mackenzie et al. 1981). Aromatization is not a reversible reaction, unlike isomerization. Initially there are no triaromatic steroidal hydrocarbons, but during catagenesis C-ring monoaromatics tend to be converted into triaromatics. In humic coals aromatization of cycloalkanes in general increases throughout coalification and is virtually complete towards the end of the bituminous coal stage. Although the distribution of 4-desmethyl C-ring monoaromatic steroids can be quite complicated, the loss of methyl groups during aromatization leads to a relatively simple distribution of triaromatics: mainly

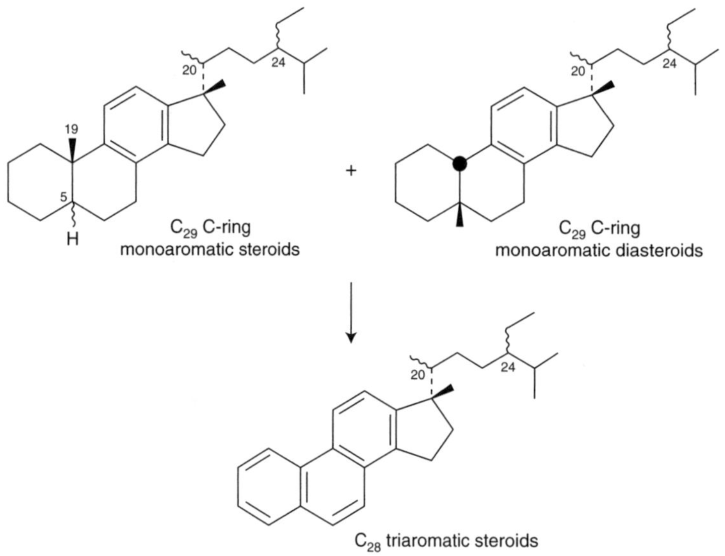

Fig. 5.36 Aromatization of C_{29} C-ring monoaromatic regular steroids and diasteroids.

C_{20}, C_{21} and 20S/R epimers of C_{26}–C_{28} (Fig. 4.25; Riolo et al. 1986).

5.5.3 Enrichment of short-chain hydrocarbons and cracking processes

(a) Steroids

As maturity progresses through the oil window there is an apparent increase in the abundance of steroidal hydrocarbons with short relative to long alkyl chains (e.g. C_{21} and C_{22} alkanes, and C_{20} and C_{21} triaromatics). This appears to reflect the greater resistance of the short-chained components towards thermal degradation rather than their production by thermal cracking of the alkyl chains in the longer-chain components (Mackenzie 1984). Although there is not a direct product–precursor relationship, the ratio of short-chain to long-chain components appears to be a useful indicator of maturity in closely related samples (Wingert & Pomerantz 1986). However, the ratio in immature sediments can vary significantly under the influence of changing sources of organic matter and depositional environments.

(b) Porphyrins

Porphyrin distributions in oils and ancient sediments are dominated by C_{27}–C_{33} DPEP and etio species complexed with Ni^{2+} and VO^{2+}. With increasing temperature the ratio DPEP:etio for vanadyl porphyrins decreases. This may in part be due to a cracking process involving cleavage of the isocyclic ring (i.e. the non-pyrrole five-membered ring) of the DPEPs to yield etio porphyrins (Fig. 5.11), but is more likely to result from a combination of generation of etioporphyrins from kerogen and the generally faster degradation rate for DPEP than etio porphyrins with increasing temperature (Barwise & Roberts 1984; Baker & Louda 1986).

5.5.4 Methyl group migration in aromatic hydrocarbons

The mobility of methyl groups on lignin and other moieties during the thermal degradation of kerogen and coal affords the possibility of migration of methyl groups from less to more thermodynamically stable positions in aromatic compounds (see Sections 4.3.2c and 5.4.3). Dimethylnaphthalenes (DMNs) and methylphenanthrenes (MPs) are usually relatively abundant components of bitumens and oils and demonstrate the application of changes in the position of methylation in estimating maturity. However, the following principles apply to all methylated aromatic compounds (Radke 1987).

There are ten possible isomers of dimethylnaphthalene (DMN): 1,2, 1,3, 1,4, 1,5, 1,6, 1,7, 1,8, 2,3, 2,6 and 2,7; all other arrangements effectively duplicate these ten isomers. As maturity increases, the thermodynamically more stable isomers become dominant in bitumens. The stability is determined by the degree of steric interaction (i.e. electrostatic repulsion caused by close proximity) between each methyl group and the substituents on the immediately adjacent C atoms of the ring system (either an H atom or another methyl group). The greater the space between these substituents, the lower the steric interaction and the greater the stability. The relative stabilities can be seen by reference to Fig. 5.37, bearing in mind that methyl groups are bulkier than hydrogen atoms. A methyl group at any of the β positions (C-2, C-3, C-6 or C-7) is further from its neighbours than one at an α position (C-1, C-4, C-5 or C-8). So for DMNs, the most stable arrange-

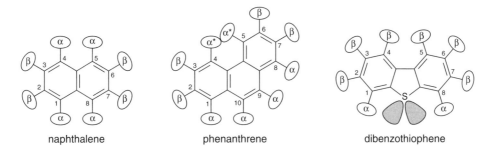

Fig. 5.37 Structures, numbering schemes and steric interactions of methylated naphthalenes, phenanthrenes and dibenzothiophenes (steric constraints in dibenzothiophenes include the two lone-pairs of electrons on the S atom, shown as half-tone lobes).

ment is a methyl on C-2 and the other on C-6 or C-7. The least stable arrangement is where the methyl groups are in adjacent α positions (1,8DMN). It is not surprising, therefore, that 1,8DMN is rarely detectable in oils (Alexander et al. 1984b). The next least stable isomers are 1,4DMN and 1,5DMN.

Similarly, there are five possible isomers of methylphenanthrene (MP): 1, 2, 3, 4 and 9 (Fig. 5.37). A methyl group at C-4 (α* position) results in intense steric interactions and is not thermodynamically favoured. Therefore, 4MP is at most a minor component in oils (<1% total MPs; Garrigues & Ewald 1983) and can be discounted. Of the other four isomers, 1MP and 9MP have methyl groups in α positions, where there is some steric interaction with H atoms on adjacent carbons, whereas in 2MP and 3MP the methyl groups are further from these hydrogens (in β positions) and so their steric interactions are smaller. Consequently, the 1 and 9 isomers are slightly less thermodynamically stable than their 2 and 3 counterparts, with the result that 2MP and 3MP become dominant as maturity increases. As well as migration within a particular aromatic compound, it has been suggested that methyl groups can migrate between aromatic moieties, such as during the methylation and subsequent demethylation that phenanthrene appears to experience with increasing maturity (Radke & Welte 1983).

More recent studies suggest that the distributions of methylated naphthalenes and phenanthrenes reflect the dilution of the components in the bitumen inherited from diagenesis by addition of components with different distributions generated from kerogen (Radke et al. 1990; Killops et al. 2001). The effects of this dilution process are probably more noticeable in coals because their larger adsorption capacity tends to delay oil expulsion to a greater extent than in marine shales.

The steric considerations proposed above to account for the differing thermodynamic stabilities of the various DMN and MP isomers can also be applied to methyldibenzothiophene (MDBT) isomers, as shown in Fig. 5.37. The least stable isomer appears to be 1MDBT, in which the methyl-group hydrogens are close to the valency electron pairs of the sulphur atom. The relative concentrations of this isomer are low above a vitrinite reflectance of $c.0.8\%$ R_o.

5.6 Palaeotemperature and age measurement

5.6.1 Introduction

From the previous section it may be deduced that the extent of isomerization reactions can potentially provide an idea of the thermal history or the age of a sediment, provided there has not been sufficient time for all isomerization reactions to yield an equilibrium mixture of isomers. The thermal histories of Recent sediments are often relatively simple, providing the opportunity for reasonably accurate estimates of temperature (if samples can be dated) or age (if independent measurements of temperature are available) to be obtained from isomerization reactions. Isomerization of amino acids has been used in studying the sequence of glacial and interglacial episodes during the Quaternary, which is examined in Section 5.6.2. However, isomerization reactions are not the only useful parameters, and we have seen in Section 5.1.3 how organisms can adjust the degree of unsaturation in lipids to maintain their fluidity as temperature fluctuates. As an example of this type of palaeothermometer, the degree of unsaturation in long-chain alkenones is discussed in Section 5.6.3. The further back into geological history we travel, the more difficult it is to model burial and thermal histories accurately, so isomerization and other reactions, such as vitrinite reflectance and the degree of kerogen degradation (as examined in Section 5.6.4), are used to provide more general estimates of maturity.

5.6.2 Amino acid epimerization

As noted in Section 3.3.3a, amino acids can be preserved in proteinaceous materials associated with resistant mineral coatings, such as molluscan shells and foraminiferal tests. The degree of isomerization of these amino acids has been successfully used to estimate the ages or burial temperatures of Quaternary sedimentary strata (Bada & Schroeder 1975; Goodfriend et al. 2001). This application is possible because living organisms synthesize proteins from the L form of amino acids (see Section 2.3.1), but after death isomerization occurs, eventually resulting in a racemic mixture (see Section 2.1.3). Racemization occurs via proton abstraction and the formation of a trigonal planar carbanion intermediate (cf. Fig. 5.5), and at equilibrium the rates of forward and backward reactions are equal, so the concentrations of the L and D enantiomers are also equal. The neutral amino acid isoleucine is frequently used because of its relative stability (Section 3.3.3a). It has two stereogenic carbon centres, and of the possible isomerizations, the epimerization of L-isoleucine to D-alloisoleucine is the major reaction pathway and is used for age/temperature estimation (Fig. 5.38). At equilibrium, the rates of forward and backward reactions are not equal, which results in the ratio of D-alloisoleucine : L-isoleucine being

Chemical stratigraphic concepts and tools | 213

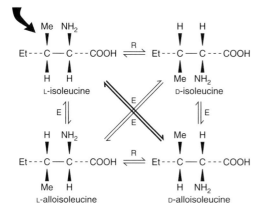

Fig. 5.38 Epimerization (E) and racemization (R) of isoleucine (major pathway is epimerization of L-isoleucine and D-alloisoleucine; large arrow indicates biogenic input).

ward reaction). The value of the constant is determined from A/I values in modern shells (usually <0.03).

The rate of epimerization (or racemization) depends upon the stability of the carbanion intermediate: the more stable it is the greater the rate of isomerization. Stability is controlled by the electrostatic influences of the groups bonded to the α-C atom. In carbonate tests the pH is slightly alkaline (see Box 3.4), so neutral amino acids exist as the protonated form together with the corresponding zwitterion (the left and central structures in Fig. 2.11a, respectively). The NH_3^+ group formed by protonation is electron-withdrawing and so stabilizes the carbanion, leading to faster isomerization, whereas a COO^- group is electron-donating and so destabilizes the carbanion, resulting in slower isomerization. The balance between these contrasting influences, and hence the rate of isomerization of an amino acid in a peptide chain, is modified by the other amino acids attached to its amine and carboxyl groups. There are five possible environments for amino acids in peptide chains, as shown in Fig. 5.39. Epimerization/racemization is slower for a C-terminal than a N-terminal unit (Smith & Sol 1980), because the former can form a COO^- group while the electron-withdrawing power of the amino group is diminished by its involvement in bonding to another amino acid (Mitterer & Kriausakul 1984). For a N-terminal unit the carboxyl group is involved in bonding to another amino acid, and so has a weakened electron-donating effect, whereas the protonated amine group stabilizes the carbanion intermediate of the isomerization process.

The observed relative rates of isoleucine epimerization in heating experiments are: N-terminal and diketopiperazine groups ≫ C-terminal and interior groups ≈ free isoleucine (in aqueous solution). However, the relative extents of epimerization observed in real samples is diketopiperazine > N-terminal > C-terminal > free ≫ interior (Mitterer & Kriausakul 1984). The apparent dilemma of why C-terminal and free isoleucine can exhibit extensive epimerization but their rates of epimerization are slow is explained by the conversion of some of the more highly epimerized isoleucine from quickly (N-terminal and diketopiperazine) to slowly (C-terminal and free isoleucine) epimerizing units via diketopiperazine and hydrolysis (Fig. 5.39; Kriausakul & Mitterer 1983).

The rate of hydrolysis is important, because if it were rapid it would yield the rate curve predicted by the simple first order kinetics in Eqn 5.12 for the epimerization of free aqueous isoleucine. If hydrolysis were slow, the initial part of the rate curve would also be similar to that

c.1.3. At first sight the application of racemization or epimerization to estimating ages or palaeotemperatures seems to be relatively simple, through the application of first order kinetics (Box 5.4). However, the potential loss of racemized amino acids by aqueous leaching has been noted (Section 3.3.3a), and there are several other complicating factors.

Normally the degree of epimerization is determined for the total amino acids, which include both the free amino acids and those bound in peptides and proteins. The free amino acids might be expected to be the most susceptible to epimerization, so the rate at which hydrolysis frees individual amino acids from peptide chains is also an important consideration. The hydrolysis rate of each peptide bond depends on factors such as protein chain length (the bonds are generally more labile in larger chains) and the nature of the adjacent amino acids (Kriausakul & Mitterer 1978). Because the sequence of amino acids is genetically determined, it is not surprising that the hydrolysis rates of proteins in calcified structures exhibit taxonomic variability. For the L-isoleucine:D-alloisoleucine epimerization the first order kinetics yield the following equation (Bada & Schroeder 1972):

$$\log_e\left[\frac{1+(A/I)}{1-K'(A/I)}\right] - \text{const} = (1+K')k_f t \quad \text{[Eqn 5.12]}$$

where A = D-alloisoleucine, I = L-isoleucine, k_f is the rate of the forward reaction (conversion of L-isoleucine to D-alloisoleucine), and K' is the reciprocal of the equilibrium constant (i.e. 1/1.3; it is equal to the ratio of the rate constant for the reverse reaction to that of the for-

Box 5.4 | Reaction rates

The speed, or rate, of a reaction involving a single chemical species (a unimolecular reaction) depends on the concentration of the species involved. It is also dependent on temperature, and so the rate at a particular temperature is proportional to some power (n) of the concentration:

$$\text{rate} = k[Y]^n \quad \text{[Eqn 5.2]}$$

where [Y] = concentration of the chemical species Y, and k = the rate constant. The value of n (the reaction order) can only be determined experimentally. Isomerization reactions and kerogen degradation can be approximated by reactions in which n = 1, i.e. they are first-order reactions. The rate constant, k, is related to the activation energy (E_{act}; see Box 2.6) and temperature:

$$k = Ae^{-(E_{act}/RT)} \quad \text{[Eqn 5.3]}$$

where A = Arrhenius constant (also known as the frequency factor, because its units are time^{-1}), E_{act} = activation energy, R = universal gas constant (8.314 kJ mol^{-1} K^{-1}) and T = absolute temperature (in Kelvin).

From Eqn 5.3 reaction rates can be seen to increase as temperature rises. For most reactions near room temperature (300 K) E_{act} = c.50 kJ mol^{-1}, so a 10 °C rise in temperature results in an approximate doubling of the reaction rate. However, the exponential dependence of reaction rate on temperature means that the rate increases by successively smaller amounts for every 10 °C interval as temperature rises so that, for example, the increase in rate is only c.1.4 for a 10 °C rise in the region of 200 °C.

The rate of a first order reaction can be represented by the change in concentration of the starting material with time, which yields the following differential equation:

$$\frac{dx}{dt} = k(a - x) \quad \text{[Eqn 5.4]}$$

where a = initial concentration and x = change in concentration. Integration then gives:

$$\log_e\left(\frac{a}{a-x}\right) = kt \quad \text{[Eqn 5.5]}$$

which represents exponential decay (e.g. radioactive decay). The time taken for half the material to disappear (i.e. x = a/2), no matter what the initial concentration, is constant. It is related to the rate constant and is termed the **half-life** ($t_{1/2}$):

$$t_{1/2} = \frac{\log_e 2}{k} \quad \text{[Eqn 5.6]}$$

A reversible reaction, such as epimerization, can be described by:

$$\underset{\text{reactant}}{Y} \underset{K_b}{\overset{k_f}{\rightleftharpoons}} \underset{\text{product}}{Z} \quad \text{[Eqn 5.7]}$$

where k_f and k_b = rate constants of forward and backward reactions, respectively. As the forward reaction progresses, the concentration of the reactant ([Y]) decreases and that of the product ([Z]) increases, so the rate of the forward reaction slows, while that of the backward reaction increases. The rate of the forward reaction (disappearance of Y) can be represented by:

$$\frac{-d[Y]}{dt} = k_f[Y] - k_b[Z] \quad \text{[Eqn 5.8]}$$

Writing this in the same form as Eqn 5.4 yields:

$$\frac{dx}{dt} = \left(\frac{K+1}{K}\right)k_f(m - x) \quad \text{[Eqn 5.9]}$$

where m = $k_f[Y_0]/(k_f + k_b)$, $[Y_0]$ = initial reactant concentration, x = change in concentration, K = equilibrium constant (k_f/k_b) and initially there is no product (i.e. [Z] = 0). Integration then gives:

$$\log_e\left(\frac{m}{m-x}\right) = \left(\frac{K+1}{K}\right)k_f \quad \text{[Eqn 5.10]}$$

It is difficult to obtain absolute concentrations, but it is possible to use the ratio of reactant and product:

$$\frac{[Z]}{[Y]+[Z]} = \left(\frac{K}{K+1}\right)(1 - e^{-(K+1)k_b t}) \quad \text{[Eqn 5.11]}$$

In all the above equations the rate of a reaction is seen to be temperature dependent, and the equations only apply to isothermal conditions. In geological systems, complex burial histories are usually involved, with variable rates of heating and possibly also intervals of cooling. For such systems the rate equations require integration with respect to both temperature and time, and it is assumed that A values are little affected by temperature change (Lewis 1993).

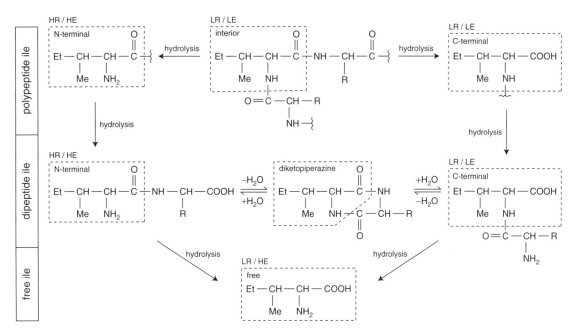

Fig. 5.39 Possible variations in molecular environment for isoleucine (ile; after Mitterer & Kriausakul 1984). With increasing diagenesis, hydrolytic cleavage will move interior amino acids to terminal positions in peptide chains and ultimately to free amino acids. The N-terminal unit has a free amino group, whereas the C-terminal analogue has a free carboxyl group. These two varieties of terminal groups are interconverted via the formation of diketopiperazines. HR = high epimerization rate; LR = low epimerization rate; HE = high epimerization extent; LE = low epimerization extent.

predicted for epimerization of free isoleucine because most of the isoleucine would be in the slowly epimerizing interior positions, with the epimerization rate being effectively controlled by the small amounts of free isoleucine. However, the rate of hydrolysis generally lies between these extremes (and depends upon the strength of the peptide bond between isoleucine and its neighbouring amino acids), with isoleucine remaining in terminal positions long enough for a significant degree of epimerization to occur before free isoleucine is formed. This helps to explain the appearance of the rate curves for natural epimerization of isoleucine, such as that in Fig. 5.40. During the early part of the rate curve, hydrolysis rapidly produces a mixture of smaller peptides from the proteins, containing some isoleucine in N-terminal and diketopiperazine units, which epimerize rapidly. Further hydrolysis of these peptides is slower, ultimately releasing free isoleucine. As the proportion of free isoleucine builds up, its slower rate of epimerization becomes the rate determining factor. The resulting non-linear rate curves can, therefore, be interpreted as consecutive, first order rate lines, as shown in Fig. 5.40,

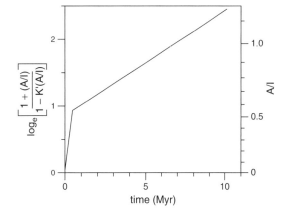

Fig. 5.40 Isoleucine epimerization in fossil foraminiferans (*Globorotalia* spp.) from deep-sea sediments interpreted as sequential first order reactions with different rates (A = D-alloisoleucine, I = L-isoleucine; after Müller 1984).

representing these two epimerization processes (Mitterer & Kriausakul 1984), with the intersection occurring around an A/I value of 0.6. More detailed analysis sometimes reveals additional linear sections in the rate curve, which may be attributable to leaching of free amino acids (affecting the early part of the curve, A/I <0.6) and to decreasing rates of protein hydrolysis with increasing age (Müller 1984; Kimber & Griffin 1987). The rate of epimerization/racemization of isoleucine and other amino acids is dependent upon organism genus (Müller 1984; Kimber & Griffin 1987) because of the influence of peptide hydrolysis rate, as mentioned above.

Despite the above complications, epimerization and racemization of amino acids have been used for dating Quaternary deposits (Coleman et al. 1987; Smart & Frances 1991). Relative ages can be obtained or, using a suitable calibration (e.g. ^{14}C dating; Box 5.5), estimates of actual ages can be obtained. The simplest dating application is in deep-sea sedimentary cores, because the thermal environment is stable and uniform over large expanses of the ocean. The extent of isomerization in individual species of benthonic foraminiferans can be calibrated against radiometric ages to provide a dating tool (Fig. 5.41a; Müller 1984), and data for two species can be combined to generate a concordia curve (Fig. 5.41b; Wehmiller 1993), providing greater confidence in the dating procedure.

In coastal settings, marine molluscs are frequently used for dating maxima in relative sea level (i.e. continental ice-volume minima) during the warm periods of interglacials and interstadials (the latter are smaller scale warming episodes during glacial periods; Kaufman & Miller 1992). The variations in temperature during successive warming and cooling episodes make absolute dating difficult, but calibration of epimerization values against other age data (e.g. oxygen isotopic stages; Box 5.6) can provide a means of identifying specific sea-level maxima at various locations (e.g. Shackleton 1987a).

Although the main application of amino-acid isomerization is in dating strata, differences in mean temperature between horizons have been estimated, yielding information on Quaternary climatic changes (e.g. Miller et al. 1987). This application requires independent dating and the use of approximate first order kinetics to describe the isomerization.

5.6.3 Degree of unsaturation in long-chain ketones

Although the ratio of $^{18}O:^{16}O$ in carbonate tests of marine invertebrates can be used to examine sea surface temperatures, the extent of continental glaciation, the balance between precipitation and evaporation, the carbonate compensation depth and the diagenetic alteration of carbonate all influence whether an estimation can be made and its accuracy (Boxes 5.6, 5.7; Marshall 1992). There is also a reliable organic geochemical palaeothermometer based on the distributions of long-chain (C_{37}–C_{39}) unsaturated ketones (Brassell et al.

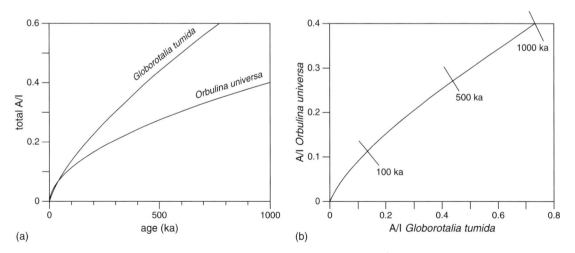

Fig. 5.41 (a) Isoleucine epimerization in total hydrolysates from fossil foraminiferans *Globorotalia tumida* and *Orbulina universa* in eastern Atlantic deep-sea sediment core 13519-2; (b) concordia curve based on (a) (A = D-alloisoleucine, I = L-isoleucine; after Müller 1984).

Box 5.5 | Radiocarbon dating

^{14}C is an unstable isotope of carbon that is formed in the upper atmosphere by neutron bombardment of ^{14}N, but subsequently decays back to ^{14}N by electron loss:

$$^{14}N + {}^{1}n \rightarrow {}^{14}C + {}^{1}p \quad [\text{Eqn 5.13}]$$

$$^{14}C \rightarrow {}^{14}N + e \quad [\text{Eqn 5.14}]$$

In the atmosphere the ^{14}C is rapidly oxidized to CO_2, which is then incorporated into organic matter by photosynthesis, and subsequently enters other organisms. The amount of ^{14}C in living organisms is reflected by the atmospheric content (it is not quite in equilibrium, due to isotopic fractionation favouring the lighter isotopes, but this can be allowed for using the $\delta^{13}C$ value; see Box 1.3), but upon death the amount of ^{14}C declines by radioactive decay, as represented by Eqn 5.14. The amount of ^{14}C relative to stable C isotopes provides a measure of the age of organic material. The half-life of ^{14}C was first estimated as 5568 yr (Libby et al. 1949), so that after about ten half-lives, or 55–60 kyr, there is so little ^{14}C left that the precision of the method decreases and age estimations become unreliable, requiring the use of other techniques (e.g. radioisotopes with longer half-lives).

The concentration of ^{14}C in a sample is compared to that in a standard (generally oxalic acid I or II), which permits the conventional radiocarbon age to be determined using first order kinetics for a half-life of 5568 yr, and correcting for any isotopic fractionation (Stuiver & Polach 1977). Ages before present (BP) relate to the arbitrary zero age of 1950. The conventional radiocarbon age assumes that the sizes of all ^{14}C reservoirs have remained constant throughout time (at 1950 values, before nuclear testing introduced a new source of ^{14}C), but this is not so. Variations in the atmospheric production of ^{14}C have occurred, related to changes in the intensity of cosmic radiation, and the exchange of $^{14}CO_2$ between the atmospheric and oceanic reservoirs has also varied. A further inaccuracy derives from the fact that the half-life is actually $c.5730$ yr. Radiocarbon ages consequently require calibration to provide accurate calendar ages, and this has been successfully achieved back to $c.11$ ka by counting tree rings (dendrochronology), as shown in Fig. 5.42. Radiocarbon age increasingly underestimates true calendar age as the latter increases.

Radiocarbon ages for marine organisms are typically 400 yr greater than for terrestrial organisms of the same calendar age (Stuiver & Braziunas 1993), as a result of the exchange rate between marine bicarbonate and atmospheric CO_2, and also the mixing of upwelling old deep water with surface water (Mangerud 1972).

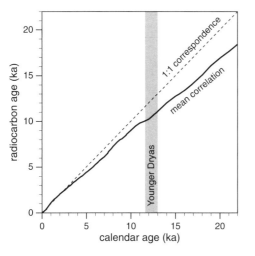
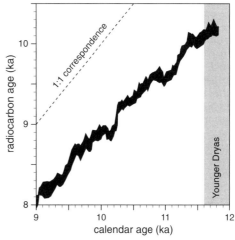

Fig. 5.42 (a) Deviation of radiocarbon age from true calendar age (based on dendrochronology); (b) detailed variation and error range of age calibration for the period following the Younger Dryas stadial (after Roberts 1998; based on OxCal calibration). See Box 6.5 for discussion of the Younger Dryas.

Box 5.6 | Oxygen isotopic record in carbonate tests

Fractionation of oxygen isotopes

Oxygen exists as three stable isotopes, ^{16}O, ^{17}O and ^{18}O (see Box 1.3 and Table 1.1), all of which are present in the water of the oceans and so take part in the equilibrium reaction involving carbonate shown in Eqn 3.8. Organisms that secrete calcium carbonate tests incorporate the isotopes of oxygen broadly in the ratio in which they occur in ambient seawater, although there is also an isotopic fractionation, which varies with temperature (the colder the water, the greater is the proportion of ^{18}O relative to ^{16}O) and species. The ratio of the two most abundant isotopes can be expressed by:

$$\delta^{18}O(‰) = [(^{18}O/^{16}O)_{sample}/(^{18}O/^{16}O)_{standard} - 1] \times 10^3$$
[Eqn 5.15]

The $\delta^{18}O$ value of the ambient water used to form carbonate tests depends upon a number of factors (Faure 1986). Water containing ^{16}O evaporates more readily than that containing ^{18}O, an effect that is more pronounced in cold than warm water because the partial pressure of $H_2{}^{18}O$ increases faster than that of $H_2{}^{16}O$ with increasing temperature. Similarly, during precipitation, $H_2{}^{18}O$ condenses more rapidly than $H_2{}^{16}O$, and the effect is more pronounced as temperature increases (a 1 °C rise causes a $\delta^{18}O$ increase of $c.0.7‰$ in snow relative to its source vapour; Dansgaard 1964). These **Rayleigh fractionation** processes lead to polar ice being enriched in ^{16}O compared with carbonates. There are various contributions to the final isotopic signature in rain and snow: the isotopic composition of surface water in the source region of moisture; the temperature during evaporation and condensation; and the amount of precipitation that has occurred en route from the source region of the moisture (Craig 1965). The lower the ambient temperature during precipitation and the further the moisture has travelled, the lower is the ratio of $^{18}O:^{16}O$ in rain and snow. For example, the $\delta^{18}O$ values for rain in the tropics, ice in Greenland and ice at the South Pole are, respectively, of the order of −30, −50 and 0‰. The deuterium content of precipitation behaves similarly to ^{18}O content, with δD increasing by 5–6‰ for every °C rise in temperature (Dansgaard 1964).

During periods of glaciation the ^{18}O content of seawater increases as more ^{16}O becomes locked up in continental ice-caps at high latitudes (Kahn et al. 1981). The $\delta^{18}O$ value of carbonate in the preserved shells of marine organisms can give a measure of the amount of water locked up in ice. The remains of benthonic foraminiferans are usually chosen, because

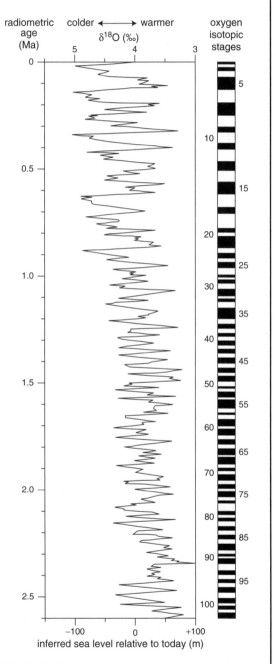

Fig. 5.43 Oxygen isotopic record and stages from sedimentary carbonate at ODP site 677 (eastern equatorial Pacific Ocean; after Shackleton et al. 1995).

Continued

bottom water temperatures in the Quaternary oceans were fairly constant and so temperature-dependent variations in $\delta^{18}O$ are minimal. A change in $\delta^{18}O$ of 0.1‰ corresponds to a sea-level change of c.100 m during the last glaciation, due to the change in amount of seawater trapped as ice on continental areas (Chappell & Shackleton 1986). Terraces in coral reefs can provide a means of calibrating $\delta^{18}O$ values with sea-level changes (Aharon & Chappell 1986).

Data from planktonic foraminiferans can provide information on sea-surface temperature (Epstein et al. 1953). This information can be more difficult to interpret because it is superimposed on the ice-volume effect and because sea-surface temperature patterns and the factors that affect them can be quite complex. A major factor is the balance between evaporation and precipitation, which affects salinity. All other conditions being constant, a fall of 1 °C results in an increase in $\delta^{18}O$ of c.0.2‰. If independent estimation of surface-water temperature is possible (e.g. $U^{K'}_{37}$), $\delta^{18}O$ values for planktonic foraminiferans can be used to evaluate salinity variations, which can provide information on climate, such as rainfall patterns in the Amazon Basin (Maslin & Burns 2000) and the intensity of the Indian monsoon (Rostek et al. 1993; Kudrass et al. 2001).

Oxygen isotopic stages

The oxygen isotopic record for the past 2.6 Myr preserved in carbonates at ODP (Ocean Drilling Programme) site 677 (eastern equatorial Pacific; Shackleton et al. 1995) is shown in Fig. 5.43, and reveals cycles of high and low $\delta^{18}O$ values. The high values correspond to relatively abundant ^{18}O, characteristic of colder periods with greater volumes of land ice. Although the detail of such records varies with location, the overall patterns are constant, reflecting the global nature of the climatic variations, and can be used as a dating tool. Odd numbered stages represent warm periods (interglacials and interstadials) and even numbered stages cold periods (glacials and stadials).

Box 5.7 | **Lysocline and carbonate compensation depth**

Carbonate tests are not always preserved in marine sediments. Shallow marine waters are generally supersaturated in calcium carbonate, which enables various organisms to precipitate carbonate tests. With increasing depth (i.e. pressure) in the water column, the degree of carbonate saturation decreases and the sinking carbonate tests of dead organisms begin dissolving. The **lysocline** marks the depth range over which there is a rapid increase in the undersaturation of carbonate, resulting in a dramatic increase in the dissolution rate.

Below the lysocline a depth is reached at which the rate of supply of carbonate particles equals the rate of dissolution, termed the **carbonate compensation depth** (**CCD**). Below the CCD no carbonate is deposited. The CCD varies with crystalline form (polymorph) of calcium carbonate, as well as the chemical composition and temperature of seawater, and the last two properties vary with ocean locality. For example, the CCD for calcite it is c.4500 m in the Atlantic and c.3000 m in the Pacific, but it is significantly shallower for aragonite. The dissolved CO_2 concentration (and hence the atmospheric partial pressure of CO_2; see Box 1.11) influences the CCD, so it is not surprising that the CCD has varied over geological time. It was much shallower from the Cretaceous to Eocene, and was probably only 1–2 km during the Cambrian and Devonian as a result of high atmospheric pCO_2 (Briggs & Crowther 2001).

1986). These compounds appear to be present only in a few living algae, the haptophytes, but are found in marine and lacustrine sediments throughout the world, probably reflecting blooms (Marlowe et al. 1984). The major contemporary source is the ubiquitous coccolithophore *Emiliania huxleyi*, which first appeared during the late Pleistocene (c.250 ka), but the compounds are also characteristic components in the morphologically related Gephyrocapsaceae, the emergence of which dates back to the Eocene (c.45 Ma; Marlowe et al. 1990). The haptophyte lineage can be traced back to the Cretaceous (see coccolithophores; Fig. 1.12).

Emiliania huxleyi and other haptophytes have been found to increase the degree of unsaturation of their long-chain ketones as temperature decreases, resulting in a lowering of the melting points of these lipids and so enabling the phytoplankton to maintain cellular fluidity and function in colder climates (as noted for fatty acids in Section 2.4.1a). These long-chain ketones seem to be more stable than most unsaturated lipids and can survive diagenesis (Prahl et al. 1989; Sikes et al. 1991), possibly aided by the unusual E (i.e. *trans*) configuration of the double bonds (Rechka & Maxwell 1988). However, under certain oxic conditions it is possible for the more highly unsaturated components to suffer a greater degree of degradation, leading to an overestimation of palaeotemperature (Hoefs et al. 1998; Gong & Hollander 1999).

The degree of unsaturation of long-chain ketones in *Emiliania huxleyi* can be evaluated from the concentrations of the dominant di- and triunsaturated C_{37} components ($C_{37:2}$ and $C_{37:3}$ respectively; see Fig. 5.1):

$$U^{K'}_{37} = \left(\frac{[C_{37:2}]}{[C_{37:2}]+[C_{37:3}]} \right) \quad \text{[Eqn 5.16]}$$

Values of $U^{K'}_{37}$ derived from Eqn 5.16 have been found to increase with sea-surface temperature (SST), and although there is some biogeographical variation, the following linear calibration is generally applied:

$$U^{K'}_{37} = 0.034(SST) + 0.039 \quad \text{[Eqn 5.17]}$$

This temperature calibration (Fig. 5.44) is derived from a range of culture experiments, sediment-trap data and sedimentary cores calibrated against $\delta^{18}O$ values for planktonic species of the foraminiferan *Globigerina* (Prahl & Wakeham 1987; Prahl et al. 1988). The $U^{K'}_{37}$ parameter can be measured to an accuracy of 0.02 units, potentially allowing temperature to be determined to within 0.5 °C.

Variations have been noted in the temperature calibration between different genera of haptophytes, although the impact on open-marine Quaternary temperature studies is likely to be limited given the dominance of *Emiliania huxleyi* in pelagic settings (*Gephyrocapsa oceanica* can be important in coastal waters; Sawada et al. 1996; Versteegh et al. 2001). Despite episodes of major changes in the composition of populations of *Gephyrocapsa* and *Emiliania* in the central North Atlantic during the past 290 kyr, the close correlation of $U^{K'}_{37}$ and $\delta^{18}O$ surface-temperature records indicates that the assemblage changes did not affect $U^{K'}_{37}$ values (Villanueva et al. 2002). In laboratory cultures, *Isochrysis galbana* has provided evidence of sensitivity of $U^{K'}_{37}$ values towards light and phosphate (but not nitrate) limitation, equivalent to a temperature decrease of 5–7 °C at 16 °C, which suggests that there is not a simple temperature control on alkenone unsaturation, at least in this species (Versteegh et al. 2001). Although algae usually experience growth-limiting conditions in natural environments, such factors would have a minor influence on $U^{K'}_{37}$ (Popp et al. 1998a), particularly if the bulk of the alkenones results from blooms.

The linear calibration has been found to fail at low temperature (<5 °C) in the Southern Ocean (Sikes & Volkman 1993) and at both high (>21 °C) and low (<12 °C) extremes in laboratory cultures (Conte et al. 1998). These deviations may reflect the temperature limits of the alkenone regulation (Conte et al. 1998), and suggest that a sigmoidal curve may be more appropriate (Fig. 5.44; Sikes & Volkman 1993). It has been proposed that errors in estimated palaeo-SST values are likely to be small, with the biogeographical variations mostly deriving from differences in population genetics and physiological status (alkenone production falls in the late log growth phase; see Box 5.8; Conte et al. 1998). However, in a review of global oceanic surficial sediment data (60° N to 60° S) Müller et al. (1998) obtained a good linear correlation over an annual mean sea-surface temperature range of 0–29 °C, identical within error limits to that in Eqn 5.17 (0.033(SST) + 0.44). This suggests that under natural conditions a reasonably linear

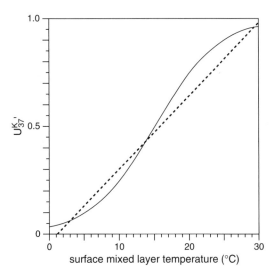

Fig. 5.44 Linear (0.034(SST) + 0.039; after Prahl et al. 1988) and sigmoidal ($1/(1 + e^{-0.22(SST-15.1)})$; after Sikes & Volkman 1993) temperature calibrations of $U^{K'}_{37}$.

Box 5.8 | Microbial growth phases

Microbial growth is generally described in terms of cell numbers, although an increase in the mass of the cell population also usually occurs. In laboratory culture, bacteria exhibit a growth curve that can be divided into four main phases: lag; exponential; stationary; and death (Fig. 5.45).

The **lag phase** can be observed upon a change in environment, and in bacteria it can be observed when nutrient supply increases. For example, when passing from a stationary phase to an exponential phase a lag may be observed before division resumes. Although cell numbers do not increase during the lag phase, cell mass may. The length of the lag phase is a function of how rapidly the microbial population acclimatizes to the new environmental conditions (Black 1996).

The **exponential (or logarithmic) phase** is characterized by uninterrupted division cycles, such that the population doubles at regular intervals (the generation time). There is usually no change in average bacterial cell mass, although the mass of individual cells increases before rapidly decreasing during cell division (by binary fission).

The **stationary phase** classically represents the stage at which the rate of cell division equals the rate of cell death, so the number of viable cells remains constant. This phase usually occurs when the cell concentration reaches a sufficient size that some property of the environment restricts growth rate, often a nutrient where phytoplankton are concerned. Physiological changes can occur, including adaptations that promote cell survival through periods of limited growth. Some bacteria (e.g. species of *Bacillus* and *Clostridium*) form endospores, which represent a dormant state that is resistant towards levels of desiccation, heat, chemical or radiation exposure that would prove lethal to the non-endospore forms.

The **death phase** (involving exponential/logarithmic decline) begins when cell deaths exceed births (i.e. the viable-cell count declines).

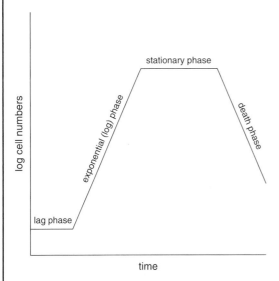

Fig. 5.45 Microbial growth phases.

relationship does exist, and that growth rate and nutrient availability do not significantly affect the temperature calibration.

Confidence in the $U_{37}^{K'}$ parameter is gained from good correlations with planktonic foraminiferal $\delta^{18}O$ records, an example of which is shown in Fig. 5.46.

5.7 Maturity of ancient sedimentary organic matter

5.7.1 Bulk compositional indicators of maturity

Macerals are affected by increasing maturity in a variety of ways, and we have already looked at changes in the reflectance of vitrinite and the fluorescence of liptinite in Box 4.1. Perhaps the simplest optical property to monitor is the change in colour, observed under transmitted light, of spores and pollen grains with increasing temperature (in a similar way to the charring of toasted bread). The carbonization of these palynomorphs results in a colour change from yellow in immature samples, through shades of orange/yellow-brown during diagenesis and brown during catagenesis, to black in metagenesis. Standardization is required to remove the subjectivity of assessing colour change, and colour charts can be used to assign numerical values, as in the **thermal alteration index** (TAI; Staplin 1969).

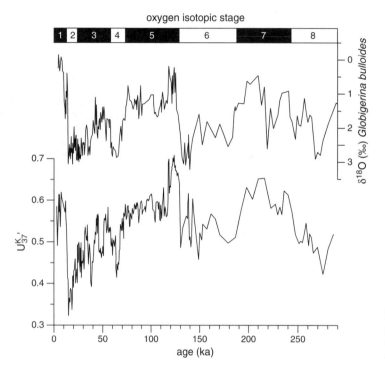

Fig. 5.46 Comparison of $U_{37}^{K'}$ and planktonic foraminiferal $\delta^{18}O$ profiles in a gravity core from the central North Atlantic (after Villanueva et al. 2002).

The behaviour of bulk organic material under pyrolysis can also provide maturity information. Among the changes that can provide such information are the amount of bitumen (a measurement of the thermal breakdown of kerogen) relative to remaining kerogen, and the temperature required to achieve the maximum rate of conversion of remaining kerogen into bitumen. A standardized method, Rock-Eval, is routinely applied in petroleum exploration (see Box 5.9). When evaluating maturity it is advisable to use a range of molecular, optical and pyrolytic parameters because any maturity indicator may be adversely affected by one factor or another (e.g. migration or source-related effects) and so provide misleading information when considered in isolation.

5.7.2 Molecular maturity parameters

(a) Light hydrocarbons

The light hydrocarbons of the gasoline range (C_4–C_{10}) account for c.30% of a crude oil, and so any information that can be gained from these components is likely to be reasonably representative of the bulk oil. The C_7 components are often selected for analysis because the number of isomers is manageable (i.e. virtually all can be resolved by gas chromatography) but sufficiently large to provide several ratios for correlation (Philippi 1981) and other purposes (see Section 4.5.3a and Fig. 4.24).

There is some conjecture about the origin of the light hydrocarbons. Thermal degradation of hydrocarbons previously generated from kerogen together with acidic clay-catalysed rearrangements have been proposed (Kissin 1987, 1990; Thompson 1979), although there is evidence that hydrocarbons are quite stable under catagenetic conditions (Mango 1991).

In general, it appears that the ratio of acyclic alkanes to cycloalkanes increases with increasing thermal maturity (Philippi 1975; Thompson 1979, 1983). To account for this behaviour, it has been proposed that higher polycyclic alkanes undergo cracking to form the cyclohexanes and cyclopentanes (Thompson 1979), which in turn undergo ring-opening with increasing temperature to produce acyclic alkanes (Thompson 1983). Two parameters based on these assumptions are the *heptane value* and *isoheptane value*, which both increase with increasing maturity:

Box 5.9 | Rock-Eval and related pyrolytic assessment of maturity

Method

There have been many variations on the general theme of bulk pyrolysis, including several generations of Rock-Eval instruments (Espitalié et al. 1977, 1985; Lafargue et al. 1998). As a general guide, a sample of powdered rock is heated rapidly to a maximum of 300 °C in an inert atmosphere, which vaporizes the bitumen already present, allowing it and any gas to be quantified by a flame-ionization detector, as the S_1 parameter. The temperature is then raised progressively ($c.25$ °C min^{-1}) to 600 °C, converting all the remaining petroleum potential of the kerogen into bitumen and gas, which is measured as the S_2 parameter. The temperature at which the maximum rate of hydrocarbon generation occurs during the S_2 measurement is also recorded, as the T_{max} parameter. In the latest instruments the CO and CO_2 evolved are also measured throughout the pyrolysis by an infrared (IR) detector. In earlier instruments only CO_2 was measured, over a narrow temperature range (using a thermal conductivity detector), providing an underestimate as the S_3 parameter. Prior treatment of the sample to remove carbonate provides more reliable estimates of oxygen content. The units of S_1 and S_2 are mg of hydrocarbons per g of rock (or kg per tonne of rock, i.e. parts per mil). The residual carbon is recorded in some instruments by the S_4 parameter, although usually total organic carbon (TOC) is independently evaluated.

Interpretation

With increasing maturity during catagenesis, hydrocarbons are generated in increasing quantity, so the S_2 measurement decreases while S_1 increases (in the absence of migration). Hence $S_1/(S_1 + S_2)$, the transformation ratio (or production index, PI), increases with increasing maturity. The value of T_{max} also increases with increasing maturity, reflecting the increasing thermal energy that is required to break the remaining bonds in kerogen associated with hydrocarbon generation. It is possible, therefore, to use T_{max} as a maturity parameter, but it is affected by the type of organic matter, because the thermal energy required to break the different types of bonds present varies (see Section 5.1.4). There are various guides in the literature to the interpretation of Rock-Eval data and the limitations of the method (e.g. Peters 1986; Killops et al. 1998).

A pseudo-van Krevelen diagram can be constructed from a plot of S_2/TOC (termed the hydrogen index, or HI) versus S_3/TOC (the oxygen index, OI). Although the underlying assumption that H content is proportional to hydrocarbon yield and O content is proportional to CO_2 can lead to significant errors in kerogen-type assessment, HI does provide an indication of the likely quantitative importance of oil among the total hydrocarbons generated. A source rock sample just approaching the onset of petroleum generation is generally considered to have only gas potential if its HI value is ≤ 150 mg g^{-1}, but to be mainly oil-prone if HI >300 mg g^{-1}.

heptane value = n-heptane/(n-heptane
+ cyclohexane + 2-methylhexane
+ 2,3-dimethylpentane
+ 1,1-dimethylcyclopentane + 3-methylhexane
+ cis-1,3-dimethylcyclopentane
+ trans-1,3-dimethylcyclopentane
+ 3-ethylpentane   [Eqn 5.18]
+ trans-1,2-dimethylcyclopentane
+ methylcyclohexane
+ 2,2,4-trimethylpentane
+ 2,2-dimethylhexane
+ 1,1,3-trimethylcyclopentane
+ 2,2,3,3-tetramethylbutane)

isoheptane value
= (2-methylhexane + 3-methylhexane)/
(cis-1,3-dimethylcyclopentane   [Eqn 5.19]
+ trans-1,3-dimethylcyclopentane
+ trans-1,2-dimethylcyclopentane)

Mango (1990a, b) has suggested that the various branched and cyclic C_7 components are formed entirely from n-heptane by steady-state, catalytic rearrangements involving transition metals and a cyclopropyl intermediate. Under such circumstances, 2,4-dimethylpentane (2,4DMP) and 2,3-dimethylpentane (2,3DMP) would be produced by the opening of different bonds in the cyclopropyl intermediate, which have slightly different energies, and so the ratio of these two components should be sensitive to temperature (Mango 1997):

$$\text{maximum temperature } (°C) = 140 + 15[\log_e(2,4\text{DMP}/2,3\text{DMP})] \quad \text{[Eqn 5.20]}$$

For an oil, the maximum temperature can generally be assumed to equate to the temperature of generation/expulsion of the hydrocarbons.

The mode of formation of light hydrocarbons clearly has implications for their application as maturity indicators, and Mango's model is not universally accepted (ten Haven 1996). The distribution of light hydrocarbons is significantly affected by variations in the original contributions to kerogen (Hayes 1991; Odden et al. 1998), so maturity differences inferred from light hydrocarbon distributions must be treated with caution.

(b) Carbon preference index

The carbon preference index (CPI) is a numerical means of representing the odd-over-even predominance in n-alkanes in a particular carbon-number range. It is often used as a maturity measurement where C_{25}–C_{33} n-alkanes from higher plant waxes are present (Bray & Evans 1961):

$$\text{CPI} = \frac{1}{2}\left[\frac{(C_{25}+C_{27}+C_{29}+C_{31}+C_{33})}{(C_{24}+C_{26}+C_{28}+C_{30}+C_{32})} + \frac{(C_{25}+C_{27}+C_{29}+C_{31}+C_{33})}{(C_{26}+C_{28}+C_{30}+C_{32}+C_{34})}\right] \quad \text{[Eqn 5.21]}$$

This form of the CPI calculation sometimes does not give a value of 1 when there is a smooth distribution of n-alkanes with no apparent C-number preference. An alternative formula, CPI2, has been proposed to overcome this problem (Marzi et al. 1993):

$$\text{CPI2} = \frac{1}{2}\left[\frac{(C_{25}+C_{27}+C_{29}+C_{31})}{(C_{26}+C_{28}+C_{30}+C_{32})} + \frac{(C_{27}+C_{29}+C_{31}+C_{33})}{(C_{26}+C_{28}+C_{30}+C_{32})}\right] \quad \text{[Eqn 5.22]}$$

Values of CPI for immature higher plant contributions are often $\gg 1.0$ but approach 1.0 with increasing maturity. This is a result of two factors: (a) the random cleavage of alkyl chains in the kerogen matrix producing n-alkanes (in the requisite carbon-number range) with no odd or even predominance, which dilute the original OEP of the higher plant contribution; (b) progressive loss of the high-OEP component during the expulsion of hydrocarbons. Consequently, the CPI values for oils are usually $c.1.0$. EOPs (i.e. CPI values <1.0) are relatively rare, but have been reported among the shorter-chain n-alkanes ($<C_{23}$) in sediments of various ages and from a variety of depositional environments, but without corresponding distributions among other straight-chain lipid components. Under such circumstances, the EOP seemingly does not derive from reduction of fatty acids/alcohols with C-number preservation, but is probably the result of direct microbial biosynthesis (Grimalt & Albaigés 1987).

(c) Biomarker transformations

The transformations reviewed in Section 5.5 are potential maturity indicators. The most useful reactions are those in which only one of the pair of components is present initially in immature sediments, so that the extent of the transformation can be attributed entirely to thermal maturation (the kinetics of the transformation are also simpler; Box 5.4). Such reactions include isomerization of pristane at C-6 and C-10, of steranes at C-20 and of hopanes at C-22, and also the aromatization of C-ring monoaromatic steroidal hydrocarbons. A number of molecular maturity parameters are shown in Fig. 5.47, together with some bulk maturity measurements. The correlation of values is approximate and varies with the type of organic matter present, its potential for generating petroleum and its heating rate.

It can be seen from Fig. 5.47 that most of the biomarker hydrocarbon maturity indicators operate at relatively low maturities and have reached their end points before the end of the oil generation window, except for the ratio of short-chain to long-chain triaromatic steroids. This limits their use in petroleum exploration mainly to the assessment of whether potential source rocks have reached the onset of catagenesis. The relative abundances of some families of biomarkers can be used to extend the effective range of maturity indications. For example, 17α-diahopanes are predicted to be more thermodynamically stable than the 18α-neohopanes, which are in turn more stable than the 17α-hopanes, so diahopane abundance increases towards the end of the oil window (Moldowan et al. 1991; Killops et al. 1998). Similarly, cheilanthanes appear to be more thermally resistant and are generated from kerogen at higher temperatures than the hopanes, so their relative abundance tends to increase towards the end of the oil window too (Peters et al. 1990; Farrimond et al. 1999). In coaly source rocks the abundance of C_{30} relative to C_{31} 17α-hopanes increases during catagenesis (Fig. 5.48; Killops et al. 1998).

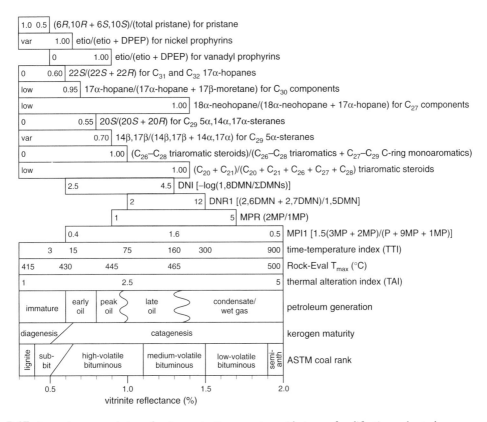

Fig. 5.47 Approximate correlation of various maturity parameters with stages of coalification and petroleum generation.

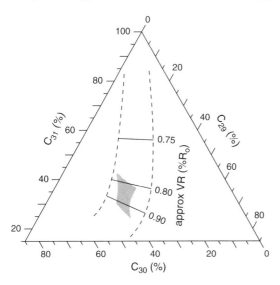

Fig. 5.48 Ternary plot of relative abundance of C_{29}, C_{30} and C_{31} 17α-hopanes (22S + 22R) in New Zealand coals (shaded area shows province of coal-sourced oils in Taranaki Basin; after Killops et al. 1998). VR = vitrinite reflectance.

All such maturity-related changes indicate the caution that needs to be applied when using biomarkers as an aid to correlating oils with their sources. The potential problems are not restricted to the higher-maturity end of catagenesis. For example, 28,30-dinorhopanes appear to derive from the original free bitumen in source rocks rather than from kerogen breakdown, so their concentration declines rapidly with increasing maturity during catagenesis (Peters & Moldowan 1993). The relative abundance of oleanane, an indicator of angiospermous contributions, increases to a maximum at the start of the oil window and thereafter remains relatively constant (Ekweozor & Telnaes 1990). The initial low abundance may be partly due to the relative stability of oleanenes during late diagenesis/early catagenesis.

(d) Methyl group isomerization in aromatic hydrocarbons

Various molecular maturity parameters, based on the positions of methylation of aromatic compounds, have been developed from studies of coals and type III kerogens (Section 5.5.4) and can also be applied, with caution, to type II kerogens. A few of the many possible parameters are examined here (Radke 1987; Kvalheim et al. 1987).

The dimethylnaphthalene index (DNI) is based on the decrease in the ratio of the least thermodynamically stable isomer, 1,8-dimethylnaphthalene (1,8DMN), to the total amount of all DMNs with increasing maturity (Alexander et al. 1984b). Amounts of 1,8DMN are usually very small in relation to other DMNs, and DMNs can be lost during the isolation procedure because they are quite volatile. However, these drawbacks are ameliorated by the use of a logarithmic scale:

$$\text{DNI} = -\log_{10}\left(\frac{1,8\text{DMN}}{\Sigma \text{DMNs}}\right) \quad \text{[Eqn 5.23]}$$

Comparison with other molecular maturity indicators suggests that DNI probably operates across the oil window, as shown in Fig. 5.47.

The ratio of 2,6DMN and 2,7DMN, which both have the most thermodynamically stable ββ-methyl groups (Fig. 5.37), to 1,5DMN, which has the less stable αα substitution pattern, is the basis for another maturity parameter, DNR1 (dimethylnaphthalene ratio 1). Values of this ratio appear to be variable below $c.1.0\%$ R_o (which generally marks the onset of oil expulsion), but then increase linearly from 2 to $c.12$ by 1.5% R_o (Fig. 5.47; Radke et al. 1984).

The ratio 2-methylphenanthrene to 1-methylphenanthrene (2MP:1MP), known as the methylphenanthrene ratio (MPR), compares β and α methyl substituents (Fig. 5.37). Like DNR1, MPR is variable below $c.1.0\%$ R_o for coals, but then increases linearly from a value of $c.1$ to $c.5$ by 1.7% R_o (Fig. 5.47; Radke et al. 1984). In comparison, the methylphenanthrene index (MPI1)—the first of the methyl-aromatic maturity parameters to be developed—is more complex, being based on the relative abundances of phenanthrene (P) and its four major methyl homologues (Radke & Welte 1983):

$$\text{MPI1} = \left[\frac{1.5(2\text{MP} + 3\text{MP})}{(\text{P} + 1\text{MP} + 9\text{MP})}\right] \quad \text{[Eqn 5.24]}$$

This parameter incorporates the potential for methylation of P at lower maturities, demethylation of MPs to yield P at higher maturities and isomerization of MPs. The result of these combined processes is that MPI1 increases to $c.1.6$ at the end of the oil window (vitrinite reflectance 1.3%) and thereafter decreases. MPI1 has been calibrated against vitrinite reflectance (Fig. 5.49)

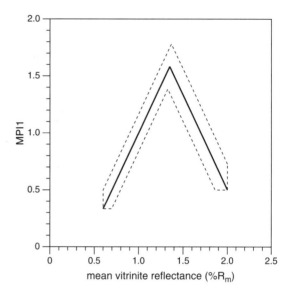

Fig. 5.49 Correlation of methylphenanthrene index (MPI1) with vitrinite reflectance for bituminous coals and type III kerogens (envelope shows error range; after Radke & Welte 1983).

and it is possible to obtain a calculated value of vitrinite reflectance (R_c) from MPI1:

$$\%R_c = 0.60(\text{MPI1}) + 0.40 \quad \text{(for } R_o < 1.35\%)$$
$$\text{[Eqn 5.25a]}$$
$$\%R_c = -0.60(\text{MPI1}) + 2.30 \quad \text{(for } R_o > 1.35\%)$$
$$\text{[Eqn 5.25b]}$$

It can be seen that a single MPI1 value can correspond to two different levels of maturity, but it is usually obvious from other maturity indicators (e.g. MPR <2.65 indicates that the lower maturity applies; Radke et al. 1986) whether the lower or higher maturity level is correct. Changes in the type of organic matter present down a stratigraphic section (facies changes; see Box 1.1) can adversely affect the maturity inferences of MPI1. However, this parameter does give useful comparative maturity data for bitumens from related source rocks over a maturity range associated with the oil window and the wet gas zone. The MPI1 and %R_c values for oils should reflect those of the source rock at the maturity level associated with the main phase of oil expulsion.

Differing thermodynamic stabilities in methyldibenzothiophenes have also been used for maturity estimations. The ratio that shows the greatest maturity-related changes is 4MDBT/1MDBT (the methyldibenzothio-

Table 5.4 Estimates of activation energies (E_{act}) and frequency factors (A) for some biomarker reactions (k_f and k_b = rates of forward and backward reactions respectively; after Mackenzie & McKenzie 1983; Abbott et al. 1985)

reaction	E_{act} (kJ mol^{-1})	A (s^{-1})	k_f/k_b
20R to 20S sterane isomerization	91	6×10^{-3}	1.174
22R to 22S hopane isomerization	91	1.6×10^{-2}	1.564
6R,10S to 6R,10R + 6S,10S pristane isomerization	120	2.1×10^{7}	1
C-ring monoaromatic steroid aromatization	200	1.8×10^{14}	∞

phene ratio, MDR), but it is dependent upon the type of organic matter present (Radke et al. 1986).

As noted in Section 5.5.4, the distributions of methylated naphthalenes and phenanthrenes in bitumens are affected by the addition of components from the thermal breakdown of kerogen that have different distributions from those in the bitumen inherited from diagenesis (Radke et al. 1990). In particular, the beginning of the linear increases in MPR and DNR1 seems to coincide with the onset of oil expulsion from oil-prone coals (Killops et al. 2001). The effects of the dilution and expulsion processes are more noticeable in coals, because their relatively large adsorption capacity tends to delay oil expulsion to a greater extent than occurs in marine shales. Although thermodynamically controlled isomerization and methyl-group migration may not, therefore, occur to the extent previously thought necessary to account for the changes in various methyl-aromatic maturity parameters, the utility of the maturity parameters is not greatly diminished.

The various methyl-aromatic ratios are likely to be adversely affected by water washing (see Section 4.5.6d) because α-type isomers are more water-soluble than their β counterparts. However, the ratios are useful for highly biodegraded oils (see Section 4.5.6c) because of the resistance of the compounds towards biodegradation.

5.7.3 Modelling kerogen maturation

(a) Temperature and time

Measuring maturity is not as straightforward as may at first be thought, because it is dependent upon the parameter selected (e.g. steroid aromatization, sterane isomerization). Each parameter involves a particular reaction or group of reactions, the rates of which are governed by characteristic activation energies and frequency factors (Table 5.4; see Box 5.4). In terms of petroleum generation, the extent of kerogen transformation is the most obvious choice, but different kerogen types involve the cleavage of different types of

Table 5.5 Activation energy distribution for kinetic modelling of vitrinite reflectance (EASY%R_o; A = 10^{13} s^{-1}; after Burnham & Sweeney 1989; Sweeney & Burnham 1990)

E_{act} (kcal mol^{-1})	stoichiometric fraction
34	0.0353
36	0.0353
38	0.0471
40	0.0471
42	0.0588
44	0.0588
46	0.0706
48	0.0471
50	0.0471
52	0.0824
54	0.0706
56	0.0706
58	0.0706
60	0.0588
62	0.0588
64	0.0471
66	0.0353
68	0.0235
70	0.0235
72	0.0118

chemical bonds, requiring different amounts of energy (e.g. Tegelaar & Noble 1994). Consequently, a kerogen with plentiful relatively weak bonds, such as type II-S kerogen, appears more mature, in terms of kerogen transformation, than a type III kerogen that has experienced an identical burial/thermal history. Often vitrinite reflectance is used as a maturity reference frame, and kinetic parameters have been derived to enable the accuracy of burial histories to be assessed against measured values (Table 5.5; Burnham & Sweeney 1989; Sweeney & Burnham 1990).

If a sedimentary horizon experiences uplift, its temperature decreases and maturation reactions slow down and may be effectively frozen, because the dominant control is temperature (Huang 1996). The importance of temperature, and particularly the maximum temperature experienced, can be seen from the fact that most oil is generated from kerogen over a relatively narrow window of $c.100–150\,°C$ (Mackenzie & Quigley 1988). Because of this, the term thermal maturity is often used. The general relationship between temperature and depth of burial is shown in Fig. 4.20, based on an average value for the geothermal gradient of $30\,°C\,km^{-1}$ (see Box 4.4). In basins where the temperature gradients are higher or lower, the depths for the peak generation of oil would be correspondingly shallower or deeper. Geothermal gradients can range from 10 to $80\,°C\,km^{-1}$. The lowest values are found in convergent plate margins and cratons (Precambrian shields) where orogenic (mountain building) events are of Palaeozoic or greater age (surface heat flow $c.20–50\,mW\,m^{-2}$). High geothermal gradients and surface heat flows ($c.100–250\,mW\,m^{-2}$) are typical along mid-ocean ridges and in rifted intracratonic areas or where the crust is thin. Geothermal gradients generally fall in the range $25–45\,°C\,km^{-1}$, and given a burial rate range of $10–500\,m\,Myr^{-1}$, heating rates are $0.25–22.5\,°C\,Myr^{-1}$, although they mostly fall in the range $0.3–15\,°C\,Myr^{-1}$.

In reconstructing the thermal history of the kerogen in a sedimentary rock it must be taken into account that geothermal gradients can vary significantly with depth and time during the life of a basin. Gradients increase when the basin is affected by orogenic and/or magmatic events. For example, the Tertiary Pannonian Basin (Central Europe) has been affected by the Carpathes alpine orogeny; the associated high heat flow and geothermal gradient ($c.50\,°C\,km^{-1}$) have resulted in source rocks of only Pliocene age reaching sufficiently high temperatures to generate oil.

Stable continental margins are areas where the geothermal gradient has decreased with time. Initially, sediments were deposited in areas of very high heat flows, close to a spreading ridge, as the margin developed upon rifting of the continent. However, with continued sea-floor spreading the distance between these marginal sediments and the spreading ridge increased and so heat flow and geothermal gradient decreased (Royden et al. 1980). Such conditions were prevalent during the Cretaceous period, when a particularly active phase of sea-floor spreading was commencing and continental land masses, providing copious sediments, were close to the spreading ridges (e.g. the Cretaceous organic-rich deposits off the Atlantic coast of South America). Similarly, failed rifts, such as the North Sea, experienced high heat flows initially, as extension brought hot asthenosphere to shallow depth beneath the thinned crust (see Box 1.2), but as extension waned the heat flow and geothermal gradient declined (Cornford 1998).

On stable platforms geothermal gradients have remained fairly constant because there has been little or no tectonic or magmatic activity for a very long time. The geothermal gradient is usually in the range $25–35\,°C\,km^{-1}$ for basins in such areas (e.g. Paris Basin, Palaeozoic basins on Precambrian basement in Australia and Mesozoic basins on Palaeozoic or older basement in west Canada).

Temperature and, to a lesser extent, time control petroleum generation. The higher the temperature experienced by kerogen the less time is required for oil generation. The amount of time needed increases exponentially with decreasing temperature. For example, if $c.200\,Myr$ are required at $60\,°C$, only $c.10\,Myr$ would be needed at $100\,°C$. The most dramatic evidence of petroleum generation from young sedimentary organic matter at elevated temperatures is provided by oceanic hydrothermal systems (Simoneit 1990). For example, petroleum-like hydrocarbons have been detected in hydrothermal vent mounds in the Guaymas Basin (California) and appear to originate from sediments ^{14}C-dated at $<5\,ka$. The vent waters have recorded temperatures of $c.200\,°C$, but are believed to be expelled at $c.315\,°C$ and a pressure of $c.20\,MPa$ (Simoneit et al. 1986b; Didyk & Simoneit 1989). Igneous intrusions can also cause rapid thermal evolution of sedimentary organic matter (e.g. Murchison & Raymond 1989).

(b) Kerogen transformation kinetics

If the thermal history of a potential petroleum source rock can be pieced together (Box 5.10), the amount and timing of petroleum generation can be modelled, provided appropriate kinetic parameters for the kerogen are available. One of the earliest and simple pseudo-kinetic models is the time–temperature index (TTI; Waples 1980). It was based on the observation that the rate of a chemical reaction, such as bond cleavage during hydrocarbon formation from kerogen, approximately doubles for every $10\,°C$ rise in temperature (see Box 5.4). Numerical values reflecting the rate doubling for $10°$ intervals are simply multiplied by the time spent in each of these intervals (Table 5.6), and oil generation is considered to occur over a particular TTI range (15–160; see Fig. 5.47). However, the TTI model tends to overestimate the influence of time and underestimate the effects of temperature in petroleum formation (it is best suited to average heat flow conditions) and fails to take into account compositional variations of kerogen

Box 5.10 | Thermal history modelling

The maturity of a sedimentary horizon depends on its thermal history, a topic that is beyond the scope of this book, but that has been reviewed in a number of texts (e.g. Barker 1996). The more deeply the horizon is buried, the greater the temperature it experiences, although it is unlikely to have experienced a uniform increase in temperature. Thermal history can be modelled using average geothermal gradients, although a more accurate method is to estimate heat flow variations over time. The heat input at the bottom of the crustal unit being modelled must be estimated; it represents the heat flowing out from the Earth's core (termed the reduced heat flow). Radiogenic heat production in the upper crust (mainly from ^{238}U, ^{235}U,

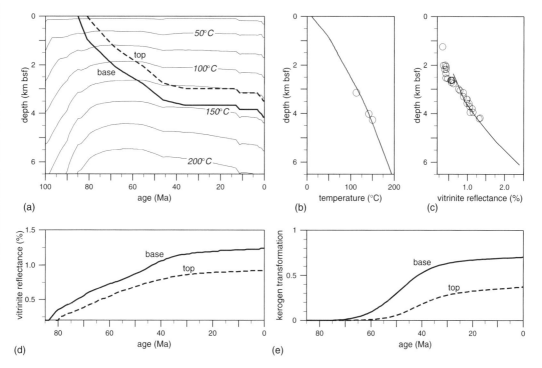

Fig. 5.50 A simple example of thermal and kinetic modelling based on a Cretaceous petroleum source rock in an exploration well (Tara-1, Great South Basin, New Zealand; after Killops et al. 2003). The thermal model (a) is constrained by a present-day surface heat flow of $61\,mW\,m^{-2}$ (a similar value being used for pre-sedimentation conditions), by measured bottom-hole temperatures (b) and by measured vitrinite reflectance (c). Vitrinite reflectance for the top and bottom of the kerogen-rich interval (d) is modelled using EASY%R_o (after Sweeney & Burnham 1990) and bulk kerogen transformation (e) is modelled using a discrete-E_{act} distribution and single A value determined for a representative kerogen sample. (Depths bsf = below sea floor.) The initial rise in isotherms is caused by the rise in hot asthenosphere accompanying extension (c.100–80 Ma, β factor = 1.7), which is modelled by a rise in basal lithospheric heat flow from 40 to $55\,mW\,m^{-2}$, followed by a decline to $35\,mW\,m^{-2}$ upon cessation of rifting. The resulting subsidence creates room for the rapid accumulation of sediments in the basin, and over a period of c.35 Myr the top of the source rock has been buried to a depth of c.3 km (corresponding to a heating rate of $c.3\,°C\,Myr^{-1}$). Following a thermal lag period after the end of rifting, isotherms gradually sink, sedimentation rates decline and the source rock experiences little change in temperature.

Continued

^{232}Th and ^{40}K), which depends upon the lithologies present, together with any igneous heat sources, must also be taken into consideration (Deming & Chapman 1989; Armstrong & Chapman 1998). The upward flow of heat depends upon the thermal conductivities of the various lithological units present, which in turn depend upon matrix conductivity (i.e. the lithology without porosity), porosity and fluid conductivity (i.e. the brines in pores). In typical rift basins (see Box 3.11), a higher heat flow is applied at the base of the crustal unit being modelled during active extension, which is then allowed to decay over time to represent the subsequent thermal relaxation (and subsidence). A change in heat flow takes time to propagate through the stratigraphic column (i.e. transient heat flow), and depends upon the thermal conductivity and heat capacity of each stratigraphic unit. Such disturbances of heat flow include an increase in basal heat flow, the influence of an igneous intrusion or the cold-blanket effect of the rapid deposition of a thick layer of sediment.

To build up an accurate thermal history requires the lithology and the age of the base of each stratigraphic unit to be known, so a depth versus age plot can be constructed, to which is subsequently applied the thermal model. Compaction (porosity decrease) with increasing overburden and episodes of uplift/erosion are important considerations. Allowances for any changes in the sedimentation rate within a unit also help to refine the burial history. The physical characteristics required for thermal modelling (e.g. thermal conductivity and porosity) can be measured or estimated from published data. Water depth has little influence on thermal maturity, other than controlling the temperature at the surface of the sedimentary pile.

Calibration of the thermal model is possible using present-day bottom-hole temperatures from exploration wells (Deming & Chapman 1989) and by the comparison of modelled and measured vitrinite reflectance (see Section 5.7.3), an example of which is shown in Fig. 5.50. There are other measurements that can provide time–temperature constraints, such as fission-track analysis (Naeser 1993; Gleadow & Brown 1999), homogenization temperatures of fluid inclusions (Roedder 1984) and clay transformations (Hoffman & Hower 1979).

Table 5.6 Calculation of TTI values (after Waples 1980). TTI = $\Sigma(\delta t_i)(r^n_i)$, where t = time spent (Myr) in 10 °C temperature interval i and r^n = the temperature factor associated with that interval, as below

temperature interval i (°C)	temperature factor r^n
10–20	2^{-9}
20–30	2^{-8}
30–40	2^{-7}
40–50	2^{-6}
50–60	2^{-5}
60–70	2^{-4}
70–80	2^{-3}
80–90	2^{-2}
90–100	2^{-1}
100–110	1
110–120	2
120–130	2^2
130–140	2^3
140–150	2^4
150–160	2^5

that significantly affect the temperature associated with the oil window. The TTI is little used today when computer power can be applied to more sophisticated kinetic models (Wood 1988).

The simplest first order kinetics are derived from the rate at which the petroleum-forming component of kerogen disappears in a series of experiments in which the kerogen is heated at different rates. Even so, there is no exact solution of the resulting Arrhenius equations (see Box 5.4). Usually, a single frequency factor (A value) is chosen and the equations are solved (by regression analysis) for a distribution of activation energies. A degree of optimization of the A value is possible in some computer programs; an increase in the value of A speeds up a reaction, which can be roughly compensated by an increase in the mean value of the E_{act} distribution. The distribution of activation energies represents the variety of bonds that are broken during kerogen degradation, which ideally should be characterized by different pairs of A and E_{act} values. The kinetic parameters for vitrinite maturation shown in Table 5.5 are an example of this type of approximation. Examples for different kinds of kerogen (Tegelaar & Noble 1994) are compared in Fig. 5.51 and can be represented by Fig. 5.52a. The dominance of algaenan in the type I kerogen, comprising long alkyl chains with a restricted range of C–C bond energies and hence a narrow E_{act} distribution, is reflected in the rapid transformation of the kerogen once sufficient thermal energy is available. Coal and type III kerogens have a much wider range of E_{act} values. The

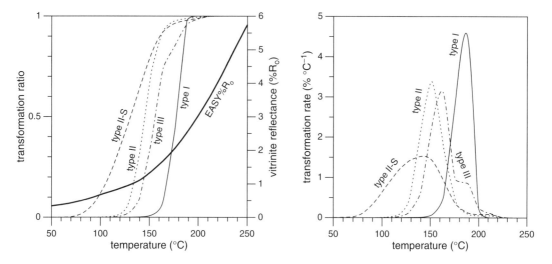

Fig. 5.51 Kerogen transformation at a constant heating rate of $3\,°C\,Myr^{-1}$, based on single A plus discrete E_{act} distributions (after Tegelaar & Noble 1994) for tasmanites (type I, Cretaceous, North Slope, Alaska), Monterey shale (type II-S, Miocene, Ventura, California), Kimmeridge Clay (type II, Jurassic, North Sea) and Manville Formation (type III, Cretaceous, Alberta). Modelled vitrinite reflectance based on EASY%R_o (after Sweeney & Burnham 1990; Table 5.5).

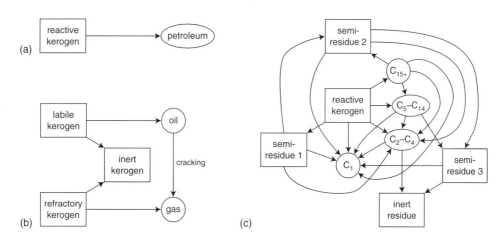

Fig. 5.52 Examples of reaction networks of varying complexity that can be used to model kerogen transformation.

presence of weak C–S and S–S bonds in type II-S kerogen accounts for the earlier onset of transformation compared with standard type II kerogen. An example of the application of simple bulk coal transformation kinetics is shown in Fig. 5.50.

Kinetic models can be further refined by incorporating functions that describe in more detail the different types of reactions involved in petroleum generation. For example, by monitoring the products of kerogen transformation, separate sets of kinetic parameters can be obtained for oil and gas (Pepper & Corvi 1995a), and kinetics for the cracking of oil to gas at higher temperatures can also be incorporated (Pepper & Dodd 1995), as shown in the slightly more complicated scheme in Fig. 5.52b. Yet more complexity can be introduced by dividing the oil into various fractions (Espitalié et al. 1988). One possible reaction network of this type is shown in Fig. 5.52c, each reaction having its own A value and E_{act} distribution. Nevertheless, these refinements are still a long way from accurately representing the chemical complexity of kerogen.

Kinetic models can be incorporated into basin models, which take into account expulsion efficiencies and migration (see Section 4.5.5), and are best calibrated against available kerogen transformation data obtained from exploration wells (e.g. Ungerer 1990, 1993; Pepper & Corvi 1995b). A problem in studying the kinetics of kerogen transformation is whether laboratory simulations, which have to be carried out in a reasonable time (i.e. $c.10^{10}$ times faster than in nature), accurately reflect *in situ* processes. There is the potential for significant error when extrapolating laboratory-scale reactions over short time periods at high heating rates to geological conditions (Snowdon 1979; Jarvie 1991; Espitalié et al. 1993; Pepper & Corvi 1995a). The mineral matrix can also affect the derived kinetic parameters (Dembicki 1992; Huizinga et al. 1987).

A major concern is which type of pyrolysis best represents natural maturation. Open-system pyrolysis, based on the Rock-Eval method (Box 5.9), is frequently used for convenience and appears to be able to predict natural petroleum systems (Burnham 1998). In this method kerogen samples, preferably of a maturity immediately preceding oil generation, are heated at different rates and the petroleum-like products escape from the system via a suitable detector. It is also possible to determine kinetic parameters using closed-system pyrolysis, whereby kerogen samples are sealed into tubes and series of experiments are carried out heating isothermally for varying periods of time. The products are confined until the end of each experiment. Open-system pyrolysis generates a range of *n*-alkenes as well as *n*-alkanes, due to the lack of available hydrogen (cf. Fig. 4.27). Water can be added as a source of hydrogen during closed-system pyrolysis (hydrous pyrolysis), resulting in a more natural distribution of hydrocarbons (Lewan 1997; Burnham 1998).

Comparison with natural systems suggests that kinetic parameters from open and closed pyrolysis can predict natural maturation of type II kerogens with similar accuracy, but coal and type III kerogens are not well modelled by data from open-system pyrolysis (Burnham et al. 1995; Schenk & Horsfield 1993, 1997, 1998). The discrepancy has been attributed to solid-state aromatization reactions that occur under slow, natural, heating rates in coaly material, primarily during the high- to medium-volatile bituminous rank range (vitrinite reflectance $c.0.6–1.5\%$). This aromatization competes with petroleum-generating reactions in natural systems, giving rise to new bonds with higher activation energies, but does not occur to a significant extent during rapid, artificial pyrolysis (Schenk & Horsfield 1998). Open-system kinetic parameters for coaly material may, therefore, lead to inaccurate predictions of the timing and extent of petroleum generation (Schenk & Horsfield 1997; Killops et al. 2002). Coals seem to undergo structural rearrangement from relatively low maturities, because pyrolytic measurements of oil potential give values that increase right up to the apparent onset of oil generation (Killops et al. 1998). Source rocks initially behave as closed systems, providing ample opportunity for maturation products to undergo secondary reactions (Ungerer 1990; Ritter et al. 1995), but once petroleum expulsion occurs a source rock may be better represented by open-system kinetics. The relatively late expulsion threshold observed for coals is likely to be a significant factor.

(c) Effect of geothermal gradient on molecular maturity parameters

The relationships between the various maturity indicators in Fig. 5.47 is intended as a guide only, because the greater temperature dependence of some processes will alter their relative operational range in basins with higher than average heat flows, as shown for sterane isomerization at C-20 and steroid aromatization in Fig. 5.53. Similarly, vitrinite reflectance values may underestimate maturity, in terms of the onset of catagenesis, in rapidly subsiding basins with high geothermal gradients such as the Mahakam Delta, Indonesia (Radke 1987).

The kinetic data in Table 5.4 show that steroidal aromatization is much more sensitive to temperature than the isomerization reactions (i.e. it has larger E_{act} and A values). This is more obvious from the plot of $\log_e(k)$ versus $1/T$ (i.e. a log plot of the Arrhenius equation) for these steroid and hopane reactions in Fig. 5.54a. Hopane isomerization is faster than sterane isomerization at all temperatures. In most oils the level of maturity is such that equilibrium has been reached in the isomerization at C-22 in hopanes, while isomerization at C-20 in steranes may not quite have reached equilibrium.

At temperatures above $c.100\,°C$ steroidal aromatization occurs at an increasingly faster rate than sterane isomerization. As a result, in young extensional basins in which there is a high heat flow, such as the Pannonian Basin in Hungary, aromatization can be virtually complete while sterane isomerization is at an early stage (Fig. 5.53a). In older basins with more moderate heat flows, such as the East Shetland Basin (northern North Sea; Fig. 5.53b), the steroid isomerization and

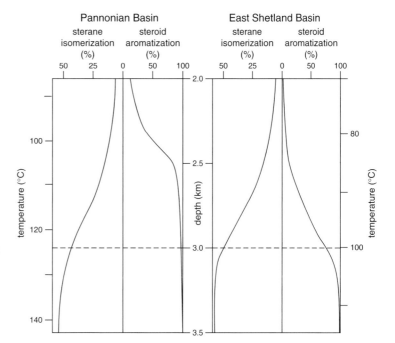

Fig. 5.53 Plots of depth versus isomerization of steranes at C-20 and aromatization of C-ring monoaromatic steroids in the Pannonian Basin (Pliocene deposits, geothermal gradient $\geq 50\,°C\,km^{-1}$) and East Shetland Basin (Jurassic, geothermal gradient $c.30\,°C\,km^{-1}$). The broken line represents the approximate depth of onset of oil generation. (Data after Mackenzie & McKenzie 1983; Mackenzie 1984.)

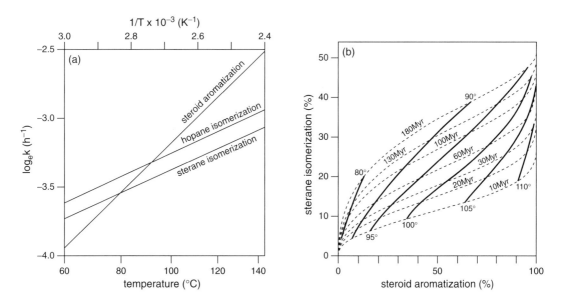

Fig. 5.54 (a) Arrhenius plots for isomerization of steranes at C-20, isomerization of hopanes at C-22 and aromatization of C-ring monoaromatic steroids, showing the greater temperature dependence of the aromatization reaction. (b) Plot of the extent of aromatization of C-ring monoaromatic steroids versus sterane isomerization at C-20 (as given in ratios in Fig. 5.47) for a basin undergoing 50% extension (i.e. β factor = 1.5). The point plotted gives the elapsed time between extension and subsequent uplift (broken lines), and the maximum temperature (°C) prior to uplift (solid lines), assuming that the cooling upon uplift freezes both reactions. (After Mackenzie et al. 1984.)

aromatization reactions may proceed at approximately equivalent rates (Mackenzie & McKenzie 1983). This behaviour is useful in assessing the maximum temperatures experienced by sedimentary material in different types of basins and the time periods over which these temperatures persisted (e.g. Beaumont et al. 1985). One approach is shown in Fig. 5.54b for extensional basins in which the β factor is 1.5 (see Box 3.11), and is based on the assumption that crustal stretching typically leads to higher temperatures and is followed by uplift, usually resulting in a decrease in temperature. It can be considered that the uplift effectively freezes the steroidal isomerization and aromatization reactions. Therefore, the position at which the steroidal isomerization and aromatization values for a particular sedimentary rock sample plots in Fig. 5.54b gives the maximum temperature experienced prior to uplift and the time spent at that temperature (i.e. the time between extension and uplift). For example, a sterane isomerization value of 30% and an aromatization value of 50% suggest that a maximum temperature of 90 °C was experienced for c.130 Myr.

5.8 Isotopic palaeontology

5.8.1 Isotopic fractionation

In Chapter 1 it was noted that the most common pathway for photosynthesis, the C_3 path (Box 1.10), results in a comparatively low value for the $^{13}C:^{12}C$ ratio of the fixed carbon (conventionally represented by the $\delta^{13}C$ notation; Box 1.3). This preference for light isotopes is a characteristic of life processes, and the extent of the depletion of the heavier isotope (or isotopic fractionation; see Boxes 1.3 and 5.11) in sedimentary organic matter can provide clues about the source organisms, their trophic relationships and the environments in which they lived. However, simply consider-

---

Box 5.11 | Isotopic fractionation

The isotopic fractionation factor (α) between two compounds (A and B) is defined as (Hoefs 1997):

$$\alpha_{A-B} = \frac{R_A}{R_B} \quad \text{[Eqn 5.26]}$$

where R = isotope ratio (e.g. $^{13}C/^{12}C$). However, by analogy with the δ notation for isotope ratio, isotopic fractionations are often given in terms of Δ values, which have units of permil:

$$\Delta_{A-B}(‰) = \left(\frac{R_A - R_B}{R_B}\right) \times 10^3 \quad \text{[Eqn 5.27a]}$$

which can also be written as:

$$\Delta_{A-B}(‰) = \left(\frac{R_A}{R_B} - 1\right) \times 10^3 \quad \text{[Eqn 5.27b]}$$

and in terms of δ values this becomes:

$$\Delta_{A-B}(‰) = \left[\frac{(\delta_A + 1000)}{(\delta_B + 1000)} - 1\right] \times 10^3 \quad \text{[Eqn 5.27c]}$$

To avoid any possible confusion with the use of Δ to denote differences in δ values, isotopic fractionation is often given the symbol ε and, as an example, the C isotopic fractionation associated with phytoplanktonic photosynthesis can be represented by (Hayes 1993):

$$\varepsilon_{\text{photosynthesis}}(‰) = \left[\frac{(\delta^{13}C_{CO_2} + 1000)}{(\delta^{13}C_{\text{biomass}} + 1000)} - 1\right] \times 10^3$$

[Eqn 5.28]

This expression, perhaps counter-intuitively, has the product in the numerator and the precursor in the denominator, but by convention it provides a positive value. Values of ε_{A-B} are approximately equal to $\delta_A - \delta_B$, although δ_B values widely different from zero can potentially lead to significant error in the estimation of ε using this approximation. For example, if $\delta^{13}C_A = -27‰$ and $\delta^{13}C_B = -59‰$, $\varepsilon_{A-B} = 34‰$, but $\delta^{13}C_A - \delta^{13}C_B = 32‰$. Transposing the values of A and B produces a different value for ε ($\varepsilon_{B-A} = -32.9‰$ in this example), so $\varepsilon_{\text{photosynthesis}}$ in Eqn 5.28 is more clearly written as $\varepsilon_{CO_2 - \text{biomass}}$.

The isotopic fractionation (ε) between reactant and product is a kinetic isotope effect, and can be represented by the rate constants for the two isotopes:

$$\varepsilon = \left(\frac{k_{12_C}}{k_{13_C}} - 1\right) \times 10^3 \quad \text{[Eqn 5.29]}$$

ing the $\delta^{13}C$ value of a material can be misleading, because it depends upon several factors (Hayes 1993):
- the isotopic composition of the primary C source used;
- the isotopic effect associated with C assimilation by a particular organism;
- the isotopic effect associated with the organism's metabolic and biosynthetic processes;
- cellular C budgets involved in competing reaction pathways.

It is important to consider these factors when interpreting isotopic fractionation, which apply to all elements exhibiting fractionation, not just carbon.

Subareal plants use atmospheric (gaseous) CO_2 ($CO_2(g)$) as their photosynthetic carbon source, which has a mean $\delta^{13}C$ value of $c.-7‰$. Subaquatic plants use the dissolved (aqueous) CO_2 ($CO_2(aq)$), which is at one end of the series of equilibria shown in Eqn 3.8. Both these assimilatory processes are accompanied by isotopic fractionations, as discussed in Section 5.8.2. In marine environments there are also C isotopic fractionations associated with the formation of calcium carbonate tests (using bicarbonate) by some organisms; that for formation of calcite is different from that for aragonite. The overall fractionation, $\varepsilon_{calcite-CO_2(aq)}$, is large and temperature dependent (Fig. 5.55; Mook et al. 1974; Morse & Mackenzie 1990), primarily because of the equilibrium between dissolved CO_2 and bicarbonate (Freeman & Hayes 1992). The fractionation in Fig. 5.55 assumes that equilibrium is attained between dissolved CO_2 and calcite, but that does not necessarily occur because of various metabolic processes, and the calcite preserved in sediments can have a slightly different $\delta^{13}C$ value from that predicted by the equilibrium fractionation relationship (Spero et al. 1991, 1997). Post-depositional diagenetic alteration of carbonates can alter the $\delta^{13}C$ value significantly by exchange with the CO_2 produced from oxidation of organic matter (Irwin et al. 1977).

For heterotrophes, significant isotopic effects during assimilation appear to be restricted to the smallest molecules (Hayes 1993). The $\delta^{13}C$ value of acetate, from which lipids are derived, is clearly important. So too are branch points in the flow of carbon through cells, because there must be a mass balance in the cellular carbon budget. This can be examined by considering the simple system in Fig. 5.56, in which a substrate (s) enters a cell at a rate of ϕ_r mol s^{-1}, providing an instantaneous internal source of reactant (r), some of which is used to form a product (p) and the remainder is unused (u, which can pass to other reaction sites to make other products). Because the product is enriched in the lighter isotope, the residual reactant must be depleted relative to the external substrate (i.e. $\delta^{13}C_r > \delta^{13}C_s$). This can be seen from the fact that the total number of C atoms of each isotope entering and leaving the cell must be conserved, so:

$$\phi_r(\delta^{13}C_s) = \phi_u(\delta^{13}C_u) + \phi_p(\delta^{13}C_p) \quad \text{[Eqn 5.30]}$$

If the fractional flow of carbon at the branch point to product p is represented by f (i.e. ϕ_p/ϕ_s):

$$\delta^{13}C_u = \delta^{13}C_s + f\varepsilon \quad \text{[Eqn 5.31a]}$$

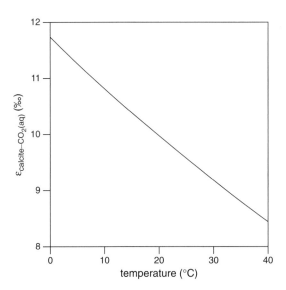

Fig. 5.55 Influence of temperature on the equilibrium isotopic fractionation between calcite and dissolved CO_2 (after Mook et al. 1974; Morse & Mackenzie 1990).

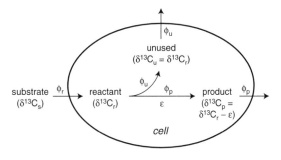

Fig. 5.56 A representation of cellular carbon budget involving a simple branch point, in which ϕ represents carbon flow (mol s^{-1}) and ε is the kinetic isotope effect (after Hayes 1993).

and

$$\delta^{13}C_p = \delta^{13}C_s - (1-f)\varepsilon \quad \text{[Eqn 5.31b]}$$

Equations 5.31a, b give the effect of the carbon flow at a branch point on the $\delta^{13}C$ values of the products, as shown in Fig. 5.57.

A summary of the isotopic fractionations associated with important processes involved in assimilation and dissimilation of carbon is presented in Table 5.7, and the various processes are considered in the following sections.

Our knowledge of the fractionation of hydrogen isotopes is at a relatively early stage, although it appears that δD values depend upon hydrogenation reactions as well as the isotopic composition of precursors and the biosynthetic fractionation and exchange of hydrogen (Sessions et al. 1999, 2002). Fractionations with respect to water in the growth medium are of the order of 100–250‰ for n-alkyl lipids, but can be higher for polyisoprenoidal lipids (c.150–350‰; Sessions et al. 1999). The δD of phytol is generally 50‰ lower than that of other isoprenoids, which may reflect different biosynthetic pathways (i.e. mevalonate and alternative pathways; Section 2.4.3). These fractionations are much higher for hydrogen than for isotopes of other atoms because the mass ratio of $^2H:^1H$ is so large.

5.8.2 Autotrophic pathways

In subaerial C_3 plants substrate and reactant (s and r, respectively, in Fig. 5.56) for photosynthesis are both gaseous (atmospheric) CO_2, which flows through the Calvin cycle (the dark reactions of photosynthesis; see Box 1.10) to yield simple carbohydrates (p), which are in turn the source of various metabolic intermediates. The source of the intracellular (kinetic) isotopic fractionation during C fixation is the enzyme rubisco (D-ribulose 1,5-diphosphate carboxylase/oxygenase). There is also an isotopic fractionation resulting from the passage of CO_2 into the cell. Passive diffusion of CO_2, at a rate ϕ_r, favours ^{12}C, but the fractionation is small

Table 5.7 Typical carbon isotopic fractionations associated with major C-cycling processes (equilibria are represented by =; data after de Leeuw et al. 1995)

process	maximum fractionation (‰)
aqueous equilibria	
atmospheric CO_2 = dissolved CO_2	0.7
dissolved CO_2 = bicarbonate	9
bicarbonate = calcite	2
terrestrial autotrophy	
CO_2(g) assimilation by C_3 plants	22
CO_2(g) assimilation by C_4 plants	4
CO_2(g) assimilation by CAM plants	15
aquatic autotrophy	
CO_2(aq) assimilation by C_3 plants	25
CO_2(aq) assimilation by Chlorobiaceae*	12.5
HCO_3^-(aq) assimilation by phytoplankton	12
CO_2(aq) assimilation by chemotrophes†	25
heterotrophy	
herbivore biomass‡	1
methanogen biomass (cf. CO_2)	40
CH_4 from methanogenesis§	70
methanotrophe biomass (cf. CH_4)	20
CO_2 from methanotrophy	25

*Chlorobiaceae photosynthesis is via reversal of the citric acid cycle; †chemotrophes include sulphur and ammonium oxidizing bacteria (see Box 1.7); ‡each additional heterotrophic step involves a similar fractionation; §methanogenesis generally leads to a fractionation of c.40‰ in freshwater conditions but c.70‰ under marine conditions.

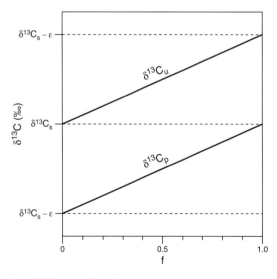

Fig. 5.57 The effect of variation in C flow at a branch point (fractional yield, f) on isotopic composition of products (after Hayes 1993). $\delta^{13}C_p = \delta^{13}C_u - \varepsilon$.

(ε_r = 4.4‰ for CO_2 diffusion into stomata of subaerial C_3 plants, and 0.7‰ for CO_2 diffusion in water at 25 °C) in comparison to that involving rubisco (ε_p = c.27‰; Farquhar et al. 1982).

$$\varepsilon_{photosynthesis} = \varepsilon_r + (concn_r/concn_s) \cdot (\varepsilon_p - \varepsilon_r)$$
$$\approx \delta^{13}C_s - \delta^{13}C_p$$
[Eqn 5.32]

From Eqn 5.32 it can be seen that $\varepsilon_{photosynthesis}$ potentially varies over a wide range, between ε_r and ε_p. The bulk $\delta^{13}C$ value for C_3 plants therefore reflects $\varepsilon_{photosynthesis}$, and the typical range for extant organisms is shown in Table 5.8, together with ranges for other types of autotrophes.

Some subaerial plants use a different pathway, involving carboxylation of phosphoenol pyruvic acid (PEP) instead of ribulose diphosphate (the CO_2 is actually transformed into bicarbonate before incorporation), which subsequently forms a C_4 compound, oxaloacetic acid, instead of PGA (Fig. 1.7). Consequently such plants are termed C_4 plants. The C_4 path is a relatively recent evolutionary development of particular advantage in hot dry climates (see Box 1.10). The PEP cycle effectively transfers CO_2 to the Calvin cycle, and each cycle confers an isotopic fractionation. Some plants, the CAM plants (see Box 1.10), can use the combined PEP–Calvin cycle path (with some leakage of CO_2 out of the cell between the cycles) or just the Calvin cycle. The effects of these pathways on the overall isotopic fraction are reflected in the $\delta^{13}C$ values in Table 5.8.

In subaqueous plants and chemoautotrophes the substrate (s, Fig. 5.56) and reactant (r) for carbon fixation is normally dissolved CO_2. However, when aqueous concentrations of CO_2 are low (e.g. during blooms), phytoplankton can actively pump bicarbonate across the cell membrane (Burns & Beardall 1987; Pancost et al. 1997), which can lead to $\varepsilon_{photosynthesis}$ values < 0‰ (Deuser 1970; Raven et al. 1987). The bicarbonate is transformed back into CO_2 (the reactant, r) before incorporation into the Calvin cycle. The bicarbonate pump places a greater energy drain on cells, so ideally no bicarbonate should be lost in producing the reactant (i.e. $\delta^{13}C_p = \delta^{13}C_r$). However, inevitably there is some leakage from the cell, which controls the isotopic variation observed among products. The overall photosynthetic fractionation can be represented by:

$$\varepsilon_{photosynthesis} = \delta^{13}C_s - \delta^{13}C_p$$
$$= \varepsilon_{CO_2(aq)-bicarbonate} + \varepsilon_{bicarbonate\ pump} + L(\varepsilon_p - \varepsilon_r)$$
[Eqn 5.33]

where L = fraction of bicarbonate that leaks away.

In acetogens, methanogens and Chlorobiaceae (green photosynthetic bacteria) the substrate is CO_2, but C fixation can occur at three different enzymatic sites, at least, involving CO_2 or HCO_3^- as the reactant (House et al. 2003). This can lead to very large isotopic contrasts between molecules from the same cell. These organisms can fix C using acetyl coenzyme A or by reversing the citric acid cycle (Fig. 2.10; Sirevag et al. 1977; Wood et al. 1986). Among the largest fractionations are those associated with methanogenic bacteria (Table 5.7), so the resulting methane exhibits very low $\delta^{13}C$ values and methanogenic biomass can also be isotopically light (Table 5.8). The use of the reverse citric acid cycle by the Chlorobiaceae causes the $\delta^{13}C$ of isorenieratene to be >10‰ higher than corresponding algal lipids in some Mediterranean sapropels (Passier et al. 1999).

All the above discussion of isotopic fractionation during carbon fixation is much simplified; there are many potential complications, such as variants of rubisco (Raven & Johnston 1991; Guy et al. 1993). Stomatal activity, and hence CO_2 uptake, can vary between terrestrial plant species and with changing environmental conditions, which can produce significant shifts in $\delta^{13}C$ values (Lockheart et al. 1997). The partial pressure of atmospheric CO_2 (see Box 1.11) may affect phytoplanktonic $\varepsilon_{photosynthesis}$, through its control of surface-water CO_2 concentrations (Rau et al. 1992; Pancost et al. 1997). Under conditions of abundant CO_2, mass transport into cells is unlikely to limit the rate of photosynthesis, so the large isotopic fractionation of rubisco should dominate the overall fractionation. There are many other factors that can influence $\varepsilon_{photosynthesis}$

Table 5.8 Isotopic composition of major autotrophes and methanogens (data after Schidlowski 1988)

organism	$\delta^{13}C$ (‰)
higher plants	
C_3	−23 to −34
C_4	−6 to −23
CAM	−11 to −33
algae	−8 to −35
cyanobacteria	−3 to −27
photosynthetic bacteria	
green	−9 to −21
purple	−26 to −36
red	−19 to −28
methanogenic bacteria	+6 to −41

(Goericke et al. 1994; Laws et al. 1997; Pancost et al. 1997; Rau et al. 1997). For modern phytoplankton $\varepsilon_{photosynthesis} \leq 25‰$ (Popp et al. 1998b) and is generally 8–18‰ (Bidigare et al. 1997). The influence of specific growth rate (μ day^{-1}), cell volume:surface area ratio (V/S μm) and dissolved CO_2 concentration (c μmol kg^{-1}) can be represented by the following relationship (Popp et al. 1998b):

$$\varepsilon_{photosynthesis} = 25.3 - 182(\mu/c)(V/S) \quad \text{[Eqn 5.34]}$$

5.8.3 Biosynthetic pathways

Nearly all sedimentary biomarkers are derived from lipids, the most obvious exceptions being porphyrins. In prokaryotes the substrate derived from the Calvin cycle for all subsequent biosynthesis is a simple C_6 carbohydrate (Fig. 1.7), each C atom of which generally has the same $\delta^{13}C$ value (although there may be exceptions in eukaryotes; Rossmann et al. 1991; Hayes 1993). However, the acetyl-CoA system and reverse citric acid cycle produce C_2 monomers in which one C atom is derived from the methyl group and the other from the carboxyl group of the acetyl unit, which are produced at different reaction sites. There is an isotopic effect that appears to arise from the reaction between pyruvate and acetyl-CoA (catalysed by pyruvate dehydrogenase; Melzer & Schmidt 1987), and results in a depletion of ^{13}C at the carbonyl group of acetate relative to the unaltered isotopic composition of the methyl group. This has implications for the bulk isotopic composition of compounds derived from acetate units, which contain one of each type of C atom, and those derived from isoprene units, which comprise three methyl-derived C and two carboxyl-derived C atoms. For heterotrophes utilizing carbohydrates or autotrophes producing carbohydrates via the Calvin cycle (from which acetate is generated via acetyl-CoA or the reverse citric acid cycle):

$$\delta^{13}C_{acetogenic\ lipids} = \frac{\delta^{13}C_{methyl} + \delta^{13}C_{carboxyl}}{2} \quad \text{[Eqn 5.35a]}$$

$$\delta^{13}C_{isoprenoidal\ lipids} = \frac{3(\delta^{13}C_{methyl}) + 2(\delta^{13}C_{carboxyl})}{5} \quad \text{[Eqn 5.35b]}$$

where

$$\delta^{13}C_{methyl} = \delta^{13}C_{carbohydrate} \quad \text{[Eqn 5.35c]}$$

$$\delta^{13}C_{carboxyl} = \delta^{13}C_{carbohydrate} - (1-f)\varepsilon_{pyruvate\ dehydrogenase} \quad \text{[Eqn 5.35d]}$$

and f = fraction of carbon flowing to acetyl-CoA.

The pathways of carbon flow in eukaryotes are more complex than in prokaryotes, mainly because the C_2 units produced within mitochondria can be exported (e.g. for lipid biosynthesis) only as part of a citrate unit produced in the citric acid cycle. The branch point in carbon flow to either isoprene or acetate can lead to additional isotopic fractionation. In general, it seems that n-alkyl lipids in a particular organism are depleted in ^{13}C by c. 1.5‰ relative to isoprenoids produced from the same substrate (Hayes 1993). In higher plants, the phenolic precursors of lignin derive from glucose (Fig. 2.29), so it is not surprising that the carbon isotopic composition of lignin reflects the major photosynthetic pathway involved (C_3 or C_4; Benner et al. 1987).

5.8.4 Heterotrophy

The bulk carbon isotopic signature of heterotrophes reflects the signature of their main substrates, modified by the signature of carbon outputs. The main outputs are respired CO_2, faecal carbon and CH_4. The CO_2 is mainly from the citric acid cycle (involving decarboxylation of organic acids), which results in the evolved CO_2 having a light signature, so the $\delta^{13}C$ value of the organism is increased. Typically, then, animals have $\delta^{13}C$ values c. 1‰ greater than their food source (DeNiro & Epstein 1978; Hayes 1993).

Bacterial heterotrophy may yield greater isotopic shifts than eukaryotic heterotrophy (Coffin et al. 1990; Pancost & Sinninghe Damsté 2003). Methanogens in the digestive systems of zooplankton generate isotopically light CH_4, which probably accounts for zooplanktonic faecal pellets being enriched in ^{13}C by larger amounts than are consistent with aerobic respiration (Fischer 1991). Methane-oxidizing bacteria (methanotrophes; see Section 3.3.2b) utilize methanogenic methane that is already isotopically very light, and they impose additional fractionation during carbon assimilation (Table 5.7; Whiticar 1999).

Combined carbon and nitrogen isotopic compositions can be useful in assessing the diets of ancient peoples, particularly where conventional archaeological evidence is sparse. The most suitable materials for $\delta^{15}N$ measurements are those in which the proteins are most resistant towards alteration. The reliability of data from bone collagen has been questioned (Sillen et al. 1989; Macko & Engel 1991), but human hair appears to be suitable, because the hydrophobic proteins in α-keratins are not easily degraded over thousands of years (see Section 3.3.3a; Lubec et al. 1987). The extent of amino-acid degradation can be assessed by hydrolysing the proteins and examining the degree of loss of the less stable amino acids such as serine and threonine. Whereas bone collagen reflects diet over a period of $c.10$ yr, human hair represents a much shorter time period, although there is good correlation of isotopic data between the two substrates (DeNiro & Epstein 1978, 1981).

Carbon isotopic composition reflects the major C-fixation pathway of plants (average $\delta^{13}C$ value for C_3 plants = $c.-26.5‰$, and for $C_4 = c.-12.5‰$) and other sources (e.g. terrestrial herbivores and marine animals; $\delta^{13}C = c.-20‰$ for the latter). Nitrogen isotopic compositions show similar trends between average $\delta^{15}N$ values of 0‰ for terrestrial plants and +6‰ for marine organisms. Animals exhibit an enrichment of ^{15}N relative to their food source of $c.3‰$ (DeNiro & Epstein 1981; Macko & Engel 1991), compared to the +1‰ shift observed for $\delta^{13}C$ values (Ostrom & Fry 1993). The isotopic values obtained from hair can be compared with those for modern food sources from the hinterland of the archaeological site to obtain indications of ancient diets (Chisholm et al. 1982). Modern populations show little variation in $\delta^{13}C$ values, and it is not possible to distinguish vegans from ovo-lacto vegetarians and omnivores. The variable availability of C_4 plants—as both primary foods and animal feeds—also hampers the identification of such dietary distinctions.

However, nitrogen isotopic data are more useful, with vegans ($\delta^{15}N$ $c.7‰$) being readily distinguished from ovo-lacto vegetarians and omnivores (which share similar $\delta^{15}N$ values of $c.3‰$).

The Neolithic Ice Man from Oetztaler in the Alps (5200 yr old) appears to have been a vegan, when the isotopic composition of his hair is compared with that of the goat hair found with him and local grass, and applying the expected degree of isotopic fractionation (Fig. 5.58a; Macko et al. 1999a, b). Teeth wear is consistent with this proposition, and the $\delta^{13}C$ value suggests C_3 plants were the main food source. A further example of this application is shown in Fig. 5.58b, based on data for Chinchorro mummies from Arica, Chile (5000 to 800 yr old), together with authentic dried foods (Macko et al. 1999b). The food comprises C_3 plants, C_4 plants (maize) and fish, and the mixing lines shown between these end members are shifted by the appropriate assimilatory fractionation effect. The hair samples lie on the mixing line between terrestrial plants and marine sources, with mummies from a coastal site plotting closest to the marine end member. Sulphur isotopic composition (which appears not to exhibit significant dietary fractionation; MacAvoy et al. 1998) supports these indications, with the marine samples showing the highest $\delta^{34}S$ values (similar to the authentic fish food at $c.15‰$).

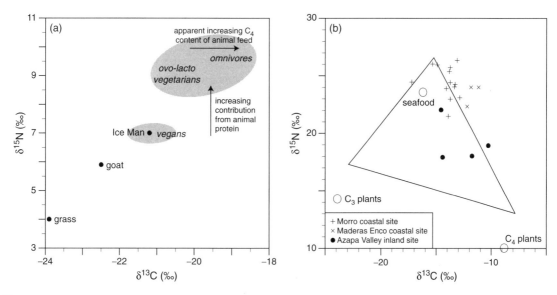

Fig. 5.58 Carbon and nitrogen isotopic evaluation of ancient human diets relative to authentic foodstuffs (after Macko et al. 1999a, b). (a) Ice Man from the Oetztaler Alps (5200 yr old) compared to modern vegans, ovo-lacto vegetarians and omnivores. (b) Chinchorro mummies from the Arica region of Chile (5000 to 800 yr old); the triangle represents end-member contributions after allowing for assimilatory fractionations.

5.8.5 Sedimentary isotopic record

Estimates of $\varepsilon_{photosynthesis}$ can be obtained from estimates of $\delta^{13}C_{photosynthesis}$ (i.e. the bulk value for phytoplankton), which in turn can be derived from measured $\delta^{13}C_{porphyrins}$ and the known difference between $\delta^{13}C_{porphyrins}$ and $\delta^{13}C_{photosynthesis}$ in modern ecosystems. This approach can provide information on the main types of primary producers and their environments. Comparison of $\delta^{13}C_{photosynthesis}$ with $\delta^{13}C$ values for total sedimentary organic carbon can then enable an assessment of the degree of reworking of primary production and the overall trophic structure of the ecosystem (Hayes 1993). There are many approximations involved in such models, so considerable caution is required in their application.

The major fractionations influencing the isotopic signatures of marine sediments are summarized in Fig. 5.59 for the most common scenario, in which C_3 photosynthesis dominates autotrophy. The value of $\delta^{13}C$ for dissolved inorganic C (DIC) varies with depth, but surface-water rather than average values are most important when evaluating $\varepsilon_{bicarbonate-calcite}$ for planktonic tests (Shackleton 1987b; Kroopnick 1985). The fractionation between calcite and bicarbonate is small and little affected by temperature (c.0.01‰ decrease per °C rise), and is generally c.1.2‰. Consequently, the fractionation between dissolved CO_2 and calcite is mainly controlled by $\varepsilon_{bicarbonate-CO_2(aq)}$ (c.0.12‰ decrease per °C rise), with $\varepsilon_{calcite-CO_2(aq)}$ being c.7‰ in warm waters and c.10‰ in cold (Hayes et al. 1999). Under aerobic conditions, herbivory and biodegradation (i.e. respiration processes) impose an isotopic fractionation (ε_{het}) of c.1.5‰ on the proportion of primary production that survives to form kerogen.

In ancient sediments, only the $\delta^{13}C$ values of calcitic tests and kerogen can be measured, but the difference between them ($\varepsilon_{sedimentary}$) bears the signature of the prevailing environmental conditions and primary production:

$$\varepsilon_{sedimentary} = \varepsilon_{photosynthesis} + \varepsilon_{calcite-CO_2(aq)} - \varepsilon_{het} \quad [\text{Eqn } 5.36]$$

The application of $\varepsilon_{sedimentary}$ in assessing palaeoenvironments is examined further in Section 6.2.1.

Comparison of the isotopic compositions of kerogen samples can be made using plots of $\delta^{13}C$ versus δD, which can show organofacies variation (Schoell 1984a, b; Whiticar 1990; see Box 1.1). With increasing maturity and the onset of hydrocarbon generation, $\delta^{13}C$ and δD values increase. This is because $^{12}C-^{12}C$ bonds are more easily broken than $^{12}C-^{13}C$ bonds (see Section 4.5.4a). The expelled hydrocarbons are isotopically light relative to the kerogen, and so the latter becomes increasingly heavier with advancing maturity (Galimov 1980).

Differences in $\delta^{15}N$ values for the various reservoirs of nitrogen used for assimilation are generally preserved in plant material. For example, a $\delta^{15}N$ value of 7–10‰ is typical for dissolved nitrate and phytoplankton, whereas c.0‰ is characteristic of atmospheric N_2 and C_3 land plants (Meyers 1997). These values are subsequently conserved in sedimentary organic matter (Peters et al. 1978). However, the $\delta^{15}N$ of phytoplankton increases as nitrate concentration decreases, as a result of both decreasing discrimination and denitrification of NO_3^-(aq) in O_2-depleted waters; the latter releases isotopically light N_2 into the atmosphere and leaves the residual nitrate enriched in ^{15}N (Cline & Kaplan 1975; see Box 3.9).

Although bulk carbon-isotopic measurements can give information on sources of sedimentary organic matter, potentially more information can be gained from the isotopic signatures of individual biomarkers (Freeman et al. 1990). For instance, it should be possible

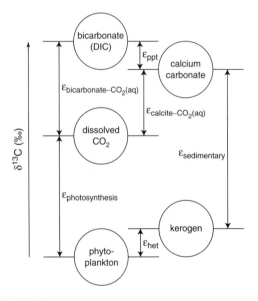

Fig. 5.59 Summary of factors affecting the observed fractionation between marine carbonate and kerogen ($\varepsilon_{sedimentary}$), assuming autotrophy is dominated by C_3 photosynthesis (after Hayes et al. 1999). Calcite test precipitation is represented by ε_{ppt} ($\varepsilon_{bicarbonate(aq)-calcite}$); heterotrophic reworking of primary production is represented by ε_{het} ($\varepsilon_{kerogen-phytoplankton}$); n.b. the effect of thermal alteration on kerogen $\delta^{13}C$ is not included here.

to identify the source of a particular compound with more certainty and to evaluate its metabolic pathways. Isolating sufficient amounts of single compounds has been an obstacle to such measurements in the past, but with the advent of improved technology these measurements are possible from relatively small amounts of components, extending the application of this technique to biomarkers. Among the first compounds to be studied were *n*-alkanes, which are generally abundant in sedimentary organic matter. For example, in one study of Recent sediments from a freshwater lake (Ellesmere, UK) C_{25}-C_{33} *n*-alkanes were found to have $\delta^{13}C$ values of <−30‰, consistent with an origin from tree-leaf waxes, whereas shorter-chain components were isotopically heavier (−20 to −22‰) and probably attributable to an algal source (Rieley et al. 1991). It is sometimes possible to distinguish a variety of sources of *n*-alkanes from their C-number ranges and isotopic signatures, including cyanobacteria and chemosynthetic bacteria (Collister et al. 1994). The $\delta^{13}C$ record of higher plant derived C_{25}-C_{33} *n*-alkanes has been used to examine the rise of C_4 grasses, which have significantly higher $\delta^{13}C$ values than C_3 vegetation (see Table 5.7), at *c*.6 Ma ago (Cerling et al. 1993; Freeman & Colarusso 2001), and the climatic (particularly aridity) induced variation in dominance between the two plant groups during glacial–interglacial cycles (Huang et al. 2001). Similarly, individual lignin-derived phenols have been used to follow Quaternary vegetational changes (Huang et al. 1999), the relative contributions of C_3 and C_4 plants being determined from the average isotopic composition of their phenols (−34 and −14‰, respectively; Goñi & Eglinton 1996):

$$\delta^{13}C \text{ lignin phenol} = (-34)(\text{proportion of } C_3 \text{ plants})$$
$$+ (-14)(\text{proportion of } C_4 \text{ plants})$$
[Eqn 5.37]

Such molecular studies offer some advantages over the pollen record when reconstructing local climate, because pollen distributions are strongly influenced by aeolian transport and the amount of pollen produced by different species, whereas leaf and wood components are more likely to be deposited *in situ* (although some aquatic transport is possible).

Compound-specific $\delta^{13}C$ analysis can be applied to the whole range of compound classes found in sediments; some examples from the Eocene Messel Shale (Germany) are presented in Table 5.9. The $\delta^{13}C$ value for the total kerogen in this shale is −28.21‰, while that for the solvent-extractable organic matter (bitumen) is −29.72‰, the latter value reflecting the general relative

Table 5.9 Carbon isotopic composition of some compounds in the Eocene Messel Shale (data after Freeman et al. 1990)

compound	$\delta^{13}C$ (‰)
pristane	−25.4
phytane	−31.8
17β-30-norhopane	−65.3
17β-hopane	−35.2

lightness of lipid-derived components. The differing values of $\delta^{13}C$ for pristane and phytane indicate that they do not share the same source. The value for pristane is compatible with an algal origin, but the considerably lighter isotopic signature of phytane probably reflects a phytanyl ether input from methanogenic bacteria. The value of $\delta^{13}C$ for 17β-hopane is characteristic of the hopanoids in general in Messel Shale, which probably originate from chemosynthetic bacteria. The intense depletion in ^{13}C of 30-nor-17β-hopane suggests that it does not share the same source as the other hopanoids but derives from a precursor already exhibiting a light isotopic signature, probably originating from methylotrophic bacteria. The hopanoids of methylotrophes, such as *Methylomonas*, are dominated by a C_{35} pentahydroxyamine which may degrade to the norhopane (Freeman et al. 1990). Isotopic analysis of individual porphyrins in the Messel Shale has permitted assessment of the relative contributions from algal and bacterial chlorophylls (Hayes et al. 1987).

The analysis of isotopic signatures of individual biomarkers can be a powerful tool in examining the composition of microbial communities and their influence on carbon cycling, and it is the only tool available for ancient sediments, because bacteria leave no other evidence (Pancost & Sinninghe Damsté 2003). For example, it appears that a consortium of anaerobic bacteria—eubacteria as well as archaebacteria—are involved in the oxidation of methane (i.e. methanotrophy) in marine sediments (Boetius et al. 2000). Some non-isoprenoidal lipids (e.g. *sn*-1-alkylglycerolmonoethers), which cannot be attributed to archaebacteria, share similar low $\delta^{13}C$ values of *c*.−100‰ with phytyl-containing lipids (e.g. archaeol, *sn*-2-hydroxyarchaeol and dihydrophytol; see Section 2.4.1b) that are unambiguously derived from methanogens (Hinrichs et al. 1999, 2000). The similar and extreme isotopic fractionation of the two compound groups suggests a trophic relationship, with the non-isoprenoids deriving from eubacteria, possibly related to thermophilic sulphate reducers.

As noted in Section 5.8.1, the δD values of individual biomarkers reflect the δD value of the growth water and so can provide useful climatic information, if the fractionation processes can be evaluated. For example, δD for precipitation in the tropics is c.0‰, whereas over mountains at high latitude it is <−150‰. High positive values correspond to intense evaporation, which increases the abundance of ^{2}H in the residual water. However, ^{2}H enrichment decreases as humidity increases, so values approaching +100‰ are found only in modern deserts (where relative humidity is low; Confiantini 1986). There are complications in terrestrial plants arising from evapotranspiration from leaves, which enriches ^{2}H in leaf water and hence photosynthetic products. However, algal sterols appear to be consistently depleted in ^{2}H by c.201‰ compared to growth water (whether marine or fresh)—i.e. $\varepsilon_{water-algal\ sterol}$ = 201‰—and so their δD values are potentially useful indicators of climatic conditions and the balance between precipitation and evaporation (Sauer et al. 2001).

5.8.6 Isotopic distributions in petroleum

(a) Oils

There are various problems associated with correlating oils and their parent kerogens from bulk isotopic data: an oil is only isotopically related to the oil-generating fraction of kerogen; there are maturity-related changes in isotopic composition; and migrated oil often comprises only a proportion of the generated bitumen and is comparatively depleted in the heavier and more polar components. For a given maturity, a bitumen is often c.0.5–1.5‰ lighter than its source kerogen, and the related oil is c.0.0–1.5‰ lighter again (Peters & Moldowan 1993). Oils of similar maturity that differ by >2‰ are usually not genetically related.

The carbon isotopic compositions of the main fractions of petroleum—saturates, aromatics, resins and asphaltenes (Box 4.2)—behave reasonably conservatively (Galimov 1973), although all fractions usually become slightly isotopically heavier as maturity increases (Chung et al. 1981). Consequently, oils and source-rock bitumens can be correlated using plots of two or more components (Fuex 1977; Stahl 1978; Peters & Moldowan 1993). For example, δ^{13}C versus δD plots for aromatic fractions of oils can be useful because the aromatics appear least affected by thermal processes and biodegradation (Schoell 1984a; Peters et al. 1986). Biodegradation is likely to result in the saturates fraction becoming isotopically heavier as the relatively light *n*-alkanes are removed. Cross plots of δ^{13}C values from saturates and aromatics are frequently used for correlation of oils and bitumens (Fuex 1977), and it has been suggested that marine and terrestrial origins can be distinguished by such plots (Fig. 5.60; Sofer 1984, 1988), although the differentiation is not always reliable (Peters et al. 1986). As noted for sediments (Section 5.8.5), compound-specific isotopic ratios provide a more reliable means of correlating oils, and branched, cyclic and aromatic members of the gasoline range appear particularly useful (Whiticar & Snowdon 1999).

Sulphur isotopic ratios have been successfully used to correlate crude oils (Orr 1974; Gaffney et al. 1980; Thode 1981). In general δ^{34}S values do not appear to be much affected by processes like migration, biodegradation or water washing (Hirner & Robinson 1989), although they can be affected by S incorporated into oils in reservoirs at high temperatures (Orr 1974). With respect to organic-S, genetically related oils and source rocks should exhibit δ^{34}S values within 2‰ (Orr 1986). In the absence of biodegradation there may be a general increase in δ^{34}S values with increasing polarity of oil and bitumen fractions, similar to that noted for δ^{13}C values above (Hirner et al. 1984).

Although the isotopic fractionation between pyritic, native and organically bound sulphur is complex (Krouse 1977), the δ^{34}S values of organically bound S in kerogen and oils can provide some useful information about depositional environments. The relative abun-

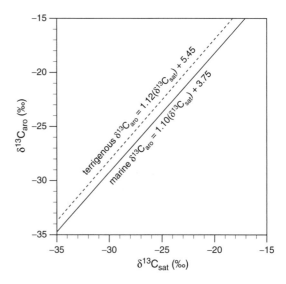

Fig. 5.60 Oil correlation and source-related plot based on δ^{13}C values for saturated and aromatic hydrocarbon fractions (after Sofer 1984).

Chemical stratigraphic concepts and tools | 243

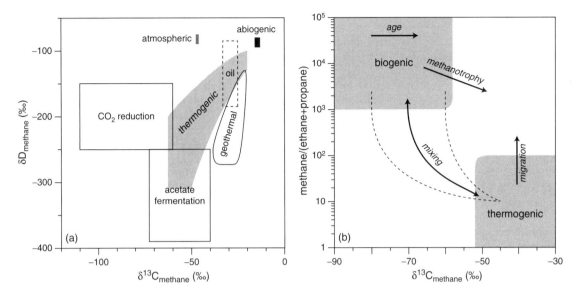

Fig. 5.61 Methane sources based on isotopic composition and relative abundance (abiogenic = derived from mantle CO_2; after Schoell 1984a, b, 1988; Whiticar 1990).

dance of the sulphate supply available for bacterial sulphate reduction appears to be the key factor (Chambers 1982). In an open system, with plentiful sulphate, a high degree of isotopic fractionation is possible and kerogen tends to have low $\delta^{34}S$ values. In open marine environments, seawater provides a plentiful supply of sulphate, presently with a $\delta^{34}S$ range of +19 to +20‰, and isotopic fractionation can give rise to values in the region of −10‰ for kerogens (and their associated oils). In comparison, the $\delta^{34}S$ value of fresh water is +1 to +7‰. It might be expected that bitumens and oils from coals would reflect the freshwater range of $\delta^{34}S$, but values close to that of ambient seawater can be observed (Hirner & Robinson 1989). The reason is that, in closed systems with a restricted supply of sulphate, sulphate-reducer activity is limited by the rate of bacterial sulphide oxidation, so there may be little or no isotopic fraction relative to the substrate. Fresh water contains little sulphate, so only a small amount of seawater, with its high sulphate content, needs to infiltrate peats during early diagenesis to overwhelm the freshwater $\delta^{34}S$ signature. Coal deposition is generally terminated by marine transgression, and if only a limited supply of sulphate is able to enter the peat, sulphate reducers effectively utilize the entire supply, conferring a $\delta^{34}S$ signature on the sulphide incorporated into the kerogen structure that is similar to the ambient marine value (Killops et al. 1994).

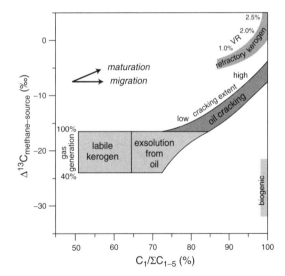

Fig. 5.62 Maturity-related variations in the relative abundance and isotopic composition of methane (after Clayton 1991). The effects of increasing maturation and migration are shown by arrows; the maturity of labile kerogen is represented by the degree of gas generation, and of refractory kerogen by vitrinite reflectance.

(b) Gases

There is only a limited number of different components present in petroleum gases, so isotopic composition is an important tool for identifying the origins of various accumulations. Sources of methane can be discriminated by use of a plot of $\delta^{13}C$ versus δD (Schoell 1984a, b, 1988; Whiticar 1990), as shown in Fig. 5.61a. Atmospheric methane and the main field for oils are included for comparison. The methane formed by thermal breakdown of kerogen is termed thermogenic, and there is some overlap at the lightest isotopic extremes with bacterial sources (methanogenesis via acetate fermentation or CO_2 reduction) and at the heavier extremes with geothermal sources (i.e. gas generated by pyrolysis of organic matter during magmatism). Conditions within a sediment can cycle (e.g. on a seasonal basis) between those favouring fermentation (warm) and those favouring reduction (cold and acetate-depleted), producing biogenic methane of a mixed signature (Schoell 1988). Bacterial oxidation can significantly alter the isotopic composition of biogenic methane because the isotopically lighter molecules are preferentially used, and the residual methane can become isotopically indistinguishable from thermogenic methane (Coleman et al. 1981). Mantle-derived CO_2 can be converted by abiogenic processes into CH_4 (there is really no such thing as mantle methane because the gas is not thermodynamically stable and only CO_2 exists under the prevailing conditions in the mantle; Giggenbach et al. 1993), which is labelled abiogenic in Fig. 5.61a. Examples of this type of methane have been obtained from sediment-free, mid-ocean ridge, hydrothermal systems.

Plots of $\delta^{13}C_{methane}$ versus methane/(ethane + propane) (Fig. 5.61b) have been used successfully for correlation purposes and to distinguish between thermogenic and biogenic sources of gas seeps in the Gulf of Mexico and the Po valley (Claypool & Kvendvolden 1983; Schoell 1984a). The molecular composition parameter in these plots makes use of the fact that biogenic gas is almost entirely methane (i.e. dry gas), whereas the gas produced by thermogenic decomposition of kerogen contains significantly greater amounts of hydrocarbons $>C_1$ (i.e. wet gas; see Section 4.5.2). The effects of mixing thermogenic and biogenic gases are shown in Fig. 5.61b, together with the depletion in $>C_1$ hydrocarbons that occurs during migration of thermogenic

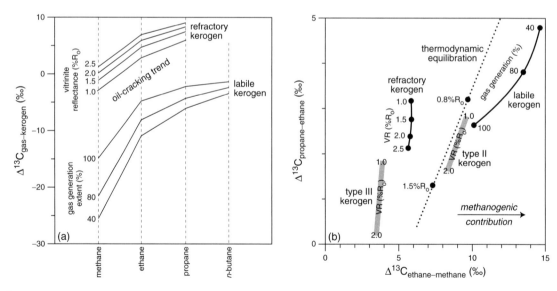

Fig. 5.63 Maturity-related variations in isotopic compositions of gaseous hydrocarbons (a) relative to source kerogen (after Clayton 1991), (b) relative to one another. In (a) trends for gases from oil cracking lie between those for labile (polymethylenic) and refractory (lignin-derived) kerogen. Trends in (b) are based on various models: kinetic isotope effect applied to labile and refractory kerogens (derived from (a); after Clayton 1991), thermodynamic equilibration of isotopic distributions (after James 1983; Schoell 1984a), empirical fractionation from pyrolysis studies of types II and III kerogen (after Berner & Faber 1996). The influence of contributions from methanogens is shown by an arrow; VR = vitrinite reflectance. Inferred source maturities are given as vitrinite reflectance values except for labile kerogen, for which degree of gas generation is used.

gas, and the depletion in $^{12}CH_4$ that occurs as a biogenic gas accumulation ages and the isotopically lighter molecules are preferentially lost by diffusion. Migration does not appear to affect the isotopic signature of gaseous hydrocarbons (Schoell 1984a, b).

If the $\delta^{13}C$ value of a source kerogen can be measured or estimated, more detailed information about the maturity of thermogenic gases from that source and their migration history can be obtained, as demonstrated in Fig. 5.62 (Clayton 1991). Two sources of methane are considered: refractory kerogen, which is a source of methane only; and labile kerogen, which yields mainly oil (see Section 4.5.2). The maturity trends for gas from labile kerogen and oil cracking in Fig. 5.62 merge because some of the former gas is present in solution in oil during the initial phase of oil cracking.

The $\delta^{13}C$ values of gaseous hydrocarbons generated from a common source increase with increasing C number, although the difference between successive *n*-alkanes decreases because the influence of the kinetic isotope effect diminishes with increasing C number (Fig. 5.63a; Clayton 1991). As the maturity of the source increases, its $\delta^{13}C$ value increases (because it evolves isotopically light hydrocarbons), the $\delta^{13}C$ values of thermogenic gases increase, but the C-number-related difference in $\delta^{13}C$ values decreases (Fig. 5.63a; James 1983; Clayton 1991). Cross plots of differences in $\delta^{13}C$ values between pairs of gaseous alkanes can potentially reveal maturity-related trends where the $\delta^{13}C$ value of the source kerogen is unknown and cannot be estimated, as shown in Fig. 5.63b. The trends in Fig. 5.63b are based on a number of models, which show some differences between related kerogens (although both types II and III kerogens may contain mixtures of labile and refractory kerogens; see Section 4.5.2), but reasonable agreement with regard to the differentiation of labile and refractory sources. The trend based on thermodynamic equilibration of isotopic distributions between all the gaseous alkanes (after James 1983; Schoell 1984a) appears to correlate most closely with type II kerogen, although full thermodynamic equilibrium (involving exchange of C isotopes between molecules) is thought unlikely (Clayton 1991). Contributions from biogenic methane should be discernible because of the influence of the isotopically light methanogenic signature (arrow, Fig. 5.63b), but oil-cracking will tend to blur the distinction between the primary kerogen sources of gas, although the plot can still be a useful correlation tool.

6 The carbon cycle and climate

6.1 Global carbon cycle

6.1.1 Carbon reservoirs and fluxes

In preceding chapters we have examined the production and fate of carbon-containing compounds, which provide a background for considering the patterns of carbon flow through the Earth system. A much simplified summary of the carbon cycle is shown in Fig. 6.1, which gives an idea of the sizes of the various compartments (or **reservoirs**) in which carbon is located, the exchange rates (or **fluxes**) between these reservoirs and the main forms in which carbon exists in each reservoir. The carbon cycle in Fig. 6.1 is in a steady state, i.e. inputs and outputs from each reservoir are in equilibrium, and under such circumstances the residence time of carbon in each reservoir can be calculated by dividing the reservoir size by the total flux either in or out of the reservoir (at equilibrium, these two total fluxes are equal). A time just before the onset of the industrial revolution is chosen in order to avoid the imbalance to the system resulting from exploitation of fossil fuels (which is considered in Section 6.5.1). All the reservoir and flux values are approximate, because many cannot be measured directly and so are inferred from other measurements; estimates are constantly being revised. Nevertheless, it can be seen that by far the largest reservoir of carbon, accounting for $c.99.9\%$ of the total, is sedimentary rock, mainly in the form of carbonates.

It is convenient to consider the carbon cycle as comprising two subcycles. The larger ($c.75$ Pt of C) involves sedimentary rocks and residence times of millions of years, which may be thought of as the geochemical subcycle. The smaller ($c.40$ Tt of C) incorporates dissolved and non-living particulate organic matter in water bodies (primarily the oceans) and soils, together with the biota biomass. It involves biological recycling and residence times of up to a hundred years or so only, and may be thought of as the biochemical subcycle. These two subcycles are linked by a small two-way flux. That from biochemical to geochemical subcycles is the rate of incorporation of carbon into sedimentary rocks, estimated at $c.0.3$ Gt yr^{-1}, with a carbonate-C to kerogen-C ratio of $c.4:1$ (and similar to the mean ratio for all sedimentary rocks; Fig. 6.1). In a steady state there is an equal flux in the opposite direction, primarily through erosion and weathering of sedimentary rocks. However, human exploitation of fossil fuels has greatly augmented this flux from the geochemical to the biochemical subcycle.

While the larger, geochemical, subcycle is quantitatively the most important ($c.99.95\%$ of total carbon), all the organic matter and a large part of the carbonate in this cycle has originated from the biochemical subcycle (i.e. from living tissues). As was discussed in Section 1.4.2, the quantity of free oxygen in the atmosphere is controlled by the amount of reduced (organic) carbon compounds preserved in sedimentary rocks, which is a result of the link between the biochemical and geochemical subcycles.

6.1.2 Biochemical subcycle

In the atmosphere carbon exists mainly as carbon dioxide (Fig. 6.1), which is taken up by plants during photosynthesis. The amount of solar energy captured by plants during photosynthesis is referred to as **gross primary production** and can be measured by the amount of carbon dioxide that has been fixed. As noted in Section 3.2.3, some of the gross primary production is used to provide the energy needed for the performance of normal biochemical processes, which are collectively termed respiration. Respiration can be viewed as the 'burning' of organic compounds to release stored energy, and because it releases carbon dioxide back into the atmosphere it is effectively the opposite of photosynthesis. The part of gross primary production that is

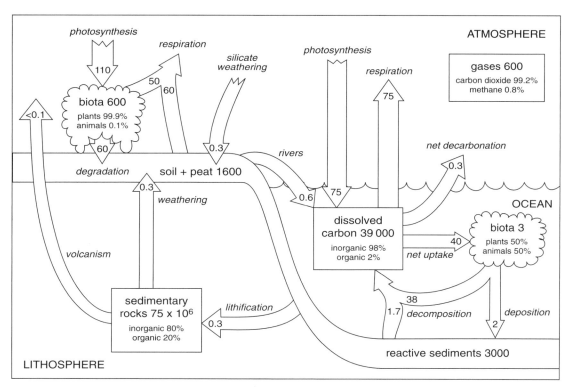

Fig. 6.1 Simplified summary of the preindustrial global carbon cycle, showing approximate sizes of the main reservoirs (variously shaped boxes) and annual fluxes (arrows) in Gt (10^{15} g) of carbon (after several sources, including Bolin et al. 1979, 1983; Kempe 1979; Mopper & Degens 1979; De Vooys 1979; Siegenthaler & Sarmiento 1993; Sundquist 1993; Arthur 2000; Falkowski et al. 2000). Reactive sediments are capable of exchanging material with the water column.

not respired, but is available for growth and reproduction, is called **net primary production**.

From Fig. 6.1 it can be seen that the annual net primary production for land plants and marine plants is similar (c.60 and 40 Gt, respectively), although the biomass of terrestrial plants is much greater than that of marine plants. This is an important demonstration of the fact that biomass is not necessarily a guide to productivity. There is another difference between the marine and terrestrial parts of the biochemical subcycle: the residence time of C in the main reservoirs. From Fig. 6.1 it can be seen that the residence time of carbon in the terrestrial biota is c.5.5 years (i.e. 600/110 yr), and c.26 years (1600/60.6 yr) in soil organic matter. In contrast, the residence time of C in marine phytoplanktonic biomass is only c.2 weeks (1.5/40 yr), but c.338 years (39 000/115.3 yr) in oceanic dissolved carbon.

(a) Terrestrial component of the biochemical subcycle

Net primary productivity on land varies greatly between biomes (Section 3.2.8, Table 3.2). The relatively large biomass of terrestrial plants is due to the storage of organic matter as woody material during life spans that are considerably longer than those of the unicellular photosynthesizers in aquatic environments. The detritus from terrestrial plants makes a large annual contribution to the organic matter in soils, which in turn supports a considerable decomposer community. The latter respires some $60\,\text{Gt}\,\text{C}\,\text{yr}^{-1}$, mostly as CO_2 but accompanied by CH_4. The residence time of C in soil of c.25 yr is an approximate mean: the most readily decomposed organic matter in fresh litter being decomposed within a couple of years, whereas the more recalcitrant material can survive and contribute to peat formation (Section 3.3.3).

Organic matter is quite efficiently recycled in soils, and so preservation of organic matter is confined mostly to peat formation in moorland bogs and low-lying swamps, but the size of this carbon reservoir (c.160 Gt) is not known with any accuracy. While the present annual accumulation rate of peat is low compared with that of organic-rich marine sediments, the presence of large coal deposits in the sedimentary record suggests that peat formation was of much greater importance in the past (see Section 4.6.1).

Freshwater environments contribute to the terrestrial biochemical subcycle, but the present-day net primary production in such environments is relatively minor at $<1\,Gt\,yr^{-1}$. Carbon-rich deposits are being formed under certain conditions in some lakes, but lakes contain only a very small volume of water compared with the oceans and so it is in marine sediments that most of the global preservation of organic carbon is occurring. The river run-off component of c.0.6 Gt yr^{-1} in Fig. 6.1 represents dissolved and particulate organic carbon together with inorganic C derived from rock weathering. Fluvial particulates contribute little to the open ocean, because they are mainly deposited in estuaries and deltas as a result of slackening currents and the flocculation caused by salinity differences.

(b) Marine component of the biochemical subcycle

On a global basis, macroscopic, multicellular algae (seaweeds) make only a minor contribution to marine primary production; 95% is accounted for by the phytoplankton. Phytoplankton are short-lived compared with terrestrial plants, especially trees, and do not need to produce supportive structural tissue, which is mostly photosynthetically inactive. Virtually the whole of the net primary production of phytoplankton is directed towards reproduction and growth, but much of this is grazed by herbivorous zooplankton. As a consequence of this efficient grazing, phytoplanktonic biomass is low and the ratio of animal to plant biomass in the oceans is greater than on land.

The flow of carbon in the marine part of the biochemical subcycle involves buffering by the large reservoir of dissolved C (39 Tt), most of which is in the form of dissolved inorganic carbon (DIC). Whereas terrestrial plants take up carbon dioxide directly from the atmosphere in gaseous form, aquatic plants utilize the CO_2 dissolved in water, and it is for this reason that the photosynthesis and respiration flux arrows in Fig. 6.1 do not point directly to marine biota. A dynamic equilibrium exists whereby CO_2 molecules are constantly exchanging between the atmosphere and oceans (where it forms part of the DIC). In solution, the CO_2 is involved in a sequence of reversible reactions that favour the bicarbonate ion (see Box 3.12). Consequently, at constant temperature, increasing the level of CO_2 in the atmosphere causes more of the gas to dissolve in the oceans, resulting in an increase in all the chemical species of carbon in Eqn 3.8, once the dynamic equilibria have been re-established.

The presence of the permanent thermocline effectively divides the marine carbon cycle into surface-water and deep-water compartments, as in Fig. 6.2. The deep compartment contains most of the DIC, and its residence time can be seen to be c.1000 years (37 000/37 yr), which equates to the rate of turnover of the oceanic water column. The organic remains of organisms and faecal pellets represent a particulate organic carbon (POC) reservoir of c.30 Gt (Mopper & Degens 1979). Dissolved organic carbon (DOC; c.700 Gt; Hedges 1992), therefore, accounts for some 95% of the total marine organic carbon. On average, only c.10%

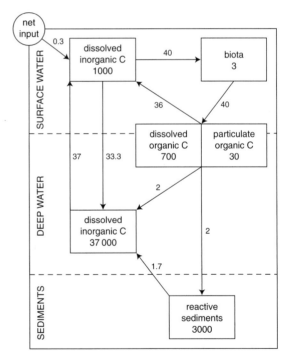

Fig. 6.2 Simplified summary of the marine carbon cycle, showing approximate sizes of reservoirs (boxes) and annual fluxes (arrows) in Gt of carbon (sources as for Fig. 6.1). Approximately 10% of DOC and POC are within surface waters. Exchanges involving DOC and POC are poorly constrained and so are not shown.

of net primary production escapes from the surface compartment, of which at most half reaches the surface sediments (Fig. 6.2). Some of the recycled POC is converted to DOC rather than being completely respired to CO_2. The small amount of POC that reaches the sediment is mostly recycled by detrital-feeding benthonic organisms. Carbonate tends to dissolve and, in addition to the CO_2 and DOC produced by decomposition processes, can be leached out of the upper layers of sediments (the reactive sediments in Fig. 6.1) into pore waters, which are able to exchange with water overlying the sediments. With increasing sedimentation this exchange process ceases and long-term preservation of carbon can occur as sediments consolidate and lithify.

Although carbon is chiefly assimilated by phytoplankton as CO_2, a significant amount of bicarbonate-C is removed from seawater by those producing tests of calcium carbonate:

$$Ca^{2+}(aq) + 2HCO_3^-(aq) \rightarrow CaCO_3(s) + CO_2 + H_2O$$
[Eqn 6.1]

The removal of bicarbonate from the pool of DIC causes a decrease in total dissolved C and a decrease in alkalinity (the loss of every Ca^{2+} is matched by two HCO_3^-, so the ionic balance is maintained, but at a lower bicarbonate contribution to total alkalinity; see Box 3.12). The decline in alkalinity is double that of dissolved C: while two negative charges are lost, only one C atom is removed from the DIC because one of the bicarbonate C atoms is converted to a molecule of dissolved CO_2 (Fig. 6.3a). Although photosynthesis effectively removes dissolved C, more CO_2 enters surface waters from the atmosphere. Movement of CO_2 into and out of surface waters does not affect alkalinity *per se* (see Box 3.12). The warm and cold surface waters in Fig. 6.3a have similar alkalinities, but their different dissolved C content is due to the declining solubility of CO_2 with increasing temperature (as for all gases) and the balance between photosynthesis and respiration. The trend from cold surface water to deep Pacific water reflects the transport of water towards the deep Pacific and the associated increasing contributions from dissolution of $CaCO_3$ and respiration of organic matter. The solubility of $CaCO_3$ increases with pressure (depth), and increasing the DIC concentration raises the acidity of water, further increasing carbonate dissolution.

The influence of DIC concentration and alkalinity on the partial pressure of CO_2 in surface waters (pCO_2; see Box 1.11) is shown in Fig. 6.3b. When the input of $CaCO_3$ (e.g. from dissolution) is greater than its output (e.g. by test secretion and ultimately burial in sediments), DIC concentration and alkalinity increase and pCO_2 decreases slightly. The opposite behaviour is observed when $CaCO_3$ input is less than its output. The effects of exportation of net primary production are also shown for contrasting regimes in Fig. 6.3b. The export of organic matter alone ($C_{org}/C_{CaCO_3} = \infty$, Fig. 6.3b) is representative of high-nutrient polar surface waters,

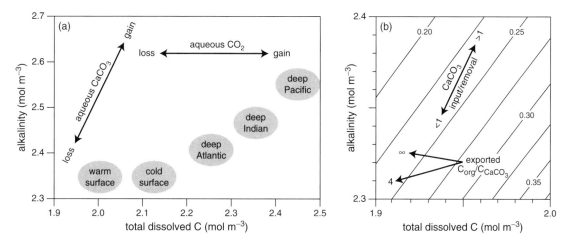

Fig. 6.3 Total dissolved C (DIC + DOC) versus alkalinity plots showing (a) relationships between different oceanic water bodies and the effects of net changes in dissolved $CaCO_3$ and CO_2 (after Broecker 1974); (b) variation in pCO_2 (shown by diagonal isobars in units of atmospheres) for surface water of 35‰ salinity at 20°C, and the effects of changes in dissolved $CaCO_3$ and export of organic matter (after Sigman & Boyle 2000).

where production is dominated by organic-walled and siliceous phytoplankton, and it involves a slight increase in alkalinity owing to the extraction of the nutrient nitrate (and hence dissociation of additional carbonic acid to restore the ionic balance; Sigman & Boyle 2000). Export of organic-C and carbonate-C in a ratio of $c.4:1$ is characteristic of warm low-latitude surface waters, where calcareous microalgae are important among the phytoplankton, and it results in decreasing alkalinity. From Fig. 6.3 it can be seen that the export of organic matter out of surface waters results in a decrease in pCO_2, so there is a net drawdown of CO_2 from the atmosphere until equilibrium is re-established for the ocean–atmosphere CO_2 exchange. Consequently, the transfer of carbon from surface waters to the deep ocean via phytoplanktonic production is known as the **biological pump**. The export of $CaCO_3$ exerts a relatively limited influence on the drawdown of CO_2 (Fig. 6.3b).

Although the CO_2 exchange between atmosphere and oceans is shown to be in equilibrium in Fig. 6.3, there are latitudinal variations. The major polar areas of deep-water formation (Antarctic Bottom Water and North Atlantic Deep Water; Box 3.2) correspond to net sinks of atmospheric CO_2, because the gas is particularly soluble in the cold waters that sink in these regions. Regions where deep water upwells, such as in the equatorial belt, are net sources of CO_2 to the atmosphere, because of their high DIC content and the increase in temperature at the surface in low latitudes. Other regions can be net sinks or sources, depending upon the patterns of seasonal primary productivity (e.g. Antarctic Divergence; Box 3.5).

6.1.3 Geochemical subcycle

Despite the large quantity of carbon in the geochemical subcycle, the average carbon content of the Earth's crust is <0.5% (by weight). The amount of carbon in sedimentary rocks can be calculated from estimates of the total volume of different types of sedimentary material and their average carbon contents (Kempe 1979). Such approximations give the value of $c.75$ Pt C for the sedimentary rock carbon reservoir in Fig. 6.1. Although some 5% of marine net primary production enters sediments, the greater part of it is recycled, and only a small proportion is preserved in sedimentary rocks. From Fig. 6.1 it can be seen that the approximate residence time of C in the reactive surface sediments is 1.5 kyr (3000/2 yr), but once lithification occurs (see Box 1.1) the residence time increases to $c.250$ Myr (75×10^6/0.3 yr). The latter residence time is of the order of tectonic cycling of lithospheric plates, so it is not surprising that the major processes involved include volcanism and the weathering of uplifted and exposed sedimentary rocks. The role of human exploitation of fossil fuels is considered in Section 6.5.

At first sight, the geochemical subcycle appears to play a minor role, but the fact that it involves such a large reservoir of C indicates its importance over geological time scales. It is slightly more complex than indicated in Fig. 6.1; a more detailed representation is given in Fig. 6.4, with the relatively small fluxes shown in kt yr^{-1}.

The major sink for atmospheric CO_2 involves the subaerial weathering of carbonates and silicates in sedimentary rocks. Physical weathering (e.g. freeze-thaw and tectonic action) increases the surface area available for chemical weathering by the carbonic acid formed upon dissolution of CO_2 in rain water:

$$CaCO_3(s) + CO_2 + H_2O \rightarrow Ca^{2+}(aq) + 2HCO_3^-(aq)$$
[Eqn 6.2]

$$CaSiO_3(s) + 2CO_2 + H_2O \rightarrow Ca^{2+}(aq) + 2HCO_3^-(aq) + SiO_2$$
[Eqn 6.3]

Although simple calcium-containing minerals are shown in Eqn 6.2, other cations are involved (e.g. dolomite, $CaMg(CO_3)_2$; albite, $NaAlSi_3O_8$; anorthite, $CaAl_2Si_2O_8$). When the bicarbonate produced reaches the oceans it may be precipitated (mainly by calcareous organisms) as calcium carbonate, as represented by Eqn 6.1. This reaction is the reverse of Eqn 6.2, so carbonate weathering and subsequent precipitation and long-term preservation of calcium carbonate do not result in a net change in atmospheric CO_2 levels. In contrast, silicate weathering followed by precipitation of calcium carbonate releases back to the atmosphere only half the amount of CO_2 originally drawn down during the weathering process, as can be seen by adding Eqns 6.1 and 6.3:

$$CaSiO_3(s) + CO_2 \rightarrow CaCO_3(s) + SiO_2 \quad \text{[Eqn 6.4]}$$

Silicate weathering resulting from the dissolution of ambient atmospheric CO_2 would be slow under the low carbonic acid concentrations produced at present-day atmospheric CO_2 levels. However, much higher CO_2 concentrations are found in soil pore-waters, due to the respiration of plant roots and the microbial oxidation of organic matter. Organic acids from root exudates and microbial activity also contribute to the weathering process (Berner 1992).

The weathering of kerogen in sedimentary rocks may be considered as equivalent to aerobic respiration

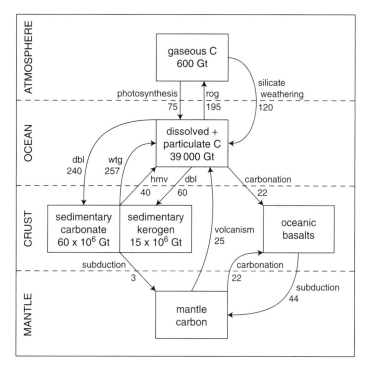

Fig. 6.4 Simplified summary of the geochemical carbon subcycle, showing main reservoirs and approximate sizes of annual fluxes (arrows) in kt of carbon (after Arthur 2000). dbl = deposition, burial + lithification; hmv = hydrothermal metamorphism + volcanism; rog = respiration + outgassing; wtg = weathering.

(Eqn 1.27; Keller & Bacon 1998), and the opposite of photosynthesis, which can be represented simply by:

$$CH_2O + O_2 \underset{\text{photosynthesis \& burial}}{\overset{\text{kerogen weathering}}{\rightleftharpoons}} CO_2 + H_2O \quad \text{[Eqn 6.5]}$$

The oxidation of coal and kerogen appears to be fast on geological time scales, so uplift and erosion rates are the limiting factors, as for silicate weathering (Chang & Berner 1999). However, some DOC and solid oxidation products are formed as well as CO_2, so the process of kerogen weathering is not as simple as implied by Eqn 6.5 and the arrow to dissolved C in Fig. 6.4. As can be seen from Eqn 6.5, the oxidative weathering of kerogen has an impact on atmospheric O_2 levels, as does pyrite weathering (Eqn 1.26); both processes consume oxygen.

Oceanic basalts can be carbonated by the CO_2 escaping from the mantle and by dissolved CO_2 during hydrothermal circulation in the vicinity of mid-ocean ridges, as shown in Fig. 6.4. Volcanism leads to CO_2 from the mantle dissolving in seawater or escaping directly to the atmosphere (only the first route is shown in Fig. 6.4 for simplicity). Volcanism, metamorphism and deep diagenesis can all release CO_2 by reactions that are effectively the reverse of Eqn 6.4.

There are various feedback mechanisms that can control some of the processes in Fig. 6.4, some of which are summarized in Fig. 6.5 in terms of the overall controls on atmospheric CO_2 and O_2 levels. Any combination of pathways forming a closed loop can be followed in Fig. 6.5, and the overall feedback is negative (i.e. stabilizing) if there is an odd number of negative feedback pathways in the loop. The loop *ahrc*, which stabilizes atmospheric O_2 levels by the negative feedback involving phosphate availability to phytoplankton, has already been examined simply in Fig. 3.10 (Box 3.6; Holland 1994; van Cappellen & Ingall 1996). Other important negative feedbacks are represented by the following loops:

1 *blg*: weathering of silicates draws down CO_2 from the atmosphere, causing a decrease in greenhouse warming (see Box 6.1), lower temperatures and less atmospheric moisture, and hence slower weathering rates (Berner & Caldeira 1997).

2 *ntg*: plant growth draws down CO_2, accelerates silicate weathering, decreasing atmospheric CO_2 levels and slowing plant growth (Volk 1987).

3 *dec*: higher atmospheric O_2 levels lead to more vegetation fires, which limit terrestrial net primary production and reduce O_2 levels (Kump 1988), although the

252 | Chapter 6

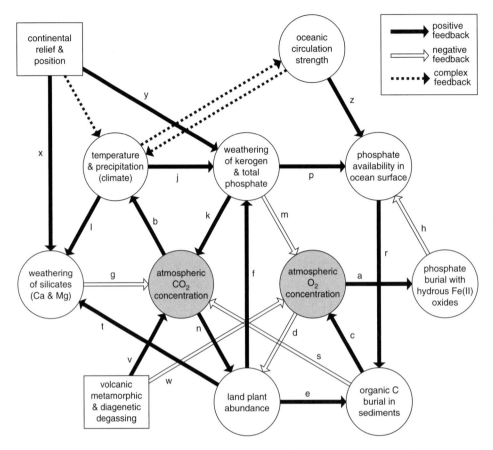

Fig. 6.5 Important feedback mechanisms operating in the long-term carbon cycle (after Berner 1999). Letters refer to feedbacks in various loops discussed in the text. The weathering of silicates implies subsequent precipitation and long-term burial of carbonate. Processes in boxes are regulated by tectonic activity and are not subject to feedbacks from the other processes depicted.

Box 6.1 | Greenhouse effect

The mean surface temperature of the Earth is primarily the result of a balance between the energy it receives from the Sun and the energy it radiates back into space. There is also a contribution from heat flow from the Earth's interior ($c.8 \times 10^{-2} W m^{-2}$), but this amounts to c.0.02% of the Sun's contribution. The high surface temperature of the Sun (equivalent to a black body temperature of c.6000 K) means that solar energy is confined to the short wavelength region of the electromagnetic spectrum, with a maximum in the visible region (Fig. 6.6). Of the solar energy incident at the top of the atmosphere (or **insolation**, which totals $c.343 W m^{-2}$), some is reflected back into space by clouds, atmospheric dust and the Earth's surface (land, plants and water), but the remainder heats up the Earth's surface ($c.240 W m^{-2}$). The reflected portion is c.30% (Trenberth et al. 1996), so the average reflectivity of the Earth, its **albedo**, is 0.3. Ice is particularly reflective, with an albedo of up to 0.9.

The Earth is warmer than surrounding space, so it radiates energy back into space, at wavelengths dictated by its black-body temperature (c.255 K, or −18°C). The relatively low surface temperature

Continued

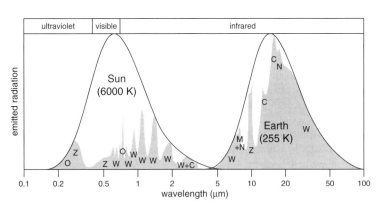

Fig. 6.6 Relative variation in energy radiated by the Sun (black body at 6000 K) and Earth (black body at 255 K). Half-tone shows effective absorption by atmospheric components (C = CO_2, M = CH_4, N = N_2O, O = O_2, W = H_2O, Z = O_3; after Mitchell 1989).

means that the emitted radiation is at longer wavelengths than the incoming radiation, in the infrared region of the electromagnetic spectrum (like heat emission from an electric fire). An exact balance between this back radiation and absorbed insolation would maintain a constant mean surface temperature at the Earth's surface.

Although the gases in the atmosphere cause only minor attenuation of the incoming energy at short wavelength, they can absorb significant amounts of the longer-wavelength outgoing energy. This is because infrared wavelengths correspond to the resonant energies associated with vibration (stretching, compression and rotation) of chemical bonds, each type of bond having its own characteristic resonant frequencies. The energy absorbed by gas molecules is re-emitted, but equally in all directions, so roughly half of this energy ($c.148\,W\,m^{-2}$) is trapped and travels back to the Earth's surface. The trapping of this back radiation by molecules of various gases is so effective that only a narrow window ($c.8$–$12\,\mu m$; Fig. 6.6) exists by which radiation can escape to space. This effect has been likened to that caused by the glass in a greenhouse, and it causes the effective average surface temperature of the Earth to be 15 °C ($c.288\,K$).

Quantitatively the most important greenhouse gas is water vapour, followed by carbon dioxide. Methane, ozone and nitrous oxide all have greater warming potentials than CO_2 on a molecule to molecule basis (see Box 6.4), but their overall greenhouse effects are smaller because of their low concentrations. Although absorption of radiation by CO_2 appears to be almost 100% efficient, an increase in atmospheric CO_2 results in an increase in trapped radiation because band broadening occurs (a widening of the adsorption wavelength window shown for CO_2 in Fig. 6.6), causing an increase in the Earth's surface temperature. As temperature rises so does the level of back radiation from the Earth's surface so that, if atmospheric CO_2 levels were allowed to stabilize at a constant higher value, an equilibrium would once again be established between incident and emitted energy fluxes, but at a higher mean surface temperature. The instantaneous heating caused by a doubling of atmospheric CO_2 concentration from 300 to 600 ppm is estimated to be 4.0–$4.5\,W\,m^{-2}$ (Cess et al. 1993) but once the Earth system has adjusted, the heating effect (or **radiative forcing**) is $c.3.5$–$4.0\,W\,m^{-2}$, for which the associated temperature rise has been modelled at 1.5–$4.5\,°C$ (Harvey 2000; Houghton et al. 2001). Based on observed increases in ocean-water temperature (see Section 6.5.1) the temperature increase caused by such a doubling of atmospheric CO_2 concentration could be towards the higher end of this range ($c.3\,°C$; Kerr 2000).

---

charcoal formed from woody tissue is resistant to decay and so may limit the extent of the negative feedback.

4 *dfprc*: increased O_2 levels cause more vegetation fires, decreasing the standing crop, so leading to less weathering of phosphate minerals, a decreased supply of phosphate to the oceans, decreased marine net primary production, less organic C burial and so lower O_2 production (Berner 1998, 1999).

5 *bjprs*: higher atmospheric CO_2 levels result in a warmer and wetter (owing to increased evaporation) climate, which increases weathering of phosphate minerals, increasing phosphate availability in the oceans, increasing

net primary production and hence burial of organic C, which reduces atmospheric CO_2 levels (Berner 1999).

An even number of (or zero) negative feedback steps results in an overall positive feedback loop, examples of which include:

1 *dfm*: higher atmospheric O_2 levels result in more fires and hence fewer land plants, leading to less weathering of organic matter and hence lower O_2 drawdown (Berner 1999).
2 *bjk*: higher atmospheric CO_2 levels result in greenhouse warming, increasing precipitation, which increases weathering of organic matter and production of CO_2 (Berner 1999).

The potential negative feedbacks tend to stabilize atmospheric CO_2 and O_2 concentrations in the long term. However, the processes represented by boxes in Fig. 6.5 are the result of tectonic activity and so are not controlled by feedbacks. They have the potential to disrupt the carbon cycle severely, as considered in Section 6.3. As indicated in Fig. 6.5, oceanic circulation can influence the C cycle in complex ways via its influence on climate, through its impact on heat distribution over the surface of the Earth and the rate and net direction of CO_2 exchange with the atmosphere. Similarly, continental relief and distribution have an effect on the climate through their influence on winds and currents. All these factors are investigated in Section 6.3.1.

6.2 Changes in carbon reservoirs over geological time

6.2.1 Sedimentary preservation of organic carbon

In Section 1.4.2 it was seen how the oxygen content of the atmosphere is linked to primary production and the long-term burial of organic matter in sedimentary rocks (see Eqn 6.5). However, the amount of carbon in the kerogen reservoir cannot always have been $c.15 \times 10^6$ Gt, as shown for the preindustrial steady state represented in Fig. 6.1. The presence of kerogen is attributable to the appearance and evolution of life. We can estimate the amount of organic C in sedimentary rocks and investigate the importance of photosynthesis throughout the history of life on Earth by examining the $\delta^{13}C$ record of kerogen and carbonates. The mean isotopic signature for the Earth as a whole can be assumed to be that inherited from the parent solar nebula, which is represented by the isotopic signature of the mantle ($\delta^{13}C$ $c.-5‰$) in Fig. 6.7. The different isotopic signatures of the various surficial (atmospheric, hydrospheric and lithospheric) reservoirs of carbon are attributable to a range of isotopic fractionation processes.

As reviewed in Section 5.8, biological processes discriminate in favour of the lighter isotope, with C_3 photosynthesis exerting the major influence on the isotopic signature of organic matter. As a result, the average $\delta^{13}C$ value for organic matter (living and dead) in the crust is $c.-25‰$. The two surficial reservoirs of the CO_2 used in photosynthesis (atmosphere and ocean) are correspondingly depleted in the lighter isotope. The atmospheric reservoir, used by terrestrial plants, has a $\delta^{13}C$ value of $-7‰$. The oceanic reservoir, used by phytoplankton, is dissolved CO_2, which rapidly equilibrates with bicarbonate ($\delta^{13}C$ $1‰$), the isotopic signature of which is broadly preserved in sedimentary carbonate ($\delta^{13}C$ $c.0‰$). All the carbon in the surficial reservoirs is ultimately derived from outgassing from the mantle (Fig. 6.7), so on average it must retain the primordial $\delta^{13}C$ signature of $c.-5‰$:

$$\delta^{13}C_{primordial} = (f_{CO_2})(\delta^{13}C_{CO_2}) + (f_{HCO_3^-})(\delta^{13}C_{HCO_3^-}) + (f_{CO_3^{2-}})(\delta^{13}C_{CO_3^{2-}}) + (f_{org})(\delta^{13}C_{org}) \quad \text{[Eqn 6.6]}$$

where f is the fraction of C in each of the surficial reservoirs.

From Eqn 6.6 it can be seen that the geological record of $\delta^{13}C$ for kerogens and carbonates reflects changes in the sizes of C reservoirs, in addition to changes in the relative importance of the various fractionation pro-cesses (such as C_3 photosynthesis). Because the kerogen and carbonate in sedimentary rocks dominate the shallow reservoirs (>99.9%), the isotopic mass balance in Eqn 6.6 can be written simply as:

$$\delta^{13}C_{primordial} = f_{organic}\delta^{13}C_{organic} + (1 - f_{organic})\delta^{13}C_{carbonate} \quad \text{[Eqn 6.7]}$$

where $f_{organic}$ is the fraction of C in kerogen (relative to kerogen + carbonate). Inserting the approximate $\delta^{13}C$ values from Fig. 6.7 into Eqn 6.7 gives a value of 20% for $f_{organic}$.

The carbon isotopic records of sedimentary carbonate ($+0.5‰ \pm 2.5‰$) and organic matter (mainly $-26‰ \pm 7‰$) appear fairly constant for at least the past 3.5 Gyr (Fig. 6.8), and possibly longer, if allowance is made for metamorphism affecting the $\delta^{13}C$ values of the 3.8 Gyr old samples from Isua, Greenland (carbonate-C becomes slightly lighter and organic-C heavier; Schidlowski 1988). The carbon isotopic record has not been entirely uniform throughout geological time. Some excursions have occurred, which can be attributed to natural perturbations that were followed by a relatively

The carbon cycle and climate | **255**

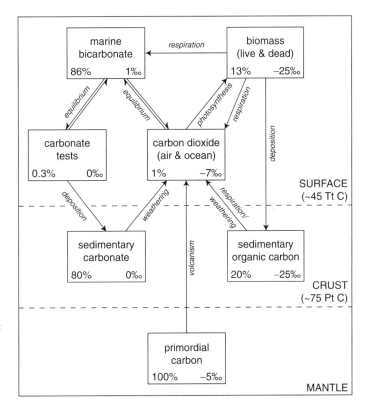

Fig. 6.7 Much simplified carbon cycle showing exchanges between the surface, crust and mantle compartments. Approximate relative sizes of reservoirs within each compartment are shown together with average $\delta^{13}C$ values (after Schidlowski 1988; Hoefs 1997).

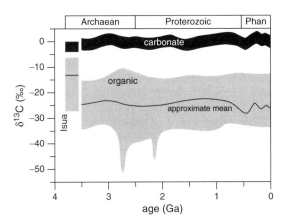

Fig. 6.8 Variation in carbon isotopic composition of carbonate and kerogen over the past 3.8 Gyr (after Schidlowski 1988). The samples from Isua, Greenland, have experienced metamorphism.

rapid return to the mean isotopic levels described above. For example, unusually heavy carbonates (having $\delta^{13}C$ values >+5‰) have been recorded for two periods in the Precambrian (2.7 and 1.8–1.9 Ga; not shown in Fig. 6.8), coinciding with the two most powerful stages of magmatism and granitization (Rhodesian and Belomorian) in the Earth's history (Galimov 1976). The heightened tectonic activity caused injection into the atmosphere and ocean of huge quantities of metamorphogenic CO_2, relatively rich in ^{13}C, which was incorporated in biogenic carbonates until levels once again returned to their equilibrium values. Enhanced methanotrophic activity may be involved in two negative spikes in $\delta^{13}C_{organic}$ at *c.*2.7 and 2.1 Ga (Schidlowski 1988).

The residence time of carbon in the DIC reservoir (mainly bicarbonate) is relatively short (*c.*1 kyr from Fig. 6.2), so any changes in processes affecting the $\delta^{13}C$ signature of bicarbonate (such as the amount of net marine primary production) would be preserved in the sedimentary carbonate record. In view of this, it might be thought that the broadly constant $\delta^{13}C$ records for kerogen and carbonate indicate that $f_{organic}$ has remained at 0.2 for the last *c.*4 Gyr and that net photosynthetic production rapidly reached a limiting level and then remained more or less constant (Schidlowski 1988). However, the data in Fig. 6.8 do not take into account any thermally induced increase in the $\delta^{13}C$ values of kerogen due to the loss of isotopically light volatiles (Section 4.5.4a). This effect is more likely to have been experienced by Archaean and Proterozoic than Phanerozoic samples, but it can be taken into account. As can be seen from Fig. 4.15, the atomic H/C ratio of a kerogen decreases with increasing thermal maturity, and can be correlated with the increase in $\delta^{13}C$ (McKirdy & Powell 1974):

$$\delta^{13}C_{original} = \delta^{13}C_{measured} - 4.05 + 3.05r \\ - 0.785/r - 0.0165/r^2 + 0.000879/r^3 \quad [Eqn\ 6.8]$$

where r = measured atomic H/C ratio. Applying this correction to Proterozoic data, after discarding data for samples that have experienced the most extreme thermal alteration, provides the plot in Fig. 6.9b, which is significantly different from the plot of the raw data (Fig. 6.9a).

Clearly, $\delta^{13}C_{organic}$ increased during the Proterozoic. Equation 6.7 can be rearranged to:

$$f_{organic} = \frac{\delta^{13}C_{primordial} - \delta^{13}C_{carbonate}}{\delta^{13}C_{organic} - \delta^{13}C_{carbonate}} \\ = \frac{\delta^{13}C_{carbonate} + 5}{\delta^{13}C_{carbonate} - \delta^{13}C_{organic}} \quad [Eqn\ 6.9]$$

The $\delta^{13}C_{carbonate}$ record has remained fairly constant, and $\delta^{13}C_{carbonate}$ values do not appear to be affected by post-depositional alteration (Veizer et al. 1992). So if it is assumed that $\delta^{13}C_{organic}$ has likewise remained reasonably constant (i.e. controlled largely by C_3 photosynthesis), the changes in ($\delta^{13}C_{carbonate} - \delta^{13}C_{organic}$), which we can call $\varepsilon_{sedimentary}$, are primarily due to changes in $f_{organic}$ (Des Marais et al. 1992). The resulting changes in $f_{organic}$ are shown in Fig. 6.9c. If it is assumed that the value of $f_{organic}$ at 2.6 Ga can be applied to all previously buried carbon, and that the crust contained c.8.2 Pt $C_{organic}$ at 2.6 Ga (based on data from Schidlowski 1988), changes in crustal $C_{organic}$ can be estimated as shown in Fig. 6.9d (Des Marais et al. 1992).

There appear to have been two main increases in $f_{organic}$ and the total amount of organic C buried, at c.2.1–1.8 and c.1.0–0.7 Ga. The likely agent is tectonic activity, because these two periods correspond to major episodes of rifting and orogeny, resulting in enhanced erosion and weathering of rocks, which delivers more nutrients and clastic material to the oceans. The combined effect is to promote marine primary production and increase the rate of burial of organic matter. During the earliest Proterozoic, $f_{organic}$ is predicted to be low (c.0.1; Fig. 6.9c), which is consistent with the low amounts of organic matter present in early Proterozoic stromatolitic carbonates (Des Marais 1991). This may reflect efficient remineralization of organic carbon similar to that seen in modern microbial mats (Des Marais et al. 1992). Major rifting commenced at 2.2 – 2.1 Ga, producing basins suitable for large-scale preservation of organic matter (see Box 3.11). There is evidence for enhanced continental erosion and silicate weathering in the form of a sustained rise in seawater $^{87}Sr/^{86}Sr$ (Veizer & Compston 1976; Box 6.2). The associated drawdown of CO_2 and climatic cooling may account for the glaciations of the time (Harland 1983). The opening and closing of major ocean basins seem to have reached maximum intensity at 2.1 – 1.7 Ga, and the predicted high rate of organic burial may be reflected in the formation of ^{13}C-rich carbonates (Baker & Fallick 1989). With the increase in crustal organic carbon content, the atmosphere would have become more oxidizing, possibly accounting for increased iron(III) retention in palaeosols (Holland & Beukes 1990) and the occurrence of extensive redbeds (Eriksson & Cheney 1992), as discussed in Section 1.4.2. Tectonic and orogenic activity declined at 1.7 – 1.2 Ga, but then increased at 1.2 – 0.9 Ga (Grenvillian orogeny) during the assembly of the supercontinent Rodinia (see Section 6.3.2). The general trend in burial of organic carbon appears to parallel these major fluctuations in tectonic activity (Fig. 6.9d).

The declining difference in $\delta^{13}C$ values between carbonate and kerogen during the Proterozoic (Fig. 6.9b) suggests a decreasing isotopic discrimination during photosynthesis as a consequence of falling atmospheric (and hence dissolved) CO_2 concentration (see Section 5.8.2). The largest declines in $\varepsilon_{sedimentary}$ coincide with the most intense phases of global rifting and orogeny, consistent with enhanced drawdown of atmospheric CO_2 during silicate weathering (Berner et al. 1983).

Major changes in $f_{organic}$ appear to have occurred during the Neoproterozoic (c.900–545 Ma), but since the Early Devonian, when vascular plants had completed their colonization of the land, this parameter appears to have remained at c.0.25 (Fig. 6.10d; Hayes et al. 1999). Two sharp drops in $f_{organic}$ can be seen in Fig. 6.10d during the Sturtian and Varangerian ice ages. As can be seen in Section 6.3.2, primary productivity was probably very low during these glaciations, but the cold episodes were terminated by pronounced warming, which caused the precipitation of carbonates directly from solution (Fairchild 1993; see Box 1.1).

The value of $\varepsilon_{sedimentary}$ varies significantly in Fig. 6.10c, the implications of which can be evaluated using the much simplified expression in Eqn 5.36. The interplay between $\varepsilon_{sedimentary}$, $\varepsilon_{photosynthesis}$ (allowing for heterotrophic alteration, ε_{het}) and temperature (which has

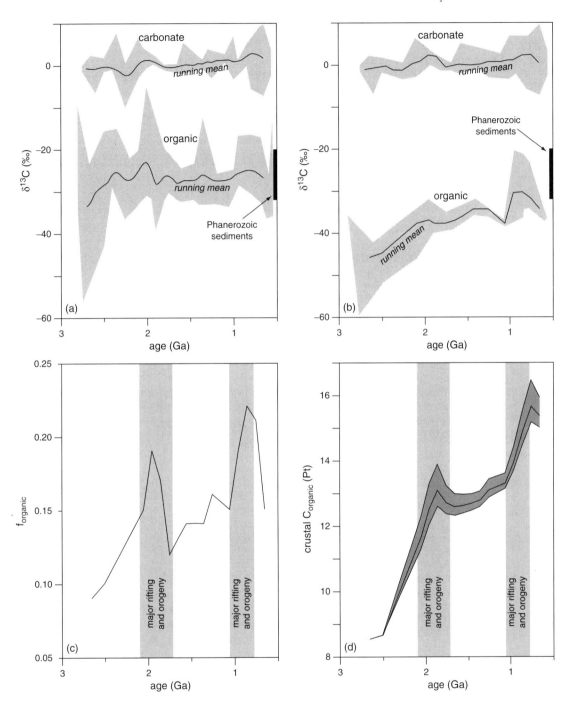

Fig. 6.9 Estimation of organic carbon burial during the Proterozoic (after Des Marais et al. 1992). (a) $\delta^{13}C$ values for carbonates and kerogens; (b) kerogen $\delta^{13}C$ values corrected for thermal evolution; (c) fraction of carbon in kerogen (relative to carbonate); (d) mass of carbon in kerogen. In (a) and (b) running means are calculated for successive 100 Myr increments, and $\delta^{13}C$ ranges are shown by half-tone bands. In (d) the effects of varying sediment half-life are represented; the upper limit of the range is for a half-life of 300 Myr and the lower 500 Myr.

Box 6.2 | Strontium isotope record of continental weathering

The ratio of two of the stable isotopes of strontium, $^{87}Sr/^{86}Sr$, can be used to determine the relative importance of weathering of continental rocks. The oceans are the ultimate repository of the products of weathering, and strontium ions can replace calcium ions during the formation of calcite, so carbonate tests provide a record of variation in the $^{87}Sr/^{86}Sr$ of seawater over geological time (Burke et al. 1982; Koepnick et al. 1988). The $^{87}Sr/^{86}Sr$ ratio of seawater can be considered to represent mixing of two end members with characteristic $^{87}Sr/^{86}Sr$ values (Palmer & Edmond 1989). That with the lower value of 0.703 is derived from the mantle and enters the oceans primarily via the hydrothermal circulation through oceanic crust. The higher value of c.0.715 derives from the weathering of continental crust and is transported to the oceans via rivers. Continental crust contains greater concentrations of the radionuclide ^{87}Rb than oceanic crust, the decay of which yields ^{87}Sr. Continental crust does not have uniform composition, and its $^{87}Sr/^{86}Sr$ ratio can vary between 0.71 and 0.72. About 75% of the Sr input to the oceans derives from weathering of ancient marine carbonates, which bear the isotopic signature of the oceans prevailing at the time of their original deposition (Brass 1976).

Changes in the $^{87}Sr/^{86}Sr$ ratio of marine carbonates reflect variations in the relative rates of continental weathering and hydrothermal activity. Consequently, independent evidence of tectonic activity (i.e. the total length and spreading rates of mid-ocean ridges) is required for accurate assessment of continental weathering rates. The $^{87}Sr/^{86}Sr$ ratio decreased at the beginning of the Jurassic, coincident with the onset of rifting of the supercontinent Pangaea and increased seafloor spreading (see Fig. 6.14a, b). In contrast, the $^{87}Sr/^{86}Sr$ ratio has risen over the past 25 Myr, and is apparently attributable primarily to the uplifting and weathering of the Himalayas (Edmond 1992).

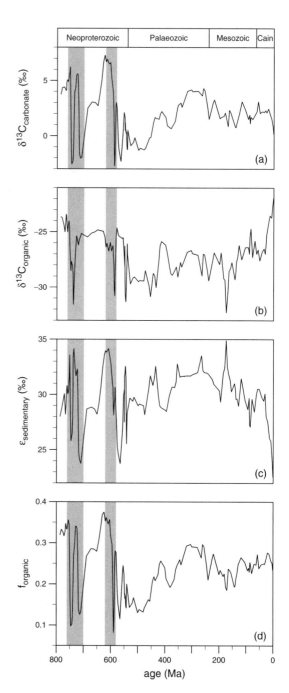

Fig. 6.10 Averaged $\delta^{13}C$ record of (a) carbonates and (b) kerogen, together with (c) associated $\varepsilon_{carbonate-organic}$ fractionation ($\varepsilon_{sedimentary}$), and (d) estimated proportion of C in kerogen ($f_{organic}$) for the past 800 Myr (after Hayes et al. 1999). The half-tone areas denote the Sturtian (c.760–700 Ma) and Varangerian (c.620–580 Ma) ice ages.

a significant influence on $\varepsilon_{calcite-CO_{2}(aq)}$; Section 5.8.5) can be represented by the plot in Fig. 6.11. As described in Section 5.8.2, the maximum value for $\varepsilon_{photosynthesis} - \varepsilon_{het}$ is c.25‰, assuming that chemoautotrophy is insignificant, and so few data points are expected to lie above the uppermost, inclined solid line

The carbon cycle and climate | 259

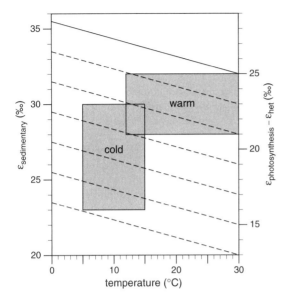

Fig. 6.11 Likely relationships between $\varepsilon_{carbonate-organic}$ fractionation ($\varepsilon_{sedimentary}$) and surface-water temperature (which controls $\varepsilon_{calcite-CO2(aq)}$) for varying photosynthetic fractionation and allowing for heterotrophic fractionation ($\varepsilon_{photosynthesis} - \varepsilon_{het}$), as described in Section 5.8.5 and Fig. 5.59 (after Hayes et al. 1999). The half-tone areas define contrasting climatic conditions. Any chemotrophic contributions to kerogen are ignored.

(passing through 25‰) on the $\varepsilon_{photosynthesis}-\varepsilon_{het}$ axis. During the period 800–750 Ma and most of the Phanerozoic up to the Neogene (excluding the Carboniferous–Permian and mid-Jurassic), $\varepsilon_{sedimentary}$ is within the range 28–32‰ in Fig. 6.10c, corresponding to the warm zone in Fig. 6.11. Conditions were generally warm and likely to result in significant photosynthetic fractionation, because slow growth (low μ), high CO_2 concentrations (high c) and small phytoplankton (low V/S) result in maximal $\varepsilon_{photosynthesis}$ values in Eqn 5.34. In contrast, $\varepsilon_{sedimentary}$ values down to 23‰ are generally associated with cooler periods (cold zone in Fig. 6.11), such as during the Sturtian and Varangerian ice ages, and in the Neogene cooling.

Just prior to each of the Varangerian and Sturtian glaciations, $\varepsilon_{sedimentary}$ rises to >32‰. These periods were associated with flourishing sulphur-oxidizing bacteria (Canfield & Teske 1996) and generally low oxygen levels in the deep ocean (Canfield 1998). A large contribution from chemosynthetic bacteria to the sedimentary organic matter is likely to have resulted, the influence of which upon $\varepsilon_{sedimentary}$ is shown in Fig. 6.12. Degradation of sinking organic detritus in the water column leads to a build-up of the CO_2 respired by degraders in the region of the chemocline (see Section 3.4.3b), where chemosynthesizers congregate. This CO_2 is enriched in ^{12}C relative to its source, phyto-

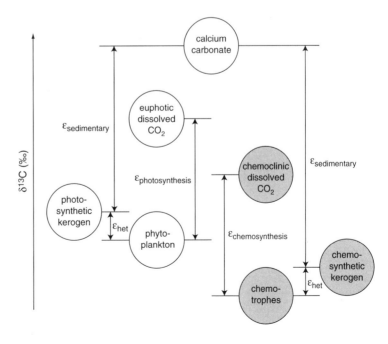

Fig. 6.12 Comparison of the effects of chemotrophy (shown by half-tone) and C_3 phototrophy on the observed fractionation between marine carbonate and kerogen ($\varepsilon_{sedimentary}$; after Hayes et al. 1999). Heterotrophic reworking of primary production is represented by ε_{het}.

planktonic remains (DeNiro & Epstein 1978), which is itself isotopically light compared with the dissolved CO_2 in the euphotic zone above. Because $\varepsilon_{chemosynthesis} \approx \varepsilon_{photosynthesis}$ (Table 5.7) and the CO_2 source for the chemosynthesizers has a lower $\delta^{13}C$ value than that for the photosynthesizers, the kerogen derived from chemosynthesizers is isotopically lighter than that from the photosynthesizers (Fry et al. 1991). Consequently, a significant chemosynthetic contribution to the kerogen increases the value of $\varepsilon_{sedimentary}$. This effect can be allowed for in Eqn 5.36 by modifying the fractionation term that incorporates reworking of organic matter (ε_{het}), which may require negative values (the other terms are little affected).

6.2.2 Atmospheric levels of carbon dioxide

In Section 1.4.2 the increase in atmospheric O_2 levels during the Proterozoic was described (Fig. 1.8). Because this was a consequence of photosynthesis, it is to be expected that atmospheric CO_2 levels correspondingly declined. In addition, the long-term burial of carbonates must also draw down atmospheric CO_2. However, from Eqn 6.1 it can be seen that carbonate precipitation acutally liberates CO_2 locally, in the short term, until the ocean–atmosphere system readjusts to compensate for the depleted total dissolved C in the affected surface waters. Estimated changes in atmospheric CO_2 concentration since the formation of the Earth are shown in Fig. 6.13 (Kasting 1993). If all the carbon present in the crust now had been in the atmosphere as CO_2, it would have exerted a partial pressure of 6–8 MPa (60–80 bars; see Box 1.11). If a more modest 15% existed in the Hadean atmosphere, the partial pressure would have been $c.1$ MPa, and climatic modelling suggests an associated mean surface temperature of $c.85\,°C$ (Kasting 1993). After the heavy bombardment period (Section 1.3.1) a weakly reducing atmosphere, comprising mainly CO_2 and N_2 (but with traces of CO, H_2 and reduced sulphur gases), would have contained $c.0.02$ MPa ($c.600$ times present atmospheric level (PAL)) of CO_2.

Some constraints on CO_2 levels can be applied during climatic modelling in order to reproduce changes from unglaciated to glaciated conditions (i.e. whether the mean surface temperature was above or below $0\,°C$) and to permit oceans to be present from at least the beginning of the Archaean (i.e. $<100\,°C$; see Section 1.3.1). For modelling purposes, it has been assumed that CO_2 and H_2O vapour were the major greenhouse gases and that changes in cloud cover (which influence albedo and hence the Sun's warming influence; Box 6.1) were insignificant (Kasting 1993). Other factors that must be taken into consideration include the lack of continental crust on the young Earth, which affects the planet's albedo, and the fact that the Sun was some 30% less luminous at 4.5 Ga (Newman & Rood 1977; Gilliland 1989). All the preceding factors constrain the upper and lower limits of the CO_2 content of the early atmosphere in Fig. 6.13. A dimmer Sun would result in lower temperatures and hence less silicate weathering, allowing CO_2 from volcanic activity to accumulate in the atmosphere, increasing greenhouse warming. This

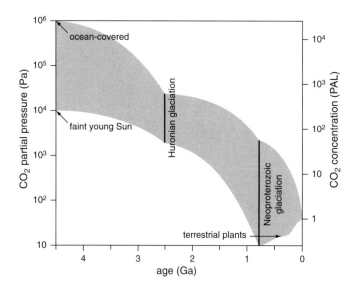

Fig. 6.13 Estimated atmospheric CO_2 concentration throughout the Earth's life (after Kasting 1993). Glaciations assume a mean surface temperature range of 5–20 °C; CO_2 levels at other times are set to maintain ice-free conditions (minimum of 0 °C) with CO_2 offsetting the decrease in solar energy with age (at 4.5 Ga it was $c.30\%$ weaker than at present) and with no change in albedo.

provides a stabilizing negative feedback. There is no geological evidence for glaciations prior to c.2.5 Ga, suggesting temperatures were ≥0 °C, but atmospheric CO_2 levels probably fell to 1–10 PAL near the end of the Proterozoic (Fig. 6.13).

There are no direct means of determining palaeo-CO_2 levels, but support for the modelled changes is provided by the $\delta^{13}C$ values of marine kerogens (Des Marais et al. 1992). It is possible that CO_2 levels could have been lower than those shown in Fig. 6.13 if greenhouse contributions from methane are considered (Kasting 1993). Methanogens may have dominated the mineralization of organic matter during the Archaean, potentially releasing large amounts of methane to the atmosphere (Pavlov et al. 2001b; Habicht et al. 2002), a topic to which we return in Section 6.3.2. Palaeosol evidence has been used to suggest that atmospheric CO_2 levels must have been <100 PAL between 2.75 and 2.2 Ga, which is lower than indicated in Fig. 6.13 and may be consistent with the presence of other greenhouse gases (Rye et al. 1995). However, once O_2 became abundant in the atmosphere between c.1.0 and 0.6 Ga (see Section 1.4.2), concentrations of reducing gases (e.g. methane and ammonia) would have fallen dramatically.

Over geological periods, atmospheric CO_2 levels are controlled by the geochemical cycle. Various models have been constructed to permit changes in atmospheric CO_2 concentrations to be estimated throughout the Earth's history. They are constructed using estimates (often based on proxy measurements) of carbon reservoir and flux sizes for successive time intervals, generally working back from the present (and the only precisely known) atmospheric CO_2 concentration. Modelling is most accurate for the Phanerozoic, because of its plentiful and relatively well preserved sedimentary rock record. One such model is GEOCARB, which has been refined over time, and we consider here the latest version (GEOCARB III; Berner & Kothavala 2001) in some detail. Most of the other models give comparable results to those of GEOCARB in Fig. 6.14c (e.g. Wallmann 2001b; Hansen & Wallmann 2003), although there are exceptions (e.g. Rothman 2002; Fig. 6.14c). GEOCARB balances the major sinks for CO_2—burial of organic matter, and silicate weathering combined with subsequent burial of carbonates (Eqns 6.1–6.4)—with the major sources of the gas—oxidative weathering of kerogen and the thermal breakdown of both kerogen and carbonates (including volcanic emissions). Climate is an important variable and its effects are incorporated into GEOCARB in the form of a general circulation model (GCM; Kiehl et al. 1998; Harvey 2000). The feedbacks within the GEOCARB model are critical because, as seen from Fig. 6.5, temperature and atmospheric CO_2 levels are linked. For example, weathering of C-containing rocks affects CO_2 levels and hence temperature, but is itself influenced by temperature through the rate of weathering and palaeogeography (the elevation of land and its location relative to warm, humid zones).

In the construction of the GEOCARB model, two or more independent measurements of geological processes were used where possible in order to constrain estimates of reservoirs and fluxes. For example, the rates of burial of kerogen and carbonates were estimated from volumes of sedimentary rocks of given age ranges as well as from $\delta^{13}C$ mass balance calculations. Similarly, weathering was estimated from the amounts of terrigenous sediments for each time period as well as by interpretation of strontium isotopic ratios (Box 6.2). Among the least well constrained parameters are those related to weathering caused by plants, which was modelled to start at 380–350 Ma, with angiosperms dominating the process by 80 Ma (Berner 1998). The rate of plant-induced weathering varies with atmospheric CO_2 levels and may be greater for angiosperms than gymnosperms (Volk 1989), but these influences are poorly quantified at present. Other factors that influence modelled CO_2 levels significantly are land uplift (leading to erosion and weathering) and volcanic degassing rates (particularly during increased tectonic activity in the Cretaceous; Larson 1995). All these factors generate the envelope surrounding the main trend in Fig. 6.14c. However, it appears that CO_2 concentrations were very high during the Palaeozoic, fell sharply during the Devonian (when plants colonized the land), rose moderately during the Mesozoic and then gradually declined throughout the late Mesozoic and Cainozoic.

The model used time-averaged data over 10 Myr or greater intervals, so abrupt changes in CO_2 concentrations are not represented. The model would clearly benefit from improved understanding of plant-induced weathering and a more detailed description of the palaeogeographic influence on weathering (Otto-Bleisner 1995). For example, the centre of the supercontinent of Pangaea (see Fig. 6.19) was dry, but there were continental fragments within the tropical precipitation zone where weathering would have been enhanced (see Section 6.3.3 and Fig. 3.12). It might be expected that an increase in the weathering rate of continental rocks would cause $^{87}Sr/^{86}Sr$ to rise and atmospheric CO_2 concentration to decrease. Conversely, an increase in mid-ocean ridge volcanicity would cause $^{87}Sr/^{86}Sr$ to fall and atmospheric CO_2 concentration to increase. Clearly, there is no simple relationship of this

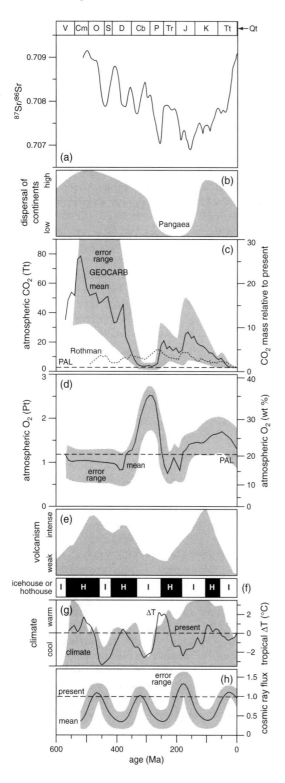

type between the strontium isotopic record and CO_2 concentrations in Fig. 6.14a, c. However, the Late Ordovician–Early Silurian and Carboniferous–Permian glaciations followed periods of enhanced weathering, indicated by elevated $^{87}Sr/^{86}Sr$ in Fig. 6.14a, as might be expected (Raymo 1991). The Quaternary glaciation has also followed a significant (and continuing) rise in $^{87}Sr/^{86}Sr$ (see Box 6.2).

The rate of chemical weathering appears to have been relatively constant since the mid-Proterozoic (Holland 1984), which suggests that the absence of plant-induced weathering prior to land plants becoming widespread in the Devonian must have been compensated for by higher levels of atmospheric CO_2 (Berner 1990, 1991). However, the degree of elevation of atmospheric CO_2 levels may have been offset by contributions of the gas from microbial respiration in pre-Silurian soils (Keller & Wood 1993), and by the fact that temperature may be a more important control on silicate weathering than soil CO_2 levels (Brady & Carroll 1994).

In Fig. 6.14c, CO_2 concentrations show a generally positive correlation with volcanic activity (Fig. 6.14e), the major source of the gas, and a negative correlation with oxygen (Fig. 6.14d), which is a proxy for the major sink of CO_2 (relating to production and burial of organic matter). Atmospheric O_2 concentrations during the Phanerozoic have been modelled using a similar approach to that for CO_2 (Berner & Canfield 1989; Berner 2001), involving the balance between the rate of O_2 generation (from estimates of the burial of organic matter and pyrite) and the rate of O_2 consumption (from estimates of the oxidative weathering of kerogen and pyrite) (see Eqns 1.26 and 6.5). Confidence in the model is gained from the reasonable agreement

Fig. 6.14 A comparison of Phanerozoic climatic changes with variations in some potentially related parameters: (a) marine Sr isotopic ratio (after Veizer et al. 1999; McArthur et al. 2001); (b) continental dispersal (after Veevers 1990); (c) atmospheric CO_2 levels (solid line and error range from GEOCARB III after Berner & Kothavala 2001; broken line after Rothman 2002); (d) atmospheric O_2 levels (after Berner & Canfield 1989; Berner 2001); (e) volcanism (after Fischer 1984); (f) icehouse versus hothouse (after Frakes et al. 1992); (g) climate (half-tone trend after Allegre & Schneider 1994; Crowley & Berner 2001; variation in mean tropical sea-surface temperature (ΔT) after Shaviv & Veizer 2003); (h) cosmic ray flux (after Shaviv & Veizer 2003). Shaded areas in (c), (d) and (h) represent confidence limits; PAL = present atmospheric level.

between independent evaluations of reservoir and flux sizes (e.g. burial rates for successive time intervals estimated either from rock volumes combined with C and S content (Berner & Canfield 1989) or from C and S isotopic mass balance calculations (Berner 2001)). The model's predictions appear particularly sensitive to the mean C concentration assumed for coal deposits (Berner & Canfield 1989).

6.3 Palaeoclimatic variations

6.3.1 Factors affecting climate

(a) Solar energy variations

In the preceding sections we have touched upon some of the factors that affect climate, but there are many others, which are summarized in Table 6.1. The amount of energy reaching the surface of the Earth is a key parameter, and is primarily contributed by the Sun (Box 6.1). However, the Sun's luminosity has been increasing steadily (some 30% since the Earth's formation; Gough 1981), and it also varies throughout the sunspot cycle (but by <1% of its mean luminosity; Eddy et al. 1982; Hoffert et al. 1988). In addition, there are variations in the Earth's orbit, collectively termed Milankovich cycles (Box 6.3), which primarily affect seasonal contrasts. The period over which each factor varies is important in terms of its climatic influence. For example, sunspot activity exhibits $c.11$ and 90 year cycles (Friis-Christensen & Lassen 1991), whereas the orbital eccentricity cycle has a major period of $c.100$ kyr. For a link to exist between a particular astronomical cycle and a climatic oscillation, the two cycles must share the same periodicity.

Modelling suggests that a change in insolation (i.e. radiative forcing) of $1\,W\,m^{-2}$ causes a mean surface temperature change of $c.0.27\,°C$ (Harvey 2000), so the relatively small changes in total insolation over time periods of 100 Myr or less seem incapable, on their own, of causing the large changes between hothouse and icehouse worlds that have occurred throughout the Phanerozoic (Fig. 6.14f). However, other factors can amplify the astronomical energy variations, such as the latitudinal distribution of continents and oceans and their varying albedos (Barron et al. 1980; Zachos et al. 2001). The expansion and retreat of polar ice caps is subject to a strong positive feedback because of the high albedo of ice.

(b) Lithospheric and hydrospheric processes

Low latitudes receive more solar energy than high latitudes over a year, because of the greater angle of incidence at high latitude, which presents a greater effective thickness of atmosphere to be penetrated and spreads out the radiation over a larger surface area. There is a net deficit between incoming short-wave radiation and outgoing long-wave radiation above $c.60°$ latitude, but a net surplus at lower latitudes. This difference in energy supply drives the meridional circulation patterns of the oceans and atmosphere that redistribute the energy (with a net transfer polewards). The atmosphere accounts for the greater part of this transfer above $30°$ latitude, but the oceans are the more important heat-transfer agent in the tropics. The position of the continents can significantly affect atmospheric and oceanic heat transfer and hence the effective temperature gradient between equator and poles (Barnes 1999). For example, a concentration of land at the poles provides suitable conditions for polar ice caps to form, insulated against the relatively warming influence of the oceans during the winter (Worsley & Kidder 1991; Crowley et al. 1993). An example of this regime is the

Table 6.1 Factors influencing climate and their associated time scales (after Wilson et al. 2000)

factor	time scale (yr)
astronomical	
solar luminosity	10, 100, 10^9
orbital (Milankovich) cycles	$2 \times 10^4, 4 \times 10^4, 10^5$
lithospheric	
tectonic processes	$10^6 - 10^8$
isostatic response to ice	$10^4 - 10^5$
volcanic emissions (e.g. CO_2)	$1 - 10^8$
cryospheric	
ice-sheet life	$10^2 - 10^5$
hydrospheric	
sea-level change (land ice)	$10^4 - 10^5$
deep-water circulation	10^3
surface-water circulation	10–100
atmospheric	
long-term composition	$10^7 - 10^8$
glacial-interglacial	$10^3 - 10^5$
circulation	1–10
methane hydrate destabilization	1–10
biospheric	
marine primary production	10–100
peat-forming environments	100
anthropogenic	
fossil fuel burning, land use	10–100

Box 6.3 | Milankovich cycles

The variations in the Earth's orbit around the Sun can be grouped into three major cycles—eccentricity, obliquity and precession—each associated with one or more characteristic frequencies (Hays et al. 1976; Berger et al. 1984).

Eccentricity cycle
At present, the orbit is almost circular, but it has been much more elliptical in the past. The eccentricity cycle has periods of 95, 123 and 413 kyr, the first two combining to impose an approximately 100 kyr cycle between successive eccentricity maxima (or minima). The weaker 413 kyr cycle is associated with how the degree of eccentricity varies over successive maxima. Eccentricity affects the mean distance of the Earth from the Sun, so it is the only one of the three cycles that causes a variation in the total insolation. It also contributes to seasonal variations; at maximum eccentricity the difference in insolation between perihelion and aphelion (the orbital points of smallest and greatest distance between Earth and Sun, respectively) has been c.30% during the past 5 Myr (Wilson et al. 2000).

Obliquity cycle
The tilt of the axis of rotation of the Earth relative to the plane of its orbit varies over a period of 41 kyr. It has oscillated between 21.8° and 24.4° over the past 800 kyr or so. This variation determines how insolation is distributed over the face of the Earth, and hence the strength of seasonal variations, but does not affect the total insolation.

Precession cycle
The rotational axis of the Earth gradually precesses like that of a spinning top, and the elliptical orbit of the Earth also precesses. The periods of these two precessions are 19 and 23 kyr, respectively, yielding an average of c.21 kyr. Precession controls the distribution of insolation over the Earth's surface by varying the timing of the seasons relative to the perihelion, but, like obliquity, does not affect the total insolation.

Insolation variation
The combined effect of the Milankovich cycles on summer insolation at 65° N over the past 600 kyr is shown in Fig. 6.15. The maximum variation in insolation is c.20% of the mean value (equivalent to a shift from 65° to 77° N). The total annual variation in insolation caused by the eccentricity cycle is only 0.03%, so the maximum change in the Sun's radiation reaching the Earth's surface is c.7 W m^{-2}. This could cause a temperature change of c.5°C (using a radiative forcing factor of c.0.8° W^{-1} m^2).

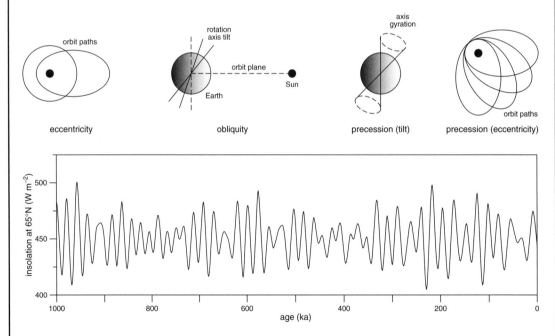

Fig. 6.15 Summary of the orbital characteristics underlying the Milankovich cycles and their combined effect upon insolation at 65° N during summer in the northern hemisphere (after Berger et al. 1984).

presence of a large land mass at the south pole during the Late Carboniferous, when there was extensive glaciation (see Section 6.3.3). A supply of moisture is the key to ice-sheet growth, and requires a suitable regime of oceanic currents and related atmospheric circulation to transport warm moist air from low-latitude evaporation zones to high-latitude precipitation sites (Zachos et al. 2001). Continental reconfiguration operates on the time-scale of ocean-basin growth and destruction (i.e. hundreds of millions of years).

Major orogenies cause uplift, increasing the amount of land above the snow line, and increasing silicate weathering, which draws CO_2 out of the atmosphere. Both processes have the effect of promoting climatic cooling (Raymo & Ruddiman 1992), although latitudinal effects are important (e.g. faster weathering rates in warm humid zones, and greater winter cooling of land masses at high latitudes). The circulation of surface and deep waters in the ocean and their rate of exchange can have a significant impact on the meridional redistribution of heat, as well as the exchange of energy and CO_2 between atmosphere and ocean. The oceanic and atmospheric circulation systems operate over significantly shorter periods than lithospheric processes (Table 6.1). As the Earth warms, sea level rises because of the melting of ice sheets and the thermal expansion of surface waters (the latter tends to have a positive feedback on ice-sheet melting).

(c) Greenhouse gases and aerosols

Volcanoes produce lava, pyroclastic material (ranging from sizable rocks to fine ash and dust) and volatiles (comprising gases, including water vapour, CO_2, SO_2, HCl and CO). Their eruption styles depend upon the volatiles content of the magma. A low volatiles content results in the relatively gentle effusion typical of basaltic lava (e.g. at mid-ocean ridges). However, in magmas of higher silica content the viscosity is greater, trapping volatiles, which expand explosively when pressure falls in the rising magma. Such magmas produce more acidic lavas (e.g. rhyolites and andesites) and are typically associated with subduction zones; they are also likely to have higher water content than basaltic magmas (from the water associated with subducting sediments). Eruptions of basaltic magma are not confined to mid-ocean ridges, under considerable depths of water that prevent volatiles entering the atmosphere directly. Unusually big plumes of hot rising mantle (superplumes) can cause large-scale eruptions termed **flood basalts**, under the oceans or within the continents, as occurred during the Cretaceous (see Section 6.3.4). Subaerial volcanism leads to volatiles entering the atmosphere directly, which can result in a major climatic impact.

Increased tectonic activity may result in more CO_2 reaching the atmosphere from increased volcanism, causing greenhouse warming (Williams et al. 1992; Box 6.1). Subaerial volcanoes also emit sulphur dioxide into the atmosphere, which is oxidized to sulphate and forms aerosols that reflect insolation, leading to cooling (Rampino 1991; Gagan & Chivas 1995; Ramanathan et al. 2001). Airborne volcanic ash has a similar cooling effect, but it also absorbs and reradiates long-wave radiation (a greenhouse warming effect). Overall, the cooling effect may dominate, although the climatic implications are complex (Kerr 1993). The largest known eruption of the late Quaternary, involving the Toba super-volcano (Sumatra) 73.5 kyr ago, appears to have resulted in pronounced cooling (Rampino & Self 1992; Zielinski et al. 1996).

The products of subaerial, effusive volcanism are usually confined to the lowest layer of the atmosphere (the troposphere; see Box 7.1 and Fig. 7.1), and the sulphate aerosols and ash last for a few months only, because they are washed out by rain. In contrast, the products of explosive volcanism (such as the Toba eruption) can be propelled higher into the atmosphere and reach the stratosphere in significant amounts. The volcanic volatiles have a more widespread and longer-lasting climatic influence in the relatively dry stratosphere, because removal by wash-out is slow (Skelton et al. 2003). The initial cooling effects of sulphur dioxide and ash during large-scale subaerial volcanism may be superseded by longer-term greenhouse warming from the associated CO_2 emissions.

The most important greenhouse gas at present is not carbon dioxide but water vapour, simply because there is so much of it in the atmosphere (Box 6.1). Volcanoes emit large amounts of water vapour ($c.1\,\text{Tt yr}^{-1}$; Skelton et al. 2003), but even so this flux is minor compared to evaporation from the oceans and evapotranspiration from plants ($c.0.25\%$; see Fig. 3.12). In a warmer world, such as during the Cretaceous, the atmosphere can hold more water vapour. However, the extent of the warming caused by extra atmospheric water vapour is difficult to predict because clouds also exhibit an albedo effect, and the balance between the greenhouse and albedo effects varies with cloud type and altitude (Lovelock & Whitfield 1982).

In Fig. 6.14f it can be seen that the sedimentological indications of hothouse versus icehouse conditions (e.g. Frakes et al. 1992) correlate reasonably well with variations in the mean surface-water temperature for tropical oceans estimated from carbonate $\delta^{18}O$ values, and exhibit a periodicity of $c.135$ Myr (Fig. 6.14g; Shaviv & Veizer 2003). Clearly, atmospheric CO_2 concentration does not vary in the same way, and so is not an obvious candidate for the driving force behind this climatic oscillation (Veizer et al. 2000), although it is undoubtedly an important agent in climatic change. In contrast, the cosmic ray flux has a similar periodicity to the major hothouse/icehouse oscillation, as can be seen from Fig. 6.14h. A physical basis for this apparent empirical correlation has yet to be established. Maxima in the cosmic ray flux are related to the passage of our Solar System through the spiral arms of the galaxy, and it has been postulated that these periods correlate with more low-altitude clouds, which in turn may exert a net cooling effect because of their high albedo (Shaviv & Veizer 2003).

In addition to CO_2 there are two other C-containing gases that merit our attention, one of which exerts a cooling effect (dimethyl sulphide) and the other a greenhouse effect (methane); they are discussed in the following subsections.

(d) Dimethyl sulphide

There is a net flux of dimethyl sulphide (DMS, CH_3-S-CH_3) from ocean to atmosphere (Kiene & Bates 1990). The precursor of DMS, dimethylsulphoniopropionate (DMSP) is produced by certain phytoplankton, and DMS is liberated mainly during zooplanktonic grazing (Dacey & Wakeham 1986), resulting in enrichment of DMS in surface waters. DMS transfer to the atmosphere is highest in the tropics, partly due to the latitudinal distribution of the phytoplanktonic groups that produce most DMS (e.g. coccolithophores; Charlson et al. 1987). DMS is quantitatively the most important volatile sulphur compound involved in air–sea transfer processes at present, its flux amounting to $c.30–50$ Mt yr^{-1} (Liss 1983). This corresponds to $c.15–25$ Mt yr^{-1} of S, which is estimated to account for nearly half the total biogenic input of sulphur to the atmosphere, and is of comparable size to volcanic emissions (Bolin et al. 1983). Once in the atmosphere, DMS undergoes oxidation and some of the products, such as sulphur dioxide and methanesulphonic acid (MSA), form sulphate aerosols and cloud condensation nuclei (CCN; Ayers & Gras 1991). A negative feedback to greenhouse warming is likely: the initial warming stimulates higher productivity among DMS-producing phytoplankton, resulting in more sulphate aerosols and increased cloud cover, which in turn causes a decrease in incident solar radiation levels (through increased albedo), offsetting the warming. However, the lower light levels caused by increased atmospheric albedo would provide a negative feedback on phytoplanktonic production, reducing both CO_2 uptake and amelioration of the greenhouse effect. It is not clear, at present, whether these opposing effects are likely to offset or reinforce global warming, or whether they have any significant effect at all. For example, without increased cloud cover from DMS production the Earth's surface would be warmer, promoting more evaporation of water, so cloud cover would increase by a different mechanism. However, the combined effects of the negative feedbacks involving DMS may help to maintain an equitable climate for life, as proposed in the Gaia theory (Lovelock et al. 1972).

(e) Methane

As noted in Section 4.6.2a, two major reservoirs of methane are the methane hydrates in permafrost and deep-sea sediments, resulting from the activity of methanogenic bacteria. Although they do not appear to make a major contribution to atmospheric methane at present, their potential for rapidly releasing large amounts of methane is important. However, it is difficult to incorporate methane into climatic models. The continental (i.e. permafrost) and marine hydrates do not necessarily respond in the same way towards global warming, because of the different temperature and pressure regimes involved (Fig. 4.32). Pressure changes are small for continental hydrates, so temperature fluctuations have the greatest influence on stability. In contrast, temperature variation at depth in the oceans is relatively restricted, so pressure variations are likely to govern the stability of marine hydrates. A positive feedback can be postulated for continental methane hydrates as temperature increases (and pressure decreases slightly owing to melting; Fig. 6.16a). In contrast, provided deep-water temperatures do not rise too much, the sea-level mediated pressure variations, caused by the changing size of ice sheets on land, may have a negative, stabilizing feedback (Fig. 6.16b, c; Kvenvolden 1998). However, hydrates on polar continental shelves are likely to be more sensitive to climatic change during transgression by a relatively warmer polar ocean. The associated pressure increase imposed by an increase in water depth of $c.100$ m would not offset the destabilizing effect of a temperature rise of $\geq 10\,°C$ (see Fig. 4.32).

The major sink for atmospheric methane is oxidation by highly reactive hydroxyl radicals (·OH; formed by photolytic dissociation of water molecules in the atmosphere; Crutzen 1988; Eisele et al. 1997). The life-time of methane in the atmosphere is, consequently, relatively short and unlikely to have exceeded 100 years

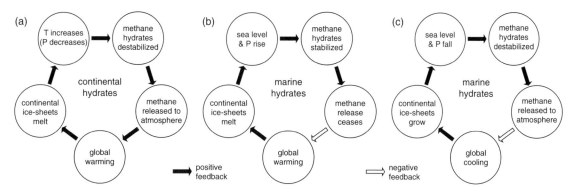

Fig. 6.16 Feedbacks between climatic change and methane hydrate stability: (a) positive greenhouse warming feedback for continental hydrates; (b) and (c) negative feedbacks for marine hydrates (after Kvenvolden 1998).

even during times of potentially higher atmospheric concentrations in the Archaean (Pavlov & Kasting 2002). High levels of methane in the Archaean could have aided the oxidation of the young Earth, through photolysis of methane and escape of the resulting hydrogen atoms (which are reducing agents) to space (Catling et al. 2001).

6.3.2 Neoproterozoic glaciations

During the Proterozoic there was a major glaciation at $c.2.3$ Ga (Huronian) in the Palaeoproterozoic (Evans et al. 1997), but then there seems to have been a long period of nearly 1500 Myr before an episode of at least two glaciations between 760 and 580 Ma in the Neoproterozoic (Kennedy et al. 1998; Hoffman & Schrag 2000). These ancient ice ages are different from their Phanerozoic counterparts in two major ways: (a) the glaciation extended into the tropics at sea level, possibly resulting in an almost completely iced-up 'Snowball Earth'; (b) the glacial deposits are capped by abiotic carbonates (<5 m thick) with particularly low $\delta^{13}C$ values (down to −5‰; Kaufman 1997). The Palaeoproterozoic glaciation seems to have followed the fragmentation of a supercontinent, and similarly the Neoproterozoic glaciations followed on from the splitting up of the supercontinent Rodinia. Precise reconstructions of Rodinia are difficult to accomplish, but the supercontinent started assembling $c.1100$ Myr ago and began disintegrating at $c.850$–800 Ma (Dalziel 1997; Hoffman et al. 1998; Meert & Powell 2000; Torsvik 2003; Meert & Torsvik 2003). The first episode of Neoproterozoic glaciation (Sturtian, $c.760$–700 Ma) accompanied the opening of the Pacific Ocean, while the second (Varangerian or Marinoan, $c.620$–580 Ma) was associated with the opening of the Iapetus Ocean (Hoffman et al. 1998).

The continental fragments appear to have been largely confined within the tropics following the disintegration of Rodinia, but despite the ability of oceanic currents to transfer heat to the the polar regions with this arrangement of continents, extensive glaciation was able to occur. Once ice had spread to ≤30° latitude, the Earth's albedo increased rapidly (due to higher insolation at low latitude, resulting from the Sun's energy being spread over a smaller area of the Earth's surface than at high latitude), and a runaway positive feedback occurred. However, the Earth escaped from the grip of ice and even appears to have rebounded into a very warm period at the end of each glaciation, because the inorganic precipitation of the cap carbonates would have required warm surface waters. Various explanations have been proposed for the mechanisms of the glacial–greenhouse transitions and the formation of the characteristic, isotopically light, cap carbonates (e.g. Kaufman et al. 1997; Hoffman et al. 1998; Kennedy et al. 2001).

In the original Snowball Earth model (Kirschvink 1992; Hoffman & Schrag 2000), it was thought that the distribution of continental fragments promoted silicate weathering, leading to a reduction in atmospheric CO_2 and the formation of polar ice packs. The ice-albedo feedback then plunged the Earth into a glaciation, with an average temperature of $c.-50\,°C$ and an average ice thickness of >1 km; only the Earth's internal heat preventing complete freezing of the oceans. Carbonate $\delta^{13}C$ values were proposed to have fallen from ≥5‰ to as low as −5‰ just before glaciation, remained similar in the cap carbonates and subsequently recovered to $c.0$‰.

The low $\delta^{13}C$ values were believed to reflect the DIC in the oceans approaching the isotopic signature of mantle CO_2 that continued to enter the oceans and atmosphere (this isotopic equilibration would take >10^5 yr) and were taken as an indication, together with a paucity of kerogen in the immediately post-glacial deposits, of an almost complete collapse of primary production. Further support for this contention appears to be provided by the presence of iron-rich deposits among the glacial debris, which indicate low oxygen levels in the oceans, due to the ice cover. Dissolved Fe(II) would have built up in the oceans as hydrothermal activity continued, but would have deposited as Fe(III) when the ice melted and dissolved O_2 levels rose again. The duration of the negative $\delta^{13}C$ excursion was estimated to be of the order of 10 Myr.

During the glaciation, the prevailing cold dry air is assumed to have prevented the growth of glaciers on the continents. Melting of the ice was accomplished in the model by the build-up of volcanic CO_2 in the atmosphere, which was possible because of the greatly diminished rates of CO_2 drawdown by both primary production and silicate weathering under the dry conditions. It has been estimated that c.0.12 bar (i.e. c.120 000 ppm or c.350 PAL) of CO_2 would be required to achieve the thaw (Caldeira & Kasting 1992a), the accumulation of which would take c.4 Myr at the present-day volcanic emission rate (c.0.24 Gt yr^{-1}). This is a minimum time, because it assumes no air–sea exchange, and does not allow for the c.6% lower solar luminosity or decreased pelagic carbonate deposition (which would result in lower CO_2 content of volcanic emissions at convergent plate margins). A likely maximum time is 30 Myr. The transition to a greenhouse climate appears to have been rapid, and the proposed atmospheric CO_2 concentration suggests the temperature approached 50°C.

During the glaciation, hydrothermal activity and low-temperature basalt alteration would have increased the Ca:Mg ratio of seawater, but the river supply of alkalinity (i.e. bicarbonate) would have been drastically diminished. Carbon dioxide continued to enter the oceans from spreading-ridge volcanism, so the net effect was carbonate dissolution (from Eqn 3.9 and Box 3.12, increasing the amount of CO_2(aq) causes the equilibrium to shift in favour of bicarbonate, lowering carbonate concentration and so encouraging dissolution of previously precipitated carbonate). As melting progressed and temperature increased, evaporation of water would have contributed to an accelerating greenhouse effect, and further carbonate dissolution would have been likely initially. However, silicate weathering on unglaciated continental fragments would have increased rapidly under the high temperatures and vigorous hydrological cycle, providing bicarbonate to the oceans and resulting in the precipitation of the cap carbonates with an isotopic signature dictated by the mantle (and average crustal) value. As life and kerogen formation recovered, the isotopic fractionation of primary production resulted in the $\delta^{13}C$ value of carbonates increasing once more, over c.10^5 yr. A reduction in atmospheric CO_2 levels from 12 kPa to typical Proterozoic levels of c.0.1 kPa corresponds to a transfer of c.250 Tt C (or some 8×10^5 km^3 of carbonate), sufficient to account for the cap carbonates.

A summary of the isotopic excursion for the Snowball Earth model is shown in Fig. 6.17. Some objections have been raised to the model, including the difficulty of climatic models achieving freezing of tropical oceans under such large amounts of atmospheric CO_2 (Hyde et al. 2000). In addition, an implausibly high rate of silicate weathering would be required for the deposition of the cap carbonates (Kennedy et al. 2001). Deposition in c.10^4 yr would require silicate weathering rates three to four orders of magnitude greater than those preceding the CO_2 build-up, for which there is no evidence in the Sr isotopic record (Jacobsen & Kaufman 1999). These difficulties can be overcome by a model incorporating destabilization of methane hydrates in terrestrial permafrost during post-glacial warming and the associated marine transgression (Jacobsen 2001; Kennedy et al. 2001). In this model the continents can be glaciated, as

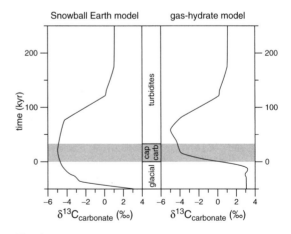

Fig. 6.17 Carbonate isotopic excursions during Neoproterozoic glaciations predicted by the Snowball Earth and gas-hydrate models (after Jacobsen 2001). Zero on the time scale corresponds to onset of cap-carbonate deposition.

predicted by climatic models, because there is no need for conditions favouring rapid silicate weathering.

The methane-hydrate model is shown in Fig. 6.17. It fits data suggesting that the greater part of the negative $\delta^{13}C$ excursion occurs in the cap carbonate, over a relatively short time (Kennedy et al. 2001). The deposition of cap carbonates can be accommodated by large-scale consumption of the released methane by anaerobic methanotrophes in a consortium with sulphate reducers (see Section 3.3.2b). The by-products of methane oxidation are passed on to the sulphate reducers (Werne et al. 2002), from which isotopically light bicarbonate can be generated (Boetius et al. 2000):

$$2CH_2O + SO_4^{2-} \rightarrow H_2S + 2HCO_3^- \quad [Eqn\ 6.10]$$

This stimulates the precipitation of the cap carbonates. Assuming the DIC (mainly bicarbonate) reservoir was $c.10$ Tt (as at present) with a $\delta^{13}C$ value of 0‰, and that the methane had a mean $\delta^{13}C$ value of −60‰, the minimum amount of methane required to achieve the required isotopic excursion is $c.5$ Tt, which is about the estimated quantity of the cap carbonates. There also seem to be sedimentary structures consistent with methane venting (Kennedy et al. 2001).

It is possible that the Neoproterozoic icehouse to greenhouse transitions provided the stimulus for the radiation of metazoa that followed during the Cambrian explosion (Hoffman & Schrag 2000). However, the recent discovery of microbial mat communities from the Sturtian glaciation, which are apparently identical to preglacial communities, raises a question about the severity of the glaciation (Corsetti et al. 2003). It has been suggested that the glaciations were more like those of the Pleistocene (see Section 6.3.5b), although the ages of the sections upon which this suggestion has been based are poorly constrained and may not correspond to the original Snowball Earth sections (Leather et al. 2002). A major difference between the Neoproterozoic and the Phanerozoic is the absence during the former of planktonic carbonate-secreting organisms and the associated, probably stabilizing, effect of pelagic carbonate deposition (Ridgwell et al. 2003).

The absence of glaciations during the greater part of the Proterozoic suggests a strong greenhouse effect, which has previously been attributed to high levels of CO_2 in the atmosphere (Kasting 1993; see Section 6.2.2). However, methanogens flourished during the early stages of life's evolution, and methane is likely to have been an important greenhouse gas while atmospheric O_2 levels were low and the major sink for methane (oxidation by hydroxyl radicals) was suppressed (Pavlov et al. 2001a; see Sections 6.2.2 and 6.3.2). If O_2 levels were lower than 1 PAL from 2.3 to 0.75 Ga, as discussed in Section 1.4.2, methane would still have been a major atmospheric component even in the late Proterozoic (Pavlov et al. 2003). Destruction of methane by methanotrophy is likely to have been suppressed because the aerobic pathway is limited when the atmospheric O_2 concentration falls below 0.45%, and the associated low levels of sulphate in the oceans (Hurtgen et al. 2002) would have inhibited the anaerobic pathway. Methanogenesis may have been important in the water column as well as in sediments. It is estimated that the production of methane was some 10–20 times modern levels (Pavlov et al. 2003). The life-time of methane in the atmosphere increases by a factor of six for a tenfold increase in its rate of input, and it also increases slightly as the amount of O_2 decreases from 1 to 0.1 PAL, but at 0.01 PAL there is an abrupt decrease in life-time because ozone shielding declines and hydroxyl radicals are formed by photolysis of water in the troposphere (see Fig. 6.18a).

Methane would probably have exerted a positive feedback on surface temperature, with higher temperatures promoting rapid growth of methanogens. However, an increase in dissolved O_2 and hence sulphate concentrations in the oceans just before 750 Ma would have caused a significant decrease in methane production, resulting in an atmospheric CH_4 level similar to today's, possibly triggering the late Neoproterozoic glaciations. Figure 6.18b demonstrates the modelled effect of varying atmospheric methane concentration on temperature for the Palaeoproterozoic (2.3 Ga) and the Neoproterozoic (700 Ma), allowing for variation in solar luminosity (Gough 1981), but at a constant CO_2 level of 1 PAL. To achieve the present-day mean surface temperature would have required $c.16$ ppm of methane at 700 Ma but $c.380$ ppm at 2.3 Ga (Pavlov et al. 2003).

6.3.3 Carboniferous–Permian icehouse

From Fig. 6.14f it can be seen that climatic oscillations between hothouse and icehouse conditions have continued through the Phanerozoic. Two major episodes of glaciation can be discerned spanning the Ordovician–Silurian and the Carboniferous–Permian boundaries. It has been suggested from studies of the geological water cycle and the associated variations of $\delta^{18}O$ in marine carbonates that there may have been three icehouse–greenhouse cycles, with an additional cold period near the Jurassic–Cretaceous boundary and greenhouse maxima occurring near the Cambrian–

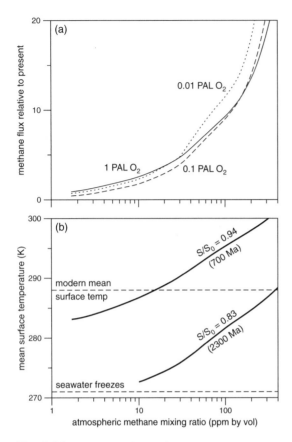

Fig. 6.18 (a) Variation of atmospheric CH_4 concentration with input flux (from methanogenesis) for varying O_2 levels (controlling oxidative destruction). (b) Influence of atmospheric methane on mean surface temperature (assuming constant O_2 and CO_2 levels, but different solar luminosity relative to present (S/S_0)) during the Palaeoproterozoic (2300 Ma) and Neoproterozoic (700 Ma). (After Pavlov et al. 2003.)

Ordovician, Devonian–Carboniferous and Permian–Triassic boundaries (Wallmann 2001a).

The major Carboniferous–Permian icehouse episode at c.300 Ma is associated with elevated atmospheric oxygen concentrations (Fig. 6.14d) and so has been attributed to elevated levels of photosynthesis, which caused atmospheric O_2 levels to increase while CO_2 was drawn out of the atmosphere and locked away in the geochemical C cycle (Berner 1990; Crowell 1999). At the time, the land mass was assembled into a supercontinent, Pangaea, which formed a continuous meridional belt almost from pole to pole (Fig. 6.19). In the equatorial belt, where the areas now represented by Europe and North America lay, the warm and humid climate during the Carboniferous was ideal for the growth of extensive coal-forming rain-forests, dominated by lycopsids. This zone was bounded by deserts, and the interior of the supercontinent would undoubtedly have been dry (Crowley 1994; see Section 1.5.5). In low-latitude shelf areas with high evaporation to precipitation ratios (see Fig. 3.12) extensive evaporite deposits were formed. The Carboniferous–Permian transition was marked by a change in both peat-forming environments and the dominant plants involved (Gastaldo et al. 1996). The southern ice cap advanced towards the equator during the Early Permian, but during warmer interglacial episodes coal-forming environments also developed in the swamps of the southern temperate regions (e.g. Australia; Fig. 6.19), dominated by deciduous glossopterids.

Of course, factors other than high terrestrial productivity may have contributed to the drawdown of CO_2 and the onset of glaciation, such as elevated silicate weathering rates coupled to carbonate burial in the oceans. As can be seen from Fig. 6.14a, the Sr isotopic ratio was quite high, consistent with an elevated contribution from continental weathering. The assemblage of Pangaea involved considerable orogeny (e.g. Carboniferous–Permian Alleghenian and Hercynian), which would have increased the potential for weathering (Klein 1994). In addition, solar luminosity was 3–5% lower than at present (Gough 1981), which would have aided the cooling caused by decreasing atmospheric CO_2 levels. Prior to 325 Ma (the Mississippian–Pennsylvanian boundary) there was a subtropical oceanic gateway between the Panthalassa and Tethys oceans (Fig. 6.19), the influence of which on oceanic and atmospheric circulation kept the central parts of Gondwana cool and arid. However, the closure of this gateway by the collision of Euramerica and Gondwana during the final assembly of Pangaea greatly enhanced poleward transport of atmospheric moisture, contributing to the growth of the southern polar ice sheet (Saltzman 2003). Pangaea gradually moved northwards during the Permian, the glaciers retreated and temperatures rose.

The C isotopic record provides support for the greatly diminished CO_2 and elevated O_2 atmospheric concentrations (Beerling et al. 2002). The evidence arises from the fact that the enzyme rubisco catalyses both photosynthetic carboxylation and photorespirational oxidation. Consequently, CO_2 and O_2 compete for acceptor molecules of the enzyme. So if pCO_2 rises the oxygenase activity of rubisco is suppressed, and conversely if pO_2 rises the carboxylase activity of rubisco is

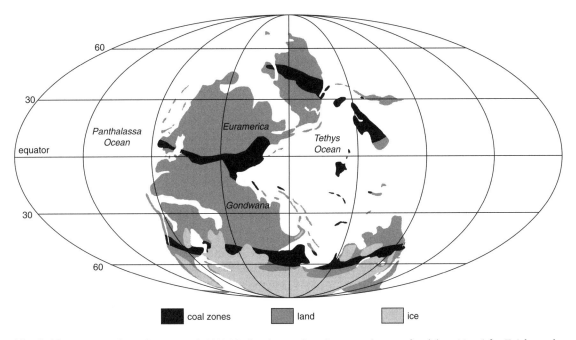

Fig. 6.19 Pangaea in the Early Permian (*c*.280 Ma), showing southern ice cap and zones of coal deposition (after Zeigler et al. 2001; Scotese 2003).

suppressed. In addition, the extent of O_2 inhibition of photosynthesis decreases as pCO_2 rises. At *c*.300 Ma, O_2 levels have been modelled at 35% and CO_2 at *c*.0.03%, resulting in a pO_2/pCO_2 ratio of ≥1000, which should have favoured the oxygenation role of rubisco and caused a decrease in the uptake of CO_2 from substomatal cavities in leaves. A decline in substomatal CO_2 uptake causes a rise in the substomatal partial pressure of CO_2 relative to atmospheric pCO_2 (see Box 1.11). The effect of this on carbon isotopic fractionation during photosynthesis can be seen by applying the appropriate parameters to Eqn 5.32 (Section 5.8.2):

$$\varepsilon_{photosynthesis} = \varepsilon_{diffusion} + (pCO_{2\,substomatal}/pCO_{2\,atmospheric})$$
$$\cdot (\varepsilon_{rubisco} - \varepsilon_{diffusion})$$

[Eqn 6.11]

A rise in ($pCO_{2\,substomatal}/pCO_{2\,atmospheric}$) causes an increase in $\varepsilon_{photosynthesis}$, other factors remaining constant (Berner et al. 2000). The value of $\varepsilon_{photosynthesis}$ can be obtained from Eqn 5.28 (Box 5.11), using measured values of fossil-plant material for $\delta^{13}C_{plant}$ and by assuming that $\delta^{13}C_{air}$ is 7‰ greater than corresponding marine $\delta^{13}C_{carbonate}$ values. This simple approach gives the trend for $\varepsilon_{photosynthesis}$ shown in Fig. 6.20, although it does

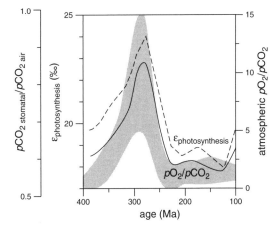

Fig. 6.20 Estimated carbon isotopic fractionation during photosynthesis (broken line) and derived atmospheric O_2/CO_2 ratio (solid line) in comparison with mass-balance model of atmospheric O_2/CO_2 ratio (half-tone band) spanning the Carboniferous–Permian glaciation (after Beerling et al. 2002).

ignore other factors that affect $\varepsilon_{photosynthesis}$ (e.g. humidity and temperature; Farquhar et al. 1989). The $\varepsilon_{photosynthesis}$ values can then be used to predict pO_2/pCO_2 variations, which are seen in Fig. 6.20 to be in reasonable agreement with the trend obtained from the independent models described in Section 6.2.2 (and shown in Fig. 6.14c,d; Berner et al. 2000; Berner & Kothavala 2001).

6.3.4 Cretaceous hothouse

During the Cretaceous the arrangement of the continents was approaching that of the present day, although an equatorial seaway, the remnant of the Tethys Ocean, was closing and the Atlantic was in the early stages of opening (Fig. 6.21). The climate warmed through the Early Cretaceous, and from c.120 to 80 Ma (Aptian to mid-Campanian) greenhouse conditions existed, with arid belts in mid-latitudes (Chumakov et al. 1995). By the Late Cretaceous there was probably no polar ice at sea level (Spicer & Corfield 1992). It had been thought that the meridional temperature gradient was low, but diagenetic alteration of carbonates may have led to misleading indications of tropical temperatures from $\delta^{18}O$ data, and more recent data from well preserved samples suggest that tropical sea-surface temperatures reached 28–32 °C (rather than 15–23 °C; Pearson et al. 2001; Norris et al. 2002). Towards the end of the Cretaceous (Maastrichtian) the climate cooled (Barrera 1994). Maximal temperature and sea level were reached near the Cenomanian–Turonian boundary (c.93.5 Ma; Jenkyns et al. 1994; Gale 2000). The sea level was c.200 m higher than at present, only c.60 m of which can be accounted for by complete melting of ice, so the other 140 m is attributable to other causes, primarily the volume of ocean basins. Spreading ridges were particularly active 125–80 Myr ago, during the break-up of Pangaea, and occupied a greater proportion of the ocean basins than presently (Gurnis 1990). New oceanic crust was produced 50–75% faster than at present (Larson 1991a).

Because of low-latitude aridity, there appears to have been no recognizable tropical rain-forests. The moisture from high evaporation rates at low latitude was transported to high latitude, where the moist and temperate conditions, with few frosts, provided favourable conditions for the growth of forests near both poles, and significant deposits of coal were formed (see Fig. 3.14; Saward 1992; Parrish et al. 1982). The relatively warm

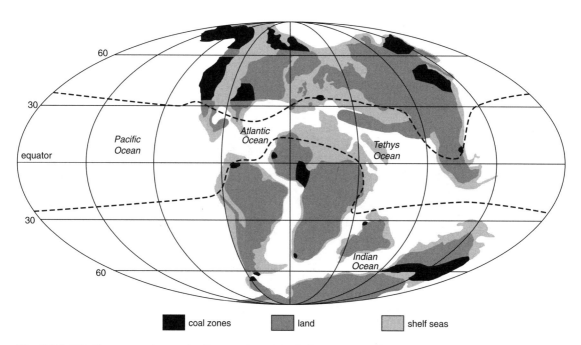

Fig. 6.21 Mid-Cretaceous palaeography (Cenomanian, c.95 Ma), showing zones of coal deposition and shallow seas on flooded continental shelves (after Parrish et al. 1982; Scotese 2003). Carbonate platforms developed between the broken lines in warm shallow waters (after Sohl 1987).

climate but shortage of light in the winter months at high latitudes would have presented the potential problem of excessive photorespiration, but this was solved by the dominant gymnosperms being deciduous (Spicer & Corfield 1992). Within the tropics there was a large area of shallow water where carbonate deposition occurred (via calcareous algae, benthonic foraminiferans, bivalves and gastropods). The geographical limits of this warm-water, platform carbonate deposition during the Cretaceous are shown in Fig. 6.21. Major carbonate deposition also occurred at mid-latitudes, in the form of extensive chalk deposits. They were produced from the remains of coccolithophores and other calcareous plankton, deposited in waters up to c.200 m deep on the flooded continental margins (Skelton et al. 2003; Fig. 6.21).

It might be supposed that the pronounced biological activity and deposition of large amounts of carbonate and coal would have caused atmospheric CO_2 levels to decline and result in an icehouse like that of the Carboniferous–Permian. However, the high levels of tectonic activity appear to have more than compensated for the drawdown of CO_2, and it is estimated that the atmosphere contained up to ten times as much of the gas as at present (Berner 1990), which is reflected in the low $\delta^{13}C$ values of marine organic matter (Dean et al. 1986). The dominant contributor may have been a mantle superplume under the Pacific (Larson 1991a, b, 1995). It appears to have fed dramatic eruptions of basalt that formed large submarine plateaux (e.g. Ontong-Java Plateau, c.122 Ma) as well as continental flood basalts (e.g. Deccan Traps, India, c.65 Ma). The continental flood-basalt eruptions are likely to have had a more direct effect on climate because they evolved CO_2 directly into the atmosphere. Both types of magmatism vented large amounts of CO_2 and basalt over relatively short periods: for example, c.1 Myr for the Ontong-Java province and periodic episodes over c.5 Myr for the Deccan Traps. The Ontong-Java and Deccan eruptions produced similar amounts of basalt (c.10^6 km^3), equating to c.1–10 Tt CO_2. The estimated rate for the Deccan Traps during each of the sporadic eruptions is 0.3 Gt CO_2 yr^{-1} which, although a significant amount in terms of natural fluxes, is only c.1.5% of present-day industrial emissions (Skelton et al. 2003). It may have caused atmospheric levels to rise by 75 ppm, resulting in a temperature increase of 1–2 °C, but this rate of CO_2 supply is insufficient to account for the scale of global warming during the Cretaceous. However, the CO_2 emissions associated with the elevated tectonic plate spreading rates, which lasted for c.40 Myr, may have produced the observed greenhouse warming (Skelton et al. 2003).

The effects of increased and reduced CO_2 emissions are modelled in Fig. 6.22, at a constant rate of silicate weathering but in the absence of feedback mechanisms, for simplicity (Berner & Caldeira 1997). Two initial atmospheric CO_2 levels are shown, corresponding to preindustrial (280 ppm) and Cretaceous (2800 ppm) concentrations, and the rate of CO_2 input is allowed to increase or decrease by 25%. Starting at the Cretaceous high atmospheric concentration but with the reduced rate of CO_2 supply, the Earth is plunged into an icehouse after <2 Myr as the gas is drawn down by silicate weathering. In contrast, at the increased rate of CO_2 supply for preindustrial initial conditions, atmospheric levels of the gas reach those of the Cretaceous within c.1 Myr and after c.40 Myr would result in temperatures too great to sustain life. That such runaway greenhouse warming did not happen during the Cretaceous indicates the importance of negative feedback controls on atmospheric CO_2 levels.

(a) *Cretaceous anoxic events*

During the Cretaceous there were episodes (generally ≤1 Myr long) of widespread deposition of organic-rich black shales on outer shelves and within basins throughout the Tethyan Ocean (which was closing) and the young Atlantic (in the early stages of opening), which

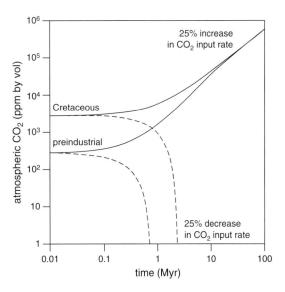

Fig. 6.22 Modelled effects of varying rate of volcanic CO_2 input to atmosphere in absence of feedbacks (after Berner & Caldeira 1997).

have been termed **oceanic anoxic events** (OAEs; Fig. 6.23a; Jenkyns 1980). OAEs are not confined to the Cretaceous; for example, there was one in the Early Jurassic. All the OAEs have cyclic bedding that could reflect Milankovich cyclicity (Kuhnt et al. 1997; Galeotti et al. 2003). The black shales and contemporaneous carbonates are characterized by positive $\delta^{13}C$ excursions, which commonly persist after the end of the OAEs (Jenkyns 1999). Although the isotopic excursions are not observed in the generalized global $\delta^{13}C_{carbonate}$ record in Fig. 6.23, they are readily recognized in Tethyan sections (Fig. 6.23b). Among the OAEs, the most widely distributed are OAE1a (the C isotopic excursion is actually a doublet; Weissert et al. 1998; Jenkyns & Wilson 1999) in the early Aptian (at $c.120.5$ Ma, called the Selli Event) and OAE2 at the Cenomanian–Turonian boundary (at $c.93.5$ Ma, called the Bonarelli Event). The other OAEs are more regional in geographical extent (Leckie et al. 2002). The organic matter in OAEs 1a, 2 and 3 is mostly of marine origin, but that in OAEs 1c and 1d is predominantly terrestrial (Erbacher et al. 1996). Each C isotopic excursion implies an increase in the burial rate of organic matter (Jenkyns & Clayton 1997). The associated black shales have exceptionally high organic C contents, up to 30%, which are considerably greater than the highest recorded in modern sediments ($c.15\%$) and of much larger geographical extent (Robinson et al. 2002), so their deposition, over periods spanning hundreds of thousands of years, appears to have involved unusual conditions.

The precise causes of the OAEs have yet to be established, but a number of factors probably contributed. Under the warm conditions, the volume of cold oxygen-rich waters sinking at the poles would have been limited. A more likely source of dense deep water would have been high-salinity brines produced by evaporation at low latitudes (see Fig. 3.12) in basins with restricted deep-water connections to the main oceans (Barron & Peterson 1989). Such locations were present in the young Atlantic Ocean and various parts of the Tethys Ocean. Some cooling of the water by migration to higher latitudes during winter may have been needed to achieve the necessary density increase for bottom-water

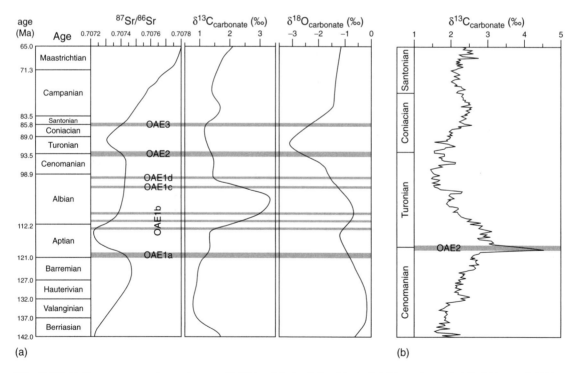

Fig. 6.23 (a) Cretaceous oceanic anoxic events (OAEs) and their relationship to global isotopic trends (after Veizer et al. 1999; Jones & Jenkyns 2001; Leckie et al. 2002); (b) $\delta^{13}C_{carbonate}$ excursion associated with OAE2 in the Chalk of East Kent, UK (after Jenkyns et al. 1994).

formation (Brady et al. 1998). The deep-water circulation would then have been driven primarily by contrasts in salinity (halothermal) rather than temperature (thermohaline). Because of the higher temperatures in their source regions, the Cretaceous warm-saline deep waters (WSDWs) would have had lower oxygen concentrations than modern deep waters, although the oceanic circulation was probably not sluggish and enough oxygen would have been present to maintain oxidizing conditions at the sea floor, especially in the large Pacific Ocean, most of the time. However, conditions were such that oxygen starvation could be triggered in some areas, resulting in the OAEs.

It is generally assumed that increased marine primary production was a major factor in triggering the OAEs (Leckie et al. 2002). From $\delta^{18}O$ data, the OAEs appear to be associated with warmer (and wetter) episodes (Fig. 6.23), so a climatic influence on oceanic nutrient availability may have been the trigger for sea-floor anoxia (Pedersen & Calvert 1990). Under conditions of relatively low oxygen availability, oxygen demand during degradation of sinking organic detritus could have exceeded supply when productivity was high. An intensified and expanded oxygen-minimum layer (over a depth range of up to 3000 m; see Section 3.3.4a) in areas of high productivity would explain the deposition of organic-rich shales where the zone intercepted the sea floor: on large areas of continental shelf and slope (and particularly in restricted basins), and also on seamounts and the flanks of spreading ridges (Jenkyns 1980). Biomarker evidence for anoxicity is present in many of the black shales (e.g. 28,30-dinorhopane; Farrimond et al. 1990; see Section 5.4.3d). The elevated rates of organic C burial seem to have drawn down a significant proportion of atmospheric CO_2, because the OAEs are succeeded by rapid climatic cooling, suggesting lower atmospheric CO_2 levels.

The OAEs coincided with marine transgressions, and the associated release of nutrients from the flooded continental margins is likely to have helped to fuel primary production (Jenkyns 1999). Increased tectonic activity would have raised volcanic emissions of CO_2, which could also have contributed to higher productivity in a number of ways associated with the resulting climatic warming:
• acceleration of the hydrological cycle, increasing the rate of continental weathering and so increasing the nutrient flux to the oceans, fuelling higher primary production (Weissert 1989);
• enhanced atmospheric circulation, increasing wind strength, which intensified oceanic current divergence and upwelling (Berner et al. 1983; Pedersen & Calvert 1990; see Boxes 3.2 and 3.5), and more efficient recycling of nutrients into the euphotic zone;
• higher evaporation rates at low latitudes increased production of WSDW, which would have to be counter-balanced by increased upwelling, again increasing the supply of nutrients to surface waters (Arthur et al. 1987).

Evidence for increased hydrothermal circulation associated with elevated rates of oceanic crust production appears to be provided by short negative excursions in the $^{87}Sr/^{86}Sr$ record coincident with the Cretaceous (and Early Jurassic) OAEs (Jones & Jenkyns 2001). Greater hydrothermal activity could have increased the iron supply to the oceans, increasing primary production in areas where the nutrient had previously been biolimiting (Sinton & Duncan 1997; see Box 3.7). However, the OAEs do not persist throughout the periods spanned by the Sr isotopic excursions (which last 5–13 Myr), so other factors must have been important.

High productivity *per se* is not the entire story. Although primary production among the usual algal phytoplankton may have been stimulated initially, intensifying the mid-water oxygen-minimum layer (OML), it appears there was a shift in the dominant members of the phytoplanktonic community from algae to bacteria (Kuypers et al. 2001). Low $\delta^{15}N$ values for marine organic matter in black shales from the eastern North Atlantic suggest that nitrogen-fixing cyanobacteria were dominant (Rau et al. 1987). In addition, isorenieratane and other biomarkers derived from isorenieratene are generally abundant, and are unambiguously derived from the green sulphur bacteria (see Section 5.3.3c; Kuypers et al. 2002). These phototrophic bacteria are obligate anaerobes and require free sulphide (i.e. euxinic conditions) in order to fix CO_2 in the euphotic zone. In addition, nitrogen-fixing cyanobacteria compete best under conditions of low dissolved oxygen content in the euphotic zone (see Box 3.9). So it seems that an intensified OML may have extended upwards into the euphotic zone and aided the establishment there of euxinic conditions, leading to bacterial domination of primary production, which was sustained over wide areas for prolonged periods. Clearly, nitrogen limitation would have been irrelevant to nitrogen-fixing cyanobacteria, and phosphate recycling from organic matter is accelerated under dysaerobic/anaerobic conditions (Filippelli et al. 2003). The increased freshwater run-off from land, which appears to correlate with black shale deposition (Erbacher et al. 2001; Galeotti et al. 2003), is likely to have aided stratification in the upper water column and development of conditions favouring high bacterial productivity. Mediterranean sapropels from the Neogene–Quaternary may represent more modern analogues of the Cretaceous black shales, although favourable conditions for their deposition were of shorter duration (Rohling 1994; Menzel et al. 2002; Meyers & Negri 2003).

6.3.5 Quaternary ice ages

The climate has cooled considerably since the start of the Neogene. Changes in oceanic currents related to movement of continental plates appear to have been one of several important factors. Antarctica split away from South America at $c.35$ Ma, opening the Drake Passage and allowing the circumpolar current to isolate the southern continent, leading to increased cooling and development of a major ice cap. Another major factor is the enhanced weathering resulting from the onset of major orogenies in the Tertiary, such as the uplift of the Tibetan Plateau following India's collision with Asia and continued northward motion (Raymo 1991). The increased rate of chemical weathering of silicates, reflected in the rising $^{87}Sr/^{86}Sr$ ratio in Fig. 6.14a, together with the associated deposition of marine carbonates, would have drawn down considerable amounts of CO_2 from the atmosphere, causing cooling.

The 'modern' ice ages began 2.6 Myr ago, at the start of the Pliocene, when northern polar glaciation intensified. This approximately coincided with the final closure of the seaway between North and South America, which promoted the formation of North Atlantic Deep Water (NADW; see Box 3.2, Fig. 3.2). Because this deep-water formation process draws warm, saline, surface water up from low latitudes, a change in the rate of NADW formation alters heat transfer to high northern latitudes and influences the extent of the northern glaciers. We are presently in an interglacial period (the Holocene) that has lasted for about half a Milankovich precessional cycle ($c.20$ kyr). The thermal maximum of this interglacial episode was 4–8 kyr ago, when the mean surface temperature was $c.1$ °C higher and the climate was wetter than today, and lakes and vegetation existed in the Sahara (Wilson et al. 2000). Superimposed on the long-term glacial–interglacial climatic changes are smaller-scale periods of relative warming during glacials (**interstadials**) and cooling during interglacials (**stadials**). Opinion is divided over whether reorganization of thermohaline circulation or the equatorial atmosphere–ocean system is the cause of the major climatic swings during the last glacial (Broecker 2003).

A major source of information on temperature changes during the Quaternary ice ages is the $\delta^{18}O$ record from pelagic marine carbonates, but δD and $\delta^{18}O$ data from ice cores can also be used in the same way (Jouzel et al. 1994; see Box 5.6 and Fig. 6.24). Data have been obtained from various ice cores from Greenland (dating back to $c.100$ ka) and Antarctica ($c.400$ ka): the Greenland cores are thicker and more obviously layered, making dating generally easier than for the Antarctic cores. Ice cores can provide additional information: they contain small air bubbles that provide a record of atmospheric composition dating from the complete isolation of the bubbles from atmospheric exchange (Blunier & Brook 2001). Concentrations of the greenhouse gases CO_2 and CH_4 can be assessed directly from ice cores (Fig. 6.25), although it is thought that some CO_2 can be produced by reactions involving mineral impurities in the ice, particularly in Greenland cores (Monnin et al. 2001). Methane concentrations can be used to correlate ages between ice cores, because it can be assumed that atmospheric exchange occurs rapidly between the hemispheres (within $c.1$ yr; Blunier & Brook 2001). Based on the adjusted age estimations, it appears that warming in Antarctica slightly preceded that in the northern hemisphere during the last glacial–interglacial transition, and appears to be related to changes in the thermohaline circulation (Broecker 1998; Peterson et al. 2000; Shackleton 2001).

During the last glacial maximum ($c.18$ ka) the mean surface temperature was some 5 °C lower than at present, but cooling of this extent cannot have been the result of CO_2 drawdown alone, because all the CO_2 would have had to be removed from the atmosphere. The $c.30\%$ reduction in atmospheric CO_2 concentration during the last glacial maximum compared to the Holocene preindustrial level (Archer & Maier-Reimer

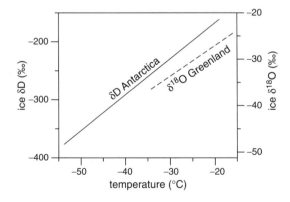

Fig. 6.24 Relationship between isotopic composition of polar ice and annual average surface temperature (after Jouzel et al. 1994).

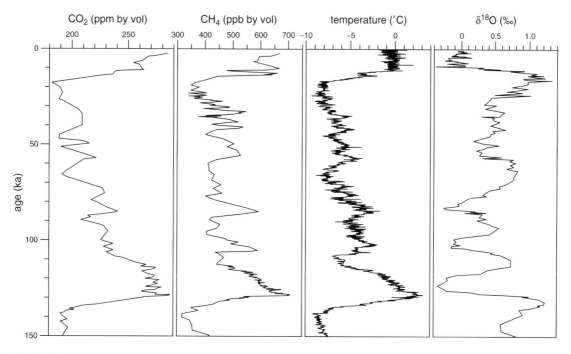

Fig. 6.25 Records of atmospheric CO_2 and CH_4 concentrations, air temperature just above the Antarctic inversion layer (derived from water δD values) and water $\delta^{18}O$ values from the Vostok ice core (after Petit et al. 1999). (An atmospheric inversion layer is one in which temperature increases with altitude, rather than the usual decrease (see Fig. 7.1).)

1994) is typical of the general c.100 ppm (by vol.) difference between interglacials (260–300 ppm) and glacials (160–200 ppm; Sigman & Boyle 2000). Such a change in CO_2 concentration is estimated to cause a temperature change of 1.4°C (Kuo et al. 1990; Wilson et al. 2000; see Box 6.1), so other factors must have contributed to glacial cooling. For example, during glacials the cooler atmosphere holds less water vapour, causing a decrease in greenhouse warming. In addition, the increasing size of polar ice sheets results in cooling because of the albedo effect.

Whether ice sheets grow or contract seems to depend upon the efficiency of melting during the summer (Wilson et al. 2000), so seasonal changes in insolation are likely to be an important trigger of the ice-albedo feedback, and the 60–80° latitude belt is most sensitive to changes in summer insolation. Despite the fact that Milankovich orbital forcing (see Box 6.3) is too small on its own to cause the observed climatic fluctuations (<0.7 W m^{-2} over the past 160 kyr), it can stimulate various feedbacks. This is seemingly borne out by the observation that the periodicity of glaciations from c.2 to 1 Ma was c.41 kyr (with approximately equal periods of glacial and interglacial conditions), consistent with obliquity-cycle forcing. Over the past 800 kyr the precessional and eccentricity cycles have become more important, the reasons for which are a matter for debate (Wilson et al. 2000; Zachos et al. 2001). The dominant periodicity has been c.100 kyr, with rapid warming out of glacials but slow cooling into them, such that the interglacials represent only c.10% of the cycle. However, Milankovich cycles cannot explain the timing of all glacial–interglacial transitions (Lorius et al. 1990; Karner & Muller 2000).

Ice-sheet growth does not appear to be the primary trigger for the onset of glacial episodes, because increases in ice volume lag behind the changes in Antarctic air temperature, deep-water temperature and atmospheric CO_2 levels, all three of which are in phase with orbital eccentricity (Sowers & Bender 1995; Shackleton 2000, 2001). It may well be that the response of the carbon cycle to the orbital-eccentricity cycle is the trigger for glaciations (Raynaud et al. 1993). During glacial periods the climate is generally more arid and winds are relatively strong (due to enhanced atmospheric pressure gradients), increasing the supply of dust from the continents to the oceans where there are favourable prevailing winds (Tiedemann et al. 1994). The dust is iron-rich, and it is likely that the increased supply of the micronutrient iron resulted in higher phytoplanktonic productivity, drawing down atmospheric CO_2 (Falkowski et al. 1998), and increasing the rate of dimethyl sulphide (DMS) production (Turner et al. 1996; see Box 3.7). The increased production of DMS would have reinforced the positive cooling feedback of decreasing atmospheric CO_2 concentrations (see Section 6.3.1). The extent of cooling stimulated by iron enrichment is likely to have been limited by the strength of the Trade Winds and their capacity to deliver nutrients to the remote oceans (Henriksson et al. 2000).

Evidence for increased DMS production during glacials has been found in Antarctic (but not Arctic) ice cores (Legrand 1997). Figure 6.26 suggests that productivity in the Southern Ocean increased during glacial periods, possibly fuelled by iron-containing dust blown from Patagonian deserts (Kumar et al. 1995). In Fig. 6.26 the $\delta^{18}O$ plot from an Antarctic ice core is a proxy for temperature, and shows the contrast between the last glacial and the present interglacial stage. The iron content estimated in another ice core was greater during the glacial episode and corresponded to lower CO_2 levels in air trapped in the ice, as would be expected if increased CO_2 drawdown had occurred. This is consistent with an increase in oceanic primary productivity, as suggested by elevated levels of organic C in a marine sediment core and high concentrations of MSA (methane sulphonic acid) in an Antarctic ice core (Fig. 6.26).

The amount of CO_2 trapped in ice cores correlates with the degree of glaciation, and as we have just seen, changes in marine primary production seem to be a major cause of the variation in atmospheric CO_2 concentration. However, it is important to examine other potential causes. The terrestrial C reservoir is likely to decrease rather than increase during glacial advance over the continents, if only because of the loss of plant habitat, probably resulting in a slight increase in atmospheric CO_2 levels. The solubility of CO_2 increases as oceanic surface water cools, but this is partially offset by the reduction in solubility caused by an increase in salinity resulting from intensified sea-ice formation (via brine rejection; see Box 3.2), so there would probably have been only a small (<10 ppm) net decrease in atmospheric CO_2 during glaciations. As shown by Fig. 6.3b, a decrease in alkalinity, from export of $CaCO_3$ from surface to deep waters/sediments, causes a decrease in dissolved C, permitting more CO_2 to enter solution. This process causes a deepening of the lysocline

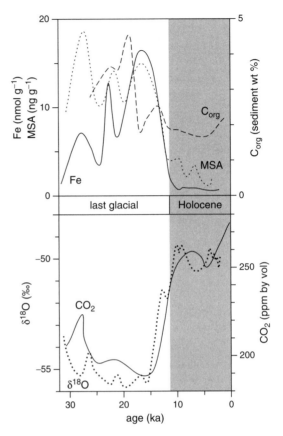

Fig. 6.26 Variations in proxy measurements for various climatic parameters in the Southern Ocean during the last glacial–interglacial cycle. MSA (methane sulphonic acid, a proxy for DMS) and $\delta^{18}O$ (proxy for temperature) in an eastern Antarctic ice core; estimated iron concentration (Fe) and CO_2 trapped in air in another Antarctic ice core (Vostok); C_{org} = organic carbon (proxy for primary productivity) in an eastern tropical Pacific sediment core (after Turner et al. 1996).

(because there is more carbonate at depth; see Box 5.7), the extent of which is directly related to the amount of CO_2 removed from the atmosphere. The depth range over which carbonates were deposited during the last glacial maximum suggests that the lysocline deepening was <1 km (Catubig et al. 1998), which limits the drawdown of CO_2 to <55 ppm (Sigman & Boyle 2000). In low-latitude oceans, nutrients are used efficiently by phytoplankton, and there seems little scope for increasing productivity significantly in order to lower CO_2 levels by an appreciable amount. It is possible that nitrogen fixation rates increased with the elevated influx of iron during glacials, but if phosphate is the limiting nutrient it would be necessary for the Redfield ratio to change (see Section 3.2.7), allowing N and C to increase relative to P, for enhanced photosynthetic drawdown of CO_2 to occur (Falkowski 1997).

The greatest potential for reduction in atmospheric CO_2 during glacials probably lies in changes in primary productivity and deep-water circulation in the Southern Ocean (Sigman & Boyle 2000). At present in this area, major nutrient utilization is low because iron is biolimiting, primary productivity is dominated by diatoms and there is strong upwelling (from mixed NADW and AAIW in Fig. 3.2b; see also Box 3.5), which vents CO_2 back into the atmosphere. However, in a cooler glacial climate, the westerlies would shift towards the equator, suppressing deep-water ventilation around Antarctica (Toggweiler et al. 1999). Despite the decrease in overall upwelling, the efficiency of nutrient utilization appears to have increased (Sigman et al. 1999)—possibly associated with dominance of non-siliceous phytoplankton—effectively increasing the drawdown of CO_2 (Sigman & Boyle 2000).

If anything, the atmospheric concentration of methane follows the air-temperature trend in Fig. 6.25 more closely than does the CO_2 record. The more direct coupling of methane to air temperature suggests that terrestrial sources of the gas were probably responsible for the fluctuations in concentration. Increased humidity in the tropics and active plant growth in high-latitude wetlands can be expected to increase methane production during interstadials (Brook et al. 1996).

6.3.6 The ultimate hothouse

The Sun is almost halfway through its $c.10$ Gyr life as a main-sequence star, converting hydrogen into helium in its core. Subsequently, it will enter a red-giant phase and expand towards the Earth over another $c.2$ Gyr (Hufnagel 1997). However, conditions for life on Earth will become increasingly hostile long before that catastrophe. As the Sun converts more hydrogen into helium, its core will get denser and hotter, the rate of thermonuclear fusion will increase and so the Sun's luminosity will continue to increase (Newman & Rood 1977). The resulting warming of the Earth is likely to increase the rate of chemical weathering of silicates, which will draw down CO_2 from the atmosphere (assuming subsequent deposition of carbonate occurs in the oceans), minimizing the temperature increase. However, there is a limit to the amount of CO_2 that can be removed from the atmosphere before another factor comes into play. Below $c.150$ ppm CO_2, C_3 photosynthesis becomes unviable, assuming that plants have a limited ability to adapt to lower CO_2 levels (Lovelock & Whitfield 1982). At this point, the rate of removal of CO_2 from the atmosphere would slow, the Earth would warm rapidly and life would eventually cease (a critical time would be when liquid water could no longer exist).

Assuming fairly conservative limits for the temperature and C requirements of life, and that volcanic CO_2 emissions and silicate weathering rates remain similar to today, it is possible to estimate the minimum amount of time left for life on Earth from the simple model represented in Fig. 6.27 (Caldeira & Kasting 1992b). Insolation will increase at $c.0.12$ W m^{-2} Myr^{-1}, so a reduction in CO_2 could maintain a constant mean surface temperature for another 0.5 Gyr, at which time the critical concentration of CO_2 of $c.150$ ppm for C_3 plants would be reached (Caldeira & Kasting 1992b). However, some C_4 plants (see Box 1.10) can survive at <10 ppm CO_2, a

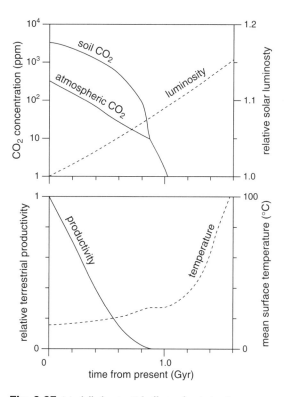

Fig. 6.27 Modelled potential effects of variation in atmospheric CO_2 concentration (as a function of weathering and primary production) and solar luminosity upon relative terrestrial primary productivity and mean surface temperature (after Caldeira & Kasting 1992b).

limit that would not be reached for 0.9 Gyr, which may extend the period of temperature buffering if these plants became dominant (Cerling et al. 1993).

When the global mean temperature exceeds 300 K (27 °C), silicate dissolution is likely to occur at a faster rate than volcanism can supply CO_2, so not all the Ca^{2+} and Mg^{2+} ions supplied to the oceans would be able to precipitate as carbonates. The global mean surface temperature would exceed 50 °C after $c.1.5$ Gyr, ignoring the possible negative feedback of increased cloud cover as more water vapour enters the atmosphere. Only prokaryotes and some protozoa can live much above this temperature. Warming is then likely to accelerate, with atmospheric water vapour levels increasing exponentially with temperature (at constant relative humidity), adding to the warming. A rise in temperature from 50 to 100 °C is likely to occur in <200 Myr, leaving hyperthermophilic archaebacteria as the only survivors. Photodissociation of water and escape of the liberated hydrogen into space will accelerate, leading to complete loss of water after $c.2.5$ Gyr. At this stage, life will certainly be eliminated. Silicate weathering will stop, so volcanic CO_2 will accumulate in the atmosphere until the Earth resembles Venus.

6.4 Isotopic excursions at period boundaries

Biotic extinctions of varying magnitudes are associated with many era boundaries (see Section 1.5.5), and there are often accompanying excursions in the carbon isotopic record. Such isotopic excursions reflect environmental events of sufficient size to have an impact on the global carbon cycle. However, it is not always easy to determine whether the excursion was the result of the extinction event (due to biotic collapse), is attributable to some other consequence of the event leading to the extinction or may even be largely unrelated to the extinction event. The problems associated with interpretation of isotopic excursions are demonstrated by three important boundary events in the following subsections.

6.4.1 Cretaceous–Tertiary boundary event

Perhaps the best known of these extinction events occurred at the Cretaceous–Tertiary boundary (KTB), which left a clear signature in the C isotopic record. As we have seen in Section 5.8, the preferential assimilation of $^{12}CO_2$ during photosynthesis leaves surface waters depleted in the lighter C isotope, but decomposition of sinking organic detritus in the underlying water column releases this isotopically light C, resulting in a gradient of decreasing $\delta^{13}C$ value with increasing depth in the main form of dissolved C, bicarbonate. The signature of this gradient is preserved in carbonate tests of planktonic and benthonic organisms such as foraminiferans (Fig. 6.28a). At the KTB the gradient rapidly fell to zero or even reversed (Fig. 6.28a), with planktonic $\delta^{13}C$ values typically decreasing by 2–3‰. This is believed to reflect a catastrophic collapse in marine primary productivity after the boundary impact event, lasting $c.1$ kyr (Zachos et al. 1989; Kump 1991; Hollander et al. 1993).

An ocean in which the export of organic matter from the surface to deep water has ceased and there is an associated zero $\delta^{13}C_{carbonate}$ gradient has been termed a 'Strangelove ocean' (Fig. 6.29; Hsü & McKenzie 1990). In such an ocean, isotopic homogeneity is likely to be reached in no more than a few hundred years, and in the longer term ($c.100$ kyr) the $\delta^{13}C$ value would be expected to approach the mean crustal value ($c.-5$‰) because of the fluvial input of bicarbonate from weathered rock (Kump 1991). The fact that $\delta^{13}C_{carbonate}$ values did not approach the crustal mean indicates that global primary production (and subsequent burial of organic C) did not shut down completely. The trend for $\delta^{13}C_{kerogen}$ derived from plankton shows a decrease of 3‰ during the Strangelove ocean period, as expected for a collapse in primary production, but then increases by 7‰, as shown in Fig. 6.28b and observed at several KTB sites (Meyers & Simoneit 1990; Hollander et al. 1993). The $\Delta\delta^{13}C_{carbonate-kerogen}$ trend gives an idea of the degree of isotopic fractionation during photosynthesis (see Section 5.8.2), and it can be seen to increase by 1.2‰ in the boundary clay layer, but decreases by 5.4‰ immediately above, before increasing to more normal values ($c.28$‰; Fig. 6.28b). The decrease has been attributed to phytoplanktonic blooms, which are known to exhibit such decreases in photosynthetic isotopic fractionation under conditions of low concentrations of dissolved CO_2, which result from demand exceeding the rate of supply from air–sea transfer, organic decomposition and bicarbonate equilibration (see Section 5.8.2). As the phytoplankton recovered after the boundary catastrophe, blooms are likely to have been stimulated by the nutrients that had accumulated in surface waters in the intervening period. Supporting evidence for such blooms is provided by monospecific calcareous nanoplanktonic assemblages commonly found in the earliest Tertiary of KTB sections (Hollander et al. 1993).

Although primary production recovered, few zooplankton appear to have been present, so bacterial

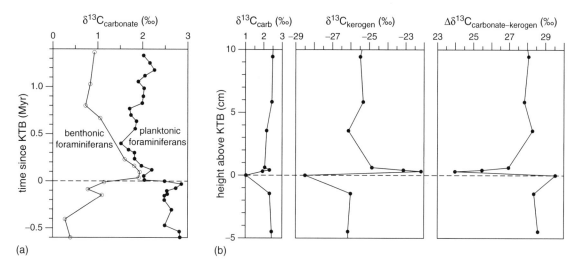

Fig. 6.28 Carbon isotopic records across the Cretaceous–Tertiary boundary at (a) Deep-Sea Drilling Project site 577 (Shatsky Rise, Pacific Ocean; after Zachos et al. 1989; Hsü & McKenzie 1990) and (b) Woodside Creek, New Zealand, where the carbonate and kerogen are dominantly of planktonic origin (after Hollander et al. 1993).

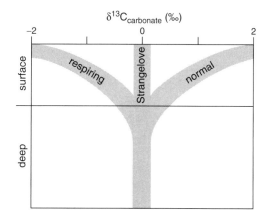

Fig. 6.29 Idealized representation of C isotopic gradients for marine carbonate under normal photosynthesizing conditions (e.g. modern ocean), after biotic collapse (Strangelove ocean) and for dominance of microbial respiration in surface waters (after Hsü & McKenzie 1990).

decomposition of organic detritus would have dominated in surface waters. Because phytoplanktonic (and particularly picoplanktonic) remains sink much more slowly down through the thermocline than zooplanktonic faecal pellets (see Section 3.3.1), it is likely that organic detritus remained in the surface waters for a considerable time and that microbial respiration there became an important control on the C isotopic gradient. The distinctive inverted (negative) gradients for $\delta^{13}C_{carbonate}$ that persisted for up to 1 Myr after the KTB in some regions is believed to reflect the influence on the overall isotopic fractionation of the dominance of respiration over photosynthesis. Not surprisingly, the term 'respiring ocean' has been applied to such conditions (Hsü & McKenzie 1990; Fig. 6.29). The fact that the C isotopic record of carbonates did not recover to pre-extinction levels for >3 Myr probably reflects the relative rate of exportation of detritus to deep water and hence the recovery period of the oceanic grazing chain (D'Hondt et al. 1998).

Atmospheric pCO_2 is likely to have increased as a result of equilibration between the air and Strangelove ocean, and the terrestrial wild fires triggered by the bolide impact (Wolbach et al. 1988). However, the periodic phytoplanktonic blooms would have drawn down significant amounts of CO_2. Consequently, it is likely that there were dramatic climatic swings during the recovery phase, until atmospheric CO_2 levels stabilized (Wolfe & Upchurch 1986).

6.4.2 Palaeocene–Eocene thermal maximum

The climate spanning the end of the Palaeocene and the start of the Eocene was warmer than at any other time during the Cainozoic (Bains et al. 1999). Within this interval there was a dramatically rapid warming—the

Palaeocene–Eocene Thermal Maximum (PETM) — at 55.5 Ma. The whole of the water column warmed by several degrees (Zachos et al. 2001). At the onset of the PETM the $\delta^{13}C$ of bulk carbonate decreased rapidly by 2–3‰ (Fig. 6.30; Kennett & Stott 1991; Bralower et al. 1997), with about two-thirds of the excursion occurring over just a few thousand years (Norris & Röhl 1999). Prior to the excursion, tropical surface waters seem to have warmed slightly, which may have been caused by the CO_2 released during a phase of circum-Caribbean volcanism, the scale of which matched the Toba eruption (Bralower et al. 1997). Volcanism in the North Atlantic igneous province increased from 56.0 to 55.55 Ma, and intensified considerably $c.50$ kyr before the PETM. The latitudinal variation in temperature trends of surface waters during this time probably reflects a change in oceanic heat transport, with the formation of warm, saline deep water becoming more important (Kennett & Stott 1991; O'Connell et al. 1996). It is possible that there were rapid changes in the relative importance of high latitude versus subtropical sources of deep water as a result of variations in continental run-off (Bice et al. 1997). A major benthonic extinction event occurred during the C isotopic excursion, with $c.35\%$ of benthonic foraminiferal species disappearing at the onset of the excursion (Kennett & Stott 1991; Kaiho et al. 1996), possibly as a result of low dissolved oxygen levels and elevated salinity (Bains et al. 1999). There were also changes among the phytoplankton during the warming leading up to and throughout the PETM (Kelly et al. 1996; Killops et al. 2000b; Zachos et al. 2001), and also in the terrestrial biota (Koch et al. 1992).

The rapidity (<10 kyr) and amplitude of the C isotopic excursion suggest methane hydrate collapse as the most likely cause (Dickens et al. 1995, 1997;

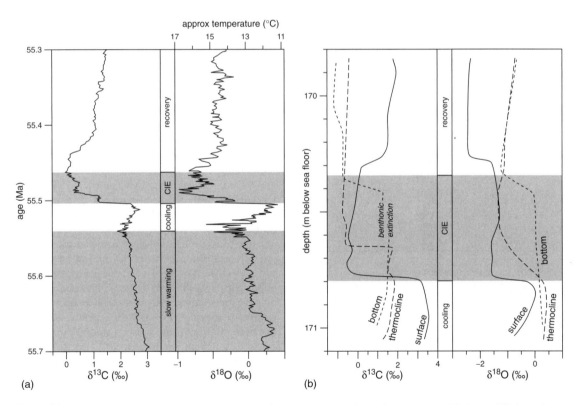

Fig. 6.30 (a) Bulk carbonate isotopic records across the Palaeocene–Eocene Thermal Maximum at ODP site 690B (Maud Rise, Weddell Sea, Antarctica; 65° 09′ S, 01° 12′ E; after Bains et al. 1999). Temperature scale after Shackleton (1984). CIE = carbon isotopic excursion. (b) Generalized trends in carbonate isotopic record for individual genera of surface (*Acarinina*) and thermocline (*Subbotina*) dwelling foraminiferans, together with single (*Nuttallides truempyi*) and multiple benthonic species from ODP site 690 (after Thomas et al. 2002).

Thomas et al. 2002). The hydrate destabilization could have occurred as a result of warming, which was initially gradual but finally became more rapid owing to the sinking of warm saline waters (bottom-water temperature increased from 11 to 15 °C; Dickens et al. 1995). Alternatively, it may have been triggered by seismic activity causing slope failure along continental margins, which removed overburden and hence lowered the pressure confining the hydrates (Bains et al. 1999). Apparent evidence for the latter has been found off the coast of Florida (Katz et al. 1999).

A biotic collapse can be ruled out as a cause of the isotopic excursion because of the proportion of the total marine and terrestrial organic C surface reservoirs (i.e. biota, soil + peat and dissolved components in Fig. 6.1) that would need to be transferred into the combined ocean/atmosphere inorganic reservoirs, which would be the only mechanism to account for an excursion over a time interval less than the oceanic C residence time (39 000/0.3 yr = 130 kyr from Fig. 6.1) but greater than the ocean mixing time (c.1 kyr). A −2‰ shift in $\delta^{13}C_{carbonate}$ would require 75–90% of the total marine and terrestrial organic C surface reservoirs to be transferred, and there would not be sufficient organic matter for a −3‰ excursion.

Similarly, the amount of volcanically emitted CO_2 needed to account for the isotopic excursion, which can be estimated by mass balance, is without precedence. For a transient input of carbon (x) we can write (Dickens et al. 1995):

$$(M_{surface} + M_x)(\delta^{13}C_{final\ surface}) = (M_x)(\delta^{13}C_x) + (M_{surface})(\delta^{13}C_{initial\ surface}) \quad [Eqn\ 6.12]$$

where surface = biota, soil + peat, dissolved and atmospheric carbon, and M = mass of carbon. $M_{surface}$ = 41.803 Tt C from Fig. 6.1, so assuming a mantle signature of −5‰ for $\delta^{13}C_x$, the isotopic excursion would require 61–92 Tt of CO_2. The corresponding emission rate of $c.6.1$–9.2 Gt yr^{-1} is improbably large compared to the estimated rates of 0.3–2.0 Mt yr^{-1} for the Deccan Traps and 0.02–0.11 Gt yr^{-1} for long-term global outgassing (Leavitt 1982; Williams et al. 1992; Arthur 2000).

Equation 6.12 can also be used to calculate the amount of methane that would account for the isotopic excursion: assuming $\delta^{13}C_x$ = −60‰, a $\delta^{13}C_{carbonate}$ excursion of −2 to −3‰ would require 1.4–2.1 Tt C, which amounts to 1.9–2.8 Tt methane. A deep-water temperature increase from 11 to 15 °C would move the upper stability depth for methane hydrates from $c.920$ to 1460 m, assuming a hydrostatic gradient of 10 kPa m^{-1} (Fig. 4.32). It is conservatively estimated that $c.14\%$ of present-day methane hydrates are located over this depth range, which corresponds to $c.2.9$ Tt (Table 4.13). So the required methane release is feasible, and its rate of release (0.29 Gt yr^{-1} over 10 kyr) is comparable to the estimated modern agricultural release of $c.3.6$ Gt yr^{-1} (Dickens et al. 1995).

There seem to be three phases to the bulk carbonate isotopic excursion in Fig. 6.30a, which might be considered to indicate a pulsed liberation of methane (Bains et al. 1999). However, high-resolution studies involving isotopic analysis of a single, surface dwelling, foraminiferal genus suggest a single, geologically instantaneous, methane release (Fig. 6.30b; Thomas et al. 2002). The apparent discrepancy between bulk and single-genus carbonate data arises because the bulk carbonate data in Fig. 6.30a represent the entire community of phytoplankton, the composition of which varied with depth and also as a result of the warming event. These variations affect the averaged isotopic signature from bulk carbonate, apparently accounting for the stepped excursion profile. Analysis of single genera from the Weddell Sea site (690) has indicated that the isotopic excursion was triggered first in the surface dwellers, then spread down to the thermocline and only later reached the benthonic community (Fig. 6.30b; Kelly 2002; Thomas et al. 2002). In addition, there appears to have been an increase in surface-water temperature of $c.2$ °C (a decrease in $\delta^{18}O$, which is not seen in the bulk carbonate data in Fig. 6.30) just before the C isotopic excursion, possibly attributable to large-scale CO_2 emissions from the North Atlantic igneous province. Data from surface-dwelling foraminiferans suggest that the sea-surface temperature rose another 4 °C at the onset of the C isotopic excursion. The recovery of $\delta^{13}C$ to pre-excursion values took $c.200$ kyr (Röhl et al. 2000), and began in surface waters while the thermoclinic and benthonic foraminiferans were still recording the isotopic excursion (Fig. 6.30b).

A single surface-water $\delta^{13}C$ excursion suggests a single rapid release of methane. It is possible, therefore, that seismic activity or some other rapid process may have been the trigger for destabilization of rich methane-hydrate deposits in a particular location, rather than the general warming of deep waters (which is unlikely to have been synchronous throughout the oceans). The apparent downward propagation of the excursion indicates that much of the methane reached the surface ocean/atmosphere. Such a large injection of C into the oceans, involving oxidation of the methane to CO_2 and subsequent equilibration with bicarbonate, would cause the lysocline and CCD to shoal (see Box 5.7), and there is evidence for the expected consequential dissolution of carbonate (Bralower et al. 1997).

6.4.3 Permo-Triassic boundary event

There was a drastic reduction in oceanic biomass at the end of the Permian, and a fern spike in the spore record demonstrates that there was catastrophic deforestation, as at the KTB (Hsü & McKenzie 1990; see Section 1.5.5). However, environmental stress on ecosystems appears to have built up over several millions of years (Erwin 1993; Knoll et al. 1996; Hallam & Wignall 1997), so identifying a single cause and linking it to the observed isotopic excursion is problematical (Bowring et al. 1998). Various causes have been postulated, such as the emplacement of the Siberian Traps, in flood-basalt eruptions that were quite explosive and emitted large amounts of dust and aerosols into the atmosphere (Campbell et al. 1992; Renne et al. 1995). This may have led initially to cooling and restricted primary production by reducing the amount of solar energy reaching the surface, but subsequently to greenhouse warming under the influence of the large quantity of CO_2 emitted. A change in ocean chemistry and reduction in oxygenation levels may also have contributed to the extinction event (Knoll et al. 1996; Isozaki 1997). Even a bolide impact has been suggested (Becker et al. 2001). Based on a relatively expanded boundary section from East Greenland, which contains both marine faunal remains and terrestrial palynomorphs, it has been suggested that the collapse in marine and terrestrial ecosystems was synchronous, beginning just before a sharp negative C isotopic excursion, and took 10–60 kyr (Twitchett et al. 2001). The final disappearance of the Permian flora seems to have taken a further few hundred thousand years, with gymnosperms being among the earliest departures.

Varied patterns of C isotopic excursions in carbonate and kerogen have been recorded in marine sequences across the Permo-Triassic boundary, ranging from a single large, sharp, negative excursion to complex patterns of rapid excursions (Magaritz et al. 1992; Bowring et al. 1998; Twitchett et al. 2001; de Wit et al. 2002). Similar patterns have been observed in terrestrial sections. In all environments the isotopic excursions appear to be superimposed on a gradual decrease in $\delta^{13}C$ during the Late Permian, which may be at least partly attributable to the isotopically light CO_2 released upon uplift and oxidation of Pangaea's extensive peat deposits (Faure et al. 1995). The main C isotopic excursions are typically $c.5‰$ (for both carbonate and kerogen), which is much greater than that at the KTB (Raup & Sepkowski 1984; Holser et al. 1989b; Magaritz et al. 1992). The multiple excursions in many sections suggest a series of events, making it difficult to define their relationship to the extinction.

An isotopic excursion of $c.5‰$ is difficult to explain by a biotic collapse of the type seen at the KTB (Magaritz 1989; Wang et al. 1994); it is more likely to be the result of destabilization of methane hydrates (Erwin 1993), which is consistent with carbon-cycle modelling. Biotic collapse and volcanic degassing may have played subsidiary roles, but atmospheric CO_2 levels would not have been excessively high, and certainly not sufficient to account for the isotopic excursion (Berner 2002). A palaeo-high-latitude methane release has been suggested on the basis of greater isotopic excursions at high latitudes (≥8‰ in carbonates and ≥10‰ in kerogen from an East Greenland section; Krull & Retallack 2000; Twitchett et al. 2001). However, the isotopic excursion occurs after the onset of the biotic collapse, so the mechanism responsible for the isotopic excursion cannot be the direct cause of the extinction event. Modelling of a coupled atmosphere–ocean system suggests that rapid injection into the ocean of methane at a rate of $0.48\,Gt\,yr^{-1}$ over 10 kyr (representing some 24–48% of the present-day, global, hydrate reservoir) would produce atmospheric and oceanic $\delta^{13}C$ excursions of the right order, which would be reflected in similar excursions in kerogen and carbonate (Fig. 6.31; de Wit et al. 2002). A −12‰ excursion for the

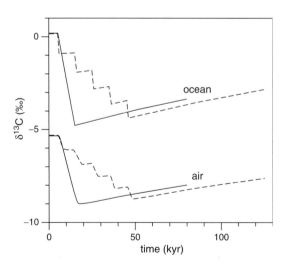

Fig. 6.31 Modelled response of a coupled ocean–atmosphere system to a single influx of CH_4 to ocean of $0.48\,Gt\,yr^{-1}$ over 10 kyr (solid lines) and five pulsed inputs of $0.96\,Gt\,yr^{-1}$ of 1 kyr duration at 10 kyr intervals (after de Wit et al. 2002).

atmosphere could be achieved by the same methane release if 5% of the gas escaped immediately to the atmosphere, which is possible given that methanotrophes may well be overwhelmed by a sudden large flux. The modelled results from five pulses of methane release, each of $0.96\,\mathrm{Gt\,yr^{-1}}$ over 1 kyr periods at intervals of 10 kyr are also shown in Fig. 6.31. The net effect of the recovery periods between each pulse is that the overall negative isotopic excursion is more gradual and slightly smaller.

6.5 Human influence on the carbon cycle

6.5.1 Deviation from the steady state system

The preindustrial carbon cycle in Fig. 6.1 is shown as a steady state system, with inputs balancing outputs for each reservoir. Human activity, particularly since the industrial revolution in the mid-1700s, has had a significant influence on the size of fluxes between some of the carbon reservoirs. Since industrialization there has been an imbalance between uptake and release of CO_2, as can be seen from the averaged data for the 1980s represented in Fig. 6.32. This imbalance is mainly the result of human (**anthropogenic**) contributions from burning fossil fuels ($5.3\,\mathrm{Gt\,C\,yr^{-1}}$) and cement manufacturing ($0.1\,\mathrm{Gt\,C\,yr^{-1}}$). The impact of agricultural practices such as deforestation ($c.2.0\,\mathrm{Gt\,C\,yr^{-1}}$) also contributes to an increased flux of CO_2 into the atmosphere from the terrestrial biota. Drying out, whether due to climatic change or agricultural drainage practice and subsequent burning of tropical peatland, is also a potentially major contributor to anthropogenic CO_2 (Sorensen 1993). The widespread fires in Indonesia during the 1997 El Niño event are estimated to have released 0.81–2.57 Gt C, mostly from peat (Siegert et al. 2001; Page et al. 2002). This contributed to the largest annual increase in atmospheric CO_2 since records began in 1957 of 6.0 Gt C (the average for the 1990s was $3.2 \pm 0.1\,\mathrm{Gt\,C\,yr^{-1}}$; Houghton et al. 2001).

Measurements of atmospheric CO_2 concentrations suggest that the gas is accumulating in the atmosphere more slowly than the rate at which it is being supplied,

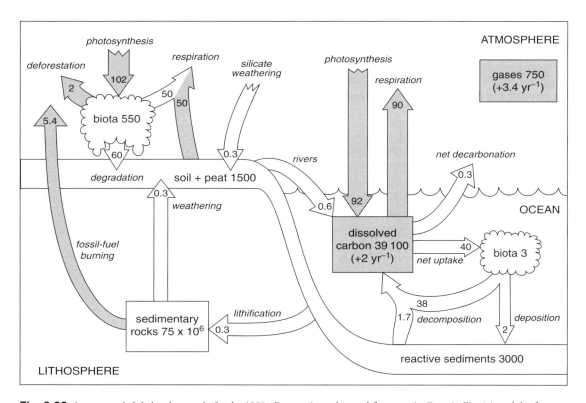

Fig. 6.32 An averaged global carbon cycle for the 1980s. Reservoirs and annual fluxes are in Gt, as in Fig. 6.1, and the flux attributed to fossil-fuel burning also contains a minor contribution from cement manufacturing ($0.1\,\mathrm{Gt\,yr^{-1}}$; after Siegenthaler & Sarmiento 1993). Major changes from the preindustrial cycle are indicated by half tone.

so the carbon cycle has adjusted to take up some of the excess anthropogenic generation. It appears that the oceans are absorbing $c.2\,Gt\,yr^{-1}$ of the gas, with a 20:80 split between surface and deep water masses. In addition, a similar amount may be taken up by terrestrial vegetation, as a result of the stimulation of photosynthesis by higher CO_2 concentrations (commonly referred to as CO_2 fertilization; Houghton et al. 1995; Keeling et al. 1996). However, the precise effect of increased CO_2 levels on the ability of terrestrial ecosystems to sequester some of the extra atmospheric C in the long term is difficult to assess (Smith & Shugart 1993; Oechel et al. 1994). In C_3 plants the carboxylase activity of rubisco increases with CO_2 concentration until the saturation limit of 800–1000 ppm is reached, after which terrestrial plants become a less efficient sink for the gas (Falkowski et al. 2000). However, net terrestrial primary production may level off at a much lower CO_2 concentration (550–650 ppm) because of nutrient limitation and higher soil microbial respiration rates as temperature increases (Falkowski et al. 2000). So although photosynthesis may be stimulated by rising CO_2 levels, so too may C transfer between root systems and soil, leading to faster C cycling but little long-term sequestration. Such limitations have been observed in grasslands (Hungate et al. 1997), but boreal forests may offer more potential as a C sink (Jarvis & Linder 2000). Reforestation in temperate latitudes may also help to take up some of the CO_2 released by tropical deforestation (Dixon et al. 1994; Fan et al. 1998).

Despite the increased uptake of CO_2 by some of the surface reservoirs, the net annual surplus input of CO_2 to the atmosphere was $c.3.4\,Gt$ during the 1980s (Siegenthaler & Sarmiento 1993; Houghton et al. 1996). As can be seen from Fig. 6.33c, the annual rate of increase of atmospheric CO_2 varies somewhat.

6.5.2 Greenhouse gas sources and fluxes

In preceding sections we have considered the main greenhouse gases, which have natural as well as anthropogenic sources: water vapour, carbon dioxide and methane. There are other naturally occurring gases that are present in the atmosphere in sufficient concentrations to influence the Earth's thermal budget, such as nitrous oxide (nitrogen(I) oxide, N_2O) and ozone (see Section 7.2.1). Nitrous oxide is produced naturally during denitrification (see Box 3.9 and Fig. 3.17), but the increased use of nitrogenous fertilizers has stimulated this production by $c.1$–$3\,Mt\,N\,yr^{-1}$ (Vitousek et al. 1997; Harvey 2000). As well as CO_2 and CH_4, carbon also enters the atmosphere as carbon monoxide (CO). Carbon monoxide is released into the air by natural processes, such as atmospheric oxidation of methane, and by anthropogenic processes, such as incomplete combustion of fossil fuels. Ultimately, however, CO, like CH_4, is converted to CO_2.

Determining the precise anthropogenic contribution of greenhouse gases that also occur naturally is difficult, but estimates are necessary if predictions are to be made of the likely associated climatic warming. Data from trapped gases in ice cores and remote atmospheric monitoring stations have provided a sound basis for estimating trends for the most abundant gases (Fig. 6.33). Carbon dioxide sources are reasonably well understood, and the major anthropogenic source over the past two centuries has been fossil-fuel combustion, particularly that related to power generation and transportation, as demonstrated by Table 6.2.

The size of the various natural sources of other gases is less well understood, making anthropogenic contributions difficult to quantify. Table 6.3 shows estimates of

Table 6.2 Representative distribution of anthropogenic sources of CO_2 emissions in the UK (after Aldhous 1990)

source	% of total emissions
power stations	31
other industries	25
road transport	17.5
domestic	14
commercial/public services	6.5
refineries	4
shipping	2

Table 6.3 Estimated changes in trace tropospheric gas concentrations and their contributions to radiation trapping in preindustrial times compared to 1985 values (after Dickinson & Cicerone 1986)

	1985		preindustrial	
	concn (ppb)	trapped heat (W m^{-2})	concn (% 1985)	trapped heat (% 1985)
carbon dioxide	345 000	~50	80	97
methane	1 700	1.7	41	65
ozone	10–100	1.3	75–100	85–100
nitrous oxide	304	1.3	94	96

the changes in tropospheric gas concentrations between preindustrial times and 1985 and their influence on climatic warming.

(a) Methane

After CO_2, CH_4 can be seen from Table 6.3 to be the next most significant contributor to global warming at present. As noted in Section 6.3.1, the major sink for atmospheric methane is oxidation by highly reactive hydroxyl radicals (·OH). The life-time of methane in the atmosphere is, consequently, relatively short at $c.12\,yr$ (Houghton et al. 2001). There is a variety of sources for methane entering the atmosphere, both natural and anthropogenic, and the fluxes are poorly constrained, so the estimates given in Table 6.4 may prove to be inaccurate for some sources. For example, estimations of emissions from wetlands range from 92 to 237 Mt $CH_4\,yr^{-1}$ (Houghton et al. 2001), and the methane generated by enteric bacteria in termites is extremely difficult to quantify (Whiticar 1990; Lelieveld et al. 1998).

Hydrothermal and mantle contributions of methane are not significant. Oceanic surface waters are oversaturated with respect to methane, due to bacterial (methanogenic) activity in localized anaerobic environments, such as the digestive tracts of zooplankton, resulting in a net flux of methane to the air. Methane is similarly produced in freshwater environments. Deep ocean waters contain much lower methane concentrations than surface waters and the methane generated within anaerobic sediments is mostly oxidized by methanotrophes. Marine and lacustrine environments as a whole do not make a large contribution to the methane flux, but natural wetlands do. The bacterial gases produced by these and other anaerobic soils are dominated by methane but contain other hydrocarbons (e.g. ethane and propane).

Some biogenic sources of methane can be considered primarily anthropogenic because of the significant influence of human activities. An example is methanogenesis in paddy fields under the anaerobic conditions that develop during flooding, although much of the methane produced appears to be oxidized by methanotrophes. The amount of methane reaching the atmosphere from this source is seasonably variable; estimates lie in the range $5–25\,mg\,m^{-2}\,h^{-1}$, and the mean gives a total annual flux of 110 Mt (Whiticar 1990). Other important anthropogenic sources of methane arise from methanogenesis in landfill sites and from burning of vegetation during deforestation/land clearance. Often the combustion of vegetation is incomplete, so that CO_2 is not the only product (e.g. CO and certain polycyclic aromatic hydrocarbons are produced; see Section 7.3.1a). There is significant production of enteric methane by ruminants, which can also be considered as anthropogenic because it arises mainly from domestic animals (chiefly cattle), although it must be remembered that this source is offset by the disappearance of large herds of wild ruminants (like bison) as a result of hunting and agricultural practices. The current ruminant input accounts for $c.15–20\%$ of the total methane flux into the atmosphere. Probably $c.2.0–2.5\%$ of natural gas is lost to the atmosphere during production, and methane is also emitted during incomplete combustion of fossil fuels.

The total annual input of methane from all sources to the atmosphere shown in Table 6.4 is 540 Mt, while the estimated output from atmosphere to sinks is 500 Mt. The potential inaccuracies in flux data can be seen by comparing the observed carbon isotopic signature of atmospheric methane of −47‰ with that calculated from the data in Table 6.4 of $c.$−54‰ (the latter is actually equivalent to −58‰ upon correcting for the kinetic isotope effect (see Box 1.3) that operates during the hydroxyl abstraction reaction). There are clearly major gaps in our understanding of the pathways of methane into and out of the atmosphere and the fluxes involved, as there are for many anthropogenic substances (see Chapter 7).

As well as raising global temperatures by the trapping of thermal radiation, increasing methane concentration in the atmosphere influences atmospheric chemistry in two further ways: first, by the increase in amounts of oxidation products (H_2O, CO and CO_2); second, by the rapid abstraction of hydroxyl radicals, which are important in a variety of oxidation processes and in ozone chemistry (see Section 7.2.1).

Table 6.4 Major sources of methane, their mean carbon isotopic signatures and estimated contribution to total methane flux entering the atmosphere (after Whiticar 1990)

source	$\delta^{13}C$ (‰)	flux into atmosphere	
		(Mt yr^{-1})	(% of total)
termites	−70.0	40	7.4
rice paddies	−63.0	110	20.4
ruminants	−60.0	80	14.8
oceans	−60.0	10	1.9
methane hydrates	−60.0	5	0.9
wetlands	−58.3	115	21.3
fresh waters	−58.0	5	0.9
landfill	−55.0	40	7.4
natural gas	−44.0	45	8.3
coal	−37.0	35	6.5
vegetation burning	−25.0	55	10.2

(b) Carbon monoxide

The chief anthropogenic source of CO is the incomplete combustion of fossil fuels, amounting to $\geq 450\,Mt\,yr^{-1}$ (Freyer 1979b). Natural sources, however, outweigh anthropogenic inputs; for example, atmospheric oxidation of CH_4 to CO, mainly by photolysis involving hydroxyl radicals, probably contributes $\geq 1500\,Mt\,yr^{-1}$. Microbial activity in ocean surface waters leads to a net transfer from sea to air of a further $c.100\,Mt\,yr^{-1}$ of CO. Despite these large inputs, atmospheric CO levels do not appear to be increasing, which implies that various sinks are removing CO as fast as it enters the atmospheric reservoir. Among these sinks are oxidation to CO_2 by soil bacteria and uptake of CO by plants, but the major sink appears to be hydroxyl radical oxidation to CO_2 (Freyer 1979b). In terms of its contribution to global warming, therefore, it is convenient to consider anthropogenic CO as simply an additional source of CO_2.

6.5.3 Atmospheric concentrations of carbon dioxide and methane

The level of CO_2 in the atmosphere prior to the industrial revolution was probably 280–290 ppm (Freyer 1979a; Houghton et al. 2001). However, earlier agricultural practices and deforestation dating back to Neolithic times (c.10 ka) have also contributed to the rise in atmospheric CO_2 concentration (Fig. 6.33a). Burning timber during deforestation returns the carbon stored in woody tissue to the atmosphere as CO_2, while the crops that replace the forests probably provide negligible long-term storage capacity. When these effects are taken into consideration, the level of atmospheric CO_2 before humans made a significant impact on the environment may have been c.260 ppm (consistent with ice-core data). By 2000, the level had risen to c.370 ppm, suggesting human activities have contributed some 110 ppm of atmospheric CO_2 (1 ppm = $c.2\,Gt\,C$; Siegenthaler & Sarmiento 1993; Houghton et al. 1996). An indication of the increase in atmospheric levels of CO_2 and CH_4 is shown in Fig. 6.33. Over the past century the anthropogenic CO_2 contribution has caused the $\delta^{13}C$ of atmospheric and surface-ocean reservoirs to rise by 1.2‰, but the isotopic composition of the deep ocean has yet to be affected because of its size and the relatively slow rate of mixing with surface waters (Joos & Bruno 1998).

Figure 6.33a shows that CO_2 and, more particularly, CH_4 levels are higher now than at any time during the Quaternary. From Fig. 6.33c it can be seen that CO_2 levels have been rising steadily over the past century, but there is a short-period cycle superimposed on the annual increase. This cycle exhibits maximum amplitude at high northern latitudes but becomes attenuated towards the equator and is hardly discernible in the southern hemisphere (inset, Fig. 6.33c; Conway et al. 1988; Keeling et al. 1996). In the northern hemisphere the cycle exhibits a maximum during the northern summer and a minimum in the winter, which is attributed to the influence of the seasonal growth of terrestrial vegetation and its preponderance in the northern hemisphere, together with incomplete mixing of the atmosphere over relatively short time scales. The lack of cyclicity in atmospheric CO_2 concentrations near the equator reflects the absence of seasonal variation in plant growth there. At the South Pole the small-amplitude cyclicity exhibits maxima during the southern summer, under the influence of the less abundant southern hemisphere flora.

Figure 6.33b shows that there is also a seasonal variation in atmospheric CH_4 concentration, but with the maxima occurring during northern hemisphere winters. This is apparently primarily attributable to the decay of higher plant detritus at the end of each growing season in the dominantly northern hemisphere terrestrial environments. In contrast to CO_2, the rate of increase of atmospheric CH_4 is declining, but the reasons are complex and relate to potential variations in the natural sources of the gas as well as its destruction by hydroxyl radicals (see Section 6.3.1).

6.5.4 Climatic change and greenhouse gases

The warming effect of each greenhouse gas is difficult to determine accurately because it depends upon a number of processes. For example, the atmospheric chemistry of many gases is complex and directly affects their life-time (adjustment time; see Box 6.4) in the atmosphere as well as their transformation into other compounds that may have their own greenhouse effect. For example, an increase in CO, which exerts a relatively small direct warming effect, causes a decrease in hydroxyl radicals (see Section 6.5.2b) and so increases the adjustment time of CH_4 (because the main route for CH_4 destruction, as for most gases containing an H atom, is via attack by ·OH; see Section 6.5.2a). In addition, CO is eventually oxidized to CO_2, which also causes warming (Isaksen & Hov 1987). It is possible to compare the warming effects of individual greenhouse gases with an equal amount of CO_2 over a given time period, using a parameter termed the global warming potential (GWP; Box 6.4). From Table 6.5 it can be seen

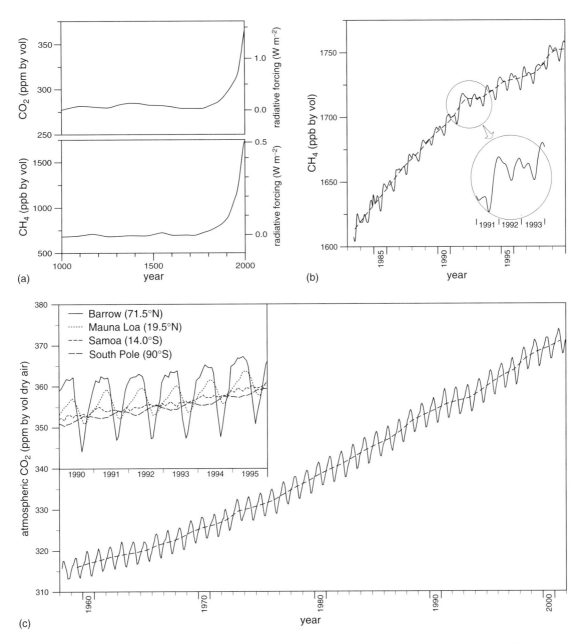

Fig. 6.33 (a) Trends in atmospheric CO_2 and CH_4 concentrations for the past 2 kyr from ice-core data (after Stauffer et al. 1985; Siegenthaler et al. 1988; Blunier et al. 1993, 1995; Neftel et al. 1994; Barnola et al. 1995; Etheridge et al. 1996, 1998; Chappellaz et al. 1997; Dlugokency et al. 1998; Houghton et al. 2001). (b) Globally averaged monthly CH_4 atmospheric concentrations (annual means shown by broken line; magnified section shows mid-year decline; after Dlugokency et al. 1998). (c) Mid-tropospheric CO_2 levels at Mauna Loa (Hawaii) and, in inset, at a number of sites over 1990–5 (after Keeling & Whorf 2002). Mauna Loa is favourable for analysing CO_2 concentrations in undisturbed air because local influences of vegetation and human activity are minimal and any effects of volcanic activity can be corrected (see Keeling et al. 1982; Bacastow et al. 1985). There was a 17.4% increase in the mean annual CO_2 concentration (shown by broken line) from 316.0 ppm (by vol.) of dry air in 1959 to 370.9 ppm in 2001; the largest increase in annual growth rate (2.87 ppm) occurred in 1997–8.

Box 6.4 | Warming potential of trace gases

The effects of increasing atmospheric concentrations of greenhouse gases on warming potential are not the same for all gases. Radiative forcing (see Box 6.1) increases approximately linearly with ozone concentration, with the square root of methane and nitrous oxide concentration, but only with the logarithm of CO_2 concentration. Hence a doubling of CO_2, regardless of initial concentration, will result in an instantaneous direct warming of $c.4.0–4.5\,W\,m^{-2}$ (see Box 6.1).

The data in Table 6.3 indicate the warming (i.e. trapped radiation) attributable to a given concentration of a trace gas at a particular time, but the overall effect of a given amount of a particular gas on global warming also depends on how long the elevated atmospheric levels persist (the **adjustment time**), which is a function of its atmospheric chemistry. In addition, various chemical transformations and feedbacks can amplify or attenuate the immediate warming effect. The adjustment time is not necessarily the same as the steady-state residence time. For example, the atmospheric residence time for CO_2 is $c.4\,yr$ (750/194.9 yr) from Fig. 6.32, but recycling processes within soils and ocean surface waters rapidly return a large proportion of the gas to the atmosphere, so it is generally considered meaningless to quote a single adjustment time for the gas (it is nominally $c.150\,yr$; Houghton et al. 2001; Fuglestvedt et al. 2003). The

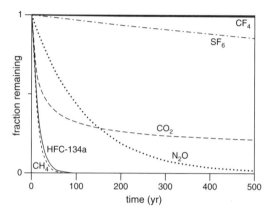

Fig. 6.34 Modelled decay rates of pulses of various greenhouse gases (after Fuglestvedt et al. 2003).

Table 6.5 Direct global warming potential (GWP) of tropospheric gases (molecule for molecule) relative to CO_2 (after Highwood & Shine 2000; Jain et al. 2000; Houghton et al. 2001)

gas	1998 concn (ppb vol.)	1998 radiative forcing ($W\,m^{-2}$)	radiative efficiency ($W\,m^{-2}\,ppb^{-1}$)	adjustment time (yr)	direct GWP over time period		
					20 yr	100 yr	500 yr
carbon dioxide	365	1.46	variable	variable	1	1	1
methane	1745	0.48	3.7×10^{-4}	12	62	23	7
nitrous oxide	314	0.15	3.1×10^{-3}	114	275	296	156
sulphur hexafluoride (SF_6)	0.0042	0.002	0.52	3200	15 100	22 200	32 400
carbon tetrachloride (CCl_4)	0.102	0.01	0.13	35	2700	1800	580
PFC-14 (CF_4)	0.8	0.003	0.08	>50 000	3900	5700	8900
perfluoroethane (CF_3CF_3)	0.0030	0.001	0.26	10 000	8000	11 900	18 000
CFC-11 ($CFCl_3$)	0.268	0.07	0.25	45	6300	4600	1600
CFC-12 (CF_2Cl_2)	0.533	0.17	0.32	100	10 200	10 600	5200
CFC-113 ($CF_2ClCFCl_2$)	0.084	0.03	0.30	85	6100	6000	2700
Halon 1301 (CF_3Br)	0.0025	0.001	0.32	65	7900	6900	2700
HCFC-22 (CHF_2Cl)	0.132	0.03	0.20	11.9	4800	1700	540
HCFC-141b (CH_3CCl_2F)	0.010	0.001	0.14	9.3	2100	700	220
HCFC-142b (CH_3CClF_2)	0.011	0.002	0.20	19	5200	2400	740
HFC-23 (CHF_3)	0.014	0.002	0.16	260	9400	12 000	10 000
HFC-134a (CF_3CH_2F)	0.0075	0.001	0.15	13.8	3300	1300	400
HFC-152a (CH_3CHF_2)	0.0005	0.000	0.09	1.40	410	120	37

Continued

adjustment time for CH$_4$ is greater than its atmospheric residence time, and actually increases with amount of CH$_4$ because of the feedback on ·OH concentration (8.4 yr; Houghton et al. 2001). Methane and CO can both contribute to tropospheric O$_3$, which is also a greenhouse gas. A comparison of the decay times for pulses of some greenhouse gases into the atmosphere is shown in Fig. 6.34.

The influence of variable atmospheric life-times on the warming effects of greenhouse gases can be allowed for by using a parameter termed the **global warming potential** (GWP; Lashof & Ahuja 1990; Rodhe 1990). This compares the radiative forcing of the same mass of each gas to that of CO$_2$ over a given period of time (Houghton et al. 1996). Carbon dioxide is arbitrarily assigned a GWP of 1.0. Uncertainties in GWP values are quite high, at c.±35% (Houghton et al. 2001), but they still provide one of the best means of comparing the net greenhouse effects of different gases (Fuglestvedt et al. 2003). The influence of adjustment time on GWPs is apparent in Table 6.5, with N$_2$O having a greater warming potential over a 100- than a 20-yr period, because its warming effect accumulates over its adustment time of 114 years. In contrast, methane exhibits the greatest warming effect over the 20-yr period because its effects disappear after 12 yr. The GWPs in Table 6.5 are for the direct effect of a gas; they do not include indirect effects, such as any influence on the abundance of other greenhouse gases. A number of entirely anthropogenic gases have considerably longer adjustment times and greater GWPs than naturally occurring greenhouse gases, as shown in Table 6.5. The most potent greenhouse gas known is SF$_6$. It is a very heavy, inert, non-flammable and non-toxic gas; useful properties for a range of industrial applications, particularly as an insulating gas in electrical switch-gear (Maiss & Brenninkmeijer 1998). Because of its stability and recent introduction to the environment, it has been useful for the study of mixing in the stratosphere and ocean, and has even been deliberately introduced into the ocean for this purpose (Maiss et al. 1996; Watson et al. 2000).

that CO$_2$ has the lowest direct GWP, whereas CH$_4$ has a greater warming capability. There are many entirely anthropogenic compounds that are much more potent greenhouse agents, some of which are included in Table 6.5, but fortunately they are present in the atmosphere at very low levels (they are discussed in Section 7.2.2).

In absolute terms, anthropogenic additions of CO$_2$ and CH$_4$ to the atmosphere are relatively small, and levels of the gases have varied widely throughout the Earth's history (see Section 6.2.2). However, the current rate of increase is exceptionally fast compared to all but catastrophic events such as the extinction event at the Cretaceous–Tertiary boundary and the likely destabilization of methane hydrates during the Paleocene–Eocene thermal maximum (see Section 6.4). Such rapid changes over the past 200 years have led Nobel laureate Paul Crutzen to suggest that we are no longer in the Holocene epoch but the 'Anthropocene'. There is growing consensus that the climatic warming experienced over the past 50 years is primarily attributable to human activities (Houghton et al. 2001). It would seem that, by 2005, the mean surface temperature rise that is attributable to greenhouse warming will be towards the lower end of the range predicted in the mid-1980s (1–5 °C; Dickinson & Cicerone 1986). Rather than warming the atmosphere, much of the additional heat has been taken up by the oceans, which, because of their vast volume and heat capacity, have experienced only a small temperature rise (<0.1 °C when averaged over the whole water column; Levitus et al. 2000, 2001; Barnett et al. 2001). Atmospheric concentrations of a variety of greenhouse gases have been increasing over the past two centuries, but so have anthropogenic emissions of SO$_2$ (c.100 Mt yr^{-1}; Bolin et al. 1983), particularly from coal and oil burning, which have led to increasing concentrations of sulphate aerosols and a cooling effect (Houghton et al. 2001). The effects of rising mean temperature upon climatic patterns, cloud cover and ocean-current patterns make it extremely difficult to predict local temperature changes, which may actually decrease in some areas while increasing significantly in others. Shifts in climatic zones and rainfall patterns will affect the distribution of vegetation zones and the type of agriculture that will be possible (Karl & Trenberth 1999).

A warmer atmosphere holds more water vapour, so global warming will lead to an increase in the rate of water flow through the hydrological cycle (Fig. 3.12). Climatic modelling suggests that half of the overall warming that would occur upon a doubling of atmospheric CO$_2$ concentration would result from the associated increase in water-vapour content (Hall & Manabe 1999; Schneider et al. 1999; Held & Soden

2000). Net evaporation occurs from the oceans at mid-latitudes in a high-pressure zone, and in the northern hemisphere the north-easterly Trade Winds carry the water vapour towards higher latitudes, where it is precipitated in the sub-polar low-pressure zone (Fig. 3.12). Increasing the amount of fresh water precipitated at high latitude in the North Atlantic reduces the salinity of surface water, which, together with the general warming of surface waters, will lower surface-water density and hence reduce the rate of, or even shut down, NADW formation (Broecker 1997, 1999). This could lead to cooling of north-west Europe similar to that experienced during the Younger Dryas (Box 6.5).

Reduction of anthropogenic CO_2 emissions is the most obvious means of ameliorating the effects of greenhouse gas induced climatic change. Consideration has also been given to stimulating phytoplanktonic uptake by providing a key biolimiting nutrient, iron, in areas of the open ocean (primarily the eastern equatorial Pacific and the Southern Ocean) where nitrate and phosphate levels are high but iron concentrations and hence primary productivity are low (see Section 3.2.7 and Box 3.7). Although iron-seeding experiments have proved successful, there is some doubt that a significant long-term sequestration of atmospheric CO_2 is possible (Boyd et al. 2000). Iron fertilization of the oceans is likely to fluctuate without deliberate intervention, because the aeolian flux of iron from the continents is affected by land use and the hydrological cycle (Fung et al. 2000). The flux may increase with increasing aridity or evaporation, although an increase in precipitation in some areas may decrease the flux. Even if the natural flux stimulated nitrogen fixation and hence marine primary production to the maximum, the estimated reduction in atmospheric CO_2 would be at most 40 ppm (Falkowski et al. 2000). This is not an insignificant decrease, but because it would occur over centuries it would have minimal impact on the current warming trend.

6.5.5 Eutrophication

On a localized basis, human activity can affect the carbon cycle indirectly by increasing the supply of mineral nutrients to aquatic environments. A large nutrient supply supports high levels of primary productivity by phytoplankton (algae and cyanobacteria), a process described as **eutrophication**. Under natural conditions, eutrophication is usually seasonal outside the tropics, being related to the stratification of the water column and the succession of phytoplankton (see Section 3.2.4). When nutrients are exhausted in the upper layers of water in thermally stratified lakes the phytoplanktonic bloom, which can form a substantial mat of material on the water surface, comes to an end. The cycle can only begin again when light levels are adequate and sufficient nutrients have accumulated in the surface waters. The latter is brought about by overturning of the water column when strong winds and cooling destroy the stratification (see Section 3.2.2b). However, when a constant anthropogenic supply of nutrients is available a high level of productivity can be sustained over longer periods, with seasonal variation in light levels probably becoming the limiting growth factor. It has been estimated that the flux of fixed inorganic N from land to the coastal zone, which is already causing some coastal eutrophication (Walsh 1991), will increase from $c.20$ to $c.40$ Mt N yr^{-1} by 2050 (Kroeze & Seitzinger 1998). Denitrification in the coastal zone presently prevents significant amounts of land-derived N from reaching the open ocean (Christensen et al. 1987; Seitzinger & Giblin 1996).

Important anthropogenic sources of nutrients include leachates from agricultural applications (nitrates and phosphates) and domestic waste water (which contains phosphates from cleaning agents). Degrading organic detritus can also release large amounts of nitrate and phosphate, with important environmental inputs being discharges from sewage treatment plants and run-off of rain water that has leached agricultural dung. The nutrients are then transported by rivers and streams to larger water bodies, and those in which flow is relatively restricted and nutrients can accumulate may be particularly prone to eutrophication. Still, freshwater bodies are likely candidates, but certain marine environments are also vulnerable. Densely inhabited and industrialized coastal areas in the Mediterranean, such as the Adriatic, have suffered from sporadic eutrophication (Cognetti et al. 2000). Aerobic decomposition of large amounts of detritus resulting from phytoplanktonic blooms in still bodies of fresh water can lead to severe depletion of oxygen in the water column, causing high mortality rates among fish. Eutrophication can even affect stretches of rivers, usually in the summer in temperate regions, when light intensity is high, low water levels result in a sluggish flow and oxygen diffuses only slowly into the calm, warm water.

Some of the species involved in phytoplanktonic blooms are toxic and can pose a danger to humans. Cyanobacterial blooms (e.g. *Anabaena*) resulting from eutrophication in lakes and other still waters in late summer can release toxins into the water. Although cyanobacteria occur naturally in the succession of phytoplanktonic species during a typical growth season

Box 6.5 | The Younger Dryas stadial

The present mild climate of the UK is attributable to the influence of the Gulf Stream/North Atlantic Drift, a surface current that brings warm water from the Gulf of Mexico to the north-western coast of Europe. This current is intimately coupled to the formation of North Atlantic Deep Water (NADW; see Box 3.2), because the surface water that sinks to form NADW off Greenland draws warm surface waters from the south to replace it, via the Gulf Stream/North Atlantic Drift (Fig. 6.35; Broecker 1997). Before surface water can sink its density must increase above that of the underlying water bodies, which is achieved by cooling and an increase in salinity (see Box 3.2). If the density of surface waters at high northern latitudes is lowered for any reason, NADW formation is likely to weaken or even cease, and so will the flow of the Gulf Stream/North Atlantic Drift. The switching off of the Gulf Stream is believed to be a key feature of the onset of glaciations in northern Europe during the Quaternary.

The Earth emerged from the last glacial episode some 15 kyr ago, and the warming during the subsequent interglacial period resulted in increased iceberg shedding from glaciers around the North Atlantic, and in particular the large Laurentide ice sheet covering much of North America. The resulting melt-water produced a low-density freshwater lid, effectively switching off NADW formation (Stocker 2000; Clark et al. 2001). This caused a sudden return to very cold conditions (a stadial), known as the Younger Dryas, which lasted from c.13 to 11.6 ka (see Fig. 5.42), and is reflected in changes in the position of the boundary between cold surface waters of polar origin and warm Atlantic waters, called the polar front. During most interglacials, the polar front seems to have run between Newfoundland and Iceland, but during the Younger Dryas it extended as far south as Portugal.

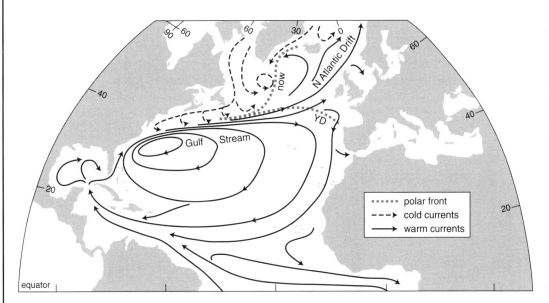

Fig. 6.35 Surface currents in the North Atlantic and the location of the polar front now and during the Younger Dryas (YD; after Ruddiman & McIntyre 1981).

(e.g. in the UK), eutrophication can cause a population explosion. The development of suitable control methods requires an understanding of the factors that affect the bloom. It is in the late summer that cyanobacteria grow vigorously, when competition with other species for light is at its lowest. The demise of other species in the phytoplanktonic succession (e.g. diatoms) is generally the result of depletion of silicate and nitrate within the epilimnion. Cyanobacteria contain gas vacuoles to prevent their sinking out of the euphotic zone and they are not affected by low nitrate levels because some species can fix nitrogen directly from the atmosphere (see Section 3.2.7). Phosphate is, therefore, usually the biolimiting nutrient for cyanobacteria, so controlling anthropogenic inputs of phosphate (and potential remobilization of sedimentary phosphate) will reduce the likelihood of a bloom. Alternatively, if the dominance of cyanobacteria can be sufficiently delayed, peak density may be diminished, so reducing potential toxicity problems. This may be possible if early summer conditions can be prolonged artificially, permitting early species in the succession, which can compete effectively against cyanobacteria, to persist for a longer period. For example, maintaining a supply of nutrients to the epilimnion, particularly silicate, allows diatom growth to continue (Reynolds et al. 1983).

7 | Anthropogenic carbon and the environment

7.1 Introduction

The preceding chapters have been concerned with the behaviour of biogenic compounds in the geosphere. An understanding of the processes that control the environmental fate of naturally occurring organic compounds can help in developing models to predict the likely behaviour of manmade (anthropogenic) components in the environment. Many of the anthropogenic organic compounds, such as CO_2 and CH_4, also occur naturally, and in sufficient quantities they can have an impact on the natural carbon cycle on a global scale, as we have seen in Chapter 6. In the following sections we examine anthropogenic inputs of compounds on a smaller scale and their environmental impact.

Some anthropogenic organic compounds, such as the insecticide DDT, do not have a natural source and so can be termed **xenobiotic**. Since the Second World War there has been a dramatic increase in the number, amount and applications of xenobiotic substances, examples of which are the CFCs discussed in Section 6.5.4. Because they have no natural source their presence in the environment is unambiguously indicative of pollution. It can be argued that although these compounds are often present at very low concentrations in the environment, their impact on the biosphere is significant because organisms have not had the opportunity to evolve in their presence, and so may lack the ability to metabolize or otherwise eliminate the chemicals. Some of these compounds are highly toxic, teratogenic (i.e. causing foetal deformities) and carcinogenic (i.e. causing cancer). In particular, the more stable, hydrophobic compounds tend to accumulate in certain tissues in higher organisms and cannot be metabolized, and because they are fat-soluble they can even be passed on to the offspring of animals (e.g. via milk for mammals and via egg yolk for birds and fish).

7.2 Halocarbons

7.2.1 Ozone depletion

Ozone is a vital component of the stratosphere because it adsorbs most of the UV radiation that is harmful to life (see Box 7.1). In contrast, it is an undesirable gas in the troposphere, the layer of the atmosphere underlying the stratosphere and in immediate contact with the Earth's surface and the biosphere, because it is toxic to organisms, being an extremely active oxidant. The burning of fossil fuels is a major source of tropospheric ozone, via photochemical dissociation of gaseous products such as nitrogen oxides (often abbreviated to NO_x). Unfortunately, little of the ozone produced in the troposphere enters the stratosphere, where it could have a beneficial action. This is mainly because it is produced near the ground and, being highly reactive, has a short residence time and is destroyed by photochemical reactions (some of which may play a part in the oxidative degradation of other gaseous pollutants) before it can reach the stratosphere. Further, the thermal stratification of the atmosphere restricts transport of ozone and other gases between the troposphere and stratosphere.

While ozone levels are increasing in the troposphere, they are decreasing in the stratosphere by the action of anthropogenic gases, particularly the xenobiotic chlorofluorocarbons (**CFCs**). CFCs have been widely used as refrigerants, blowing/propulsion agents and cleaning agents for electronic components. They have long residence times in the atmosphere (c.45 and 100 yr, respectively, for CFC-11 and CFC-12; Table 6.5) because they lack a removable hydrogen and so are not destroyed by reaction with hydroxyl radicals (Houghton et al. 2001). They can diffuse into the stratosphere, where they undergo photolytic degradation to provide chemical species that are extremely effective scavengers of ozone, in reactions that take place on the surfaces of ice crystals (Wuebles 1981; WMO 1999). The main reactive

Box 7.1 | Atmospheric stratification and the ozone layer

The atmosphere is composed of a number of layers in which temperature either decreases or increases with increasing altitude. This thermal stratification can be likened to that of the ocean (Fig. 3.1). The lowest two layers are the **troposphere** ($c.$0–15 km altitude, where weather occurs) and the **stratosphere** ($c.$15–50 km altitude). The boundary between these two layers is the tropopause (Fig. 7.1), the altitude of which is greatest at the equator ($c.$18 km) and least at the poles ($c.$8 km). With increasing altitude in the troposphere air density and temperature both decrease until the tropopause is reached; then temperature increases again in the stratosphere, until the stratopause.

The increase in temperature in the stratosphere is due to the absorption of UV radiation by the ozone present in this layer. The ozone is itself formed by the action of UV radiation on oxygen, which splits the molecule into two extremely reactive atoms. These atoms then react with further oxygen molecules to produce ozone:

$$O_2 \xrightarrow{UV} O + O \quad \text{[Eqn 7.1]}$$

$$O + O_2 \rightarrow O_3 \quad \text{[Eqn 7.2]}$$

Ozone absorbs UV radiation at wavelengths of $c.$200–300 nm (see Fig. 6.6). Wavelengths in the range 290–320 nm may cause significant damage to DNA, and those shorter than $c.$290 nm can be fatal to life.

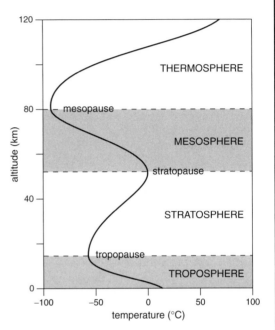

Fig. 7.1 Typical variation of temperature with altitude in the atmosphere and the resulting thermal layers. (The tropopause varies from $c.$18 km and a temperature of $c.$–80 °C at the equator to $c.$8 km and $c.$–50 °C at the poles.)

species is the chlorine atom, which behaves like a catalyst in its reaction with ozone: it is regenerated and can react with many ozone molecules before finally being consumed by other chemical species (Solomon et al. 1986). The range of reactions that can occur is complex, but the consumption of ozone can be simplified to:

$$Cl + O_3 \rightarrow ClO + O_2 \quad \text{[Eqn 7.3]}$$

$$ClO + O \rightarrow Cl + O_2 \quad \text{[Eqn 7.4]}$$

In effect, the reactions in Eqns 7.3 and 7.4 bring about the reverse of Eqns 7.1 and 7.2. Another anthropogenic gas, nitric oxide (NO), scavenges ozone in a similar way (to form nitrogen dioxide, NO_2) and is produced by the dissociation of N_2O, which can be generated from fossil fuel combustion and from the decomposition of nitrate fertilizers (i.e. the last step of denitrification; Fig. 3.17):

$$N_2O \rightarrow N_2 + O \quad \text{[Eqn 7.5]}$$

$$N_2O + O \rightarrow 2NO \quad \text{[Eqn 7.6]}$$

$$NO + O_3 \rightarrow NO_2 + O_2 \quad \text{[Eqn 7.7]}$$

$$NO_2 + O \rightarrow NO + O_2 \quad \text{[Eqn 7.8]}$$

Equations 7.7 and 7.8 together represent the catalytic conversion of ozone and the reactive atomic oxygen that is involved in the formation of ozone into molecular oxygen, so a small amount of the nitrogen oxides can do a lot of damage:

$$O_3 + O \rightarrow 2O_2 \quad \text{[Eqn 7.9]}$$

CFCs are just one class of **halocarbons**, which are compounds that contain carbon and at least one halogen atom (i.e. fluorine, chlorine, bromine or iodine). Halocarbons containing bromine (collectively called halons) have also been found to destroy ozone. So too does carbon tetrachloride (CCl_4), formerly used as a dry-cleaning agent, but its perfluorinated analogue CF_4 (PFC-14) does not. Because of the long-term effects of the build-up of stratospheric halocarbons and other gases on ozone levels, various measures have been taken to reduce emissions of these compounds (e.g. the Montreal Protocol and its amendments). Less harmful alternatives have been introduced into industrial applications (e.g. Prather & Watson 1990). Alternatives to the most commonly used CFCs (CFC-11 and CFC-12) are HCFCs (hydrochlorofluorocarbons) and HFCs (hydrofluorocarbons). These compounds degrade more rapidly in the atmosphere than CFCs because they contain hydrogen. In particular, HFCs contain no chlorine and the hydrogen fluoride that is formed upon decomposition is believed not to be detrimental to ozone.

7.2.2 Greenhouse effect

There are several xenobiotic gases that have a pronounced greenhouse effect, a number of which are also ozone depleters, as noted in the previous section and demonstrated by the global warming potential (GWP; Box 6.4) values in Table 6.5. The importance of CFC contributions to global warming, despite the relatively small quantities involved compared to anthropogenic CO_2 emissions (see Section 6.5.4), is attributable to the long atmospheric lives of the gases. As noted in Section 7.2.1, many of the xenobiotic greenhouse gases — such as the PFCs, CFCs, halons and SF_6 — do not react with hydroxyl radicals in the troposphere and so, like N_2O, reach the stratosphere before they are destroyed, mainly by UV radiation at wavelengths <240 nm (Houghton et al. 2001). For the most stable compounds, SF_6 and the PFCs, the major mode of destruction is via photolysis or ionic reactions in the mesosphere (see Fig. 7.1).

Although HFCs and HCFCs are replacing CFCs in many industrial applications, because they are less detrimental to the ozone layer and are more readily destroyed within the troposphere (see Section 7.2.1), they can still make a significant contribution to global warming over a long period, as can be seen from the GWPs in Table 6.5. HFC-23, which has one of the largest GWPs (Table 6.5), is an unintended by-product of HCFC-22 production (Houghton et al. 2001). However, the values given in Table 6.5 are for the direct warming effect, and do not take account of any ability to destroy ozone, which is itself a greenhouse gas. This major indirect effect is included in the net GWPs in Table 7.1. Despite the error margin associated with the estimates in Table 7.1, the net GWP values can be seen to be considerably lower than the direct GWPs. For brominated halocarbons the net GWPs are negative, because they so readily destroy ozone (Daniel et al. 1995).

The CFCs, PFCs, HFCs and HCFCs were not produced before the twentieth century (Butler et al. 1999). Thanks to measures to reduce emissions of ozone depleters under the Montreal Protocol, atmospheric concentrations of most CFCs and CCl_4 have begun to decrease or, in the case of CFC-12, the rate of increase has declined. In contrast, atmospheric concentrations of HFCs and HCFCs are growing quite rapidly (Fig. 7.2). Although we have called CF_4 and SF_6 xenobiotic compounds above, they have recently been found to occur naturally in fluorites, and can outgas (Harnisch & Eisenhauer 1998). However, the natural background levels are low compared to the anthropogenic contribu-

Table 7.1 Comparison of the direct and net global warming potential (GWP) of some halocarbons (after Daniel et al. 1995; Houghton et al. 2001). The net GWPs include indirect effects on ozone

gas	GWP 20-year period			GWP 100-year period		
	direct	net (min)	net (max)	direct	net (min)	net (max)
carbon tetrachloride (CCl_4)	2700	−4700	1300	1800	−3900	660
CFC-11 ($CFCl_3$)	6300	100	5000	4600	−600	3600
CFC-12 (CF_2Cl_2)	10200	7100	9600	10600	7300	9900
CFC-113 ($CF_2ClCFCl_2$)	6100	2400	5300	6000	2200	5200
halon 1301 (CF_3Br)	7900	−79000	−9100	6900	−76000	−9300
HCFC-22 (CHF_2Cl)	4800	4100	4700	1700	1400	1700
HCFC-141b (CH_3CCl_2F)	2100	180	1700	700	−5	570
HCFC-142b (CH_3CClF_2)	5200	4400	5100	2400	1900	2300

tions ($c.0.01$ ‰ for SF_6 and $c.40$‰ for CF_4; hence the non-zero pre-manufacturing concentration for CF_4 (shown as PFC-14) in Fig. 7.2).

7.3 Hydrocarbon pollution in aquatic environments

7.3.1 Fossil fuel combustion

(a) Polycyclic aromatic hydrocarbons in Recent sediments

Carbon dioxide and water are the major products from combustion of fossil fuels. However, combustion is rarely totally efficient and among the products from oil and coal are various aromatic hydrocarbons that contain several fused benzenoid rings and are hence generally referred to as polycyclic aromatic hydrocarbons (**PAHs**). They are formed by the action of the heat generated during the combustion process, but under conditions of local oxygen deficiency, so they can be considered pyrolysis products.

Distributions of pyrolytic PAHs are characterized by the dominance of the non-alkylated species shown in Fig. 7.3. Particularly abundant are the highly peri-condensed compounds—such as pyrene, the benzo-pyrenes, benzo[*ghi*]perylene and coronene—that result from extensive angular fusion of benzenoid systems. The presence of such PAH distributions in the aromatic

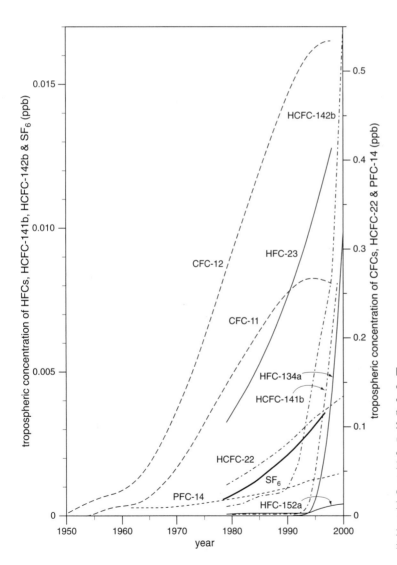

Fig. 7.2 Changes in tropospheric concentrations of some xenobiotic, ozone-depleting greenhouse gases over the past half century. HCFC, HFC and SF_6 data from Cape Grim, Tasmania (values slightly lower than global mean), other data are global mean values (after Harnisch et al. 1996, 1999; Maiss et al. 1996; Montzka et al. 1996a, b, 1999; Oram et al. 1996, 1998, 1999; Maiss & Brenninkmeijer 1998; Miller et al. 1998; Sturrock et al. 1999; Prinn et al. 2000). See Table 6.5 for fluorocarbon identifications.

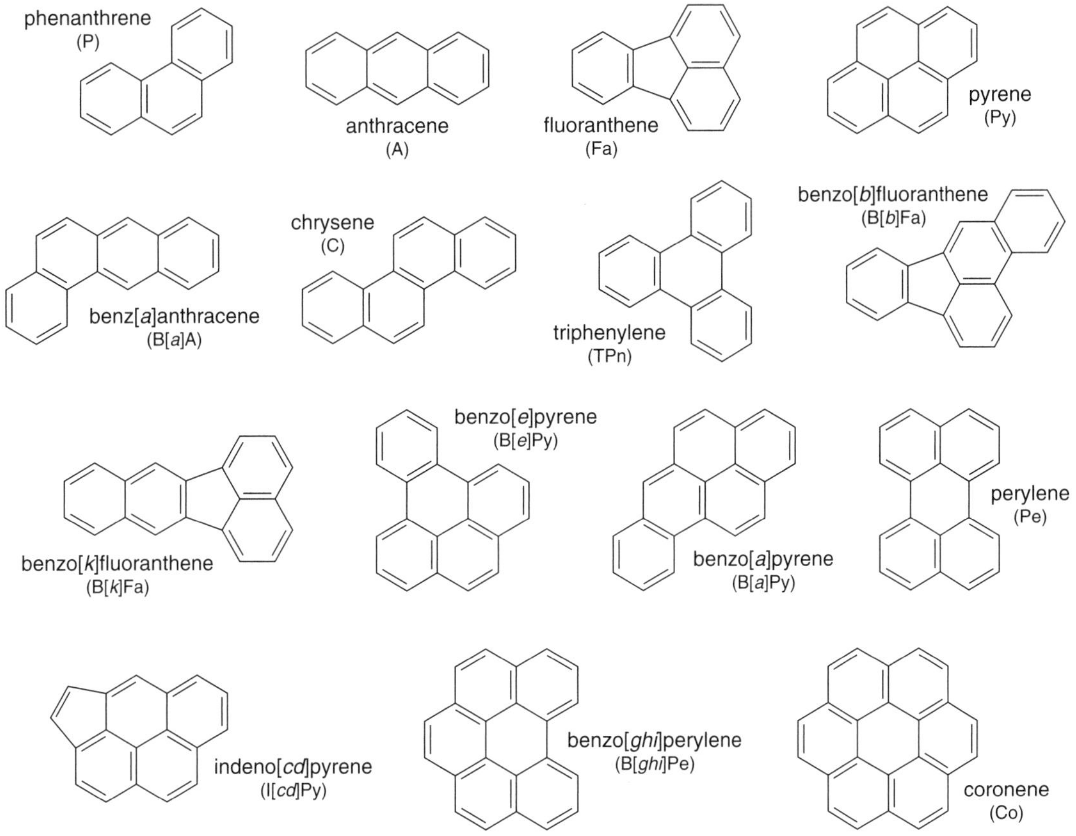

Fig. 7.3 Major pyrolytic polycyclic aromatic hydrocarbons (PAH) found in sediments (abbreviations refer to Fig. 7.4).

hydrocarbon fractions of Recent sediments is generally considered to reflect inputs from the combustion of organic matter, such as wood and fossil fuels. Pyrolytic PAHs are widespread in sediments post-dating the industrial revolution and demonstrate the extent of human influence on the environment.

Highly peri-condensed PAHs are more reactive than PAHs exhibiting lower degrees of angular fusion and so their high concentrations among pyrolysis products are attributed to rapid quenching by adsorption (through hydrogen bonding) on to particles of soot, which is itself a polycondensed PAH material. The observed distributions of pyrolytic PAHs in Recent sediments from a variety of environments are remarkably constant (Youngblood & Blumer 1975; Laflamme & Hites 1978; Wakeham et al. 1980; Gschwend & Hites 1981); an example is shown in Fig. 7.4. This uniformity suggests that the PAH distributions finally preserved in the sediment are largely protected by their association with soot particles against the actions of environmental agents during transportation and sedimentation (Prahl & Carpenter 1983). For example, although all the PAHs are virtually insoluble in water, slight differences in solubility exist that would be expected to lead to preferential leaching of some components, but this does not appear to occur. Further, some PAHs (e.g. anthracene) are more susceptible to photo-oxidation than others and might be expected to be preferentially degraded during aeolian transport and subaerial exposure in sediments, but again such losses appear to be minimal.

The immobilization of PAHs within the sedimentary matrix is important because many of these compounds are proven carcinogens and it would not be desirable for them to be readily released and subsequently incorporated into the aquatic food chain. Benzo[a]pyrene is a particularly potent carcinogen and its concentration in sediments in the Severn Estuary (UK) has been estimated at 9 ppm (based on sediment

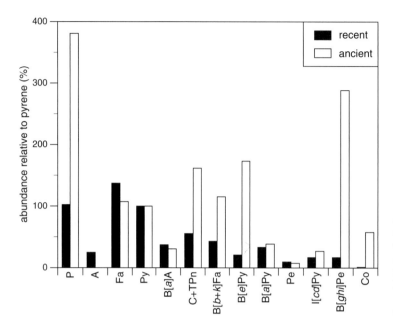

Fig. 7.4 Major pyrolytic polycyclic aromatic hydrocarbon (PAH) distributions in Recent and ancient sediments, normalized to pyrene abundance (from molecular ion responses during GCMS analysis; see Box 4.3; after Killops & Massoud 1992). See Fig. 7.3 for key to names and structures.

dry weight; Thomson & Eglinton 1978). A significant proportion of pyrolytic PAHs associated with soot particles is likely to be distributed by wind, the influence of which can extend a great distance out into the oceans (e.g. the Trade Winds). Fluvial transport of PAHs is also important for lacustrine and nearshore marine environments. Street dust and asphalt have been found to contain pyrolytic PAHs and so the disposal network for urban drainage water can influence local sedimentary concentrations of PAHs. However, in a high-energy environment such as the Severn Estuary, much surface sediment is periodically resuspended and redistributed by the complex current systems. The associated mixing is likely to contribute to the uniformity of PAH distributions.

(b) Polycyclic aromatic hydrocarbons in ancient sediments

It appears that a recognizable pyrolytic PAH fingerprint can survive over geological time-scales. Wild fires, primarily initiated by lightning strikes, are the most likely source of these PAHs in ancient sediments. Such fires have probably been a feature of terrestrial ecosystems from at least the Late Devonian (see Section 1.4.2), as suggested by the occurrence in sediments of fusinites and semifusinites (see Section 4.3.1b), the proposed products of vegetation fires (Chaloner 1989). Observed PAH distributions in ancient sediments are often like those seen in Recent sediments and are consistent with moderate to high combustion temperatures of 400–800 °C, as established for contemporary vegetation combustion (Youngblood & Blumer 1975).

Suitably high temperatures for the formation of PAHs can occasionally arise in sedimentary environments as a result of igneous activity. For example, pyrolytic PAHs associated with ash, coal and wood fragments and apparently linked with igneous activity have been noted in sedimentary material from the Midland Valley of Scotland (Murchison & Raymond 1989). Pyrolytic PAHs can, therefore, be adsorbed by charred woody remains as well as by soot. An interesting example of PAH generation by igneous activity is the pyrolytic PAHs detected in oil seeps in the Guaymas Basin, California (Kawka & Simoneit 1990). Their formation has been attributed to the *in situ* pyrolysis of contemporary organic matter (with a significant terrestrial component) at temperatures in excess of 300 °C. The heat is supplied by hydrothermal activity associated with the spreading centre in the basin.

Pyrolytic PAH distributions in ancient sediments sometimes differ from those typical of Recent sediments in exhibiting enhanced levels of the more highly peri-condensed structures, especially benzo[*e*]pyrene, benzo[*ghi*]perylene and coronene (Fig. 7.4; Killops & Massoud 1992). The reasons are as yet unknown but may reflect the effects of either different formation conditions or varying geochemical processes over geological time periods.

It can be seen that there are natural inputs of pyrolytic PAHs to the environment as there are for atmospheric carbon dioxide. However, unlike CO_2, anthropogenic PAH contributions in contemporary sediments exceed natural inputs. This increase in environmental burden may, therefore, be significant in its effects on organisms.

7.3.2 Oil spills

(a) Effects of oil pollution

Oil spills in coastal areas cause immediate and obvious problems to animals and plants from external fouling; ingestion is an added problem for animals. Unseen effects include the suppression of primary production in phytoplankton owing to reduction in light levels, the suppression of oxygen transfer from the atmosphere and the damage caused to benthonic communities when the oil sinks to the sea bed. There are also long-term effects on ecosystems related to the release of toxic components (**toxicants**) over a prolonged period as the oil breaks up and the concentration of toxicants in organisms towards the top of the food chain increases.

Once again, however, oil is a naturally produced material and there are abundant natural oil seeps. It seems likely, therefore, that ecosystems can cope to an extent with the presence of oil, providing the amount is not overwhelming. Recovery of spilt oil is the most effective response in terms of minimizing ecological effects, but weathering occurs rapidly, making recovery difficult. For example, approximately half the oil released into the Arabian Gulf during the 1991 hostilities evaporated within 24 hours. Loss of the more volatile, liquid hydrocarbons left a tarry material, rich in asphaltenes, that broke up with wave and current action and drifted to the sea floor as its density increased. Asphaltenes contain relatively polar components, including organometallics, which gradually leach into the water. These components exhibit varying toxicities, and once in solution they are readily available for ingestion by organisms. Emulsification can also occur soon after an oil spill, as wind and wave action breaks up the oil into fine droplets and disperses them within the water matrix, again making recovery of the oil extremely difficult.

Beached oil can be equally difficult to deal with. Spraying with detergent-based dispersing agents drives the oil deeper into the substrate on sandy and muddy shores and, until the recent development of new agents, the toxicity to sediment-dwelling communities of the chemicals employed was similar to that of oils. The use of dispersants is most effective while the oil spill is still at sea. Oil will degrade naturally with time under the effects of weathering and bacterial action. As well as leaching of soluble components, weathering processes involve photo-oxidation, which fragments molecules and introduces polar oxygen groups, resulting in products that are generally more water-soluble and may also be more readily metabolized by organisms (although some products may be toxic). In contrast, apolar components are generally more difficult to metabolize and can build up in certain body tissues, with possible long-term toxicity implications. However, some fungi and a variety of bacteria (e.g. *Nocardia* species; Chosson et al. 1991) are able to use oil components as a carbon source (Atlas 1981), as was noted in our consideration of in-reservoir biodegradation (Section 4.5.6c). They may be present in relatively large numbers in areas where oil pollution is a chronic problem, as in the Arabian Gulf, and provide a degree of buffering to the ecosystem against small spills. Cultures of these bacteria are now being exploited in combating major spills. Biodegradation of oil and PAHs has been observed in anoxic sediments, apparently involving sulphate reducers (Coates et al. 1997).

(b) Oil pollution monitoring

Monitoring levels of oil contamination can be difficult because only the hydrocarbons fraction is readily amenable to routine analytical techniques. The dominant components, *n*-alkanes, are biodegraded more rapidly than other components and there are also natural sources of *n*-alkanes, such as higher plants (mainly odd carbon numbers in the range C_{23}–C_{35}) and benthonic algae (in which C_{15} and C_{17} often dominate; see Section 5.4.2a). When the obvious signs of oil fouling have disappeared, oil contamination is usually confirmed by the presence of certain molecular indicators of petrogenic inputs, particularly characteristic sterane and terpane distributions (see Section 4.5.3b). Although only trace components in oils, these biomarkers are relatively resistant towards biodegradation and can be used to identify the source of an oil spill and to estimate pollution levels.

Examples of the aerobic biodegradation pathways of hydrocarbons are shown in Figs 7.5–7.7 (anaerobic biodegradation of aromatic compounds has been reviewed by Heider & Fuchs 1997a, b). Biodegradation proceeds quite rapidly in subaerial environments, eventually leading to the removal of all the major components that can be resolved by gas chromatography (see Section 4.5.3a and Box 4.3). However, another

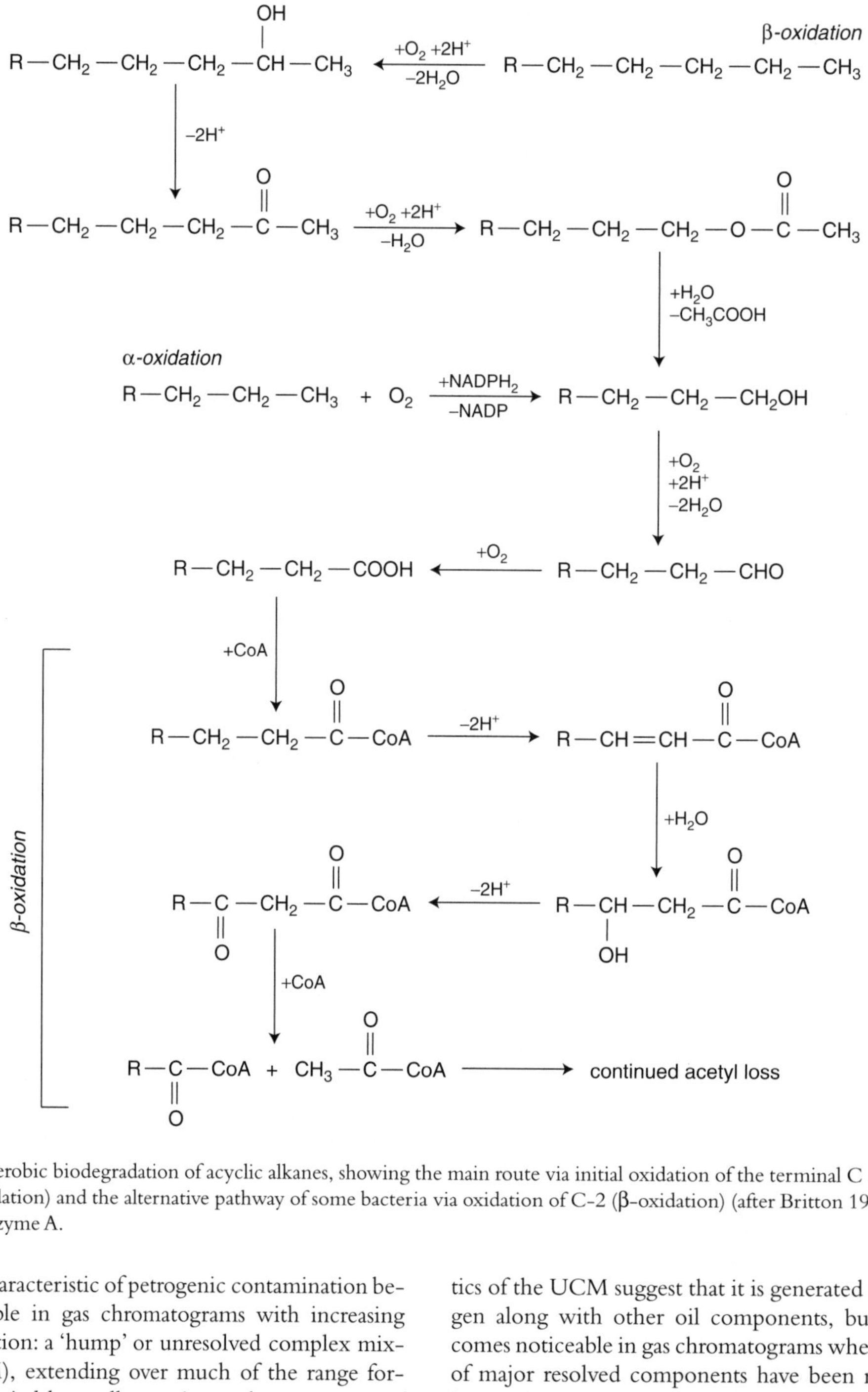

Fig. 7.5 Aerobic biodegradation of acyclic alkanes, showing the main route via initial oxidation of the terminal C atom (C-1; α-oxidation) and the alternative pathway of some bacteria via oxidation of C-2 (β-oxidation) (after Britton 1984). CoA = coenzyme A.

apparent characteristic of petrogenic contamination becomes visible in gas chromatograms with increasing biodegradation: a 'hump' or unresolved complex mixture (UCM), extending over much of the range formerly occupied by n-alkanes. A gas chromatogram of the total hydrocarbons fraction of a heavily biodegraded oil is shown in Fig. 7.8 (see Fig. 4.16 for comparison with a non-biodegraded oil). The chemical characteristics of the UCM suggest that it is generated from kerogen along with other oil components, but only becomes noticeable in gas chromatograms when the levels of major resolved components have been reduced by bacterial action (Killops & Al-Juboori 1990). The area of the UCM in a gas chromatogram can be used to determine the level of environmental contamination and is particularly useful for assessing chronic pollution.

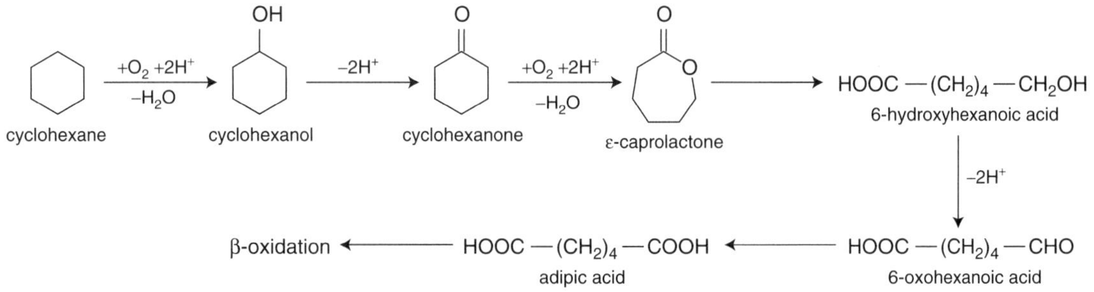

Fig. 7.6 Aerobic biodegradation of cycloalkanes (after Britton 1984).

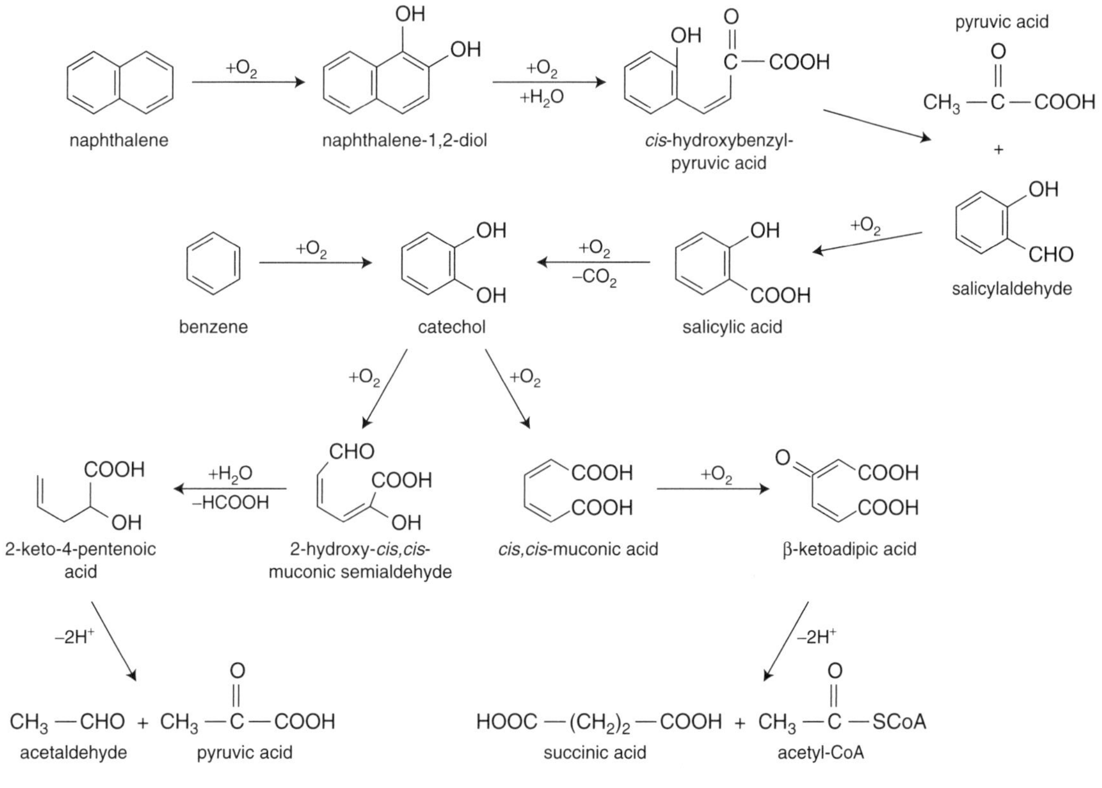

Fig. 7.7 Simplified scheme of aerobic biodegradation of aromatic hydrocarbons (after Gibson & Subramanian 1984). CoA = coenzyme A.

Chronic, low-level inputs of oil into aquatic environments can be conveniently studied using suitably sessile, filter-feeding organisms, such as mussels (e.g. *Mytilus edulis*). Mussels have been found to accumulate hydrocarbons in their tissues, and accumulation appears to continue with increasing exposure. These organisms naturally colonize oil-production platform legs in the North Sea, providing a useful means of monitoring pollution during oil production. Acyclic alkanes are also absorbed by the mussels from algal food sources, so monitoring the UCM may be the most accurate way of measuring petrogenic contamination levels. It has been

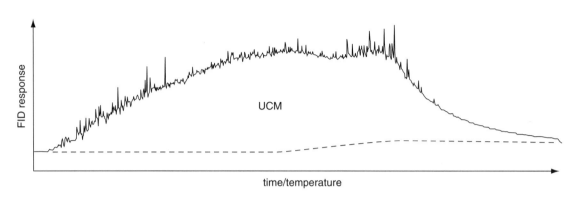

Fig. 7.8 Gas chromatogram of the total hydrocarbons fraction from a heavily biodegraded oil, showing the unresolved complex mixture (UCM). The elution range of the UCM is broadly equivalent to the *n*-alkane range shown in Fig. 4.16. The broken line shows the approximate zero signal (the elevated section towards right corresponds to the signal created by bleed of stationary phase at higher temperature; see Box 4.3).

found that exposure to the monoaromatic fraction of the UCM causes a toxic response in mussels at sublethal doses (Rowland et al. 2001). For example, feeding rates over a 24-h period declined by >40% at an aqueous UCM concentration of $200\,mg\,l^{-1}$, which corresponded to an accumulation of up to $90\,mg\,g^{-1}$ wet wt. These results are comparable to the 50% reduction in feeding rates associated with a burden of $94\,mg\,g^{-1}$ of 1-phenyldecane (Donkin et al. 1991).

In one study of the association of petrogenic inputs with various grades of sedimentary material it was found that significantly more of the UCM was associated with the clay fraction than with the sand and silt fractions (Brassell & Eglinton 1980). This behaviour is not surprising because the approximately spherical silt particles would be expected to exhibit a smaller available adsorption surface area than the clay platelets. In addition, the aluminosilicates in clay may exhibit a greater adsorption affinity for UCM components. It is important to remember that **adsorption** is a surface effect that can involve strong chemical interactions (chemisorption), whereas **absorption** is a weaker, physical interaction in which a fluid enters pores within a solid (e.g. water in a sponge). The particulate associations of oil components in the depositional environment contrast with those of pyrolytic PAHs, which remain associated with the relatively low-density particulates by which they are adsorbed at their time of formation (Prahl & Carpenter 1983; Readman et al. 1984). These differences in particulate associations may result in significant variations in the behaviour of petrogenic and pyrolytic hydrocarbons in sedimentary environments.

7.4 Endocrine-disrupting chemicals

7.4.1 Endocrine activity

In 1938 it was noted that some substituted phenols had oestrogenic activity. In particular, the drug 4,4′-dihydroxy-αβ-diethylstilbene (also known as diethylstilboestrol or DES; Fig. 7.9) was observed to be a potent oestrogen. Subsequently, various other chemicals that have been released into the environment have been discovered to affect the endocrine system of organisms to varying extents. Over the past 50 years or so there is compelling evidence for depressed sperm counts and elevated incidence of testicular and breast cancer among humans, which may be attributable to endocrine disruption. Initially interest was focused on compounds that mimic sex hormones (agonists) or block their activity (antagonists) and their consequences for reproductive health. More recently, concern has been growing about the influence of environmental compounds on other glands of the endocrine system (e.g. pituitary, thyroid and adrenal; Box 7.2) and related endocrine functions (e.g. immune and neurological). Consequently, there is interest in the fate and behaviour of potential endocrine disrupters and in the factors that may influence the effects of exposure (e.g. sensitive periods during organism development; Box 7.2). Among the abnormalities in wildlife that have been suggested to arise from endocrine disruption are: thyroid dysfunction in birds and fish; decreased fertility in birds, fish and turtles; metabolic abnormalities in birds, fish and mammals; behavioural abnormalities in birds; demasculinization and feminization of male fish and birds; and

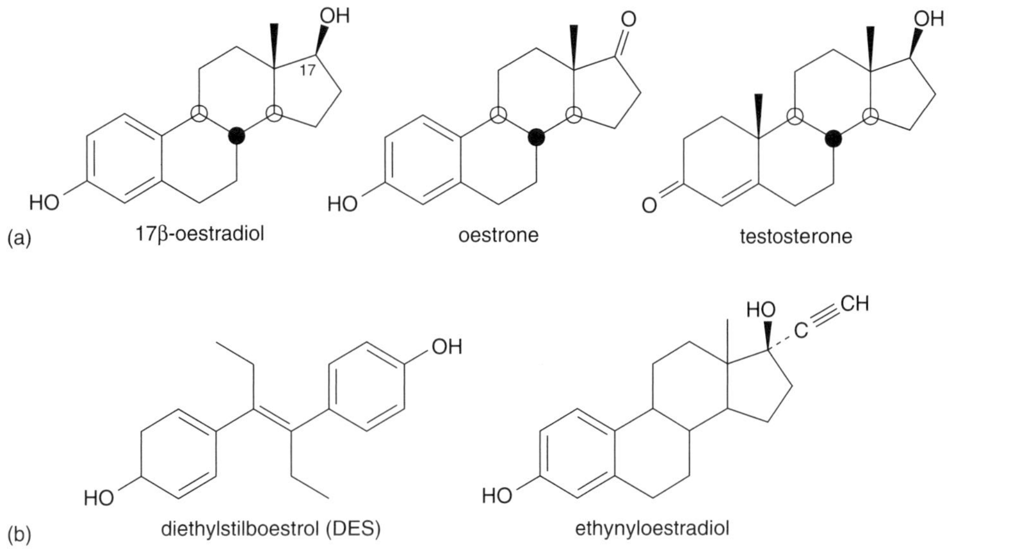

Fig. 7.9 (a) Natural and (b) synthetic human sex hormones.

compromised immune systems in birds and mammals (Colborn & Clement 1992). However, it has not yet been established for most endocrine disrupters whether the effects noted in individuals have an impact at the population level.

7.4.2 Natural and synthetic oestrogens

Our present meagre understanding of normal endocrine processes in invertebrates makes the assessment of chemical endocrine disruption in the field difficult (LeBlanc 1999). Steroid roles differ between species and sometimes sexes, and their influence may vary at different developmental stages. In most studies of invertebrates, endocrine disruption appears to involve androgenization rather than oestrogenic effects (see Box 7.2). Arthropods (crustaceans and insects), annelids and molluscs use ecdysteroids, terpenoids and vertebrate-like sex steroids for endocrine control. For example, the ecdysteroid ecdysone is naturally converted to 20-hydroxyecdysone (Fig. 7.10), which induces moulting (ecdysis) in both insect larvae and crustaceans.

Many plants produce endocrine disrupters as a defence against phytophagous (plant-eating) insects, which work by affecting growth, development or reproduction of the insects (Bergamasco & Horn 1983). One such group of compounds is the ecdysteroids, including the ubiquitous 20-hydroxyecdysone, which is found in many families of pteridophytes, gymnosperms and angiosperms. The waste products of paper pulping

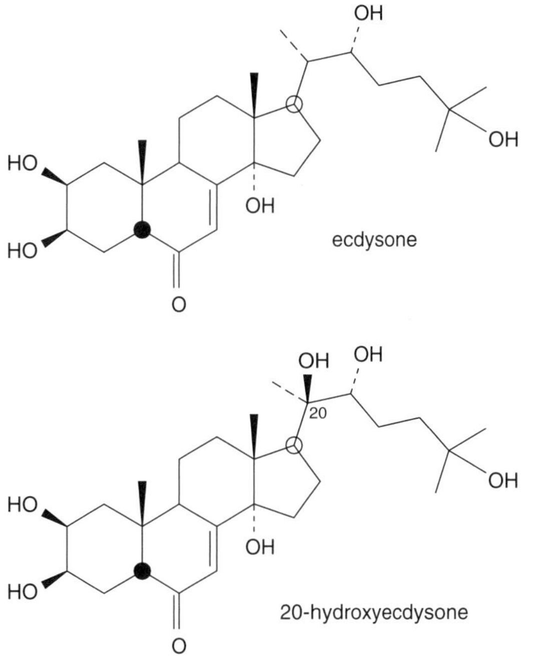

Fig. 7.10 Examples of ecdysteroids.

(Kraft-mill effluent) contain various plant sterols, primarily β-sitosterol (c.85%) and stigmasterol (c.3%) (see Fig. 2.22 for structures), which have oestrogenic activity (they are known to cause masculinization in fish),

> **Box 7.2** | **Endocrine system and its disruption**
>
> The conventional definition of the **endocrine** system is the collection of ductless glands that secrete small amounts of hormones directly into the blood of vertebrates. The hormones are transported to other parts of the body where they have a profound effect on biological processes. Perhaps the best examples are the sex hormones oestrogen and testosterone, but there are many others. Hormones are also produced by invertebrates (e.g. moulting hormones) and even by plants (e.g. growth hormones).
>
> Some xenobiotic compounds can have hormone-like effects and cause disruption of the endocrine system. A compound that mimics the action of a particular hormone is termed an **agonist**, whereas one that blocks its activity is an **antagonist**. Hormone activity is governed by receptors that recognize the shape of the molecule, so potentially other molecules with suitable structural characteristics can interact with these receptors. There are other potential modes of action of endocrine disrupters. To summarize:
> 1 They can mimic the effects of naturally occurring hormones by recognizing their binding sites (agonists).
> 2 They can reduce the effects of natural hormones by blocking their interaction with their physiological binding sites (antagonists).
> 3 They can react directly or indirectly with natural hormones.
> 4 They can alter the natural pattern of hormone synthesis.
> 5 They can alter hormone receptor levels.
>
> Compounds that mimic female sex hormones are called **oestrogenic**, and those that mimic male sex hormones are **androgenic**. Antagonistic counterparts to these compounds exist, with anti-oestrogenic and anti-androgenic properties (Colborn & Clement 1992).
>
> **Exposure timing**
> The timing of exposure of an organism to endocrine-disrupting chemicals is important. There can be organizational effects if exposure occurs early in life, especially in embryonic stages, involving permanent developmental changes. Alternatively, activational effects can be experienced in adulthood, and these are usually more transitory (Guillette et al. 1995).
>
> Activational effects include female–female pair bonding in gulls, linked to DDT (Fry & Toone 1981), and plasma vitellogenin (yolk protein) appearance in male trout, which has been attributed to effluent from sewage works (Purdom et al. 1994). Organizational effects include embryonic deformities in turtles (Bishop et al. 1991), and premature sexual maturity in salmon (Leatherland 1992), both associated with PCBs and PCDD/Fs. Many organizational effects do not become apparent until later in life, so it can be difficult to establish the direct cause, but some laboratory studies have confirmed potential associations (Guillette et al. 1995). For example, lipophilic organochlorines affect sex differentiation in birds. The compounds are concentrated in fat reserves, which are mobilized during egg production and deposited in egg yolk (Fry 1995). It would seem that species producing large, yolky eggs are especially at risk (Guillette et al. 1995). Controlled exposure of sea-bird eggs to DDT, at concentrations found in contaminated eggs in the wild, has been found to induce development of female reproductive tissues in male embryos (Fry & Toone 1981).
>
> Sublethal exposure to oil or DDT leads to adverse reproductive effects, probably through non-specific morbidity or increased stress. Disturbingly little is known about how microorganisms are affected by endocrine-disrupting chemicals, although these organisms can transform natural steroids, some of which are hormones.

so they can be termed **phytoestrogens** (i.e. plant-derived oestrogens). Some plant sterols can be converted microbially into other active compounds, including testosterone (Fig. 7.9). Other phytoestrogens include isoflavones (e.g. isoflavone; Fig. 7.11), lignans (see Section 2.5.1) and coumestans (e.g. coumestrol; Fig. 7.11), which are found in various vegetables such as beans, peas, sprouts, cabbage, spinach, soya beans, grains and hops. Resorcylic acid lactones (e.g. zearalenone, also known as mycotoxin F2; Fig. 7.11), which are fungal metabolites found in wheat *inter alia*, also display oestrogenic activity.

Phytoestrogens are generally present at low levels and have lower activity than the main human oestrogen, 17β-oestradiol (Fig. 7.9). Some can show anti-oestrogenic activity (Box 7.2), such as indole-3-carbinol (Fig. 7.11; present in cruciferous plants, e.g. sprouts and cabbage) and isoflavones. High levels of

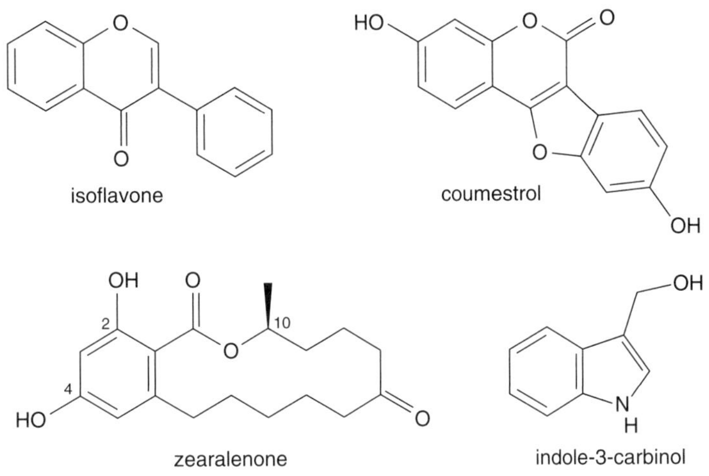

Fig. 7.11 Examples of phytoestrogens.

consumption may present symptoms of oestrogenic activity, such as anatomical changes in the cervix and uterus, infertility and aberrant sexual behaviour in ewes and reduced sperm counts in rams feeding on clover containing large amounts of isoflavones (Kaldas & Hughes 1989).

Various drugs with oestrogenic activity have found their way into the aquatic environment via the sewage system, in addition to the natural oestrogens—17β-oestradiol and oestrone (Fig. 7.9)—that are excreted by pregnant females in significant amounts as well as deriving from breakdown of the contraceptive pill. DES (Fig. 7.9) is one of the earliest xenobiotic compounds; it was used to promote growth in cattle and to prevent human miscarriages in the early 1950s to early 1970s, but was subsequently found to cause abnormalities. However, surprisingly little information is available on the environmental fate of DES. The degree to which a compound is likely to be distributed between water and organic material in the environment can be estimated from its partition coefficient in an octanol–water system (K_{ow}), because octanol appears to be a reasonable analogue for sedimentary organic matter. DES has a high $\log K_{ow}$ value of 5.07, so it is likely to adsorb strongly on to organic matter in soils and sediments. Only limited evaporation is likely, and that reaching the air is rapidly oxidized. Unlike 17β-oestradiol, DES does not bind to the sex-hormone binding globulin (SHBG) and so enters cells more freely, where it binds with oestrogen receptors.

In a key study of the effluent from sewage treatment works processing domestic sewage, the only significant oestrogenic activity was traced to the unbound forms of oestrone, 17β-oestradiol and ethynyloestradiol (Desbrow et al. 1996). The first two compounds occur naturally, but ethynyloestradiol (Fig. 7.9) is anthropogenic; the combination of the three is attributable to the human contraceptive pill, in which ethynyloestradiol is the main active ingredient. Importantly, the oestrogenic activity of the naturally occurring hormones is additive. The level of oestrone and oestradiol in sewage sludge is approximately an order of magnitude greater than that of ethynyloestradiol, which is not always detectable. However, ethynyloestradiol is considerably more potent than the two natural steroids (e.g. Purdom et al. 1994) and so may have significant environmental activity at levels below present detection limits. Ethynyloestradiol has a lower aqueous solubility than oestrone and oestradiol, so it is not as readily degraded during microbial waste-water treatment. Primary treatment (i.e. solids removal) has been found to remove 5–25% of ethynyloestradiol compared to 35–55% for the natural oestrogens, and the corresponding removal figures for combined primary and secondary (i.e. biodegradation) treatment are 20–40% and 50–70%, respectively (Tabak et al. 1981). It has been suggested that ethynyloestradiol is excreted in a conjugated form that is less active, although there is the possibility that bacterial sewage treatment results in deconjugation and release of the active form (Sumpter & Jobling 1995).

Relatively little is known about environmental concentrations and the fate and behaviour of the oestrogens in the aquatic environment. There appears to be an increasing burden of oestrogens in water bodies owing to the use of the contraceptive pill—up to 15 ng l^{-1} of ethynyloestradiol have been recorded in some UK waters (Aherne & Briggs 1989) — which seems to be the main source of oestrogenic activity in the environment in general. They can be quite persistent; for example, 17β-oestradiol can be present in poultry litter for at least a week after being spread on pasture (Nichols et al. 1997).

7.5 Environmental behaviour of selected xenobiotic compounds

7.5.1 Physicochemical properties and bioaccumulation

The fate and behaviour of chemicals in the environment depend not only upon various environmental processes and conditions, but also upon the physicochemical properties of the compounds. All these factors are summarized in Table 7.2. We have considered some physicochemical properties of oestrogenic compounds in the previous section. Xenobiotic compounds can display widely varying physicochemical parameters, as demonstrated in Table 7.3, which profoundly influence factors such as transportation and modes of introduction into organisms. The octanol–water (K_{ow}) partition coefficient represents the equilibrium concentration ratio of a compound in the two immiscible solvents, and is usually expressed as the logarithm. Values >4 suggest that the contaminant is highly **lipophilic** (i.e. has an affinity with fatty substances) and will tend to bind strongly to the organic material in soils and sediments. Values of Henry's Law constant $>10^{-4}$ indicate high volatility and suggest a relatively high potential for existence in the vapour phase. It can be seen that the degree of chlorination of individual PCBs (see Section 7.5.5) significantly affects the various physicochemical parameters in Table 7.3, so even within one class of compound environmental behaviour can vary enormously.

Table 7.2 Examples of factors affecting the environmental fate of chemicals

environmental processes
   transportation (e.g. land into groundwater)
   movement from one medium to another
     (e.g. deposition and volatilization)
   dilution in air or water
   alteration (e.g. biodegraded) to give compounds that
     are more/less toxic
   absorption and concentration by organism ingestion

environmental conditions
   temperature
   pH
   salinity
   current (wind/water) velocity
   microbial activity
   organic matter content
   sunlight intensity (photodegradation)

physicochemical properties of compounds
   vapour pressure
   aqueous solubility
   partition coefficients (solid/liquid partitioning, e.g.
     lipophilicity)
   Henry's Law constant (tendency to evaporate from
     soil/water/plant surface into air)

Table 7.3 Physicochemical constants for pollutants (mean values; after IEH 1999)

compound	water solubility (mg l^{-1} at 25°C)	vapour pressure (Pa at 25°C)	log K_{ow}*	Henry's Law constant
ethynyloestradiol	4.83	1.00×10^{-4}	4.1	–
oestrone	12.4	–	–	–
17β-oestradiol	13.0	–	–	–
TBT	10.00	–	2.2–4.4	–
DEHP	0.34	8.00×10^{-4}	5.1	4.91×10^{-4}
p,p'-DDT	3.00×10^{-3}	2.00×10^{-5}	6.0	9.53×10^{-4}
p,p'-DDE	4.00×10^{-2}	1.00×10^{-3}	5.7	3.21×10^{-3}
PCB-28†	3.70×10^{-2}	3.40×10^{-2}	5.6	1.09×10^{-2}
PCB-153†	2.06×10^{-3}	1.20×10^{-4}	6.8	7.12×10^{-3}
PCB-206†	1.10×10^{-4}	3.80×10^{-5}	7.1	5.59×10^{-3}
2,3,7,8-TCDD	1.93×10^{-5}	1.18×10^{-7}	6.8	1.35×10^{-3}

*Logarithm of octanol-water partition coefficient. †PCB numbering scheme: 28 = 2,4,4'-trichlorobiphenyl; 153 = 2,2',4,4',5,5'-hexachlorobiphenyl; 206 = 2,2',3,3',4,4',5,5',6-nonachlorobiphenyl.

In the following sections we use physicochemical parameters to assess the environmental behaviour and the tendency for bioaccumulation (Box 7.3) of the compound groups represented in Table 7.3. Several of the compounds considered exhibit varying degrees of endocrine disruption.

Many factors can control the persistence of a pollutant in a particular reservoir. Although it can be difficult to determine the exact influence of each factor, a useful overall measurement of the environmental persistence of a compound can be gained from its half-life, which is the time needed for the concentration of a contaminant in a reservoir to halve once inputs have ceased. It is analogous to radioactivity half-life (see Box 5.4).

7.5.2 Tributyl tin

One of the most thoroughly studied endocrine disrupters is tributyl tin (**TBT**). Trialkyl tin formulations (e.g. as the chloride salt; Fig. 7.12) were widely used in

---

Box 7.3 | Bioaccumulation

The term **bioaccumulation** is often used to describe the overall build-up of contaminants within the tissues of an organism. However, sometimes more specific terminology is used to reflect the different routes for absorption of contaminants, with bioconcentration used to denote passive absorption through skin or gills, while bioaccumulation is reserved for ingestion of particulates (feeding) and biomagnification is used to describe the cumulative effects over successive trophic levels in a food chain (Franke et al. 1994).

The main routes of uptake and absorption of pollutants by organisms vary according to the physicochemical properties of the compounds and the assimilatory functions of the organism. A major route for dissolved lipophilic pollutants is diffusion through the cell walls of phytoplankton and through the gills of fish (bioconcentration; LeBlanc 1995). However, passive diffusion is generally not important for hydrophobic chemicals (which have high partition coefficients; Iannuzzi et al. 1996). The **bioconcentration factor** (BCF) can be defined as:

$$BCF = \frac{\text{concentration in organism}}{\text{concentration in surrounding medium}}$$

[Eqn 7.10]

BCF varies with duration of exposure, concentration of contaminant, environmental conditions and the species of organism involved. Highly lipophilic compounds at very low exposure concentrations are likely to yield high BCFs, whereas BCF values tend to be suppressed at high concentrations. Meaningful comparison of published BCF values is difficult unless all the conditions under which they were determined were identical.

Uptake by active ingestion (to which the more restricted use of the term bioaccumulation can be applied) depends on feeding preferences, ingestion rate and absorption efficiency (Iannuzzi et al. 1996). Absorption efficiency is defined as:

$$\text{absorption efficiency} = \frac{\text{amount of substance accumulated}}{\text{total amount ingested or absorbed}}$$

[Eqn 7.11]

and is controlled by factors such as **depuration** (the ability of an organism to eliminate a contaminant from its tissues when placed in an environment free of the compound), dilution due to growth (after a single contamination event), metabolism and excretion (Gobas et al. 1988). Absorption efficiencies tend to be greatest for the most lipophilic compounds. For most organochlorine pesticides and PCBs (see Section 7.5.5) the absorption efficiency is c.40–90% (Iannuzzi et al. 1996). Lower efficiencies of c.10–30% have been suggested for PCDD/Fs (see Section 7.5.6) and some PCBs (Iannuzzi et al. 1996), although they may be higher for 2,3,7,8-TCDD (e.g. 40–50% in trout; Niimi 1996). The absorption efficiency of PCDD/Fs has been found to vary with the logarithm of the octanol-water partition coefficient ($\log K_{ow}$) in fish (Gobas et al. 1988). It is likely that depuration and metabolism are not significant expulsion routes for highly lipophilic (and hence hydrophobic) compounds.

Bioconcentration and bioaccumulation (in its more restricted definition) can operate together, and determining their precise contributions can be extremely difficult. For example, tributyl tin (TBT) is relatively soluble in water, so the dog whelk absorbs it across its gills and mantle tissue, but it can also absorb a significant proportion from its food (Section 7.5.1; Bryan et al. 1989; Langston 1996). Throughout this book bioaccumulation is used in its most general sense.

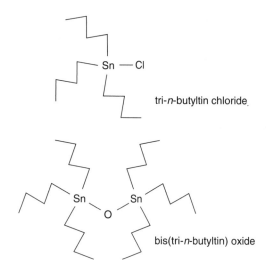

Fig. 7.12 Some TBT formulations.

anti-fouling paints in the 1970s and early 1980s; the tributyl and triphenyl compounds are the most toxic. Unfortunately, the formulations were found to affect organisms other than the target species, and legislation was introduced to restrict usage in the UK and to impose environmental quality standards (Langston 1996).

Organotins have very low volatilities, so they are unlikely to enter the atmosphere. They are reasonably water-soluble (Table 7.3) but also moderately lipophilic, with $\log K_{ow}$ values of 2.2–4.4, depending on the pH, ionic strength and nature of the medium (Fent 1996). It might be expected, therefore, that TBT can enter organisms via absorption from water and ingestion of sedimentary organic matter, and there is a reasonable amount of corroborative evidence from TBT concentrations in environmental samples. During the busy summer months of 1986–9, UK estuaries and harbours contained up to $2\,\mu g\,l^{-1}$ of TBT (Waite et al. 1991). Associated sediments in harbours generally contain several % dry wt (Law et al. 1998), while concentrations in aquatic organisms are usually <1‰ by weight, although they can reach their highest levels of a few ‰ by weight in molluscs (e.g. Langston 1996; Bryan et al. 1987). Concentrations of TBT in UK waters have generally decreased since the mid-1980s, but the rate of disappearance at some sites has declined, so the environmental quality standard of $2\,ng\,l^{-1}$ ($0.8\,ng\,l^{-1}$ as Sn) may not be achieved in the near future (Langston 1996). Interestingly, there is no environmental quality standard for sediments, other than to prevent levels of TBT from increasing, although the sedimentary reservoir is clearly likely to be a major source of TBT for ecosystems for some time.

Complexation with dissolved organic matter seems to reduce the bioavailability and hence toxicity of TBT, which may explain why BCFs (Box 7.3) are lower for algae (e.g. 1500 for *Fucus vesiculosis*) than suspension- and deposit-feeding invertebrates (e.g. 2000–20 000 for *Nereis diversicolor* and 5×10^5 for *Mya arenaria*; Langston et al. 1994; Langston 1996). In addition to physico-chemical factors and the feeding habits of organisms, organotin accumulation levels are strongly influenced by variations in the ability of organisms to metabolize the contaminant. Depuration studies (see Box 7.3) have shown that metabolism of TBT is faster in algae (half-life of 2–3 weeks) and fish (2–7 days) than molluscs (7–14 weeks) (Bryan et al. 1989; Langston et al. 1994). Degradation is relatively efficient in mammals and birds, so there is little biomagnification in top-level consumers. Among the benthonic invertebrates, gastropods and copepods seem the most sensitive to the effects of TBT absorbed via ingestion of particulate organic matter. The dog whelk (*Nucella lapillus*) can absorb about half its burden of TBT from its food, the mussel *Mytillus edulis*, but the digestive gland breaks down some of the TBT into di- and monobutyl tin (Bryan et al. 1989).

Organotins (especially in triorgano form) have a range of toxic effects in many phyla (Langston 1996). One effect is the inhibition of oxidative phosphorylation (via NADPH and NADH; see Section 2.2.2). Another is imposex in gastropods, which is the suppression of female reproductive capability by the induction of male characteristics. Imposex could arise from inhibition of cytochrome P450-dependent androgen aromatization, which is responsible for the conversion of testosterone into oestrogen. The extent of imposex in gastropods has been positively correlated with body concentrations of tin, in tributyl or dibutyl form (Bryan et al. 1987). This is, to date, the only firmly established case of endocrine disruption causing effects at the population level in the field. Imposex can be initiated in the female dog whelk at TBT concentrations as low as $1\,ng\,l^{-1}$, with accumulated levels of Sn (in tributyl form) reaching 1‰ by weight after 4 months' exposure and with 40% of the animals developing some imposex (Bryan et al. 1986). In another study, all female dog whelks were sterilized by imposex after two years' exposure at ambient aqueous concentrations of TBT of $>2\,ng\,l^{-1}$ (Gibbs et al. 1988). There have been reports from around the world of similar effects of TBT on many species of marine gastropods (Ellis & Pattisina 1990).

Degradation of tributyl tin in the environment occurs by sequential removal of alkyl chains, ultimately yielding inorganic tin (Waldock 1994). A variety of chemical and biological processes can bring about this dealkylation. Photolytic decay can occur within 2–3 days under clear conditions (Soderquist & Crosby 1980), although half-lives in the aqueous phase can be anything from days to months, depending upon temperature and water clarity. Tributyl tin should not be particularly persistent, but degradation is slowed by interaction with sediments; the contaminant can persist for years in anaerobic sediments (Waldock et al. 1990).

7.5.3 Phthalates

Phthalates are primarily used as plasticizers, to increase the pliability of plastics such as PVC. Most concern centres on the leaching of phthalates from plastics into drinking water and food. About $2.7\,\text{Mt}\,\text{yr}^{-1}$ are produced globally, of which some 50% comprises di(2-ethylhexyl)phthalate (or DEHP; Fig. 7.13). Total losses to the environment are estimated at c. $7.7\,\text{kt}\,\text{yr}^{-1}$. DEHP is a reproductive toxicant and a carcinogen in animals.

Despite the ubiquity of phthalate contamination of environmental samples, there is a general lack of data for concentrations in the various environmental compartments. Concentrations are usually quoted for DEHP, which can reach $300\,\text{ng}\,\text{m}^{-3}$ in air, tens of ppm dry wt in sediments, but are usually low in water (up to $2.5\,\mu\text{g}\,\text{l}^{-1}$). DEHP has a low vapour pressure and aqueous solubility (Table 7.3), so it will evaporate and dissolve only very slowly. The most likely release to air and water is during the production of DEHP and disposal of plastics by incineration and landfill. The high concentrations found in waste water and sewage sludges in urban/industrial areas may derive from plasticized particulates in road dust. Adsorption on to particulates, followed by atmospheric transport and subsequent wash-out in rain, may be a significant route into the environment. Photodegradation occurs rapidly in the atmosphere, and a half-life of <2 days has been reported (DoE 1991). It is estimated that c.90% of available DEHP is readily adsorbed by organic material in soil (Wams 1987), from which it is unlikely to evaporate or leach. Biodegradation of DEHP occurs fairly rapidly under aerobic conditions but very slowly anaerobically (ECPI 1995).

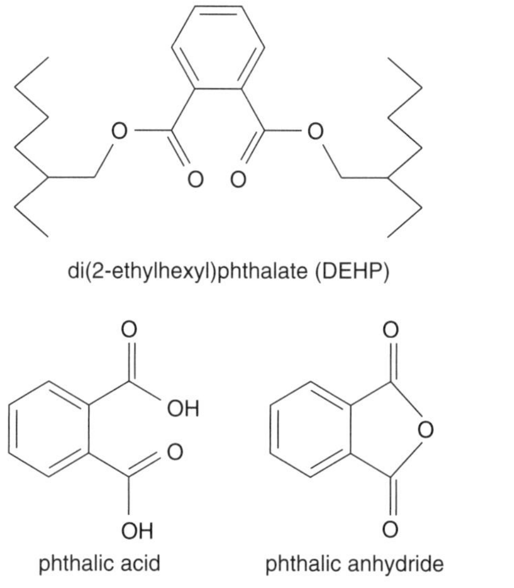

The high K_{ow} of DEHP (Table 7.3) suggests high lipophilicity and hence a potential for bioconcentration in aquatic organisms. BCFs of 10–500 in fish and aquatic invertebrates have been reported (Wofford et al. 1981). Organisms seem to be able to reduce their DEHP burden when placed in clean water (depuration), suggesting metabolic elimination (Mayer & Sanders 1973). Radiolabelled DEHP has been found to decrease to 37% after 2 days and to 17% after 7 days in a freshwater fish (*Lebistes reticulatus*), whereas c.95% remained in some invertebrates and plants (Metcalf et al. 1973). Degradation occurs via hydrolysis of the ester groups, yielding monoethylhexyl phthalate, phthalic acid, phthalic anhydride (Fig. 7.13) and a variety of polar metabolites and conjugates.

7.5.4 DDT and related compounds

The commercial insecticide **DDT** is a mixture of several isomers, the major one (75–80%) being p,p'-DDT (an abbreviation for p,p'-dichlorodiphenyltrichloroethane, which is more correctly named 1,1,1-trichloro-2,2-di(p-chlorophenyl)ethane; Fig. 7.14). It was widely used after the Second World War, and appeared to offer the ultimate means of controlling a variety of insects that were major crop pests or disease vectors. It was believed that excess DDT would be washed away, finally being diluted to negligible levels in the oceans. However, the harmful effects of the accumulation of the compound in vertebrates became apparent, and the use of DDT has virtually ceased in many countries, although it was still widely used in the UK as recently as the early 1980s (Newton et al. 1993) and continues to be used extensively in the tropics (Smith 1991). It is estimated that some 2.8 Mt had been manufactured worldwide by 1971 (Woodwell et al. 1971).

Fig. 7.13 DEHP and some of its important degradation products.

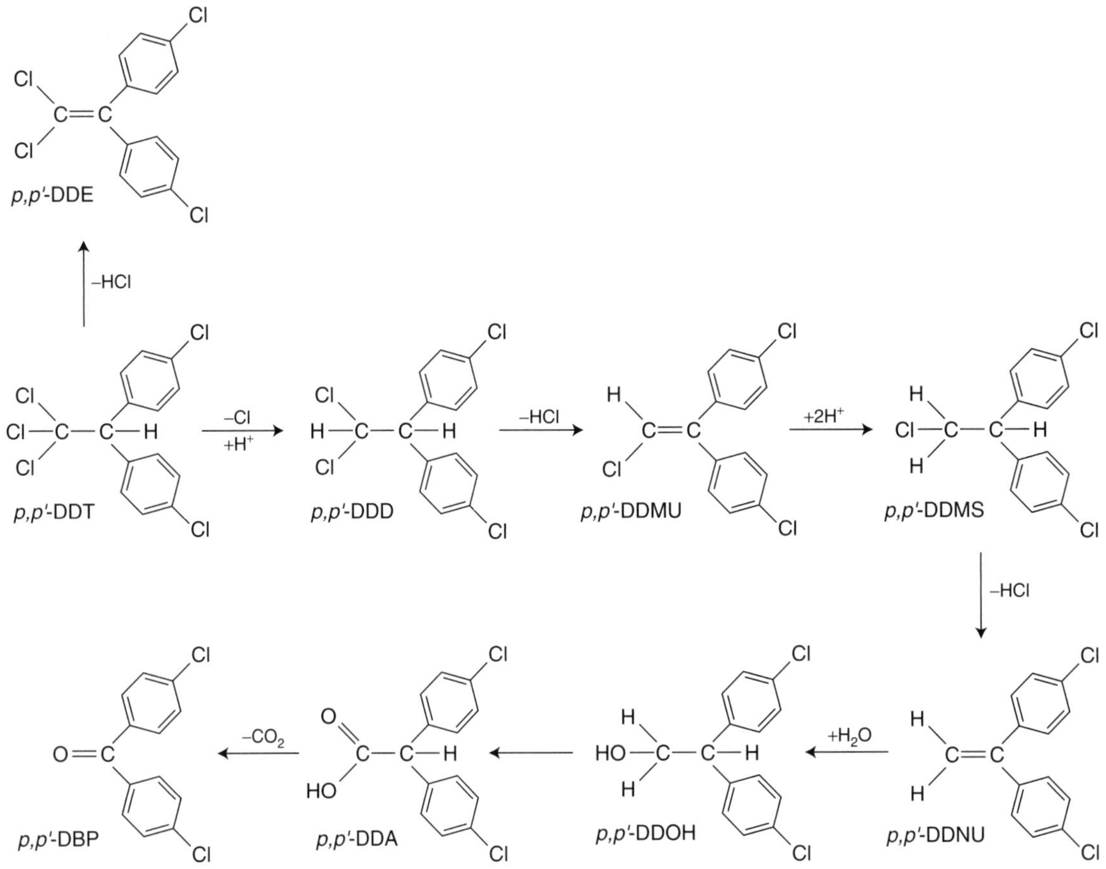

Fig. 7.14 Biodegradation of DDT by *Aerobacter aerogenes* (after Wedemeyer 1967).

DDT is hydrophobic (Table 7.3) and so its application by aqueous spraying is achieved by the use of a wettable powder formulation. DDT has a high $\log K_{ow}$ value (Table 7.3) and so is efficiently adsorbed by detritus and colloidal humic material (mainly by hydrogen bonding; see Section 2.1.2). It is concentrated in the tissue of detrital feeders and is passed up the food chain to the top consumers, generally large vertebrates, including humans, and is particularly persistent in fat. Adult vertebrates may not be killed directly by the compound, but their fecundity (ability to produce the next generation) may be drastically reduced. At each step in the food chain accumulation levels increase, and this is shown by the data in Table 7.4, which represent the effects of the careful, restricted use of DDT to control mosquitoes on Long Island, USA. However, deleterious effects are not confined to the higher trophic levels; the plankton can also be affected.

Table 7.4 Example of DDT concentration in an aquatic food chain (after Woodwell et al. 1967)

substrate	DDT residues* (ppm wet wt)
water	0.00005
plankton	0.04
silverside minnow	0.23
pickerel (predatory fish)	1.33
heron (feeds on small animals)	3.57
herring gull (scavenger)	6.00
merganser (fish-eating duck)	22.8
cormorant (feeds on larger fish)	26.4

*Total residues of DDT and related compounds DDE and DDD.

Measurements of DDT residues in soils and sediments often include the two related compounds DDD (1,1-dichloro-2,2-di(*p*-chlorophenyl)ethane; Fig. 7.14) and DDE (1,1-dichloro-2,2-di(*p*-chlorophenyl)ethylene; Fig. 7.14), which are degradation products of DDT. Where DDT is still in use, concentrations of up to 590 pg m^{-3} in air and up to 860 ng l^{-1} in seawater have been recorded (Iwata et al. 1993). Soil concentrations depend on past DDT usage, organic matter content, texture, nutrient status, pH, moisture and even temperature; all factors that affect the bioavailability and rate of biodegradation of DDT (Weber et al. 1993; Aislabie & Lloyd-Jones 1995).

The degradation of DDT to DDD and DDE is probably mainly microbially mediated, with reductive dechlorination being a major pathway, and favoured by anaerobic conditions (Fig. 7.14). Further degradation can lead to the conversion of the CCl$_3$ group of DDT into a COOH group, the product being called DDA (2,2-di(*p*-chlorophenyl)acetic acid; Fig. 7.14). DDA is polar and will dissolve in water, whereas DDD and DDE, like DDT, are practically insoluble (Table 7.3) and can persist for a long time in soils and sediments, due partly to their relative chemical stability.

p,p'-DDT and its main metabolite *p,p'*-DDE do not apparently bind to oestrogen receptors, although the latter does bind strongly with androgen receptors. In contrast, *o,p'*-DDT binds weakly to oestrogen receptors, but it comprises only 15–20% of technical DDT and is rapidly degraded in the environment (Bitman et al. 1968). DDD has been used in its own right as an insecticide but it is less toxic than DDT. The final product in Fig. 7.14, *p,p'*-dichlorodibenzophenone (*p,p'*-DBP), is also recalcitrant and has biological activity. So although DDT is biodegraded, it is not completely mineralized. Humic material is able to trap all these chloroaromatic compounds, but has the potential to release them as changes in pH and redox conditions affect the structural and chelating characteristics of the humic macromolecules (see Sections 4.2.2 and 7.6.5).

Other chlorinated pesticides, such as lindane (γ-hexachlorocyclohexane), have presented similar toxicological problems in the environment, and share the properties of accumulation in organisms and slow breakdown and elimination from tissue. There are also parallels among herbicides, such as 2,4,5-T (2,4,5-trichlorophenoxyacetic acid; Fig. 7.15), which is a defoliant and acts systemically. It persists for months in soil because the *meta* chloro group impedes hydroxylation and cleavage of the aromatic ring. Although its toxicity to animals is apparently low, the associated and extremely carcinogenic compound 2,3,7,8-tetrachlorodibenzo-*p*-dioxin (see Section 7.5.6) may be present, formed as an impurity during the manufacturing process. In contrast to 2,4,5-T, the related 2,4-D (Fig. 7.15) is biodegraded within days owing to the lack of the *meta* chloro group.

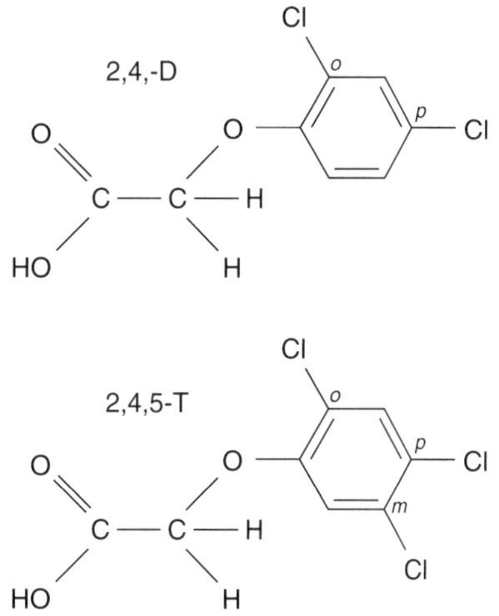

Fig. 7.15 Herbicides 2,4-D and 2,4,5-T (*o* = *ortho*, *m* = *meta*, *p* = *para* substitution positions).

7.5.5 Polychlorinated biphenyls

Polychlorinated biphenyls (**PCBs**) have a variety of industrial applications related to their high chemical and thermal stabilities, electrical resistance and low volatilities. Some of their uses are as dielectric fluids in capacitors and transformers, as lubricants and hydraulic fluids, as heat exchangers and fire retardants and as plasticizers. Commercial production began in the USA in 1929, but restrictions were imposed in the 1970s because of growing concerns about adverse biological effects, and the use of PCBs was banned in the USA in 1986. Some 37 kt were produced in 1970 (Goldberg 1976), and total production to 1980 has been estimated at 1.2 Mt, of which *c*.65% is still in use or has been deposited in landfills, *c*.4% has been destroyed and *c*.31% has been released to the environment (Tanabe 1988).

Manufacture involves the chlorination of the parent hydrocarbon, biphenyl, resulting in a mixture of individual PCBs (Fig. 7.16). For any degree of chlorination there are usually several isomers, or **congeners**, and there is a total of 209 possible congeners (Table 7.5). The properties of the congeners depend largely upon the degree of chlorination, so different commercial mixtures are manufactured by controlling the ratio of

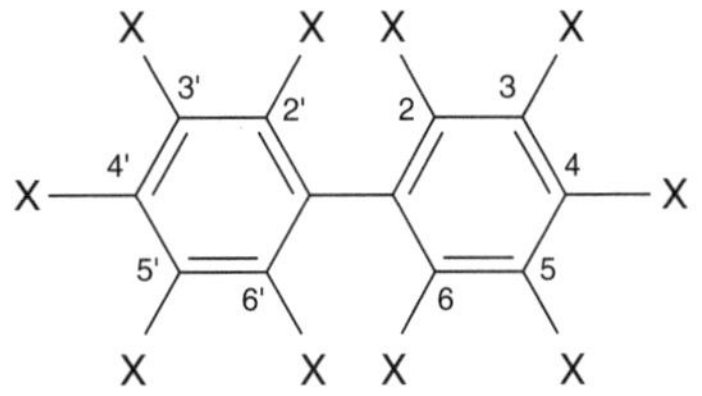

Fig. 7.16 PCB structure and numbering scheme (each X = H or Cl).

Table 7.5 Number of possible PCB congeners by degree of chlorination

PCB formula	number of congeners
$C_{12}H_9Cl$	3
$C_{12}H_8Cl_2$	12
$C_{12}H_7Cl_3$	24
$C_{12}H_6Cl_4$	42
$C_{12}H_5Cl_5$	46
$C_{12}H_4Cl_6$	42
$C_{12}H_3Cl_7$	24
$C_{12}H_2Cl_8$	12
$C_{12}HCl_9$	3
$C_{12}Cl_{10}$	1
total	209

chlorine to biphenyl. Examples of commercial PCB mixtures manufactured by Monsanto are Aroclor 1242, which comprises mainly di-, tri- and tetrachlorobiphenyls, and Aroclor 1254, which contains tetra-, penta-, hexa- and heptachlorobiphenyls (the last two digits give the wt % of Cl present).

PCBs are now widespread in the environment and can be detected in air, water, soils, sediments and organisms. They are persistent owing to their chemical inertness but bacterial degradation, at least of the less chlorinated congeners, does occur slowly. Typical environmental concentrations are 0.05–$5.0\,ng\,m^{-3}$ in air, 2–50 ppb by weight in surface soil, 0.01–$2.0\,ng\,l^{-1}$ in water, $c.10$ ppb by weight in vegetation and 5–50 ppm by weight in the blubber of marine mammals (1 ppb = 1 part in 10^9). In the more remote oceanic settings, atmospheric levels can reach $0.71\,ng\,m^{-3}$ and seawater concentrations $0.063\,ng\,l^{-1}$ (Iwata et al. 1993). Like DDT, PCBs have extremely low solubilities in water and they are lipophilic (they have high $\log K_{ow}$ values; Table 7.3), which results in progressive enrichment on passing along a food chain.

Significant accumulation can occur in the fatty tissues of animals, particularly marine mammals containing a high proportion of blubber. PCBs do not appear to be carcinogenic, but chronic exposure of organisms to low levels is probably important. Toxicity appears to be related to molecular structure and, in particular, the positions of chlorination: $3,3',4,4'$ seems especially toxic. A major difficulty in assessing the toxic effects of exposure is that higher levels of PCBs in tissue are invariably associated with higher levels of other chloroaromatics, such as DDT. Some congeners have oestrogenic properties, but at levels some million times lower than 17β-oestradiol. The oestrogenic activity seems to decrease as chlorination increases. Much greater oestrogenic activity is exhibited by some of the *p*-hydroxylated metabolites, but these compounds are more water soluble and so more easily excreted. Reproductive cycles appear to be affected by PCBs and there may also be immune system damage. For example, the reproductive failure in common seals feeding on fish from polluted coastal waters in the Wadden Sea (coastal Netherlands) has been attributed to PCBs (Reijnders 1986). A further problem is the passing on of PCBs in milk to the offspring of mammals.

Various studies have suggested that the phasing out of PCB production during the 1970s was accompanied by a decrease in environmental levels, in both substrates and organisms. However, levels in organisms appear to have remained fairly constant subsequently, suggesting that PCBs have remained biologically available. In contrast, levels of other organochlorine compounds, like DDT, continued to decline.

Individual PCB congeners exhibit different physical and chemical properties, such as the degree of solubility in water, volatility and susceptibility to bioaccumulation. The more highly chlorinated species tend to exhibit lower solubilities and volatilities. For example, the solubility of monochlorinated congeners is about 1 $mg\,l^{-1}$, whereas that of decachlorobiphenyl is around five orders of magnitude lower. In addition, the more highly chlorinated species are more resistant towards microbial degradation and are more lipophilic. However, the PCB distributions found in sediments often parallel those of particular commercial mixtures, suggesting that the congeners in a mixture tend to behave as a single phase. This can be explained by the strong association with colloidal organic (humic) material, which has the effect of reducing the availability of PCBs for microbial degradation.

In the past, landfill has been an important means for the disposal of PCBs, but degradation occurs only slowly and the partial volatility of PCBs at ambient temperatures leads to their remobilization and release into the atmosphere from both soil and water. Transfer into the air can occur by direct vaporization or by association of PCBs with aerosols. Redeposition of PCBs is brought about by wash-out in rain (**wet deposition**) or by fall-out of particulates (**dry deposition**). Atmospheric transport has resulted in the widespread occurrence of PCBs in surface sediments and soils (at ppb levels). In addition to waste burial, PCBs are most likely to be introduced into the environment by leaks or accidental spills, which is different from the mode of introduction of DDT. More recently, incineration has been used for disposal of PCBs, but unless temperatures are sufficiently high for efficient combustion even more toxic compounds are produced, collectively called dioxins, the most potent of which is TCDD (see Section 7.5.6).

(a) Aeolian transport

Aeolian transport is important for both DDT and PCBs. Although our overall knowledge of the pathways and fluxes of these substances into the marine environment is limited, detailed information has been obtained for the North American Basin, an area of c. 10^7km^2 centred on Bermuda in the North Atlantic, which has been studied since 1970. Estimates of the atmospheric inputs of DDT and PCBs to this area from the eastern USA in the early 1970s are $c.14$ and 140 t yr^{-1}, respectively (Bidleman et al. 1976). These values represent the sum of the estimates of three contributions to the air-to-sea transfer: direct transfer of the gaseous phase, wet deposition and dry deposition. Factors controlling each of these transfer processes are poorly understood, restricting the reliability of the final flux estimates, but to a first approximation it appears that the three air-to-sea transfer mechanisms make broadly equivalent contributions for both DDT and PCBs.

Data for PCB concentrations in the North American Basin in the 1970s (Harvey & Steinhauer 1976) reflect the physical processes operating in the water column and the association of PCBs with particulate organic matter. The isolation of the **surface mixed layer** (SML, $c.100$ m deep here) from the underlying water by the thermocline resulted in higher concentrations of PCBs in the SML due to the trapping of the aeolian input. Concentrations in the SML in 1972 were estimated at $c.16 \text{ ng l}^{-1}$, but by 1973 they had fallen to $c.1 \text{ ng l}^{-1}$, reflecting the reduction in industrial sources of PCBs from the eastern USA that occurred during the early 1970s. The rate of loss of PCBs from the SML during this period corresponded to a flux of c. $15 \text{ mg m}^{-2} \text{ yr}^{-1}$. This is at least two orders of magnitude greater than can be accounted for by adsorption on to sinking particulates (see Box 7.4). Similarly, advection and diffusion are not considered to make significant contributions over the time-scale involved, so another process must have been responsible for most of the PCB flux out of the SML.

PCB concentrations were found to be relatively low immediately above and below the air–sea interface, suggesting that rapid loss of PCBs from the surface microlayer of the sea was occurring. A likely cause was that a large proportion of the aeolian input of PCBs was entrained back into the atmosphere during periods when water evaporation was high (a characteristic of mid-latitudes), and subsequently transported to areas of greater precipitation/condensation at higher latitudes (see Fig. 3.12). This behaviour is consistent with the global distillation (or cold condensation) model (Simonich & Hites 1995), which is applicable to all semi-volatile compounds. It predicts that there would be higher losses of PCBs and compounds of similar volatility to the air from both soil and water in warm mid-latitudes, and that these compounds would be carried by the dominant air mass movements to cooler higher latitudes, where redeposition would occur. At higher latitudes the lower temperatures result in greater persistence of PCBs because volatilization and degradation processes are slower. Further support for this model has been obtained in the form of increasing concentrations of PCBs and other organochlorine compounds in Arctic biota in recent years (Wania & Mackay 1993), together with enhanced concentrations in high mountains, where low average temperatures mimic high-latitude environments (Blais et al. 1998).

PCBs are one of the most intensely studied groups of anthropogenic components, and this example shows how difficult it can be to predict the areas that will act as the final reservoirs for anthropogenic inputs without detailed knowledge of the environmental behaviour of the compounds in question.

7.5.6 Dioxins and related compounds

Polychlorodibenzo-*p*-dioxins (**PCDDs**) and polychlorodibenzofurans (**PCDFs**) are mostly produced by incineration, although they are also a waste product of the pulp/paper and chemical manufacturing industries (Duarte-Davidson et al. 1997). There are potentially 75 PCDD and 135 PCDF congeners (see Fig. 7.17), but PCDDs from most biological samples contain only the tetra- to octachloro congeners with 2,3,7,8 chlorination; the absence of the other congeners has been attributed to rapid metabolism and excretion (Ahlborg

> **Box 7.4** | **PCB fluxes in the North American Basin**
>
> Dissolved PCB concentrations in the Sargasso Sea, which forms part of the $10^7 km^2$ North American Basin, declined from $c.16\,ng\,l^{-1}$ in 1972 to $c.1\,ng\,l^{-1}$ in 1973 (Harvey et al. 1974b). This represents a loss of $c.15\,mg\,m^{-2}\,yr^{-1}$ from the surface mixed layer (SML), which is $c.100\,m$ deep across the basin:
>
> a volume of ocean $1\,m^2$ by $100\,m$ deep $= 100\,m^3 = 10^6$ litres
> and $15\,ng = 15 \times 10^{-6}\,mg$
>
> $$\text{so } 15\,ng\,l^{-1} = \frac{(15 \times 10^{-6}\,mg)}{(10^{-6}\,m^2)} = 15\,mg\,m^{-2}$$
>
> The bulk of the PCBs are associated with particulate matter, at a concentration factor of $c.10^6$ compared with levels of dissolved PCBs (Harvey & Steinhauer 1976). Consequently, the 1972 aqueous concentrations suggest a particulate concentration of $c.1.6\,mg\,kg^{-1}$, which is similar to that measured in phytoplankton (Harvey et al. 1974a):
>
> $$16\,ng\,l^{-1} = 16\,ng\,kg^{-1} \text{ in solution}$$
>
> which gives $(16) \times (10^6)\,ng\,kg^{-1} = 1.6\,mg\,kg^{-1}$ in particulates.
>
> To a first approximation, the sinking of particulate-associated PCBs was found to account for the levels of PCBs detected in the underlying sediments of the basin. Based on an estimated 270t of accumulated PCBs in the sediments, the fact that major production of PCBs started around 1955, and assuming the Sargasso Sea is representative of the basin as a whole, an average sedimentary accumulation rate of $c.1.6\,\mu g\,m^{-2}\,yr^{-1}$ is obtained:
>
> $$10^7\,km^2 = 10^{13}\,m^2$$
> $$270\,t = 2.7 \times 10^{14}\,\mu g$$
> accumulation time $= 1972 - 1955 = 17\,yr$
>
> $$\text{so accumulation rate} = \frac{(2.7 \times 10^{14})}{(10^{13}\,m^2) \times (17)} = 1.6\,\mu g\,m^{-2}\,yr^{-1}$$
>
> This value agrees quite well with a measured PCB deposition rate of $4.9\,\mu g\,m^{-2}\,yr^{-1}$ from sediment-trap experiments, which corresponds to a particulate deposition rate of $c.3.1\,g\,m^{-2}\,yr^{-1}$:
>
> $$\text{particulate deposition rate} = \frac{(4.9\,\mu g\,m^{-2}\,yr^{-1})}{(1.6\,mg\,kg^{-1})}$$
> $$= \frac{(4.9\,\mu g\,m^{-2}\,yr^{-1})}{(1.6\,\mu g\,g^{-1})} = 3.1\,g\,m^{-2}\,yr^{-1}$$
>
> Primary production of organic matter in the open ocean is $c.100\,g\,m^{-2}\,yr^{-1}$, but of this particulate matter $c.96\%$ is recycled within the SML (i.e. $c.4\,g\,m^{-2}\,yr^{-1}$ is lost from the SML), so the particulate loss rate from the SML of $3.1\,g\,m^{-2}\,yr^{-1}$ calculated above is reasonable. However, such a deposition rate would scavenge only $c.3.1\,\mu g\,m^{-2}\,yr^{-1}$ of PCBs from the SML. A likely maximum deposition rate for the basin, corresponding to the area under the Sargasso Sea, is $c.13\,g\,m^{-2}\,yr^{-1}$:
>
> accumulation rate $(0.5\,cm\,kyr^{-1}) \times$ area $(10^4\,cm^2)$
> $\times$ density $(2.6\,g\,cm^{-3})$
>
> This yields a PCB loss rate of $c.21\,\mu g\,m^{-2}\,yr^{-1}$:
>
> $$(13\,g\,m^{-2}\,yr^{-1}) \times (1.6\,\mu g\,g^{-1})$$
>
> which is over 100 times less than the total loss from the SML.

et al. 1992). Not surprisingly, then, attention has been focused on the congeners with chlorines at the 2,3,7,8 positions, which also seem to be the most toxic and carcinogenic to humans (Kaiser 2000). In addition, dioxins appear to have some anti-oestrogenic activity.

PCDD and PCDF concentrations in environmental samples are generally an order of magnitude lower than those of DDT and PCBs, although the same main repositories of sediments and soils are involved. Henry's Law constants are relatively high (Table 7.3) and, together with the main source, incineration, suggest that aeolian transport is an important route into the environment for these compounds. The annual flux into the environment is very small ($c.15\,kg$ in the USA according to EPA figures), but the activity of some congeners is extremely high (e.g. the LD_{50}—the median lethal dose—for the major tetrachloro congener 2,3,7,8-TCDD in guinea pigs is 0.6 ppb by weight). The normal human exposure route is via food, notably meat, milk and fish.

As for the PCBs, the lower chlorinated PCDDs and PCDFs exhibit higher aqueous solubilities and vapour

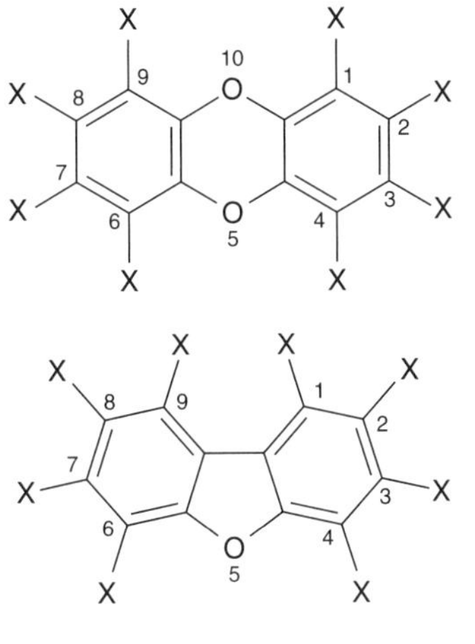

Fig. 7.17 Structures and numbering schemes of PCDDs and PCDFs.

pressures than the higher congeners, so they tend to dominate in air, whereas the higher congeners are more abundant in biotic samples, particularly at higher trophic levels (Ahlborg et al. 1992). 2,3,7,8-TCDD shares common toxic properties with PCBs (Ahlborg et al. 1994), and is similarly resistant towards degradation, has a long residence time in the environment and is subject to long-range transport and global distribution. *Dehalococcoides* spp. of bacteria can perform reductive dechlorination of PCDDs, although very slowly for the 2,3,7,8-tetrachlorinated congener (Fig. 7.18; Bunge et al. 2003).

7.6 Factors affecting the fate of anthropogenic components

7.6.1 Environmental fate of chloroaromatics

We can get an idea of the important factors controlling the environmental behaviour of anthropogenic compounds by taking a general look at the chloroaromatics discussed in the preceding sections. Their fate is governed by their low reactivity (persistence), volatility (potential for vapour-phase transport) and low solubility in water but high lipophilicity. Among PCBs, PCDDs and PCDFs, the lower chlorinated congeners have the higher vapour pressures and aqueous solubilities, but lower $\log K_{ow}$ values, and so may be more prone to atmospheric degradation. Chloroaromatics appear slow to photodegrade, in view of their global atmospheric distribution. They are readily adsorbed by sediments and soils, which provide long-term sinks and potential sources of bioavailability. The estimated half-life in soils is >10 years. Invertebrates redistribute the compounds within the upper layers of soil, but uptake by plants is inefficient, possibly reflecting the strong adsorption on soil particles.

Concentrations of chloroaromatics are usually higher in aquatic than terrestrial animals (de Voogt 1996). In fish, most of the variation in concentrations of DDT and PCBs can be explained by variations in lipid content, trophic level and trophic structure (Rowan & Rasmussen 1992), with environmental levels (in water and soil) playing a subordinate role. The ocean serves as a sink for chloroaromatics.

Bioconcentration factors are higher for fish than their invertebrate prey, by almost an order of magnitude for some Aroclor mixtures (LeBlanc 1995; Box 7.3). Biomagnification factors over several marine trophic levels are usually an order of magnitude greater for PCBs than PCDDs and PCDFs (Niimi 1996; Box 7.3). The pattern of PCBs observed in a particular organism may be controlled by metabolic processes, rather than dietary factors such as the type of prey consumed (Leonards et al. 1994).

7.6.2 General considerations

In the above sections a number of important factors that influence the environmental impact of anthropogenic substances have been examined. These include the sources, pathways and associated fluxes of inputs into the environment, the residence times and reactions of individual components and the influence of possible associations with particular types of sedimentary material on transport, biological uptake and long-term sedimentary fate. All these factors, in addition to toxicity, require consideration when assessing the environmental impact of an anthropogenic component, but often there is insufficient information about a number of them. This limits our ability to predict the effects of anthropogenic inputs, which weakens the case for the introduction of effective legislation to control such inputs at national and international scales. When an anthropogenic compound also has a natural source (i.e. is not xenobiotic), there can be considerable problems in determining both the level and the environmental impact of the anthropogenic input relative to the natural background level at a particular location.

It is worth noting that humans are not the only organisms to affect ecosystems by the input of potentially toxic materials. Phytoplanktonic succession is affected by various chemicals exuded by different species. Some

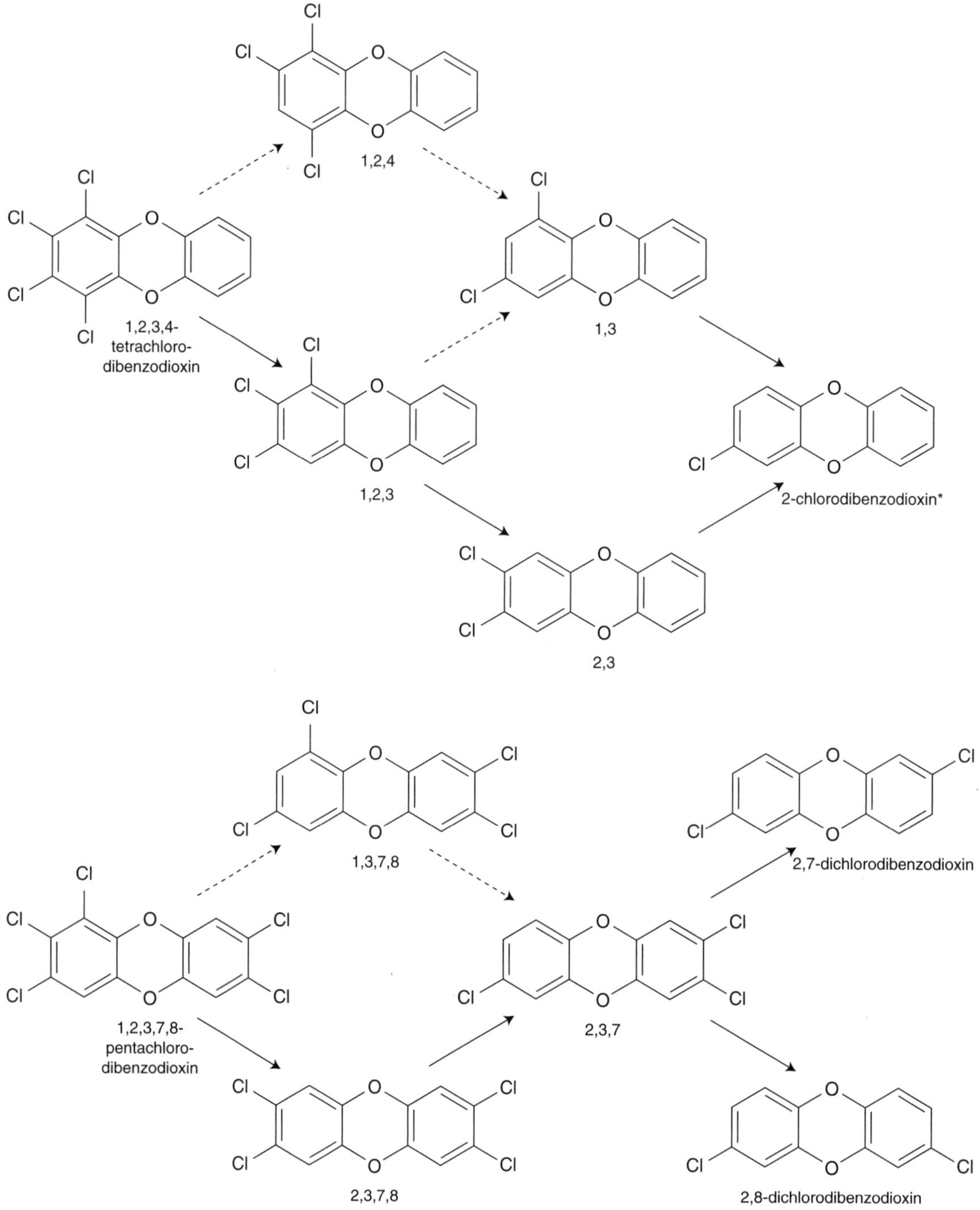

Fig. 7.18 Reductive dechlorination of PCDDs by anaerobic bacteria (broken arrows show minor routes; after Bunge et al. 2003). (*By convention, substituent positions bear the lowest possible numbers, so the loss of the Cl atom from C-1 of 1,3-dichlorodibenzodioxin yields 2- rather than 3-chlorodibenzodioxin, because the structure can be rotated to bring the chlorine atom into the C-2 position, with reference to the numbering scheme in Fig. 7.17.)

of these chemicals have been found to be essential for the growth of the next group of organisms in the succession, while others may suppress the growth of competing organisms or inhibit predation. These compounds are generally called **allelochemicals**. Bromophenols are an example of toxins with bactericidal properties, and they also offer some protection against predation. They are found in marine algae and invertebrates, particularly annelids, phoronids and hemichordates. Phlorotannins, which are biosynthesized by phytoplanktonic algae as well as macroalgae, are important in chelating metal ions from seawater (Ragan et al. 1979) and in deterring herbivore grazing (Tugwell & Branch 1992).

7.6.3 Bioavailability

Bioavailability is a major control of the extent to which a contaminant builds up in the tissues of an organism, and is determined by factors such as solubility and adsorption on to soil/sediment. Aqueous solubility is influenced by the shape, size and functional group content of a chemical (Aislabie & Lloyd-Jones 1995). Whether sorption (adsorption or absorption) is likely to increase or decrease bioavailability and degradation rates in soils is difficult to predict (Harms & Zehnder 1995). The degree and position of chlorination affects the bioavailabilities of chloroaromatic compounds. For example, the higher chlorinated PCBs, PCDDs and PCDFs tend to exhibit lower bioaccumulation rates than their lower chlorinated counterparts in the presence of sediment and soil, apparently reflecting variations in adsorption characteristics of the chloroaromatics (Larsen et al. 1992; Loonen et al. 1994).

Soil properties (**edaphic factors**) are important in determining bioavailability and include nutrient status, organic-matter content, soil structure, temperature, moisture and pH (Weber et al. 1993). Ageing is another important edaphic factor, because it appears that the amount of extractable contaminant decreases over time. For example, 8–14% of ^{14}C-labelled 1,3,6,8-TCDD was found to become unextractable in soil after about a year in one study (Muir et al. 1985). It seems that a contaminant will diffuse into kinetically remote compartments of soil over time. It is possible, therefore, to distinguish three contaminant fractions: a labile, relatively bioavailable fraction; a recalcitrant, relatively firmly bound but still extractable fraction; and a non-extractable, irreversibly bound fraction (Jones et al. 1996).

The extent to which sediment-associated contaminants can move through aquatic food webs is not well known (Suedel et al. 1994), although the trophic transfer potential is an important factor to understand if the environmental impact of a contaminant is to be assessed. It has been argued that, as an organism grows, chemical compartmentalization within it increases, so the storage capacity for lipophilic chemicals rises at the expense of elimination efficiency (LeBlanc 1995). This can give the impression of biomagnification (see Box 7.3) rather than reflecting the real cause of differences in contaminant concentrations (e.g. species-related differences in bioconcentration). The generally higher concentrations of contaminants in organisms at higher trophic levels may be at least partly attributable to higher lipid content and reduced efficiency of chemical elimination. A slight (< twofold) biomagnification between adjacent trophic levels has been suggested for compounds with BCFs >114 000 (or $\log K_{ow}$ >6.3); for example, DDT and some PCBs and PCDDs (LeBlanc 1995).

Top predators are particularly prone to the effects of bioaccumulation through the food chain. Bioaccumulation may be important for endocrine disrupters, which may be released from fat stores during critical periods, such as starvation, egg production, early pregnancy and lactation. It should be remembered that bioaccumulation is affected by metabolism and other elimination pathways. For example, DDT can be metabolized to DDE, which is even more persistent, but it can alternatively be converted to DDD, which is more rapidly eliminated.

7.6.4 Mixtures and interactions

Most toxicological research has been conducted on single compounds, but it must also be considered whether the mixtures found in the environment behave as individual components or interact in some way. To illustrate the potential effects of mixtures, we can consider endocrine disruption. Mixtures of contaminants potentially can have complementary, additive (agonistic) effects, or counteractive (antagonistic) effects. For example, complex mixtures of chloroaromatics can give rise to antagonistic effects (de Voogt 1996).

There is also the possibility of amplifying (**synergistic**) effects, where the total activity is greater than that of the sum of the component parts. Synergism is well known in the pharmaceutical industry, and it has also been suggested for anthropogenic mixtures in the environment, where the compounds share the same mode of action, such as PCB congeners (Ahlborg et al. 1992, 1994) and 17β-oestradiol plus oestrone (Environment Agency 1996). It is possible that synergism may exist between compounds with different modes of action, such that one compound might inhibit the detoxifying enzyme that would remove another contaminant, or alternatively it might activate an enzyme that increases the toxicity of another contaminant (Walker 1990). For example, relatively high levels of PCBs in sea birds are associated with abundance of a particular form of cytochrome P450 (Borlakoglu et al. 1988), which may make the birds particularly vulnerable to certain aromatic hydrocarbons in oil spills by enhancing the effects of carcinogenic metabolites (Walker & Johnston 1989). Recent developments in biochemistry and molecular biology have led to the development of assay systems that allow the effects of a range of pollutants to be investigated under field conditions (Walker 1990). However, a number of environmental assay studies have failed to yield evidence for synergism in endocrine disruption (Ashby et al. 1997; Ramamoorthy et al. 1997), so the concept remains controversial for the time being.

Some xenobiotic chemicals can affect oestrogen metabolism. Two potential pathways for the metabolism of 17β-oestradiol lead to 2-hydroxyoestrone or 16α-hydroxyoestrone; the latter is a potent oestrogen as well as being genotoxic. PAHs like benzo[a]pyrene inhibit the pathway to 2-hydroxyoestrone. Anti-oestrogenic activity is shown by 2,3,7,8-TCDD and related chloroaromatics, various combustion PAHs and some compounds in cruciferous vegetables (e.g. indole-3-carbinol). However, depending upon their concentrations, some compounds appear to have the ability to be weakly oestrogenic or anti-oestrogenic (e.g. bioflavonoids), so the overall endocrine activity of a mixture of xenobiotic chemicals can be extremely difficult to predict. Table 7.6 shows the results of a study of the estimated daily exposure of humans in terms of oestrogen equivalents (EQ) and TCDD anti-oestrogen equivalents (TEQ, where 1 TEQ is approximately equivalent to 1 EQ). The conclusion is that the major human intake of endocrine disrupters associated with oestrogen-induced response pathways derives from naturally occurring oestrogens in foods, with the exception of hormone administration.

7.6.5 Humic substances and pollutants

Humic and fulvic acids play an important role in the complexation (or chelation) and release of metals. For example, lowering the pH to a value of $c.2$ can release most of the humic-bound iron, and the binding power of Fe(II) ions with humic substances appears to be greater than that of Fe(III) ions. Carboxyl groups appear particularly important in metal complexation and, in general, retention of metallic cations by humic acids and brown coals is favoured by an increase in pH (Stevenson 1976). However, the interaction of humic material with metals is complex and for the series of metals Hg(II), Fe(III), Pb, Cu, Al, Ni, Cr(III), Cd, Zn, Co and Mn it has been found that Hg and Fe are adsorbed most efficiently, and Co and Mn least (Kerndorf & Schnitzer 1980). Within this series, the order of the relative stabilities of metal complexes again depends on pH. For example, Cu(II) complexes are more stable than those of Ni(II) at pH 3, but less stable at pH 5. These laboratory studies have been found to parallel the behaviour in natural aquatic systems, which result in metals such as Cu, Pb, Ni and Cr being enriched in humic substances in the water column compared with the underlying sediment. There may be competition between metals for active sites on humic material and so the concentration of humic material and individual metal ions is an important factor.

In addition to physically trapping organic compounds within their macromolecular structure (see Section 4.2.2), humic substances in soils and aquatic

Table 7.6 Estimated daily human exposure to oestrogenic/antioestrogenic agents (after Safe 1995)

agent	oestrogen equivalents (EQ g d^{-1})	TCDD anti-oestrogen equivalents (TEQ g d^{-1})
dietary bioflavonoids	102×10^{-6}	–
oestrogenic pesticides	2.5×10^{-12}	–
post-menopausal therapy	3.35×10^{-3}	–
birth-control pill	16.7×10^{-3}	–
morning-after pill	333×10^{-3}	–
PAHs in food	–	$1.2–5.0 \times 10^{-9}$
indole-3-carbinol*	–	$0.25–1.28 \times 10^{-9}$

*Represents active derivative of indole-3-carbinol in Brussels sprouts.

environments can form chemical bonds with various organic pollutants and their metabolites (e.g. Senesi et al. 1987; Richnow et al. 1994). The potential of humic substances to form water-soluble complexes with toxic metals and organics is an important pollution problem, because it affects the bioavailability of these toxicants and potentially increases their transportation range. Natural environmental changes (e.g. a fall in pH) can cause the release of previously complexed toxicants as the structure of humic material becomes more open (see Section 4.2.2). Similarly, chlorination of humic-rich potable waters can release or even form additional toxic compounds (e.g. chloroform). One solution is to remove the humic substances by granular activated charcoal prior to water disinfection, but this is expensive. It is possible that the addition of humic acids could be useful in removing some pollutants, upon subsequent precipitation, during waste-water treatment.

Appendix 1
SI units used in this book

parameter	SI unit	abbreviation	comment
amount of substance	mole	mol	A number, equal to the Avogadro constant ($L = 6.022 \times 10^{23}$). Chemical concentrations are often recorded as mol per litre ($1\,l = 1\,dm^3$).
energy (e.g. heat)	joule	J	Chemical energies are often measured in kilocalories (1 cal = 4.2 J). ($1\,J = 1\,N\,m = 1\,kg\,m^2\,s^{-2}$)
length	metre	m	
mass	kilogram	kg	For geological masses, the metric tonne ($1\,t = 1000\,kg = 1\,Mg$) is often more convenient.*
pressure	pascal	Pa	Mean atmospheric pressure = $1.013 \times 10^5\,Pa$, which is equivalent to 1 atmosphere (atm), 1 torr or 1.013 bar. ($1\,Pa = 1\,N\,m^{-2}$ or $1\,kg\,m^{-1}\,s^{-2}$)
temperature	degrees Celsius	°C	Thermodynamic temperature uses the absolute scale of kelvin (K). Increments are identical in the two scales, but 0°C = 273.15 K.
time	second	s	For geological time-scales, the year is more convenient and is abbreviated to yr or, when describing age, a (for annum).† (1 year = $31.56 \times 10^6\,s$)
work	watt	W	($1\,W = 1\,J\,s^{-1}$)

*The term weight is used frequently in the literature (e.g. dry wt, molecular weight), and for ease of comparison it is used throughout this text. However, more correctly the term mass should be applied (e.g. dry mass, relative molecular mass). †For example, the Proterozoic began at 2.5 Ga or 2.5 Gyr ago.

Appendix 2
SI unit prefixes

order	prefix	symbol
10^{-18}	atto	a
10^{-15}	femto	f
10^{-12}	pico	p
10^{-9}	nano	n
10^{-6}	micro	μ
10^{-3}	milli	m
10^{-2}	centi	c
10^{-1}	deci	d
10	deca	da
10^{2}	hecto	h
10^{3}	kilo	k
10^{6}	mega	M
10^{9}	giga	G
10^{12}	tera	T
10^{15}	peta	P
10^{18}	exa	E

Appendix 3
Geological time scale

Note: Epochs can be further subdivided, e.g. into Ages.

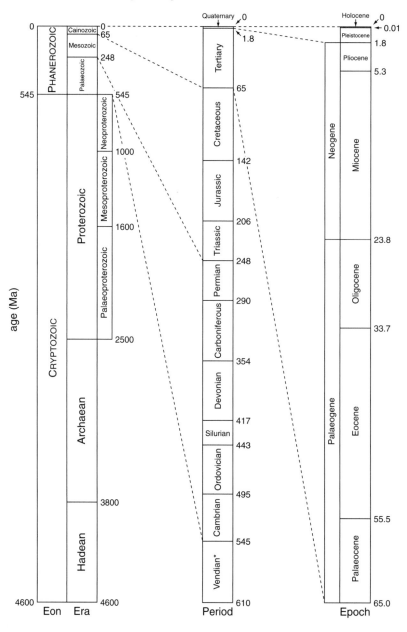

★ The International Commission on Stratigraphy has defined the Neoproterozoic periods as: Tonian (1000–850 Ma); Cryogenian (850–600 Ma); Ediacaran (600–540 Ma). The Cryogenian incorporates the 'Snowball Earth' glaciations and the Ediacaran replaces the Vendian (the latter term is adopted in this book because of its common usage in the literature).

References

Abbott G.D., Lewis C.A., Maxwell J.R. (1985) Laboratory models for aromatization and isomerization of hydrocarbons in sedimentary basins. *Nature 318*, 651–3.

Abbott G.D., Wang G.Y., Eglinton T.I., Home A.K., Petch G.S. (1990) The kinetics of sterane biological marker release and degradation processes during the hydrous pyrolysis of vitrinite kerogen. *Geochim. Cosmochim. Acta 54*, 2451–61.

Abbott G.D., Bashir F.Z., Sugden M.A. (2001) Kerogen-bound and free hopanoic acids in the Messel oil shale kerogen. *Chirality 13*, 510–16.

Adler E. (1977) Lignin chemistry—past, present and future. *Wood Sci. Technol. 11*, 69–218.

Aharon P., Chappell J. (1986) Oxygen isotopes, sea level changes and the temperature history of a coral reef environment in New Guinea over the last 10^5 years. *Palaeogeogr., Palaeoclim., Palaeoecol. 56*, 337–9.

Aherne G.W., Briggs R. (1989) The relevance of the presence of certain synthetic steroids in the aquatic environment. *J. Pharm. Pharmacol. 41*, 735–6.

Ahlborg U.G., Brouwer A., Fingerhut M.A., Jacobson J.L., Jacobson S.W. & 9 others (1992) Impact of polychlorinated dibenzo-*p*-dioxins, dibenzofurans and biphenyls on human and environmental health, with special emphasis on application of the toxic equivalence factor concept. *Eur. J. Pharmacol. Environ. Toxicol. 228*, 179–99.

Ahlborg U.G., Becking G.C., Birnbaum L.S., Brouwer A., Derks H.J.G.M. & 10 others (1994) Toxic equivalence factors for dioxin-like PCBs. Rep. on a WHO-ECEH & IPCS consultation, Dec. 1993. *Chemosphere 28*, 1049–67.

Aislabie J., Lloyd-Jones G. (1995) A review of bacterial degradation of pesticides. *Aust. J. Soil Res. 33*, 925–42.

Albaigés J. (1980) Identification and geochemical significance of long chain acyclic isoprenoid hydrocarbons in crude oils. In *Advances in Organic Geochemistry 1979* (ed. Douglas A.G., Maxwell J.R.), 19–28. Oxford: Pergamon.

Aldhous P. (1990) Taxation or regulation? *Nature 347*, 412.

Alexander R., Kagi R.I., Noble R., Volkman J.K. (1984a) Identification of some bicyclic alkanes in petroleum. *Org. Geochem. 6*, 63–70.

Alexander R., Kagi R., Sheppard P. (1984b) 1,8-Dimethylnaphthalene as an indicator of petroleum maturity. *Nature 308*, 442–3.

Alexander G., Hazai I., Grimalt J., Albaigés J. (1987) Occurrence and transformation of phyllocladanes in brown coals from Nograd Basin, Hungary. *Geochim. Cosmochim. Acta 51*, 2065–73.

Alexander R., Kralert P.G., Kagi R.I. (1992) Kinetics and mechanism of the thermal decomposition of esters in sediments. *Org. Geochem. 19*, 133–40.

Allard B., Templier J., Largeau C. (1997) Artifactual origin of mycobacterial bacteran. Formation of melanoidin-like artifact macromolecular material during the usual isolation process. *Org. Geochem. 26*, 691–703.

Allard B., Rager M.-N., Templier J. (2002) Occurrence of high molecular weight lipids (C_{80+}) in the trilaminar outer cell walls of some freshwater microalgae. A reappraisal of algaenan structure. *Org. Geochem. 33*, 789–801.

Allègre C.L., Schneider S.H. (1994) The evolution of the Earth. *Sci. Am. 271 Oct*, 44–51.

Allen P.A., Allen J.R. (1990) *Basin Analysis—Principles and Applications*. Oxford: Blackwell Scientific.

Allen P.A., Collinson J.D. (1986) Lakes. In *Sedimentary Environments and Facies* (ed. Reading H.G.), 63–94. Oxford: Blackwell Scientific.

Aller R.C. (1998) Mobile deltaic and continental shelf muds as suboxic, fluidized bed reactors. *Mar. Chem. 61*, 143–55.

Alvarez L.W., Alvarez W., Asaro F., Michel H.V. (1980) Extraterrestrial cause for the Cretaceous–Tertiary extinction. *Science 208*, 1095–108.

Aplin R.T., Cambie R.C., Rutledge P.S. (1963) The taxonomic distribution of some diterpene hydrocarbons. *Phytochem. 2*, 205–14.

Aquino Neto F.R., Trendel J.M., Restle A., Connan J., Albrecht P.A. (1983) Occurrence and formation of tricyclic and tetracyclic terpanes in sediments and petroleums. In *Advances in Organic Geochemistry 1981* (ed. Bjorøy M. et al.), 659–67. Chichester: Wiley.

Archer D., Maier-Reimer E. (1994) Effect of deep-sea sedimentary calcite preservation on atmospheric CO_2 concentrations. *Nature 367*, 260–3.

Archer B.L., Audley B.G., Cockbain E.G., McSweeney G.P. (1963) The biosynthesis of rubber, of mevalonate and isopentenyl pyrophosphate into rubber, by *Hevea brasiliensis*-latex fractions. *Biochem. J. 89*, 565–85.

Armstrong P.A., Chapman D.S. (1998) Beyond surface heat flow: an example from a tectonically active sedimentary basin. *Geology 26*, 183–6.

Armstrong R.A., Lee C., Hedges J.I., Honjo S., Wakeham S.G. (2002) A new, mechanistic model for organic carbon fluxes in the ocean based on the quantitative association of POC with ballast minerals. *Deep-Sea Res. II 49*, 219–36.

Arthur M.A. (2000) Volcanic contributions to the carbon and sulfur geochemical cycles and global change. In *Encyclopedia of Volcanoes* (ed. Houghton B.F., McNutt S.R., Rymer H., Stix J.), 1045–56. San Diego: Academic Press.

Arthur M.A., Schlanger S.O., Jenkyns H.C. (1987) The Cenomanian-Turonian Oceanic Anoxic Event, II. Palaeoceanographic controls on organic-matter production and preservation. In *Marine Petroleum Source Rocks* (ed. Brooks J., Fleet A.J.) *Geol. Soc. Spec. Publn 26*, 401–20. London: Geological Society.

Ash R.D., Knott S.F., Turner G. (1996) A 4-Gyr shock age for a martian meteorite and implications for the cratering history of Mars. *Nature 380*, 57–9.

Ashby J., Lefevre P.A., Odum J., Harris C.A., Routledge E.J., Sumpter J.P. (1997) Synergy between synthetic oestrogens? *Nature 385*, 494.

Atlas R.M. (1981) Microbial degradation of petroleum hydrocarbons: an environmental perspective. *Microbiol. Revs. 45*, 180–209.

Atlas R.M., Bartha R. (1998) *Microbial Ecology*. Menlo Park, CA: Addison Wesley Longman.

Ayers G.P., Gras J.L. (1991) Seasonal relationship between cloud condensation nuclei and aerosol methanesulphonate in marine air. *Nature 353*, 834–5.

Bacastow R.B., Keeling C.D., Whorf T.P. (1985) Seasonal amplitude increase in atmospheric CO_2 concentration at Mauna Loa, Hawaii, 1959–1982. *J. Geophys. Res. 90*, 10529–40.

Bada J.L., Mann E.H. (1980) Amino acid diagenesis in DSDP cores: kinetics and mechanisms of some reactions and their applications in geochronology and in paleotemperature and heat flow determinations. *Earth Sci. Rev. 16*, 21–57.

Bada J.L., Schroeder R.A. (1972) Racemization of isoleucine in calcareous marine sediments. *Earth Planet. Sci. Letts 15*, 1–11.

Bada J.L., Schroeder R.A. (1975) Amino acid racemization reactions and their geochemical implications. *Naturewissenschaften 62*, 71–9.

Bailey J., Chrysostomou A., Hough J.H., Gledhill T.M., McCall A., Clark S., Ménard F., Tamura M. (1998) Circular polarization in star-formation regions: implications for biomolecular homochirality. *Science 281*, 672–4.

Bains S., Corfield R.M., Norris R.D. (1999) Mechanisms of climate change at the end of the Paleocene. *Science 285*, 724–7.

Baker A.J., Fallick A.E. (1989) Evidence from Lewisian limestones for isotopically heavy carbon in two-thousand-million-year-old water. *Nature 337*, 352–4.

Baker E.W., Louda J.W. (1986) Porphyrins in the geological record. In *Biological Markers in the Sedimentary Record* (ed. Johns R.B.), 125–225. Amsterdam: Elsevier.

Ballantine J.A., Lavis A., Morris R.J. (1979) Sterols of the phytoplankton—effects of illumination and growth stage. *Phytochem. 18*, 1459–66.

Barbeau K., Rue E.L., Bruland K.W., Butler A. (2001) Photochemical cycling of iron in the surface ocean mediated by microbial iron(III)-binding ligands. *Nature 413*, 409–13.

Barker C. (1996) *Thermal Modelling of Petroleum Generation: Theory and Applications*. Amsterdam: Elsevier.

Barnes C.R. (1999) Paleoceanography and paleoclimatology: an Earth system perspective. *Chem. Geol. 161*, 17–35.

Barnes R.S.K., Hughes R.N. (1988) *An Introduction to Marine Ecology*. London: Blackwell Scientific.

Barnett T.P., Pierce D.W., Schnur R. (2001) Detection of anthropogenic climate change in the world's oceans. *Science 292*, 270–4.

Barnola J.M., Anklin M., Porcheron J., Raynaud D., Schwander J., Stauffer B. (1995) CO_2 evolution during the last millennium as recorded by Antarctic and Greenland ice. *Tellus Ser. B 47*, 264–72.

Barrera E. (1994) Global environmental changes preceding the Cretaceous–Tertiary boundary: early–late Maastrichtian transition. *Geology 22*, 877–80.

Barron E.J., Peterson W.H. (1989) Model simulations of the Cretaceous ocean circulation. *Science 244*, 684–6.

Barron E.J., Sloan J.L., Harrison C.G.A. (1980) Potential significance of land–sea distributions and surface-albedo variations as a climatic forcing factor. *Palaeogeogr. Palaeoclim. Palaeoecol. 30*, 17–40.

Barth T., Borgund A.E., Hopland A.L., Graue A. (1988) Volatile organic acids produced during kerogen maturation—amounts, composition and role in migration of oil. *Org. Geochem. 13*, 461–5.

Barton D., Nakanishi K. (eds) (1999) *Comprehensive Natural Product Chemistry*. Oxford: Pergamon.

Barwise A.J.G., Roberts I. (1984) Diagenetic and catagenetic pathways for porphyrins in sediments. *Org. Geochem. 6*, 167–76.

Bates A.L., Hatcher P.G. (1989) Solid-state ^{13}C NMR studies of a large fossil gymnosperm from the Yallourn Open Cut, Latrobe Valley, Australia. *Org. Geochem. 14*, 609–17.

Beaumont C., Boutilier R., Mackenzie A.S., Rullkötter J. (1985) Isomerization and aromatization of hydrocarbons and the paleothermometry and burial history of Alberta Foreland Basin. *Am. Assoc. Pet. Geol. Bull. 69*, 546–66.

Becker L., Poreda R.J., Junt A.G., Bunch T.E., Rampino M. (2001) Impact event at the Permian–Triassic Boundary: evidence from extra-terrestrial noble gasses in fullerenes. *Science 291*, 1530–3.

Beerling D.J., Lake J.A., Berner R.A., Hickey L.J., Taylor D.W., Roger D.L. (2002) Carbon isotope evidence implying high O_2/CO_2 ratios in the Permo-Carboniferous atmosphere. *Geochim. Cosmochim. Acta 66*, 3757–67.

Begon M., Mortimer M. (1986) *Population Ecology*. Oxford: Blackwell Scientific.

Béhar F., Vandenbroucke M. (1987) Chemical modelling of kerogen. *Org. Geochem. 11*, 15–24.

Behrenfeld M.J., Bale A.J., Kolber Z.S., Aiken J., Falkowski P.G. (1996) Confirmation of iron limitation of phytoplankton photosynthesis in the equatorial Pacific Ocean. *Nature* 383, 508–11.

Belt S.T., MasséG., Allard W.G., Robert J.-M., Rowland S.J. (2001) C_{25} highly branched isoprenoid alkenes in planktonic diatoms of the *Pleurosigma* genus. *Org. Geochem.* 32, 1271–5.

Bend S.L. (1992) The origin, formation and petrographic composition of coal. *Fuel* 71, 851–70.

Benner R., Fogel M.L., Sprague E.K., Hodson R.E. (1987) Depletion of ^{13}C in lignin and its implications for stable carbon isotope studies. *Nature* 329, 708–10.

Bennett K.D. (1983) Devensian late-glacial and Flandrian vegetational history at Hockham Mere, Norfolk, England. I. Pollen percentages and concentrations. *New Phytologist* 95, 457–87.

Benton M.J. (1995) Diversification and extinction in the history of life. *Science* 268, 52–8.

Bergamasco R., Horn D.H.S. (1983) Distribution and role of insect hormones in plants. In *Endocrinology of Insects* (ed. Downer R.G.H., Laufer H.), 627–54. New York: Alan R. Liss.

Berger A., Imbrie J., Hays J., Kukla G., Saltzman B. (eds) (1984) *Milankovitch and Climate*. Boston: Reidel.

Berner R.A. (1984) Sedimentary pyrite formation: an update. *Geochim. Cosmochim. Acta* 48, 605–15.

Berner R.A. (1989) Biogeochemical cycles of carbon and sulfur and their effect on atmospheric oxygen over Phanerozoic time. *Palaeogeogr. Palaeoclimatol. Palaeoecol.* 73, 97–122.

Berner R.A. (1990) Atmospheric carbon dioxide levels over Phanerozoic time. *Science* 249, 1382–6.

Berner R.A. (1991) A model for atmospheric CO_2 over Phanerozoic time. *Am. J. Sci.* 291, 339–76.

Berner R.A. (1992) Weathering, plants, and the long-term carbon cycle. *Geochim. Cosmochim. Acta* 56, 3225–31.

Berner R.A. (1998) The carbon cycle and CO_2 over Phanerozoic time: the role of land plants. *Roy. Soc. Phil. Trans. B* 353, 75–82.

Berner R.A. (1999) A new look at the long-term carbon cycle. *GSA Today* 9(11), 1–6.

Berner R.A. (2001) Modeling atmospheric O_2 over Phanerozoic time. *Geochim. Cosmochim. Acta* 65, 685–94.

Berner R.A. (2002) Examination of hypotheses for the Permo-Triassic boundary extinction by carbon cycle modelling. *Proc. Natl Acad. Sci. USA* 99, 4173–7.

Berner R.A., Caldeira K. (1997) The need for mass balance and feedback in the geochemical carbon cycle. *Geology* 25, 955–6.

Berner R.A., Canfield D.E. (1989) A new model of atmospheric oxygen over Phanerozoic time. *Am. J. Sci.* 289, 333–61.

Berner U., Faber E. (1996) Empirical carbon isotope/maturity relationships for gases from algal kerogens and terrigenous organic matter, based on dry, open-system pyrolysis. *Org. Geochem.* 24, 947–55.

Berner R.A., Kothavala Z. (2001) GEOCARB III: a revised model of atmospheric CO_2 over Phanerozoic time. *Am. J. Sci.* 301, 182–204.

Berner R.A., Lasaga A.C., Garrels R.M. (1983) The carbonate-silicate geochemical cycle and its effects on atmospheric carbon dioxide over the past 100 million years. *Am. J. Sci.* 283, 641–83.

Berner R.A., Petsch S.T., Lake J.A., Beerling D.J., Popp B.N., & 6 others (2000) Isotope fractionation and atmospheric oxygen: implications for Phanerozoic O_2 evolution. *Science* 287, 1630–3.

Berner R.A., Beerling D.J., Dudley R., Robinson J.M., Wildman R.A. (2003) Phanerozoic atmospheric oxygen. *Ann. Rev. Earth Planet. Sci.* 31, 105–34.

Berthéas O., Metzger P., Largeau C. (1999) A high molecular weight complex lipid, aliphatic polyaldehyde tetraterpenediol polyacetal from *Botryococcus braunii* (L race). *Phytochem.* 50, 85–96.

Bice K.L., Barron E.J., Peterson W.H. (1997) Continental runoff and early Cenozoic bottom-water sources. *Geology* 25, 951–4.

Bidigare R.R., Fluegge A., Freeman K.H., Hanson K.L., Hayes J.M. & 10 others (1997) Consistent fractionation of ^{13}C in nature and in the laboratory: growth-rate effects in some haptophyte algae. *Glob. Biogeochem. Cycles* 11, 279–92.

Bidleman T.F., Ritt C.P., Olney C.E. (1976) High molecular weight chlorinated hydrocarbons in the air and sea: rules and mechanisms of air-sea transfer. In *Marine Pollutant Transfer* (ed. Windom H.L., Duce R.A.), 323–51. Boston: D.C. Heath.

Billard C., Dauguet J.-C., Maume D., Bert M. (1990) Sterols and chemotaxonomy of marine Chrysophyceae. *Bot. Mar.* 33, 225–8.

Bishop A.N., Abbott G.D. (1993) The interrelationship of biological marker maturity parameters and molecular yields during contact metamorphism. *Geochim. Cosmochim. Acta* 57, 3661–8.

Bishop C.A., Brooks R.J., Carey J.H., Ng P., Nostrum R.J., Lean D.R.S. (1991) The case for cause–effect linkage between environmental contamination and development in eggs of the common snapping turtle (*Chelydra s. serpentina*) from Ontario, Canada. *J. Toxicol. Environ. Health* 33, 521–47.

Bitman J., Cecil H.C., Harris S.J., Fries G.F. (1968) Estrogenic activity of o,p′-DDT in the mammalian uterus and avian oviduct. *Science* 162, 371–2.

Bjørlykke K. (1994) Fluid-flow processes and diagenesis in sedimentary basins. In *Geofluids: Origin, Migration and Evolution of Fluids in Sedimentary Basins* (ed. Parnell J.) *Geol. Soc. Spec. Publn.* 78, 127–40. Oxford: Blackwell Scientific.

Black J.G. (1996) *Microbiology. Principles and Applications.* Englewood Cliffs, NJ: Prentice Hall.

Blais J.M., Schindler D.W., Muire D.C.G., Kimpe L.E., Donald D.B., Rosenberg B. (1998) Accumulation of persistent organochlorine compounds in mountains of western Canada. *Nature* 395, 585–8.

Blanc P., Connan J. (1992) Origin and occurrence of 25-norhopanes: a statistical study. *Org. Geochem.* 18, 813–28.

Blank C.E. (2004) Evolutionary timing of the origins of mesophilic sulphate reduction and oxygenic photosynthesis: a phylogenomic dating approach. *Geobiol. 2*, 1–20.

Blöchl E., Rachel R., Burggraf S., Hafenbradl D., Jannasch H.W., Stetter K.O. (1997) *Pyrolobus fumarii*, gen. and sp. nov., represents a novel group of archaea, extending the upper temperature limit for life to 113°C. *Extremophiles 1*, 14–21.

Blokker P., Schouten S., van den Ende H., de Leeuw J.W., Hatcher P.G., Sinninghe Damsté J.S. (1998) Chemical structure of algaenans from the fresh water algae *Tetraedron minimum*, *Scenedesmus communis* and *Pediastrum boryanum*. *Org. Geochem. 29*, 1453–68.

Blokker P., van Bergen P., Pancost R., Collinson M.E., de Leeuw J.W., Sinninghe Damsté J.S. (2001) The chemical structure of *Gloeocapsomorpha prisca* microfossils: Implications for their origin. *Geochim. Cosmochim. Acta 65*, 885–900.

Blumer M., Guillard R.R.L., Chase T. (1971) Hydrocarbons of marine phytoplankton. *Mar. Biol. 8*, 183–9.

Blunier T., Brook E.J. (2001) Timing of millennial-scale climate change in Antarctica and Greenland during the last glacial period. *Science 291*, 109–12.

Blunier T., Chappellaz J., Schwander J., Barnola J.-M., Desperts T., Stauffer B., Raynaud D. (1993) Atmospheric methane, record from a Greenland ice core over the last 1000 years. *J. Geophys. Res. 20*, 2219–22.

Blunier T., Chappellaz J., Schwander J., Stauffer B., Raynaud D. (1995) Variations in atmospheric methane concentration during the Holocene epoch. *Nature 374*, 46–9.

Boetius A., Ravenschlag K., Schubert C.J., Rickert D., Widdel F., Gieseke A., Amann R., Jørgensen B.B., Witte U., Pfannkuche O. (2000) A marine microbial consortium apparently mediating anaerobic oxidation of methane. *Nature 407*, 623–6.

Bohacs K., Suter J. (1997) Sequence stratigraphic distribution of coaly rocks: fundamental controls and paralic examples. *Am. Assoc. Pet. Geol. Bull. 81*, 1612–39.

Bois C., Bouche P., Pelet R. (1982) Global geologic history and distribution of hydrocarbon reserves. *Am. Assoc. Pet. Geol. Bull. 66*, 1248–70.

Bolin B., Degens E.T., Duvigneaud P., Kempe S. (1979) The global biogeochemical carbon cycle. In *The Global Carbon Cycle* (ed. Bolin B., Degens E.T., Kempe S., Ketner P.) SCOPE Rep. No. 13, 1–56. Chichester: Wiley.

Bolin B., Rosswall T., Richey J.E., Freney J.R., Ivanov M.V., Rodhe H. (1983) C, N, P, and S cycles: major reservoirs and fluxes. In *The Major Biogeochemical Cycles and their Interactions* (ed. Bolin B., Cook R.B.) SCOPE Rep. No. 21, 41–65. Chichester: Wiley.

Bonnett R., Burke P.J., Czechowski F., Reszka A. (1984) Porphyrins and metalloporphyrins in coal. *Org. Geochem. 6*, 177–82.

Bordovskiy O.K. (1965) Sources of organic matter in marine basins. *Mar. Geol. 3*, 5–31.

Borgund A.E., Barth T. (1994) Generation of short-chain organic acids from crude oil by hydrous pyrolysis. *Org. Geochem. 21*, 943–52.

Borlakoglu J.T., Wilkins J.P.G., Walker C.H. (1988) Polychlorinated biphenyls in fish-eating sea birds: molecular features and metabolic interpretations. *Mar. Environ. Res. 24*, 15–19.

Botto R.E. (1987) Solid ^{13}C NMR tracer studies to probe coalification. *Energy Fuels 1*, 228–30.

Boudou J.-P., Pelet R., Letolle R. (1984) A model of the diagenetic evolution of coaly sedimentary organic matter. *Geochim. Cosmochim. Acta 48*, 1357–62.

Bowring S.A., Erwin D.H., Jin Y.G., Martin M.W., Davidek K., Wang W. (1998) U/Pb zircon geochronology and tempo of the end-Permian mass extinction. *Science 280*, 1039–45.

Boyd P.W., Watson A.J., Law C.S., Abraham E.R., Trull T. & 30 others (2000) A mesoscale phytoplankton bloom in the polar Southern Ocean stimulated by iron fertilization. *Nature 407*, 695–702.

Bradshaw S.A., O'Hara S.C.M., Corner E.D.S., Eglinton G. (1990) Dietary lipid changes during herbivory and coprophagy by the marine invertebrate *Nereis diversicolor*. *J. Mar. Biol. Assoc. UK 70*, 771–87.

Bradshaw S.A., O'Hara S.C.M., Corner E.D.S., Eglinton G. (1991) Effects on dietary lipids of the marine bivalve *Scrobicularia plana* feeding in different modes. *J. Mar. Biol. Assoc. UK 71*, 635–53.

Brady P.V., Carroll S.A. (1994) Direct effects of CO_2 and temperature on silicate weathering: possible implications for climate control. *Geochim. Cosmochim. Acta 58*, 1853–6.

Brady E.C., DeConto R.M., Thompson S.L. (1998) Deep water formation and poleward ocean heat transport in the warm climate extreme of the Cretaceous (80 Ma). *Geophys. Res. Letts 25*, 4205–8.

Bralower T.J., Thomas D.J., Zachos J.C., Hirschmann M.M., Röhl U., Sigurdsson H., Thomas E., Whitney D.L. (1997) High-resolution records of the late Paleocene thermal maximum and circum-Caribbean volcanism: Is there a causal link? *Geology 25*, 963–6.

Brasier M.D., Green O.R., Jephcoat A.P., Kleppe A.K., Van Kranendonk M.J., Lindsay J.F., Steele A., Grassineau N.V. (2002) Questioning the evidence for Earth's oldest fossils. *Nature 416*, 76–82.

Brass G.W. (1976) The variation of the marine $^{87}Sr/^{86}Sr$ ratio during Phanerozoic time: interpretation using a flux model. *Geochim. Cosmochim. Acta 40*, 721–30.

Brassell S.C. (1985) Molecular changes in sediment lipids as indicators of systematic early diagenesis. *Phil. Trans. Roy. Soc. Lond. A 315*, 57–75.

Brassell S.C., Eglinton G. (1980) Environmental chemistry—an interdisciplinary subject. Natural and pollutant organic compounds in contemporary aquatic environments. In *Environmental Chemistry* (ed. Albaigés J.), 1–22. Oxford: Pergamon.

Brassell S.C., Wardroper A.M.K., Thomson I.D., Maxwell J.R., Eglinton G. (1981) Specific acyclic isoprenoids as biological markers of methanogenic bacteria in marine sediments. *Nature 290*, 693–6.

Brassell S.C., McEvoy J., Hoffmann C.F., Lamb N.A., Peakman T.M., Maxwell J.R. (1984) Isomerisation, rearrangement and aromatisation of steroids in distinguishing early stages of diagenesis. *Org. Geochem.* 6, 11–23.

Brassell S.C., Eglinton G., Marlowe I.T., Pflaumann U., Sarnthein M. (1986) Molecular stratigraphy: a new tool for climatic assessment. *Nature* 320, 129–33.

Braun R.L., Burnham A.K. (1992) PMOD: a flexible model of oil and gas generation, cracking, and expulsion. *Org. Geochem.* 19, 161–72.

Bray E.E., Evans E.D. (1961) Distribution of *n*-paraffins as a clue to recognition of source beds. *Geochim. Cosmochim. Acta* 22, 2–15.

Briggs D.E.G. (1999) Molecular taphonomy of animal and plant cuticles: selective preservation and diagenesis. *Phil. Trans. Roy. Soc. Lond. B.* 354, 7–17.

Briggs D.E.G., Crowther P.R. (eds) (2001) *Palaeobiology II*. Oxford: Blackwell Science.

Briggs D.E.G., Evershed R.P., Stankiewicz B.A. (1998) The molecular preservation of fossil arthropod cuticles. *Ancient Biomols* 2, 135–46.

Britton G. (1983) *The Biochemistry of Natural Products*. Cambridge: Cambridge University Press.

Britton L.N. (1984) Microbial degradation of aliphatic hydrocarbons. In *Microbial Degradation of Organic Compounds* (ed. Gibson D.T.), 89–129. New York: Marcel Dekker.

Brocks J.J., Logan G.A., Buick R., Summons R.E. (1999) Archean molecular fossils and the early rise of eukaryotes. *Science* 285, 1033–6.

Broecker W.S. (1974) *Chemical Oceanography*. New York: Harcourt Brace.

Broecker W.S. (1997) Thermohaline circulation, the Achilles Heel of our climate system: will man-made CO_2 upset the current balance? *Science* 278, 1582–8.

Broecker W.S. (1998) Paleocean circulation during the last deglaciation: a bipolar seesaw? *Paleoceanogr.* 13, 119–21.

Broecker W.S. (1999) What if the conveyor were to shut down? Reflections on a possible outcome of the great global experiment. *GSA Today*, Jan., 1–7.

Broecker W.S. (2003) Does the trigger for abrupt climate change reside in the ocean or in the atmosphere? *Science* 300, 1519–22.

Brook E.J., Sowers T., Orchardo J. (1996) Rapid variations in atmospheric methane concentration during the past 110,000 years. *Science* 273, 1087–91.

Brooks J., Cornford C., Archer R. (1987) The role of hydrocarbon source rocks in petroleum exploration. In *Marine Petroleum Source Rocks* (ed. Brooks J., Fleet A.J.) *Geol. Soc. Spec. Publn* 26, 17–46. Oxford: Blackwell Scientific.

Bryan G.W., Gibbs P.E., Humberstone L.G., Burt G.R. (1986) The decline of the gastropod *Nucella lapillus* around southwest England: evidence for the effect of tributyl tin from antifouling paints. *J. Mar. Biol. Assoc. UK* 66, 611–40.

Bryan G.W., Gibbs P.E., Humberstone L.G., Burt G.R. (1987) Copper, zinc and organotin as long-term factors governing the distribution of organisms in the Fal estuary in southwest England, UK. *Estuaries* 10, 208–19.

Bryan G.W., Gibbs P.E., Humberstone L.G., Burt G.R. (1989) Uptake and transformation of ^{14}C-labelled tributyltin chloride by the dog-whelk, *Nucella lapillus*—importance of absorption from the diet. *Mar. Environ. Res.* 28, 241–5.

Buick R. (1992) The antiquity of oxygenic photosynthesis: evidence from stromatolites in sulphate-deficient Archaean lakes. *Science* 255, 74–7.

Bu'lock J.D., de Rose M., Gambacorta A. (1981) Isoprenoid biosynthesis in Archaebacteria. In *Biosynthesis of Isoprenoid Compounds, Vol. 1* (ed. Porter J.W., Spurgeon S.L.), 159–89. Chichester: Wiley.

Bungard R.A., Ruban A.V., Hibberd J.M., Press M.C., Horton P., Scholes J.D. (1999) Unusual carotenoid composition and a new type of xanthophyll cycle in plants. *Proc. Natl Acad. Sci. USA* 96, 1135–9.

Bunge M., Adrian L., Kraus A., Opel M., Lorenz W.G., Andreesen J.R., Görisch H., Lechner U. (2003) Reductive dehalogenation of chlorinated dioxins by an anaerobic bacterium. *Nature* 421, 357–60.

Burke W.H., Denison R.E., Hetherington E.A., Koepnick R.B., Nelson H.F., Otto J.B. (1982) Variation in seawater $^{87}Sr/^{86}Sr$ throughout Phanerozoic time. *Geology* 10, 516–19.

Burnham A.K. (1998) Comments on 'Experiments on the role of water in petroleum formation' by M.D. Lewan. *Geochim. Cosmochim. Acta* 62, 2207–10.

Burnham A.K., Sweeney J.J. (1989) A chemical kinetic model of vitrinite maturation and reflectance. *Geochim. Cosmochim. Acta* 53, 2649–57.

Burnham A.K., Schmidt B.J., Braun R.L. (1995) A test of the parallel reaction model using kinetic measurements on hydrous pyrolysis residues. *Org. Geochem.* 23, 931–9.

Burnham A.K., Gregg H.R., Ward R.L., Knauss K.G., Copenhaver S.A., Reynolds J.G., Sanborn R. (1997) Decomposition kinetics and mechanism of *n*-hexadecane-1,2-$^{13}C_2$ and dodec-1-ene-1,2-$^{13}C_2$ doped in petroleum and *n*-hexadecane. *Geochim. Cosmochim. Acta* 61, 3725–37.

Burns B.D., Beardall J. (1987) Utilization of inorganic carbon by marine microalgae. *J. Exp. Mar. Biol. Ecol.* 107, 75–86.

Butler J.H., Battle M., Bender M.L., Montzka S.A., Clarke A.D., Saltzman E.S., Sucher C.M., Severinghaus J.P., Elkins J.W. (1999) A record of atmospheric halocarbons during the twentieth century from polar firn air. *Nature* 399, 749–55.

Caldeira K., Kasting J.F. (1992a) Susceptibility of the early Earth to irreversible glaciation caused by carbon dioxide clouds. *Nature* 359, 226–8.

Caldeira K., Kasting J.F. (1992b) The life span of the biosphere revisited. *Nature* 360, 721–3.

Campbell I.H., Czamanske G.K., Fedorenko V.A., Hill R.I., Stepanov V. (1992) Synchronism of the Siberian Traps and the Permian–Triassic boundary. *Science* 258, 1760–3.

Canfield D.E. (1994) Factors influencing organic carbon preservation in marine sediments. *Chem. Geol.* 114, 315–29.

Canfield D.E. (1998) A new model for Proterozoic ocean chemistry. *Nature* 396, 450–3.

Canfield D.E., Des Marais D.J. (1993) Biogeochemical cycles of carbon, sulfur, and free oxygen in a microbial mat. *Geochim. Cosmochim. Acta* 57, 3971–84.

Canfield D.E., Teske A. (1996) Later Proterozoic rise in atmospheric oxygen concentration inferred from phylogenetic and sulphur-isotope studies. *Nature 382*, 127–32.

Canfield D.E., Thamdrup B. (1994) The production of ^{34}S-depleted sulfide during bacterial disproportionation of elemental sulfur. *Science 266*, 1973–5.

Cardoso J.N., Eglinton G. (1983) The use of hydroxyacids as geochemical indicators. *Geochim. Cosmochim. Acta 47*, 723–30.

Carlile M.J., Watkinson S.C., Gooday G.W. (2001) *The Fungi*. San Diego: Academic Press.

Carlson D.J., Ducklow H.W. (1995) Dissolved organic carbon in the upper ocean of the central equatorial Pacific Ocean, 1992: daily and finescale vertical variations. *Deep-Sea Res. 42*, 639–50.

Carlson R.M.K., Teerman S.C., Moldowan J.M., Jacobson S.R., Chan E.I., Dorrough K.S., Seetoo W.C., Mertani B. (1993) High temperature gas chromatography of high-wax oils. *Proc. Indonesian Petroleum Assoc., 29th Ann. Convention*, 483–504.

Carpenter S.R., Elser M.M., Elser J.J. (1986) Chlorophyll production, degradation and sedimentation: implications for paleolimnology. *Limnol. Oceanogr. 31*, 112–24.

Cassani F., Eglinton G. (1986) Organic geochemistry of Venezuelan extra-heavy oils. I. Pyrolysis of asphaltenes: a technique for the correlation and maturity evaluation of crude oils. *Chem. Geol. 56*, 167–83.

Catling D.C., Zahnle K.J., McKay C.P. (2001) Biogenic methane, hydrogen escape, and the irreversible oxidation of early Earth. *Science 293*, 839–42.

Catubig N.R., Archer D.E., Francois R., deMenocal P., Howard W., Yu E.-F. (1998) Global deep-sea burial rate of calcium carbonate during the last glacial maximum. *Paleoceanogr. 13*, 298–310.

Cerling T.E., Wang Y., Quade J. (1993) Expansion of C4 ecosystems as an indicator of global ecological change in the late Miocene. *Nature 361*, 344–5.

Cess R.D., Zhang M.-H., Potter G.L., Barker H.W., Coleman R.A. & 25 others (1993) Uncertainties in carbon dioxide radiative forcing in atmospheric general circulation models. *Science 262*, 1252–5.

Chaffee A.L., Johns R.B. (1983) Polycyclic aromatic hydrocarbons in Australian coals. I. Angularly fused pentacyclic tri- and tetra-aromatic components of Victorian brown coal. *Geochim. Cosmochim. Acta 47*, 2141–55.

Chaffee A.L., Strachan M.G., Johns R.B. (1984) Polycyclic aromatic hydrocarbons in Australian coals. II. Novel tetracyclic components from Victorian brown coal. *Geochim. Cosmochim. Acta 48*, 2037–43.

Chalansonnet S., Largeau C., Casadevall E., Berkaloff C., Peniguel G., Couderc R. (1988) Cyanobacterial resistant biopolymers. Geochemical implications of the properties of *Schizothrix sp.* resistant material. *Org. Geochem. 13*, 1003–10.

Chaloner W.G. (1989) Fossil charcoal as an indicator of palaeoatmospheric oxygen level. *J. Geol. Soc. Lond. 146*, 171–4.

Chambers L.A. (1982) Sulfur isotope study of a modern intertidal environment, and the interpretation of ancient sulfides. *Geochim. Cosmochim. Acta 46*, 721–8.

Chang S., Berner R.A. (1999) Coal weathering and the geochemical carbon cycle. *Geochim. Cosmochim. Acta 63*, 3301–10.

Chappell J., Shackleton N.J. (1986) Oxygen isotopes and sea level. *Nature 324*, 137–40.

Chappellaz J., Blunier T., Kints S., Dällenbach A., Barnola J.-M., Schwander J., Raynaud D., Stauffer B. (1997) Changes in the atmospheric CH_4 gradient between Greenland and Antarctica during the Holocene. *J. Geophys. Res. 102*, 15987–99.

Charlson R.J., Lovelock J.E., Andreae M.O., Warren S.G. (1987) Oceanic phytoplankton, atmospheric sulphur, cloud albedo and climate. *Nature 326*, 655–61.

Chavez F.P., Buck K.R., Coale K.H., Martin J.J., Ditullio G.R., Welschmeyer N.A., Jacobson A.C., Barber R.T. (1991) Growth rates, grazing, sinking and iron limitation of equatorial Pacific phytoplankton. *Limnol. Oceanogr. 36*, 1816–33.

Chester R. (2000) *Marine Geochemistry*. Oxford: Blackwell Science.

Chicarelli M.I., Maxwell J.R. (1986) A novel fossil porphyrin with a fused ring system: evidence for water column transformation of chlorophyll. *Tetr. Letts 27*, 4653–4.

Chisholm S.W., Morel F.M.M. (eds) (1991) *What Controls Phytoplankton Production in Nutrient-rich Areas of the Open Sea? Limnol. Oceanogr. 36*.

Chisholm B.S., Nelson D.E., Schwarcz H.P. (1982) Stable carbon isotope ratios as a measure of marine versus terrestrial protein in ancient diets. *Science 216*, 1131–2.

Chosson P., Lannau C., Connan J., Dessort D. (1991) Biodegradation of refractory hydrocarbon biomarkers from petroleum under laboratory conditions. *Nature 351*, 640–2.

Christensen J.P., Murray J.W., Devol A.H., Codispoti L.A. (1987) Denitrification in continental shelf sediments has major impact on the oceanic nitrogen budget. *Glob. Biogeochem. Cycles 1*, 97–116.

Chumakov N.M., Zharkov M.A., Herman A.B., Doludenko M.P., Kalandadze N.N., Lebedev E.L., Ponomarenko A.G., Rautian A.S. (1995) Climatic Belts of the MidCretaeceous Time. *Strat. Geol. Correl. 3*, 241–60.

Chung H.M., Brand S.W., Grizzle P.L. (1981) Carbon isotope geochemistry of Paleozoic oils from Big Horn Basin. *Geochim. Cosmochim. Acta 45*, 1803–15.

Chyba C., Sagan C. (1992) Endogenous production, exogenous delivery and impact-shock synthesis of organic molecules: an inventory for the origins of life. *Nature 355*, 125–32.

Chyba C.F., Thomas P.J., Brookshaw L., Sagan C. (1990) Cometary delivery of organic molecules to the early Earth. *Science 249*, 366–73.

Clark P.U., Marshall S.J., Clarke G.K.C., Hostetler S.W., Licciardi J.M., Teller J.T. (2001) Freshwater forcing of abrupt climate change during the last glaciation. *Science 293*, 283–7.

Claypool A.E., Kvenvolden K.A. (1983) Methane and other hydrocarbon gases in marine sediments. *Ann. Rev. Earth Planet. Sci. 11*, 299–327.

Clayton C. (1991) Carbon isotope fractionation during natural gas generation from kerogen. *Mar. Pet. Geol. 8*, 232–40.

Cline J.D., Kaplan I.R. (1975) Isotopic fractionation of dissolved nitrate during denitrification in the Eastern Tropical North Pacific. *Mar. Chem. 3*, 271–99.

Cloud P. (1973) Paleoecological significance of the banded-iron formation. *Econ. Geol. 68*, 1135–43.

Coale K.H., Johnson K.S., Fitzwater S.E., Gordon R.M., Tanner S. & 14 others (1996) A massive phytoplankton bloom induced by an ecosystem-scale iron fertilization experiment in the equatorial Pacific Ocean. *Nature 383*, 495–501.

Coates J.D., Woodward J., Allen J., Philp P., Lovley D. (1997) Anaerobic degradation of polycyclic aromatic hydrocarbons and alkanes in petroleum-contaminated harbor-sediments. *Appl. Environ. Microbiol. 63*, 3589–93.

Coffin R.B., Velinsky D.J., Devereux R., Price W.A., Cifuentes L.A. (1990) Stable carbon isotope analysis of nucleic acids to trace sources of dissolved substrates used by estuarine bacteria. *Appl. Environ. Microbiol. 56*, 2012–20.

Cognetti G., Lardicci C., Abbiati M., Castelli A. (2000) The Adriatic Sea and the Tyrrhenian Sea. In *Seas at the Millennium: An Environmental Evaluation. Vol. 1 Regional Chapters: Europe, The Americas and West Africa* (ed. Sheppard C.R.C.), 267–84. Amsterdam: Pergamon.

Cohen B.A., Swindle T.D., Kring D.A. (2000) Support for the lunar cataclysm hypothesis from lunar meteorite impact melt ages. *Science 290*, 1754–6.

Colborn T., Clement C. (eds.) (1992) *Advances in Modern Environmental Toxicology Vol. 21, Chemically-induced Alterations in Sexual and Functional Development: the Wildlife/Human Connection*. Princeton, NJ: Princeton Scientific.

Coleman D.D., Risatti J.B., Schoell M. (1981) Fractionation of carbon and hydrogen isotopes by methane-oxidizing bacteria. *Geochim. Cosmochim. Acta 45*, 1033–7.

Coleman S.M., Pierce K.L., Birkeland P.W. (1987) Suggested terminology for Quaternary dating methods. *Quat. Res. 28*, 314–19.

Colling A., Dise N., Francis P., Harris N., Wilson C. (1997) *The Dynamic Earth*. Milton Keynes: Open University.

Collins M.J., Westbroek P., Muyzer G., de Leeuw J.W. (1992) Experimental evidence for condensation reactions between sugars and proteins in carbonate skeletons. *Geochim. Cosmochim. Acta 56*, 1539–44.

Collister J.W., Lichtfouse E., Hieshima G., Hayes J.M. (1994) Partial resolution of sources of n-alkanes in the saline portion of the Parachute Creek Member, Green River Formation (Piceance Creek Basin, Colorado). *Org. Geochem. 21*, 645–59.

Confiantini R. (1986) Environmental isotopes in lake studies. In *Handbook of Environmental Isotope Geochemistry, Vol. 2. The Terrestrial Environment, B* (ed. Fritz P., Fontes J.C.), 113–68. Amsterdam: Elsevier.

Connan J. (1984) Biodegradation of crude oils in reservoirs. In *Advances in Petroleum Geochemistry, Vol. 1* (ed. Brooks J., Welte D.), 299–335. New York: Academic Press.

Connan J., Bouroullec J., Dessort D., Albrecht P. (1986) The microbial input in carbonate-anhydrite facies of a sabkha palaeoenvironment from Guatemala: A molecular approach. *Org. Geochem. 10*, 29–50.

Connan J., Lacrampe-Coulombe G., Magot M. (1996) Origin of gases in reservoirs. In *Proc. of the 1995 Int. Gas Res. Conf., Vol. 1* (ed. Dolenc D.), 21–62. Rockville: Government Institutes.

Conte M.H., Thompson A., Lesley D., Harris R.P. (1998) Genetic and physiological influences on the alkenone/alkenoate versus growth temperature relationship in *Emiliania huxleyi* and *Gephyrocapsa oceanica*. *Geochim. Cosmochim. Acta 62*, 51–68.

Conway T.J., Tans P., Waterman L.S., Thoning K.W., Masarie K.A., Gammon R.H. (1988) Atmospheric carbon dioxide measurements in the remote global troposphere. *Tellus 40B*, 81–115.

Corliss J.B. (1990) Hot springs and the origin of life. *Nature 347*, 624.

Cornford C. (1998) Source rocks and hydrocarbons of the North Sea. In *Petroleum Geology of the North Sea* (ed. Glennie K.W.), 376–462. Oxford: Blackwell Scientific.

Corsetti F.A., Awramik S.M., Pierce D. (2003) A complex microbiota from snowball Earth times: Microfossils from the Neoproterozoic Kingston Peak Formation, Death Valley, USA. *Proc. Natl Acad. Sci. USA 100*, 4399–404.

Cowie G.L., Hedges J.I. (1984) Carbohydrate sources in a coastal marine environment. *Geochim. Cosmochim. Acta 48*, 2075–87.

Cowie G.L., Hedges J.I. (1996) Digestion and alteration of the biochemical constituents of a diatom (*Thalassiosira weissflogii*) ingested by an herbivorous zooplankton (*Calanus pacificus*). *Limnol. Oceanogr. 41*, 581–94.

Cox C.B., Moore P.D. (1993) *Biogeography: an Ecological and Evolutionary Approach*. Oxford: Blackwell Scientific.

Craig H. (1965) Isotopic variations in meteoric waters. *Science 133*, 1702–3.

Crane P.R., Lidgard S. (1989) Angiosperm diversification and paleolatitudinal gradients in Cretaceous floristic diversity. *Science 246*, 675–8.

Cranwell P.A. (1982) Lipids of aquatic sediments and sedimenting particulates. *Prog. Lipid Res. 21*, 271–308.

Crowell J.C. (1999) Pre-Mesozoic ice ages: Their bearing on understanding the climate system. *Geol. Soc. Am. Mem. 192*. Boulder: Geol. Soc. America.

Crowley T.J. (1994) Pangean climates. In *Pangea: Paleoclimate, Tectonics and Sedimentation During Accretion, Zenith and Breakup of a Supercontinent* (ed. Klein G.D.) *Geol. Soc. Am. Spec. Pap. 288*, 25–40. Boulder: Geol. Soc. America.

Crowley T.J., Berner R.A. (2001) CO_2 and climate change. *Science 292*, 870–2.

Crowley T.J., Baum S.K., Kim K.Y. (1993) GCM sensitivity experiment with pole-centred supercontinents. *J. Geophys. Res. 98*, 8793–800.

Crutzen P.J. (1988) Variability in atmosphere-chemical systems. In *Scales and Global Change* (ed. Rosswall T., Woodmansee R.G., Risser P.G.) *SCOPE Rep. 35*, 81–108. Chichester: Wiley.

Dacey J.W.H., Wakeham S.G. (1986) Oceanic dimethylsulphide: production during zooplankton grazing on phytoplankton. *Science 233*, 1314–16.

Dahl J.J., Moldowan M.J., Sundararaman P. (1993) Relationship of biomarker distribution to depositional environment: Phosphoria Formation, Montana, USA. *Org. Geochem. 20*, 1001–17.

Dalziel I.W.D. (1997) Overview: Neoproterozoic–Paleozoic geography and tectonics: Review, hypothesis, environmental speculation. *Geol. Soc. Am. Bull. 109*, 16–42.

Daniel J.S., Solomon S., Albritton D. (1995) On the evaluation of halocarbon radiative forcing and global warming potentials. *J. Geophys. Res. 100*, 1271–85.

Dansgaard W. (1964) Stable isotopes in precipitation. *Tellus 4*, 437–68.

Davin L.B., Wang H., Crowell A.L., Bedgar D.L., Martin D.M., Sarkanen S., Lewis N.G. (1997) Stereoselective biomolecular phenoxy radical coupling by an auxiliary (dirigent) protein without an active center. *Science 275*, 362–6.

Dean W.E., Gorhan E. (1998) Magnitude and significance of carbon burial in lakes, reservoirs, and peatlands. *Geology 26*, 535–8.

Dean W.E., Arthur M.A., Claypool G.E. (1986) Depletion of ^{13}C in Cretaceous marine organic matter: Source, diagenetic, or environmental signal? *Mar. Geol. 70*, 119–57.

De Grande S.M.B., Aquino Neto F.R., Mello M.R. (1993) Extended tricyclic terpanes in sediments and petroleums. *Org. Geochem. 20*, 1039–47.

De Jong E., De Vries F.P., Field J.A., Van De Zwan R.P., De Bont J.A.M. (1992) Isolation of basidiomycetes with high peroxidative activity. *Mycol. Res. 96*, 1098–104.

de las Heras F.X., Grimalt J.O., Albaigés J. (1991) Novel C-ring cleaved triterpenoid-derived aromatic hydrocarbons in Tertiary brown coals. *Geochim. Cosmochim. Acta 55*, 3379–85.

de Leeuw J.W., Baas M. (1986) Early-stage diagenesis of steroids. In *Biological Markers in the Sedimentary Record* (ed. Johns R.B.), 101–23. Amsterdam: Elsevier.

de Leeuw J.W., Cox H.C., van Grass G., van de Meer F.W., Peakman T.M., Baas J.M.A., van de Graaf B. (1989) Limited double bond isomerisation and selective hydrogenation of sterenes during early diagenesis. *Geochim. Cosmochim. Acta 53*, 903–9.

de Leeuw J.W., Frewin N.L., van Bergen P.F., Sinninghe Damsté J.S., Collinson M.E. (1995) Organic carbon as a palaeoenvironmental indicator in the marine realm. In *Marine Palaeoenvironmental Analysis from Fossils* (ed. Bosence D.W.J., Allison P.A.) *Geol. Soc. Spec. Publ. 83*, 43–71. London: Geological Society.

Demaison G.J., Moore G.T. (1980) Anoxic environments and oil source bed genesis. *Am. Assoc. Pet. Geol. Bull. 64*, 1179–209.

Dembicki H. Jr. (1992) The effects of the mineral matrix on the determination of kinetic parameters using modified Rock Eval pyrolysis. *Org. Geochem. 18*, 531–9.

Deming D., Chapman D.S. (1989) Thermal histories and hydrocarbon generation: example from Utah-Wyoming thrust belt. *Am. Assoc. Pet. Geol. Bull. 73*, 1455–71.

DeNiro M.J., Epstein S. (1978) Influence of diet on the distribution of carbon isotopes in animals. *Geochim. Cosmochim. Acta 42*, 495–506.

DeNiro M.J., Epstein S. (1981) Influence of diet on the distribution of nitrogen isotopes in animals. *Geochim. Cosmochim. Acta 45*, 341–51.

Department of the Environment (1991) *Environmental Hazard Assessment: Di-(2-ethylhexyl)phthalate*. London: HMSO.

Derenne S., Largeau C., Casadevall E., Connan J. (1988) Comparison of torbanites of various origins and evolutionary stages. Bacterial contribution to their formation. Cause of the lack of botryococcane in bitumens. *Org. Geochem. 12*, 43–59.

Derenne S., Metzger P., Largeau C., Van Bergen P.F., Gatellier J.P., Sinninghe Damsté J.S., de Leeuw J.W., Berkaloff C. (1992) Similar morphological and chemical variations of *Gloeocapsomorpha prisca* in Ordovician sediments and cultured *Botryococcus braunii* as a response to changes in salinity. *Org. Geochem. 19*, 299–313.

Derenne S., Largeau C., Behar F. (1994) Low polarity pyrolysis products of Permian to Recent *Botryococcus*-rich sediments: first evidence for the contribution of an isoprenoid algaenan to kerogen formation. *Geochim. Cosmochim. Acta 58*, 3703–11.

Dereppe J.-M., Moreaux C., Debyser Y. (1980) Investigation of marine and terrestrial humic substances by ^1H and ^{13}C nuclear magnetic resonance and infrared spectroscopy. *Org. Geochem. 2*, 117–24.

Derry L.A., Kaufman A.J., Jacobsen S.B. (1992) Sedimentary cycling and environmental change in the Late Proterozoic: evidence from stable and radiogenic isotopes. *Geochim. Cosmochim. Acta 56*, 1317–29.

Desbrow C., Routledge E., Sheehan D., Waldock M., Sumpter J. (1996) The identification and assessment of oestrogen substances in sewage treatment works effluents (P2-i490/7). London: MAFF Lab. & Brunel University, Environment Agency.

Des Marais D.J. (1991) Microbial mats, stromatolites and the rise of oxygen in the Precambrian atmosphere. *Palaeogeogr. Palaeoclimatol. Palaeoecol. Glob. Planet. Change Sect. 97*, 93–6.

Des Marais D.J., Strauss H., Summons R.E., Hayes J.M. (1992) Carbon isotopic evidence for the stepwise oxidation of the Proterozoic environment. *Nature 359*, 605–9.

Deuser W.G. (1970) Isotopic evidence for diminishing supply of available carbon during diatom bloom in the Black Sea. *Nature 225*, 1069–71.

Devol A.H. (2003) Solution to a marine mystery. *Nature 422*, 575–6.

de Voogt P. (1996) Ecotoxicology of chlorinated aromatic hydrocarbons. In *Chlorinated Organic Micropollutants* (ed. Hester R.E., Harrison R.M.), 89–112. Cambridge: Royal Society of Chemistry.

De Vooys C.G.N. (1979) Primary production in aquatic environments. In *The Global Carbon Cycle* (ed. Bolin B., Degens

E.T., Kempe S., Ketner P.) *SCOPE Rep. No. 13*, 259–92. Chichester: Wiley.

de Wit M.J., Ghosh J.G., de Villiers S., Rakotosolofo N., Alexander J., Tripathi A., Looy C. (2002) Multiple organic carbon isotope reversals across the Permo-Triassic boundary of terrestrial Gondwana sequences: clues to extinction patterns and delayed ecosystem recovery. *J. Geol. 110*, 227–46.

D'Hondt S., Donaghay P., Zachos J.C., Luttenberg D., Lindinger M. (1998) Organic carbon fluxes and ecological recovery from the Cretaceous–Tertiary mass extinction. *Science 282*, 276–9.

Dickens G.R., O'Neil J.R., Rea D.K., Owen R.M. (1995) Dissociation of oceanic methane hydrate as a cause of the carbon isotopic excursion at the end of the Paleocene. *Paleoceanogr. 10*, 965–71.

Dickens G.R., Castillo M.M., Walker J.C.G. (1997) A blast of gas in the latest Paleocene: Simulating first-order effects of massive dissociation of oceanic methane hydrate. *Geology 25*, 258–62.

Dickinson R.E., Cicerone R.J. (1986) Future global warming from atmospheric trace gases. *Nature 319*, 109–15.

Didyk B.M., Simoneit B.R.T. (1989) Hydrothermal oil of Guaymas Basin and implications for petroleum formation mechanisms. *Nature 342*, 65–9.

Didyk B.M., Simoneit B.R.T., Brassell S.C., Eglinton G. (1978) Organic geochemical indicators of palaeoenvironmental conditions of sedimentation. *Nature 272*, 216–22.

Dieckmann V., Schenck H.J., Horsfield B., Welte D.H. (1998) Kinetics of petroleum generation and cracking by programmed-temperature closed-system pyrolysis of Toarcian shales. *Fuel 77*, 23–31.

Disnar J.R., Harouna M. (1994) Biological origin of tetracyclic diterpanes, *n*-alkanes and other biomarkers found in Lower Carboniferous Gondwana coals (Niger). *Org. Geochem. 21*, 143–52.

Dixon R.K., Brown S., Houghton R.A., Solomon A.M., Trexler M.C., Wisniewski J. (1994) Carbon pools and flux of global forest ecosystems. *Science 263*, 185–90.

Dlugokency E.J., Masarie K.A., Lang P.M., Tans P.P. (1998) Continuing decline in the growth of the atmospheric methane burden. *Nature 393*, 447–50.

Dominé F., Bounaceur R., Scacchi G., Marquaire P.-M., Dessort D., Pradier B., Brevart O. (2002) Up to what temperature is petroleum stable? New insights from a 5200 free radical reactions model. *Org. Geochem. 33*, 1487–99.

Donkin P., Widdows J., Evans S.V., Brinsley M.D. (1991) QSARs for the sublethal responses of marine mussels (*Mytilus edulis*). *Sci. Tot. Environ. 109/110*, 461–76.

Doolittle W.F. (1998) A paradigm gets shifty. *Nature 392*, 15–16.

Doolittle W.F. (1999) Phylogenetic classification and the universal tree. *Science 284*, 2124–8.

Douglas A.G., Sinninghe Damsté J.S., Fowler M.G., Eglinton T.I., de Leeuw J.W. (1991) Unique distributions of hydrocarbons and sulphur compounds released by flash pyrolysis from the fossilised alga *Gloeocapsomorpha prisca*., a major constituent in one of four Ordovician kerogens. *Geochim. Cosmochim. Acta 55*, 275–91.

Duarte-Davidson R., Sewart A., Alcock R.E., Cousins I.T., Jones K.C. (1997) Exploring the balance between sources, deposition, and the environmental burden of PCDD/Fs in the UK terrestrial environment: an aid to identifying uncertainties and research needs. *Environ. Sci. Technol. 31*, 1–11.

Duch M.V., Grant D.M. (1970) Carbon-13 chemical shift studies of the 1,4-polybutadienes and the 1,4-polyisoprenes. *Macromolecules 3*, 165–74.

Dudley R. (1998) Atmospheric oxygen, giant Paleozoic insects and the evolution of aerial locomotor performance. *J. Exp. Biol. 201*, 1043–50.

Dugdale R.C., Wilkerson F.P. (1992) Nutrient limitation of new production. In *Primary Productivity and Biogeochemical Cycles in the Sea* (ed. Falkowski P.G., Woodhead A.D.), 107–22. New York: Plenum Press.

Dugdale R.C., Wilkerson F.P., Minas H.J. (1995) The role of a silicate pump in driving new production. *Deep-Sea Res. 42*, 697–719.

Durand B. (ed.) (1980) *Kerogen–Insoluble Organic Matter from Sedimentary Rocks*. Paris: Editions Technip.

Durand B. (1985) Diagenetic modification of kerogens. *Phil. Trans. Roy. Soc. Lond. A 315*, 77–90.

Durand B., Paratte M., Bertrand P. (1983) Le potentiel en huile des charbons: une approche géochemique. *Rev. Inst. Fr. Pét. 38*, 709–21.

ECPI (1995) *Assessment of the Release, Occurrence and Possible Effects of Plasticisers in the Environment. Phthalate esters used in PVC* (draft report). European Council for Plasticisers and Intermediates.

Eddy J.A., Gilliland R.L., Hoyt D.V. (1982) Changes in the solar constant and climatic effects. *Nature 300*, 689–93.

Edmond J.M. (1992) Himalayan tectonics, weathering processes, and the strontium isotope record in marine limestones. *Science 258*, 1594–7.

Eglinton G., Hamilton R.J. (1967) Leaf epicuticular waxes. *Science 156*, 1322–35.

Eisele F.L., Mount G.H., Tanner D., Jefferson A., Shetter R., Harder J.M., Williams E.J. (1997) Understanding the production and interconversion of the hydroxyl radical during the Tropospheric OH Photochemistry Experiment. *J. Geophys. Res. 102*, 6457–65.

Eisenreich W., Schwarz M., Cartayrade A., Arigoni D., Zenk M., Bacher A. (1998) The deoxyxylulose phosphate pathway of terpenoid biosynthesis in plants and microorganisms. *Chem. Biol. 5*, R221–3.

Ekweozor C.M., Telnaes N. (1990) Oleanane parameter: verification by quantitative study of the biomarker occurrence in sediments of the Niger delta. *Org. Geochem. 16*, 401–13.

Ekweozor C.M., Okogun J.I., Ekong D.E.U., Maxwell J.R. (1979) Preliminary organic geochemical studies of samples from the Niger delta (Nigeria). I. Analyses of crude oils for triterpanes. *Chem. Geol. 27*, 11–28.

Elliott T. (1986) Deltas. In *Sedimentary Environments and Facies* (ed. Reading H.G.), 113–54. Oxford: Blackwell Scientific.

Ellis D.V., Pattisina L.A. (1990) Widespread neogastropod imposex: a biological indicator of global TBT contamination? *Mar. Pollut. Bull. 21*, 248–53.

Emerson S., Hedges J.I. (1988) Processes controlling the organic carbon content of open ocean sediments. *Paleoceanogr. 3*, 621–34.

Engel M.H., Macko S.A. (1986) Stable isotope evaluation of the origins of amino acids in fossils. *Nature 323*, 531–3.

Engel M.H., Macko S.A. (1997) Isotopic evidence for extraterrestrial non-racemic amino acids in the Murchison meteorite. *Nature 389*, 265–8.

Engel M.H., Macko S.A. (2001) The stereochemistry of amino acids in the Murchison meteorite. *Precambr. Res. 106*, 35–45.

Engel M.H., Nagy B. (1982) Distribution and enantiomeric composition of amino acids in the Murchison meteorite. *Nature 296*, 837–40.

Engel M.H., Macko S.A., Silfer J.A. (1990) Carbon isotope composition of individual amino acids in the Murchison meteorite. *Nature 348*, 47–9.

England W.E., Fleet A.J. (eds) (1991) *Petroleum Migration. Geol. Soc. Spec. Publn 59.* Oxford: Blackwell Scientific.

England W.A., Mackenzie A.S., Mann D.M., Quigley T.M. (1987) The movement and entrapment of petroleum fluids in the subsurface. *J. Geol. Soc. Lond. 144*, 327–47.

Ensminger A., van Dorsselaer A., Spyckerelle Ch., Albrecht P., Ourisson G. (1973) Pentacyclic triterpenes of the hopane type as ubiquitous geochemical markers: origin and significance. In *Advances in Organic Geochemistry 1973* (ed. Tissot B., Bienner F.), 245–60. Paris: Technip.

Environment Agency (1996) *The Identification and Assessment of Oestrogenic Substances in Sewage Treatment Works Effluents. R&D Technical Summary P38.* London: Environment Agency.

Epstein S., Buchsbaum R., Lowenstam H., Urey H.C. (1953) Revised carbonate–water isotopic temperature scale. *Bull. Geol. Soc. Am. 64*, 1315–26.

Epstein S., Krishnamurthy R.V., Cronin J.R., Pizzarello S., Yuen G.U. (1987) Unusual stable isotope ratios in amino acids and carboxylic acid extracts from the Murchison meteorite. *Nature 326*, 477–9.

Erbacher J., Thurow J., Littke R. (1996) Evolution patterns of radiolaria and organic variations: a new approach to identify sea-level changes in mid-Cretaceous pelagic environments. *Geology 24*, 499–502.

Erbacher J., Huber B.T., Norris R.D., Markey M. (2001) Increased thermohaline stratification as a possible cause for an ocean anoxic event in the Cretaceous period. *Nature 409*, 325–9.

Ericksson P.G., Cheney E.S. (1992) Evidence for the transition to an oxygen-rich atmosphere during the evolution of red beds in the Lower Proterozoic sequences of southern Africa. *Precambr. Res. 54*, 257–69.

Erlich H.L. (1995) *Geomicrobiology.* New York: Marcel Dekker.

Erwin D.H. (1993) *The Great Paleozoic Crisis.* New York: Columbia University Press.

Erwin D.H. (1994) The Permo-Triassic extinction. *Nature 367*, 231–6.

Eshet Y., Rampino M.R., Visscher H. (1995) Fungal event and palynological record of ecological crisis and recovery across the Permian-Triassic boundary. *Geology 23*, 967–70.

Espitalié J., Laporte J.L., Madec M., Marquis F., Leplat P., Paulet J., Boutefeu, A. (1977) Méthode rapide de caractérisation des roches mères, de leur potentiel pétrolier et de leur degré d'evolution. *Rev. Inst. Fr. Pét. 32*, 23–42.

Espitalié J., Deroo G., Marquis F. (1985) La pyrolyse Rock-Eval et ses applications. Partie 1. *Rev. Inst. Fr. Pét. 40*, 563–79.

Espitalié J., Ungerer P., Irwin I., Marquis F. (1988) Primary cracking of kerogens. Experimenting and modeling C_1, C_2–C_5, C_6–C_{15} and C_{15+} classes of hydrocarbons formed. *Org. Geochem. 13*, 893–9.

Espitalié J., Marquis F., Drouet S. (1993) Critical study of kinetic modelling parameters. In *Basin Modelling: Advances and Applications* (ed. Doré A.G. et al.) *Norwegian Petroleum Soc. Spec. Publn 3*, 233–42. Amsterdam: Elsevier.

Etheridge D.M., Steele L.P., Langenfelds R.L., Francey R.J., Barnola J.M., Morgan V.I. (1996) Natural and anthropogenic changes in atmospheric CO_2 over the last 1000 years from air in Antarctic ice and firn. *J. Geophys. Res. 101*, 4115–28.

Etheridge D.M., Steele L.P., Francey R.J., Langenfelds R.L. (1998) Atmospheric methane between 1000 A.D. and present: Evidence of anthropogenic emissions and climate variability. *J. Geophys. Res. 103*, 15979–93.

Eugster H.P., Kelts K. (1983) Lacustrine chemical sediments. In *Chemical Sediments and Geomorphology* (ed. Goudie A.S., Pye K.) 321–8. London: Academic Press.

Evans D.A., Beukes N.J., Kirschvink J.L. (1997) Low-latitude glaciation in the Palaeoproterozoic era. *Nature 386*, 262–6.

Fairchild I.J. (1993) Balmy shores and icy wastes: the paradox of carbonates associated with glacial deposits in Neoproterozoic times. In *Sedimentology Review 1* (ed. Wright V.P.), 1–16. Oxford: Blackwell.

Falkowski P.G. (1997) Evolution of the nitrogen cycle and its influence on the biological sequestration of CO_2 in the ocean. *Nature 387*, 272–5.

Falkowski P.G., Barber R.T., Smetacek V. (1998) Biogeochemical controls and feedbacks on ocean primary production. *Science 281*, 200–6.

Falkowski P., Scholes R.J., Boyle E., Canadell J., Canfield D. & 12 others (2000) The global carbon cycle: a test of our knowledge of Earth as a system. *Science 290*, 291–6.

Fan S., Gloor M., Mahlman J., Pacala S., Sarmiento J., Takahashi T., Tans P. (1998) A large terrestrial carbon sink in North America implied by atmospheric and oceanic carbon dioxide data and models. *Science 282*, 442–6.

Fang H., Yongchaun S., Sitian L., Qiming A. (1995) Overpressure retardation of organic matter and petroleum generation: a case study from the Yinggehai and Qwiongdongnan basins, South China Sea. *Am. Assoc. Pet. Geol. Bull. 79*, 551–62.

FAO (1972) *Atlas of the Living Resources of the Seas.* Rome: Food & Agriculture Organization, UN Department of Fisheries.

Farquhar G.D., O'Leary M.H., Berry J.A. (1982) On the relationship between carbon isotope discrimination and the intercellular carbon dioxide concentration in leaves. *Aust. J. Plant Physiol.* **9**, 121–37.

Farquhar G.D., Ehleringer J.R., Hubrick K.T. (1989) Carbon isotope discrimination and photosynthesis. *Ann. Rev. Plant Physiol. Plant Mol. Biol.* **40**, 503–37.

Farquhar J., Wing B.A., McKeegan K.D., Harris J.W., Cartigny P., Thiemens M.H. (2002) Mass-independent sulfur of inclusions in diamond and sulfur recycling on early Earth. *Science* **298**, 2369–72.

Farrimond P., Eglinton G., Brassell S.C., Jenkyns H.C. (1990) The Cenomanian/Turonian anoxic event in Europe: an organic geochemical study. *Mar. Pet. Geol.* **7**, 75–98.

Farrimond P., Bevan J.C., Bishop A.N. (1996) Hopanoid hydrocarbon maturation by an igneous intrusion. *Org. Geochem.* **25**, 149–64.

Farrimond P., Taylor A., Telnæs N. (1998) Biomarker maturity parameters: the role of generation and thermal degradation. *Org. Geochem.* **29**, 1181–97.

Farrimond P., Bevan J.C., Bishop A.N. (1999) Tricyclic terpane maturity parameters: response to heating by an igneous intrusion. *Org. Geochem.* **30**, 1011–19.

Faulon J.L., Carlson G.A., Hatcher P.G. (1994) A three-dimensional model for lignocellulose from gymnospermous wood. *Org. Geochem.* **21**, 1169–80.

Faure G. (1986) *Principles of Stable Isotope Geology.* New York: Wiley.

Faure K., de Wit M.J., Willis J.P. (1995) Late Permian global coal hiatus linked to ^{13}C-depleted CO_2 flux into the atmosphere during the final consolidation of Pangea. *Geology* **23**, 507–10.

Fenchel T. (2001) Marine bugs and carbon flow. *Science* **292**, 2444–5.

Fenchel T., Finlay B.J. (1995) *Ecology and Evolution in Anoxic Worlds.* Oxford: Oxford University Press.

Fenchel T.M., Jørgensen B.B. (1977) Detritus food chains of aquatic ecosystems: the role of bacteria. In *Advances in Microbial Ecology, Vol. 1* (ed. Alexander M.), 1–57. New York: Plenum Press.

Fenchel T., King G.M., Blackburn T.H. (1998) *Bacterial Biogeochemistry: the Ecophysiology of Mineral Cycling.* San Diego: Academic Press.

Fent K. (1996) Ecotoxicology of organotin compounds. *Crit. Rev. Ecotoxicol.* **26**, 1–117.

Filippelli G.M., Sierro F.J., Flores J.A., Vazquez A., Utrilla R., Perez-Fogado M., Latimer J.C. (2003) A sediment–nutrient–oxygen feedback responsible for productivity variations in Late Miocene sapropel sequences of the western Mediterranean. *Palaeogeogr. Palaeoclimatol. Palaeoecol.* **190**, 335–48.

Filley T.R., Cody G.D., Goodell B., Jellison J., Noser C., Ostrofsky A. (2002) Lignin demethylation and polysaccharide decomposition in spruce sapwood degraded by brown rot fungi. *Org. Geochem.* **33**, 111–24.

Finar I.L., Finar A.L. (1998) *Organic Chemistry, Vol. 2. Stereochemistry and the Chemistry of Natural Products.* New York: Addison Wesley.

Finean J.B., Coleman R., Michell R.H. (1984) *Membranes and Their Cellular Functions.* Oxford: Blackwell Scientific.

Fischer A.G. (1984) The two Phanerozoic supercycles. In *Catastrophes in Earth History* (ed. Berggren W.A., Van Couvering J.A.), 129–50. Princeton, NJ: Princeton University Press.

Fischer G. (1991) Stable carbon isotope ratios of plankton carbon and sinking organic matter from the Atlantic sector of the Southern Ocean. *Mar. Chem.* **35**, 581–96.

Fischer F., Zillig W., Stetter K.O., Schreiber G. (1983) Chemolithoautotrophic metabolism of anaerobic extremely thermophilic archaebacteria. *Nature* **301**, 511–13.

Fisher S.J., Alexander R., Kagi R.I., Oliver G.A. (1998) Aromatic hydrocarbons as indicators of biodegradation in North Western Australian reservoirs. In *The Sedimentary Basins of Western Australia 2. Proc. of Petroleum Explor. Soc. of Australia Symp., Perth* (ed. Purcell P.G., Purcell R.R.), 185–94.

Fortey R. (1997) *Life: An Unauthorised Biography.* London: Flamingo.

Fossey J., Lefort D., Sorba J. (1995) *Free Radicals in Organic Chemistry.* Paris: Masson.

Fowler M.G. (1992) The influence of *Gloeocapsomorpha prisca* on the organic geochemistry of oils and organic-rich rocks of Late Ordovician age from Canada. In *Early Organic Evolution: Implications for Mineral and Energy Resources* (ed. Schidlowski M., Golubic S., Kimberley M.M., McKirdy D.M., Trudinger P.A.), 336–56. Berlin: Springer-Verlag.

Frakes L.A., Francis E., Syktus J.I. (1992) *Climate Modes of the Phanerozoic.* Cambridge: Cambridge University Press.

Francois R. (1987) A study of sulphur enrichment in the humic fraction of marine sediments during early diagenesis. *Geochim. Cosmochim. Acta* **51**, 17–27.

Franke C., Studinger G., Berger G., Bohling M., Bruckamann U., Cohors-Fresenborg D., Jöhncke U. (1994) The assessment of bioaccumulation. *Chemosphere* **29**, 1501–14.

Freeman K.H., Colarusso L.A. (2001) Molecular and isotopic records of C_4 grassland expansion in the late Miocene. *Geochim. Cosmochim. Acta* **65**, 1439–54.

Freeman K.H., Hayes J.M. (1992) Fractionation of carbon isotopes by phytoplankton and estimates of ancient CO_2 levels. *Glob. Biogeochem. Cycles* **6**, 185–98.

Freeman K.H., Hayes J.M., Trendel J.-M., Albrecht P. (1990) Evidence from carbon isotope measurements for diverse origins of sedimentary hydrocarbons. *Nature* **343**, 254–6.

Freyer H.-D. (1979a) Variations in the atmospheric CO_2 content. In *The Global Carbon Cycle* (ed. Bolin B., Degens E.T., Kempe S., Ketner P.) *SCOPE Rep. No. 13*, 79–96. Chichester: Wiley.

Freyer H.-D. (1979b) Atmospheric cycles of trace gases containing carbon. In *The Global Carbon Cycle* (ed. Bolin B., Degens E.T., Kempe S., Ketner P.) *SCOPE Rep. No. 13*, 101–28. Chichester: Wiley.

Friis-Christensen E., Lassen K. (1991) Length of the solar cycle: an indicator of solar activity closely associated with climate. *Science* **254**, 698–700.

Fry D.M. (1995) Reproductive effects in birds exposed to pesticides and industrial chemicals. *Environ. Health Perspect.* **103** (suppl. 7), 165–71.

Fry D.M., Toone C.K. (1981) DDT-induced feminization of gull embryos. *Science* 213, 922–4.

Fry B., Jannasch H., Molyneaux S.J., Wirsen C.O., Muramoto J.A., King S. (1991) Stable isotope studies of the carbon, nitrogen and sulfur cycles in the Black Sea and the Cariaco Trench. *Deep-Sea Res. 38(S2)*, S1003–19.

Fuex A.N. (1977) The use of stable carbon isotopes in hydrocarbon exploration. *J. Geochem. Explor.* 7, 155–88.

Fuglestvedt J.S., Berntsen T.K., Godal O., Sausen R., Shine K.P., Skodvin T. (2003) Metrics of climate change: assessing radiative forcing and emission indices. *Clim. Change* 58, 267–331.

Fung I.Y., Meyn S.K., Tegen I., Doney S.C., John J.G., Bishop J.K.B. (2000) Iron supply and demand in the upper ocean. *Glob. Biogeochem. Cycles* 14, 281–95.

Gacesa P. (1988) Alginates. *Carbohydr. Polym.* 8, 161–82.

Gaffney J.S., Premuzic E.T., Manowitz B. (1980) On the usefulness of sulfur isotope ratios in crude oil correlations. *Geochim. Cosmochim. Acta* 44, 135–9.

Gagan M.K., Chivas A.R. (1995) Oxygen isotopes in western Australian coral reveal Pinatubo aerosol-induced cooling in the Western Pacific Warm Pool. *Geophys. Res. Letts* 22, 1069–72.

Gagosian R.B., Farrington J.W. (1978) Sterenes in surface sediments from the southwest African shelf and slope. *Geochim. Cosmochim. Acta* 42, 1091–101.

Gagosian R.B., Smith S.O., Nigrelli G.E. (1982) Vertical transport of steroid alcohols and ketones measured in a sediment trap experiment in the equatorial Atlantic Ocean. *Geochim. Cosmochim. Acta* 46, 1163–72.

Gale A.S. (2000) The Cretaceous world. In *Biotic Response to Global Change: The Last 145 Million Years* (ed. Culver S.J., Rawson P.F.), 4–19. Cambridge: Cambridge University Press.

Galeotti S., Sprovieri M., Coccioni R., Bellanca A., Neri R. (2003) Orbitally modulated black shale deposition in the upper Albian Amadeus Segment (central Italy): a multi-proxy reconstruction. *Palaeogeogr. Palaeoclimatol. Palaeoecol.* 190, 441–58.

Galimov E.M. (1973) *Carbon Isotopes in Oil-Gas Geology.* Moscow: Nedra; Washington, DC: NASA (1974 translation from Russian).

Galimov E.M. (1976) Variations of the carbon cycle at present and in the geological past. In *Environmental Geochemistry, Vol. 1* (ed. Nriagu J.O.), 3–11. Ann Arbor, MI: Ann Arbor Science.

Galimov E.M. (1980) C^{13}/C^{12} in kerogen. In *Kerogen, Insoluble Organic Matter from Sedimentary Rocks* (ed. Durand B.). Paris: Editions Technip.

Garrigues P., Ewald M. (1983) Natural occurrence of 4-methyl-phenanthrene in petroleums and recent marine sediments. *Org. Geochem.* 5, 53–6.

Gastaldo R.A., DiMichele W.A., Pfefferkorn H.W. (1996) Out of the Icehouse into the Greenhouse: a Late Paleozoic analog for modern global vegetational change. *GSA Today* 6, 1–10.

Gautier D.L. (ed.) (1986) *Roles of Organic Matter in Sediment Diagenesis.* SEPM Spec. Publn 38. Tulsa, OK: Society of Economic Paleontologists & Mineralogists.

Gelin F., Volkman J.K., Largeau C., Derenne S., Sinninghe Damsté J.S., de Leeuw J.W. (1999) Distribution of aliphatic, nonhydrolyzable biopolymers in marine microalgae. *Org. Geochem.* 30, 147–59.

Gélinas Y., Baldock J.A., Hedges J.I. (2001) Organic carbon composition of marine sediments: effect of oxygen exposure on oil generation potential. *Science* 294, 145–8.

George S.C., Boreham C.J., Minifie S.A., Teerman S.C. (2002) The effect of minor to moderate biodegradation on C_5 to C_9 hydrocarbons in crude oils. *Org. Geochem.* 33, 1293–317.

Gibbs P.E., Pascoe P.L., Burt G.R. (1988) Sex change in the female dog whelk, *Nucella lapillus*, induced by tributyltin from antifouling paints. *J. Mar. Biol. Assoc. UK* 68, 715–31.

Gibson D.T., Subramanian V. (1984) Microbial degradation of aromatic hydrocarbons. In *Microbial Degradation of Organic Compounds* (ed. Gibson D.T.), 181–252. New York: Marcel Dekker.

Giggenbach W.F. (1997) Relative importance of thermodynamic and kinetic processes in governing the chemical and isotopic composition of carbon gases in high-heatflow sedimentary basins. *Geochim. Cosmochim. Acta* 61, 3763–85.

Giggenbach W.F., Sano Y., Wakita H. (1993) Isotopic composition of helium, and CO_2 and CH_4 contents in gases produced along the New Zealand part of a convergent plate boundary. *Geochim. Cosmochim. Acta* 57, 3427–55.

Gilliland R.L. (1989) Solar evolution. *Glob. Planet. Change* 1, 35–55.

Gleadow A.J.W., Brown R.W. (1999) Fission track thermochronology and the long-term denudational response to tectonics. In *Geomorphology and Global Tectonics* (ed. Summerfield M.A.), 57–75. Chichester: Wiley.

Gluyas J., Swarbrick R. (2004) *Petroleum Geoscience.* Oxford: Blackwell.

Goad L.J., Lenton J.R., Knapp F.F., Goodwin T.W. (1974) Phytosterol side chain biosynthesis. *Lipids* 9, 582–95.

Gobas F.A.P.C., Muir D.C.G., Mackay D. (1988) Dynamics of dietary bioaccumulation and faecal elimination of hydrophobic organic chemicals in fish. *Chemosphere* 17, 943–62.

Goericke R., Montoya J.P., Fry B. (1994) Physiology of isotope fractionation in algae and cyanobacteria. In *Stable Isotopes in Ecology* (ed. Lajtha K., Michener B.), Chapter 9. Oxford: Blackwell.

Gold T. (1992) The deep, hot biosphere. *Proc. Natl Acad. Sci. USA* 89, 6045–9.

Goldberg E.D. (1976) *The Health of the Oceans.* Paris: UNESCO.

Gong C., Hollander D.J. (1999) Evidence for differential degradation of alkenones under contrasting bottom water oxygen conditions: implication for paleotemperature reconstruction. *Geochim. Cosmochim. Acta* 63, 405–11.

Goñi M.A., Eglinton T.I. (1996) Stable carbon isotopic analyses of lignin-derived CuO oxidation products by isotope ratio monitoring-gas chromatography-mass spectrometry (irm-GC-MS). *Org. Geochem.* 24, 601–15.

Goodfriend G.A., Collins M.J., Fogel M.L., Macko S.A., Wehmiller J.F. (eds) (2001) *Perspectives in Amino Acid and Protein Geochemistry*. Oxford: Oxford University Press.

Goossens H., de Leeuw J.W., Schenck P.A., Brassell S.C. (1984) Tocopherols as likely precursors of pristane in ancient sediments and crude oils. *Nature 312*, 440–2.

Goossens H., Rijpstra I.C., Düren R.R., de Leeuw J.W., Schenck P.A. (1986) Bacterial contribution to sedimentary organic matter; a comparative study of lipid moieties in bacteria and Recent sediments. *Org. Geochem.* 10, 683–96.

Goossens H., de Lange F., de Leeuw J.W., Schenk P.A. (1988a) The Pristane Formation Index, a molecular maturity parameter. Confirmation in samples from the Paris Basin. *Geochim. Cosmochim. Acta 52*, 2439–44.

Goossens H., Due A., de Leeuw J.W., van de Graaf B., Schenck P.A. (1988b) The Pristane Formation Index, a molecular maturity parameter. A simple method to assess maturity by pyrolysis/evaporation–gas chromatography of unextracted samples. *Geochim. Cosmochim. Acta 52*, 1189–93.

Goth K., de Leeuw J.W., Püttmann W., Tegelaar E.W. (1988) Origin of Messel Oil Shale kerogen. *Nature 336*, 759–61.

Gough D.O. (1981) Solar interior structure and luminosity variations. *Solar Physics 74*, 21–34.

Gould S.J. (1991) *Wonderful Life. The Burgess Shale and the Nature of History*. London: Penguin Books.

Grady M.M. (1999) Meteorites and microfossils from Mars. *Geoscientist 9*, 4–7.

Graham J.B., Dudley R., Aguilkar N.M., Gans C. (1995) Implications of the late Paleozoic oxygen pulse for physiology and evolution. *Nature 375*, 117–20.

Granger J., Price N.M. (1999) Importance of siderophores in iron nutrition of heterotrophic marine bacteria. *Limnol. Oceanogr.* 44, 541–55.

Grantham P.J., Wakefield L.L. (1988) Variations in the sterane carbon number distributions of marine source rock derived crude oils through geological time *Org. Geochem.* 12, 61–73.

Grantham P.J., Posthuma J., DeGroot K. (1980) Variation and significance of the C_{27} and C_{28} triterpane content of a North Sea core and various North Sea crude oils. In *Advances in Organic Geochemistry 1979* (ed. Douglas A.G., Maxwell J.R.), 29–38. New York: Pergamon Press.

Grantham P.J., Posthuma J., Baak A. (1983) Triterpanes in a number of Far-Eastern crude oils. In *Advances in Organic Geochemistry 1981* (ed. Bjorøy M. et al.), 675–83. Chichester: Wiley.

Greenberg J.M. (1997) Prebiotic chiral molecules created in interstellar dust and preserved in comets, comet dust and meteorites: an exogenous source of life's origins. *Proc. SPIE 3111*, 226–37.

Greiner A.C., Spyckerelle C., Albrecht P. (1976) Aromatic hydrocarbons from geological sources–I. New naturally occurring phenanthrene and chrysene derivatives. *Tetrahedron 32*, 257–60.

Grice K., Schouten S., Nissenbaum A., Charrach J., Sinninghe Damsté J.S. (1998) A remarkable paradox: Sulfurised freshwater algal (*Botryococcus braunii*) lipids in an ancient hypersaline euxinic ecosystem. *Org. Geochem.* 28, 195–216.

Grice K., Alexander R., Kagi R.I. (2000) Diamondoid hydrocarbon ratios as indicators of biodegradation in Australian crude oils. *Org. Geochem.* 31, 67–73.

Grimalt J., Albaigés J. (1987) Sources and occurrence of C_{12}–C_{22} n-alkane distributions with even carbon-number preference in sedimentary environments. *Geochim. Cosmochim. Acta 51*, 1379–84.

Grossi V., Hirschler A., Raphel D., Rontani J.-F., de Leeuw J.W., Bertrand J.-C. (1998) Biotransformation pathways of phytol in Recent anoxic sediments. *Org. Geochem.* 29, 845–61.

Gschwend P.M., Hites R.A. (1981) Fluxes of polycyclic aromatic hydrocarbons to marine and lacustrine sediments in the northern United States. *Geochim. Cosmochim. Acta 45*, 2359–67.

Guillette L.J. Jr, Crain D.A., Rooney A.A., Pickford D.B. (1995) Organization versus activation: the role of endocrine-disrupting contaminants (EDCs) during embryonic development in wildlife. *Environ. Health Perspect. 103 (suppl. 7)*, 157–64.

Gurnis M. (1990) Ridge spreading, subduction, and sea level fluctuations. *Science 250*, 970–2.

Guy R.D., Fogel M.L., Berry J.A. (1993) Photosynthetic fractionation of the stable isotopes of oxygen and carbon. *Plant Physiol. 101*, 37–47.

Habicht K.S., Gade M., Thamdrup B., Berg P., Canfield D.E. (2002) Calibration of sulfate levels in the Archean Ocean. *Science 298*, 2372–8.

Hall A., Manabe S. (1999) The role of water vapor feedback in unperturbed climate variability and global warming. *J. Clim. 12*, 2327–46.

Hallam A., Wignall P.B. (1997) *Mass Extinctions and Their Aftermath*. Oxford: Oxford University Press.

Halliday A.N. (2000) Terrestrial accretion rates and the origin of the Moon. *Earth Planet. Sci. Letts 176*, 17–30.

Halpern H.I. (1995) Development and applications of light-hydrocarbon-based star diagrams. *Am. Assoc. Pet. Geol. Bull. 79*, 801–15.

Hammel K.E., Kalyamaraman B., Kirk T.K. (1986) Substrate free radicals are intermediates in ligninase catalysis. *Proc. Natl Acad. Sci. USA 83*, 3708–12.

Han J., Calvin M. (1969) Hydrocarbon distribution of algae and bacteria, and microbiological activity in sediments. *Proc. Natl Acad. Sci. USA 64*, 436–43.

Han T.M., Runnegar B. (1992) Megascopic eukaryotic algae from the 2.1-billion-year-old Negaunee Iron-Formation, Michigan. *Science 257*, 232–5.

Hansen K.W., Wallmann K. (2003) Cretaceous and Cenozoic evolution of seawater composition, atmospheric O_2 and CO_2: a model perspective. *Am. J. Sci. 303*, 94–148.

Hanson R.S., Hanson T.E. (1996) Methanotrophic bacteria. *Microbiol. Rev. 60*, 439–71.

Harland W.B. (1983) The Proterozoic glacial record. In *Proterozoic Geology: Selected Papers from an International Proterozoic*

Symposium (ed. Medaris L.G. Jr. et al.) *Geol. Soc. Am. Memoir 161*, 279–88.

Harms H., Zehnder A.J.B. (1995) Bioavailability of sorbed 3-chlorodibenzofuran. *Appl. Environ. Microbiol. 61*, 27–33.

Harnisch J., Eisenhaeuer A. (1998) Natural CF_4 and SF_6 on Earth. *Geophys. Res. Letts. 25*, 2401–4.

Harnisch J., Borchers R., Fabian P., Maiss M. (1996) Tropospheric trends for CF_4 and CF_3CF_3 since 1982 derived from SF_6 dated stratospheric air. *Geophys. Res. Letts 23*, 1099–102.

Harnisch J., Borchers R., Fabian P., Maiss M. (1999) CF_4 and the age of mesospheric and polar vortex air. *Geophys. Res. Letts 26*, 295–8.

Harradine P.J., Harris P.G., Head R.N., Harris R.P., Maxwell J.R. (1996) Steryl chlorin esters are formed by zooplankton herbivory. *Geochim. Cosmochim. Acta 60*, 2265–70.

Hartgers W.A., Lòpez J.F., Sinninghe Damsté J.S., Reiss C., Maxwell J.R., Grimalt J.O. (1997) Sulfur-binding in recent environments: II. Speciation of sulfur and iron and implications for the occurrence of organo-sulfur compounds. *Geochim. Cosmochim. Acta 61*, 4769–88.

Hartnett H.E., Keil R.G., Hedges J.I., Devol A.H. (1998) Influence of oxygen exposure time on organic carbon preservation in continental margin sediments. *Nature 391*, 572–4.

Harvey J.G. (1982) *Atmosphere and Ocean*. London: Artemis Press.

Harvey L.D.D. (2000) *Global Warming: The Hard Science*. Harlow: Pearson Education.

Harvey H.R., McManus G.B. (1991) Marine ciliates as a widespread source of tetrahymanol and hopan-3β-ol in sediments. *Geochim. Cosmochim. Acta 55*, 3387–90.

Harvey G.R., Steinhauer W.G. (1976) Biogeochemistry of PCB and DDT in the North Atlantic. In *Environmental Biogeochemistry, Vol. 1* (ed. Nriagu J.O.), 203–21. Ann Arbor, MI: Ann Arbor Science.

Harvey G.R., Miklas H.P., Bowen V.T., Steinhaeuer W.G. (1974a) Observations on the distribution of chlorinated hydrocarbons in Atlantic Ocean organisms. *J. Mar. Res. 32*, 103–18.

Harvey G.R., Steinhaeuer W.G., Miklas H.P. (1974b) Decline of PCB concentrations in North Atlantic surface waters. *Nature 252*, 387–8.

Harvey H.R., Eglinton G., O'Hara S.C.M., Corner E.D.S. (1987) Biotransformation and assimilation of dietary lipids by *Calanus* feeding on a dinoflagellate. *Geochim. Cosmochim. Acta 51*, 3031–40.

Harwood J.L., Russell N.J. (1984) *Lipids in Plants and Microbes*. London: George Allen & Unwin.

Hatcher P.G. (1988) Dipolar dephasing ^{13}C NMR studies of decomposed wood and coalified xylem tissue: defunctionalization of lignin structural units during coalification. *Energy Fuels 2*, 48–58.

Hatcher P.G. (1990) Chemical structural models for coalified wood (vitrinite) in low rank coal. *Org. Geochem. 16*, 959–68.

Hatcher P.G., Clifford D.J. (1997) The organic geochemistry of coal: from plant materials to coal. *Org. Geochem. 27*, 251–74.

Hatcher P.G., Rowan R., Mattingly M.A. (1980) ^{1}H and ^{13}C NMR of marine humic acids. *Org. Geochem. 2*, 77–85.

Hatcher P.G., Maciel G.E., Dennis L.W. (1981) Aliphatic structure of humic acids; a clue to their origin. *Org. Geochem. 3*, 43–8.

Hatcher P.G., Spiker E.C., Szeverenyi N.M., Maciel B.E. (1983) Selective preservation: the origin of petroleum-forming aquatic kerogen. *Nature 305*, 498–501.

Hatcher P.G., Lerch H.E. III, Bates A.L., Verheyen T.V. (1989a) Solid-state ^{13}C nuclear magnetic resonance studies of coalified gymnosperm xylem tissue from Australian brown coals. *Org. Geochem. 14*, 145–55.

Hatcher P.G., Wilson M.A., Vassallo A.M., Lerch H.E. III (1989b) Studies of angiospermous wood in Australian brown coal by nuclear magnetic resonance and analytical pyrolysis: new insights into the early coalification process. *Int. J. Coal Geol. 13*, 99–126.

Hatcher P.G., Faulon J.-P., Wenzel K.A., Cody G.D. (1992) A structural model for lignin-derived vitrinite from high-volatile bituminous coal (coalified wood). *Energy Fuels 6*, 813–20.

Hayatsu R., Botto R.E., Scott R.G., McBeth R.L., Winans R.E. (1987) Thermal catalytic transformation of pentacyclic triterpenoids: alteration of geochemical fossils during coalification. *Org. Geochem. 11*, 245–50.

Hayes J.M. (1991) Stability of petroleum. *Nature 352*, 108–9.

Hayes J.M. (1993) Factors controlling ^{13}C contents of sedimentary organic compounds: Principles and evidence. *Mar. Geol. 113*, 111–25.

Hayes J.M., Strauss H., Kaufman A.J. (1999) The abundance of ^{13}C in marine organic matter and isotopic fractionation in the global biogeochemical cycle of carbon during the past 800 Ma. *Chem. Geol. 161*, 103–25.

Hayes J.M., Takigiku R., Ocampo R., Callot H.J., Albrecht P. (1987) Isotopic compositions and probable origins of organic molecules in the Eocene Messel shale. *Nature 329*, 48–53.

Hayes M.H.B., MacCarthy P., Malcolm R.L., Swift R.S. (eds) (1989) *Humic Substances II. In Search of Structure*. Chichester: Wiley.

Hays J.D., Imbrie J., Shackleton N.J. (1976) Variations in the Earth's orbit: pacemaker of the Ice Ages. *Science 194*, 1121–32.

Heath D.J., Lewis C.A., Rowland S.J. (1997) The use of high temperature gas chromatography to study the biodegradation of high molecular weight hydrocarbons. *Org. Geochem. 26*, 769–85.

Hecky R.E., Kilham P. (1988) Nutrient limitation of phytoplankton in freshwater and marine environments: a review of recent evidence on the effects of enrichment. *Limnol. Oceanogr. 33*, 796–822.

Hedges J.I. (1978) The formation and clay mineral reactions of melanoidins. *Geochim. Cosmochim. Acta 42*, 69–72.

Hedges J.I. (1992) Global biogeochemical cycles: progress and problems. *Mar. Chem. 39*, 67–93.

Hedges J.I., Ertel J.R. (1982) Characterisation of lignin by gas capillary chromatography of cupric oxide oxidation products. *Anal. Chem. 54*, 174–8.

Hedges J.I., Keil R.G. (1995) Sedimentary organic matter preservation: an assessment and speculative synthesis. *Mar. Chem.* 49, 81–115. This article should be read in conjunction with the following comments by Pedersen T.I. (117–19), Berner R.A. (121–2), Mayer L.M. (123–6), Henrichs S.M. (127–36), and response by Hedges J.I. & Keil R.G. (137–9).

Hedges J.I., Mann D.C. (1979) The characterisation of plant tissues by their lignin oxidation products. *Geochim. Cosmochim. Acta* 43, 1803–7.

Hedges J.I., Cowie G.L., Ertel J.R., Barbour R.J., Hatcher P.G. (1985) Degradation of carbohydrates and lignins in buried woods. *Geochim. Cosmochim. Acta* 49, 701–11.

Hedges J.I., Eglinton G., Hatcher P.G., Kirchman D.L., Arnosti C. & 7 others (2000) The molecularly-uncharacterized component of nonliving organic matter in natural environments. *Org. Geochem.* 31, 945–58.

Hedges J.I., Baldock J.A., Gélinas Y., Lee C., Peterson M.L., Wakeham S.G. (2001) Evidence for non-selective preservation of organic matter in sinking marine particles. *Nature* 409, 801–4.

Heider J., Fuchs G. (1997a) Anaerobic metabolism of aromatic compounds. *Eur. J. Biochem.* 243, 577–96.

Heider J., Fuchs G. (1997b) Microbial anaerobic metabolism. *Anaerobe* 3, 1–22.

Heinrich A.K. (1962) The life histories of plankton animals and seasonal cycles of plankton communities in the oceans. *J. Cons.* 27, 15–24.

Held I.M., Soden B.J. (2000) Water vapor feedback and global warming. *Ann. Rev. Energy Env.* 25, 441–75.

Henrichs S.M. (1992) Early diagenesis of organic matter in marine sediments: progress and perplexity. *Mar. Chem.* 39, 119–49.

Henriksson A.S., Sarnthein M., Eglinton G., Poynter J. (2000) Dimethylsulfide production variations over the past 200 ky in the equatorial Atlantic: a first estimate. *Geology* 28, 499–502.

Heredy L.A., Wender I. (1980) Model structure for a bituminous coal. *Am. Chem. Soc. Div. Fuel Chem. Prepr.* 25, 38–45.

Highwood E.J., Shine K.P. (2000) Radiative forcing and global warming potentials of 11 halogenated compounds. *J. Quant. Spectrosc. Radiat. Transf.* 66, 169–83.

Higuchi T. (ed.) (1985) *Biosynthesis and Biodegradation of Wood*. San Diego: Academic Press.

Hines M.E., Banta G.T., Giblin A.E., Hobbie J.E. (1994) Acetate concentrations and oxidation in salt-marsh sediments. *Limnol. Oceanogr.* 39, 140–8.

Hinrichs K.-U., Hayes J.M., Sylva S.P., Brewer P.G., De Long E.F. (1999) Methane-consuming archaebacteria in marine sediments. *Nature* 398, 802–5.

Hinrichs K.-W., Summons R.E., Orphan V., Sylva S.P., Hayes J.M. (2000) Molecular and isotopic analysis of anaerobic methane-oxidizing communities in marine sediments. *Org. Geochem.* 31, 1685–701.

Hirner A.V., Robinson B.W. (1989) Stable isotope geochemistry of crude oils and of possible source rocks from New Zealand—2: sulfur. *Appl. Geochem.* 4, 121–30.

Hirner A.V., Graf W., Treibs R., Melzer A.N., Hahn-Weinheimer P. (1984) Stable sulfur and nitrogen isotopic compositions of crude oil fractions from southern Germany. *Geochim. Cosmochim. Acta* 48, 2179–86.

Hoefs J. (1997) *Stable Isotope Geochemistry*. Berlin: Springer.

Hoefs M.J.L., Sinninghe Damsté J.S., de Leeuw J.W. (1995) A novel C_{35} highly branched isoprenoid polyene in Recent Indian Ocean sediments. *Org. Geochem.* 23, 263–7.

Hoefs M.J.L., Versteegh G.J.M., Rijpstra W.I.C., de Leeuw J.W., Sinninghe Damsté J.S. (1998) Postdepositional oxic degradation of alkenones: implications for the measurement of palaeo sea surface temperatures. *Paleoceanogr.* 13, 42–9.

Hoehler T.M., Alperin M.J., Alpert D.B., Martens C.S. (1994) Field and laboratory studies of methane oxidation in an anoxic marine sediment: evidence for a methanogen-sulfate reducer consortium. *Glob. Biogeochem. Cycles* 8, 451–63.

Hoffert M.I., Frei A., Narayanan V.K. (1988) Application of solar max acrim data to analysis of solar-driven climatic variability on earth. *Clim. Change* 13, 267–86.

Hoffman P.F., Schrag D.P. (2000) Snowball Earth. *Sci. Am.*, Jan., 50–7.

Hoffman P.F., Kaufman A.J., Halverson G.P., Schrag D.P. (1998) A Neoproterozoic snowball Earth. *Science* 281, 1342–6.

Hoffmann J., Hower J. (1979) Clay mineral assemblages as low grade metamorphic geothermometers: applications to the thrust faulted disturbed belt of Montana, USA. In *Aspects of Diagenesis* (ed. Scholle P.A., Schluger P.R.), *SEPM Spec. Publn 26*, 5579. Tulsa, OK: Society of Economic Paleontologists & Mineralogists.

Hoffmann C.F., Foster C.B., Powell T.G., Summons R.E. (1987) Hydrocarbon biomarkers from Ordovician sediments and the fossil alga *Gloeocapsomorpha prisca* Zalessky 1917. *Geochim. Cosmochim. Acta* 51, 2681–97.

Holba A.G., Tegelaar E.W., Huizinga B.J., Moldowan J.M., Singletary M.S., McCaffrey M.A., Dzou L.I.P. (1998) 24-Norcholestanes as age-sensitive molecular fossils. *Geology* 26, 783–6.

Holba A.G., Tegelaar E., Ellis B.L., Singletary M.S., Albrecht P. (2000) Tetracyclic polyprenoids: Indicators of freshwater (lacustrine) algal input. *Geology* 28, 251–4.

Holland H.D. (1984) *The Chemical Evolution of the Atmosphere and Oceans*. Princeton, NJ: Princeton University Press.

Holland H.D. (1994) The phosphate–oxygen connection. *Eos (Trans., Am. Geophys. Union)* 75, OS96.

Holland H.D., Beukes N.J. (1990) A paleoweathering profile from Griqualand West, South Africa: evidence for a dramatic rise in atmospheric oxygen between 2.2 and 1.9 BYBP. *Am. J. Sci. A290*, 1–34.

Hollander D.J., McKenzie J.A., Hsü K.J. (1993) Carbon isotope evidence for unusual plankton blooms and fluctuations of surface water CO_2 in 'Strangelove Ocean' after terminal Cretaceous event. *Palaeogeogr. Palaeoclim. Palaeoecol.* 104, 229–37.

Holloway P.J. (1982) The chemical constitution of plant cutins. In *The Plant Cuticle* (ed. Cutler D.F., Alvin K.L., Price C.E.), 45–85. Linnean Society. London: Academic Press.

Holm N.G., Charlou J.L. (2001) Initial indicators of abiotic formation of hydrocarbons in the Rainbow ultramafic hydrothermal system, Mid-Atlantic Ridge. *Earth Planet. Sci. Letts 191*, 1–8.

Holmes S. (1983) *Outline of Plant Classification*. London: Longman.

Holser W.T., Maynard J.B., Cruikshank K.M. (1989a) Modelling the natural cycle of sulphur through Phanerozoic time. In *Evolution of the Global Biogeochemical Sulphur Cycle* (ed. Brumblecombe P., Lein A.Y.), 21–56. New York: Wiley.

Holser W.T., Schönlaub H.P., Attrep M. Jr., Boeckelmann K., Klein P., & 10 others (1989b) A unique geochemical record at the Permian/Triassic boundary. *Nature 337*, 39–44.

Horgan J. (1991) In the beginning . . . *Scientific Am. 264, Feb.*, 100–9.

Horita J., Berndt M.E. (1999) Abiogenic methane formation and isotopic fractionation under hydrothermal conditions. *Science 285*, 1055–7.

Horsfield B. (1989) Practical criteria for classifying kerogens: some observations from pyrolysis–gas chromatography. *Geochim. Cosmochim. Acta 53*, 891–901.

Houghton J.T., Meira Filho L.G., Bruce J., Hoesung Lee, Callander B.A., Haites E., Harris N., Maskell K. (1995) *Climate Change 1994: Radiative Forcing of Climate Change*. Cambridge: Cambridge University Press.

Houghton J.T., Meira Filho L.G., Callander B.A., Harris N., Kattenberg A., Maskell K. (1996) *Climate Change 1995: The Science of Climate Change*. Cambridge: Cambridge University Press.

Houghton J.T., Ding Y., Griggs D.J., Noguer M., van der Linden P.J., Dai X., Maskell K., Johnson C.A. (eds.) (2001) *Climate Change 2001: The Scientific Basis*. Cambridge: Cambridge University Press.

House C.H., Schopf J.W., Stetter K.O. (2003) Carbon isotopic fractionation by Archaeans and other thermophilic prokaryotes. *Org. Geochem. 34*, 345–56.

Hsü K.J., McKenzie J.A. (1990) Carbon-isotope anomalies at era boundaries; global catastrophes and their ultimate cause. *Geol. Soc. Am. Spec. Pap. 247*, 61–70.

Huang W.-L. (1996) Experimental study of vitrinite maturation: effects of temperature, time, pressure, water, and hydrogen index. *Org. Geochem. 24*, 233–41.

Huang W.-Y., Meinschein W.G. (1979) Sterols as ecological indicators. *Geochim. Cosmochim. Acta 43*, 739–45.

Huang Y., Freeman K.H., Eglinton T.I., Street-Perrott F.A. (1999) $\delta^{13}C$ analyses of individual lignin phenols in Quaternary lake sediments: a novel proxy for deciphering past terrestrial vegetation changes. *Geology 27*, 471–4.

Huang Y., Street-Perrott F.A., Metcalfe S.E., Brenner M., Moreland M., Freeman K.H. (2001) Climate change as the dominant control on glacial–interglacial variations in C_3 and C_4 plant abundance. *Science 293*, 1647–51.

Hufnagel B. (1997) Biography of a Star: Our Sun's Birth, Life, and Death. San Francisco: Astronomical Society of the Pacific (http://www.astrosociety.org/education/publications/tnl/39/39.html).

Hughes W.B., Holba A.G., Dzou L.I.P. (1995) The ratios of dibenzothiophene to phenanthrene and pristane to phytane as indicators of depositional environment and lithology of petroleum source rocks. *Geochim. Cosmochim. Acta 59*, 3581–98.

Huizinga B.J., Tannenbaum E., Kaplan I.R. (1987) The role of minerals in the thermal alteration of organic matter—IV. Generation of n-alkanes, acyclic isoprenoids and alkenes in laboratory experiments. *Geochim. Cosmochim. Acta 51*, 1083–97.

Hungate B.A., Holland E.A., Jackson R.B., Chaplin F.S., Mooney H.A., Field C.B. (1997) The fate of carbon in grasslands under carbon dioxide enrichment. *Nature 388*, 576–9.

Hunten D.M. (1993) Atmospheric evolution of the terrestrial planets. *Science 259*, 915–20.

Hurtgen M.T., Arthur M.A., Suits N.S., Kaufman A.J. (2002) The sulfur isotopic composition of Neoproterozoic seawater sulfate: implications for a snowball Earth? *Earth Planet. Sci. Letts 203*, 413–29.

Hussler G., Albrecht P. (1983) C_{27}–C_{29} Monoaromatic anthrasteroid hydrocarbons in Cretaceous black shales. *Nature 304*, 262–3.

Hussler G., Chappe B., Wehrung P., Albrecht P. (1981) C_{27}–C_{29} ring A monoaromatic steroid hydrocarbons in Cretaceous black shales. *Nature 294*, 556–8.

Hussler G., Connan J., Albrecht P. (1984) Novel families of tetra- and hexacyclic aromatic hopanoids predominant in carbonate rocks and crude oils. *Org. Geochem. 6*, 39–49.

Hutchins D.A., Witter A.E., Butler A., Luther G.W. (1999) Competition among marine phytoplankton for different chelated iron species. *Nature 400*, 858–61.

Hutchinson G.E., Löffler H. (1956) The thermal classification of lakes. *Proc. Natl Acad. Sci. USA 42*, 84–6.

Hwang J., Druffel E.R.M. (2003) Lipid-like material as the source of uncharacterized organic carbon in the ocean? *Science 299*, 881–4.

Hyde W.T., Crowley T.J., Baum S.K., Peltier W.R. (2000) Neoproterozoic 'snowball Earth' simulations with a coupled climate/ice-sheet model. *Nature 405*, 425–9.

Iannuzzi T.J., Harrington N.W., Shear N.M., Curry C.L., Carson-Lynch H., Henning M.H., Su S.H., Raabe D.E. (1996) Distribution of key exposure factors controlling the uptake of xenobiotic chemicals in an estuarine food web. *Environ. Toxicol. Chem. 15*, 1979–92.

Ibata K., Kageyu A., Takigawa T., Okada M, Nishida T., Mizuno M., Tanaka Y. (1984) Polyprenols from conifers: multiplicity in chain length distribution. *Phytochem. 23*, 2517–21.

IEH (1999) *IEH Assessment on the Ecological Significance of Endocrine Disruption: Effects on Reproductive Function and Consequences for Natural Populations (Assessment A4)*. Leicester: MRC Inst. Environment & Health.

Irwin H., Hurst A. (1983) Applications of geochemistry to sandstone reservoir studies. In *Petroleum Geochemistry of Europe* (ed. Brooks J.) *Geol. Soc. Spec. Publn 12*, 127–46. Oxford: Blackwell Scientific.

Irwin H., Curtis C., Coleman M. (1977) Isotopic evidence for source of diagenetic carbonates formed during burial of organic-rich sediments. *Nature 269*, 209–13.

Isaksen I.S.A., Hov O. (1987) Calculations of trends in the tropospheric concentrations of O_3, OH, CO, CH_4 and NO_x. *Tellus 39B*, 271–83.

Ishiwatari R. (1985) Geochemistry of humic substances. In *Humic Substances in Soil, Sediment, and Water: Geochemistry, Isolation, and Characterization* (ed. McKnight D.M.), 147–80. New York: Wiley.

Ishiwatari R., Fukushima K. (1979) Generation of unsaturated and aromatic hydrocarbons by thermal alteration of young kerogen. *Geochim. Cosmochim. Acta 43*, 1343–9.

Isozaki Y. (1997) Permo-Triassic boundary superanoxia and stratified superocean: records from the lost deep sea. *Science 276*, 235–8.

Iwata H., Tanabe S., Sakal N., Tatsukawa R. (1993) Distribution of persistent organochlorines in the oceanic air and surface seawater and the role of ocean on their global transport and fate. *Environ. Sci. Toxicol. 27*, 1080–98.

Jacobsen S.B. (2001) Gas hydrates and deglaciations. *Nature 412*, 691–3.

Jacobsen S.B., Kauffman A.J. (1999) The Sr, C and O isotopic evolution of Neoproterozoic seawater. *Chem. Geol. 161*, 37–57.

Jahnke L., Klein H.P. (1979) Effects of low levels of oxygen on *Sacchromyces cerevisial*. *Origins Life 9*, 329–34.

Jain A.K., Briegleb B.P., Minschwaner K., Wuebbles D.J. (2000) Radiative forcings and global warming potentials of 39 greenhouse gases. *J. Geophys. Res. 105*, 20773–90.

James A.T. (1983) Correlation of natural gas by use of carbon isotope distribution between hydrocarbon components. *Am. Assoc. Pet. Geol. Bull. 67*, 1176–91.

Jannasch H.W., Mottl M.J. (1985) Geomicrobiology of deep-sea hydrothermal vents. *Science 229*, 717–25.

Jarvie D.M. (1991) Factors affecting Rock-Eval derived kinetic parameters. *Chem. Geol. 93*, 79–99.

Jarvis P., Linder S. (2000) Constraints to growth of boreal forests. *Nature 405*, 904–5.

Jenkyns H.C. (1980) Cretaceous anoxic events:from continents to oceans. *J. Geol. Soc., Lond. 137*, 171–88.

Jenkyns H.C. (1999) Mesozoic anoxic events and palaeoclimate. *Zentr. Geol. Paläontol. 1997*, 943–9.

Jenkyns H.C., Clayton C.J. (1997) Lower Jurassic epicontinental carbonates and mudstones from England and Wales: chemostratigraphic signals and the early Toarcian anoxic event. *Sedimentology 44*, 687–706.

Jenkyns H.C., Wilson P.A. (1999) Stratigraphy, paleoceanography, and evolution of Cretaceous Pacific guyots: relics from a greenhouse Earth. *Am. J. Sci. 299*, 341–92.

Jenkyns H.C., Gale A.S., Corfield R.M. (1994) Carbon- and oxygen-isotope stratigraphy of the English Chalk and Italian Scaglia and its palaeoclimatic significance. *Geol. Mag. 131*, 1–31.

Jiamo F., Guoying S. (1989) Biological marker composition of typical source rocks and related crude oils of terrestrial origin in The People's Republic of China: a review. *Appl. Geochem. 4*, 13–22.

Jiamo F., Guoying S., Pingan P., Brassell S.C., Eglinton G., Jiyang J. (1986) Peculiarities of salt lake sediments as potential source rocks in China. *Org. Geochem. 10*, 119–26.

Jiang Z.S., Fowler M.G. (1986) Carotenoid-derived alkanes in oils from northwestern China. *Org. Geochem. 10*, 831–9.

Jobson A.M., Cook F.D., Westlake D.W.S. (1979) Interaction of aerobic and anaerobic bacteria in petroleum biodegradation. *Chem. Geol. 24*, 355–65.

Johns R.B., Nichols P.D., Perry G.J. (1979) Fatty acid composition of ten marine algae from Australian waters. *Phytochem. 18*, 799–802.

Johns R.B., Gillan F.T., Volkman J.K. (1980) Early diagenesis of phytyl esters in a contemporary temperate intertidal sediment. *Geochim. Cosmochim. Acta 44*, 183–8.

Jones R.W. (1987) Organic facies. In *Advances in Petroleum Geochemistry Vol. 2* (ed. Brooks J., Welte D.), 1–90. London: Academic Press.

Jones C.E., Jenkyns H.C. (2001) Seawater strontium isotopes, oceanic anoxic events, and seafloor hydrothermal activity in the Jurassic and Cretaceous. *Am. J. Sci. 301*, 112–49.

Jones D.M., Douglas A.G., Parkes R.J., Taylor J., Giger W., Schaffner C. (1983) The recognition of biodegraded petroleum-derived aromatic hydrocarbons in recent marine sediments. *Mar. Pollut. Bull. 14*, 103–8.

Jones K.C., Alcock R.E., Johnson D.L., Northcott G.L., Semple K.T., Woolgar P.J. (1996) Organic chemicals in contaminated land: analysis, significance and research priorities. *Land Contam. Reclam. 4*, 189–97.

Joos F., Bruno M. (1998) Long-term variability of the terrestrial and oceanic carbon sinks and the budgets of the carbon isotopes ^{13}C and ^{14}C. *Glob. Biogeochem. Cycles 12*, 277–95.

Jörgensen B.B. (1982) Mineralisation of organic matter in the sea bed: the role of sulphate reduction. *Nature 296*, 643–5.

Jörgensen B.B. (1983a) The microbial sulphur cycle. In *Microbial Geochemistry* (ed. Krumbein W.E.), 91–124. Oxford: Blackwell Scientific.

Jörgensen B.B. (1983b) Processes at the sediment–water interface. In *The Major Biogeochemical Cycles and their Interactions* (ed. Bolin B., Cook R.B.) SCOPE Rep. No. 21, 477–515. Chichester: Wiley.

Jouzel J., Lorius C., Johnsen S., Grootes P. (1994) Climate instabilities: Greenland and Antarctic records. *Compte Rondue Acad. Sci. Paris Ser. II*, 65–77.

Kaeberlein T., Lewis K., Epstein S.S. (2002) Isolating 'uncultivable' microorganisms in pure culture in a simulated natural environment. *Science 296*, 1127–29.

Kahn M.I., Tadamichi O., Ku T. (1981) Paleotemperatures and the glacially induced changes in the oxygen-isotope composition of sea water during late Pleistocene and Holocene time in Tanner Basin, California. *Geology 9*, 485–90.

Kaiho K., Arinobu T., Ishiwatari R., Morgans H., Okada H. & 8 others (1996) Latest Paleocene benthic foraminiferal extinction and environmental changes at Tawanui, New Zealand. *Paleoceanogr. 11*, 447–65.

Kaiser J. (2000) Just how bad is dioxin? *Science 288*, 1941–4.

Kaiser W.R., Hamilton D.S., Scott A.R., Tyler R., Finley R.J. (1994) Geological and hydrological producibility of coalbed methane. *J. Geol. Soc. Lond. 151*, 417–20.

Kaldas R.S., Hughes G.L. (1989) Reproductive and general metabolic effects of phytoestrogens in mammals. *Reprod. Toxicol. 3*, 81–9.

Kannenberg E.L., Poralla K. (1999) Hopanoid biosynthesis and function in bacteria. *Naturwissenschaflen 86*, 168–76.

Karl T.R., Trenberth K.E. (1999) The human impact on climate. *Sci. Am., Dec.*, 62–7.

Karner D.B., Muller R.A. (2000) A causality problem for Milankovitch. *Science 288*, 2143–4.

Kashefi K., Lovley D.R. (2003) Extending the upper temperature limit for life. *Science 301*, 934.

Kasting J.F. (1987) Theoretical constraints on oxygen and carbon dioxide concentrations in the Precambrian atmosphere. *Precambr. Res. 34*, 205–29.

Kasting J.F. (1991) Box models for the evolution of atmospheric oxygen: an update. *Palaeogeogr. Palaeoclimatol. Palaeoecol. Glob. Planet. Change Sect. 97*, 125–31.

Kasting J.F. (1993) Earth's early atmosphere. *Science 259*, 920–6.

Katz M.E., Pak D.K., Dickens G.R., Miller K.G. (1999) The source and fate of massive carbon input during the latest Paleocene thermal maximum. *Science 286*, 1531–3.

Kaufman A.J. (1997) An ice age in the tropics. *Nature 386*, 227–8.

Kaufman D.S., Miller G.H. (1992) Overview of amino acid geochronology. *Comp. Biochem. Physiol. 102B*, 199–204.

Kaufman A.J., Knoll A.H., Narbonne G.M. (1997) Isotopes, ice ages, and terminal Proterozoic Earth history. *Proc. Natl Acad. Sci. USA 94*, 6600–5.

Kawka O.E., Simoneit B.R.T. (1990) Polycyclic aromatic hydrocarbons in hydrothermal petroleums from the Guaymas Basin spreading center. *Appl. Geochem. 5*, 17–27.

Keeling C.D., Whorf T.P. (2002) Atmospheric CO_2 records from sites in the SIO air sampling network. In *Trends: A Compendium of Data on Global Change*. Oak Ridge: Carbon Dioxide Information Analysis Center, Oak Ridge National Lab., US Department of Energy.

Keeling C.D., Bacastow R.B., Whorf T.P. (1982) Measurements of the concentration of carbon dioxide at Mauna Loa Observatory, Hawaii. In *Carbon Dioxide Review: 1982* (ed. Clark W.C.). New York: Oxford University Press.

Keeling C.D., Chin J.F.S., Whorf T.P. (1996) Increased activity of northern vegetation inferred from atmospheric CO_2 measurements. *Nature 382*, 146–9.

Keely B.J., Prowse W.G., Maxwell J.R. (1990) The Treibs hypothesis: an evaluation based on structural studies. *Energy Fuels 4*, 628–34.

Keil R.G., Montluçon D.B., Prahl F.G., Hedges J.I. (1994) Sorptive preservation of labile organic matter in marine sediments. *Nature 370*, 549–52.

Keller C.K., Bacon D.H. (1998) Soil respiration and georespiration distinguished by transport analyses of vadose CO_2, $^{13}CO_2$, and $^{14}CO_2$. *Glob. Biogeochem. Cycles 12*, 361–72.

Keller C.K., Wood B.D. (1993) Possibility of chemical weathering before the advent of vascular land plants. *Nature 364*, 223–5.

Kelly D.C. (2002) Response of Antarctic (ODP Site 690) planktonic foraminifera to the Paleocene–Eocene thermal maximum: faunal evidence for ocean/climate change. *Paleoceanogr. 17*, 1071, doi: 10.1029/2002PA000761.

Kelly D.C., Bralower T.J., Zachos J.C., Permoli Silva I., Thomas E. (1996) Rapid diversification of planktonic foraminifera in the tropical Pacific (ODP site 865) during the late Paleocene thermal maximum. *Geology 24*, 423–6.

Kelts K. (1988) Environments of deposition of lacustrine petroleum source rocks: an introduction. In *Lacustrine Petroleum Source Rocks* (ed. Fleet A.J., Kelts K., Talbot M.R.) *Geol. Soc. Spec. Publn 40*, 3–26. Oxford: Blackwell Scientific.

Kempe S. (1979) Carbon in the rock cycle. In *The Global Carbon Cycle* (ed. Bolin B., Degens E.T., Kempe S., Ketner P.) SCOPE Rep. No. 13, 343–77. Chichester: Wiley.

Kennedy J.F. (ed.) (1988) *Carbohydrate Chemistry*. Oxford: Oxford University Press.

Kennedy M.J., Runnegar B., Prave A.R., Hoffmann K.-H., Arthur M.A. (1998) Two or four Neoproterozoic glaciations? *Geology 26*, 1059–63.

Kennedy M.J., Christie-Blick N., Sohl L.E. (2001) Are Proterozoic cap carbonates and isotopic excursions a record of gas hydrate destabilization following Earth's coldest intervals? *Geology 29*, 443–6.

Kennett J.P., Stott L.D. (1991) Abrupt deep-sea warming, palaeoceanographic changes and benthic extinctions at the end of the Palaeocene. *Nature 353*, 225–9.

Kenney J.F., Kutcherov V.A., Bendeliani N.A., Alekseev V.A. (2002) The evolution of multicomponent systems at high pressures: VI. The thermodynamic stability of the hydrogen-carbon system: the genesis of hydrocarbons and the origin of petroleum. *Proc. Natl Acad. Sci. USA 99*, 10976–81.

Kerndorf H., Schnitzer M. (1980) Sorption of metals on humic acid. *Geochim. Cosmochim. Acta 44*, 1701–8.

Kerr R.A. (1993) Volcanoes may warm locally while cooling globally. *Science 260*, 1232.

Kerr R.A. (2000) Globe's 'missing warming' found in the ocean. *Science 287*, 2126–7.

Kiehl J.T., Hack J.J., Bonan G., Boville B.A., Williamson D., Rasch P. (1998) The National Center for Atmospheric Research Community Climate Model: CCM3. *J. Clim. 11*, 1151–78.

Kiene R.P., Bates T.S. (1990) Biological removal of dimethyl sulphide from sea water. *Nature 345*, 702–5.

Killops S.D. (1991) Novel aromatic hydrocarbons of probable bacterial origin in a Jurassic lacustrine sequence. *Org. Geochem. 17*, 25–36.

Killops S.D., Al-Juboori M.A.H.A. (1990) Characterisation of the unresolved complex mixture (UCM) in the gas chromatograms of biodegraded petroleums. *Org. Geochem. 15*, 147–60.

Killops S.D., Frewin N.L. (1994) Triterpenoid diagenesis and cuticular preservation. *Org. Geochem. 21*, 1193–209.

Killops S.D., Howell V.J. (1991) Complex series of pentacyclic triterpanes in a lacustrine sourced oil from Korea Bay Basin. *Chem. Geol.* 91, 65–79.

Killops S.D., Massoud M.S. (1992) Polycyclic aromatic hydrocarbons of pyrolytic origin in ancient sediments—evidence for Jurassic vegetation fires. *Org. Geochem.* 18, 1–7.

Killops S.D., Woolhouse A.D., Weston R.J., Cook R.A. (1994) A geochemical appraisal of oil generation in the Taranaki Basin, New Zealand. *Am. Assoc. Pet. Geol. Bull.* 78, 1560–85.

Killops S.D., Raine J.I., Woolhouse A.D., Weston R.J. (1995) Chemostratigraphic evidence of higher-plant evolution in the Taranaki Basin, New Zealand. *Org. Geochem.* 23, 429–45.

Killops S.D., Allis R.G., Funnell R.H. (1996) Carbon dioxide generation from coals in Taranaki Basin, New Zealand: implications for petroleum migration in southeast Asian Tertiary basins. *Am. Assoc. Pet. Geol. Bull.* 80, 545–69.

Killops S.D., Funnell R.H., Suggate R.P., Sykes R., Peters K.E., Walters C., Woolhouse A.D., Weston R.J., Boudou J.-P. (1998) Predicting generation and expulsion of paraffinic oil from vitrinite-rich coals. *Org. Geochem.* 29, 1–21.

Killops S.D., Carlson R.M.K., Peters K.E. (2000a) High-temperature evidence for the early formation of C_{40+} n-alkanes in coals. *Org. Geochem.* 31, 589–97.

Killops S.D., Hollis C.J., Morgans H.E.G., Sutherland R., Field B.D., Leckie D.A. (2000b) Paleoceanographic significance of Late Paleocene dysaerobia at the shelf/slope break around New Zealand. *Palaeogeogr. Palaeoclim. Palaeoecol.* 156, 51–70.

Killops S., Walker P., Wavrek D. (2001) Maturity-related variations in the bitumen compositions of coals from Tara-1 and Toko-1 wells. *NZ J. Geol. Geophys.* 44, 157–69.

Killops S., Jarvie D., Sykes R., Funnell R. (2002) Maturity-related variation in the bulk-transformation kinetics of a suite of compositionally related New Zealand coals. *Mar. Pet. Geol.* 19, 1151–68.

Killops S., Cook R., Raine I., Weston R., Woolhouse A. (2003) A tentative New Zealand chemostratigraphy for the Jurassic–Cretaceous based on terrestrial plant biomarkers. *NZ J. Geol. Geophys.* 46, 63–77.

Kimber R.W.L., Griffin C.V. (1987) Further evidence of the complexity of the racemization process in fossil shells with implications for amino acid racemization dating. *Geochim. Cosmochim. Acta* 51, 839–46.

Kirk T.K. (1984) Degradation of lignin. In *Microbial Degradation of Organic Compounds* (ed. Gibson D.T.), 399–437. New York: Dekker.

Kirk T.K., Farrell R.L. (1987) Enzymatic 'combustion': the microbial degradation of lignin. *Ann. Rev. Microbiol.* 41, 465–505.

Kirschvink J.L. (1992) Late-Proterozoic low-latitude global glaciation: the snowball earth. In *The Proterozoic Biosphere* (ed. Schopf J.W., Klein C.), 51–2. New York: Cambridge University Press.

Kissin Y.V. (1987) Catagenesis and composition of petroleum: origin of n-alkanes and isoalkanes in petroleum crudes. *Geochim. Cosmochim. Acta* 51, 2445–57.

Kissin Y.V. (1990) Catagenesis of light cycloalkanes in petroleum. *Org. Geochem.* 15, 575–94.

Kleeman G., Poralla K., Englert G., Kjøsen H., Liaaen-Jensen S., Neunlist S., Rohmer M. (1990) Tetrahymanol from the phototrophic bacterium *Rhodopseudomonas palustris*: first report of a gammacerane triterpene from a prokaryote. *J. Gen. Microbiol.* 136, 2551–3.

Klein G.D. (ed.) (1994) *Pangea: Paleoclimate, Tectonics and Sedimentation During Accretion, Zenith and Break-up of a Supercontinent*. Geol. Soc. Am. Spec. Pap. 288. Boulder: Geological Society of America.

Klemme H.D. (1980) Petroleum basins–classification and characteristics. *J. Pet. Geol.* 3, 187–207.

Klemme H.D., Ulmishek G.F. (1991) Effective petroleum source rocks of the world: stratigraphic distribution and controlling depositional factors. *Am. Assoc. Pet. Geol. Bull.* 75, 1809–51.

Klok J., Cox H.C., Baas M., Schuyl P.J.W., de Leeuw J.W., Schenck P.A. (1984a) Carbohydrates in recent marine sediments—I. Origin and significance of deoxy- and O-methyl-monosaccharides. *Org. Geochem.* 7, 73–84.

Klok J., Baas M., Cox H.C., de Leeuw J.W., Schenck P.A. (1984b) Loliolides and dihydroactinidiolide in a recent marine sediment probably indicate a major transformation pathway of carotenoids. *Tetr. Letts* 25, 5577–80.

Knights B.A., Brown A.C., Conway E., Middleditch B.S. (1970) Hydrocarbons from the green form of the freshwater alga *Botryococcus braunii*. *Phytochem.* 9, 1317–24.

Knoll A.H., Hayes J.M., Kaufman A.J., Swett K., Lambert I.B. (1986) Secular variation in carbon isotope ratios from Upper Proterozoic successions of Svalbard and East Greenland. *Nature* 321, 832–8.

Knoll A.H., Bambach R.K., Canfield D.E., Grotzinger J.P. (1996) Comparative Earth history and Late Permian mass extinction. *Science* 273, 452–7.

Koblenz-Mishke O.I.V., Volkonsky V.V., Kabanova J.G. (1970) Planktonic primary production of the world oceans. In *Symposium on Scientific Exploration of the South Pacific* (ed. Wooster W.S.), 183–93. Washington, DC: National Acadamy of Sciences.

Koch P.L., Zachos J.C., Gingerich P.D. (1992) Correlation between isotope records in marine and continental carbon reservoirs near the Palaeocene/Eocene boundary. *Nature* 359, 319–22.

Koepnick R.B., Denison R.R., Dahl D.A. (1988) The Cenozoic seawater $^{87}Sr/^{86}Sr$ curve: data review and implications for correlation of marine strata. *Paleoceanogr.* 3, 743–56.

Kögel-Knabner I. (2000) Analytical approaches for characterizing soil organic matter. *Org. Geochem.* 31, 609–25.

Kohnen M.E.L., Sinninghe Damsté J.S., ten Haven H.L., de Leeuw J.W. (1989) Early incorporation of polysulphides in sedimentary organic matter. *Nature* 341, 640–1.

Kohnen M.E.L., Sinninghe Damsté J.S., de Leeuw J.W. (1991a) Biases from natural sulphurization in palaeoenvironmental reconstruction based on hydrocarbon biomarker distributions. *Nature* 349, 775–8.

Kohnen M.E.L., Sinninghe Damsté J.S., Koch-van Dalen A.C., de Leeuw J.W. (1991b) Di- or polysulphide-bound biomarkers in sulphur-rich geomacromolecules as revealed by selective chemolysis. *Geochim. Cosmochim. Acta 55*, 1375–94.

Kohnen M.E.L., Sinninghe Damsté J.S., ten Haven H.L., Koch-van Dalen A.C., Schouten S., de Leeuw J.W. (1991c) Identification and geochemical significance of cyclic di- and trisulphides with linear and acyclic isoprenoid carbon skeletons in immature sediments. *Geochim. Cosmochim. Acta 55*, 3685–95.

Kohnen M.E.L., Schouten S., Sinninghe Damsté J.S., de Leeuw J.W., Merritt D.A., Hayes J.M. (1992) Recognition of paleobiochemicals by a combined molecular sulfur and isotope geochemical approach. *Science 256*, 358–62.

Kolber Z.S., Plumley F.G., Lang A.S., Beatty J.T., Blankenship R.E., VanDover C.L., Vetriani C., Koblizek M., Rathgeber C., Falkowski P.G. (2001) Contribution of aerobic photoheterotrophic bacteria to the carbon cycle in the ocean. *Science 292*, 2492–5.

Konhauser K.O., Hamade T., Raiswell R., Morris R.C., Ferris F.G., Southam G., Canfield D.E. (2002) Could bacteria have formed the Precambrian banded iron formations? *Geology 30*, 1079–82.

Kriausakul N., Mitterer R.M. (1978) Isoleucine epimerization in peptides and proteins: Kinetic factors and application to fossil proteins. *Science 20*, 1011–14.

Kriausakul N., Mitterer R.M. (1983) Epimerization of COOH-terminal isoleucine in fossil dipeptides. *Geochim. Cosmochim. Acta 47*, 963–6.

Kroeze C., Seitzinger S. (1998) Nitrogen inputs to rivers, estuaries and continental shelves and related nitrous oxide emissions in 1990 and 2050. *Nutr. Cycl. Agroecosys. 52*, 195–212.

Kroopnick P. (1985) The distribution of ^{13}C in ΣCO_2 in the world oceans. *Deep-Sea Res. 32*, 57–84.

Krouse H.R. (1977) Sulfur isotope studies and their role in petroleum exploration. *J. Geochem. Explor. 7*, 189–211.

Krouse H.R., Viau C.A., Eliuk L.S., Ueda A., Halas S. (1988) Chemical and isotopic evidence of thermochemical sulfate reduction by light hydrocarbon gases in deep carbonate reservoirs. *Nature 333*, 415–19.

Kruge M.A., Stankiewicz B.A., Crelling J.C., Montanari A., Bensley D.F. (1994) Fossil charcoal in Cretaceous–Tertiary boundary strata: evidence for catastrophic firestorm and megawave. *Geochim. Cosmochim. Acta 58*, 1393–7.

Krull E.S., Retallack G.J. (2000) δ^{13}C depth profile from paleosols across the Permian–Triassic boundary; evidence for methane release. *Geol. Soc. Am. Bull. 112*, 1459–72.

Krumholz L.R. (2000) Microbial communities in the deep subsurface. *Hydrogeol. J. 8*, 4–10.

Kudrass H.R., Hofmann A., Doose H., Emeis K., Erlenkeuser H. (2001) Modulation and amplification of climatic changes in the Northern Hemisphere by the Indian summer monsoon during the past 80 ky. *Geology 29*, 63–6.

Kuenen J.G., Robertson L.A., Gemerden H.V. (1985) Microbial interactions among aerobic and anaerobic sulfur-oxidising bacteria. *Adv. Microbial. Ecol. 8*, 1–59.

Kuhnt W., Nederbragt A., Leine L. (1997) Cyclicity of Cenomanian–Turonian organic-carbon-rich sediments in the Tarfaya Atlantic Coastal Basin (Morocco). *Cret. Res. 18*, 587–601.

Kumar N., Anderson R.F., Mortlock R.A., Froelich P.N., Kubik P., Dittrich-Hannen B., Suter M. (1995) Increased biological productivity and export production in the glacial Southern Ocean. *Nature 378*, 675–80.

Kump L.R. (1988) Terrestrial feedback in atmospheric oxygen regulation by fire and phosphorus. *Nature 335*, 152–4.

Kump L.R. (1991) Interpreting carbon-isotope excursions: Strangelove oceans. *Geology 19*, 299–302.

Kuo C., Lindberg C., Thomson D.J. (1990) Coherence established atmospheric carbon dioxide and global temperature. *Nature 343*, 709–14.

Kuypers M.M.M., Blokker P., Erbacher J., Kinkel H., Pancost R.D., Schouten S., Sinninghe Damsté J.S. (2001) Massive expansion of marine archaea during a mid-Cretaceous oceanic anoxic event. *Science 293*, 92–4.

Kuypers M.M.M., Pancost R.D., Nijenhuis I.A., Sinninghe Damsté J.S. (2002) Enhanced productivity led to increased organic carbon burial in the euxinic North Atlantic basin during the late Cenomanian oceanic anoxic event. *Paleoceanogr. 17*, 1051, doi:10.1029/222PA000569.

Kvalheim O.M., Christy A.A., Telnaes N., Bjørseth A. (1987) Maturity determination of organic matter in coals using the methylphenanthrene distribution. *Geochim. Cosmochim. Acta 51*, 1883–8.

Kvenvolden K.A. (1988) Methane hydrate—a major reservoir of carbon in the shallow geosphere? *Chem. Geol. 71*, 41–51.

Kvenvolden K.A. (1998) A primer on the geological occurrence of gas hydrate. In *Gas Hydrates: Relevance to World Margin Stability and Climate Change* (ed. Henriet J.-P., Mienert J.) *Geol. Soc. Spec. Publn 137*, 9–30. London: Geological Society.

Lafargue E., Marquis F., Pillot D. (1998) Rock-Eval 6 applications in hydrocarbon exploration, production, and soil contamination studies. *Rev. Inst. Fr. Pét. 53*, 421–37.

Laflamme R.E., Hites R.A. (1978) The global distribution of polycyclic aromatic hydrocarbons in recent sediments. *Geochim. Cosmochim. Acta 42*, 289–303.

Laflamme R.E., Hites R.A. (1979) Tetra- and pentacyclic, naturally-occurring, aromatic hydrocarbons in recent sediments. *Geochim. Cosmochim. Acta 43*, 1687–91.

Lalli C.M., Parsons T.R. (1997) *Biological Oceanography, an Introduction.* Oxford: Butterworth-Heinemann.

Lancet M.S., Anders E. (1970) Carbon isotope fractionation in Fischer-Tropsch synthesis and in meteorites. *Science 170*, 980–2.

Langston W.J. (1996) Recent developments in TBT ecotoxicology, *TEN 3*, 179–87.

Langston W.J., Bryan G.W., Burt G.R., Pope N.D. (1994) *Effects of Sediment Metals on Estuarine Benthis Organisms*, R&D note 203. Bristol: National Rivers Authority.

Largeau C., Derenne S., Casadevall E., Kadouri A., Sellier N. (1986) Pyrolysis of immature Torbanite and of the resistant biopolymer (PRB A) isolated from extant alga *Botryococcus braunii*. Mechanism of formation and structure of Torbanite. *Org. Geochem.* 10, 1023–32.

Larsen T., Pelusio H.K., Paya-Perez A. (1992) Bioavailability of polychlorinated biphenyl congeners in the soil to earthworm (*L. rubellus*) system. *Int. J. Environ. Anal. Chem.* 4, 149–62.

Larsen J.W., Parikh H., Michels R. (2002) Changes in the cross-link density of Paris Basin Toarcian kerogen during maturation. *Org. Geochem.* 33, 1143–52.

Larson R.L. (1991a) Latest pulse of Earth: evidence for a mid-Cretaceous superplume. *Geology* 19, 547–50.

Larson R.L. (1991b) Geological consequences of superplumes. *Geology* 19, 963–6.

Larson R.L. (1995) The mid-Cretaceous superplume episode. *Sci. Am.*, Feb., 66–70.

Larter S.R., Senftle J.T. (1985) Improved kerogen typing for petroleum source rock analysis. *Nature* 318, 277–80.

Lasaga A.C., Ohmoto H. (2002) The oxygen geochemical cycle: dynamics and stability. *Geochim. Cosmochim. Acta* 66, 361–81.

Lashof D.A., Ahuja D.R. (1990) Relative contributions of greenhouse gas emissions to global warming. *Nature* 344, 529–31.

Law B.E., Rice D.D. (eds.) (1993) *Hydrocarbons from Coal. AAPG Studies in Geology 38*. Tulsa, OK: American Association of Petroleum Geologists.

Law R.J., Blake S.J., Jones B.R., Rogan E. (1998) Organotin compounds in liver tissue of Harbour Porpoises (*Phocoena phocoena*) and Grey Seals (*Halichoerus grypus*) from the coastal waters of England and Wales. *Mar. Pollut. Bull.* 36, 241–7.

Laws E.A., Bidigare R.R., Popp B.N. (1997) Effect of growth rate and CO_2 concentration on carbon isotope fractionation by the marine diatom *Phaeodactylum tricornutum*. *Limnol. Oceanogr.* 42, 1152–60.

Leather J., Allen P.A., Brasier M.D., Cozzi A. (2002) Neoproterozoic snowball Earth under scrutiny: evidence from the Fiq glaciation of Oman. *Geology* 30, 891–4.

Leatherland J.F. (1992) Endocrine and reproductive function in Great Lakes salmon. In *Chemically-Induced Alterations in Sexual and Functional Development: the Wildlife/Human Connection* (ed. Colborn T., Clement C.), 129–46. Princeton, NJ: Princeton Scientific Publishing.

Leavitt S.W. (1982) Annual volcanic carbon dioxide emission: An estimate from eruption chronologies. *Environ. Geol.* 4, 15–21.

LeBlanc G.A. (1995) Trophic-level differences in the bioconcentration of chemicals: implications in assessing environmental biomagnification. *Environ. Sci. Technol.* 29, 154–60.

LeBlanc G.A. (1999) Steroid hormone regulated processes in invertebrates and their susceptibility to environmental endocrine disruption. In *Environmental Endocrine Disrupters: an Evolutionary Perspective* (ed. Guillette L.J., Crain A.). London: Taylor & Francis.

Leckie R.M., Bralower T.J., Cashman R. (2002) Oceanic anoxic events and plankton evolution: biotic response to tectonic forcing during the mid-Cretaceous. *Paleoceanogr.* 17, 10.1029/2001PA000623.

Lee C. (1994) Kitty litter for carbon control? *Nature* 370, 503–4.

Lee C., Henrichs S.M. (1993) How the nature of dissolved organic matter might affect the analysis of dissolved organic carbon. *Mar. Chem.* 41, 105–20.

Lee C., Wakeham S.G. (1992) Organic matter in the water column: future research challenges. *Mar. Chem.* 39, 95–118.

Lee R.F., Hirota J., Barnett A.M. (1971) Distribution and importance of wax esters in marine copepods and other zooplankton. *Deep-Sea Res.* 18, 1147–65.

Legrand M. (1997) Ice core records of atmospheric sulphur. *Roy. Soc. Lond. Phil. Trans. B 352*, 241–50.

Leif R.N., Simoneit B.R.T. (2000) The role of alkenes produced during hydrous pyrolysis of a shale. *Org. Geochem.* 31, 1189–208.

Lelieveld J., Crutzen P.L., Dentener F.J. (1998) Changing concentration, lifetime and climate forcing of atmospheric methane. *Tellus 50B*, 128–50.

Lenton T.M., Watson A.J. (2000) Redfield revisited 2. What regulates the oxygen content of the atmosphere? *Glob. Biogeochem. Cycles* 14, 249–68.

Leonards P.E.G., van Hattum B., Cofino W.P., Brinkman U.A.T. (1994) Occurrence of non-ortho-, mono-ortho- and di-ortho- substituted PCB congeners in different organs and tissues of polecats (*Mustela putorius* L.) from the Netherlands. *Environ. Toxicol. Chem.* 13, 129–42.

Levitus S., Antonov J.I., Boyer T.P., Stephens C. (2000) Warming of the world ocean. *Science* 287, 2225–9.

Levitus S., Antonov J.I., Wang J., Delworth T.L., Dixon K.W., Broccoli A.J. (2001) Anthropogenic warming of Earth's climate system. *Science* 292, 267–70.

Levorsen A.I. (1967) *Geology of Petroleum*. San Francisco: Freeman.

Lewan M.D. (1984) Factors controlling the proportionality of vanadium to nickel in crude oils. *Geochim. Cosmochim. Acta* 48, 2231–8.

Lewan M.D. (1997) Experiments on the role of water in petroleum formation. *Geochim. Cosmochim. Acta* 61, 3691–723.

Lewis C.A. (1993) The kinetics of biomarker reactions. In *Organic Geochemistry* (ed. Engel M.H., Macko S.A.), 491–510. New York: Plenum Press.

Leythaeuser D., Schaefer R.G., Weiner B. (1979) Generation of low molecular weight hydrocarbons from organic matter in source beds as a function of temperature and facies. *Chem. Geol.* 25, 95–108.

L'Haridon S., Reysenbach A.-L., Glénat P., Prieur D., Jeanthon C. (1995) Hot subterranean biosphere in a continental oil reservoir. *Nature* 377, 223–4.

Libby W.F., Anderson E.C., Arnold J.R. (1949) Age determination by radiocarbon content: world-wide assay of natural radiocarbon. *Science* 109, 227–8.

Liss P.S. (1983) The exchange of biogeochemically important gases across the air–sea interface. In *The Major Biogeochemical*

Cycles and Their Interactions (ed. Bolin B., Cook R.B.), SCOPE Rep. No. 21, 411–26. Chichester: Wiley.

Lockheart M.J., van Bergen P.F., Evershed R.P. (1997) Variations in the stable carbon isotope compositions of individual lipids from the leaves of modern angiosperms: implications for the study of higher land plant-derived sedimentary organic matter. *Org. Geochem. 26*, 137–53.

Lohr M., Wilhelm C. (1999) Algae displaying the diadinoxanthin cycle also possess the violaxanthin cycle. *Proc. Natl Acad. Sci. USA 96*, 8784–9.

Loonen H., Parsons J.R., Govers H.A.J. (1994) Effect of sediment on the bioaccumulation of a complex mixture of polychlorinated dibenzo-p-dioxins (PCDDs) and polychlorinated dibenzofurans (PCDFs) by fish. *Chemosphere 28*, 1433–46.

Looy C.V., Brugman W.A., Dilcher D.L., Visscher H. (1999) The delayed resurgence of equatorial forests after the Permian-Triassic ecologic crisis. *Proc. Natl Acad. Sci. USA 96*, 13857–62.

Lorius C., Jouzel J., Raynaud D., Hansen J., Le Treut H. (1990) The ice-core record: climate sensitivity and future greenhouse warming. *Nature 347*, 139–45.

Louda J.W., Li J., Liu L., Winfree M.N., Baker E.W. (1998) Chlorophyll degradation during senescence and death. *Org. Geochem. 29*, 1233–51.

Louda J.W., Loitz J.W., Rudnick D.T., Baker E.W. (2000) Early diagenetic alteration of chlorophyll-a and bacteriochlorophyll-a in a contemporaneous marl ecosystem. *Org. Geochem. 31*, 1561–80.

Louda J.W., Liu L., Baker E.W. (2002) Senescence- and death-related alteration of chlorophylls and carotenoids in marine phytoplankton. *Org. Geochem. 33*, 1635–53.

Lovelock J.E., Whitfield M. (1982) Life span of the biosphere. *Nature 296*, 561–3.

Lovelock J.E., Maggs R.J., Rasmussen R.A. (1972) Atmospheric sulphur in the natural sulphur cycle. *Nature 237*, 452–3.

Lubec G., Nauer G., Seifert K., Strouhal E., Portecler H., Szilvassy I., Teschler M. (1987) Structural stability of hair 4000 years old. *J. Arch. Sci. 14*, 113–20.

McArthur J.M., Howarth R.J., Bailey T.R. (2001) Strontium isotope stratigraphy: LOWESS version 3: best fit to the marine Sr-isotope curve for 0–509 Ma and accompanying look-up table for deriving numerical age. *J. Geol. 109*, 155–70.

MacAvoy S.E., Macko S.A., Garman G.C. (1998) Tracing marine biomass into tidal freshwater ecosystems using stable sulphur isotopes. *Naturwissenschaften 85*, 544–6.

McCarthy M., Hedges J.I., Benner R. (1998) Major bacterial contribution to marine dissolved organic nitrogen. *Science 281*, 231–4.

McCartney J.T., Teichmüller M. (1972) Classification of coals according to degree of coalification by reflectance of the vitrinite component. *Fuel 51*, 64–8.

MacGregor E.A., Greenwood C.T. (1980) *Polymers in Nature*. Chichester: Wiley.

Machel H.G., Krouse H.R., Sassen R. (1995) Products and distinguishing criteria of bacterial and thermochemical sulfate reduction. *Appl. Geochem. 10*, 373–89.

McKay D.S., Gibson E.K. Jr, Thomas-Keprta K.L., Vali H., Romanek C.S., Clemett S.J., Chillier X.D.F., Maechling C.M., Zare R.N. (1996) Search for past life on Mars: possible relic biogenic activity in Martian meteorite ALH84001. *Science 273*, 924–30.

Mackenzie A.S. (1984) Applications of biological markers in petroleum geochemistry. In *Advances in Petroleum Geochemistry, Vol. 1* (ed. Brooks J., Welte D.), 115–214. New York: Academic Press.

Mackenzie A.S., McKenzie D. (1983) Isomerization and aromatization of hydrocarbons in sedimentary basins formed by extension. *Geol. Soc. Mag. 120*, 417–528.

Mackenzie A.S., Quigley T.M. (1988) Principles of geochemical prospect appraisal. *Am. Assoc. Pet. Geol. Bull. 72*, 399–415.

Mackenzie A.S., Hoffmann C.F., Maxwell J.R. (1981) Molecular parameters of maturation in the Toarcian shales, Paris Basin, France — III. Changes in the aromatic steroid hydrocarbons. *Geochim. Cosmochim. Acta 45*, 1345–55.

Mackenzie A.S., Brassell S.C., Eglinton G., Maxwell J.R. (1982) Chemical fossils: the geological fate of steroids. *Science 217*, 491–504.

Mackenzie A.S., Beaumont C., McKenzie D.P. (1984) Estimation of the kinetics of geochemical reactions with geophysical models of sedimentary basins and applications. *Org. Geochem. 6*, 875–84.

Mackenzie A.S., Leythaeuser D., Muller P., Quigley T.M., Radke M. (1988) The movement of hydrocarbons in shales. *Nature 331*, 63–5.

McKinney D.E., Hatcher P.G. (1996) Characterization of peatified and coalified wood by tetramethylammonium hydroxide (TMAH) thermochemolysis. *Int. J. Coal Geol. 32*, 217–28.

McKinney D.E., Bortiatynski J.M., Carson D.M., Clifford D.J., de Leeuw J.W., Hatcher P.G. (1996) Tetramethylammonium hydroxide (TMAH) thermochemolysis of the aliphatic biopolymer cutan: insights into the chemical structure. *Org. Geochem. 24*, 641–50.

McKinney D.E., Béhar F., Hatcher P.G. (1998) Reaction kinetics and n-alkane product profiles from the thermal degradation of ^{13}C labelled n-C_{25} in 2 dissimilar oils as determined by SIM/GC/MS. *Org. Geochem. 29*, 119–36.

McKirdy D.M., Powell T.G. (1974) Metamorphic alteration of carbon isotopic composition in ancient sedimentary organic matter: new evidence from Australia and South Africa. *Geology 2*, 591–5.

McKirdy D.M., Cox R.E., Volkman J.K., Howell V.J. (1986) Botryococcane in a new class of Australian non-marine crude oils. *Nature 320*, 57–9.

Macko S.A., Engel M.H. (1991) Assessment of indigeneity in fossil organic matter, amino acids and stable isotopes. *Phil. Trans. Roy. Soc. Lond. B 333*, 367–74.

Macko S.A., Lubec G., Teschler-Nicola M., Andrusevich V., Engel H.M. (1999a) The Ice Man's diet as reflected by the stable nitrogen and carbon isotopic composition of hair. *FASEB J. 13*, 559–62.

Macko S.A., Engel M.H., Andrusevich V., Lubec G., O'Connell T.C., Hedges R.E.M. (1999b) Documenting the diet in ancient human populations through stable isotope analysis of hair. *Phil. Trans. Roy. Soc. Lond. B 354*, 65–76.

McNair H.M., Miller J.M. (1997) *Basic Gas Chromatography*. New York: Wiley.

McNeil R.L., BeMent W.O. (1996) Thermal stability of hydrocarbons: laboratory criteria and field examples. *Energy Fuels 10*, 60–7.

Magaritz M. (1989) ^{13}C minima follow extinction events: A clue to faunal radiation. *Geology 17*, 337–40.

Magaritz M., Krishnamurthy R.V., Holser W.T. (1992) Parallel trends in organic and inorganic carbon isotopes across the Permian/Triassic boundary. *Am. J. Sci. 292*, 727–39.

Maher K.A., Stevenson D.J. (1988) Impact frustration of the origin of life. *Nature 331*, 612–14.

Maiss M., Brenninkmeijer C.A.M. (1998) Atmospheric SF_6: trends, sources and prospects. *Environ. Sci. Technol. 32*, 3077–86.

Maiss M., Steele L.P., Francey R.J., Fraser P.J., Langenfelds R.L., Trivett N.B.A., Levin I. (1996) Sulfur hexafluoride — a powerful new atmospheric tracer. *Atmos. Env. 30*, 1621–9.

Mangerud J. (1972) Radiocarbon dating of marine shells, including a discussion of apparent ages of Recent shells from Norway. *Boreas 1*, 143–72.

Mango F.D. (1990a) The origin of light cycloalkanes in petroleum. *Geochim. Cosmochim. Acta 54*, 23–7.

Mango F.D. (1990b) The origin of light hydrocarbons in petroleum: A kinetic test of the steady-state catalytic hypothesis. *Geochim. Cosmochim. Acta 54*, 1315–23.

Mango F.D. (1991) The stability of hydrocarbons under the time-temperature conditions of petroleum genesis. *Nature 352*, 146–8.

Mango F.D. (1997) The light hydrocarbons in petroleum: a critical review. *Org. Geochem. 26*, 417–40.

Manzano B.K., Fowler M.G., Machel H.G. (1997) The influence of thermochemical sulphate reduction on hydrocarbon composition in Nisku reservoirs, Brazeau river area, Alberta, Canada. *Org. Geochem. 27*, 507–21.

Marlowe I.T., Green J.C., Neal A.C., Brassell S.C., Eglinton G., Course P.A. (1984) Long-chain (n-C_{37}-C_{39}) alkenones in the Prymnesiophyceae. Distribution of alkenones and other lipids and their taxonomic significance. *Br. Phycol. J. 19*, 203–16.

Marlowe I.T., Brassell S.C., Eglinton G., Green J.C. (1990) Long-chain alkenones and alkyl alkenoates and the fossil coccolith record of marine sediments. *Chem. Geol. 88*, 349–75.

Marshall J.D. (1992) Climatic and oceanographic isotopic signals from the carbonate rock record and their preservation. *Geol. Mag. 129*, 143–60.

Martin M. (1986) Effects of steeper Archean geothermal gradient on geochemistry of subduction-zone magmas. *Geology 14*, 753–6.

Martin W., Russell M.J. (2003) On the origin of cells: an hypothesis for the evolutionary transitions from abiotic geochemistry to chemoautotrophic prokaryotes, and from prokaryotes to nucleated cells. *Phil. Trans. Roy. Soc. Lond. 358*, 59–85.

Martin J.H., Knauer G.A., Karl D.M., Broenkow W.W. (1987) VERTEX: carbon cycling in the northeast Pacific. *Deep-Sea Res. 34*, 267–85.

Martin J.H., Gordon R.M., Fitzwater S.E. (1991) The case for iron. *Limnol. Oceanogr. 36*, 1793–802.

Martin J.H., Coale K.H., Johnson K.S., Fitzwater S.E. and 40 others (1994) Testing the iron hypothesis in ecosystems of the equatorial Pacific Ocean. *Nature 321*, 61–3.

Martinez J.S., Zhang G.P., Holt P.D., Jung H.-T., Carrano C.J., Haygood M.G., Butler A. (2000) Self-assembling amphiphilic siderophores from marine bacteria. *Science 287*, 1245–7.

Marzi R., Torkelson B.E., Olson R.K. (1993) A revised carbon preference index. *Org. Geochem. 20*, 1303–6.

Masiello C.A., Druffel E.R.M. (1998) Black carbon in deep-sea sediments. *Science 280*, 1911–13.

Maslin M.A., Burns S.J. (2000) Reconstruction of the Amazon Basin effective moisture availability over the past 14,000 years. *Science 290*, 2285–7.

Matile P., Hortensteiner S., Thomas H., Krautler B. (1996) Chlorophyll breakdown in senescent leaves. *Plant Physiol. 112*, 1403–9.

Matsumoto G.I., Nagashima H. (1984) Occurrence of 3-hydroxy acids in microalgae and cyanobacteria and their geochemical significance. *Geochim. Cosmochim. Acta 48*, 1683–7.

Maxwell J.R., Douglas A.G., Eglinton G., McCormick A. (1968) The Botryococcenes — hydrocarbons of novel structure from the alga *Botryococcus braunii*, Kutzing. *Phytochem. 7*, 2157–71.

Maxwell J.R., Mackenzie A.S., Volkman J.K. (1980) Configuration at C-24 in steranes and sterols. *Nature 286*, 694–7.

Mayer L.M. (1994) Surface area control of organic carbon accumulation in continental shelf sediments. *Geochim. Cosmochim. Acta 58*, 1271–84.

Mayer J.F.L., Sanders H.O. (1973) Toxicology of phthalic acid esters in aquatic organisms. *Environ. Health Perspect. 3*, 153–7.

Meert J.G., Powell C.McA. (2000) Assembly and break-up of Rodinia: introduction to the special volume. *Precambr. Res. 110*, 1–8.

Meert J.G., Torsvik T.H. (2003) The making and unmaking of a supercontinent: Rodinia revisted. *Tectonophysics 375*, 261–88.

Mello M.R., Telnaes N., Gaglianone P.C., Chicarelli M.I., Brassell S.C., Maxwell J.R. (1988a) Organic geochemical characterisation of depositional palaeoenvironments of source rocks and oils in Brazilian marginal basins. *Org. Geochem. 13*, 31–45.

Mello M.R., Gaglianone P.C., Brassell S.C., Maxwell J.R. (1988b) Geochemical and biological marker assessment of depositional environments using Brazilian offshore oils. *Mar. Pet. Geol. 5*, 205–23.

Melzer E., Schmidt H.-L. (1987). Carbon isotope effects on the pyruvate dehydrogenase reaction and their importance for relative carbon-13 depletion in lipids. *J. Biol. Chem.* 262, 8159–64.

Menzel D., Hopmans E.C., van Bergen P.F., de Leeuw J.W., Sinninghe Damsté J.S. (2002) Development of photic zone euxinia in the eastern Mediterranean Basin during deposition of Pliocene sapropels. *Mar. Geol.* 189, 215–26.

Metcalf R.L., Booth G.M., Schuth C.K., Hansen D.J., Lu P.-Y. (1973) Uptake and fate of di-2-ethylhexyl phthalate in aquatic organisms and in a model ecosystem. *Environ. Health Perspect.* 4, 27–34.

Metzger P., Casadevall E. (1987) Lycopadiene, a tetraterpenoid hydrocarbon from new strains of the green alga *Botryococcus braunii*. *Tetr. Letts* 28, 3931–4.

Metzger P., Berkaloff C., Cassadevall E., Couté A. (1985) Alkadiene- and botryococcene producing races of wild strains of *Botryococcus braunii*. *Phytochem.* 24, 2305–12.

Meulbroek P. (2002) Equations of state in exploration. *Org. Geochem.* 33, 613–34.

Meyers P.A. (1997) Organic geochemical proxies of paleoceanographic, paleolimnologic, and paleoclimatic processes. *Org. Geochem.* 27, 213–50.

Meyers P.A., Negri A. (2003) Introduction to 'Paleoclimatic and Paleoceanographic Records in Mediterranean Sapropels and Mesozoic Black Shales'. *Palaeogeogr. Palaeoclimatol., Palaeoecol.* 190, 1–8.

Meyers P.A., Simoneit B.R.T. (1990) Global comparison of organic matter in sediments across the Cretaceous/Tertiary boundary. *Org. Geochem.* 16, 641–8.

Michels R., Landais P., Philp R.P., Torkelson B.E. (1995) Influence of pressure and the presence of water on the evolution of the residual kerogen during confined, hydrous pyrolysis, and high-pressure hydrous pyrolysis of Woodford Shale. *Energy Fuels* 9, 204–15.

Miki K., Renganathan V., Gold M.H. (1986) Mechanism of β-aryl ether dimeric lignin model compound oxidation by lignin peroxidase of *Phanerochaete chrysosporium*. *Biochemistry* 25, 4790–6.

Miller S.L., Urey H.C. (1959) Organic compound synthesis on the primitive earth conditions. *Science* 130, 245–51.

Miller G.H., Jull A.J.T., Linick T., Sutherland D., Sejrup H.P., Brigham J.K., Bowen D.Q., Mangerud J. (1987) Racemization-derived late Devensian temperature reduction in Scotland. *Nature* 326, 593–5.

Miller B., Huang J., Weiss R., Prinn R., Fraser P. (1998) Atmospheric trend and lifetime of chlorodifluoromethane (HCFC-22) and the global tropospheric OH concentration. *J. Geophys. Res.* 103, 13237–48.

Mills J.S., White R., Gough L.J. (1984/5) The chemical composition of Baltic amber. *Chem. Geol.* 47, 15–39.

Mitchell J.F.B. (1989) The 'greenhouse' effect and climate change. *Rev. Geophys.* 27, 115–39.

Mitterer R.M. (1993) The diagenesis of proteins and amino acids in fossil shells. In *Organic Geochemistry* (ed. Engel M.H., Macko S.A.), 739–53. New York: Plenum Press.

Mitterer R.M., Kriausakul N. (1984) Comparison of rates and degrees of isoleucine epimerization in dipeptides and tripeptides. *Org. Geochem.* 7, 91–8.

Mojzsis S., Arrhenius G., McKeegan K.D., Harrison T.M., Nutman A.P., Firend C.R.L. (1996) Evidence for life on Earth before 3,800 million years ago. *Nature* 385, 55–9.

Moldowan J.M., Seifert W.K. (1979) Head-to-head linked isoprenoid hydrocarbons in petroleum. *Science* 204, 169–71.

Moldowan J.M., Seifert W.K. (1980) First discovery of Botryococcane in petroleum. *J. Chem. Soc., Chem. Commun.*, 912–14.

Moldowan J.M., Seifert W.K., Gallegos E.J. (1985) Relationship between petroleum composition and depositional environment of petroleum source rocks. *Am. Assoc. Pet. Geol. Bull.* 69, 1255–68.

Moldowan J.M., Fago F.J., Lee C.Y., Jacobson S.R., Watt D.S., Slougui N.-E., Jeganathan A., Young D.C. (1990) Sedimentary 24-n-propylcholestanes, molecular fossils diagnostic of marine algae. *Science* 247, 309–12.

Moldowan J.M., Fago F.J., Carlson R.M.K., Young D.C., Van Duyne G., Clardy J., Schoell M., Pillinger C.T., Watt D.S. (1991) Rearranged hopanes in sediments and petroleum. *Geochim. Cosmochim. Acta* 55, 3333–53.

Moldowan J.M., Dahl J., Jacobson S.R., Huizinga B.J., Fago F.J., Shetty R., Watt D.S., Peters K.E. (1996) Chemostratigraphic reconstruction of biofacies: molecular evidence linking cyst-forming dinoflagellates with pre-Triassic ancestors. *Geology* 24, 159–62.

Monnin E., Indermühle A., Dällenbach A., Flückiger J., Stauffer B., Stocker T.F., Raynaud D., Barnola J.-M. (2001) Atmospheric CO_2 concentrations over the last glacial termination. *Science* 291, 112–14.

Montzka S.A., Butler J.H., Meyers R.C., Thompson T.M., Swanson T.H., Clarke A.D., Lock L.T., Elkins J.W. (1996a) Decline in the tropospheric abundance of halogen from halocarbons: implications for stratospheric ozone depletion. *Science* 272, 1318–22.

Montzka S.A., Meyers R.C., Butler J.H., Elkins J.W., Lock L.T., Clarke A.D., Goldstein A.H. (1996b) Observations of HFC-134a in the remote troposphere. *Geophys. Res. Letts* 23, 169–72.

Montzka S.A., Butler J.H., Elkins J.W., Thompson T.M., Clarke A.D., Lock L.T. (1999) Present and future trends in the atmospheric burden of ozone-depleting halogens. *Nature* 398, 690–4.

Mook W.G., Bommerson J.C., Staberman W.H. (1974) Carbon isotope fractionation between dissolved bicarbonate and gaseous carbon dioxide. *Earth Planet. Sci. Letts* 22, 169–76.

Mopper K., Degens E.T. (1979) Organic carbon in the ocean: nature and cycling. In *The Global Carbon Cycle* (ed. Bolin B., Degens E.T., Kempe S., Ketner P.), SCOPE Rep. No. 13, 293–316. Chichester: Wiley.

Mopper K., Zhou X., Kieber R.J., Kieber D.J., Sikorski R.J., Jones R.D. (1991) Photochemical degradation of dissolved organic carbon and its impact on the oceanic carbon cycle. *Nature* 353, 60–2.

Morris R.C., Horwitz R.C. (1983) The origin of the iron-formation-rich Haersley Group of Western Australia—deposition on a platform. *Precambr. Res. 21*, 273–97.

Morrison R.T., Boyd R.N. (2001) *Organic Chemistry*. San Francisco: Benjamin Cummings.

Morse J.W., Mackenzie F.T. (1990) *Geochemistry of Sedimentary Carbonates*. Developments in Sedimentology 48. Amsterdam: Elsevier.

Mösle B., Collinson M.E., Finch P., Stankiewicz B.A., Scott A.C., Wilson R. (1998) Factors influencing the preservation of plant cuticles: a comparison of morphology and chemical composition of modern and fossil examples. *Org. Geochem. 29*, 1369–80.

Muir D.C.G., Yarechewski A.L., Corbet R.L., Webster G.R.B., Smith A.E. (1985) Laboratory and field studies on the fate of 1,3,6,8-tetrachlorodibenzo-*p*-dioxin in soil and sediments. *J. Agric. Food Chem. 33*, 518–23.

Müller P.J. (1984) Isoleucine epimerization in Quaternary planktonic foraminifera: effects of diagenetic hydrolysis and leaching, and Atlantic-Pacific intercore correlations. *'Meteor' Forschungsergeb. Reihe C 38*, 25–47.

Müller P.J., Suess E. (1979) Productivity, sedimentation rate, and sedimentary organic matter in the oceans. *Deep-Sea Res. 26*, 1347–62.

Müller P.J., Kirst G., Ruhland G., von Storch I., Rosell-Melé A. (1998) Calibration of the alkenone paleotemperature index based on core-tops from the eastern South Atlantic and the global ocean (60°N–60°S). *Geochim. Cosmochim. Acta 62*, 1757–72.

Müller P., Li X.-P., Niyogi K.K. (2001) Non-photochemical quenching. A response to excess light energy. *Plant Physiol. 125*, 1558–66.

Murchison D.G., Raymond A.C. (1989) Igneous activity and organic maturation in the Midland Valley of Scotland. *Int. J. Coal Geol. 14*, 47–82.

Murray J.W., Grundmanis V. (1980) Oxygen consumption in pelagic marine sediments. *Science 209*, 1527–30.

Naeser N.D. (1993) Apatite fission-track analysis in sedimentary basins—a critical appraisal. In *Basin Modeling: Advances and Applications* (ed. Dore A.G. et al.) *Norwegian Petroleum Soc. Spec. Publn 3*, 147–60. Amsterdam: Elsevier.

Neal A.C., Prahl F.G., Eglinton G., O'Hara S.C.M., Corner E.D.S. (1986) Lipid changes during a planktonic feeding sequence involving unicellular algae, *Elminius nauplii* and adult *Calanus. J. Mar. Biol. Assoc. UK 66*, 1–13.

Nealson K.H., Saffarini D. (1994) Iron and manganese in anaerobic respiration: environmental significance, physiology, and regulation. *Ann. Rev. Microbiol. 48*, 311–43.

Nedwell D.B. (1984) The input and mineralisation of organic carbon in anaerobic aquatic sediments. *Adv. Microb. Ecol. 7*, 93–131.

Neftel A., Friedli H., Moor E., Lötscher H., Oeschger H., Siegenthaler U., Stauffer B. (1994) Historical CO_2 record from the Siple station ice core. In *Trends '93: A Compendium of Data on Global Change* (ed. Boden T.A., Kaiser D.P., Sepanski R.J., Stoss F.W.), 11–14. Oak Ridge: Carbon Dioxide Information Analysis Center, Oak Ridge National Lab., US Department of Energy.

Newman M.J., Rood R.T. (1977) Implications of solar evolution for the Earth's early atmosphere. *Science 198*, 1035–67.

Newton I., Wyllie I., Asher A.J. (1993) Long-term trends in organochlorine and mercury residues in some predatory birds in Britain. *Environ. Pollut. 79*, 143–51.

Nichols D.J., Jarzen D.M., Orth C.J., Oliver P.Q. (1986) Palynology and iridium anomalies at Cretaceous–Tertiary boundary, south-central Saskatchewan. *Science 231*, 714–17.

Nichols D.J., Daniel T.C., Moore P.A. Jr, Edwards D.R., Pote D.H. (1997) Runoff of estrogen hormone 17beta-estradiol from poultry litter applied to pasture. *J. Environ. Qual. 26*, 1002–6.

Niimi A.J. (1996) Evaluation of PCBs and PCDD/Fs retention by aquatic organisms. *Sci. Tot. Environ. 192*, 123–50.

Nimz H. (1974) Beech lignin-proposal of a constitutional scheme. *Angew. Chem., Int. Ed. Engl. 13*, 313–21.

Nip M., Tegelaar E.W., de Leeuw J.W., Schenck P.A., Holloway P.J. (1986a) A new non-saponifiable highly aliphatic and resistant biopolymer in plant cuticles: evidence from pyrolysis and ^{13}C NMR analysis of present day and fossil plants. *Naturwissenschaften 73*, 579–85.

Nip M., Tegelaar E.W., Brinkhuis H., de Leeuw J.W., Schenck P.A., Holloway P.J. (1986b) Analysis of modern and fossil plant cuticles by Curie-point Py-GC and Curie-point Py-GC-MS: recognition of a new highly aliphatic and resistant biopolymer. *Org. Geochem. 10*, 769–78.

Nissenbaum A., Kaplan I.R. (1972) Chemical and isotopic evidence for *in situ* origin of marine humic substances. *Limnol. Oceanogr. 17*, 570–82.

Noble R.A., Alexander R.A., Kagi R.I., Knox J. (1985) Tetracyclic diterpenoid hydrocarbons in some Australian coals, sediments and crude oils. *Geochim. Cosmochim. Acta 49*, 2141–7.

Noble R.A., Alexander R.A., Kagi R.I., Knox J. (1986) Identification of some diterpenoid hydrocarbons in petroleum. *Org. Geochem. 10*, 825–9.

Norris R.D., Röhl U. (1999) Carbon cycling and chronology of climate warming during the Palaeocene/Eocene transition. *Nature 401*, 775–8.

Norris R.D., Bice K.L., Magno E.A., Wilson P.A. (2002) Jiggling the tropical thermostat in the Cretaceous hothouse. *Geology 30*, 299–302.

Nytoft H.P., Bojesen-Koefoed J.A. (2001) $17\alpha,21\alpha(H)$-hopanes: natural and synthetic. *Org. Geochem. 32*, 841–56.

Nytoft H.P., Bojesen-Koefoed J.A., Christiansen F.G. (2000) C_{26} and C_{28}-C_{34} 28-norhopanes in sediments and petroleum. *Org. Geochem. 31*, 25–39.

O'Connell S., Chandler M.A., Ruedy R. (1996) Implications for the creation of warm saline deep water: Late Paleocene reconstructions and global climate model simulations. *Geol. Soc. Am. Bull. 108*, 270–84.

Odden W., Patience R.L., van Graas G.W. (1998) Application of light hydrocarbons (C_4-C_{13}) to oil/source rock correlations: a study of the light hydrocarbon compositions of source rocks and test fluids from offshore Mid-Norway. *Org. Geochem. 28*, 823–47.

Odier E., Monties B. (1983) Absence of microbial mineralization of lignin in anaerobic enrichment cultures. *Appl. Environ. Microbiol. 46*, 661–5.

Oechel W.C., Cowles S., Grulke N., Hastings S.J., Lawrence B., Prudhomme T., Riechers G., Strain B., Tissue D., Vourlitis G. (1994). Transient nature of CO_2 fertilization in Arctic tundra. *Nature 371*, 500–3.

Ogawa H., Amagai Y., Koike I., Kaiser K., Benner R. (2001) Production of refractory dissolved organic matter by bacteria. *Science 292*, 917–20.

Oram D.E., Reeves C.E., Sturges W.T., Penkett S.A., Fraser P.J., Langenfelds R.L. (1996) Recent tropospheric growth rate and distribution of HFC-134a (CF_3CH_2F). *Geophys. Res. Letts 23*, 1949–52.

Oram D.E., Sturges W.T., Penkett S.A., Lee J.M., Fraser P.J., McCulloch A., Engel A. (1998) Atmospheric measurements and emissions of HFC-23 (CHF_3). *Geophys. Res. Letts 25*, 35–8.

Oram D.E., Sturges W.T., Penkett S.A., Fraser P.J. (1999) Tropospheric abundance and growth rates of radiatively-active halocarbon trace gases and estimates of global emissions. In *IUGG 99: Abstracts, Birmingham (England)*. Abstract MI02/W/12-A4. International Union of Geodesy and Geophysics.

Orphan V.J., House C.H., Hinrichs K.-U., McKeegan K.D., De Long E.F. (2001) Methane-consuming archaea revealed by directly coupled isotopic and phylogenetic analysis. *Science 293*, 484–7.

Orr W.L. (1974) Changes in sulfur content and isotopic ratios of sulfur during petroleum maturation. Study of Big Horn Basin Paleozoic oils. *Am. Assoc. Pet. Geol. Bull. 50*, 2295–318.

Orr W.L. (1986) Kerogen/asphaltene/sulphur relationships in sulphur-rich Monterey oils. *Org. Geochem. 10*, 499–516.

Osborne M.J., Swarbrick R.E. (1997) Mechanisms for generating overpressure in sedimentary basins: a reevaluation. *Am. Assoc. Pet. Geol. Bull. 81*, 1023–41.

Ostrom P.H., Fry B. (1993) Sources and cycling of organic matter within modern and prehistoric food webs. In *Organic Geochemistry* (ed. Engel M.H., Macko S.A.), 785–94. New York: Plenum Press.

Otto A., Simoneit B.R.T. (2001) Chemosystematics and diagenesis of terpenoids in fossil conifer species and sediment from the Eocene Zeitz formation, Saxony, Germany. *Geochim. Cosmochim. Acta 65*, 3505–27.

Otto-Bliesner B.L. (1995) Continental drift, runoff and weathering feedbacks: implications from climate model experiments. *J. Geophys. Res. 100*, 11537–48.

Ourisson G., Nakatani Y. (1994) The terpenoid theory of the origin of cellular life: the evolution of terpenoids to cholesterol. *Chem. Biol. 1*, 11–23.

Ourisson G., Albrecht P., Rohmer M. (1979) The hopanoids. *Pure Appl. Chem. 51*, 709–29.

Ourisson G., Albrecht P., Rohmer M. (1982) Predictive microbial biochemistry from molecular fossils to prokaryote membranes. *Trends Biochem. Sci. 7*, 236–9.

Ourisson G., Rohmer M., Poralla K. (1987) Prokaryotic hopanoids and other polyterpenoid sterol surrogates. *Ann. Rev. Microbiol. 41*, 301–33.

Page S.E., Siegert F., Rieley J.O., Boehm H.-D.V., Jaya A., Limin S. (2002) The amount of carbon released from peat and forest fires in Indonesia during 1997. *Nature 420*, 61–5.

Palmer M.R., Edmond J.M. (1989) The strontium isotope budget of the modern ocean. *Earth Planet. Sci. Letts 92*, 11–26.

Pancost R.D., Sinninghe Damsté J.S. (2003) Carbon isotopic compositions of prokaryotic lipids as tracers of carbon cycling in diverse settings. *Chem. Geol. 195*, 29–58.

Pancost R.D., Freeman K.H., Wakeham S.G., Robertson C.Y. (1997) Controls on carbon isotope fractionation by diatoms in the Peru upwelling region. *Geochim. Cosmochim. Acta 61*, 4983–91.

Parkes R.J. (1987) Analysis of microbial communities within sediments using biomarkers. In *Ecology of Microbial Communities, SGM Symp. Ser. 41*, 147–77. Cambridge: Cambridge University Press.

Parkes R.J. (2000) A case of bacterial immortality? *Nature 407*, 844–5.

Parkes R.J., Taylor J. (1983) The relationship between fatty acid distributions and bacterial respiratory types in contemporary marine sediments. *Est. Coast. Shelf Sci. 16*, 173–89.

Parkes R.J., Cragg B.A., Fry J.C., Herbert R.A., Wimpenny J.W.T. (1990) Bacterial biomass and activity in deep sediment layers from the Peru margin. *Phil. Trans. Roy. Soc. Lond. A 331*, 139–53.

Parkes R.J., Cragg B.A., Getliff J.M., Harvey S.M., Fry J.C., Lewis C.A., Rowland S.J. (1993) A quantitative study of microbial decomposition of biopolymers in Recent sediments from the Peru Margin. *Mar. Geol. 113*, 55–66.

Parkes R.J., Cragg B.A., Wellsbury P. (1999) Recent studies in bacterial populations and processes in subseafloor sediments: a review. *Hydrogeol. J. 8*, 11–28.

Parrish J.T., Ziegler A.M., Scotese C.R. (1982) Rainfall patterns and the distribution of coals and evaporites in the Mesozoic and Cenozoic. *Palaeogeogr. Palaeoclim. Palaeoecol. 40*, 67–101.

Parsons T.R., Takahashi M., Hargrave B. (1977) *Biological Oceanographic Processes*. New York: Pergamon Press.

Partensky F., Hess W.R., Vaulot D. (1999) *Prochlorococcus*, a marine photosynthetic prokaryote of global significance. *Microbiol. Molec. Biol. Revs. 63*, 106–27.

Passier H.F., Bosch H.-J., Nijenhuis I.A., Lourens L.J., Böttcher M.E., Leenders A., Sinninghe Damsté J.S., de Lange G.J., de Leeuw J.W. (1999) Sulphidic Mediterranean surface waters during Pliocene sapropel formation. *Nature 397*, 146–9.

Patience R.L., Mann A.L., Poplett I.J.F. (1992) Determination of molecular structure of kerogens using ^{13}C NMR spectroscopy: II. The effects of thermal maturation on kerogens from marine sediments. *Geochim. Cosmochim. Acta 56*, 2725–42.

Pavlov A.A., Kasting J.F. (2002) Mass-independent fractionation of sulfur isotopes in Archean sediments: strong evidence for an anoxic Archean atmosphere. *Astrobiol.* 2, 27–41.

Pavlov A.A., Kasting J.F., Brown L.L. (2001a) UV-shielding of NH_3 and O_2 by organic hazes in the Archean atmosphere. *J. Geophys. Res. 106*, 23267–87.

Pavlov A.A., Kasting J.F., Eigenbrode J.L., Freeman K.H. (2001b) Organic haze in Earth's early atmosphere: source of low-^{13}C Late Archaean kerogens? *Geology 29*, 1003–6.

Pavlov A.A., Hurtgen M.T., Kasting J.F., Arthur M.A. (2003) Methane-rich Proterozoic atmosphere? *Geology 31*, 87–90.

Peakman T.M., Maxwell J.R. (1988) Early diagenetic pathways of steroid alkenes. *Org. Geochem. 13*, 583–92.

Peakman T.M., ten Haven H.L., Rechka J.R., de Leeuw J.W., Maxwell J.R. (1989) Occurrence of (20R)- and (20S)-$\Delta^{8,14}$ and Δ^{14} 5α(H)-sterenes and the origin of 5α(H),14β(H),17β(H)-steranes in an immature sediment. *Geochim. Cosmochim. Acta 53*, 2001–9.

Peakman T.M., Jervoise A., Wolff G.A., Maxwell J.R. (1992) Acid-catalyzed rearrangements of steroid alkenes. Part 4. An initial reinvestigation of the backbone rearrangement of 4-methylcholest-4-ene. In *Biological Markers in Sediments and Petroleum* (ed. Moldowan J.M., Albrecht P., Philp R.P.), 58–74. Englewood Cliffs, NJ: Prentice Hall.

Pearson P.N., Ditchfield P.W., Singano J., Harcourt-Brown K.G., Nicholas C.J., Olsson R.K., Shackleton N.J., Hall M.A. (2001) Warm tropical sea surface temperatures in the Late Cretaceous and Eocene Epochs. *Nature 413*, 481–7.

Pedersen T.F., Calvert S.E. (1990) Anoxia vs. productivity: what controls the formation of organic-carbon-rich sediments and sedimentary rocks? *Am. Assoc. Pet. Geol. Bull. 74*, 454–66.

Pennisi E. (1998) Genome data shake tree of life. *Science 280*, 672–4.

Penny D., Poole A. (1999) The nature of the last universal common ancestor. *Curr. Opin. Genet. Dev. 9*, 672–7.

Pepper A.S., Corvi P.J. (1995a) Simple kinetic models of petroleum formation. Part I: oil and gas generation from kerogen. *Mar. Pet. Geol. 12*, 291–319.

Pepper A.S., Corvi P.J. (1995b) Simple kinetic models of petroleum formation. Part III: Modelling an open system. *Mar. Pet. Geol. 12*, 417–52.

Pepper A.S., Dodd T.A. (1995) Simple kinetic models of petroleum formation. Part II: oil-gas cracking. *Mar. Pet. Geol. 12*, 321–40.

Perry G.J., Volkman J.K., Johns R.B., Bavor H.J. (1979) Fatty acids of bacterial origin in contemporary marine sediments. *Geochim. Cosmochim. Acta 43*, 1715–25.

Peters K.E. (1986) Guidelines for evaluating petroleum source rock using programmed pyrolysis. *Am. Assoc. Pet. Geol. Bull. 70*, 318–29.

Peters K.E., Moldowan J.M. (1993) *The Biomarker Guide: Interpreting Molecular Fossils in Petroleum and Ancient Sediments*. Englewood Cliffs, NJ: Prentice Hall.

Peters K.E., Sweeney R.E., Kaplan I.R. (1978) Correlation of carbon and nitrogen stable isotope ratios in sedimentary organic matter. *Limnol. Oceanogr. 23*, 598–604.

Peters K.E., Moldowan J.M., Schoell M., Hempkins W.B. (1986) Petroleum isotopic and biomarker composition related to source rock organic matter and depositional environment. *Org. Geochem. 10*, 17–27.

Peters K.E., Moldowan J.M., Sundararaman P. (1990) Effects of hydrous pyrolysis on biomarker thermal maturity parameters: Monterey Phosphatic and Siliceous members. *Org. Geochem. 15*, 249–65.

Peterson L.C., Haug G.H., Hughen K.A., Röhl U. (2000) Rapid changes in the hydrologic cycle of the tropical Atlantic during the last glacial. *Science 290*, 1947–51.

Petit J.R., Jouzel J., Raynaud D., Barkov N.I., Barnola J.-M. & 14 others (1999) Climate and atmospheric history of the past 420,000 years from the Vostok ice core, Antarctica. *Nature 399*, 429–36.

Petsch S.T., Berner R.A. (1998) Coupling the geochemical cycles of C, P, Fe, and S: the effect on atmospheric O_2 and the isotopic records of carbon and sulfur. *Am. J. Sci. 29*, 246–62.

Philp R.P. (1994) High temperature gas chromatography for the analysis of fossil fuels: a review. *J. High Res. Chromatogr. 17*, 398–406.

Philp R.P., Gilbert T.D. (1986) Biomarker distributions in Australian oils predominantly derived from terrigenous source material. *Org. Geochem. 10*, 73–84.

Philippi G.T. (1975) The deep subsurface temperature controlled origin of the gaseous and gasoline-range hydrocarbons of petroleum. *Geochim. Cosmochim. Acta 39*, 1353–73.

Phillippi G.T. (1981) Correlation of crude oils with their oil source formation, using high resolution GLC C_6-C_7 component analysis. *Geochim. Cosmochim. Acta 45*, 1495–513.

Popp B.N., Kenig F., Wakeham S.G., Laws E.A., Bidigare R.R. (1998a) Does growth rate affect ketone unsaturation and intracellular carbon isotopic variability in *Emiliania huxleyi*? *Paleoceanogr. 13*, 35–41.

Popp B.N., Laws E.A., Bidigare R.R., Dore J.E., Hanson K.L., Wakeham S.G. (1998b) Effect of phytoplankton cell geometry on carbon isotopic fractionation. *Geochim. Cosmochim. Acta 62*, 69–77.

Powell T.G. (1986) Petroleum geochemistry and depositional setting of lacustrine source rocks. *Mar. Pet. Geol. 3*, 200–19.

Powell T.G., Boreham C.J., Smyth M., Russell N., Cook A.C. (1991) Petroleum source rock assessment in non-marine sequences: pyrolysis and petrographic analysis of Australian coals and carbonaceous shales. *Org. Geochem. 17*, 375–94.

Prahl F.G., Carpenter R. (1983) Polycyclic aromatic hydrocarbons (PAH)-phase associations in Washington coastal sediment. *Geochim. Cosmochim. Acta 47*, 1013–23.

Prahl F.G., Wakeham S.G. (1987) Calibration of unsaturation patterns in long-chain ketone compositions for palaeotemperature assessment. *Nature 330*, 367–76.

Prahl F.G., Eglinton G., Corner E.D.S., O'Hara S.C.M., Forsberg T.E.V. (1984a) Changes in plant lipids during passage through the gut of *Calanus*. *J. Mar. Biol. Assoc. UK 64*, 317–34.

Prahl F.G., Eglinton G., Corner E.D.S., O'Hara S.C.M. (1984b) Copepod fecal pellets as a source of dihydrophytol in marine sediments. *Science 224*, 1235–7.

Prahl F.G., Muehlhausen L.A., Zahnle D.L. (1988) Further evaluation of long-chain alkenones as indicators of paleoceanographic conditions. *Geochim. Cosmochim. Acta 52*, 2303–10.

Prahl F.G., de Lange G.J., Lyle M., Sparrow M.A. (1989) Post-depositional stability of long-chain alkenones under contrasting redox conditions. *Nature 341*, 434–7.

Prather M.J., Watson R.T. (1990) Stratospheric ozone depletion and future levels of atmospheric chlorine and bromine. *Nature 344*, 729–34.

Price L.C., Barker C.E. (1985) Suppression of vitrinite reflectance in amorphous rich kerogen—a major unrecognised problem. *J. Pet. Geol. 8*, 59–84.

Price L.C., Wenger L.M. (1992) The influence of pressure on petroleum generation and maturation as suggested by aqueous pyrolysis. *Org. Geochem. 19*, 141–59.

Prinn R.G., Weiss R.F., Fraser P.J., Simmonds P.G., Cunnold D.M., & 12 others (2000) A history of chemically and radiatively important gases in air deduced from ALE/GAGE/AGAGE. *J. Geophys. Res. 105*, 17751–92.

Purdom C.E., Hardiman P.A., Bye V.J., Eno N.C., Tyler C.R., Sumpter J.P. (1994) Estrogenic effects of effluents from sewage-treatment works. *Chem. Ecol. 8*, 275–85.

Püttmann W., Villar H. (1987) Occurrence and geochemical significance of 1,2,5,6-tetramethylnaphthalene. *Geochim. Cosmochim. Acta 51*, 3023–9.

Quirk M.M., Wardroper A.M.K., Wheatley R.E., Maxwell J.R. (1984) Extended hopanoids in peat environments. *Chem. Geol. 42*, 25–43.

Radke M. (1987) Organic geochemistry of aromatic hydrocarbons. In *Advances in Petroleum Geochemistry, Vol. 2* (ed. Brooks J., Welte D.), 141–207. New York: Academic Press.

Radke M., Welte D.H. (1983) The methylphenanthrene index (MPI): a maturity parameter based on aromatic hydrocarbons. In *Advances in Organic Geochemistry 1981* (ed. Bjorøy et al.), 504–12. Chichester: Wiley.

Radke M., Welte D.H., Willsch H. (1986) Maturity parameters based on aromatic hydrocarbons: Influence of the organic matter type. *Org. Geochem. 10*, 51–63.

Radke M., Leythaeuser D., Teichmüller M. (1984) Relationship between rank and composition of aromatic hydrocarbons for coals of different origins. *Org. Geochem. 6*, 423–30.

Radke M., Willsch H., Teichmüller M. (1990) Generation and distribution of aromatic hydrocarbons in coals of low rank. *Org. Geochem. 15*, 539–64.

Ragan M.A., Smidsrø O., Larsen B. (1979) Chelation of divalent metal ions by brown algal polyphenols. *Mar. Chem. 7*, 265–71.

Ramamoorthy K., Wang F., Chen I.-C., Safe S., Norris J.D., McDonnell D.P., Gaido K.W., Biochinfuso W.P., Korach K.S. (1997) Potency of combined estrogenic pesticides. *Science 275*, 405–6.

Ramanathan V., Crutzen P.J., Kiehl J.T., Rosenfeld D. (2001) Aerosols, climate, and the hydrological cycle. *Science 294*, 2119–24.

Rampino M.R. (1991) Volcanism, climate change and the geologic record. *SEPM Spec. Publn 45*, 9–18. Tulsa, OK: Society of Economic Paleontologists and Mineralogists.

Rampino M.R., Self S. (1992) Volcanic winter and accelerated glaciation following the Toba super-eruption. *Nature 359*, 50–2.

Ransom B., Shea K.F., Burkett P.J., Bennett R.H., Baerwald R. (1998) Comparison of pelagic and nepheloid layer marine snow: implications for carbon cycling. *Mar. Geol. 150*, 39–50.

Rasmussen B., Buick R. (1999) Redox state of the Archean atmosphere: evidence from detrital heavy minerals in ca. 3250–2750 Ma sandstones from the Pilbara Craton, Australia. *Geology 27*, 115–18.

Rasyid U., Johnson W.D., Wilson M.A., Hanna J.V. (1992) Changes in organic structural group composition of humic and fulvic acids with depth in sediments from similar geographical but different depositional environments. *Org. Geochem. 18*, 521–9.

Rau G.H., Arthur M.A., Dean W.E. (1987) $^{15}N/^{14}N$ variations in Cretaceous Atlantic sedimentary sequences: implication for past changes in marine nitrogen biogeochemistry. *Earth Planet. Sci. Letts 82*, 269–79.

Rau G.H., Takahashi T., Des Marais D.J., Repeta D.J., Martin J.H. (1992) The relationship between $\delta^{13}C$ of organic matter and $[CO_{2(aq)}]$ in ocean surface water: data from a JGOFS site in the northeast Atlantic Ocean and a model. *Geochim. Cosmochim. Acta 56*, 1413–19.

Rau G.H., Riebesell U., Wolf-Gladrow D. (1997) CO_{2aq}-dependent photosynthetic ^{13}C fractionation in the ocean: a model versus measurements. *Glob. Biogeochem. Cycles 11*, 267–78.

Raup D.M., Sepkowski J.J. (1984) Periodicity of extinctions in the geologic past. *Proc. Natl Acad. Sci. USA 81*, 801–5.

Raven J.A., Johnston A.M. (1991) Mechanisms of inorganic-carbon acquisition in marine phytoplankton and their implications for the use of other resources. *Limnol. Oceanogr. 36*, 1701–14.

Raven J.A., Macfarlane J.J., Griffiths H. (1987) The application of carbon isotope discrimination techniques. In *Plant Life in Aquatic and Amphibious Habitats* (ed. Crawford R.M.M.). BES Special Symposium. Oxford: Blackwell Scientific.

Raymo M.E. (1991) Geochemical evidence supporting T. C. Chamberlin's theory of glaciation. *Geology 19*, 344–7.

Raymo M.E., Ruddiman W.F. (1992) Tectonic forcing of late Cenozoic climate. *Nature 359*, 117–22.

Raymont J.E.G. (ed.) (1983) *Plankton and Productivity in the Oceans*. Oxford: Pergamon.

Raynaud D., Jouzel J., Barnola J.M., Chappellaz J., Delmas R.J., Lorius C. (1993) The ice record of greenhouse gases. *Science 259*, 926–34.

Reading H.G. (1986) *Sedimentary Environments and Facies*. Oxford: Blackwell Scientific.

Readman J.W., Mantoura R.F.C., Rhead M.M. (1984) The

physico-chemical speciation of polycyclic aromatic hydrocarbons (PAH) in aquatic systems. *Fres. Z. Anal. Chem. 319*, 126–31.

Rechka J.A., Maxwell J.R. (1988) Unusual long chain ketones of algal origin. *Tetr. Letts 29*, 2599–600.

Rechka J.A., Cox H.C., Peakman T.M., de Leeuw J.W., Maxwell J.R. (1992) A reinvestigation of aspects of the early diagenetic pathways of 4-methylsterenes based on molecular mechanics calculations and the acid catalysed isomerization of 4-methylcholest-4-ene. In *Biological Markers in Sediments and Petroleum* (ed. Moldowan J.M., Albrecht P., Philp R.P.), 42–57. Englewood Cliffs, NJ: Prentice Hall.

Reeburgh W.S. (1980) Anaerobic methane oxidation: rate depth distributions in Skan Bay sediments. *Earth Planet. Sci. Letts 47*, 345–52.

Reijnders P.J.H. (1986) Reproductive failure in common seals feeding on fish from polluted coastal waters. *Nature 324*, 456–7.

Renne P.R., Zichao Z., Richards M.A., Black M.T., Basu A.R. (1995) Synchrony and causal relations between Permian–Triassic boundary crises and Siberian flood volcanism. *Science 269*, 1413–16.

Repeta D.J. (1989) Carotenoid diagenesis in recent marine sediments: II. Degradation of fucoxanthin to loliolide. *Geochim. Cosmochim. Acta 53*, 699–707.

Repeta D.J., Gagosian R.B. (1982) Carotenoid transformation in coastal marine waters. *Nature 295*, 51–4.

Repeta D.J., Gagosian R.B. (1984) Transformation reactions and recycling of carotenoids and chlorins in the Peru upwelling region (15°S, 75°W). *Geochim. Cosmochim. Acta 48*, 1265–77.

Repeta D.J., Simpson D.J., Jørgensen B.B., Jannasch H.W. (1989) Evidence for anoxygenic photosynthesis from the distribution of bacteriochlorophylls in the Black Sea. *Nature 342*, 69–72.

Requejo A.G. (1994) Maturation of petroleum source rocks—II. Quantitative changes in extractable hydrocarbon content and composition associated with hydrocarbon generation. *Org. Geochem. 21*, 91–105.

Retallack G.J. (1995) Permian–Triassic life crisis on land. *Science 267*, 77–80.

Revil A.T., Volkman J.K., O'Leary T., Summons R.E., Boreham C.J., Banks M.R., Denwer K. (1994) Hydrocarbon biomarkers, thermal maturity, and depositional setting of tasmanite oil shales from Tasmania, Australia. *Geochim. Cosmochim. Acta 58*, 3803–22.

Revsbech N.P., Sørensen J., Blackburn T.H. (1980) Distribution of oxygen in marine sediments measured with microelectrodes. *Limnol. Oceanogr. 25*, 403–11.

Reynolds C.S. (1984) Phytoplankton periodicity: the interactions of form, function and environmental variability. *Freshwater Biol. 14*, 111–42.

Reynolds C.S., Wiseman S.W., Godfrey B.M., Butterwick C. (1983) Some effects of artificial mixing on the dynamics of phytoplankton populations in large limnetic enclosures. *J. Plankton Res. 5*, 203–34.

Richnow H.H., Seifert R., Hefter J., Kästner M., Mahro B., Michaelis W. (1994) Metabolites of xenobiotica and mineral oil constituents linked to macromolecular organic matter in polluted environments. *Org. Geochem. 22*, 671–81.

Ridgwell A.J., Kennedy M.J., Caldeira K. (2003) Carbonate deposition, climate stability, and Neoproterozoic ice ages. *Science 302*, 859–62.

Rieley G., Collier R.J., Jones D.M., Eglinton G., Eakin P.A., Fallick A.E. (1991) Sources of sedimentary lipids deduced from stable carbon-isotope analyses of individual compounds. *Nature 352*, 425–7.

Ries-Kautt M., Albrecht P. (1989) Hopane-derived triterpenoids in soils. *Chem. Geol. 76*, 143–51.

Riffé-Chalard C., Verzegnassi L., Gülaçar F.O. (2000) A new series of steryl chlorin esters: pheophorbide a steryl esters in an oxic surface sediment. *Org. Geochem. 31*, 1703–12.

Riolo J., Hussler G., Albrecht P., Connan J. (1986) Distribution of aromatic steroids in geological samples: Their evaluation as geochemical parameters. *Org. Geochem. 10*, 981–90.

Ritter U. (2003) Solubility of petroleum compounds in kerogen: implications for petroleum expulsion. *Org. Geochem. 34*, 319–26.

Ritter U., Myhr M.B., Vinge T., Aareskjold K. (1995) Experimental heating and kinetic models of source rocks: comparison of different methods. *Org. Geochem. 23*, 1–9.

Roberts N. (1998) *The Holocene*. Oxford: Blackwell.

Robin P.L., Rouxhet P.G., Durand B. (1977) Caracterisation des kérogènes et de leur évolution par spectroscopie infrarouge. In *Advances in Organic Geochemistry 1975* (ed. Campos R., Goni J.), 693–716. Madrid: ENADIMSA.

Robinson N., Eglinton G., Brassell S.C. Cranwell P.A. (1984) Dinoflagellate origin for sedimentary 4α-methylsteroids and 5α(H)-stanols. *Nature 308*, 439–42.

Robinson N., Cranwell P.A., Eglinton G., Jaworski G.H.M. (1987) Lipids of four species of freshwater dinoflagellates. *Phytochem. 26*, 411–21.

Robinson R.S., Meyers P.A., Murray R.W. (2002) Geochemical evidence of variations in delivery and deposition of sediment in Pleistocene light-dark cycles under the Benguela Current Upwelling System. *Mar. Geol. 180*, 249–70.

Robson J.N., Rowland S.J. (1986) Identification of novel widely distributed sedimentary acyclic sesterterpenoids. *Nature 324*, 561–3.

Rodhe H. (1990) A comparison of the contribution of various gases to the Greenhouse Effect. *Science 248*, 1217–19.

Roedder E. (1984) Fluid inclusions. *Rev. Mineral. 12*. Washington, DC: Mineralogical Society of America.

Rogers J., Santosh M. (2002) Configuration of Columbia, a Mesoproterozoic supercontinent. *Gondwana Res. 5*, 5–22.

Röhl U., Bralower T.J., Norris R.N., Wefer G. (2000) A new chronology for the late Paleocene thermal maximum and its environmental implications. *Geology 28*, 927–30.

Rohling E.J. (1994) Review and new aspects concerning the formation of eastern Mediterranean sapropels. *Mar. Geol. 122*, 1–28.

Rohmer M., Anding C., Ourisson G. (1980) Nonspecific biosynthesis of hopane triterpenes by a cell-free system from *Acetobacter pasteurianum*. *Eur. J. Biochem.* **112**, 541–7.

Rohmer M., Boivier-Nave P., Ourisson G. (1989) Distribution of hopanoid triterpenes in prokaryotes. *J. Genetic Microbiol.* **130**, 1137–50.

Rohmer M., Bisseret P., Neunlist S. (1992) The hopanoids, prokaryotic triterpenoids and precursors of ubiquitous molecular fossils. In *Biological Markers in Sediments and Petroleum* (ed. Moldowan J.M., Albrecht P., Philp R.P.), 1–17. Englewood Cliffs, NJ: Prentice Hall.

Rontani J.-F. (2001) Visible light-dependent degradation of lipidic phytoplanktonic components during senescence: a review. *Phytochem.* **58**, 187–202.

Rontani J.-F., Bonin P. (2000) Aerobic bacterial metabolism of phytol in seawater: effect of particulate association on an abiotic intermediate step and its biogeochemical consequences. *Org. Geochem.* **31**, 489–96.

Rontani J.-F., Volkman J.K. (2003) Phytol degradation products as biogeochemical tracers in aquatic environments. *Org. Geochem.* **34**, 1–35.

Rontani J.-F., Bonin P., Volkman J.K. (1999) Biodegradation of free phytol by bacterial communities isolated from marine sediments under aerobic and denitrifying conditions. *Appl. Environ. Microbiol.* **65**, 5484–92.

Rontani J.-F., Perrote S., Cuny P. (2000) Can a high chlorophyllase activity bias the use of the phytyldiol versus phytol ratio (CPPI) for the monitoring of chlorophyll photooxidation in seawater? *Org. Geochem.* **31**, 91–9.

Rossmann A., Butzenlechner M., Schmidt H.-L. (1991) Evidence for a nonstatistical carbon isotope distribution in natural glucose. *Plant Physiol.* **96**, 609–14.

Rostek F., Ruhland G., Bassinot F.C., Müller P.J., Labeyrie L.D., Lancelot Y., Bard E. (1993) Reconstructing sea surface temperature and salinity using $\delta^{18}O$ and alkenone records. *Nature* **364**, 319–21.

Rothman D.H. (2002) Atmospheric carbon dioxide levels for the last 500 million years. *Proc. Natl Acad. Sci. USA* **99**, 4167–71.

Rothschild L.J., Mancinelli R.L. (1990) Model of carbon fixation in microbial mats from 3500 Myr ago to the present. *Nature* **345**, 710–12.

Rowan D.J., Rasmussen J.B. (1992) Why don't Great Lakes fish reflect environmental concentrations of organic contaminants? An analysis of between-lake variability in the ecological partitioning of PCBs and DDT. *J. Great Lakes Res.* **18**, 724–41.

Rowland S.J., Robson J.N. (1990) The widespread occurrence of highly branched acyclic C_{20}, C_{25} and C_{30} hydrocarbons in Recent sediments and biota—a review. *Mar. Environ. Res.* **30**, 191–216.

Rowland S., Donkin P., Smith E., Wraige E. (2001) Aromatic hydrocarbon 'humps' in the marine environment: unrecognised toxins? *Environ. Sci. Technol.* **35**, 2640–4.

Royden L., Sclater J.G., von Herzen R.P. (1980) Continental margin subsidence and heat flow: important parameters in formation of petroleum hydrocarbons. *Am. Assoc. Pet. Geol. Bull.* **64**, 173–87.

Rubinstein I., Sieskind O., Albrecht P. (1975) Rearranged sterenes in a shale: Occurrence and simulated formation. *J. Chem. Soc., Perkin Trans. I*, 1833–6.

Rubinsztain Y., Ioselis P., Ikan R., Aizenshtat Z. (1984) Investigations of the structural units of melanoidins. *Org. Geochem.* **6**, 791–804.

Ruddiman W.F., McIntyre A. (1981) The North Atlantic Ocean during the last deglaciation. *Palaeogeogr. Palaeoclim. Palaeoecol.* **35**, 145–214.

Rueter P., Rabus R., Wilkes H., Aeckersberg F., Rainey F.A., Jannasch H.W., Widdel F. (1994) Anaerobic oxidation of hydrocarbons in crude oil by new types of sulphate-reducing bacteria. *Nature* **372**, 455–8.

Rullkötter J., Michaelis W. (1990) The structure of kerogen and related materials. A review of recent progress and future trends. *Org. Geochem.* **16**, 829–52.

Rullkötter J., Peakman T.M., ten Haven H.L. (1994) Early diagenesis of terrigenous triterpenoids and its implications for petroleum geochemistry. *Org. Geochem.* **21**, 215–33.

Runnegar B. (1991) Precambrian oxygen levels estimated from the biochemistry and physiology of early eukaryotes. *Palaeogeogr. Palaeoclim. Palaeoecol. Glob. Planet. Change Sect.* **97**, 97–111.

Russell M.J., Hall A.J. (1997) The emergence of life from iron monosulphide bubbles at a submarine hydrothermal redox and pH front. *J. Geol. Soc. Lond.* **154**, 377–402.

Rye R., Holland H.D. (1998) Paleosols and the evolution of atmospheric oxygen: a critical review. *Am. J. Sci.* **298**, 621–72.

Rye R., Kuo P.H., Holland H.D. (1995) Atmospheric carbon dioxide concentrations before 2.2 billion years ago. *Nature* **378**, 603–5.

Safe S. (1995) Environmental and dietary estrogens and human health—is there a problem? *Environ. Health Perspect.* **103**, 346–51.

Saiz-Jimenez C., de Leeuw J.W. (1986) Lignin pyrolysis products: Their structures and their significance as biomarkers. *Org. Geochem.* **10**, 869–76.

Saltzman M.R. (2003) Late Paleozoic ice age: oceanic gateway or pCO_2? *Geology* **31**, 151–4.

Sandvik E.I., Young W.A., Curry D.J. (1992) Expulsion from hydrocarbon sources: the role of organic absorption. *Org. Geochem.* **19**, 77–87.

Sarmiento J.L., Herbert T.D., Toggweiler J.R. (1988) Causes of anoxia in the world ocean. *Glob. Biogeochem. Cycles* **2**, 115–28.

Sauer P.E., Eglinton T.I., Hayes J.M., Schimmelmann A., Sessions A.L. (2001) Compound-specific D/H ratios of lipid biomarkers from sediments as a proxy for environmental and climatic conditions. *Geochim. Cosmochim. Acta* **65**, 213–22.

Sawada K., Handa N., Shiraiwa Y., Danbara A., Montani S. (1996) Long-chain alkenones and alkyl alkenoates in the coastal and pelagic sediments of the northwest North Pacific, with special reference to the reconstruction of *Emil-*

iania huxleyi and *Gephyrocapsa oceanica* ratios. *Org. Geochem.* 24, 751–64.

Saward S.A. (1992) A global view of Cretaceous vegetation patterns. In *Controls on the Distribution and Quality of Cretaceous Coals* (ed. McCabe P.J., Parrish J.T.) *Geol. Soc. Am. Spec. Pap. 267*, 17–35. Boulder: Geol. Soc. of America.

Schaeffer J. (1972) Comparison of the carbon-13 nuclear magnetic resonance of some solid cis- and trans-polyisoprenes. *Macromolecules 5*, 427–32.

Schenk H.J., Horsfield B. (1993) Kinetics of petroleum generation by programmed-temperature closed- versus open-system pyrolysis. *Geochim. Cosmochim. Acta 57*, 623–30.

Schenk H.J., Horsfield B. (1997) Kinetics of petroleum generation: applications and limitations. *Zbl. Geol. Paläont. Teil I H 11/12*, 1113–17.

Schenk H.J., Horsfield B. (1998) Using natural maturation series to evaluate the utility of parallel reaction kinetic models: an investigation of Toarcian shales and Carboniferous coals, Germany. *Org. Geochem. 29*, 137–54.

Schidlowski M. (1988) A 3800-million-year isotopic record of life from carbon in sedimentary rocks. *Nature 333*, 313–18.

Schindler S.W. (1976) Evolution of phosphorus limitation in lakes. *Science 195*, 260–2.

Schmitter J.M., Sucrow W., Arpino P.J. (1982) Occurrence of novel tetracyclic geochemical markers: 8,14-seco-hopanes in a Nigerian crude oil. *Geochim. Cosmochim. Acta 46*, 2345–50.

Schneider E.K., Kirtman B.P., Lindzen R.S. (1999) Tropospheric water vapor and climate sensitivity. *J. Atm. Sci. 36*, 1649–58.

Schnitzer M. (1978) Humic substances: chemistry and reactions. *Dev. Soil Sci. 8*, 1–64.

Schoell M. (1984a) Stable isotopes in petroleum research. In *Advances in Petroleum Geochemistry, Vol. 1* (ed. Brooks J., Welte D.), 215–45. New York: Academic Press.

Schoell M. (1984b) Recent advances in petroleum isotope geochemistry. *Org. Geochem. 6*, 645–63.

Schoell M. (1988) Multiple origins of methane in the Earth. *Chem. Geol. 71*, 1–10.

Schoell M., McCaffrey M.A., Fago F.J., Moldowan J.M. (1992) Carbon isotopic composition of 28,30-bisnorhopanes and other biological markers in a Monterey crude oil. *Geochim. Cosmochim. Acta 56*, 1391–9.

Schopf J.W. (1993) Microfossils of the Early Archean Apex Chert: new evidence of the antiquity of life. *Science 260*, 640–6.

Schopf J.W., Packer B.M. (1987) Early Archaean (3.3-billion to 3.5-billion-year old) microfossils from Warrawoona Group, Australia. *Science 237*, 70–3.

Schopf J.W., Kudryavtsev A.B., Agresti D.G., Wdowiak T.J., Czaja A.D. (2002) Laser-Raman imagery of Earth's earliest fossils. *Nature 416*, 73–6.

Schouten S., de Graaf W., Sinninghe Damsté J.S., van Driel G.B., de Leeuw J.W. (1994) Laboratory simulation of natural sulphurization II. Reaction of multi-functionalized lipids with inorganic polysulphides at low temperatures. *Org. Geochem. 22*, 825–34.

Schouten S., Schoell M., Rijpstra W.I.C., Sinninghe Damsté J.S., de Leeuw J.W. (1997) A molecular stable carbon isotope study of organic matter in immature Miocene Monterey sediments, Pismo basin. *Geochim. Cosmochim. Acta 61*, 2065–82.

Schouten S., Rijpstra W.I.C., Kok M., Hopmans E.C., Summons R.E., Volkman J.K., Sinninghe Damsté J.S. (2001) Molecular organic tracers of biogeochemical processes in a saline meromictic lake (Ace Lake). *Geochim. Cosmochim. Acta 65*, 1629–40.

Schulze M., Schaeffer P., Bernasconi S., Albrecht P. (2001) Investigation of the fate of ^{13}C-labelled phytol in sulfur-rich anaerobic lacustrine sediments. *Abstracts of 20th International Meeting on Organic Geochemistry Vol. 1*, 136–7.

Scotese C.R. (2003) Paleomap Project. http://www.scotese.com/

Scott A.C. (1989) Observations on the nature and origin of fusain. *Int. J. Coal Geol. 12*, 443–75.

Scott A.C., Fleet A.J. (1994) *Coal and Coal-bearing Strata as Oil-prone Source Rocks? Geol. Soc. Spec. Publn 77*. Oxford: Blackwell Scientific.

Seewald J.S. (1994) Evidence for metastable equilibrium between hydrocarbons under hydrothermal conditions. *Nature 370*, 285–7.

Seewald J.S., Benitez-Nelson B.C., Whelan J.K. (1998) Laboratory and theoretical constraints on the generation and composition of natural gas. *Geochim. Cosmochim. Acta 62*, 1599–617.

Seitzinger S.P., Giblin A.E. (1996) Estimating denitrification in North Atlantic continental shelf sediments. *Biogeochem. 35*, 235–60.

Selley R.C. (1997) *Elements of Petroleum Geology*. New York: Academic Press.

Senesi N., Testini C., Miano T.M. (1987) Interaction mechanisms between humic acids of different origin and nature and electron donor herbicides: a comparative IR and ESR study. *Org. Geochem. 11*, 25–30.

Sepkoski J.J. Jr (1984) A kinetic model of Phanerozoic taxonomic diversity. III. Post-Paleozoic families and mass extinctions. *Paleobiol. 10*, 246–67.

Sessions A.L., Burgoyne T.W., Schimmelmann A., Hayes J.M. (1999) Fractionation of hydrogen isotopes in lipid biosynthesis. *Org. Geochem. 30*, 1193–200.

Sessions A.L., Jahnke L.L., Schimmelmann A., Hayes J.M. (2002) Hydrogen isotope fractionation in lipids of the methane-oxidizing bacterium *Methylococcus capsulatus*. *Geochim. Cosmochim. Acta 66*, 3955–69.

Shackleton N.J. (1984) In *Fossils and Climate* (ed. Brenchley P.J.), 27–34. New York: Wiley.

Shackleton N.J. (1987a) Oxygen isotopes, ice volumes, and sea level. *Quat. Sci. Rev. 6*, 183–90.

Shackleton N.J. (1987b) The carbon isotopic record of the Cenozoic: history of organic carbon burial and of oxygen in the ocean and atmosphere. In *Marine Petroleum Source Rocks* (ed. Brooks J., Fleet A.J.) *Geol. Soc. Spec. Publn 26*, 423–34. Oxford: Blackwell.

Shackleton N.J. (2000) The 100,000-year ice-age cycle identified and found to lag temperature, carbon dioxide, and orbital eccentricity. *Science* 289, 1897–902.

Shackleton N. (2001) Climate change across the hemispheres. *Science 291*, 58–9.

Shackleton N.J., Crowhurst S., Hagelberg T., Pisias N.G., Scheider D.A. (1995) A new Late Neogene time scale: application to leg 138 sites. *Proc. Ocean Drilling Prog. 138*, 73–101.

Shaviv N.J., Veizer J. (2003) Celestial driver of Phanerozoic climate? *GSA Today 13(7)*, 4–10.

Shen Y., Knoll A.H., Walter M.R. (2003) Evidence for low sulphate and anoxia in a mid-Proterozoic marine basin. *Nature 423*, 632–5.

Shiea J., Brassell S.C., Ward D.M. (1990) Mid-chain branched mono- and dimethyl alkanes in hot spring cyanobacterial mats: a direct biogenic source for branched alkanes in ancient sediments? *Org. Geochem. 15*, 223–31.

Siegenthaler U., Sarmiento J.L. (1993) Atmospheric carbon dioxide and the ocean. *Nature 365*, 119–25.

Siegenthaler U., Friedli H., Loetscher H., Moor E., Neftel A., Oeschger H., Stauffer B. (1988) Stable-isotope ratios and concentration of CO_2 in air from polar ice cores. *Annals Glaciol. 10*, 1–6.

Siegert F., Rücker G., Hinrichs A., Hoffmann A. (2001) Increased fire impacts in logged over forests during El Niño driven fires. *Nature 414*, 437–40.

Sieskind O., Joly G., Albrecht P. (1979) Simulation of the geochemical transformation of sterols: Superacid effects of clay minerals. *Geochim. Cosmochim. Acta 43*, 1675–9.

Sigman D.M., Boyle E.A. (2000) Glacial/interglacial variations in atmospheric carbon dioxide. *Nature 407*, 859–69.

Sigman D., Altabet M.A., Francois R., McCorkle D.C., Gaillard J.-F. (1999) The isotopic composition of diatom-bound nitrogen in Southern Ocean sediments. *Paleoceanogr. 14*, 118–34.

Sikes E.L., Volkman J.K. (1993) Calibration of alkenone unsaturation ratios ($U^{k'}_{37}$) for paleotemperature estimation in cold polar waters. *Geochim. Cosmochim. Acta 57*, 1883–9.

Sikes E.L., Farrington J.W., Keigwin L.D. (1991) Use of the alkenone unsaturation ratio $U^{k'}_{37}$ to determine past sea surface temperatures: Core-top SST calibrations and methodology considerations. *Earth Planet. Sci. Letts 104*, 36–47.

Sillen A., Sealy J.C., Van der Merwe N.J. (1989) Chemistry and paleodietary research: no more easy answers. *Am. Antiquities 54*, 504–12.

Simoneit B.R.T. (1977) Diterpenoid compounds and other lipids in deep-sea sediments and their geochemical significance. *Geochim. Cosmochim. Acta 41*, 463–76.

Simoneit B.R.T. (1986) Cyclic terpenoids of the geosphere. In *Biological Markers in the Sedimentary Record* (ed. Johns R.B.), 43–99. Amsterdam: Elsevier.

Simoneit B.R.T. (1990) Petroleum generation, an easy and widespread process in hydrothermal systems: an overview. *Appl. Geochem. 5*, 3–15.

Simoneit B.R.T., Grimalt J.O., Wang T.G., Cox R.E., Hatcher P.G., Nissenbaum A. (1986a) Cyclic terpenoids of contemporary resinous plant detritus and of fossil woods, ambers and coals. *Org. Geochem. 10*, 877–89.

Simoneit B.R.T., Summerhayes C.P., Meyers P.A. (1986b) Sources and hydrothermal alteration of organic matter in Quaternary sediments: a synthesis of studies from the Central Gulf of California. *Mar. Pet. Geol. 3*, 282–97.

Simonich S.L., Hites R.A. (1995) Global distribution of persistent organochlorine compounds. *Science 269*, 1851–4.

Sinninghe Damsté J.J., de Leeuw J.W. (1990) Analysis, structure and geochemical significance of organically-bound sulphur in the geosphere: state of the art and future research. *Org. Geochem. 16*, 1077–101.

Sinninghe Damsté J.S., Koch-van Dalen A.C., de Leeuw J.W., Schenck P.A., Guoying S., Brassell S.C. (1987) The identification of mono-, di- and trimethyl 2-methyl-2(4,8,12-trimethyltridecyl)chromans and their occurrence in the geosphere. *Geochim. Cosmochim. Acta 51*, 2393–400.

Sinninghe Damsté J.J., Rijpstra W.I.C., Kock-van Dalen A.C., de Leeuw J.W., Schenck P.A. (1989) Quenching of labile functionalised lipids by inorganic sulphur species: evidence for the formation of sedimentary organic sulphur compounds at the early stages of diagenesis. *Geochim. Cosmochim. Acta 53*, 1343–55.

Sinninghe Damsté J.J., de las Heras X.F.C., de Leeuw J.W. (1992) Molecular analysis of sulphur-rich brown coals by flash pyrolysis-gas chromatography-mass spectrometry: the Type III-S kerogen. *J. Chromatogr. 607*, 361–76.

Sinninghe Damsté J.J., de las Heras X.F.C., van Bergen P.F., de Leeuw J.W. (1993a) Characterization of Tertiary Catalan lacustrine oil shales: discovery of extremely organic sulphur-rich Type I kerogens. *Geochim. Cosmochim. Acta 57*, 389–415.

Sinninghe Damsté J.S., Keeley B.J., Betts S.E., Baas M., Maxwell J.R., de Leeuw J.W. (1993b) Variations in abundances and distributions of isoprenoid chromans and long-chain alkylbenzenes in sediments of the Mulhouse Basin: a molecular sedimentary record of palaeosalinity. *Org. Geochem. 20*, 1201–15.

Sinninghe Damsté J.S., Wakeham S.G., Kohnen M.E.L., Hayes J.M., de Leeuw J.W. (1993c) A 6,000-year sedimentary molecular record of chemocline excursions in the Black Sea. *Nature 362*, 827–9.

Sinninghe Damsté J.S., Kenig F., Koopmans M.P., Köster J., Schouten S., Hayes J.M., de Leeuw J.W. (1995) Evidence for gammacerane as an indicator of water column stratification. *Geochim. Cosmochim. Acta 59*, 1895–900.

Sinninghe Damsté J.S., Schouten S., van Duin A.C.T. (2001) Isorenieratene derivatives in sediments: possible controls on their distribution. *Geochim. Cosmochim. Acta 65*, 1557–71.

Sinninghe Damsté J.S., Rijpstra W.I.C., Schouten S., Fuerst J.A., Jetten M.S.M., Strous M. (2004) The occurence of hopanoids in planctomycetes: implications for the sedimentary biomarker record. *Org. Geochem. 35*, 561–6.

Sinton C.W., Duncan R.A. (1997) Potential links between ocean plateau volcanism and global ocean anoxia at the Cenomanian–Turonian boundary. *Econ. Geol. 92*, 836–42.

Sirevag R., Buchanan B.B., Berry J.A., Troughton J.H. (1977)

Mechanisms of CO_2 fixation in bacterial photosynthesis studied by the carbon isotope technique. *Arch. Microbiol.* 112, 35–8.

Skelton P.W., Spicer R.A., Kelley S.P., Gilmour I. (2003) *The Cretaceous World*. Cambridge: Cambridge University Press.

Skerrat J.H., Nichols P.D., Bowman J.P., Sly L.I. (1992) Occurrence and significance of long-chain (ω-1)-hydroxy fatty acids in methane-utilizing bacteria. *Org. Geochem.* 18, 189–94.

Sloan E.D. (1990) *Clathrate Hydrates of Natural Gas*. New York: Marcel Dekker.

Smart P.L., Frances P.D. (1991) *Quaternary Dating Methods—A User's Guide*. Technical Guide 4. Cambridge: Quaternary Research Association.

Smith A.G. (1991) Chlorinated hydrocarbon insecticides. In *The Handbook of Pesticide Toxicology Vol. 2: Classes of Pesticides*. San Diego: Academic Press.

Smith M.B., March J. (2000) *March's Advanced Organic Chemistry*. New York: Wiley.

Smith T.M., Shugart H.H. (1993) The transient response of terrestrial carbon storage to a perturbed climate. *Nature* 361, 523–6.

Smith G.G., Sol B.S. (1980) Racemization of amino acids in dipeptides shows COOH > NH_2 for nonsterically hindered residues. *Science* 207, 765–7.

Snowdon L.R. (1979) Errors in extrapolation of experimental kinetic parameters to organic geochemical systems. *Am. Assoc. Pet. Geol. Bull.* 63, 1128–34.

Snyder L.E. (1997) Detection of large interstellar molecules with radio interferometers. *Proc. SPIE* 3111, 296–304.

Soderquist C.J., Crosby D.G. (1980) Degradation of triphenyltin hydroxide in water. *J. Agric. Food Chem.* 28, 111–17.

Sofer Z. (1984) Stable carbon isotope compositions of crude oils: application to source depositional environments and petroleum alteration. *Am. Assoc. Pet. Geol. Bull.* 68, 31–49.

Sofer Z. (1988) Biomarkers and carbon isotopes of oils in the Jurassic Smackover Trend of the Gulf Coast States, USA. *Org. Geochem.* 12, 421–32.

Sohl N.F. (1987) Presidential address—Cretaceous gastropods: contrasts between Tethys and the temperate provinces. *J. Paleontol.* 61, 1085–111.

Solomon S., Garcia R.R., Rowland F.S., Wuebbles D.J. (1986) On the depletion of Antarctic ozone. *Nature* 321, 755–8.

Sorensen K.W. (1993) Indonesian peat swamp forests and their role as a carbon sink. *Chemosphere* 27, 1065–82.

Sorokin J. (1966) On the trophic role of chemosynthesis and bacterial biosynthesis in water bodies. In *Primary Productivity in Aquatic Environments* (ed. Goldman C.R.), 189–205. Berkeley: University of California Press.

Sowers T., Bender M. (1995) Climate records covering the last deglaciation. *Science* 269, 210–14.

Spero H., Lerche I., Williams D.F. (1991) Opening the carbon isotopes 'vital effect' black box, 2. Quantitative model for interpreting foraminiferal carbon isotope data. *Paleoceanogr.* 6, 639–55.

Spero H.J., Bijma J., Lea D.W., Bemis B.E. (1997) Effect of seawater carbonate concentration on foraminiferal carbon and oxygen isotopes. *Nature* 390, 497–500.

Spicer R.A., Corfield R.M. (1992) A review of terrestrial and marine climates in the Cretaceous with implications for modelling the 'Greenhouse Earth'. *Geol. Mag.* 129, 169–80.

Spooner N., Harvey H.R., Pearce G.E.S., Eckardt C.B., Maxwell J.R. (1994) Biological defunctionalisation of chlorophyll in the aquatic environment II: action of endogenous algal enzymes and aerobic bacteria. *Org. Geochem.* 22, 773–80.

Stach E., Mackowsky M.Th., Teichmüller M., Taylor G.H., Chandra D., Teichmüller R. (1982) *Stach's Textbook of Coal Petrology*. Berlin: Gebrüder Borntraeger.

Stahl W.J. (1978) Source rock-crude oil correlation by isotopic type-curves. *Geochim. Cosmochim. Acta* 42, 1573–7.

Stainforth J.G., Reinders J.E.A. (1990) Primary migration of hydrocarbons by diffusion through organic matter networks, and its effect on oil and gas generation. *Org. Geochem.* 16, 61–74.

Stankiewicz B.A., Briggs D.E.G., Michels R., Collinson M.E., Flannery M.B., Evershed R.P. (2000) Alternative origin of aliphatic polymer in kerogen. *Geology* 28, 559–62.

Staplin F.L. (1969) Sedimentary organic matter, organic metamorphism, and oil and gas occurrence. *Can. Pet. Geol. Bull.* 17, 47–66.

Stauffer B., Fischer G., Neftel A., Oeschger H. (1985) Increase of atmospheric methane recorded in Antarctic ice core. *Science* 229, 1386–8.

Stetter K.O., Huber R., Blöchl E., Kurr M., Eden R.D., Fielder M., Cash H., Vance I. (1993) Hyperthermophilic archaea are thriving in deep North Sea and Alaskan oil reservoirs. *Nature* 365, 743–5.

Stevenson F.J. (1976) Binding of metal ions by humic acids. In *Environmental Biogeochemistry. Vol. 2: Metals Transfer and Ecological Mass Balances* (ed. Nriagu J.O.), 519–40. Ann Arbor, MI: Ann Arbor Science.

Stewart W.N., Rothwell G.W. (1993) *Paleobotany and the Evolution of Plants*. Cambridge: Cambridge University Press.

Stocker T.F. (2000) Past and future reorganizations in the climate system. *Quat. Sci. Rev.* 19, 301–19.

Stout S.A., Boon J.P. (1994) Structural characterization of the organic polymers comprising a lignite's matrix and megafossils. *Org. Geochem.* 21, 953–70.

Stout S.A., Boon J.P., Spackman W. (1988) Molecular aspects of the peatification and early coalification of angiosperm and gymnosperm woods. *Geochim. Cosmochim. Acta* 52, 405–14.

Stowe K.S. (1979) *Ocean Science*. New York: Wiley.

Strachan M.G., Alexander R., Kagi R.I. (1988) Trimethylnaphthalenes in crude oils and sediments: effects of source and maturity. *Geochim. Cosmochim. Acta* 52, 1255–64.

Stuiver M., Braziunas T.F. (1993) Modelling atmospheric ^{14}C influences and ^{14}C ages of marine samples to 10000 BC. *Radiocarbon* 35, 137–91.

Stuiver M., Polach H.A. (1977) Discussion: reporting of ^{14}C data. *Radiocarbon* 19, 355–63.

Sturrock G.A., O'Doherty S., Simmonds P.G., Fraser P.J.

(1999) In situ GC-MS measurements of the CFC replacement chemicals and other halocarbon species: the AGAGE program at Cape Grim, Tasmania. *Proc. Australian Symp. on Analytical Science, Melbourne, July 1999*, 45–8. Melbourne: Royal Australian Chemical Institute.

Stryer L. (1988) *Biochemistry*. New York: W.H. Freeman & Co.

Subroto E.A., Alexander R., Kagi R. (1991) 30-Norhopanes: their occurrence in sediments and crude oils. *Chem. Geol. 93*, 179–92.

Suedel B.C., Boraczek J.A., Peddicord R.K., Clifford P.A., Dillon T.M. (1994) Trophic transfer and biomagnification potential of contaminants in aquatic ecosystems. *Rev. Environ. Contam. Toxicol. 136*, 21–89.

Suess E. (1980) Particulate organic carbon flux in the oceans — surface productivity and oxygen utilization. *Nature 288*, 260–3.

Summerhayes C.P. (1983) Sedimentation of organic matter in upwelling regimes. In *Coastal Upwelling — Its Sediment Record. Part B: Sedimentary Records of Ancient Coastal Upwelling* (ed. Thiede J., Suess E.), 29–72. New York: Plenum Press.

Summons R.E., Powell T.G. (1987) Identification of aryl isoprenoids in source rocks and crude oils: biological markers for the green sulphur bacteria. *Geochim. Cosmochim. Acta 51*, 557–66.

Summons R.E., Volkman J.K., Boreham C.J. (1987) Dinosterane and other steroidal hydrocarbons of dinoflagellate origin in sediments and petroleum. *Geochim. Cosmochim. Acta 51*, 3075–82.

Summons R.E., Thomas J., Maxwell J.R., Boreham C.J. (1992) Secular and environmental constraints on the occurrence of dinosterane in sediments. *Geochim. Cosmochim. Acta 56*, 2437–44.

Summons R., Jahnke L., Hope J.M., Logan G.A. (1999) 2-Methylhopanoids as biomarkers for cyanobacterial oxygenic photosynthesis. *Nature 400*, 554–6.

Summons R., Jahnke L., Cullings K., Logan G. (2002) Absence of sterol biosynthesis in cyanobacteria. Poster abstract, Second Astrobiology Sci. Conf., NASA AMES Research Center, Moffett Field.

Sumpter J.P., Jobling S. (1995) Vitellogenesis as a biomarker for estrogenic contamination of the aquatic environment. *Environ. Health Perspect. 103 (suppl. 7)*, 173–8.

Sunda W.G., Huntsman S.A. (1995) Iron uptake and growth limitation in oceanic and coastal phytoplankton. *Mar. Chem. 50*, 189–206.

Sunda W.G., Kieber D.J. (1994) Oxidation of humic substances by manganese oxides yields low-molecular-weight organic substrates. *Nature 367*, 62–4.

Sundquist E.T. (1993) The global carbon dioxide budget. *Science 259*, 934–41.

Suzuki Y. (1993) On the measurement of DOC and DON in seawater. *Mar. Chem. 41*, 287–8.

Sweeney J.J., Burnham A.K. (1990) Evaluation of a simple model of vitrinite reflectance based on chemical kinetics. *Am. Assoc. Pet. Geol. Bull. 74*, 1559–70.

Tabak H.H., Bloomhuff R.N., Bunch R.L. (1981) Steroid hormones as water pollutants II. Studies on the persistence and stability of natural urinary and synthetic ovulation-inhibiting hormones in untreated and treated waste waters. *Dev. Ind. Microbiol. 22*, 497–519.

Talbot H.M., Head R.N., Harris R.P., Maxwell J.R. (1999) Steryl esters of phaeophorbide b: a sedimentary sink for chlorophyll b. *Org. Geochem. 30*, 1403–10.

Talbot M.R. (1988) The origins of lacustrine oil source rocks: evidence from the lakes of tropical Africa. In *Lacustrine Petroleum Source Rocks* (ed. Fleet A.J. et al.), *Geol. Soc. Spec. Publn 40*, 3–26. Oxford: Blackwell.

Tan Y.L., Heit M. (1981) Biogenic and abiogenic polynuclear aromatic hydrocarbons in sediments from two remote Adirondack lakes. *Geochim. Cosmochim. Acta 45*, 2267–79.

Tanabe S. (1988) PCB problems in the future: foresight from current knowledge. *Environ. Pollut. 50*, 163–77.

Taylor J., Parkes R.J. (1983) The cellular fatty acids of the sulphate-reducing bacteria, *Desulfobacter* sp., *Desulfobulbus* sp. and *Desulfovibrio desulfuricans*. *J. Gen. Microbiol. 129*, 3303–9.

Tegelaar E.W., Noble R.A. (1994) Kinetics of hydrocarbon generation as a function of the molecular structure of kerogen as revealed by pyrolysis–gas chromatography. *Org. Geochem. 22*, 543–74.

Tegelaar E.M., de Leeuw J.W., Derenne S., Largeau C. (1989) A reappraisal of kerogen formation. *Geochim. Cosmochim. Acta 53*, 3103–6.

Tegelaar E.M., Kerp H., Visscher H., Schenck P.A., de Leeuw J.W. (1991) Bias of the paleobotanical record as a consequence of variations in the chemical composition of higher vascular plant cuticles. *Palaeobiol. 17*, 133–44.

Tegelaar E.W., Hollman G., van der Vegt P., de Leeuw J.W., Holloway P.J. (1995) Chemical characterization of the periderm tissue of some angiosperm species: recognition of an insoluble, non-hydrolyzable aliphatic biomacromolecule (Suberan). *Org. Geochem. 23*, 239–51.

Teichmüller M. (1974) Uber neue Macerale der Liptinit-Gruppe und die Enstehung von Micrinit. *Fortschr. Geol. Westf. 24*, 37–64.

Templeton W. (1969) *An Introduction to the Chemistry of Terpenoids and Steroids*. London: Butterworth.

ten Haven H.L. (1996) Applications and limitations of Mango's light hydrocarbon parameters in petroleum correlation studies. *Org. Geochem. 24*, 957–76.

ten Haven H.L., Rullkötter J. (1988) The diagenetic fate of taraxer-14-ene and oleanane isomers. *Geochim. Cosmochim. Acta 52*, 2543–8.

ten Haven H.L., de Leeuw J.W., Schenck P.A. (1985) Organic geochemical studies of a Messinian evaporitic basin, northern Apennines (Italy) I: hydrocarbon biological markers for a hypersaline environment. *Geochim. Cosmochim. Acta 49*, 2181–91.

ten Haven H.L., de Leeuw J.W., Rullkötter J., Sinninghe Damsté J.J. (1987) Restricted utility of the pristane/phytane ratio as a palaeoenvironmental indicator. *Nature 330*, 641–3.

ten Haven H.L., Rohmer M., Rullkötter J., Bisseret P. (1989) Tetrahymanol, the most likely precursor of gammacerane, occurs ubiquitously in marine sediments. *Geochim. Cosmochim Acta 53*, 3073–9.

ten Haven H.L., Peakman T.M., Rullkötter J. (1992) Δ^2-Triterpenes: early intermediates in the diagenesis of terrigenous triterpenoids. *Geochim. Cosmochim. Acta 56*, 1993–2000.

Thode H.G. (1981) Sulfur isotope ratios in petroleum research and exploration: Williston Basin. *Am. Assoc. Pet. Geol. Bull. 65*, 1527–35.

Thomas D.J., Zachos J.C., Bralower T.J., Thomas E., Bohaty S. (2002) Warming the fuel for the fire: evidence for the thermal dissociation of methane hydrate during the Paleocene–Eocene thermal maximum. *Geology 30*, 1067–70.

Thompson K.F.M. (1979) Light hydrocarbons in subsurface sediments. *Geochim. Cosmochim. Acta 43*, 657–72.

Thompson K.F.M. (1983) Classification and thermal history of petroleum based on light hydrocarbons. *Geochim. Cosmochim. Acta 47*, 303–16.

Thompson K.F.M. (1987) Fractionated aromatic petroleums and the generation of gas-condensates. *Org. Geochem. 11*, 573–90.

Thompson J.E., Kutateladze T.G., Schuster M.C., Venegas F.D., Messmore J.M., Raines R.T. (1995) Limits to catalysis by ribonuclease A. *Bioorg. Chem. 23*, 471–81.

Thomson S., Eglinton G. (1978) Composition and sources of pollutant hydrocarbons in the Severn Estuary. *Mar. Pollut. Bull. 9*, 133–6.

Tiedemann R., Sarnthein M., Shackleton N.J. (1994) Astronomic time scale for the Pliocene Atlantic $\delta^{18}O$ and dust flux records of Ocean Drilling Program Site 659. *Paleoceanogr. 9*, 619–38.

Tien M., Kirk T.K. (1983) Lignin-degrading enzyme from the hymenomycete *Phanerochaete chrysosporium* Burds. *Science 221*, 661–3.

Tien M., Kirk T.K. (1984) Lignin-degrading enzyme from *Phanerochaete chrysosporium*. Purification, characterization, and catalytic properties of a unique H_2O_2-requiring oxygenase. *Proc. Natl Acad. Sci. USA 81*, 2280–4.

Tissot B. (1979) Effects on prolific petroleum source rocks and major coal deposits caused by sea-level changes. *Nature 277*, 463–5.

Tissot B., Espitalié J. (1975) L'évolution thermique de la matière organique des sédiments: application d'une simulation mathématique. *Rev. Inst. Fr. Pét. 30*, 743–77.

Tissot B.P., Welte D.H. (1984) *Petroleum Formation and Occurrence*. Berlin: Springer-Verlag.

Tissot B., Durand B., Espitalié J., Combaz A. (1974) Influence of the nature and diagenesis of organic matter in formation of petroleum. *Am. Assoc. Pet. Geol. Bull. 58*, 499–506.

Toggweiler J.R., Carson S., Bjornsson H. (1999) Response of the ACC and the Antarctic pycnocline to a meridional shift in the southern hemisphere westerlies. *Eos 80*, OS286.

Torsvik T.H. (2003) The Rodinia jigsaw puzzle. *Science 300*, 1379–81.

Trenberth K.E., Houghton J.T., Meira Filho L.G. (1996) The climate system, an overview. In *Climate Change 1995: The Science of Climate Change* (ed. Houghton J.T., Filho L.G.F., Callander B.A., Harris N., Kattenberg A., Maskell K.), 51–64. Cambridge: Cambridge University Press.

Tugwell S., Branch G.M. (1992) Effects of herbivore gut surfactants on kelp polyphenol defenses. *Ecology 73*, 205–15.

Turner S.M., Nightingale P.D., Spokes L.J., Liddicoat M.I., Liss P.S. (1996) Increased dimethyl sulphide concentrations in sea water from *in situ* iron enrichment. *Nature 383*, 513–17.

Twitchett R.J., Looy C.V., Morante R., Visscher H., Wignall P.B. (2001) Rapid and synchronous collapse of marine and terrestrial ecosystems during the end-Permian biotic crisis. *Geology 29*, 351–4.

Tyrrell T. (1999) The relative influences of nitrogen and phosphorus on oceanic primary production. *Nature 400*, 525–31.

Tyson R.V. (1995) *Sedimentary Organic Matter — Organic Facies and Palynofacies*. London: Chapman & Hall.

Ungerer P. (1990) State of the art of research in kinetic modelling of oil formation and expulsion. *Org. Geochem. 16*, 1–25.

Ungerer P. (1993) Modelling of petroleum generation and expulsion — an update to recent reviews. In *Basin Modelling: Advances and Applications* (ed. Doré A.G., Auguston J.H., Hermanrud C., Stewart D.J., Sylta O.), *Norwegian Petroleum Soc. Spec. Publn 3*, 219–32. Amsterdam: Elsevier.

Vail P.R., Mitchum R.M., Thompson S. (1978) Seismic stratigraphy and global changes of sea level, part 4: global cycles of relative changes of sea level. *AAPG Memoir 26*, 83–97. Tulsa, OK: American Association of Petroleum Geologists.

Vajda V., Raine J.I., Hollis C.J. (2001) Indication of global deforestation at the Cretaceous–Tertiary boundary by New Zealand fern spike. *Science 294*, 1700–2.

Valentine J.W., Moores E.M. (1970) Plate-tectonic regulation of faunal diversity and sea level: a model. *Nature 228*, 657–9.

Valley J.W., Peck W.H., King E.M., Wilde S.A. (2002) A cool early Earth. *Geology 30*, 351–4.

van Aarssen B.G.K., Cox H.C., Hoogendoorn P., de Leeuw J.W. (1990) A cadinene biopolymer in fossil and extant dammar resins as a source for cadinanes and bicadinanes in crude oils from South East Asia. *Geochim. Cosmochim. Acta 54*, 3021–31.

van Aarssen B.G.K., Hessels J.K.C., Abbink O.A., de Leeuw J.W. (1992) The occurrence of polycyclic sesqui-, tri-, and oligoterpenoids derived from a resinous polymeric cadinene in crude oils from southeast Asia. *Geochim. Cosmochim. Acta 56*, 1231–46.

van Cappellen P., Ingall E.D. (1996) Redox stabilization of the atmosphere and oceans by phosphorus-limited marine productivity. *Science 271*, 493–6.

Vandenbroucke M., Béhar F., Rudkiewicz J.L. (1999) Kinetic modelling of petroleum formation and cracking: implications from the high pressure/high temperature Elgin Field (UK, North Sea). *Org. Geochem. 30*, 1105–25.

van Gemerden H. (1993) Microbial mats: a joint venture. *Mar. Geol. 113*, 3–25.

van Graas G., de Lange F., de Leeuw J.W., Schenck P.A. (1982) De-A-steroid ketones and de-A-aromatic steroid hydrocarbons in shale indicate a novel diagenetic pathway. *Nature* 299, 437–9.

van Heemst J.D.H., Peulvé S., de Leeuw J.W. (1996) Novel algal polyphenolic biomacromolecules as significant contributors to resistant fractions of marine dissolved and particulate organic matter. *Org. Geochem.* 24, 629–40.

van Kaam-Peters H.M.E., Köster J., van der Gaast S.J., Dekker M., de Leeuw J.W., Sinninghe Damsté J.S. (1998) The effect of clay minerals on diasterane/sterane ratios. *Geochim. Cosmochim. Acta* 62, 2923–9.

Van Valen L. (1971) The history and stability of atmospheric oxygen. *Science* 171, 439–43.

van Zuilen M.A., Lepland A., Arrhenius G. (2002) Reassessing the evidence for the earliest traces of life. *Nature* 418, 627–30.

Veevers J.J. (1990) Tectonic-climatic supercycle in the billion-year plate tectonic eon: Permian Pangaean icehouse alternates with Cretaceous dispersed-continents greenhouse. *Sed. Geol.* 68, 1–16.

Veizer J., Compston W. (1976) ^{87}Sr/^{86}Sr in Precambrian carbonates as an index of crustal evolution. *Geochim. Cosmochim. Acta* 40, 905–14.

Veizer J., Clayton R.N., Hinton R.W. (1992) Geochemistry of Precambrian carbonates: IV. Early Paleoproterozoic (2.25 ± 0.25 Ga) seawater. *Geochim. Cosmochim. Acta* 56, 875–85.

Veizer J., Ala D., Azmy K., Bruckschen P., Buhl D. & 10 others (1999) ^{87}Sr/^{86}Sr, δ^{13}C and δ^{18}O evolution of Phanerozoic seawater. *Chem. Geol.* 161, 59–88.

Veizer J., Godderis Y., François L.M. (2000) Evidence for decoupling of atmospheric CO_2 and global climate during the Phanerozoic eon. *Nature* 408, 698–701.

Venkatesan M.I. (1989) Tetrahymanol: its widespread occurrence and geochemical significance. *Geochim. Cosmochim. Acta* 53, 3095–101.

Venrick E.L. (1982) Phytoplankton in an oligotrophic ocean: observations and questions. *Ecol. Monogr.* 52, 129–54.

Versteegh G.J.M., Riegman R., de Leeuw J.W., Jansen J.H.F. (2001) $U_{37}^{k'}$ values for *Isochrysis galbana* as a function of culture temperature, light intensity and nutrient concentrations. *Org. Geochem.* 32, 785–94.

Villanueva J., Flores J.A., Grimalt J.O. (2002) A detailed comparison of the $U_{37}^{k'}$ and coccolith records over the past 290 kyears: implications to the alkenone paleotemperature method. *Org. Geochem.* 33, 897–905.

Visscher H., Brinkhuis H., Dilcher D.L., Elsik W.C., Eschet Y., Looy C.V., Rampino M.R., Traverse A. (1996) The terminal Paleozoic fungal event: evidence of terrestrial ecosystem destabilization and collapse. *Proc. Natl Acad. Sci. USA* 93, 2155–8.

Vitousek P.M., Mooney H.A., Lubchenco J., Melillo J. (1997) Human domination of Earth's ecosystems. *Science* 277, 494–9.

Volk T. (1987) Feedbacks between weathering and atmospheric CO_2 over the past 100 million years. *Am. J. Sci.* 287, 763–79.

Volk T. (1989) Rise of angiosperms as a factor in long-term climatic cooling. *Geology* 17, 107–10.

Volkman J.K. (1986) A review of sterol markers for marine and terrigenous organic matter. *Org. Geochem.* 9, 83–99.

Volkman J.K., Johns R.B. (1977) The geochemical significance of positional isomers of unsaturated acids from an intertidal zone sediment. *Nature* 267, 693–4.

Volkman J.K., Alexander R., Kagi R.I., Rowland S.J., Sheppard P.N. (1984) Biodegradation of aromatic hydrocarbons in crude oils from the Barrow Sub-basin of Western Australia. *Org. Geochem.* 6, 619–32.

Volkman J.K., Barrett S.M., Dunstan G.A. (1994) C_{25} and C_{30} highly branched isoprenoid alkenes in laboratory cultures of two marine diatoms. *Org. Geochem.* 21, 407–13.

Volkman J.K., Barrett S.M., Blackburn S.I., Mansour M.P., Sikes E.L., Gelin F. (1998) Microalgal biomarkers: A review of recent developments. *Org. Geochem.* 29, 1163–79.

Waite M.E., Evans K.E., Thain J.E., Smith D.J., Milton S.M. (1991) Reductions in TBT concentrations in UK estuaries following legislation in 1986 and 1987. *Mar. Environ. Res.* 32, 89–111.

Wakeham S.G. (1987) Steroid geochemistry in the oxygen minimum zone of the eastern tropical North Pacific Ocean. *Geochim. Cosmochim. Acta* 51, 3051–69.

Wakeham S.G. (1989) Reduction of sterols to stanols in particulate matter at oxic-anoxic boundaries in sea water. *Nature* 342, 787–90.

Wakeham S.G., Schaffner C., Giger W. (1980) Polycyclic aromatic hydrocarbons in Recent lake sediments. 1. Compounds having anthropogenic origins. *Geochim. Cosmochim. Acta* 44, 403–13.

Wakeham S.G., Lee C., Farrington J.W., Gagosian R.B. (1984a) Bio-geochemistry of particulate matter in the oceans: results from sediment trap experiments. *Deep-Sea Res.* 31, 509–28.

Wakeham S.G., Gagosian R.B., Farrington J.W., Canuel E.A. (1984b) Sterenes in suspended matter in the eastern tropical North Pacific. *Nature* 308, 840–3.

Wakeham S.G., Lee C., Hedges J.I., Hernes P.J., Peterson M.L. (1997) Molecular indicators of diagenetic status in marine organic matter. *Geochim. Cosmochim. Acta* 61, 5363–9.

Wakeham S.G., Peterson M.L., Hedges J.I., Lee C. (2002) Lipid biomarker fluxes in the Arabian Sea, with comparison to the equatorial Pacific Ocean. *Deep-Sea Res. II* 49, 2265–301.

Waldock M.J. (1994) Organometallic compounds in the aquatic environment. In *Handbook of Ecotoxicology* (ed. Calow P.), 106–29. Oxford: Blackwell Scientific.

Waldock M.J., Thain J.E., Smith D.J., Milton S.M. (1990) The degradation of TBT in estuarine sediments. In *Proc. 3rd Int. Organotin Conf., Monaco*.

Walker C.H. (1990) Persistent pollutants in fish-eating sea birds—bioaccumulation, metabolism and effects. *Aquat. Toxicol.* 17, 293–324.

Walker C.H., Johnston G.O. (1989) Interactive effects of

pollutants at the toxicokinetic level—implications for the marine environment. *Mar. Environ. Res.* 28, 521–5.

Wallmann K. (2001a) The geological water cycle and the evolution of marine $\delta^{18}O$ values. *Geochim. Cosmochim. Acta* 65, 2469–85.

Wallmann K. (2001b) Controls on the Cretaceous and Cenozoic evolution of seawater composition, atmospheric CO_2 and climate. *Geochim. Cosmochim. Acta* 65, 3005–25.

Walsh J.J. (1991) Importance of continental margins in the marine biogeochemical cycling of carbon and nitrogen. *Nature* 350, 53–5.

Wams T.J. (1987) Diethylhexylphthalate as an environmental contaminant—a review. *Sci. Tot. Environ.* 66, 1–16.

Wang K., Geldsetzer H.H.J., Krouse H.R. (1994) Permian–Triassic extinction: organic $\delta^{13}C$ evidence from British Columbia, Canada. *Geology* 22, 580–4.

Wania F., Mackay D. (1993) Global fractionation and cold condensation of low volatility organochlorine compounds in polar regions. *Ambio* 22, 10–18.

Waples D.W. (1980) Time and temperature in petroleum formation: application of Lopatin's method to petroleum exploration. *Am. Assoc. Pet. Geol. Bull.* 64, 916–26.

Waples D.W. (1984) Thermal models for oil generation. In *Advances in Petroleum Geochemistry, Vol. 1* (ed. Brooks J., Welte D.), 7–67. New York: Academic Press.

Ward C.R. (1984) *Coal Geology and Coal Technology*. Melbourne: Blackwell Scientific.

Wardroper A.M.K., Hoffmann C.F., Maxwell J.R., Barwise A.J.G., Goodwin N.S., Park P.J.D. (1984) Crude oil biodegradation under simulated and natural conditions—II. Aromatic steroid hydrocarbons. *Org. Geochem.* 6, 605–17.

Watson A.J., Lovelock J.E., Margulis L. (1978) Methanogenesis, fires and regulation of atmospheric oxygen. *Biosystems* 10, 293–8.

Watson A.J., Bakker D.C.E., Ridgwell A.J., Boyd P.W., Law C.S. (2000) Effect of iron supply on Southern Ocean CO_2 uptake and implications for glacial atmospheric CO_2. *Nature* 407, 730–3.

Weber J.B., Best J.A., Gonese J.U. (1993) Bioavailability of sorbed organic chemicals. In *Sorption and Degradation of Pesticides and Organic Chemicals in Soils* (ed. Linn D.M., Carski T.H., Brusseau M.L., Chang F.-H.), 153–96. Wisconsin: Soil Society of America.

Wedemeyer G. (1967) Dechlorination of 1,1,1-trichloro-2,2-bis(*p*-chlorophenyl)ethane by *Aerobacter aerogenes*. *Appl. Microbiol.* 15, 569–74.

Wefer G., Fisher G. (1993) Seasonal patterns of vertical particle flux in equatorial and coastal upwelling areas of the eastern Atlantic. *Deep-Sea Res.* 40, 1613–45.

Wehmiller J.F. (1993) Applications of organic geochemistry for Quaternary research: aminostratigraphy and aminochronology. In *Organic Geochemistry* (ed. Engel M.H., Macko S.A.), 755–83. New York: Plenum Press.

Weiss R.F. (1974) Carbon dioxide in water and seawater: the solubility of a non-ideal gas. *Mar. Chem.* 2, 203–15.

Weissert H. (1989) C-isotope stratigraphy, a monitor of paleoenvironmental change: a case study from the Early Cretaceous. *Surv. Geophys.* 10, 1–61.

Weissert H., Lini A., Föllmi K.B., Kuhn O. (1998) Correlation of Early Cretaceous isotope stratigraphy and platform drowning events: a possible link? *Palaeogeogr. Palaeoclim. Palaeoecol.* 137, 189–203.

Wellman C.H., Osterloff P.L., Mohiuddin U. (2003) Fragments of the earliest land plants. *Nature* 425, 282–5.

Wellsbury P., Goodman K., Barth T., Cragg B.A., Barnes S.P., Parkes R.J. (1997) Deep marine biosphere fuelled by increasing organic matter availability during burial and heating. *Nature* 388, 573–6.

Werne J.P., Baas M., Sinninghe Damsté J.S. (2002) Molecular isotopic tracing of carbon flow and trophic relationships in a methane-supported benthic microbial community. *Limnol. Oceanogr.* 47, 1694–701.

Weston R.J., Philp R.P., Sheppard C.M., Woolhouse A.D. (1989) Sesquiterpanes, diterpanes and other higher terpanes in oils from the Taranaki Basin of New Zealand. *Org. Geochem.* 14, 405–21.

Whelan J.K. (1977) Amino acids in a surface sediment core of the Atlantic abyssal plain. *Geochim. Cosmochim. Acta* 41, 803–10.

White C.M., Lee M.L. (1980) Identification and geochemical significance of some aromatic components of coal. *Geochim. Cosmochim. Acta* 44, 1825–32.

Whitehurst D.D., Mitchel T.O., Fatcasiu M. (1980) *Coal Liquefaction*. New York: Academic Press.

Whiticar M.J. (1990) A geochemical perspective of natural gas and atmospheric methane. *Org. Geochem.* 16, 531–47.

Whiticar M.J. (1999) Carbon and hydrogen isotope systematics of bacterial formation and oxidation of methane. *Chem. Geol.* 161, 291–314.

Whiticar M.J., Snowdon L.R. (1999) Geochemical characterization of selected Western Canada oils by C_5-C_8 compound specific isotope correlation (CSIC). *Org. Geochem.* 30, 1127–61.

Whitman W.B., Coleman D.C., Weibe W.J. (1998) Prokaryotes: the unseen majority. *Proc. Natl Acad. Sci. USA* 95, 6578–83.

Whittaker R.H. (1975) *Communities and Ecosystems*. New York: Macmillan.

Whittaker R.H., Likens G.E. (1975) The biosphere and man. In *Primary Productivity of the Biosphere* (ed. Lieth H., Whittaker R.H.). *Ecol. Stud.* 14, 305–28. Berlin: Springer-Verlag.

Wickramasinghe N.C., Hoyle F., Wallis D.H. (1997) Spectroscopic evidence for panspermia. *Proc. Int. Soc. Optic. Engin.* 3111, 282–96.

Wiechert U.H. (2002) Earth's early atmosphere. *Science* 298, 2341–2.

Wildman R.A. Jr, Hickey L.J., Dickinson M.B., Berner R.A., Robinson J.M., Dietrich M., Essenhigh R.H., Wildman, C.B. (2004) Burning of forest materials under late Paleozoic high atmospheric oxygen levels. *Geology* 32, 457–60.

Wilkes H., Boreham C., Harms G., Zengler K., Rabus R. (2000) Anaerobic degradation and carbon isotopic fractionation of alkylbenzenes in crude oil by sulphate-reducing bacteria. *Org. Geochem. 31*, 101–15.

Wilkins R.W.T., Wilmshurst J.R., Russell N.J., Hladky G., Ellacott M.V., Buckingham C. (1992) Fluorescence alteration and the suppression of vitrinite reflectance. *Org. Geochem. 18*, 629–40.

Williams S.N., Schaefer S.J., Calvache M.L., Lopez D. (1992) Global carbon dioxide emission to the atmosphere by volcanoes. *Geochim. Cosmochim. Acta 56*, 1765–70.

Wilson R.C.L., Drury S.A., Chapman J.L. (2000) *The Great Ice Age*. London: Routledge.

Winfrey M.R., Ward D.M. (1983) Substrates for sulfate reduction and methane production in intertidal sediments. *Appl. Environ. Microbiol. 45*, 193–9.

Wingert W.S., Pomerantz M. (1986) Structure and significance of some twenty-one and twenty-two carbon petroleum steranes. *Geochim. Cosmochim. Acta 50*, 2763–9.

WMO (1999) *Scientific Assessment of Ozone Depletion: 1998*. Global Ozone Research & Monitoring Project, *Rep. 44*. Geneva: World Meteorological Organization.

Woese C.R., Wolfe R.S. (eds) (1985) *The Bacteria. Vol. VIII, Archaebacteria*. New York: Academic Press.

Woese C.R., Kandler O., Wheelis M.L. (1990) Towards a natural system of organisms: proposal for the domains Archaea, Bacteria and Eucarya. *Proc. Natl Acad. Sci. USA 87*, 4576–9.

Wofford H.W., Wilsey C.D., Neff G.S., Giam C.S., Neff J.M. (1981) Bioaccumulation and metabolism of phthalate esters by oysters, brown shrimp and sheepshead minnows. *Excotoxicol. Environ. Saf. 5*, 202–10.

Wolbach W.S., Gilmour I., Anders E., Orth C.J., Brooks R.R. (1988) Global fires at the Cretaceous/Tertiary boundary. *Nature 334*, 665–9.

Wolfe J.A., Upchurch G.R. (1986) Vegetation, climatic and floral changes at the Cretaceous-Tertiary boundary. *Nature 324*, 148–52.

Wolfenden R., Lu X., Young G. (1998) Spontaneous hydrolysis of glycosides. *J. Am. Chem. Soc. 120*, 6814–15.

Wolff G.A., Lamb N.A., Maxwell J.R. (1986) The origin and fate of 4-methyl steroids—II. Dehydration of stanols and occurrence of C_{30} 4-methyl steranes. *Org. Geochem. 10*, 965–74.

Wood D.A. (1988) Relationship between thermal maturity indices calculated using Arrhenius equation and Lopatin method: implications for petroleum exploration. *Bull. Am. Assoc. Pet. Geol. 72*, 115–34.

Wood H.G., Ragsdale S.W., Pezaka E. (1986) The acetyl-CoA pathway of autotrophic growth. *FEMS Microbiol. Rev. 39*, 325–62.

Woodall J., Boxall J.G., Forde B.G., Pearson J. (1996) Changing perspectives in plant nitrogen metabolism: the central role of glutamine synthetase. *Science Progress 79*, 1–26.

Woodwell G.M., Wurster C.F., Isaacson P.A. (1967) DDT residues in an east coast estuary: a case of biological concentration of a persistent insecticide. *Science 156*, 821–4.

Woodwell G.W., Craig P.P., Johnson H.A. (1971) DDT in the biosphere: where does it go? *Science 174*, 1101–7.

Woolhouse A.D., Oung J.-N., Philp R.P., Weston R.J. (1992) Triterpanes and ring-A degraded triterpanes as biomarkers characteristic of Tertiary oils derived from predominantly higher plant sources. *Org. Geochem. 18*, 23–31.

World Energy Council (2002) *WEC Survey of Energy Resources*. Available at www.worldenergy.org.

Worsley T.R., Kidder D.L. (1991) First-order coupling of paleogeography and CO_2, with global surface temperature and its latitudinal contrast. *Geology 19*, 1161–4.

Wu J., Sunda W., Boyle E.A., Karl D.M. (2000) Phosphate depletion in the western north Atlantic Ocean. *Science 289*, 759–62.

Wuebbles D.J. (1981) *The Relative Efficiency of a Number of Halocarbons for Destroying Stratospheric Ozone*. UCID-18924. Livermore: Lawrence Livermore National Laboratory.

Youngblood W.M., Blumer M. (1975) Polycyclic aromatic hydrocarbons in the environment: homologous series in soils and recent marine sediments. *Geochim. Cosmochim. Acta 39*, 1303–14.

Zachos J.C., Arthur M.A., Dean W.E. (1989) Geochemical evidence for suppression of pelagic marine productivity at the Cretaceous/Tertiary boundary. *Nature 337*, 61–4.

Zachos J., Pagani M., Sloan L., Thomas E., Billups K. (2001) Trends, rhythms, and aberrations in global climate 65 Ma to present. *Science 292*, 686–93.

Ziegler F., Rees A., Rowley D. (2001) The Paleogeographic Atlas Project. University of Chicago. http://pgap.uchicago.edu/

Zielinski G.A., Mayewski P.A., Meeker D.L., Whitlow S., Twickler M.S., Taylor K. (1996) Potential atmospheric impact of the Toba mega-eruption ~71,000 years ago. *Geophys. Res. Letts 23*, 837–40.

Index

Italic entries refer to figures and tables; <u>underlined</u> entries refer to boxes; **bold** entries refer to technical terms emboldened in the text, i.e. where they are first explained.

abietadiene   *190*
abietane   201, *202*
abietatetraenoic acid   *190*
abietatriene   *190*
abietic acid   *51*, 52, 188, *190*
abietoids   202
abiogenic/abiotic chemical evolution   7, 10–13, <u>10</u>, 15
absorption   *304*, *308*, 310, 319
  efficiency   <u>309</u>
abyssal plains   79, *79*, 106, 108, **115**
Acarinina   282
accommodation space   <u>107</u>, 112, 162
acetaldehyde   93, *303*
acetate/acetic acid   15, <u>42</u>, 49, 50, 64, 67, *94*, 94–5, 98–9, 162, 244, *302*
  $\delta^{13}C$ values   235
acetoacetyl unit   *45*
Acetobacterium   95
acetogens   *see* bacteria, acetogenic
acetogenesis   *94*
acetomycetes   40
acetone   43
acetosyringone   *173*
acetovanillone   *173*
acetylmuramic acid   *39*, 40
acetyl unit   40, 44, *45*, 238, *302–3*
  $\delta^{13}C$ values   238
acidity (pH)   40, <u>81</u>, <u>98</u>, 104, <u>111</u>, *111*, 113, 127, 162, 203, <u>208</u>, *208*, 213, 222, *308*, 310, 313, 319
  and bacterial activity   98, 100–1, 127
  and carbonate deposition   110
  and humic substance properties   121, 320, 321
  and metal complexation   <u>208</u>
  and nutrient availability   80
  and soil leaching   119
acidophiles   *see* bacteria, acidophilic
acid rain   <u>150</u>
acids   **42**
  alkanoic   44, 167, 169–70, 179
  alkenoic   44, 168–70
  carboxylic   31, <u>42</u>, 44, 191

  inorganic   <u>42</u>, 136
  organic   <u>42</u>, 161–2, 191, 238, 250
  polyhydroxy aromatic   64
  volatile   95
  weak   <u>111</u>
  see also fatty acids
acritarchs   **25**, *26*, 162, *163*
actinomycetes   94, 101, 117
activation energy   *see* energy, activation
adamantane   160, *160*
adenine   68, 68–9
adenosine
  diphosphate (ADP)   <u>17</u>, *17*, 80, 96
  triphosphate (ATP)   <u>17</u>, *17*, 38, <u>43</u>, 46, 49, 50, 67, 68, 79, 80, 96, 205
adipic acid   *303*
adjustment time   288, **290**, *290*
adsorption   104, 106, 108, 156, 161, 182, 212, 227, 299, **304**, 307, 311–12, 315, 317, 319–20
advection   105, 176, 315
Aerobacter aerogenes   312
aerobes   19, 21–2, 99
  see also bacteria, aerobic
aerobia   *see* respiration, aerobic
aerosols   265–6, 284, 291, 315
agar   40, 101
agathalene   *195*, 196
agathic acid   *51*, 52
Agave americana   133, *133*
agonist   *304*, **306**, 319
agriculture   283, 291, 292
air–sea
  interface   315
  transfer/exchange   250, 254, 260, 265–6, 268, 278, 280–1, 288, 301, 315
alanine   8, *8*, 9, 40, *41*, 47, 106
albedo   **252**, 260, 263, 265–7, 277
Albian   274
albite   250
Alcaligenes   <u>96</u>
alcohols   <u>10</u>, 31, <u>37</u>, <u>42</u>, 44, 55, 95, 191
  fatty   94
  steroidal   *see* stanols; stenols; sterols

aldehydes   5, 31, 35, <u>37</u>, 47, <u>60</u>
aldohexoses   35–6, *118*, 172
aldopentoses   <u>37</u>, 172
aldopyranoses   37
aldoses   35, <u>37</u>
aldosterone   55
algae   12, 20, 21, 27, 73, *73*, <u>89</u>, *91*, <u>96</u>, *122*, 166, 197, 205, 220, 292, 310
  benthonic   25, <u>59</u>, 301
  blue-green   *see* cyanobacteria
  brackish water   167
  brown   27, 40, 57, *57*, <u>59</u>, 61, 167, 170
  calcareous   250, 273
  chemical composition   38, 40, 44, 47, 49, 56, 56–7, 59–60, 61, 62, 69, 133, 167, 169, 170–2, 177, 186, 205, 219, 237, 241–2, *303*, 319
  contribution to sedimentary organic matter   109, 121–3, 133, 136, 138, 166, 186, 204, 241
  coralline   73
  epiphytic   73, 86
  evolution   27
  freshwater   133, 138, 167, 170, 172, 208
  golden   61, 84, *85*, *171*, 186, 204
  green   25, 27, 40, 57, <u>59</u>, 61, 78, 84, *85*, 133, 162, *163*, 167, 169–70, *171*, 172, 208
  hypersaline   205
  isotopic composition   *237*, 237, 241
  macroscopic/macrophytic   *see* algae, multicellular
  marine   101, 170–2, 204, 319
  micro   170, 208, 250
  multicellular   *26*, 27, <u>59</u>, 71, *73*, 167, 169–70, 248, 319
  red   27, 56–7, <u>59</u>, 62, *85*, 170
  unicellular   4, 20, 21, 24, 25, *26*, 71, 86, 169, 171, 275; *see also* phytoplankton
  yellow-brown   61, 84, *85*, 133, *171*, 169–70, 172, 219–20
  yellow-green   *85*, 133, *171*
algaenan   **49**, 108, 133, *133*, 137–9, 151, 197, *198*, 230
alginic acid   *39*, 40

alginite   122, 123, *124*, 127, *133*, 137, *142*
aliphatic structures   **30**, *31*, *34*, 44, 49, 59, 69, 119, 152; *see also* alkanes; alkenes
   in coal   128, 131, 137
   in humic material   121
   in kerogen   136–7, *138*, 139, 144–6, 148, 155
alkadienes   198
alkalis   <u>42</u>, 119
alkalinity   <u>**111**</u>, 248, *249*, 250, 268, 278
alkaloids   67
alkanes   **30**, 121, 131, 144, 153, <u>154</u>, 160–1, 175, 205
   acyclic   30, 121, <u>128</u>, 138, 149, <u>150</u>, *151*, 152, 197, 222, *302*, 303
   anteiso-   149, *151*, 160, 197
   biodegradation   302–3
   branched   47, *146*, 149, 151–2, 156
   δ^{13}C values   146, 241, 245
   cyclic/cyclo-   30, <u>128</u>, 131, *146*, 149, <u>150</u>, *151*, 151–2, 156, 160, 200, 222, *303*
   iso-   149, 151, *151*, 153, 160, 197
   isoprenoidal   49, 133, 197–200, 204–5
      acyclic   138, 151, *160*, 197, 207
      head-to-head   197, *199*
      highly branched   199, *199*
   isomerization   207
   methyl branched   <u>154</u>, *154*, 197, *see also* alkanes, *iso*; alkanes, *anteiso*
   mid-chain   197
   normal   30, 47, 139, <u>141</u>, *141*, 146, *146*, 148–9, *151*, 151–3, <u>154</u>, *154*, 156, *160*, 160, 162, 167, 180, 197, 204–5, 224, 232, 241–2, 245, 301, 302, *304*
   in organisms   47, 167, 197, 224, 241
   in petroleum and bitumen   142, 144, 146, 148, 180, 197–200, 204–5, 224, 241; *see also* biomarkers
   source indicators   180, 197–200, 224, 241
alkanols   47, 167
alkatrienes   198
alkenes   **30**, *31*, <u>31</u>, 52, 153, <u>154</u>, 175, 191
   isoprenoidal   49, 136, 198–200
   normal   232
alkenones
   long-chain unsaturation   *see* ketones, long-chain unsaturated
alkenylresorcinols   197
alkoxyacetic acid   102
alkylation   129
5-*n*-alkylbenzene-1,3-diol   *see n*-alkylresorcinol units
alkylbenzenes   151, *151*, 160
alkylbenzothiophenes   151
alkyl chains/units   31, 32, <u>34</u>, 46, <u>55</u>, <u>60</u>, 120, 131, 133, *136*, 139, 142, 144–5, 148, *151*, 151–2, <u>154</u>, 155, 160, 171–2, 183, *184*, 185, 192, *198*, 200, 211, 230, 236, 238, 241; *see also* methyl group; propyl units
   thermal cracking   148, 153, <u>154</u>, 192, 197, 211
alkylcyclohexanes   151, 160, 197
alkylcyclopentanes   151

alkyldibenzothiophenes   151
alkylglycerolmonoethers   241
alkylnaphthalenes   151, *151*, 160
alkylnaphthobenzothiophenes   151
alkylperhydronaphthalenes   151, *151*
alkylphenanthrenes   151, *151*, 202
alkylphenols   129
n-alkylresorcinol units   197, *198*
1-alkyl-2,3,6-trimethylbenzenes   183, 185
allelochemicals   319
allochthonous inputs   69, 109–10, 112, 116, 122, 137
alloisoleucine   212–13, *213*
   isoleucine ratio   212–13, *215–16*, 216
alluvial fans   110
alterobactin A   88
Alteromonas luteoviolacea   88
altitude   <u>296</u>, *296*
alumina   140
aluminosilicates   <u>1</u>, 250, 304
amides   31, <u>42</u>, 93
amines   31, <u>42</u>
   methylated   94, 95, 98
amino acids   8, <u>9</u>, 40–1, **40**–3, 49, 79, 80, 92, 94, 94, <u>95</u>, <u>97</u>, 118, 120–1, <u>126</u>, *134*, 167, 176, 213
   abiotic   7–8, <u>9</u>, 10–11
   branched   170
   classification   41, 100, 212–13
   and dating   212–13, 216
   degradation   <u>9</u>, 100, 106, 238–9
   diketopiperazine   213, *215*, 215
   epimerization   212–16, *213*, <u>214</u>, *215*–16, 216
   extraterrestrial   8, <u>9</u>
   free   42, 213, *215*, 216
   isotopic composition   <u>9</u>
   leaching   100, 213, 216
   and mean temperature   212–13, 216
   occurrence and function   42–3
   position in proteins   213, *215*, 215
   racemization   10, 212–13, 216
   in sedimentary organic matter   100, 117
   sequence in proteins   10–11, 42, 68, 212–13, 215
   stereoisomerism   32, 40
   structure   40, *41*
aminobacteriohopanetriol   52, *53*
aminobutyric acid   8, *9*, 106
aminoisobutyric acid   <u>9</u>
ammination   96
ammonia   5, *5*, 7, 80, <u>95</u>–6, 127, 261
ammonification   95, <u>96</u>–7
ammonium   80, 80, 94, <u>95</u>–6, *95*, 96, *134*, 213
   carboxylate   41
amorphinite   142
amphipathicity/amphilicity   10, *10*, <u>10</u>, 89
amylase   100
amylopectin   37, 38
amylose   37, 38, 100
amyrin
   α   52, *53*, 192, *194*, 201–2
   β   52, *53*, 189, *191*, 195, 196, 202–3

Anabaena   88, 292
anaerobes   *see* bacteria, anaerobes
anaerobia   *see* respiration, anaerobic
Andes   204
androgen   305, **306**, 310, 313
angiosperms   16, **26**, **27**, 52, 59–60, 64, *73*, 167, 172–4, *173*–4, 196, 201–3, *202*, 225, 261, 305
anhydrite   97, <u>98</u>, <u>145</u>, 161
Animalia/animals   **12**, *12*, 14, <u>21</u>, *21*, 24–5, 56, 61, *239*, 239, 247, 248, 287, 301, *312*
   chemical composition   38, 41–2, 44–6, 53, 55, 60, 169, 238–9, 295, 310–11, 313–14, 317
anions   **60**, <u>72</u>, <u>111</u>, **208**
annelids   305, 319
anomerism   35, **37**
anorthite   250
anoxicity   6, <u>18</u>, 18–21, <u>22</u>, 71, 83, <u>87</u>–<u>8</u>, 94, *95*, <u>96</u>, *97*, <u>98</u>, 99, 102, 104–6, *106*, 108–10, *110*, 112–13, 115–16, 127, 139, 159, 162, 177, 179–80, *180*–1, 182, *185*, 185–6, *187*, 205–7, 275
antagonist   304, **306**, 319
Antarctica   276, *281*
Antarctic
   ice cores   *see* ice, cores, Antarctic
   inversion   277
   polar front   <u>85</u>
antheraxanthin   57, *61*
anthocyanins   65
anthracene   299, *299*, 300
anthracite   **125**, *125*, <u>126</u>, 127, *128*, 132, 162, *163*
   meta-   125
   semi-   *125*, *128*, 225
anthraquinones   67
anthrasteroids, monoaromatic   187
Anthropocene   291
anthropogenics   **285**, 286–8, 291–2, 294–321
anti-androgens   **306**
antibiotics   49
antibody   43
anticlines   <u>159</u>
anticyclones   <u>76</u>
antidesiccants   38
anti-oestrogen   **306**, 316, 320
   equivalents (TEQ)   320, *320*
aphelion   <u>264</u>
Apex chert   7
API gravity   *see* oil, gravity, API
Aptian   272, 274, *274*
aquachelins   88, <u>89</u>
arabinogalactans   38
arabinose   *36*, 38, 40, 172, *173*, 175
arachidic acid   44
arachidonic acid   44, <u>45</u>
aragonite   <u>219</u>, 235
Archaea   *see* archaebacteria
Archaean   6, 7, 15, 18–19, 25, **26**, **27**, <u>255</u>, 256, 260–1, 267, 324
archaebacteria   **12**, *14*, **14**, 15, 22, **26**, <u>95</u>, **206**, 241

chemical composition 40, 47, 48, 136,
  166–7, 197, 200, 203, 207, 241
halophilic 14, 14, 47, 167, 205, 207
hyperthermophilic 14, 98, 99, 117, 159, 280
isotopic composition 241
methane oxidizing/methanotrophic 99, 269
methanogenic 14, 14–15, 21, 47, 62, 94, 98,
  99–100, 112, 167, 197, 203, 205, 207,
  236–7, 237–8, 241, 261, 266, 269, 287
thermoacidophilic 11, 14, 14, 47, 98, 100,
  167, 197
archaeol 47, 241
Archaeopteris 27
Arctic Circle 90
arginine 41
argon 20
Arica 239
Arochlors see PCB
aromaticity 119–21, 129, 131, 142, 144
aromatic structures 30, 31, 34, 60, 69, 101–2,
  119, 152, 160, 175, 192, 211, 225, 298,
  313
  in coal and peat 126, 123, 127–9, 131–2
  in humic material 119–21
  in kerogen 126, 133, 137, 139, 142, 144, 144,
    146, 153
  nitrogen-containing 123, 137
  oxygen-containing 137
  sulphur-containing 123, 137, 149, 151, 152,
    155
aromatics 140, 142, 146, 150, 151, 161, 242,
  242, 304
aromatization
  of biomarkers/lipids 135, 175, 181, 182,
    186–8, 194, 196, 210, 210–11, 224, 225,
    227, 233, 310
  during coalification 131–2
  during kerogen formation/maturation 136,
    144, 153, 155
  as a maturity indicator 210, 225, 227, 233
  during petroleum formation 155
  solid-state 232
Arrhenius
  constant (A) 214, 227, 227, 229, 230, 231,
    232
  equation 230, 232
arthropods 38, 101, 305
ascomycetes 93, 101, 179
Ascophyllum 79
asparagine 41
aspartic acid 8, 9, 41, 101
Aspergillus 93
asphalt 300
asphaltenes 128, 135, 135, 148–9, 155–6, 159,
  208, 242, 301; see also NSO compounds
  precipitation 128, 158–9
  pyrolysis 148
assimilatory processes 4, 17–18, 22, 80, 89,
  92–3, 94, 95–6, 95, 97, 97, 104, 117–18,
  235–6, 238–40, 249, 280, 309
astaxanthin 57, 58, 59, 185
asteroids 5
asthenosphere 2, 2, 107, 107, 228, 229

Astronomical Unit (AU) 5
atmosphere 2, 80, 90, 97, 217, 247, 251, 252,
  254, 255, 263, 265, 284–5
chemistry 8, 97, 266, 287–8, 290–1, 295–6
circulation 75–6, 76, 263, 265, 270, 275
composition 5, 7, 16, 252, 253, 263, 266–7,
  269, 271, 272, 276–7, 280, 284–8, 287,
  289–90, 290–1, 291, 297, 307, 310–11,
  313–15, 317; see also individual gases
density 296
evolution 5, 16, 19–23, 260–2
mixing 276, 288
moisture content 260
ocean coupling 276, 284
redox conditions 5, 7–8, 15, 19, 22, 256,
  260
stratification 295, 296, 296
temperature 277, 296, 296
  meridional gradient 263
trace gases 297
atoms 3, 4, 30, 32, 43, 296
atomic
  mass 4
  number 4
autochthonous inputs 69, 70, 109–10, 112,
  116, 122, 139
autolysis 93
autotrophes 13, 16, 25, 85, 237, 238
autotrophy 17–18, 23, 236, 240
Avogadro number/constant 322
axial substituents 35, 35, 37, 54
Azotobacter 95

Bacillariophyceae see diatoms
Bacillus 40, 94, 96, 96, 101, 221
bacteria 7, 12–13, 14, 21, 38, 53, 89, 100–1,
  179, 206, 221, 238, 288, 301, 302; see also
  archaebacteria
acetogenic 95, 98–9, 237
aerobic 18, 84, 93–4, 98–9, 101, 159, 167,
  169, 170, 182
  facultative 169
  obligate 18, 94, 98, 98
aerotolerant see oxygen-tolerant
ammonium oxidizing 236
anaerobic 13, 15–16, 18, 18, 21, 47, 94–5, 95,
  99, 159, 167, 169, 182, 241, 318
  facultative 94, 96, 96, 98, 101
  obligate 94, 96, 98, 101, 275
autotrophic 24
biofilms 108
biomass 116, 237
and C cycle 94, 92
chemical composition 40, 47, 48, 49, 52,
  55–7, 60–2, 69, 116, 133, 136, 167–70,
  172, 180, 196, 7, 200–1, 204–5, 241
chemolithotrophic see lithotrophic
chemosynthetic 13, 14–15, 18, 24, 96, 97–8,
  99, 116, 241, 259, 259–60
communities 91, 99–100
competition 96, 98–9
consortia 99, 101, 112, 241, 269
contribution to sedimentary organic matter

24, 69, 99, 109–10, 121, 127, 137, 139,
  166, 170, 203, 205, 241, 259–60, 275
degradation of organic matter see
  biodegradation
denitrifying 94, 96, 96, 99
evolution 14, 25
Gram-negative 14, 40, 47, 170
Gram-positive 14, 40, 47, 98, 170
green 236, 237
  non-sulphur 14, 18
  sulphur 16, 18, 61, 97–8, 112, 116, 184,
    237, 275
heterotrophic 24, 55, 88, 89, 91, 93–4, 99,
  112, 115, 133
iron-oxidizing 20
iron-reducing 14
isotopic composition 241
lithotrophic 13, 99
marine 84
metabolism see respiration
methanogenic see archaebacteria,
  methanogenic
methanotrophic 99, 112, 236, 238, 241, 285,
  287; see also archaebacteria,
  methanotrophic
methylotrophic 98, 99, 170, 201, 241
nitrifying 96
nitrate reducing 94, 96, 98
nitrite reducing 94
nitrogen fixing 80, 95, 201; see also
  cyanobacteria, nitrogen-fixing
non-photosynthetic 56, 57
nutrient cycling 74, 88–9, 94–5, 95–6, 97,
  97–8
oxygen-tolerant 18, 94
photoheterotrophic 18, 84, 85, 91
photosynthetic/phototrophic 18, 21, 25,
  56–7, 61, 95, 97, 99, 112, 116, 167, 169,
  184, 205, 236–7, 275
  anoxygenic 18, 18, 99
planktonic 170, 275
purple 14, 112, 237
  non-sulphur 18, 167
  sulphur 16, 18, 97, 184
red 237
reductive dechlorinators 317
sulphate reducing 20, 21, 22, 94, 95, 96,
  98–9, 112, 135, 159, 168–70, 203–5, 241,
  269, 301
sulphur oxidizing 22, 112, 236, 259
thermophilic 99, 117, 159, 241
Bacteria see eubacteria
bactericides 319
bacteriochlorophylls see chlorophylls, bacterial
bacteriohopanepolyols see hopanoids, polyols
bacteriohopanoids see hopanoids
bacterioruberin 167, 168
bamboo 174
banded iron formations (BIFs) 16, 19, 19–20
Bangiophyceae/bangiophytes 85, 171
barium 80
bark 64, 67, 122
Barremian 274

Barrow 289
basalt 251, 265, 268, 273
    flood 265, 284
        continental 265, 273
        marine 273
    oceanic 251, 265
base **42**, 68
basidiomycetes 93, 100–2
Basin
    Amazon 219
    Baltic Sea 116
    Black Sea 116
    Douala 137
    East Shetland 232, *233*
    Great South *229*
    Guaymas 228, 300
    Lake Maracaibo 116
    Mahakam 232
    Mediterranean Sea 116
    North American 315, 316
    North Sea 107, 228
    Pannonian 228, 232, *233*
    Paris 148, *137*, 228
    Persian Gulf 116
    and Range Province 107
    Red Sea 116
    Songliao 164
    Taranaki *225*
    Uinta 208
basins 106, 107, 116, 148, 203, 228, 232, *233*, 234, 256
    Brazilian marginal offshore 203
    classification 107
    enclosed 116
    evolution 203
    fore-arc 107
    foreland 107
    formation 107
    intracratonic sag 109
    lacustrine 109, 112
    marine 107
        marginal 19, 106, 115
    Mesozoic of west Canada 228
    modelling 232
    oceanic 79, 83, 256, 265
        volume 83, 107, 162, 272
    Palaeozoic of Australia 228
    restricted 105, 115–16, 274–5
    rift 107, 109, 230
    silled 105, 106, 116
    strike-slip 107
    Tertiary 228
    transtensional 107
beaches 110
Beggiatoa 98
Belomorian 255
benthonic organisms/communities **25**, 27, 84, 93, 179, 249, 280, 282, 301, 310
benzene 30, *31*, 67, 152, 203, 298, *303*
benzanthracene *299–300*
benzofluoranthenes *299–300*
benzohopanes 193, *195*, 196
benzoperylene 298, 300, *299–300*

benzopyrenes 298–300, *299–300*, 320
benzothiophene 152
Berriasian 274
betulin 53
beyerane 201, *202*
β factor *see* extension, β factor
bicadinanes *199*, 203
bicarbonate 4, 72, 95, 111, *111*, 161, 217, 235, *236*, 237, 248–50, *255*, 255, 268, 280
    δ^{13}C values *240*, 255, 269, *284*
        gradient 280
    equilibrium with CO_2/carbonate 218, 235, 254, 268, 280, 283
    pump 237
bicycloalkylporphyrins 180, *182*
BIFs *see* banded iron formations
Big Bang 2
bile acids 55
bioaccumulation 309, **309**, *312*, 314, 319
bioavailability 310, 313, 317, **319**, 321
bioconcentration 309, 311, 319
    factor (BCF) **309**, 311, 317, 319
biodegradation 23, 49, 67, 71, 80, 91–2, 93, 96, 100–1, 104–6, 108–9, 113, 116–19, 121–2, 127, 132, *134*, 135, 137, 179, 182–3, 196, 204–5, 238, 242, 247, *247*, 249, 253, 260, 280–1, *285*, 288, 292, 301, *304*, *308*, 310, *312*, 319; *see also* mineralization
    aerobic 22, 93–5, *94*, 99, 108–9, 114, 127, 182, 240, 275, 292, 301, *302–3*, 311
    anaerobic *94*, 94–102, 96, 97, 113, 116, 146, 162, 186, 301, 311, *318*
    bacterial 22, 24, 100, 104, 113, 117, 128, 179, 188, 191
    fungal 24, 64, 100, 102, 117, 179
    at molecular level 160, 160, 174–9, 185, 192, 204, 227, 301, *302–3*, 307, 311, 313–14, 317, 319
    oil *see* oil, biodegradation
    scale 160, *160*
    in water column 91
biological pump 250
biomacromolecules *see* macromolecules, bio-; polymers
biomagnification 309, 310, 317, 319
biomarkers **135**, 166–7, 176
    acyclic isoprenoidal 182, *184*, 197–200, *199*, 203, 205, 207; *see also* squalane; squalene
    age indicators 200–2, *201*
    in bitumen/oil 153, *153*, 193, 200–6, *201*, 210, 222, 232, 301
    biodegradation 160
    catagenesis *183*, 187, *191*
    in coals *202*, 210
    correlation of oils and source rocks 197, 200, 222
    diagenesis 175, 197, *206*, *209*; *see also* lipids, diagenesis
    environmental indicators 196–7, 203–7, 275
    generation versus destruction 210
    isotopic composition 241, 241–2
    in kerogen/sediments 146, *202*, 210, 238

    maturity indicators 224–5, *225*, *227*, *233*
        temperature dependence *233*
    microbial 168
    source indicators 167, *168*, 169–72, *171*, 179, *184*, *191*, *195*, 196–203, *199*, *202*, 241, 275
    sulphurization 136, 200, 207
    thermal stability *see* thermodynamic stability
biomass 14, 23–4, 71, 81, 83, 112, 116, 197, 237, 246–8, *247–8*, 253, *255*, 284, *285*
biomes **87**, 89, **90–1**, 238, 247, 291
biopolymers *see* macromolecules, bio-
biosphere *16*, 295
biota 283
    Arctic 315
    marine *247*, 248
    terrestrial *247*, 247, 282, 285
biotic collapse *see* extinction events
bioturbation **93**, 102, 104, 110, *110*, 115, 317
biphenyl 313–14
    polychlorinated *see* PCB
birds 295, 304–5
    bioaccumulation 306, 310, *312*, 320
bitter principles 52
bitumen *122*, 123, 128, 127, *134*, 135, 148, 156, *163*, 197, 223, 242
    abundance 222
    composition 128, 135, 139, *140*, 149, 211–12, 225–7; *see also* biomarkers in bitumen/oil
    formation 131, 134, 159, 161
    isotopic composition 241–3
    pyro- 128
bituminite *122*, 123
bivalves 273
black body *see* radiation, black body
block rotation 107
bloom *see* phytoplankton blooms
bogs *see* mires
boiling point 141
bolide impacts 13, *29*, 260, 280–1
bombardment *see* bolide impacts; Late Heavy Bombardment
Bonarelli Event 274
bond 30–2, 47, 121, 175, 232, 321
    cleavage 102, 117, 128–9, 131, 135, 142, 155, 166, 177, *179*, 182, *183*, 192–3, 197, 200, 211, *215*, 223, 224, 227–8, 230, 313
    conjugated 30, *33*, 45
    coordinate *see* dative
    covalent **30**, 31
    dative 30
    delocalized 33
    double 30, 32, 33, 44, 45, 52, *52*, 55, 56, 136, *179*, 185–6, 196, 198, 220
        migration 136, 175, 186, 189, 193, *209*
        position 169–70, 172, 177, 188, 192
    energy 222, 223, 227, 230
    geometry 32
    glycosidic **37**
    hydrogen **31**, 42, 68, *69*, *120*, 121, 156, 299, 312
    peptide *see* peptide linkage

rotation   35, <u>253</u>
single   30–2, <u>33</u>, 35
spatial arrangement   <u>33</u>, *33*
strength   2, 30, <u>33</u>, *43*, 155, 215, 227, 231, 240
triple   30, 32
vibration   <u>253</u>
bone   42, 100, 239
borate   <u>111</u>
Borneo   137
boron   80
botryococcane   197–9, 203
botryococcenes   198, *199*
Botryococcus braunii   133, *142*, 138, 167, 197–9, 203
brine   <u>230</u>
rejection   <u>75</u>, 278
bromide   72, <u>111</u>
bromine   80, 297
bromphenols   319
Brussels sprouts   *320*
Bryophyta/bryophytes   <u>13</u>, *26*, 27, 113, 170; *see also* mosses
Bryopsida   *see* mosses
buffer   <u>111</u>, 248, 280, 301
buoyancy   157–8, <u>159</u>
burial history/rate   212, <u>214</u>, 227–8, <u>230</u>, 234, *229*, 263, 274
butane
normal   35, *35*, 149, *244*
butyrate   95
butyryl unit   44, *45*

cadalene   188, *190*
cadinene   *51*, 60
caesium   80
Cainozoic   26, *26*, *28*, *201*, 261, *258*, *281*, 324
Calanus   177
helgolandicus   177
pacificus   176
calcite   2, 110, 161, <u>219</u>, 235, *235*–6, <u>258</u>
magnesium content   112
calcium   72, 80, *79*, <u>111</u>, 249–50, <u>258</u>, 268, 280
carbonate   2, 110, <u>111</u>, 112, <u>219</u>, *240*, 249, *249*, 250, *259*, 278; *see also* aragonite; calcite; carbonate polymorphs   <u>219</u>
sulphate   <u>97</u>, <u>98</u>; *see also* anhydrite; gypsum
California   231
Calvin cycle   <u>17</u>, *17*, 236–8
Cambrian   26, *28*, 115, *163*, <u>219</u>, *262*, 324
explosion   22, 25, 28, 269
fauna   25, *28*, 28
Cameroon   137
Campanian   272, *274*
campesterol   *57*, 171–2
camphor   *51*
Canyon Diablo troilite (CDT)   *4*
capillary
column   <u>140</u>
forces   *see* pressure, capillary
cap
carbonate   *see* carbonate, cap
rock   <u>159</u>

Cape Grim   298
caprolactone   303
carbanions   <u>60</u>, 212–13
carbocations   <u>60</u>, 128–9, 153, <u>154</u>, *154*
carbohydrates   15–16, <u>17</u>, *17*, 18, 30, **35**–40, *39*–*40*, 43–4, 49, <u>54</u>, 69, *94*, 117–18, *118*, 120–1, 127, 167, 236, 238
composition   35–8
degradation/diagenesis   117–18, 127, 133, 174–5
occurrence and function   38–40, 45–6
as source indicators   40, 169, 172, 174; *see also* monosaccharides
stereoisomerism   36–7
carbon   1, 2, 3, <u>4</u>, <u>33</u>, 80, 146, *147*
amorphous   132
aromatic   *see* aromaticity
black   92
crustal   106, 132, *163*, *164*, 250, 254, *255*, 256, 260, 280
dating   *see* radiocarbon dating
dissolved   <u>111</u>, 246–9, *247*, 249, *251*, 251, 260, 278, 280, 283, *285*, 288
inorganic (DIC)   <u>111</u>, *240*, 240, *247*–9, 248–50, 255, 268–9
organic (DOC)   *91*, <u>111</u>, *116*, 247–9, 248–9, 251
export from surface water   250
fixed   *see* coal, fixed carbon
fixation   *4*, *12*, <u>13</u>, <u>17</u>–<u>18</u>, *59*, 71, 79, 110, 234, 237, 239, 246, 248, 275; *see also* photosynthesis; chemosynthesis
gases   146, *147*, 251
inorganic   248, 283
isotopes   *4*, *4*, <u>217</u>, 234, 236, 319
isotopic
composition   *4*, 6–7, <u>9</u>, *16*, 92, 146, 155, 160–1, 165, 167, <u>217</u>, <u>234</u>, 234–42, *235*, *237*, *239*, 241–4, 244–5, 254–5, *255*, *257*–8, *259*, 260, 263, 267–71, *271*, 273, 274, 280–4, *287*, 287–8
excursions   254–5, *268*, 268–9, *274*, *274*, 280–5, *281*–2, *284*
kerogen:carbonate ratio   246, 254, 256, *257*–8
mantle/primordial   *251*, 254, *255*, 268, 283
marine   255
organic   25, 86, *108*, *106*, <u>111</u>, *116*, 132, *146*, *174*, *174*, *255*, 274, 278, 283; *see also* organic matter
burial/preservation   109, *109*, *116*, 248–9, 253–60, *257*, 275, 280
fluxes   102
freshwater   119
ratio to sulphur   108
sedimentary   2, 71, *106*, 108, 162–5, 174, 246, 254–60, 278
sources   *116*, *174*, 238
total (TOC)   176, <u>223</u>
oxidation   <u>217</u>
particulate organic (POC)   92, 246, *248*, 248–9, *251*
preference index (CPI)   224

reduced   24
residual   *see* Rock–Eval, residual carbon
carbonaceous chondrites   **5**, 8
carbonate   4, 92, 110, <u>111</u>, *111*, 112, *164*, *181*, *204*, <u>223</u>, 246, 251–2, *255*, 255–6, <u>258</u>, 271, 273, 274, 276, *281*; *see also* calcium carbonate; magnesium carbonate; calcite; dolomite
abiogenic   110, 267
burial/preservation   249, 260–1, 270
cap   267–9, *268*
δ^{13}C values   161, *235*, 240, 235, 240–2, *255*, 254, *255*, *255*, *256*, *257*, *258*, *259*, 267, *268*, 268, 269, 271, 274, *274*, 280, 281, 282, 283, *281*, *281*, *282*, 284, *284*
marine gradient   281, *281*
compensation depth (CCD)   216, **219**, 283
δ^{18}O values   216, *218*, <u>218</u>–19, 220–1, *222*, 265, 269, 272, *274*, 275–6, *282*, 283
diagenetic alteration   216, 272
dissolution/solubility   98, 162, <u>219</u>, 249, 268, 284
equilibrium with bicarbonate   110, <u>218</u>, 235, 268
export from surface water   250, 278
to kerogen C-content ratio   246, 254, 256, *257*–8
pelagic   268–9, 276
platforms   *272*, 273
Precambrian   255
precipitation   162, 256, 260, 267–8, 279–80
reservoir rocks   <u>159</u>
source rocks   135, *204*
thermal decomposition   261
weathering   250
carbonation   *247*, 251, *251*
carbon cycle   23, 86, <u>87</u>, *91*, *94*, *99*, *236*, 241, 246–54, *247*, *252*, *255*, 277, 280, *285*, 286, 292, 295
biochemical subcycle   246–50
geochemical subcycle   250–4, *251*, 261, 270
human influence   285–94, *285*
marine   248–50, *248*, 283
microbial pathways   *91*, *94*
modelling   261, 284
preindustrial   *247*, 285
reservoirs and fluxes   92, 163, **246**, 247–8, *248*, 250, *251*, 254–62, *255*, 261, 263, 269, 278, 283, *285*, 285–6, 288
steady-state and deviations   246, 254, 285–6
terrestrial   247–8, 278, 283
variations over time   254–62
carbon dioxide   1, *4*, 44, *147*, <u>157</u>, 156, 162, *164*, 165, <u>223</u>
aqueous (dissolved)   72, 104, 110, <u>111</u>, 162, <u>217</u>, <u>219</u>, *235*–6, 235, 237–8, *240*, 248–51, *249*, 254, *255*, 256, 259–60, *259*, 265, 268, 278, 280, 286, <u>290</u>
equilibrium with bicarbonate/carbonate   <u>218</u>, 235, 254, 268, 283
atmospheric   5–6, *16*, 23, *23*, 27, <u>111</u>, <u>217</u>, 235–7, *236*, 246, *247*, 248–51, *252*–3,

atmospheric (*cont.*) 253, 253–6, *255*, 259–61, 265, 269, 275, 280, 284–6, *286*, *289*, *290*, 291; *see also* atmosphere, gases
  concentration variation 29, 253, *260*, 260–3, *262*, 267–8, 270–1, *271*, 273, *273*, 276–9, *277–9*, 281, 288, *290*, 291
  seasonal plant growth 288
δ^{13}C values 235, 238, *240*, 255, 259–60, 271, *284*
drawdown *see* sinks
fertilization 286
greenhouse warming 251, 261, *263*, 265, 269, 273, 284, 286–7, 291
in natural gas 149, 162
mantle/primordial 99, *243*, 244, 251, 268
partial pressure 20, *219*, 237, 249–50, 260, *260*, 270–2, *271*, 281
reduction 7, 15, 18, 23, 95, 169, 244; *see also* carbon, fixation; chemosynthesis; photosynthesis
sinks *94*, 99, *134*, 250, 256, 260–2, 265–6, 268, 270–1, 273, 275–6, 278–9, 281, 285–6; *see also* acetogenesis; chemosynthesis; methanogenesis; photosynthesis; production, primary
soil 250, 262, 279, 286, *290*, 292
sources 5, *23*, 73, *94*, 135, 145, 250–1, 254, 260–1, 273, 279, 285–8, 298
  anthropogenic *263*, 273, *286*, 285–6, 288, 291–2, 295, 297, 301
  biogenic 23, *40*, 81, *94*, 96, 98, 104, 110, 127, *134*, 142, 146, 238, 247, 249, 259, 262, 288
  thermogenic 131, *147*, 155, 161, 255
  volcanic 260, 262, 265, 268, 273, 275, 279–80, 282–4
stomatal 271, *271*
carbon disulphide 128
carbonic acid 111, *111*, 250
carbon monoxide 5, 5, *5*, *23*, *223*, 260, 291
  sinks 286, 288
  sources 265, 286–8
carbon tetrachloride 290, 297, *297*
carbon tetrafluoride *see* PFC-14
Carboniferous 22, *26*, 27, *28*, 29, 70, 112, 114, 138–9, 162, *163*, 201, *262*, 270, 324
  – Permian
    boundary 22
    icehouse 29, 269–73
carboxylase 236, 270, 286
carboxylation 45, 237, 270
carcinogens 295, 299, 311, 313–14, 316, 320
carnivores 13, 21, 24, 25
carotane 182–3, *184–5*, 205
carotene 56, 57, *58*, *185*, 185, 205
carotenoids 11, 50, 56–7, *58–9*, 59, 79, 167–8, 205
  acyclic 57, 185
  aromatic 57, 116, 182, 184–5
  degradation/diagenesis 179, *180*, 182–5, *184*, 205
  epoxide-containing 57, *180*, 185, 205
  oxygen-containing 185

ring-numbering scheme 58
saturated 182–3
sources 182, 184, 205
carrier rock 159, 156–8, 162
carvone 32, *34*
catabolic processes 177
catagenesis 117, 135, *140*, 144–9, 152–3, 154, 155, 196, 197, 203, 221, 223, *225*, 232
  of coal 125, 144, 162
  early 225
  of kerogen 126, 141–2, *143–4*, 144–6, *147*, 148, 155, 162
  late 225
  molecular
    changes 166, 175, 186, 188, 197, 207, 210
    indicators 183, 187, 191, 224
catalysis 11, 13, 42–43, 43, *43*, 117, 154, 192, 223, 238, 270, 296
  by mineral surfaces 13, 117, 169, 203, 207; *see also* clay, catalysed reactions
catalysts 43
catechols 101, 128–9, *129–30*, 303
catecholate group 88, 89
cations 35, **60**, 72, 111, 102, 182, 250, 320
  alkylammonium 106
C_1 compounds *94*, 95, 98
CCD *see* carbonate compensation depth
cedrene 51
cell 1, 7, 10, 11, 14–15, 68, 71–2, 93, 100, *122*, 176–77, 220, 221, *235*, 235–8, 307
  division 168
  membranes 4, 10, 10–11, *11*, 46, 46, 48, 53, 55, 89, 96, 167–9, 237
  bacterial 38, 40, 47, 52
  fungal 38
  plant 38, 47
  rigidifiers 10, 47, 55, 167, 205, 207
  structure 10, 78
  volume:surface area ratio 69, 86, 89, 238, 259
  walls 38, *48*, 117, 133, 176, 197, 309
  bacterial 40, 47, 133, 174
  fungal 47
  plant 38, 47, 62, 101, 174
cellulase 100
cellulose 37, 38, 42, 62, 69, 100, 101, 118, *124*, 127, 174, 175
cement
  calcite 1
  manufacturing 285, *285*
  quartz 1
cementation 1
Cennomanian 272, 274
 – Turonian boundary 272, 274
Ceratium furcoides 172
CFC (chlorofluorocarbon) 295, 297
  11 *290*, 297, 297–8
  12 *290*, 297, 297–8
  113 *290*, 297
  photolysis 295
chair conformation 35, *35*, 37, *37*, 54, *54*, *160*

chalk 2, 273, *274*
charcoal 19, 22–3, 114–15, 123, 253, 321
charring 92, 123, 221, 300
cheilanthanes 153, 160, 200, 224
  and marine incursions 204
  numbering system 153
chelation/chelators 88, **89**, 89, 101, 121, 123, *181*, *183*, 208, 310, 313, 319–20
chemisorption 304
chemoautotrophes *see* bacteria, chemosynthetic
chemocline 116, 259, *259*
chemosynthesis 10–13, 15, 18, 71, *94*, 96–8, *97*, *116*, *236*, 258, *259*
chemosynthesizers *see* bacteria, chemosynthetic
chemotaxonomy *see* molecular palaeontology
chemotrophes *see* bacteria, chemosynthetic
chemotrophy *see* chemosynthesis
chert 19
Chinchorro mummies 239, *239*
chirality 32, 207
chitin 38, 100–1
chitinase 100
chloride 72, *72*, 111, 165
chlorine 46, 296–7
chlorins 183
chloroaromatics 314, 319; *see also* organochlorines
  chlorination degree/position 319
  environmental fate 313, 317
chlorobactene 168, 184
Chlorobiaceae/Chlorobium *see* bacteria, green sulphur
2-chlorodibenzodioxin *318*
chlorofluorocarbons *see* CFC
chloroform 43, 321
chloromonads 85
chlorophyllase 177, 182
chlorophyllides 177, *178*
chlorophylls 17, 17, 18, 30, 52, 56, 59, *59*, 63, 60–1, 78–9, *80*, 83–4, 86, 167–8, 176, *178*, *181*, 182, 206, *206*, 241
  bacterial 49, *63*, 60–1, 116, 167, 177, 182, 241
  compositional variation in organisms 60–1
  degradation/diagenesis 136, 138, 177–82, *178*, *181*, *206*, 206–7
  ring-numbering scheme 63
Chlorophyta/Chlorophyceae/chlorophytes *see* algae, green
chloroplasts 15, 47, 85, 167–8
cholestane 55
cholestanol 54, 55, 186
cholestenol 55
cholesterol 10, 49, 53, 55, 57, 171–2, 175–7
cholic acid 55
choline 46, *80*
chordate *12*
chromans 199, 205
  ratio 205
Chromatium *see* bacteria, purple sulphur
chromatogram 141, *141*, 160, 302

Index | 369

chromatography   128, 140–1, 149, 152
   gas   139, 140–1, 152, 172, 222, 301
   –liquid   140
   liquid   140
chromophore   64
chromosomes   12
chrysanthemic acid   49, *51*
chrysene   151, *299–300*
Chrysophyceae/chrysophytes   *see* algae, golden
chytridiomycetes   93
ciliates   14, 91, 99, 205
cinnamic acid   64, *65*
cinnamyl lignin units   173–4, *173–4*
citric acid   38, 238
   cycle   38, *40*, 40, 46, 98, 238
      reverse   *236*, 237–8
Cladium   *see* grass, saw-
Cladosporium   93
clarain   123
clarite   123
class (taxonomic)   12, 26–7, 84, *85*, 86, 167, 200
clastics   105, 109–10, *113*, 114, 159, 256
clathrate   165; *see also* methane, hydrate
claws   42
clay   43, 92–3, 104, 280, 304
   catalysed reactions   153, 169, 175, 186, 188, 207, 222
      Kimmeridge   231
   minerals   204
      transformation   230
climate   29, 72–4, 110, 113–16, 220, 241–2, *252*, 259, 261, *262–3*, 270
   arid   87, 109, 112, 114–16, 203, 237, 241, 261, 270, 272, 277, 292
   change   23, 27–9, *29*, 70, 219, 254, 256, 263, 265–81, *267*, *278*, *282*, 283–6, 288, 291–2, *293*
      Quaternary   216, 220, 241
   factors influencing   263–7
      periodicity   263, 266, 277
   modelling   253, 260–1, *262*, 266, 268–9, 279
   temperate   27, 87, *113*, 270, 286
   tropical   87, 110, *113*, 261, 270, 272
   warm/wet   253, 261, 265, 270, 273, 275–6
   zones   87, 89
climax population   89
Clostridia/*Clostridium*   40, 94–5, 95, 101, 221
cloud   252, 260, 265–6, 280, 291
   condensation nuclei (CCN)   266
   interstellar   3
   molecular   3, 5
clubmosses   *see* lycopsids
CNO cycle   3, *3*
coal   2, 64, 71, 102, 113–14, *113*, 122–32, *164*, 204, 287, 300; *see also* fossil fuel
   adsorption capacity   131, 137, 156, 212, 227
   ash   113, 114, 123, *124*, 128, *164*, 204
   bituminous   122, **125**, 122, 126, 131, 162, *163*, 210, *226*
      high-volatile   125, 127–9, 131, *225*, 232
      low-volatile   125, 127–8, 131, *225*
      medium-volatile   125, 127–8, 131, *225*, 232

boghead   **122**, *127*; *see also* torbanite
brown   118–19, 126, 121, 123, **125**, *127*, 127–8, 131, 139, 142, 162, *163*, 196, 320
burning   291, 298
calorific value   125
cannel   **122**, *127*
carbon content   123, **128**, 131, 263
classification   122, 125
composition   122–5, *124*, *128*, 133, 137, 201, 204, 225–6
density   *124*, 131
deposits   27, 29, 162, *163*, 248, 263, *271–2*, 272
detrital   102
evolution/maturation   125, 127, 162, 230–2
fixed carbon   *124–5*
formation   16, 22, 113, 125, 127–32, 135, 141, 243
   major periods   114, 139, 162, 273
gas-prone   142
groundmass   127
hard   *see* coal, bituminous
high-H   139, *140*, 142
humic   **122**, 123, 125, *127*, 127, 131, 137, 142, 152, 157, 182, *183*, 202, *202*, 210
hydrocarbon generation   203, 224, 231
lithotypes   123
low-H   139, *140*, 142
macerals   *see* macerals
metal chelation   123, 130, 320
microlithotypes   123
mineral content   *see* ash
nitrogen content   123, 131
oil-prone   142, 146, *225*, 227, 232, 243
oxidation   251
oxygen content   123, *124*, 125, *128*, 128–9, 131
petrography/petrology   *122*, 122–3, 136
porosity   131
proximate analysis   123, *124*
rank   122, **125**, 125, 126, 127, *127*, 129–32, *130*, 142, 210, *225*, 232
reserves   *163*, 163
sapropelic   **122**, 125, *127*, 127, 137, 142, 182
seams   102
specific energy   *124*
structural evolution/maturation   128–32, 141, 211
sub-bituminous   122, **125**, *125*, 127, *127–30*, 129–31, 163, *225*
sulphur
   content   112, 123, 131, 139
   forms   *124*
swamps   *see* mires
trapping small molecules   *see* coal, adsorption capacity
ultimate analysis   123, *124*
volatiles   123, *124–5*, 130–1
water/moisture content   123, *124–5*, 131, *164*

coalification   **125**, *129*, 130–1, 133, 203, 210, *225*
   biochemical stage   **125**, 127–8
   geochemical stage   **125**, 131–2, 175
coals
   Asian   *163*
   Antarctic   29, 163
   Australian   29, *163*
   Carboniferous   22, 27, 29, 112, 114, 122, 139, 162
   Chinese   *163*
   Cretaceous   27, 112, 139, 162, 164
   Devonian   27, 29, 112
   European   122, 139, *163*
   Far Eastern   *163*
   former Soviet Union   *163*
   Gondwanan   *163*
   Holocene   113
   Indian   29
   Jurassic   162
   Laurasian   29
   New Zealand   139, *225*
   North American   *163*, 165
   Permian   29, 114, 162
   South American   29
   South-east Asian   114, *163*
   Tertiary   27, 114, 139, 162, 164
   Westphalian   122
coccolithophores   **26**, *26*, 84, *85*, 91, 136, 162, *163*, 266, 273
   chemical composition   172, 200, 219
coccoliths   79
coenzyme   44, 80, 97
   A   *40*, *45*, *50*, 67–8, *68*, *302–3*
      acetyl   38, 44, 46, 49, 237–8
      malonyl   44
   Q   *see* ubiquinones
collagen   42, 100, 238–9
collinite   **122**, *122*, 124
colloids   92, 119, 122, 312, 314
columbin   *51*, 52
combustion   22–3, 113–14, 123, 153, 253–4, *263*, *285*, 285–8, *287*, 291, 295–6, 298–300; *see also* incineration; wildfires
   temperature   300, 315
comets   5, 5, 7
   composition   5
common ancestor   12, 13–15
communic acid   *51*, 52, 60
compensation
   depth   **78**, *79*
   light intensity   78
competition
   biotic   28, 70, 72, 86, 89, 89, 96, 98–9, 275, 294, 319
   chemical   208, 186–8, 190, 192, 196, 235, 270, 320
complexation   *see* chelation
compression   107, 148, 157; *see also* sediments, compaction
Coniacian   274
concordia curves   216, *216*
condensate   **148**, 150, *163*, *225*

condensation   10, **37**, 41–2, 49, 62, 64–5,
    100–1, 106, *118*, 118–19, 121, 127–9,
    129, 131–2, *134*, 135
  cold   315
conessine   56
configuration   *see* stereoisomerism,
    configurational
conformation   *see* stereoisomerism,
    conformational
congeners   313–17, *314*, 319
coniferaldehyde   174
conifers   27, 29, 52, *91*, 127, 133, 172
coniferyl alcohol   62, 64–5, 174
conjugate   307, 311
continental
  arrangement/dispersal   29, 29, 70, 83, 107,
    162, *163*, 228, 252, 254, 261, *262*, 263,
    265, 267–8, 270, *271–2*, 272
  boundary currents   85, *85*
  crust   *see* crust, continental
  drift   *see* tectonic activity
  margin   83, 92, **115**, 162, 165, 228, 273, 275,
    283
  plates   *see* tectonic plates
  relief   254, 261, 265
  rise   79, *105*
  shelf/slope   73, 79, 85, *105*, 105, **115**, *272*,
    275; *see also* environments, shelf/slope
  stable platforms   *see* craton
contraceptive pill   307, *320*
convection   **2**, 78
convergence   76, 83, 85
conveyor (oceanic)   75
Cooksonia   27
copepods   **27**, 81, 91–2, 171, 176, 179, 310
  chemical composition   *69*, 69, 177
copper   102, *181*, 182, 320
coprophagy   177, 179
Cordaites   29
core (of Earth)   **2**, 5, 229
  inner   *2*, 2
  outer   *2*, 2
Coriolis effect   76, 76, 85, *85*
cork   122
coronene   298, *299–300*, 300
cortisone   55
corynebacteria/*Corynebacterium*   47, 94
cosmic ray flux   262, 265; *see also* radiation,
    cosmic
coumarate/coumaric acid   62, 64, *65*, 67,
    173
coumaryl alcohol   62, 64, 65
coumestan   65, 67, 306
coumestrol   *25*, 65, 306
CPI   *see* carbon preference index
cracking   *see* oil, cracking
craton   164, 228
  Pilbara   201
Cretaceous   24, *26*, 26–8, *28*, 71, 107, 112, 115,
    133, *137*, 139, 162, *163*, 201, 219, 219,
    229, 231, *262*, 272–3, 273–4, 324
  chalk   274
  climate   265, 272–6

oceanic anoxic event   *see* oceans, anoxic
    events
  polar forests   272
  rifting/tectonism   228, 261, 265, 272–3, 275
  – Tertiary boundary   28, 280–1, *281*, 284,
    291
cresols   129
critical
  depth   **78**, *79*
  point   *157*, 158
  pressure   *see* pressure, critical
  temperature   *see* temperature, critical
crust   **2**, **4**, 107, *107*, 146, 203, 229–30, *251*,
    255, 268
  continental   6, *16*, 83, *258*, 260
  heat production   **4**, 145
  oceanic   272, 275
  oxidation   16
  stretching/thinning   *see* extension
crustaceans   27, 38, 57, 171, 177, 305
Cryogenian   324
cryptomonads/Cryptophyceae   85, *171*
Cryptozoic   324
current   106, 110, 115, 248, 254, 300, 301;
    see also ocean, circulation
  Benguela   84, 85
  boundary   83, 85, *85*
  circumpolar   276
  Peru   84, 85
cutan   **49**, *133–4*, 133–4, 139, 145–6, 151
cuticles   27, 133
  arthropod   101, 134
  plant   47, 49, 69, 101, 118, 123, *133–4*, 139,
    179
cutin   **49**, 134, 170, 180
cutinase   100
cutinite   122, 123, *124*, 133
cyanidin chloride   65, *67*
cyanobacteria/Cyanophyceae/Cyanophyta   12,
    14, 16, 17–18, **18**, 25–6, 26, 70–1, 84, 85,
    89, *91*, 112, 162, *163*, *237*, 205, 292,
    294
  chemical composition   18, 47, 55, 57, 60, 62,
    133, 167, 169–70, 172, 197, 201, 205
  contribution to sedimentary organic matter
    133
  colonial   18, 112
  freshwater   88, 292
  hypersaline   72, 112, 205
  isotopic composition   241
  mats   *see* microbial mats
  nitrogen-fixing   79, 87, 95, 86, 275, 294
  non-colonial   18, 21, 84
cycads   27
cycloartenol   53, *56*
cyclohexane   151, 222–3, *303*
cyclohexanol   303
cyclohexanone   303
cyclones   76
cyclopentadiene   31
cyclopentane   151, 222
cyclopentanyl unit   47
cyclophaeophorbides   *181–2*

cypress   114–15; *see also* swamp cypress
cysts   25
cysteine   *41*, 41–2, 62, 97
cytochromes   43, 62, 80, 99, 182, *183*
  P450   310, 320
cytosine   68, *69*

2,4-D   313, *313*
Dalton's law   20
dating methods   212–16, 217, 219, 276
DBP   *312*, 313
DDA   *312*, 313
DDD   *312*, 313, 319
DDE   *308*, *312*, 313, 319
DDMS   *312*
DDMU   *312*
DDNU   *312*
DDOH   *312*
DDT   295, 306, *308*, 311–16, *312*
  bioaccumulation   306, *312*, 317, 319
  degradation products   *312*, 313, 319
dealkylation   155, 311
de-A-lupane   202, 201
deamination   62, *65*
de-A-oleanane   201, *202*
de-asphalting of oil   *see* asphaltenes,
    precipitation
de-A-ursane   201, *202*
decarbonation   247, 285
decarboxylation   44, 47, 49, **50**, 53, 125, 131,
    140, 142, *147*, 167, 175, 180, *183*, 197,
    207, 238, *303*, 312
decarboxymethylation   182, *184*
decomposers   **13**, *25*, 71, 80, 93, 95, 99, 101,
    110, 113, 117, 119, 121, 127, 247, 259; *see
    also* bacteria, heterotrophic; fungi
decomposition   *see* biodegradation
de-E-hopanes   *see* secohopanes
deep hot biosphere   14
deforestation   284–8, *285*
defunctionalization   127, 131–2, 135, 175, 182,
    188, 197
degradation
  biological   *see* biodegradation
  chemical   179, 311, 315, 317, 319
  thermal   *see* maturation
Dehalococcoides   317
DEHP   *see* phthalate, diethylhexyl-
dehydration   27, 42, 44, **45**, 62, 89, 100, *118*,
    125, 131, *140*, 169, 175, 180, *181*, 186,
    188, 190, 192, 207, *209*, 215, *302–3*
dehydroabietane   190
dehydroabietic acid   190
dehydroabietin   190
dehydrogenase   18, 238
dehydrogenation   45, 62, 81, 125, 144
dehydroxylation   129
deltas
  Amazon   115
  Mahakam   113–15, *137*
  Mississippi   114–15
  Niger   114
  see also environments, deltaic

demethylation   101–2, 127, 129, *129*, 131, 212, 226
dendrochronology   <u>217</u>, *217*
denitrification   94–5, 95–6, <u>96</u>, 99, 240, 286, 292, 296
  zone   *94*, 96, 99
denitrifiers   *see* bacteria, denitrifying
density   3, 27, 71, <u>75–6</u>, *124*, 131, <u>145</u>, <u>157</u>, *164*, 157–8, 161, 292, <u>293</u>, <u>296</u>, 301, 304; *see also* water, density
  flow   *see* sediment, density flow
  relative   <u>150</u>
deoxomesopyrophaeophorbides   *181*
deoxophylloerythrin (DPE)   *181*
deoxygalactose   35
deoxygenation   35
deoxymannose   35
deoxyribonucleic acid   *see* DNA
deoxyribose   67
deoxyxylulose
  pathway   49, *50*
  phosphate   49, *50*
depolymerization   102, 127
deposit feeders   *see* detritivores
deposition
  dry   **315**
  wet   265, 311, **315**
depositional environments   *see* environments
depuration   295, **309**, 310–11, 313, 319
DES   *see* diethylstilboestrol
desaturase   45, 169
desert   *73*, 83, *90*, 204, 242, 270
  Patagonian   278
  Sahara   276
13-desethyletioporphyrin   *183*
desferrioxamine B   88
desiccation   <u>1</u>, 27, 49, 100, 112, <u>221</u>
4-desmethylsteranes   *see* steranes, regular
4-desmethylsterols   *see* sterols, regular
desmocollinite   124
desmosterol   57
Desulfobacter   96, 98, 170
Desulfobulbus   98, 168–9
Desulfotomaculum   96
Desulfovibrio   96, 98, 100
  desulfuricans   170
Desulfuromonas acetoxidans   98
desulphuration   <u>97</u>, *97*
desulphydrase   <u>97</u>
detector
  infrared   <u>223</u>
  flame ionization (FID)   <u>141</u>, *141*, <u>223</u>, *304*
  limits   307
  mass spectrometer   <u>141</u>
detritivores   **13**, 24, *25*, 71, 93, 104, 127, 249, 310, 312
detritus   **13**, <u>18</u>, 19, 24, <u>89</u>, 91, 93–4, 99, 106, 108, 112, 114–16, 127, 238, 246–8, 281, 288, 292
  adsorption of pollutants   312
  comminution   24, 93, 104, 127
  preservation/degradation   25, 80

sinking   80, 86, 91, 106, 116, 175–7, 179, 259, 275, 280–1, 315
  sources   24, 27, 92–3, 99, 109–10, 112, *113*, 114–15, 119, 121–3, 127, 132, 136–7, 139, 169, 175–7, 260
deuterium   3; *see also* hydrogen, isotopic composition
Devonian   26, 27–9, *28*, 70, 112, 115, 162, *163*, <u>219</u>, 256, 261–2, *262*, 300, 324
diacylgalactosylglycerol   47, *48*
diadinochrome   *185*
diadinoxanthin   57, *61*, 179, *185*, 185
diagenesis   106, **117**–19, 135, 139, *140*, 149, <u>166</u>, 200, <u>208</u>, 216, 221, *225*, 251, *252*, 272; *see also* under individual compounds
  anoxic   108
  – catagenesis boundary   117, 125, 141
  coal/peat   122–3, 125, <u>126</u>, 127–32, *147*, 162
  early   180, 186, 188, 191, 197, 204, 205–6, 243
  humic material   119, *147*
  kerogen   <u>126</u>, 132–6, 141–2, *143*–4, 146, *147*, 148, 162
  late   182, 188, 196, 225
  maceral changes   122, 124–5, <u>126</u>, 127
  mid   182
  molecular changes   166, 174–96, *178*–85, *187, 189*–95, 197–8, 203, *206*, 207–9, *209*, 212, *215*, 227, 306
  oxygen availability   206
  sedimentary   93–100, 132, 177, *178*, 180–1, 179–96
  water column   89, 91–3, 132, *178*, 180, 186, 206
diahopanes   196, *196*, 200, 224
dialkylthiacyclohexanes   *136*
dialkylthiacyclopentanes   *136*
dialkylthiophenes   *136*
diamantane   160, *160*
diamond   2, 146, 160
diamondoids   **160**, *160*
diasteranes   *153*, *160*, 160, **186**, *187*, 196
  clay influence   186, 204
  ratio to steranes   204
diasterenes   **186**–7, *187*
diastereomers   *32*, *34*, <u>37</u>, *37*, <u>141</u>
diasteroids   <u>55</u>, *175*, 204
  aromatic   *187*, *210*, 210
diatoms   **26**, *26*, *61*, *78*, *79*, *80*, *84*, *85*, *86*, 92, 112, 162, 176, 279
  benthonic   *52*, 73, *79*, 84
  chemical composition   69, 169, 170–2, *171*, 179, 200
  planktonic   *52*, *52*, *163*, 294
diatomsterol   *see* epibrassicasterol
diatoxanthin   57, *61*, 179, *185*, 185
dibenzofuran   131, 152
dibenzothiophene   152, *211*
  – phenanthrene ratio   *204*, 205
DIC   *see* carbon, dissolved inorganic
dichlorodibenzodioxins   *318*
p,p'-dichloro-2,2-dibenzophenone   *see* DBP

1,1-dichloro-2,2-di(*p*-chlorophenyl)acetic acid   *see* DDA
1,1-dichloro-2,2-di(*p*-chlorophenyl)ethane   *see* DDD
1,1-dichloro-2,2-di(*p*-chlorophenyl)ethylene   *see* DDE
p,p'-dichlorodiphenyltrichloroethane   *see* DDT
Dickinsonia   *19*, 22
diesel fuel   <u>150</u>
diethylhexylphthalate   *see* phthalate, diethylhexyl-
diethylstilboestrol (DES)   304, *305*, 307
diffusion   **4**, <u>93</u>, 98, 105, <u>141</u>, 157, 161, 236–7, 292, <u>309</u>, 315, 319
  isotopic fractionation   *see* isotopic fractionation, diffusion
  thermally activated   156
digitonin   55
diglycerides   44
dihomohopanoic acid   192
dihydroactinidiolide   *185*, 185
dihydrophytol   *see* phytanol
dihydroxyacetone phosphate   35, *36*
dihydroxybenzenes   *see* catechols
dihydroxydiethylstilbene   *see* diethylstilboestrol
diketopiperazines   *see* amino acids, diketopiperazines
7,8-dimethylchroman   205
dimethylcyclopentanes   *152*, *203*, 223
1,2-dimethylcyclohexane   35, *35*
dimethylhexane   223
dimethylnaphthalene   211–12, 226
  1,8-   212, 226
  index (DNI)   *225*, 226
  isomerization   225
  ratio (DNR1)   *225*, 226–7
2,6-dimethyloctane   34
dimethylpentanes   *152*, *203*, 223
1,7-dimethylphenanthrene   *see* pimanthrene
2,4-dimethylphenol   129
1,8-dimethylpicene   202, *202*
dimethyl sulphide   *see* sulphide, dimethyl
dimethyl sulphoniopropionate (DMSP)   266
1,4a-dimethyl-7-*iso*propyl-1,2,3,4-tetrahydrophenanthrene   *190*
dinoflagellates/Dinophyceae   **26**, *26*, 56, *61*, *78*, 84, *85*, 86, *163*, 172
  chemical composition   69, 133, *171*, 167, 169–70, 179, 188, 200
  freshwater   172
28,30-dinorhopane   *199*, 206, 225, 275
dinosaurs   28
dinosterane   *199*, 200
dinosterol   57, *171*, 172
dioxins   *see* polychlorodibenzo-*p*-dioxins
dipeptides   41
diploptene   *52*, *53*, 192–3, *194*, 201
diplopterol   52–3, *53*, 201
dipterocarps   60, 114
disaccharides   *37*, 38
dissimilatory processes   **96**, *96*, 236
diterpanes   49, 201–3, *202*
  isomerization   209

diterpenes 49
diterpenoids 49, *49–51*, 52, 60, *133*
　　aromatic/aromatization 202, 210
　　diagenesis 188, *190*
　　source indicators 201–3
division (taxonomic) 12, 12–13, *14*, 166–7
divergence 75, 83, *85*, 85, 275
　　Antarctic 85, 250
DMS *see* dimethylsulphide
DNA (deoxyribonucleic acid) 10–11, 12, 15, 68, *69*, *100*, 100, 296
DOC *see* carbon, dissolved organic
dodecanoic acid 44
dog whelk 309, 310
Doldrums 75; *see also* ITCZ
dolomite 2, *16*, 250
DOM *see* organic matter, dissolved
domains (taxonomic) 12, 14
downwelling 75–6, 83, *85*, 105
drainage area 158
Drake Passage 276
drimane 192, *193*, 196
drimenol 51, 192, *193*, *195*, 196
durain 123
durite 123
dust 252
　　continental 86, 278
　　interplanetary 7, *7*, 8
　　interstellar 5, 5
　　road 300, 311
　　volcanic 265, 284
dysaerobia *see* respiration, dysaerobic

Earth 293
　　composition 2, 254
　　evolution 5, 6, 12, 15, *16*, 23, 29, 260, 267, 279–80
　　layers 2, *2*
　　magnetic field 2, *29*
　　orbit 263, 264, *264*; *see also* Milankovich cycles
　　rotation 75
　　Snowball 267–9, *268*, 324
　　structure 2
　　surface temperature 6, 165, *91*, 252–3, *252–3*, *260*, 260–1, *263*, *267*, 268–70, *270*, 272–3, *276*, 276–7, *279*, 279–80, *282*, 287, 291
　　meridional gradient 263, 272, 315
　　system 246
earthquakes 2
EASY %R$_o$ *see* vitrinite reflectance, maturation kinetics
ecdysis 305
ecdysone 305, *305*
ecdysteroids 305, *305*
ecological niches 22, 28, 70, 72
ecosystems 28–9, 72–3, 197, 205, 240, 284, 301, 310, 317
　　aquatic/marine 24, 80, 84, 284
　　terrestrial 24, 70, 78, 80, 278, 284, 286, 300
edaphic factors 87, 100, 313, **319**

Ediacaran 324
　　fauna 22, 25
eggs 295, 306
E$_h$ *see* redox potential
eicosanoic acid 44
eicos-5,8,11,14-tetraenoic acid 44
electron 3, 4, 6, 17–18, 42, 141, 175
　　acceptors 13, 94–5, 99–100
　　carrier system 17
　　delocalised 33, 60
　　density 31
　　donors 98
　　lone pairs *211*
　　releasing/donating group 60, 213
　　transfer 43, 62, 67, 81, 102
　　unpaired 154
　　valency 30, 60, 212
　　withdrawing group 60, 102, 213
electronegativity 31
electrophiles 60, 129
electrostatic interactions 30, 31, 211, 213
element 2, 3, 4, 123
　　biointermediate 80
　　biolimiting 80, 87, 88–9
　　essential 74, *80*; *see also* nutrients, biolimiting
　　non-biolimiting 80
ellagic acid 65, 67
El Niño 285
elution 140, 139
Emiliania huxleyi 136, 172, 219, 220
emoldin *67*, 67
empirical formula 15
emulsification 301
enantiomers 32, 34, 37, 100, 175, 212
endocrine
　　activity 304–5, 320
　　disruption 304–8, 306, 309–10, 319–20
　　system 304, 306
endospores 221
energy 3, 40, 79; *see also* thermodynamic stability
　　activation 43, *43*, *175*, 176, 214, *227*, 227, *229*, 230, *231*, 232
　　bond 223
　　cellular 1, 2, 15, 22–3, 40, 46, 62, 95–6, 98, 237
　　changes in reactions 43, *43*, 176
　　chemical 13, 17–18, 22, 24, 71
　　content of coal *124*–5
　　equivalence of fossil fuels *163*
　　generation 286
　　kinetic 20, 93
　　latitudinal distribution *see* heat, global distribution/transfer
　　light *see* light
　　reserve 15, 38, 81
　　solar *see* light
　　sources 13, 45–6, 99, 150, 159, 263
　　specific *124*
　　storage 17, 24, 45, 117, 167–9, 172, 174, 246
　　thermal 117
　　transfer through ecosystems 13, 24, *25*, 112
　　units 322

Enteromorpha 59
envelope conformation 37, *37*, 54
environment 1, 72, 93, 119, 197, 240, 280, 284, *308*, 309
　　acidic 81, 101
　　alkaline 81, 101
　　anoxic *see* anoxicity
　　aquatic 13, 16, 24, 21, 69, 74, 80, 93, 95, 132, 247, 292, 298–304, 307, 320–1
　　arid 29, 167
　　brackish 115, 138, 204
　　change 89, 321
　　coastal 24, 29, 79, *105*, 106, *108*, 116, 162, 204, 216, 220, 292, 301
　　deep-sea *108*, 108, 165, 176
　　deltaic 106, *105–6*, *107*, *110*, 110, 113–16, 159, 204, *204*, 248
　　depositional 70–1, 106, 109, 112, 197, 203–7, *204*, 208, 242, 304
　　dysoxic 206
　　epicontinental 115
　　estuarine 72, *73*, 78–9, 84, *105*, *171*, 171, 248, 299, 310
　　eutrophic **83**, 94, 114
　　euxinic 21, 116, 275
　　extreme 14, 22, 98
　　floodplain 113, 204
　　fluvial *73*, 73, 86, *204*
　　freshwater 29, 72, 91, 93, *105*, 138, 203–5, *236*, 248, 287
　　high energy 74, 77, 77–8, 83, *105*, 277, 292, 300, 310
　　hypersaline *see* hypersalinity
　　indicators *see* source/environmental indicators
　　intertidal *73*, 115
　　lacustrine 24, 73, 104, *105*, 109–10, 112, 115, 138, *171*, 171, *203–4*, 203–5, 287, 300; *see also* lakes
　　lagoonal *105*, 109, 114, 138
　　latitudinal variation 28–9, 69, 72, 78, 84, 87, 100, 113, 115, 242, 250, 270, 272, 277, 279, 315
　　litoral 79, 84, *110*
　　low energy 106, 122, 138
　　marine 24, 70, 72, 74, 97, 91, 99, 104, 110, 115–16, 138, 139, *171*, 171, *203*, *204*, *208*, *236*, 201, 203–4, 205, 206, 235, 287, 292, 315
　　marginal/nearshore 113, 83, 92, 162, 165, 201, 204, 300
　　open/offshore 72, 88, *105*, 106, 201, 204, 220, 243, 292, 314
　　neritic 79
　　nutrient-rich *see* environment, eutrophic; nutrients, supply
　　oligotrophic **83**, 105, 114
　　oxic/oxidising *see* oxicity
　　paralic **113**
　　peat-forming *see* mires, coal/peat-forming
　　pelagic 79, *106*, 220
　　polar 165, 249, 266
　　reducing *see* anoxicity

saline   29, 203–4
sedimentary   2, *16*, 24, 70, 100, 135, 167, 180, *205*, 300
shelf/slope/shallow sea   92, 99, *105–6*, 115–16, *164*, 165, 266, 273
shallow water   18–19, 25, 72, 112, 115, 122, 138, 273
subaerial   20, 301
suboxic   <u>**6**</u>
swamp   71, 73, *105*, 105, 115, 162
temperate   27, 73–4, 113, 270, 292
terrestrial/terrigenous   24, 27, 70, 74, 78, 93, 100, 121, *164*, *171*, 171, 203, *203*, 288
tropical   29, 74, 89, *90*, 113–14, <u>218</u>, 242, 263, 270, 273, 311
upwelling   *see* upwelling
enzymes   14, 35, 40, **43**, <u>43</u>, 43–7, *45*, 49, *50*, 53, *56*, *61*, 62, 72, 78, 96, 100, 102, 128, 169, 177, 179, 236–7, 270, 320; *see also* under individual enzymes
extracellular   <u>13</u>, 93, 100–1, 104, 117, 160
hydrolytic   93, 100–1, 102, 106, 117, 177, 179
phenol-oxidizing   102
Eocene   137, 138, 219, <u>219</u>, *241*, 241, 324
Eon   324
EOP   *see* even-over-odd predominance
epiabietic acid   188
epibrassicasterol   55, *57*, 172
epilimnion   74, **77**, 78, 83, 294
epimerization   **32**, 175, *175*, 213, *213*, <u>214</u>, *215–16*, 215–16
epimers   **32**, 172, 185, 188, 210–1
epiphytes   73, **86**, 101
Epoch   324
equations of state   158
equatorial substituents   **35**, *35*, 37, *37*
Equator   <u>75–6</u>, <u>85</u>, *85*, *90*, 263, *271*, *272*, <u>296</u>
equilibrium   246, 250, 255
chemical/thermodynamic   36, <u>111</u>, *111*, 145–6, *147*, <u>147</u>, *157*, <u>176</u>, 186, 188, <u>214</u>, <u>217–18</u>, 20–9, 212, 235, *235–6*, *244*, 245, 308
constant   213, <u>214</u>
dynamic   175, <u>176</u>, 246, 248
isotopic   268
thermal   <u>253</u>
vapour–liquid   161–2
Equisetum   27
Era   *26*, 280, 324
ergosterol   *57*, 171
erosion   71, 105, <u>107</u>, 112, *113*, 144, 148, <u>230</u>, 246, 251, 256, 261
erythrose phosphate   65
Escherichia   94, <u>96</u>, 96
esters   31, <u>42</u>, 47, 49, 57, 59, 65, 67, 207
glyceryl   44
steryl   44, **47**, <u>166</u>, 171, 177, *178*, 186
wax   44, **47**, 69, 101, <u>166</u>, 169
ethane   149, 161, *164*, 165, *243–4*, 244, 287
ethanoic acid   *see* acetic acid
ethanol   5, <u>42</u>, 94, 95
ethers   31, 47, 49, 120, 129, 167
aryl   128, 131

biphytanyl   47, *48*, 167, 197
phytanyl   47, *206*, 241
24-ethylcholesta-5,22E-diene-3b-ol   <u>55</u>
ethylcyclopentane   *152*, *203*
ethylpentane   *152*, *203*, 223
ethynyloestradiol   *305*, 307, *308*
etioporphyrins   *see* porphyrins, etio
eubacteria   <u>12</u>, *14*, **14**, 15, *26*, 40, 47, *48*, 49, 52, 55, 93, <u>95</u>, <u>98</u>, 99, 166, 169–70, 200, 241
eudesmane   *190*, 188
eudesmol   *51*
Euglenophyceae   *see* flagellates, green
Eukarya/eukaryotes   <u>12</u>, *14*, 14, *16*, 22, 55, <u>89</u>, 167, 238
unicellular   <u>12</u>, 20, 25
multicellular   <u>12</u>, 15, 20
Eumycota   *see* fungi
euphausiids   179
Euphorbia   60
euphotic zone   **78**, 80–1, 83, 86, 89, 92, 104, 116, 176–7, 179, *259*, 260, 275, 294
depth limit   78–9, 84, 105
Euramerica   270, *271*
Eustigmatophyceae/eustigmatophytes   *85*, 133, 169–72, *171*
eutrophication   *292*, 294
evaporation   72, <u>75</u>, 110, 112, 116, 216, <u>218–19</u>, 203, 242, 253, 265–6, 268, 270, 292, 301, 307, *308*, 311, 315
zones   76, *90*, 265, 272, 274–5, 292, 315
evaporites   *see* sediments, evaporitic
evaporative fractionation   161–2
evapotranspiration   72, *90*, 242, 265
even-over-odd predominance (EOP)   47, 49, **167**–70, 197, 205, 224
Everglades   114
exine   123
exinite   123, *126*
extension   83, <u>107</u>, *107*, 109, 148, 203, 228, *229*, <u>230</u>, 232, *233*, 234
β factor   <u>107</u>, *107*, *229*, *233*, 234
extinction events   28, 28–9, 280–2, *282*, 284, 291
exudates   113–14, 250
exudatinite   *122*, 123

facies   **1**, 112, 226
organo-   **1**, 109, 240
faecal pellets   <u>13</u>, 238, 248
family (taxonomic)   *12*, 28, 28–9, 84, 86, 167
Faraday constant   81
farnesol   49, *51*
farnesyl unit   *51*, *63*
fat   *40*, 40, **44**–6, *306*, 312, 319
compound solubility in   *see* lipophilicity
fatty acids   *40*, <u>42</u>, **44**, <u>45</u>, 46–7, *48*, 49, 55, 59, 101, 118, 121, *134*, 135, <u>166</u>, 177, 197, 220, 224
anteiso   *168*, 168, 170
biosynthesis   40, 44, 169–70
bound   169
common   44
cycloalkyl   *168*, 170

diagenesis   179–80
dicarboxylic   49, *170*, 179–80
free   169
hydroxy   49, 169–71, 179–80
internally branched   170
iso   *168*, 168, 170
long-chain   *94*, 94
methyl-branched   *168*, 169
nomenclature   44, *44*, <u>45</u>
polyunsaturated (PUFAs)   44, <u>45</u>, 106, 169–70, 177, 179
saturated   *see* acids, alkanoic
short-chain   *94*, 95
source indicators   44, 167, *168*, 169–71, 179
unsaturated   *see* acids, alkenoic
volatile   95
fatty acid synthase   44
fatty alcohols   47, *94*, 101, <u>166</u>
source indicators   167, 224
fatty diacids   49, 170, 179–80
faults   158, <u>159</u>
San Andreas   <u>107</u>
strike-slip   *107*
feedbacks   <u>87</u>, 102, 109, *109*, 251, *252*, 253–4, 261, 265–6, *267*, 269, 273, *273*, 277, 280, *290*
ice-albedo   263, 267, 277–8
feldspar   162
fermentation   15, 22, *94*, **94**, 98, 244
fern   27–8, 52, *53*
spike   28, 284
ferredoxins   80
ferrireductase   <u>89</u>
fertilizers   286, 296
ferulate/ferulic acid   *173*
fetch   78, 109
fibrinogen   100
FID   *see* detector, flame ionization
filter feeders   <u>13</u>, 93, 303, 310
fish   239, 292, 295
bioaccumulation   <u>309</u>, 310–11, *312*, 316–17
endocrine disruption   304–5, *306*
fission
heterolytic   **60**
homolytic   **60**
track analysis   <u>230</u>
flagellates   *12*, *14*, 84
green   84, *85*, *171*
Flavobacterium   *14*, 94
flavone   65
flavonoids   **65**, 67
bio-   320, *320*
flavonol   65
flexural loading   <u>107</u>, *107*
flocculation   248
fluoranthene   *299–300*
fluorescence   *123*, 123–5, 177
fluorine   80, 297
fluorinite   *122*, 123
fluorites   297
fluvial inputs   *see* transportation, aquatic
folds   <u>159</u>

food chains 24, 301, *309*, *312*, 312, 314, 319
   aquatic/marine 299, 317, 319
   detrital **13**, 24–5, *25*, 104
   grazing **13**, 24–5, *25*, 281
foraminiferans **26**, *26–8*, 91, 212, *215–16*, 280, 282
   benthonic 216, **218**, 273, *281*, 282–3
   planktonic **219**, 220–1, *222*, *281*, 283
   thermoclinic 283
forests 27, 29, 73, *90–1*, 114–15, 288
   boreal 87, 89, *90–1*, 286
   polar 272
   rain 71, 73, *73*, 89, *90–1*, 270, 272
   temperate 87, 89, *90–1*
formaldehyde 5, **10**, 98
formate 15, 95, 98, 177
Formation
   Green River 109
   Kimmeridge Clay *144*, 156–7, *231*
   Manville *231*
formaldehyde 5
formic acid *see* formate
formyloxobilanes 177, *179*
fossil 100, 133, 197
   assemblages *1*
   fuel 2, **150**, 246, 250
      calorific value 150
      combustion 113–14, 153, *263*, *285*, 285–8, 291, 295–6, 298–300
      distribution *163*
      reserves *163*, 162–5
      S and N content 150
   record 7, 14, 20, 22, 24–8, 89
   soils *see* palaeosols
frequency factor *see* Arrhenius constant
friedelin 53
fronts 85, **293**, *293*
fructans 38
fructofuranose 37
fructose 35, *36–7*, **37**, 37–8
fruits 38, 45, 69, 127
fucoids **59**, *73*
fucose 35, *36*, *172*, *173*
fucosterol 57, 200
fucoxanthin 58, 179, *180*, 185, 205
fucoxanthinol 179, *180*
Fucus vesiculosis 310
fulvic acids **119**, 120–1, 132
   composition *119–20*, 320
   soil *119*, 121, 132
   structure *120*
fumaroles 14
functional groups 30, **31**, **45**, 56, 100, 102, 119, 135, 192, 196, 319
   alcohol *see* functional groups, hydroxyl
   alkene 31, **31**
   amido/amide **31**, **31**, 35, 41–2, 118, 137, *138*
   amino/amine **31**, **31**, 40–1, **60**, 67, **95**, *96*, 101, 119, 123, 131, 213, 215
      in amino acids and peptides 40–1, 213, 215
   aryl **31**, 131
   in carbohydrates 35–7
   carbonyl **31**, *31*, 35, **37**, 49, 67, 102, 119, *120*, *128*, 129, 131, *133*, 137, *138*, 142, *144*, 238
   aldehydic 119, 133
   ketonic 119, 123, 139
   quinoidal 119, *120*, *138*, 139; *see also* quinoidal units
   carboxyl/carboxylate **31**, *31*, 35, 40–1, **45**, 49, **60**, 62, 65, 102, 119, *120*, 121, 123, 127, *128*, 129, 131–2, 137, *138*, 139, 142, *144*, 170, 213, *215*, 238, 313, 320
   ester 119, 121, 137–9, 169, 177, 311
   in coal 123, 125, 127–9, *128*, 131
   ether **31**, 47, 119, *120*, *128*, 131, 137–8, *138*, *144*, *144*
      alkoxy 102
      methoxy **60**, 65, 101, 102, 119, *120*, 123, 127–9, *128*, 139
   in humic material 119, *120*
   hydroxyl/hydroxy **31**, 35–6, **31**, **37**, 37, 44, 47, 49, 55, **55**, **60**, 60, 62, 65, 101–2, 119, *120*, 123, 128–9, 131–2, 137, *138*, *144*, 170, 185, 188, 196
   phenolic 64, *120*, 123, 128, *128*, 131, *138*, *144*
   in kerogen 137–9, *138*, 142, 144, *144*
   in lignins and tannins 62, 64–5
   in lipids 175
   thio/thiol 5, 31, **97**, 123
fungi **12**, *26*, 56, 88, **89**, 93, **96**, 100–1, *122*, 301
   brown-rot 101
   chemical composition 38, 40, 47, 52–3, 56, 59–60, 64, 67, 127, 168–71, 180, 306
   colonial 27, 93, 101
   contribution to sedimentary organic matter 24, 110, 121, 122, 123, 127
   decomposition of organic matter *see* biodegradation
   evolution 25, 27
   imperfecti 93, 101
   soft-rot
   unicellular *see* yeasts
   white-rot 101–2
fungicides 49
furan *31*, 35, 152
furanyl unit 118, 120
furanoses 35, 37, **37**
fusain 123
fusinite *122*, 122–3, *142*, 300
   semi- *122*, 122–3, *133*, 300

Gaia theory 266
galactans 101
galactosamine *36*
galactose *36*, 38, 40, 172, *173*, 175
galacturonans 38
galacturonic acid 35, 38, 101, 175
galaxy 266
gallic acid 64
gallium *181*, 182
gamete 27
gammacerane *199*, 205
gas 2, 123, 127, 131, **145**, **150**, 156, *157–8*, 159, *159*, 161–2, *164*, 245, *247*, *285–6*; *see also* fossil fuel; gas, natural
   carrier 140
   chromatogram *304*
   chromatography 140
      mass spectrometry 141, *300*
   composition 243, 244–5
   condensate 148, *158*; *see also* condensate
   correlation with source 244–5
   density 157, 157–8, *164*
   dry **126**, *140*, *147*, **148**, 244
   fields 139, 161
   generation 148, *149*, 231, *243–4*
      temperature 148, *149*
   greenhouse **253**, 260–1, *263*, 265–7, 269, *286*, 286–8, *290*, **290–1**, 291–2, 297–8, *298*
   hydrate *see* methane hydrate
   ideal 20
   inert 140
   marsh 114
   maturation 245
   migration *164*, *243*, 244–5; *see also* petroleum, migration
   monatomic 20
   natural 149, 161, *163–4*, **223**, *231*, 287, *287*
   –oil ratio 158
   ozone-depleting 295–7, *298*
   potential 148
   present atmospheric level (PAL) *16*, **19**, *19*, 20–3, 260–1, *260*, *262*, 268–9, *270*
   pVT behaviour 20, **157**
   reducing 5, 7–8, 18, 21, 260–1
   reserves *163*, 164–5
   seeps 244
   solubility
      in oil 148, 157–8, *243*, 245
      in water 162, 249–50, 278, 283, 287
   sour 161
   source rocks 164
      Cretaceous 164
      Former Soviet Union 164
      Middle East 164
      Tertiary 164
   sources 139, *244*, 244
      anthropogenic **290**
      biogenic 245
      thermogenic 164, 245
   trace **290**, *286*, *298*
   wet **126**, *140*, *147*, **148**, 149, *225*, 226, 244
gasoline 150
   range 159, 222, 242
gastropods 273, 310
Gelbstoff *see* humic material
gels 119, *122*, 127
gene/genetics 10–12, 14–15, 213, 220
   swapping 15
general circulation model (GCM) 261
genome 15
genus *12*, 69, 100, 177, 216, 220
geobaric gradient 145–6, 158, 162, 265, 283; *see also* pressure, hydrostatic; pressure, lithostatic

GEOCARB 261, *262*
geochemical fossils *see* biomarkers
geochromatography 159
geological time scale 324
geomacromolecules/geopolymers *see* macromolecules, geo-
geosphere 117, 295
geotherm 165
geothermal
  gradient 117, 144–6, 145, 148, 158, 162, *164*, 165, 188, 228, 229, 232, *233*, 234
  and molecular maturity parameters 210
  history *see* thermal history
Geotrichum 93
Gephyrocapsa oceanica 220
Gephyrocapsaceae 219
geranyl unit *50*, *63*
gibberellic acid 52
gibberellins 52
gills *308*, 309
Ginkgo *see* ginkgoes
ginkgoes 27, 134
glacial/interglacial cycles 212, 241, 260, *263*, 276–7, *278*
glacials 212, 216, 219, 276, 278–9
  last 89, 276, *278*, 293
glaciations 29, 162, *163*, 212, 261, 268, 270
  Carboniferous–Permian 262, 265, *271*
  Huronian *260*, 267
  Late Jurassic–Early Cretaceous 269
  Late Ordovician–Early Silurian 262, 269
  Marinoan *see* Varangerian
  Neoproterozoic 29, 256, *260*, 267–9, *268*, 324
  Carboniferous–Permian 29, 269–72, 273
  Pleistocene 269
  Palaeoproterozoic 256, 267
  Phanerozoic 267
  Quaternary 29, 114, 212, 262, 276–9
  Sturtian 256, *258*, 259, 267, 269
  triggers 263, 265–7, 270, 276–9
  Varangerian 256, *258*, 259, 267
glaciers *see* ice, sheets
glands 304, 306, 310
global distillation *see* condensation, cold
global warming 266, *267*, *282*; *see also* greenhouse; hothouse
  potential (GWP) 253, 288, *290*, **291**, 291, 297, *297*
Globigerina 91, 220
Globorotalia 215
  tumida 216
globin 61
globulin 307
Gloeocapsomorpha prisca 197, *198*
  kerogen 197
glossopterids/*Glossopteris* 29, 270
glucans 38
glucopyranose 37
glucopyranosides *100*
glucosamine 35, *36*
  acetyl 40

glucose 1, 15, 18, 35, 37, 37–8, 40, 44, *45*, 47, 62, *65*, 65, 94, 100, 127, 172, 174–5, 238
  C-numbering scheme 36
  phosphate 17
  synthesis and structure 36–7, 36–7
glucuronic acid 35
glutamate/glutamic acid 40–1, *41*, 95, *96*
  synthase *96*
glutamine *41*, 95, *96*
  synthase *96*
glycans 38
glyceraldehyde 32, *34*, 34
  phosphate 35, *36*, 49, *50*
glycerides **44**–7, 52
glycerol 10, *40*, 44, 46–7, *48*
glyceryl unit *46*
glycine *8*, *9*, 9, *40*, *41*, 68, *100*
glycogen 38
glycolipids 44, **46**–7
glycolysis 38
glycoproteins 40, 43
glycosidase 100
glycosides 67, 171
  steryl 171
Gondwana 29, 270, *271*
Graham's Law 93
granitization 255
graphite 2, 7, 126, 132, 142, 146, *147*, 149
grass 28, *73*, 89, 114, 172, *239*, 239
  C_4 241
  cord- 73
  eel- 73
  saw- 113
  sea- 79
  turtle- 73
grassland 27, *73*, 89, *90*–1, 286
greenhouse *262*, 268, 272; *see also* hothouse
  effect 251, 252–3, 254, 260–1, *263*, 265–6, *267*, 268–9, 273, 280, 284, *286*, 286–8, 290–1, 291–2, 297–8
  gases *see* gas, greenhouse
  maxima 269–70
Greenland 7, 19, 276, 284, 293
  ice cores *see* ice cores, Greenland
growth 1, 13, 24, 78–9, 86–7, 98, *113*, 113, 220, 236, 247–8, 260, 269, 279, 292, 305, 307, 309, 319
  rate 52, 100, 207, 221
  specific 238
  seasonal 288
  stages 62, 86, 176, 221, *221*, 305, 319
    death **221**, *221*
    exponential/log 186, **221**, *221*
    lag 220, **221**, *221*
    stationary 186, **221**, *221*
Grypania spiralis 19, 20
guaiacyl units *see* vanillyl lignin units
guaiaretic acid 64
guanine 68, *69*
Gulf
  Arabian 301
  of Mexico 244, 293

Persian 116
  Stream 293, *293*
guluronic acid 40
gutta-percha 60, *62*
gymnosperms *26*, **27**, 52, 59, 64, 167, 172–4, *173*–4, 201–2, *202*, 261, 284, 305
  deciduous 273
gypsum 16, 98, 112, *97*
gyres **76**
  subpolar 85
  subtropical 76, 76, 83, *84*, 85, *85*

Hadean 5, 6, 260, 324
Hadley cells 76
haem **61**, 62, *64*, 102
haematite 20
haemin 182, *183*
haemoglobin 43, 61
hair 238–9
half-life *100*, **214**, 309, 310–11, 317
halite 145
Halobacterium 40
halocarbons 295–8, **297**, *297*
  brominated *see* halon
  greenhouse warming 297–8
  ozone destruction 295–7
halocline **74**, 112
halogens 60, 297
Halomonas 88
halon 297
  1301 *290*, 297
halophiles *see* bacteria, halophilic
Haptophyceae/haptophytes *see* algae, yellow-brown
hardwoods *see* angiosperms
Hauterivian 274
Hawaii 85, *289*
Haworth structure 35
H/C atomic ratio 119–20, 123–5, *124*, 127, *127*, 131, 137, 139, *140*, 142, *143*, 144, 256
HCFC (hydrochlorofluorocarbon) 297
  22 *290*, 297, 297–8
  141b *290*, 297–8
  142b *290*, 297–8
heat
  of accretion 6
  capacity 230, 291
  convection 2, 75
  flow 6, 145, 229–30, 232, 252
    basal *229*
    reduced 229
    surface 228, *229*
    transient 230
  global distribution/transfer 254, 263, 265, 267, 276, 282, 291
  internal/radiogenic 4, 6, 145, 229, 252, 267
  of reaction 43
heating rate 158, 210, 214, 224, 228, *229*, 230, *231*, 232–4
HFC 297
  23 297
Heliobacter 18

helium   2, 3, 5, 20, 140, 149, 279
hemiacetals   37
hemicelluloses   38, 62, 64, 101, 118, 127, 172, 174–5
hemichordates   319
hemiketals   37, 180, 185
Henry's Law constant   308, 316
heptane
  branched   203, 223
  normal   152, 223
  value   222, 223
heptatriaconta-15E,22E-dien-2-one   168
heptatriaconta-8E,15E,22E-dien-2-one   168
heptoses   35
herbicides   313, 313
herbivores   13, 21, 24–6, 25, 81, 91, 100, 112, 179, 236, 239
herbivory   25, 64, 70, 72, 81, 89, 116, 200, 205, 240, 319; see also phytophagy; zooplankton grazing
herbs   89, 114
heteroatoms   30, 31, 33, 128, 138, 151
heteropolysaccharides   38
heterotrophes   13, 13, 15, 22–4, 85, 117, 238
heterotrophy   55, 94, 186, 236, 240, 238–9
Hevea brasiliensis   60
hexane   43, 140
  normal   33, 33
hexachlorobiphenyl   see PCB-153
hexachlorocyclohexane   see lindane
hexaprenol   199, 200
hexoses   34, 35, 48, 118
HFC (hydrofluorocarbon)
  134a   290, 298
  152a   290, 298
  23   298
HI   see hydrogen, index
hibaene   51, 52
higher plants   12, 27, 70, 71, 72, 87, 89, 95, 109–10, 237
  chemical composition   30, 38, 40, 44, 47, 49, 53, 55–7, 56, 59–60, 61, 62, 64–5, 67, 69, 73, 101, 110, 118, 127, 133, 167–74, 173–4, 179–80, 188–9, 191–2, 196–7, 201–3, 224, 238, 301
  contribution to sedimentary organic matter   24, 69, 110, 114–15, 121–2, 137, 139, 152, 166–7, 169–70, 172, 174, 179–80, 196, 201–4, 225, 241, 247
  degradation/diagenesis   100
  evolution   27
  isotopic composition   241
  non-woody   38, 172–4, 173–4
  population variations   87
  productivity   see production, primary, terrestrial
  woody   22, 24, 38, 60, 62, 73, 167, 172–4, 173–4, 201
Himalayas   258
hispidin   64, 67
histidine   41, 101
holocellulose   101
Holocene   71, 113, 113, 166, 276, 278, 291, 324
  thermal maximum   276
Hominidae   12
homopolysaccharides   38, 40
Homo sapiens   see humans
hopanes   55, 135, 153, 153, 160, 160, 192, 193, 193, 196, 200–1, 210, 224
  C-number distributions   205, 224, 225
  C-ring opened   see secohopanes, 8,14-
  $\delta^{13}C$ values   241, 241
  demethylated   see norhopanes
  E-ring degraded   see de-A-hopane
  environmental/source indicators   200–1
  isomerization   193, 201, 225, 227, 209–10, 233
    kinetics   232
  methylated   see methylhopanes
  pentahydroxyamino   241
  ratio to steranes   204
  rearranged   see diahopanes; neohopanes
hopanoic acids   192, 193, 210
hopanoids   11, 52, 55, 136, 167, 192–3, 196, 205–6
  aromatic/aromatization   192–3, 196
  $\delta^{13}C$ values   241
  C-ring opened   see secohopanoids
  diagenesis   191–3, 192, 195, 196
  occurrence and function   52
  oxygen-containing groups   192, 196
  polyols   52, 53, 55, 191–2, 193
  stereoisomerism   192
  structural notation scheme   54–5, 54, 209
hopenes   193, 194
  rearrangement   196, 200
hormones   43, 55, 306
  growth   306
  moulting   306; see also ecdysis
  receptors   306
  sex   304–5, 305, 306, 307, 320
hornworts   12, 27
horsetails   see sphenopsids
hothouse   263, 265–6; see also greenhouse
  Cambrian–Ordovician   269–70
  Cretaceous   265, 272–6
  Devonian–Carboniferous   270
  icehouse oscillations   267, 269
  Permian–Triassic   270
  ultimate   279–80
hot spring communities   see hydrothermal communities
humans   2, 12, 12
humic acids   119, 132
  elemental composition   119, 119, 320
  freshwater   121
  functional group content   120, 320
  marine   119, 121
  soil/terrestrial   119, 120, 132
humic material/substances   119–23, 127, 132, 134, 134
  aqueous   320–1
  chlorination   321
  classification and occurrence   119
  composition and structure   119–21, 120, 127, 320–1
  formation   120–1
  freshwater   93, 119–20
  marine   93, 119–21
  metal chelation   121, 130
  pollutant associations   312–14, 320–1
  precipitation   119, 321
  soil/terrestrial   119–21, 132, 320
  trapping small molecules   121, 313, 320
humidity   242, 272, 279–80
humification   122
humin   118, 119, 121, 132
huminite   122, 124, 127–8
hump   see unresolved complex mixture
hydration   183
hydrocarbons   7, 10, 31, 47, 56, 135, 139, 142, 149, 177, 188, 192, 196, 198, 301–4, 304, 313
  aliphatic   128, 135, 135, 139, 144, 149, 159, 188, 196; see also alkanes; alkenes
  aromatic   30, 128, 135, 139, 146, 149, 150, 151, 151–2, 156, 159, 161–2, 175, 182, 188, 196, 298, 303, 320; see also aromatics
    polycyclic   see PAHs
  biodegradation   302–3
  biomarker   see biomarkers
  C_7 152, 152, 203, 203, 222–4
  deep/mantle source   146
  diamondoid   see diamondoids
  gaseous   128, 140, 142, 144–5, 148–9, 151, 244, 244–5
    isotopic composition   244–5
  gasoline-range   159, 222, 242
  generation   146–7, 147, 148, 153–5, 154, 154, 156–7, 223, 240
  light   128, 146, 148–9, 159, 161, 222, 224
    origin   223–4
  liquid   128, 140, 145, 148–9, 151, 161, 301
  in petroleum/bitumen   123, 132, 128, 135, 146, 146, 149, 150, 151–3, 151–3, 156, 158, 160–2, 232, 298–304
  polycyclic   185
  saturated   see alkanes; saturates
  solid   128
  source/environmental indicators   203
  thermal alteration   207, 211, 222; see also oil cracking
  unsaturated   see alkenes
  viscosity   150, 151
hydrochlorofluorocarbons   see HCFC
hydrodynamic flow   158, 161–2
hydrofluorocarbons   see HFC
hydrogen   2–3, 123
  atmospheric   5, 260, 267, 280
  bond   see bond, hydrogen
  chloride   5, 16, 265, 312
  fluoride   297
  fusion   2, 279
  index (HI)   223
  isotopic composition   236, 240, 242, 243, 244, 276, 276–7
  peroxide   102, 103–4, 108

and reduction 6, 15–16, 17–18, 94, 95, 98,
    98–9, 267, 295
  sulphide see sulphide, hydrogen
  transfer/exchange 18, 153, 155, *155*, 175,
    176, 188, 207, 209, 210, 212, 232
hydrogenation see reduction
hydrological cycle 90, 109–10, 112, 114, 116,
    268–9, 275, 291, 292
hydrolysis 42, 45, *94*, 94, *100*, 100–1, 117–18,
    120–1, 128–9, 133, 166, 172, 177, 179,
    182, 186, 207, 213, *215–16*, 215–16, 238,
    302–3, 311, 312
hydrophilicity 8, **10**, 10, *11*, 46, 132
hydrophobicity 8, **10**, 10, *11*, 46, 55, 160, 238,
    295, 309, 312
hydrosphere 2, 6, 109, 254, 263
hydrothermal
  activity/vents 7, 12, 14–15, 19, 21, 98, *164*,
    228, 244, *251*, 251, 258, 268, 275, 287,
    300
  communities 12, 14–15, 98
  gradient 165
hydroxamate/hydroxamic acid 88, 89
hydroxyacetophenone 173
hydroxyarchaeol 241
hydroxybenzaldehyde 173
hydroxybenzoic acid 173
hydroxybenzylpyruvic acid 303
hydroxycarboxylate/hydroxycarboxylic acid
    88, 89
hydroxycinnamate 62
20-hydroxyecdysone 305, *305*
hydroxyhexadecanoic acid 303
hydroxyhopanone 54, 55
hydroxylation 62, 65, 101, 313–14
hydroxymuconic semialdehyde 303
hydroxyoestrones 320
p-hydroxyphenyl lignin units 64, *66*, 102, *173*,
    173–4
hypersalinity 72, 72, 112, 116, 205–7
hyperthermophiles see archaebacteria,
    hyperthermophilic
hypolimnion 74, 78, 83, 94
Hypoxylon 101

ice 5, 72, 77–8, *90*, 100, 110, *164*, 165, 252,
    260; see also permafrost
  ages see galcials; glaciations
  albedo see feedback, ice-albedo
  bergs 293
  caps/polar *90*, 107, 218–19, 263, 267, 270,
    271, 272, 276, 276–7
  cores 276, *289*
    Antarctic 276, 277–8, *278*
    gas/air bubbles 276, 278, *278*, 286, 288
    Greenland 276
    Vostok see ice core, Antarctic
  crystals/structure 165, 295
  house 262, 263, 265–6, 269, 273
    Carboniferous–Permian 29, 269–73
    –hothouse oscillations 267, 269
  δD record 276, *276*
  δ^{18}O record 276, *276–8*, 278

lakes 77
Man see Neolithic Ice Man
melting/meltwater 265–6, 272, 277, 293
sea- 72, 75, 278, 293
sheets/continental 29, 107, 218–9, *263*,
    265–6, *267*, 268, 270, 277–8
  Laurentide 293
  volume 216, 219, 277
Iceland 293
igneous activity 99, 210, 228, 230, 300
imposex 310
incineration 311, 315–16
indenopyrene 299–300
Indian monsoon 219
indole-3-carbinol 306, *307*, 320, *320*
inductive effect **60**
induration 112
industrialization 246, 254, 273, 276, 285,
    287–8, 292, 299, 311
inertinite **122**, *122–4*, 123, 126, *126*, 127, 131,
    136, 139, *142*, 142, 148
inertodetrinite 122
infrared; see also radiation, long-wave
  detector 223
  spectroscopy 137, 142
ingestion 309
insecticides 49, 295, 311–13
insects 22, 38, 67, 305, 311
insolation 13, *25*, 74, 75, 77, 83, **252**, 267
  variation 263, 264, *260, 264*, 265, *270*, 277,
    279, *279*
insulin 43
interglacials 212, 216, 219, 270, 276, 278, 293
  latest see Holocene
interstadials 216, 219, **276**, 279
Intertropical Convergence Zone see ITCZ
invertebrates 13, 25, 21, 112, 118, 127, 170–1,
    179, 205, 216, 305, 306, 310–11, 317, 319
iodine 297
ions 72, 111, 248–50, **258**; see also under
    individual ions
  ferric 19, 88–9
  fragment 141
  hydrogen see protons
  hydride 154, 175
  hydroxyl 17, 81, 111
  metal 19, 121, 135, 208
  molecular 141, *300*
  negative see anions; carbanions
  positive see cations; carbocations
iron 3, 98, 99, 108, 135, 146, 161, 205, 256, 278
  chelation 89, 88, 182, 320
  cycle 13, 14, 87
  dissolved 20–1, 88–9, 268
  hydroxyoxides *80*, 87, 87, 89, 108
  as a micronutrient 19, 79, *80*, 88–9
  supply 86, 275, 278–9, 292
  nickel alloy 2, 5
  organic 61, *80*
  oxidation 6, 18–9, *22*, 16, 23, 23, 81, 256,
    268
  oxides 6, 20–1, *252*
  sulphide see pyrite

isoarborinol 53
isobars 249
*iso*butane 149
*iso*butene 10
Isochrysis galbana 220
isoflavone 65, 306–7, *307*
*iso*heptane value 222–3
isoleucine *41*, 168, *215*
  isomerization 212–13, *213*, 215–16,
    215–16
  ratio to alloisoleucine see alloisoleucine :
    isoleucine ratio
  position in peptide chains 215
isololiolide *185*, 185
isomerization 35, 136, 169, 214, **175**, 186, 189,
    190, 193, 196; see also under individual
    compounds; stereoisomerization
  of methyl groups see methyl group,
    migration
  rate 214
  and steric hindrance 176
  and thermal maturity 207–10, 212–3, 216,
    226–7
  alternatives to 209–10
isomerism 52, 170, 188–9, 193, 203, 207–12,
    311, 313; see also stereoisomerism
*iso*pentane 148–9
*iso*pentenol 10
*iso*pentenyl pyrophosphate 49, *50*
isopimarane 201, *202*
isoprene unit 10, **49**, 60, 238
  δ^{13}C values 238
isoprenoids 60, 205, 207, 238; see also alkanes,
    isoprenoidal; alkenes, isoprenoidal
  acyclic **49**, 52, *160*, 160, 183, 200, 203
  sources 197–200
  stereoisomerism 206, 207
  δ^{13}C values 238
isorenieratane 182, *184*, 185, 275
isorenieratene 58, 116, 183, 237, 275
isostasy 263
isotherms 229
isotopes 4, 4, 6, 217
  radioactive/radiogenic 4, 6
  stable 4, 8, 217
  unstable 4, 145, 217, 258
isotopic fractionation 4, 8, 20–1, 217–19, 234,
    234–6, *235–6, 239–40, 243–4*, 254, 268,
    276, 281; see also under individual
    elements/compounds
  aqueous equilibria 235–6, *240*
  autotrophic 236–8
  bicarbonate–calcite differences 240
  bicarbonate–CO_2 difference 240
  in biosynthetic pathways 235, 238
  and bond strength 155, 240
  calcite/carbonate–kerogen difference *240*,
    258–9, *280*, 281
  chemosynthetic 236
  calcite–CO_2 difference 235, *235–6*, 240,
    240, 256, 258
  chemosynthetic *259*, 260
  during diffusion 4, 161, 236–7, 245, 271

isotopic fractionation (*cont.*)
   heterotrophic   235, *236*, 238–40, *239–40*,
      256, 258–60, *259*
   human dietary   238–9, *239*
   and mass balance   *235–6*, 254, 261, 263, 283
   methanogenic   236–7
   methanotrophic/methylotrophic   *236*, 238,
      241
   during petroleum generation   155, 240, 242,
      244–5
   photosynthetic   4, 234, 236–7, *236–7*, *240*,
      240, 256, 258–60, *259*, 271, 271–2, 280
   source indications   234
   during sulphate reduction   22, 161, 241, 243
   temperature dependence   218–19, 235, 240,
      256, 272
isovaline   *8*, *9*
Isua   19, 254, *255*
ITCZ (Intertropical Convergence Zone)   75,
   90

Jupiter   5
Jurassic   *26*, *28*, 115, 156, 162, *163*, 231, *233*,
   258, *262*, 274–5, 324

kaolinite   186
kaurane   201, *202*, 209
kaurene   *51*, 52
kelp   *59*, 73, *73*, 79
keratin   42, 100, 117, 238
kerogen   118, 121, **132**–44, *134–5*, 145, 156,
   162, *164*, 164, 166, 196, 223, *240*, *251*,
   254, *255*, *259*, 268, *281*
   amorphous   133, 136–7
   Archaean   256
   Campins   138
   – carbonate C ratio   246, 254, 256, *257–8*
   classification   137–9, 141, *142*; *see also*
      kerogen, type
   chemical composition   *133*, 133, 136–7,
      137–8, 142, 144, 144–6, 152, 155, *158*,
      166, 182, *202*, *206*, 225–6, 232, 256, 260;
      see also kerogen, type
   formation   118–19, 132–6, *134*, 139, 147,
      197, 201, 240, 268
   H-rich   122
   humic   137, *147*
   inert   **148**, *149*, *231*
   isotopic composition   146, 155, *240*, 240–3,
      245, 254–6, *255*, *257–9*, *281*, 280, 284
   labile   **148**, *149*, 157, *158*, *231*, *243–4*, 245,
      260–1
   oxidation   137, 251, 261–2
   oxygen content   137–9, 142, *144*, 144, 223
   petroleum generation   147, 149, 154, 197,
      200, 203, 207, 210–12, 222, 224, 228,
      302; *see also* petroleum generation
   Phanerozoic   256
   Proterozoic   256
   pyrolysis   *see* pyrolysis
   reactive   **148**, *231*
   refractory   **148**, *149*, *158*, *231*, *243–4*, 245
   Ribesalbes   138

sapropelic   137, *147*
S content   135–9, 144
S-rich   135, 137, 139, 152, 155, 161
structure and maturation   126, 132–6, *137*,
   138, 141–2, *143–4*, 144–6, 148–9, 152–3,
   155, 159, 161–2, 211–12, 222, *225*, 225,
   227–8, 230, 232, 244, *255*–6, 261; *see also*
   kerogen, transformation
transformation   227, *229*, 230, *231*, 232, *240*
   kinetics   228, *229*, 230–2, *231*
   ratio   *see* transformation ratio
trapping small molecules   137, 148, 153, 156
type   137–9, 141, 157, 223, 227–8, 230, 261
   I   126, **137**, 137–8, 139, *140*, *142*, 144–6,
      146, 155, 230, *231*
   I-S   138
   II   **137**, 137–8, 139, *140*, *142*, *142–3*,
      145–6, *146*, 155, 205, 226, *231*, 231–2,
      244, 245
   II-S   **139**, 155, 227, *231*, *231*
   III   **137**, 137–8, 139, *140*, *142*, 142, 144–6,
      146, 155, *225*, *226*, 227, 230, *231*, 232,
      244, 245
   III-S   139
   IV   **137**, 139, *142*
volatiles   135, *255*
weathering   250–1, 262
kerosine   150
ketoadipic acid   *303*
ketocarboxylic acids   95
ketoglutarate   96
ketohexoses   35, *37*
ketones   *31*, 35, 37, 47, 186, 191, *192*
   long-chain, unsaturated   136, *168*, 212
   degree of unsaturation   216, 219–21
2-ketopentenoic acid   *303*
ketoses   35, 37, 38
kinetic
   isotope effect   4, 9, 161, *235*, 236, *244*, 245,
      287
   modelling   227, *229*, *231*
   parameters   214, 227, 227–8, 230–2
kingdoms (taxonomic)   12, 12, 14–15
Kraft mill effluent   *see* paper pulping
Krebs cycle   38
krill   85
Kuiper Belt   5
kukersites   198

labdane   201, *202*
labdatriene   60
laccase   102
lactate   95, 98, 169
lactobacillic acid   *168*, 170
Lactobacillus   94
Lake
   Baikal   83
   Chad   109
   Ellesmere   241
   Kivu   109
   Maracaibo   116
   Salt   109
   Tanganyika   83

lakes   73, 109–10, 112, 248, 276; *see also*
   environments, lacustrine
   anoxic   112
   classification   74, 77
   dimictic   77, 77–8, 86
   ephemeral   109, 112
   evolution   138
   eutrophic   84, 94, 106, 292
   freshwater   77, 79, 83, 99, 110, 198–9, 203,
      241
   glacial   74, 109
   high-latitude   77
   hydrologically closed   **109**, 112
   hydrologically open   77, **109**, 110
   hypersaline   205
   meromictic   83
   morphology   109
   oligomictic   83
   oligotrophic   83–4
   ox-bow   109
   pluvial   109
   polar   77
   polymictic   83
   saline   112, 201
   salt   205
   size/morphology   83
   soda   14
   stratification   *see* water, column, stratification
   stratified   83, 109–10, 112, 292
   tectonic   109
   temperate   77, 83
   tropical   77, 83–4
   volcanic   109
   warm monomictic   83
Laminaria/laminarians   73, 79
land
   fill   287, *287*, 311, 313–14
   use   263, 287, 292
lanolin   47
lanosterol   47, 53, *56*
Late Heavy Bombardment   **6**, 10
lateral transfer   *see* gene swapping
latex   59
Laurasia   29
lauric acid   *44*
LD50 test   316
leaching   20, 100, 119, 127, 213, 216, 249, 292,
   299, 301, 311
leaves   27, 47, 64, 67, 69, 101, 114, 118, *122*,
   123, 127, 139, 167, 169, 172, 179, 188,
   241–2, 271
Lebistes reticulatus   311
Le Chatelier's Principle   146, 147
lecithin   46, *46*
Lepidodendron   27, 29
leucine   *8*, *9*, *41*, 168
levopimaric acid   188
Libertella   101
Lichenes/lichens   12, 67
life   2, 10, 16, 254
   conditions for   2, 6, 10, 14, *16–17*, 17, 18,
      71–3, 79, 98, 100, *134*, 141, 266, 273,
      279–80, 295

diversity   15, 22, *28*, 28–9, 70, 72, 89, 167, 269
evolution   1, 7, 10–13, 10, 15, *16*, 20–9, *26*, 68, 70, 167, 219, 237, 254, 269
  origin   5–7, 10–13, 14
lignans   64, 306
light   16, 17, *17*, 18, 24, 61, 71, 80–1, *91*, 263, 267, *308*; see also radiation, solar; insolation; solar luminosity
  intensity   78, *78*, 167, 220, 292
  and photosynthesis   see photosynthesis, light limitation
lignin   22, **62**, 64, 69, 99, 118, 121, *129*, *133*–4, 174, 211, 244
  angiospermous   101, *124*, 128–9, *174*
  beechwood   66
  contribution to sedimentary organic matter   99, 101, 117–21, 127–31, *130*, *133*–4, 139, 142, 145, 148, 173–4
  degradation   101–2, 118, 127–9
  diagenesis/diagenetic products   64, 117, 120, 128–31, *130*, 139, 145, 148
    and morphology   64, 122, 123, 130–1
  formation   62
  grass-   174
  gymnospermous   101, *124*, 128, *130*, *174*
  linkages   101, 102, *104*, 129
    β-O-4 linkage   64, 102, *103*, 128–9, *129*, 131
  model compounds   102, *103*–4
  occurrence and function   30, 42, 62
  oxidation   *103*–4, 108, *173*–4, 173–4
  peroxidase   see ligninase
  pyrolysis   174
  structure   64, *66*, 101, 128–31, *130*, 142, 145
  units/precursors   59, 62, 64–5, 102, 167, 169, 173, 238, 241; see also p-hydroxyphenyl units; syringyl units; vanillyl units
    biosynthesis   62, 65
    δ¹³C values   238
    numbering system   66
    source indicators   169, 173–4, *174*
ligninase   102, *103*–4
lignite   122, **125**, *125*, 127–9, 127–30, *163*, *225*; see also coal, brown
  A   129
  B   129, *130*
limestone   159
limonene   *51*
lindane   313
linoleic acid   44
linolenic acid   44
lipids   43–62, *48*, 92, *94*, 101–2, 110, 118, 121, 125, 127–8, *134*, 166, 175, 197, 207, 237–8, 241, 317
  bound   40, **166**, 182
  compositional variation in organisms   30, *69*, 69, 167–9, 186, 212, 219–20, 319
  environmental influence   69, 219–21
  contribution to sedimentary organic matter   69, 132, 135, 137
  diagenesis/degradation   118, 133–5, 153, *178*–85, *187*, *189*–95, 220
  in sediments   117, 179–96
  in water column   175–9
  ether-   46–**7**, *48*, 52
  free   **166**, 182
  isotopic composition   236, 238
  occurrence and function   43–62
lipophilicity   295, 306, 307, **308**, *308*, 309, 310–11, 314, 317, 319
lipopolysaccharides   40, 47, 170
lipoproteins   44, **55**
liptinite   122–3, **123**, 124, *126*, 136–7, *142*, 142, 148
  fluorescence   123, *123*, 125, *126*, 221
liptodetrinite   122
lithification   1, 117, 247, 249–50, *285*
lithology   230
lithosphere   2, **2**, *2*, 107, *107*, 229, 247, 250, *251*, 254, 263, 265, *285*
  oceanic   107
  stretching   see extension
lithotrophes   see bacteria, lithotrophic
lithotype   123
liverworts   *12*, 27
loliolide   *180*, *185*, 185
log K_ow   see partition coefficient, octanol–water
long-chain ketones   see ketones, long-chain
lupane, A-ring degraded   see de-A-lupane
lupanoids   52
lupeol   52, *53*, 201
lutein   56, *58*, *185*, 185
lycopadiene   199
lycopane   182, *184*, 197
lycopene   57, *58*, 185, 197
lycopods/lycopsids   27, 29, 59, 270
Lyngba   112
lysine   *41*
lysis   *91*
lysocline   **219**, 278, 283
lyxo-5-hexosuluronic acid   see mannuronic acid, keto-
lyxose   172

Maastrichtian   272, *274*
macerals   **122**, *122*–4, 123–5, *126*, 127–8, 131–3, *133*, 136, 139, *142*, 145, 221
  microscopy   122, 123, 126
macrinite   *122*, 123
Macrocystis   79
macromolecules   92, 118, 120, 128–9, *134*, 135, 137, 139, 142, 197
  bio-   61, *100*, 100, 117, 132–4, *134*, 167, 203
    resistant/refractory   101, 106, 118–19, 121, 132–5, *133*
  decomposition   94, 100–2
  geo-   118–19, 132, 134, *134*
  polymethylenic   49, 108, *133*, 133–4, 148, *198*, 244
  polyphenolic/polyalkylphenolic   62, *130*, 133
  S-rich   134, *134*
magmatism   107, 228, 244, 255, 265, 273

magnesium   3, 5, 30, 61, *72*, 79, 80, 111, 177, 250, 268, 280
  carbonate   2
maiden cane   114
maize   239
malonyl unit   44
maltenes   128
Mammalia/mammals   *12*, 56, 295, 304–5, 310, 314
manganese   93, 95, 108, 182, 320
  peroxidase   102
Mango parameters   *203*
mangroves   73, *73*, 84, 105, 114–15
mannans   38, 101
mannose   *36*, 38, 40, 172, *173*, 175
mannuronic acid   40
  keto-   101
mantle   **2**, *2*, 5, 145, *251*, 254, *255*, 258, 268, 287
  hot-spots   107
  oxidation   5, 21, 267
  outgassing   5, 254, 283
  plumes/superplumes   265, 273
marine
  incursions   204, 243
  regressions   162
  transgressions   83, 112–13, *113*, **115**, 162, 203–4, 243, 266, 268, 275
marl   110
Mars   5, 7
marshes   73, 114
mass
  balance modelling   21, 235, 254, 261, 263, 271, 283
  – charge ratio   141
  extinction   see extinction events
  number   3, **4**, 236
  spectrometry   4, 141
maturation   99, **117**, 188, 196, 224
  of coal/kerogen   122, 123, 131, 136, 138–9, 142, 145–6, 148–9, 153, 196, 211–12, 221, 227–8, 230, *231*, 232, 240, 244, 255–6, 261, 300
  of petroleum   161, 242, 245
  temperature/time control   227
maturity   125, 127, 128, 129, 137, *137*–8, 139, 140, 142, 144, 145, 146, *147*, 148–9, 152, 155, 161, 187–8, 196–**7**, 208, 212, 221–8, 223, 225, 229–30, 232, *243*
  indicators/parameters   *225*
    bulk organic matter   221–2, 224
    molecular   207–12, 222–7, 225–6, 232–4, *233*
    optical   126, 221–2, 225–6, 229
    pyrolytic   222, 223, 225
  and isotopic changes   155, 240, 242, 254–5
Maud Rise   282
Mauna Loa   289
Mediterranean flora   90
melanoidins   101, **118**, *118*, 121
menthol   49, *51*
mercaptan   see thiol
mesopause   296

Index

mesophyll 204
Mesoproterozoic 324
mesopyrophaeophorbides 180, *181*
mesosphere *296*, 297
Mesozoic 26, *26*, *28*, 29, 162, *201*, *258*, 261, 324
metabolic processes 4, 6, 10–12, 43, 49, 62, 69, 94, 100, 177, 179, 235–6, 241, 295, 301, 304, 306, 310–11, 315, 317, 319–20; *see also* respiration
metabolites 311, 313–14, 319–21
metagenesis 126, *140*, **142**, 144–5, *147*, 148, 221
metal 2, 20, 99
 chelation *see* chelation, metal
 heavy 121
 oxidation state 19, 208
 transition 223
metalimnion 77
metamorphism 2, 7, 132, 142, 251, *251*–2, 254–5, *255*
metazoa 21, *16*, 21, 180, 269
meteors 5
 Perseids 5
meteorites 5, **5**, 7–8, 10, 12
 impacts 8, 28
 Murchison 8, 9, *8*, 10
meteoroids 5
methanal *see* formaldehyde
methane 5, 99, 127, 135, 146, 149, 156, 157, 161, *164*, 165, 261, 267–8, 283–4, *284*
 atmospheric 5, 7–8, 244, *247*, *253*, 261, 266–7, *267*, 269, 270, 276, 277, 279, 283, 285, *286*, 287–8, *289*–*90*, 291
 life-time 266, 269, 287–8, *290*, 291; *see also* adjustment time
 seasonal variation 288
 coal-bed 131, *136*, 165
 American 165
 greenhouse warming 253, 261, 266, 269, *286*, 286–7, *290*, 290–1, 291
 hydrate *164*, **165**, 266, 283
 continental/permafrost 165, 266, *267*, 268
 Gulf of Mexico 165
 marine sedimentary 165, 266, *267*
 stability/destabilization 165, *263*, 266, 267–8, 268–9, 282–4, 291
 isotopic composition 155, 161, 165, 237–8, *243*–4, 244–5, 269, 283, *284*, 287, 287
 oxidation 6, *94*, 98–9, 161, 241, 244, 266, 269, *270*, 283, 286–8
 oxidisers *see* bacteria, methanotrophic
 photolysis 21, 267, 288
 sinks 266, 269, 287, 291
 sources *243*–4, 244–5, *287*, 287–8
 abiogenic *243*, 244, 287
 anthropogenic 283, 287, *287*, 291, 295
 biogenic 98, *94*, *140*, 146, *147*, 148, 165, 238, *243*–4, 244, 247, 266, *270*, 279, *287*, 287
 methanogenic *see* biogenic
 thermogenic 139, *140*, 142, 144–5, *147*, 147, 148, 155, *243*, 244
 sulphonate/sulphonic acid 266, 278, *278*
Methanobacillus 98
Methanococcus 98
methanogens *see* archaebacteria, methanogenic
methanogenesis *94*, 95, 99, 146, *147*, 148, 165, 182, 205, *236*, 244–5, 269, *270*, 287
methanol 95, 140
methanotrophes *see* bacteria, methanotrophic
methanotrophy *94*, *236*, *243*, 255, 269
methionine 41, 97
methyladamantane 160
methylalkanes
 2- *see* alkanes, *iso*-
 3- *see* alkanes, *anteiso*-
methylation 62, 212, 226
8-methylchroman 205
24-methylcholesta-5,24(28)-dienol 168
methylcyclohexane *152*, 203, *203*, 223
methyldiamantane 160
4-methyldiasteranes 188, *189*
4-methyldiasterenes 188, *189*
4-methyl diasteroids
 aromatic *189*, 210
methyldibenzothiophene *211*, 212, 226
 ratio (MDR) 226–7
7-methyl-3′-ethyl-1,2-cyclopentanochrysene 194
methyl group 60, 98, 131, 142, 144, 151, 184, 188, 210–12, 226, 238
 migration *154*, 186, 196, 203, 211–12, 225–7
 1,2-shift *154*, **196**
methylheptadecanes 197
10-methylhexadecanoic acid 168
methylhexanes *152*, 223
methylhopanes 201
 2- 18, 201
 3- 55
2-methylhopanoids 201
1-methyl-7-*iso*propylphenanthrene *see* retene
methylnaphthalenes 211
15-methyl-27-nor-17-hopanes *see* diahopanes
Methylomonas 40, 98, 241
methylotrophes *see* bacteria, methylotrophic
methylotrophy 94
methylphenanthrene *211*, 211–12, 226
 index (MPI1) 225–6, **226**
 isomerization 212, *225*–6
 ratio (MPR) *225*, 226–7
methylphenols *see* cresols
methylsilicone 140
4-methylspirosterenes 188, *189*
4-methylstanols 188, *189*
4-methylstenols 188, *189*
4-methylsteradienes 188, *189*
4-methylsteranes *189*, 200
4-methylsterenes 188, *189*
 rearrangement 188, *189*
4-methylsteroids 188
 aromatic *189*, 210

4-methylsterols *171*, 188
 diagenesis 188, *189*
 reduction *189*
 sources 172
methyltetradecanoic acid *168*
mevalonate/mevalonic acid 49
 pathway 49, *50*, 236
micrinite *122*, 123
microbial
 degradation *see* biodegradation
 growth stages 86, 186, 220, 221, *221*
 mats 7, 16, 72, *110*, 112, 115, 201, 205, 256, 269
Micrococcus 96
Microcoleus 112
microfracturing 156–7
microlithotype 123
Microsporidia 14
mid-ocean ridges *see* spreading ridges
Milankovich cycles *263*, 263–4, 264, 274, 276–7
 eccentricity 263, *264*, 264, 277
 obliquity/tilt *264*, 264, 277
 precessional *264*, 264, 276–7
mineral
 inorganic 23, 71, 100, 105, 112, 114, 122–3, *135*, 136–7, 146, 156, 250
 surface 10, 43, 92, 106, 108–9, 156, 159, 161–2, 175, 188, 304
mineralization 93, *94*, 94, 99, 106, *109*, 127, *134*, 182, 256, 261, 313; *see also* biodegradation
Miocene 28, *231*, 324
mires 73, 101, **112**, *263*, 248
 boreal 114
 coal/peat-forming 24, 27, 29, 70–1, 112–14, *113*, 118, 122, 127, *142*, 162, 182, 270, 272
 evolution 114
 ombrotrophic **114**
 planar 114
 raised 113, 114
 rheotrophic **114**
 temperate *113*
 tropical *113*, 114
 types 114–15
Mississippian *163*
 – Pennsylvanian boundary 270
mitochondria 15, 238
Modern fauna *28*, 28
molecular
 palaeontology 69, 135, 166–7
 structure 30–5, 33, *33*, 38, 42, 49, 55, 62, 64, 68, 314, 319
molluscs 38, 100, 170, 212, 216, 305, 310
Monera 12
Monochrysis lutheri 186
monoglycerides 44
monosaccharides **35**, 34, *36*–7, 37–8, *39*, 117, 169, 175
 deoxygenated 172
 methoxy 172
 ring formation 37

Index

source indicators   40, 172, *173*, 174
stereoisomerism   36–7, <u>37</u>
monoterpenoids   *49*, **49**, *50–1*, 188
montmorillonite   186
Montreal Protocol   297
Moon   5–6
moors   113, 248
moretanes   **209**, *225*
mosses   <u>12</u>, 27, 101, 113–14, 170; *see also* bryophytes
moulting   *see* ecdysis
mountains   *see* orogeny
muconic acid   303
mud   *110*, 115, 122, 138, 301
   flats, tidal   27, 73
   stone   *113*, 115
murein   *39*, 40, 47, *134*
mussels   303–4, 310
mutarotation   *see* anomerism
Mya arenaria   310
mycelium   168
mycobacteria   47
mycolic acid   47
Mycophyta   *see* fungi
mycotoxin F2   *see* zearalenone
myrcene   *51*
myristic acid   *44*
Mytilus edulis   303, 310
Myxomycota   *see* slime moulds

NAD/NADP/NADPH   *see* nicotinamide adenine dinucleotide
naphthalene   67, 152, *303*
   1,2-diol   303
   methylated   212, 227
   numbering system   *211*
naphthenes   <u>128</u>, 139, *142*, 144, **149**, *150–1*; *see also* alkanes, cyclic
naphthenoaromatics   151
naphthobenzothiophene   152
natural products   2, 30, <u>31</u>, 32, <u>33</u>
neoabietic acid   188
Neogene   71, 100, *163*, 276, 324
   cooling   259, 276
neohopanes   200, 224, *225*
neohopanoids   <u>55</u>
Neolithic   288
   Ice Man   *239*, 239
Neoproterozoic   256, *258*, *260*, 268, 270, 267–9, 324
Neptune   <u>5</u>
Nereis diversicolor   310
Nernst equation   <u>81</u>
neutrons   3, <u>4</u>, <u>217</u>
Newfoundland   <u>293</u>
New Zealand   *281*
nickel   <u>2</u>, <u>5</u>, 149, *181*, 182, <u>208</u>, *225*, 320
   sulphide   *see* sulphide, nickel
   to vanadium ratio in oils   *208*
nicotinamide   68
   adenine dinucleotide (NAD)   68, **68**, 80, 310
   adenine dinucleotide phosphate (NADP)   <u>17</u>, *17*, 67, 68, *68*, *96*, 302

reduced form (NADPH)   <u>17</u>, *17*, 49, 56, *80*, *96*, *302*, 310
nitrate   *80*, 79, 80, 94, *94–5*, <u>96</u>, *134*, 250
   dissimilatory reduction   *see* denitrification
   fertilizers   296
   isotopic composition   *see* nitrogen, isotopic composition
   reducers   *see* bacteria, nitrate reducing
   reductase   <u>96</u>, 96
   reduction   94, *95*, <u>96</u>
      assimilatory   <u>96</u>
      dissimilatory   *see* denitrification
   sources/supply   86, <u>87</u>, 220, 240, 292, 294
nitric oxide   *see* nitrogen (II) oxide
nitrification   **96**, 96
nitrifiers   *see* bacteria, nitrifying
nitrite   *80*, 80, *95*, <u>96</u>
Nitrobacter   <u>96</u>
nitrogen   3, 74, *80*, 86, *95*, 95, <u>96</u>, 99, 101, 127, 149, 165, 286
   atmospheric   *4*, 16, 240, 260
   base   *46*, 67, 68
   cycle   *94–5*, <u>95–6</u>, 99
   dioxide   296
   fixation   86, <u>88</u>, <u>95</u>, *95*, 278, 292
   fixers   *see* bacteria, nitrogen-fixing; cyanobacteria, nitrogen-fixing
   inorganic   292
   isotopes   *2*, <u>217</u>
   isotopic composition   <u>9</u>, 238–40, *239*, 275
   organic   30, <u>31</u>, 40, 42, 67, 79, 93, *95*, <u>96</u>, 119, 121, 123, 137, 148–9, 151–2, 176
   (I) oxide   *95*, 127, <u>253</u>, *253*, 286, *286*, 290, <u>291</u>, 296–7
   (II) oxide   *95*, <u>147</u>, 296
   oxides (NO$_x$)   *80*, <u>96</u>, <u>150</u>, 295, 296
   ratio to phosphorus   *see* Redfield ratio
   in volcanic emissions   *5*
nitrogenase   *80*, <u>95</u>
Nitrosomonas   <u>96</u>
nitrous oxide   *see* nitrogen (I) oxide
NMR   *see* nuclear, magnetic resonance
Nocardia/nocardiae   47, <u>96</u>, 301
nonachlorobiphenyl   *see* PCB-206
norabietatriene   *190*
norcholestane   200, 204
   ratio   200
norhopanes   200
   25-   206
   30-   204–5
   δ^{13}C values   *241*, 241
norisopimaroids   *202*
North Atlantic
   Drift   87, <u>293</u>, *293*
   igneous province   282, 283
North Slope   *231*
NSO compounds in bitumen/oil   <u>128</u>, 135, 149, <u>150</u>, *150*
Nucella lapillus   *see* dog whelk
nuclear
   fusion   2–4
   magnetic resonance (NMR)   92, 120, 137, 144

nucleic acids   10–12, **68**
nucleophiles   <u>60</u>, 102
nucleotides   14, **67–8**, *68*, 80, 172
nutrients   29, **79–83**, <u>85</u>, 86, 92, 104, 112, 114
   biolimiting   79, *80*, 80, 84, *85*, 86–7, <u>221</u>, 275, 279, 292, 294
   and eutrophication   83, 86, 292
   macro-   79, 86
   micro-   19, 79, 278
   and phytoplanktonic blooms   19, 80–1, 83, 86, 280, 292, 294
   sources/supply   27, 80, 83, 86–7, 105, 114–16, 125, 162, 167, 176, 220–1, 249, 256, 275, 278–80, 286, 292, 313, 319
Nuttallides truempyi   282

O/C atomic ratio   119, 123, *124*, 125, *127*, 137, *140*, 142, *143*, 144
Ocean
   Atlantic   <u>75–6</u>, *75–6*, 81, *82*, 84, <u>85</u>, 114, 203, *216*, <u>219</u>, 220, *222*, 228, *249*, 272, 272–5, 292, *293*, 315
   Iapetus   267
   Indian   75, <u>75–6</u>, *84*, *249*, 272
   Pacific   14, <u>75–6</u>, *75–6*, 80–1, *82*, 84, 84, <u>85</u>, 86, 92, <u>218</u>, <u>219</u>, 249, *249*, 267, *272*, 273, 275, *278*, *281*, 292
   Panthalassa   270, *271*
   Southern/Antarctic   <u>75</u>, *75*, 86, 220, *278*, 278–9, 292
   Tethys   270, *271–2*, *272–4*
oceans   *16*, 79, 87, *90*, 92, *247*, 248, 250, *251*, 253, *255*, 256, <u>258</u>, *260*, 263, 270, 278, *284–5*, 287, 311, 317
   anoxic events (OAEs)   115, 273, *274*, **274–6**
   basins   *see* basins, oceanic
   circulation/currents   <u>21</u>, 29, 78, 83, 87, 105, *252*, 254, 265, 267, 270, 275–6, 291, *308*; *see also* current
      halothermal   275
      deep   74, <u>75</u>, *75*, 80, 105, <u>217</u>, *250*, 263, *274–6*, *279*, 282–3, <u>293</u>
      meridional   263, 265, 267, 282
      surface   <u>75–6</u>, *76*, <u>217</u>, *263*, <u>293</u>, *293*
      thermohaline   <u>75</u>, *75*, 115, 275–6
   equatorial   *82*
   evaporation zones   *see* evaporation zones
   floor zones   79, *79*
   heat adsorption   291
   high latitude   80, 83, 105
   low latitude   80, 278
   mid-latitude   80, 292
   mixing   283, 288, <u>291</u>
   nutrient status   86, 292; *see also* seawater, nutrient supply
   polar   83, *82*, 266
   redox conditions   *16*, 19, <u>22</u>, <u>87</u>, *87*, *109*
   respiring   281, *281*
   Strangelove   280–1, *281*
   stratification   *see* water, column, stratification
   surface mixed layer   *see* surface mixed layer
   temperate   83, 86, <u>219</u>, 291
   temperature   *see* water, temperature

oceans (*cont.*)
    tropical   74, 265, 268, 282
    trenches   79; *see also* subduction
    ventilation   105, 279
    water   *see* seawater
octanol   307–8, 310–12, 314, 317, 319
odd-over-even predominance (OEP)   47, **167**, 169–70, 180, 197, 204, 224
oestradiol   *305*, 306–8, *308*, 314, 319–20
oestrogen   304–8, **306**, 307–8, 310, 320
    activity   304–8, 313–14, *320*, 320
    equivalents (EQ)   320, *320*
    natural   305–7, 320
    synthetic   306–7
oestrone   *305*, 307, *308*, 319
OI   *see* oxygen index
oil   2, 128, 149, 156, 159, 160–2, 226; *see also* bitumen; fossil fuel
    alteration   159–62, 164; *see also* oil, biodegradation
    aromatic-asphaltic   *150*, 161, 164
    aromatic-naphthenic   *150*, 161, 164
    aromatic-intermediate   150, *150*, 161
    biodegradation   152, *160*, 159–62, 227, 242, 301–2, *304*
    burning   291, 298
    chemical composition   141, 146, 149, 150, *141*, *147*, 148–9, *150–3*, 155, 159–60, 200–1, 211–12, 301–2, 304, 320; *see also* biomarkers in bitumen/oil
    classification   150, *150*
    coal sourced   149, 152, 204, *225*
    conventional   150, 163–4, 163–4
    correlation with sources/other oils   197, 200, 222, 225, *242*, 242, 301
    cracking   *147*, 148–9, 155, 157, 161, 197, 222, 231, *231*, 242, *243*, 245
    de-asphalting   *see* asphaltenes, precipitation
    density   150, 157, *164*
    distillation fractions   150
    essential   49, *122*
    expulsion   *see* petroleum, migration, primary
    fields   139, 164
    generation   117, 126, 128, *140*, 142, 146, *146*–7, 148, *149*, 152, 155, 157, *158*, 159, 209, 224, 228, 223, 231–2, *231*, *233*; *see also* oil window
    temperature   148, *149*, 224
    gravity
        API   150, 161; *see also* oil, density
        specific   *see* oil, density
    heavy   150, 163–4, 158–9, 163–4
    isotopic composition   *160*, *242*, 242–4
    lacustrine   149, 164, 204
    light   158–9
    marine   149, 150, *150*, 164, *242*
    metals content   149, 208
    migration   *see* petroleum, migration
    paraffinic   146, 150, *150*; *see also* oil, waxy
    paraffinic-naphthenic   150, *150*, 161
    peppermint   49
    pollution   301–4, 306, 320
    potential   108, 138, 144–6, 148, 232

pour point   150
reserves   *163*, 163–4
sand   164
seeps   300–1
shale   109, 122, 133, 138, *163*, 164
solubility
    in gas   148
    in supercritical fluids   156, 157
source rocks   30, 108, 162, *163*, 164, 197, 200, *201*, 204, 204–5, 207, 208, 245
    major depositional periods   164
S content   150, 155, 161, 208
stringers   156–7
terrestrial/terrigenous   150, *150*, 204, *242*
thermal alteration/stability   *see* oil, cracking
uses   150
vegetable   52
water washing   161, 227, 242
waxy   146, 149; *see also* oil, paraffinic
weathering   301
window   126, *147*, **148**, 211, 224–6, 228;
    see also oil generation
oils
    African   *163*
    American   *163*, 164
    Asian   146, *163*
    Australian   146, *163*, 199
    Chinese   164, 205
    Cretaceous   200
    Former Soviet Union   *163*, 164
    Indonesian   *163*
    Jurassic and younger   *201*
    Mahakam Delta   204
    Mesozoic   200
    Middle Eastern   *163*, 164
    New Zealand   146
    Nigerian   204, 208
    North Sea   *163*
    Omani   200, *201*
    Palaeozoic   164, 200
    post-Palaeozoic   204
    Proterozoic   200
    Saudi Arabian   208
    South-east Asian   203
    Sumatran   199
    Tertiary   200, 203
    Triassic and older   200, *201*
    Uinta Basin   208
    Venezuelan   *163*, 164, 208
okenone   57, *58*, 184–5
oleanane   190, *191*, 201, 209–10, 225
    A-ring degraded   *see* de-A-oleanane
oleanenes   189, *191*, 225
oleanoids   52
oleate/oleic acid   44, 169
Oligocene   28, 324
oligomers   203
oligosaccharides   8
O-methylmannose   172
O-methylxylose   172
omnivores   *239*, 239
onoceranes   192
onocerin   53, 192

oomycetes   93
Oort cloud   5
ooze
    calcareous   91
    siliceous   135
optical activity/isomers   *see* stereoisomerism, optical
Orbulina universa   216
order
    reaction   *see* reaction, order
    taxonomic   *12*, 27, 133, 204
Ordovician   24, *26*, 27–8, *28*, 115, *163*, 197, *262*, 324
organelles   47, 168
organic matter   13, *23*, 24, 123, 217, 244, 254, *308*, 316; *see also* kerogen; soil, organic matter
    abiotic   5, 7–8, 10–13, *16*
    amorphous   123–4, 137
    aqueous   92, 314, 317, 319
    burial   *23*, 23, 71, *87*, 93, 104–6, 108, *109*, 117, 123, 127, 131–2, 141, 145, 148, 251, *251*–2, 261–2, 274
    decomposition   *see* biodegradation; mineralization
    dissolved (DOM)   92–3, 106, 119–20, 310
    endogenous   7, 7
    exogenous   **7**, 7
    export from surface water   80, 86, 91, 92, 102, 106, 176, 179, 249, *249*, 280–1
    lacustrine   24
    marine   24, 92, 106, 120–1, 139, 172, 246, 273–5
    maturation   *see* maturation
    oxidation   *22*, 23, 87, 93, 95, 99, 105, *109*, 112–15, 119, 129, 137, 139, 250–1, 254, 284
    particulate (POM)   92, 176–7, 179, 186, 310
    preservation   *see* preservation, organic matter
    sedimentary   1, *16*, *23*, 23–4, 27, 29, 69–70, 72, 92–3, 99, 101, 105–6, *108*, 108, 117–19, 121–2, 132, *135*, 135, 137, 139, 172, 174, 177, 196, *208*, 221–2, 228, 234, 241, 259, 307–8, 310–11, 313–7
    isotopic record   240–2, 273, 275, 283
    major contributors   23–9, 30, 69, 71, 109, 117, 166, 202, 241, 260
    monolayer coverage   106, 108–9
    sources   24, *26*, 69, 92, 109, 115, 119, 137, 155, 167, 174, 177, 211, 240
    terrestrial/terrigenous   24, 92, 106, 121, 132, 172, 274; *see also* soil, organic matter
    thermal alteration   *see* maturation
    type   29, 223, 224, 226–7; *see also* kerogen type
        labile   93, 99, 106, 117, 174, 176, 179, 213, 247, 319
        refractory   92–3, 99, 106, 117–18, 121, 127, 130, 174, 176, 179, 247, 253, 313–14, 317, 319
    uncharacterized   92
    weathering   *23*, 23, 144, *247*, 250–1, *251*–2, 254, *285*

organic sulphur compounds (OSCs)   10, 197, 203–5, 207; see also sulphur, incorporation into organic matter
organochlorines   314–15; see also chloroaromatics
   bioaccumulation   306, 309
organometallic compounds   301
organotins   see tin, organo-
orogeny   107, 228, 256, 257, 265, 270
   Alleghenian   270
   Carpathes alpine   228
   Grenvillian   256
   Hercynian   270
   Palaeozoic and older   228
   Tertiary   276
Oscillatoria   95, 112
oxalic acid   217
oxaloacetic acid   38, 237
OxCal   217
oxic–anoxic boundary   18, 22, 81, 98, 179, 206
oxicity   6, 19–21, 22, 88–9, 93, 95, 97, 97, 102, 106, 108, 110, 121, 127, 180, 207, 220; see also oxygen depletion; sediment, oxygenation; water, oxygenation
oxidase   101–2
oxidation   1, 6, 22, 23, 38, 45–6, 49, 57, 68, 78, 81, 87, 95, 95–6, 97, 93–4, 98, 100, 102, 108, 109, 170, 177, 179, 182, 183, 185, 185, 206, 217, 251, 266–7, 270, 287, 302–3, 310; see also under individual components; respiration
   α   46, 302
   β   46, 170, 179, 182, 184, 302–3
   bacterial   46, 95, 96, 97, 97–8, 98, 170, 179, 182, 185, 185, 244, 250, 302–3
   biosynthetic   53, 56
   chemical   119–20, 137, 161, 179, 295, 307
   photo   see photo-oxidation
   state   20
oxidizing agent   81, 95, 295
oxohexanoic acid   303
oxygen   3, 81, 96, 296
   atmospheric   6, 15, 16, 19, 19–23, 23, 87, 87, 109, 109, 246, 251, 252–3, 253–4, 260–2, 262, 270–1, 270–1, 296; see also atmosphere, composition
   to carbon dioxide ratio   20, 271–2
   cycle   87
   demand   94, 104–5, 275
   depletion   77, 87, 94–5, 95, 104, 105, 106, 112, 115–16, 122, 138, 259, 268, 275, 282, 284, 292, 298
   dissolved   see water, oxygenation
   exposure time   108–9, 108–9
   generation   17, 17–18, 18, 23, 23, 253, 262; see also photosynthesis, oxygenic
   index (OI)   223
   isotopes   4, 218–19
   isotopic
      composition   6, 216, 218–19, 277–8, 218, 220–1, 222, 265, 269, 272, 275–6, 278, 283
      stages   216, 218, 219, 222

minimum layer (OML)   104, 105, 105–6, 115, 275
organic   30, 31, 45, 49, 56, 119, 123, 148–9, 151–2, 301
   in oxidation   6, 15, 271
   partial pressure   19, 20, 270–2, 271
   sinks   18, 22, 23, 23, 251, 254, 262, 269
   supply   102, 104–5, 262
   transport   43, 61
oxygenase   236, 270
ozone   19, 253, 287, 295, 296
   depletion   295–7, 298
   greenhouse warming   253, 253, 286, 286, 291, 297, 297
   layer   6, 21, 27, 269, 297

PAHs (polycyclic aromatic hydrocarbons)   151, 153, 298–301, 299, 303, 320
   in ancient sediments   300–1
   distribution variations   300, 300
   peri-condensed   298–300
   pyrolytic   287, 298–300, 300, 304
   in Recent sediments   298–300
PAL   see gas, present atmospheric level
Palaeocene   324
   – Eocene thermal maximum   281–4, 282, 291
palaeoclimatic variations   see climate, change
Palaeogene   163, 324
palaeogeography   271–2
Palaeoproterozoic   269, 270, 324
palaeosols   20, 256, 261
palaeothermometry   see temperature proxies
Palaeozoic   25, 26, 26–7, 28, 162, 201, 258, 261, 324
   fauna   28, 28
palmitate/palmitic acid   44, 44–6, 47, 59, 169
palmitoleic acid   44
palynology   201, 241, 284
palynomorph   284
   carbonization   221
Pangaea   29, 70, 163, 258, 262, 270, 271, 272, 284
   climate   29, 261, 272
panspermia   7
paper pulping   305, 315
paraffins   127, 128, 142, 149, 150–1, 154, 156
parasites   13
particulate organic carbon   see carbon, particulate organic
partition coefficient   141, 161, 162, 307–8, 308, 309
   octanol–water (K_{ow})   307–8, 310–12, 314, 317, 319
PCB (polychlorinated biphenyl)   306, 309, 313–17, 314, 316
   28   308
   153   308
   206   308
   bioaccumulation   306, 319–20
   degradation   314, 317

degree/position of chlorination   308, 313–14, 314, 319
numbering scheme   314
PCDDs   see polychlorodibenzo-*p*-dioxins
PCDFs   see polychlorodibenzofurans
PCR   see reaction, polymerase chain
PDB   see Peedee belemnite
peat   1, 104, 112–13, 115, 122, 125, 125, 127, 127, 163–4, 247, 283, 285
   burning   285
   composition/type   113–15, 127, 192, 243, 270
   contributing organisms   27, 113–14
   erosion/oxidation   284
   formation   110, 112–15, 113, 127, 247–8; see also peatification
   mires   see mires
peatification   125, 127, 130; see also peat, formation
pectins   38, 39, 101, 172, 174–5
pectinase   100
Pediastrum boryanum   133
Peedee belemnite (PDB)   4, 4
Pennsylvanian   163, 270
1,2,3,7,8-pentachlorodibenzodioxin   318
pentahomohopane   55
pentamethyleicosane
   2,6,10,14,18-   199, 205
   2,6,10,15,19-   197
pentane   128
   normal   148–9
pentoses   35, 67, 118
peptide   39, 40, 42, 100, 117, 213, 216
   linkage/bond   41, 42, 100, 215
   hydrolysis rate   213, 215–16
peptidoglycan   40, 93
perfluorocarbons   see PFC
perfluoroethane   290
peridinin   58, 185, 185
perihelion   264
Period   26, 324
   boundaries   281, 280–5
permafrost   164
Permian   26, 28, 29, 114, 162, 163, 262, 270, 271, 324
   flora   284
   – Triassic boundary event   28–9, 284–5
permeability   145, 146, 156–8, 159
peroxidase
   lignin   see ligninase
   manganese   see manganese peroxidase
perylene   299–300
pesticides   309, 313, 320
petrography/petrology   122–3
petroleum   30, 128, 145, 231; see also fossil, fuel
   accumulations   149, 158, 159, 161, 245
   alteration   159–62
   chemical composition   145, 149, 151–3, 158, 162
   isotopic composition   242–5
   exploration wells   229, 230, 232
   expulsion   see petroleum, migration, primary efficiency/threshold   232

petroleum (cont.)
  generation <u>147</u>, 144–6, *149*, 156, *158*, *225*, 227, 231–2; see also kerogen, petroleum generation
    deep origin 146
    and mineral matrix 232
    kinetics/modelling 227–8, *229–31*, 245
    volume change <u>147</u>
  migration 148, 156–9, <u>159</u>, 161, <u>223</u>, 232, 242, *243*
    distance 158
    mechanisms 156–7
    primary 144, 152–3, **156**–7, *158*, 159, 161, 212, 222, 224, 226–7
    secondary **156**–9, 162, 197
    tertiary 162
  phases see phase
  potential 157, 162, <u>223</u>, 224
  reserves *163*, 163–5
  reservoirs see reservoirs
  source rocks 102, 109, 116, <u>128</u>, 139, 149, 152, 156–7, 159, *163*, 206, <u>223</u>, 224–6, 228, *229*
    evaluation 139, 141, *142*, 222, <u>223</u>, 228–32
    lacustrine 158, 206
    major formation periods 162
    marine 200, 206
    maturity see maturity
    Ordovician 197
  traps see traps
PFC (perfluorocarbon) 297
14 *290*, 297–8, *298*
pH see acidity
phaeophorbides 177, *178*–9, *179*, *183*
Phaeophyta/phaeophytes see algae, brown
phaeophytins 177, *178*, 180, *183*
Phanerochaete chrysosporium 102
Phanerozoic 22, 28, 109, 201, *255*, 256, 259, 261–3, *262*, 269, 324
phase
  behaviour 156, <u>157</u>, 158, 161–2
  diagram <u>157</u>, *157*, 164
  mobile 140
  stationary
    chromatography <u>140</u>, *304*
    growth see growth, phases
  vapour/gas 308, 315, 317
phenanthrene 152, 212, 226–7, *299–300*
  methylated 202, 212
  numbering system *190*, 211
phenol 31, <u>33</u>, 33
phenolic compounds/units 62, 102, 120–1, 127, *129*, 129, 131, 133, *134*, 139, *142*, 169, 173, 238, 241, *304*, 319
phenoxyacetic acid 102
phenoxyethanol 102
phenyl units 31
phenylalanine 41, 62, 65
1-phenyldecane 304
phenylpropanoid
  acetate pathway 64
  units 64

phenylpyruvic acid 65
pheromones 49
phloroglucinol 133
phlorotannins 133, 319
phorbides 182
phoronids 319
phosphate <u>10</u>, 13, <u>17</u>, *46*, 46–7, *48*, 67–8, *69*, 80, 86, 99, *134*
  availability 86, <u>87</u>, 87, 105, 220, 251, *252*, 253, 275, 279, 292, 294
  ortho- 80
  pyro- <u>10</u>, 50
phosphatides see phospholipids
phosphoglyceric acid (PGA) *17*, 237
phospholipids <u>10</u>, 44, *46*, **46**–7, 55, 79, *80*, 167, 169
phosphoric acid see phosphate
phosphorite 115
phosphorus 74, *80*, 80, <u>87</u>, 99
  cycle <u>87</u>, 99
  organic 79
  pentoxide <u>12</u>
phosphorylation <u>12</u>, 49
  oxidative 38, 310
photochemical dissociation/photodegradation/photodissociation see photolysis
photolytic degradation see photolysis
photolysis 6, 19, 21, <u>89</u>, 93, 179, 266–7, 269, 280, 288, 295, 297, *308*, 311, 317
photo-oxidation <u>97</u>, 177, 299, 301
photoreductive dissociation see photodissociation
photorespiration <u>18</u>, 72, 270, 273
photosynthesis 1, <u>13</u>, 15–8, *16*, <u>17</u>, 22, 24, 35, 47, 56, *59*, 60, 62, 68, 71, <u>89</u>, *91*, 104, 110, *116*, 116, 177, 179, 235–8, 242, 246, *247*, 248–9, 251, 254–5, 266, 268, 270, 279, 280–1, *281*, *285*, 286
  anaerobic/anoxygenic <u>18</u>, *94*, 97, 116
  bacterial 60
  CO_2 limitation 27, 279–80
  C_3 pathway 28, 234, 238, 240, 254, 256, 279
  C_4 pathway 28, 238
  dark stage/reactions <u>17</u>, 17, 72, 78, 236
  inhibition phase 78, 78
  isotopic fractionation see isotopic fractionation, photosynthetic
  light limitation 27, <u>59</u>, 71–2, 74, 78, 78–9, 83–4, 86–7, 273, 284, 292, 294, 301
  light stage/reactions <u>17</u>, 17
  linear phase 78, 78
  nutrient limitation 74, 79–84, 286
  oxygenic 15, <u>17–18</u>, 17, 18–9, 22–3, *23*, 60, 71, *94*, 260
  O_2 inhibition 271
  saturation plateau 78, 78
  temperature effect <u>17–18</u>, 72, 86
photosynthesizers/phototrophes **13**, 16, <u>17–18</u>, 18, 20–1, <u>21</u>, 23–4, 27, 55, 57, <u>59</u>, 60, 70, 71, 79–80, 99, 167, 240, 247, 260
  biomass 23
  oxygenic 19, 71

phototrophy see photosynthesis
phthalates 311
  monoethylhexyl- 311
  diethylhexyl-(DEHP) *308*, 311, *311*
    degradation products 311
phthalic
  acid 311, *311*
  anhydride 311, *311*
phycobilins 62
phycocyanobilin 62, *64*
phycoerythrobilin 62, *64*
Phycophyta see algae
phyllocladane 201, *202*, 209
phyllocladene 188, *190*
phyllocladoids 202
phylloerythrin (PE) *181*
phylum <u>12</u>, *12*, 25, 166, 310
physicochemical properties/parameters 125, *308*, 308–9, <u>309</u>, 310, 314
phytadienes 182, *206*
  sulphurization 136, 207
phytane 151, 153, 205, 207
  $\delta^{13}C$ values *241*, 241
  isomerization 206
  sources 207
phytanic acid 179
phytanol 52, 179, 182, *206*, 207, 241
phytanyl unit 47, 197, 207
phytenal 182
phytenes 182, *206*, 207
phytenic acid 179, *206*, 207
phytochrome 62, *64*
phytoene 50, 56, *58*
phytoestrogens **306**–7
phytol 47, *51*, **52**, 60, *206*–7
  $\delta^{13}C$ values 236
  degradation/diagenesis 138, 182, 205–7, *206*
phytophagy 305
phytoplankton 20–1, <u>21</u>, 71–3, 78, *80*, 83–4, 86, *97*, 109–10, 175–7, 220, <u>221</u>, 248, 251, 254, 259, 266, 275, 278, 280–1
  bioaccumulation of pollutants <u>309</u>, <u>316</u>
  biomass 81, *82*, 83, 116, 177, <u>234</u>, 247–8
  blooms 19, 56, 80–1, 83, 86, 92, 102, 112, *200*, 219–20, 237, 280–1, 292, 294
  brackish 85
  calcareous 25, 28, 162, <u>219</u>, 250, 280
  chemical composition 55–7, 69, 92, 169–72, *171*, 176, 179, 186, 197, 200, 205–6, 319
  classification 83, *85*, *171*, 200
  contribution to sedimentary organic matter 24, 69, 112, 137, 139, 162, 203
  evolution *16*, *19*, 25
  freshwater 84, *85*, 115
  hypersaline 72, 112
  isotopic composition 180, <u>234</u>, *236*, 240, *240*, *259*
  marine 83–4, *85*, 170, *219*, *247*, *259*
  non-siliceous 29
  organic-walled 25–6, 162, 250
  population variations 29, 83–4, 86, *163*, 266, 275, 282–3

production/productivity   *see* production, phytoplankton
Proterozoic   201
remains and sinking   24, 69, 91–3, 112, 137–8, 175–7, 179
senescence   **177**, *178*, 179, *180*
siliceous   25–6, 162, 250
succession   86, 292, 294, 317, 319; *see also* population variations
phytosterols   **55**, 204; *see also* sterols, plant
phytyl unit   47, *63*, 177, 179, 207, 241
oxidation   184
pigments   61
accessory   56, **59**, 60, 62, 79, 167
carotenoid   *see* carotenoids
diagenesis/degradation   177–9
flavonoid   *see* flavonoids
hydroxyaromatic   67, *67*
light absorption   56, *59*, *61*, 79
photosynthetic   43–4, 62, 167, 180; *see also* pigments, accessory; pigments primary
primary   **17**, 18, 60; *see also* chlorophylls
pyro-   177
tetrapyrrole   **60**–2, 167, 177, 180–2; *see also* chlorophylls
non-chlorophyll   *64*
pillow lava   7
pimanthrene   188, 202, *202*
pimaric acid   *51*, 52, 188
pimaroids   202
pinene   51
planctomycetes   **96**, 167
planets   2, **5**
plankton   **21**, 28, 86, 91–3, *173*, 240, 280, *281*, *312*, 312
buoyancy   **21**
calcareous   26, 269, 273
chemical composition   30, 69, 137, 172, 280
classification   **21**, *21*
macro-   **21**, *21*
mega-   **21**, *21*
micro-   *21*
nano-   **21**, *21*, 26, 84, 280
pico-   *see* plankton, ultranano-
ultranano-   **21**, *21*, 26, 84, *85*, 86, **89**, 91, 281
Plantae/plants   15, 38, 46, 49, 84, 93, 99, 105, 112, 114, 189, *247*, 248, 250, **252**, 253, 288, 301
aquatic   27, 40, 114, 235, 237, 248; *see also* algae; grass; mangroves; phytoplankton
benthonic   79
C3   **17**, 72, 236, **236**–7, **239**, 239–41, 279, 286
C4   **18**, *236*–7, *237*, *239*, 239, 241, 279
CAM   **18**, *236*–7, 237
chemical composition   38, 41–2, 44–7, 49, 52–3, *53*, 55, 56, 60, 61, 62, 64, 67, 115, *124*, 167–8, 170, 192, 305–6, **306**, 311, 317; *see also* higher plants, chemical composition
colonization of land   24, 27, 70, 256, 261–2
contributions to sedimentary organic matter   174, 247, 270

cruciferous   306, 320
evolution   27
flowering   *see* angiosperms
green   14
higher   *see* higher plants
isotopic composition   237, 239, 241, 271
lower   27
marine   247; *see also* algae; phytoplankton
non-vascular   174
succession   89, 114
terrestrial/subaerial   16, 24, 27, 38, **59**, 67, 80, 87, **95**, 110, 112, 235, 239, 240, 242, 247–8, *252*, 254, *260*, 286, 288; *see also* higher plants
tissues   64, 78, 80, **95**, 101, 105, 114, 121–2, *122*, 123–4, *124*, 136, 167, 174, 204, 237, 240, 253
vascular   *see* higher plants
plasmalogens   47, *48*
plasma   3
membrane   *see* cell membrane
plasticizers   311, 313
plateau
Ontong-Java   273
submarine   273
Tibetan   276
Pleistocene   114, 219, 324
Pliocene   233, 276, 324
Poaceae   *see* grass
POC   *see* carbon, particulate organic
polar
groups/compounds   156, 159, 161, 242, 301, 313
NSO compounds   *see* NSO compounds
poles   263, 288, **296**, *296*
pollen   59, 89, 118, *122*, 123, 127, 133, 136, 221, 241
pollutants
gaseous   295–8, 308, 311, 314–17
aqueous solubility   299, 301, 307–8, *308*, 310–17, **309**, **316**
chloroaromatic   *see* chloroaromatics; organochlorines
dilution   308, **309**
environmental behaviour   295–321, *308*, **309**
factors affecting   295, *308*, 308–9, 317–21; *see also* individual compounds
hydrocarbons   298–304, **306**
associations
organic   307–8, 311–14, 317, 319, 321
particulate   299–300, 304, 311, 315, **316**, 317, 319
polyamides   42
polycadinanes   203
polycadinenes   60, *62*
polychaetes   93
polychlorinated biphenyls   *see* PCB
polychlorodibenzofurans (PCDFs)   **306**, **309**, **315**–16, *317*
bioaccumulation   **306**
chlorination degree   316–17, 319
degradation   317
numbering scheme   *317*

polychlorodibenzo-*p*-dioxins (PCDDs)   **315**–17, *317*–8
bioaccumulation   **306**, **309**, 319
chlorination degree/position   315–17, 319
degradation   317, *318*
numbering scheme   *317*
polycommunic acid   *62*
polycyclic aromatic hydrocarbons   *see* PAHs
poly-D-galacturonic acid   38, *39*
poly-D-manuronic acid   *39*
polyesters   49, 133
polyhydroxy-bacteriohopanes   *see* hopanoids, polyols
polyhydroxybenzene   59, *62*
polyhydroxyflavonol units   65
polyhydroxyphenol units   167
polymers   13, **37**, 38, 40, 49, 59, 65, 68, 106, 118–19, 129, 132, **166**, 176, 196; *see also* macromolecules
polymerization   60, 64, 100–2, 118, 134, 197, 200
polymethylchrysenes   192
polymethylnaphthalenes   *195*, 196
source indicators   203
polymethylpicenes   192, *194*, *202*, 202
polypeptides   40, **42**
polyprenoid
tetracyclic   *199*, 204
polyprenols   **10**, *50*, 60, *62*
polysaccharides   **37**, 38, *39*, 40, 47, 64, 100–1, 169, 172, 174, 176
degradation/diagenesis   117, 120, 127–8, 133, *134*, 172, 174
polysulphides   *see* sulphide, poly-
polyterpenoids   49–50, 59–60, *62*, 133
pores   1, 93, 106, 108, 156–7, **159**, 159, 304
fluid   145
water   *see* water, pore
porosity   131, **145**, **156**–8, **159**, 162, **230**
porphyrins   **60**, 179, 182, 238
$\delta^{13}C$ values   240–1
catagenesis   183
cracking   183, 211
deoxophylloerythroetio (DPEP)   *181*, 182, *183*, 211
diagenesis   *178*, *181*–3
DPEP:etio ratio   211, 225
enrichment of short-chain components
etio   182, *183*, 211
free-base   182
iron   **89**, *183*
metal chelation/metallation   *181*, 182, *183*
nickel   *181*, 182, **208**, 211, 225
reduction   183
sources   182
vanadyl   *181*–2, 182, **208**, 211, 225
potassium   **4**, 6, 72, 80, **111**, **145**, **230**
Prasinophyceae/prasinophytes   *85*, 170, *171*
Precambrian   25, *26*, 118, 201, 255
shields   *see* craton
precipitation
minerals   **1**, 19, 110, 112, 250

precipitation (*cont.*)
  water/ice 1, 71, 72, 87, 91, 112–4, 216, 218–19, 242, 250, 252, 254, 265, 269, 291–2
  zones 76, 90, 261, 265, 292, 315
preservation
  of organic matter 24–5, 68, 70–1, 92, 99, 100–2, 104–9, 117–18, 127, 130, 135, 162, 248, 250, 256
  selective 25, 119, 132, 134, *134*
pressure
  atmospheric 20, 75–6, 76, 85, 90, 277, 292; *see also* anticyclones; cyclones
  and burial *see* geobaric gradient
  capillary 156–8, 159
  critical 157, *157*
  hydrostatic 145, 157, 164, 165, 219, 228, 249, *267*, 283
  lithostatic 2, 145, 159
  over- 145, 156–7
  partial **20**
    of argon 20
    of carbon dioxide *see* carbon dioxide, partial pressure
    Dalton's Law *see* Dalton's Law
    of nitrogen 20
    of oxygen *see* oxygen, partial pressure
    of water vapour 218
  and petroleum migration 156–7
  and phase change 156, *157*, *157*, *164*
  and reactions 117, 125, 131–2, 145–6, 147
  solar 2–3
  and solubility 249
  units 322
  vapour 141, 162, *308*, 311, 316–17
primates 12
pristane 141, 151, 153, 179, 207
  δ^{13}C values *241*, 241
  isomerization 206, 225, 227, 207–8, 224
  meso- 207
  – phytane ratio 204, *204*, 207
  sources 207
pristanic acid 179
pristenes 179, 207
Prochlorophyceae/prochlorophytes **26**, 60, 71, 84, 201
production/productivity 18, 25, 70, 87, 102, 109, 112–13, 116–17, 162, 247, 250, 262, 275, 292
  bacterial 24, 275
  index *see* transformation ratio
  phytoplanktonic 24, 26, 82, 86, 89, 89, 104, 110, 116, 162, 177, 237–8, 248–50, 266, 278, 292, 301
  primary **13**, 24, 26, 71, 73, 78–9, 79, 81, *82*, 85, 86, 87, 87, 92, 102, 104–6, 108, 115–17, 132, 176, 179, *240*, 240, 250, 254, 256, 268, *279*, 279–80, 284, 292, 316
    annual 71, *84*
    aquatic 71, 73, 73, 78, 83
    factors affecting 71–4, 78–84
    gross 78, *79*, **246**

marine 24, 72–3, 83–4, 87, 248, 250, 253, 255–6, *263*, 275–6, *278*, 280, 292
  net 24, 71–2, **247**–9, 251, 253–5
  spatial variations in ocean 83
  terrestrial 24, 71–2, 79, 127, 251, 270, *279*, 286
progymnosperms 27
prokaryotes 12, 13–16, 21–2, 25, 89, 238, 280
proline 8–9, 9, 41
propane 149, 161, *164*, 243–4, *244*, 287
24-*n*-propenylcholesterol 204
24-*n*-propylcholestanes 204
propionate/propionic acid 95, 169
propyl units 64, 120, 128–9, 131, 142, 145
protective coatings/tissues 13, 47, 49, 52, 59, 67, 92, 100–1, 106, 118, 133, 135, 167, 169, 174, 179, 212
protease 100
proteins 8, 10–11, 14–15, 30, **40**–4, 46, 61–2, 64, 67–9, 80, 89, **94**, *100*, 102, *108*, 117–18, 134, *134*, 212, 215, 238, *239*, 306
  chemical composition 40–2, 212–13, 239
  diagenesis/degradation 100, 117–18, 133, 213, 216
  fibrous 42, 117
  globular 42
  integral 46, *46*
  isotopic composition 238–9, *239*
  occurrence and function 40
  peripheral 46, *46*
  structure 101
    primary 42
    secondary 42
proteobacteria 18, 98
Proterozoic 19, 20–2, 22, 25, 26, 27, 28, *201*, 201, 255, 256, *257*, 260–2, 268–9, 322, 324
Protista/protists 12
proton 4, 17–18, 40, 42, 81, 111, 102, 154, 213
  activity 81
protozoa 21, 26, 86, *91*, 91, 205, 280
prymnesiophytes *see* algae, yellow-brown
Pseudomonas 40, 94, 96, 96
psilophytes 27
Pteridophyta/pteridophytes 12, 26, 27, 305
purine 67–8, *68*
purpurin 183
pycnocline 74
pyran *31*, 35, *138*, 144
pyranoses 35, 37
pyrene 298, *299–300*
pyrethrum 49
pyridine *31*, 152
pyridyl unit 123
pyrimidine 11, 67, 68, *68*
pyrite 11, 20, 97, 98, *108*, 123, 161, 242
  burial 22, 23, 262
  isotopic composition 21
  oxidation 22, 23, 87, 262
  weathering 22, 23, 251
Pyrobaculum 98
Pyrodictium 98
Pyrolobus fumarii 14

pyrolysis 7, 119–20, 129, 137–9, *142*, 148, 164, 174, 222, 223, 232, *244*, 244, 298–300
  anhydrous 153
  closed-system 232
  hydrous 153, 210, 232
  open-system 232
pyrochlorophyllides *178*
pyrophaeophorbides 177, *178*, 180
pyrophaeophytins *178*
pyrrole *31*, 60, 152
pyruvate/pyruvic acid 15, 38, **40**, 45, 46, 49, 50, 65, 93, 95, 238, *303*
  phenyl 65
  phosphoenol (PEP) 65, 237

quartz 20
Quaternary 28, 71, 219, 241, 262, 265, 276, 288, 293, 324
quinoidal units *31*
quinones *31*, 67, *67*

racemic mixture 8, 10, **32**, 175, 212
racemization 9, 10, 100, **175**, 212–13, *213*, 216
radiation 221
  balance of Earth 263, 264, 266–7, 269–70, 273, 276–80
  black body 252, *253*
  cosmic 29, 217
  long-wave/infrared 253, *253*, 263, 265
  short-wave/UV-visible 7, 252, *253*, 263, 296
  solar 5, 266; *see also* insolation; solar luminosity
  trapping *see* greenhouse effect/gasses
  ultraviolet 5–8, 19, 21, 27, 67, 78, 295, 297
radiative
  efficiency 290
  forcing **253**, 263, *289–90*, 290–1
radicals 102, 179
  free 57, **60**, 148, 153, 154, *154*, 155
  hydrogen 175
  hydroxyl 17, 179, 266, 269, 287–8, 291, 295, 297
radioactivity 145, 214, 217, 229–30, 309, 311, 319
radiocarbon dating 216, 217, *217*, 228
radionuclide/radioisotopes *see* isotopes, unstable
radiolarians 26, **26**, 28, 80, 91
radium 80
rain *see* precipitation, water/ice
  pelagic 92, 110
rank *see* coal, rank
Raphidophyceae/raphidophytes 170, *171*, 172
rate constant *see* reaction, rate, constant
Rayleigh fractionation **218**
RDP *see* ribulose diphosphate
reaction 64, 68, 102, *118*, 146, 227, 231, *233*; *see also* under individual reactions
  acid catalysed 10, 153, 154
  addition 30, 44, 136, *136*
  alkylation *129*, 129–30
  anammox 96
  chain 154, *154*

cracking *155*
cyclization 53, 55, 131, 153, 155, *155*, 180, 185, 196
dark *see* photosynthesis
disproportionation 153, *154*
elimination 129, 141–2, 144
enzymatic 10
equilibrium *see* reaction, reversible
Fischer–Tropsch 7
  in hydrocarbon formation *154*, *154*–5
intermediate 43, 176, *175*, 212–13, 223
ionic 153, *154*, 297
ion-molecule 3
kinetics *see* reaction, rate; kinetic, parameters
Maillard 118
networks 231, *231*
order 213, 214, *215*, 215–16, 217
photochemical 295
polymerase chain 14
radical initiated 153
rate 43, 100, 212–13, 214, *215*, 215–16, 217, 224, 227–8, 230–2, *233*, 238
  constant 214, *233*, 227
  determining step 215
  temperature dependence 72, 100, 214, 223, 228, 232–4, 279, 315; *see also* temperature, and reactions
rearrangement 175, 223
redox 6, 13, 81, 94, 162
regiospecific **43**
reversible 176, 210, 214, 248; *see also* equilibrium, chemical
ring-opening 101–2, 153, *155*, 177, 182, 190, 192–3, 196, 200, 211, 222, 313
stereoselective 10, 32, 40, **43**, 64
Strecker 9
unimolecular 214
redbeds *16*, *19*, 19–**20**, 256
Redfield ratio 86, 105, 279
redox
  conditions 13, 21, **22**, *16*, *19*, 87, 205–7, **208**, *208*, 313
  indicators 205–7
  gradient 108
  potential (E$_h$) 80, 81, 203, 208, *208*
  reaction *see* reaction, redox
reducing
  agent 81, 267
  gases *see* gas, reducing
  power 17–18
  see also anoxicity
reduction **6–7**, 44, *45*, 47, 49, 62, 68, 81, 89, 94, 131, 135, 153, 169, 175, 179, 180, *181*, *183*, *185*, 187, 189, 192, *193*, 197–8, 205, 206, 208, *209*, 224, 236, 244, *302*; *see also* under individual compounds
  nitrate *see* nitrate reduction
  sulphate *see* sulphate reduction
reductive
  dechlorination 312, 313, 317, *318*
  decoupling 89
reeds 113–14

reefs
  carbonate 159
  coral 28, 73, *73*, 85, 105, 219
reflectance 123, *123*, 125, 126, *126*; *see also* vitrinite, reflectance
reforestation 286
refuge (for life) 6, 89
regression *113*
reproduction 1, 10, 13, 24, 78–9, 81, 89, 247–8, 304–5, 310–12, 314
reservoir 148, 156, 158–9, 161–2, 164, 301
  rock 156, **159**
  seal 159, 161–2
  spill point 159, 162
reservoirs
  Gulf Coast 162
  Jurassic 208
  Mesozoic 208
  stacked 162
  Tertiary 208
residence time 91–2, 104, 246–8, 250, 255, 283, 290–1, 295, 317; *see also* adjustment time
resin acids 52, *190*
resinite *122*, 123, *133*, 142
resins *122*, 127, 242
  bitumen/oil component 128, 135, *135*, 140, 149, 155–6, 159; *see also* NSO compounds
  dammar 60, 203
  gymnospermous 52, 127, 201
  plant 60, 122–3, **128**, 188, *190*
resorcylic acid lactone 306
respiration 23, 23–4, *25*, 38, 67, 79, 87, 94, 104, 238, 240, 246–50, 247, 251, 259, 262, 281, *281*, *285*, 286–8; *see also* oxidation; photorespiration
  aerobic **6**, 18, 22–3, 38, 40, 45–6, 78, 87, 99, 100, 105, *116*, 121, 182, 238, 240, 250, 269; *see also* methylotrophy
  anaerobic 6, 15, 38, 99–100, 112, 205, 269, 275, 287, 311, 313; *see also* acetogenesis; fermentation; methanogenesis; denitrification; sulphate reduction
  dysaerobic 6, 93, 275
retene 188, *190*
retention time 140–1
retinol *51*, 56
rhamnose 35, *36*, 40, 172, *173*
Rhizobium 95
Rhizophora 73
  mangle 114
Rhizopoda/rhizopods 26, 91
Rhodesian 255
Rhodophyceae/Rhodophyta *see* algae, red
Rhodospirillum 18; *see also* bacteria, purple non-sulphur
ribitol 47, *48*
ribonucleic acid *see* RNA
ribose *36*, 37, 40, 47, 53, 67, 68, 172, *173*
ribulose diphosphate (RDP) 17, *17*, 237
  carboxylase/oxygenase *see* rubisco
rice paddies 287, *287*

rift
  East African 109
  North Sea 228
rifting 109, 228, *229*, 256, 257
  Cretaceous 203
rigidifiers *see* cell membrane, rigidifiers
rimuane *202*, 201
River
  Amazon 115, 219
  Mississippi 114–15
  Niger 114
  Po 244
rivers 74, 83, 86, 109, 110, 112–13, *116*, 116, 204, *247*, 248, 258, 268, 280, 282, *285*, 292
RNA (ribonucleic acid) 10–12, *12*, 14, 14, 68, 100, 172
  world 11
Rock–Eval 222, 223, 232
  HI *see* hydrogen, index
  OI *see* oxygen, index
  PI (production index; *see* transformation ratio
  residual carbon 223
  T$_{max}$ 223, *225*
Rodinia 29, 256, 267
roots 49, 67, 80, 105, 114, *122*, 167, 250, 286
rubber 49, 59, *62*
rubidium *80*, 258
rubisco 236–7, 270–1, 286
ruminants 287, *287*

sabkhas 112, 115
salicylaldehyde *303*
salicylic acid *303*
salinity 72, *72*, 74, 75, 77, 98, 98, 100, 111, 112, 116, 138, 145, *164*, 219, 205, 207, 248, *249*, 273, 275–6, 278, 282–3, 292, *293*, 308
salmon 306
salt 75
  dome 145, 159
  lakes *see* lakes, saline
  marshes 73, 84, 105
  pans 102
Samoa 289
sand *110*, 301, 304
  bodies (reservoirs) 159
  stone 159, 157
sandaracopimaric acid 188
Santonian 274
saponification **42**
sapropels 71; *see also* kerogen, sapropelic; coal, sapropelic
  Mediterranean 237, 275–6
saprophytes 13
Sarcinochrysidales 204
sarcosine 8–9, *9*
saturated compounds **30**; *see also* aliphatic structures; saturates
saturates 128, 140, **149**, 160–1, *242*, 242
savannah 73, 90–1
Scenedesmus communis *133*

S/C atomic ratio   139
Schizophyta   *see* prokaryotes
Schizomycetes   *see* bacteria
Schizophyceae   *see* cyanobacteria
Scilly Isles   <u>85</u>
sclerotia   122
sclerotinite   *122*, 123
scrub   73, 89, *90–1*, 114
Sea
  Adriatic   292
  Azov   116
  Baltic   116
  Black   24, 71, *116*, 116
  Caspian   24
  Marmara   116
  Mediterranean   71, 116, 292
  North   144, *231*, 303
  Red   116
  Sargasso   <u>316</u>
  Wadden   314
  Weddell   *281*, 283
sea
  epeiric   115
  floor spreading   *see* spreading ridges; tectonic activity
  ice   *see* ice, sea-
  inland   90
  level   *29*, 105, 113–14, 162, 164, 216, 272
    change   <u>107</u>, 114, <u>219</u>, *263*, 265, *267*
    eustatic   **107**, 115, 162, *163*
  shelf/shallow   *29*, 83, *272*
  surface temperature (SST)   216, 220, 250, *262*, 265, 267, 272, 283, 292
seafood   239
seamounts   275
seawater   10, 13, 78, 119, 204, 248, 260, *270*; *see also* water
  alkalinity   *see* alkalinity
  composition   22, <u>72</u>, *72*, <u>111</u>, <u>218–19</u>, 242–3, 248, 256, <u>258</u>, 268, 284, 313–14, 319; *see also* seawater, dissolved gases; seawater, nutrient supply
  density   <u>75</u>, 77, <u>293</u>
  dissolved gases   15, 249, 251, 278, 283–4, 287, <u>290</u>
  nutrient supply   19, 80, 83, <u>85</u>, 86, 105, 115–16, 162, 220–1, 249, 256, 275, 278–80
  pH   <u>111</u>, *111*, <u>208</u>
  salinity   *see* salinity
  volume   83, <u>107</u>, 248, 291
seaweeds   *see* algae, macroscopic
secohopanes
  8,14-   **192**, *193*
  17,21-   200, 204
8,14-secohopanoids   <u>55</u>, 193, 196
  aromatic   *54*, 193, *195*, 196
8,14-secotriterpanes   **192**, 196
8,14-secotriterpenoids   193
  aromatic   *195*, 196
sedges   113–14
sediment   **1**, <u>13</u>, 93, <u>95</u>, <u>97</u>, 118, 182, *183*, 228, *248*, 265, 269, *299*, 310, <u>316</u>, 320

accumulation   69, 71, *106*, <u>107</u>, *113*, 105–6, *108*, 108–9, 112, 114, 116, 156, 207, *229*
ancient   30, 142, 166, **179**, 196–7, 200–2, 204–5, 221, 240–1, *300*, 300–1
anoxic   <u>96</u>, 205, 287, 301, 311
argillaceous   135, 204
authigenic   <u>1</u>
biogenic   <u>1</u>, 26, 91, 110
burial   *see* sediment, accumulation
carbonate   <u>98</u>, *110*, 115, 135, 161, 180, 193, 201, 203–5, 235; *see also* carbonate
chemical   <u>1</u>, 110
clastic   <u>1</u>, 110, 113, 135, 161; *see also* clastics
coastal   108, 228
column   <u>230</u>
compaction   <u>1</u>, 104, 112, 117, 127, <u>145</u>, 131, 141, 156
deep-sea   100, 108, <u>159</u>, *216*, 216
density flows   108, 110, <u>159</u>; *see also* turbidites
First Print
deposition   *see* sediment, accumulation; burial history
detrital   *see* clastic
evaporitic   <u>1</u>, <u>98</u>, 112, 135, <u>145</u>, 203–5, 270
fluidized bed   106
fluviatile   110, 112–13, 115
freshwater   98, 248
grading   102, 104–5, 108, 110, 115, 122, 138, 156, 176, 301, 304
Holocene   *see* sediment, Recent
inorganic   <u>1</u>
lacustrine   109, 118–19, 183, 200, 203, 205, 219
laminated   102, *110*, 110, 112, 115, 274
loading   <u>107</u>
marine   20, 25, 27, 71, 96, 106, 108, 118–19, 165, 176, 183, 199, 204–5, *208*, <u>208</u>, 219, <u>219</u>, 228, 240–1, 248, 278, 278
organic matter in   *see* organic matter, sedimentary
organic-rich   24, 71, 83, 92, 98, 102, 104–6, 108–9, 115–16, 138, 162, 205–6, 228, 248
over-burden   148, <u>230</u>, 283
oxygenation   93, 104–5, 108, 206
pelagic   **93**, 98, 108, *110*, 112, 115, 207, 268–9
Pliocene   228
pores   *see* pores
Proterozoic   205
Quaternary   212, 216
reactive   <u>87</u>, *247*–8, 249–50, *285*
Recent   134, **166**, 167, 169, 172, 185, 192–3, 197, 199–200, 205, 212, 241, 274, 298–304, *300*
shelf/slope   93, 108, 115, 165
siliceous   135
siliciclastic   *see* clastic
surface   79, 93, 104–6, 117, 127, 177, 180, 220, 249–50, 278
suspended   74, 78, 110, 115
temperature   <u>229</u>–30
terrigenous   261

transport/reworking   69, 112, 300; *see also* bioturbation
traps   176–7, 220, <u>316</u>
turbidity flows   *see* turbidites
water content   93, 117, 146, 156
water interface   93
sedimentary rock   1, <u>1</u>, 7, 23, 106, 132, 136, 155, 158, 200, 228, 234, 246, *247*, 250, 254, 261, 263, *285*
  density   <u>145</u>, 156
seeds   45, 89, 168–9
seismic activity   *see* tectonic activity
Selli Event   274
senescence   *178*, 180
serine   8, *41*, <u>97</u>, 100, 239
sesquiterpenoids   49, *49–51*, *133*, 188, *190*, 192, 196
sesterterpanes   153, 200
sesterterpenes   52
sesterterpenoids   49, 52, *50*
Severn Estuary   299
sewage   292, <u>306</u>, 307, 311
Shale
  Bakken   157
  Burgess   25
  Green River   *137*, 138
  Messel   133, *241*, 241
  Monterey   231
shales   <u>145</u>, 156, <u>159</u>, *204*, 201, 210
  black/organic-rich   25, 110, <u>145</u>, *204*, 273–6
  coaly   113
  deltaic   142
  Liassic α   *137*
  marine   25, *142*, 212, 227
  MidlandValley, Scotland   122
  oil   *see* oil, shale
  Toarcian   *137*
Shatsky Rise   *281*
shell   *see* test
shikimic acid   62, *65*
shizokinen   88
Shorea albida   114
shrubs/shrubland   73, 114
siderophores   88, **89**
silica   <u>140</u>, 265
silicate   <u>5</u>, 21, 79, *80*, 80, 86, 92, *110*, <u>111</u>
  dissolution   280
  ortho-   80
  supply   294
  weathering   *247*, 250–1, *251*–2, 256, 260–2, 265, 267–70, 273, 276, 279–80, *285*
silicoflagellates   **26**, *26*, *80*, *163*, 162, 200
silicon   79–80, *80*
silk   42, 100
silt   93, 304
Silurian   24–5, *26*, *28*, 70, 324
simonellite   *190*
sinapate/sinapic acid   *65*
sinapyl alcohol   62, 64, *65*, 174
sitosterol   55, *57*, 171, 200, 305
skin   42, *308*
slime moulds   <u>13</u>
SI units   322
  prefixes   323

SML *see* surface mixed layer
snow 110, 265; *see also* precipitation
  marine 91–2; *see also* rain, pelagic
Snowball Earth *see* Earth, snowball
softwoods *see* gymnosperms
sodium *72*, 72, *80*, 111, 165, 250
soil 13, 89, 90, 93, 95–6, *247*, 250, 283, *285*, 319
  age 262, 319
  chloroaromatic pollutants 319
  compartments 319
  conditions *see* edaphic factors
  hopanoids 191–2
  humics 118–21, *119–20*, 132, 320
  litter *see* detritus
  microbial respiration 94, 99, 262, 286–8
  organic matter 92–3, 101, 104, 118–19, 246–8, 307–8, 311, 313–17; *see also* soil, humics
solar
  energy *see* light
  evolution 5, 260, *260*, 270, 279, *279*
  luminosity 260, *260*, 263, *263*, 268–70, *270*, 279, *279*
  nebula 3, 5, 254
  system 3, 5, 5–6, 8, *16*, 266
solvent 43, 308
  extraction 156, 166, 182, 241, 319
soot 299–300
sorption 106, 108, 319
South Pole 218, *289*
source/environmental indicators
  isotopic 9, 237, *237*, *239*, 240–5, *241*, *243–4*, *255*
  molecular 167, *168*, 169–74, *171*, *173–4*, 177, 182–34, *184*, *191*, *195*, 196–207, *199*, *201–4*, 219, *243*
  optical 122–3, *122–3*, 126
source rock *see* petroleum, source rocks
space 252
  interstellar 5, 9
Spartina *see* grass, cord
species 12, 28–9, 69–70, 72, 86, 94, 167, 176–7, 200, 216, 220, 241, 294, 305, 310, 317, 319; *see also* life
  succession 86, 89; *see also* phytoplankton, succession
spectrum
  action 59, *59*
  electromagnetic 18, 59, 252
Spermatophyta/spermatophytes 12, 27, 167
sphagnan 101
Sphagnum 27, 101, 113–14, 171, 192
sphenopsids 27
spirosteranes 208
spirosterenes 186, 208, *209*
sponge 42
spores 27, 28, 59, 118, *122*, 122–3, 127, 133, 136, 221, 284
  fungal 29
sporinite 122, *124*, 133, *142*
sporophytes 29
sporopollenin 59, *133*, 134

spreading ridges 19, 79, 107, *107*, 162, 228, 244, 251, 258, 261, 265, 272, 275, 300
  rate 107, 272–3, 275
squalane 197, *199*, 205
squalene *50*, 52–3, 55, *56*, 197
  2,3-oxide *56*
SST *see* sea surface temperature
stadials 219, **276**, 293
standard temperature and pressure (STP) 157
stanols **55**, 177, 179, 186, *187*, 188
stanones 186
starch 38, 117
stars 2, 3, 8, 10, 279
stearic acid 44
stenols **55**, 175, 177, 179, 186
  Δ⁵- 177, 186, *187*
  Δ⁷- 177, 208, *209*
steradienes 186–7, *187*
steranes 20, 135, 141, 153, *160*, 160, 175, **186**, 204, 210–11, 301
  C-number distributions 200, *201*, 203
  isomerization 187, 207–8, *209*, *225*, *227*, 210, 224, 227, *233*, 234
  kinetics 232, *233*, 234
  rearranged *see* diasteranes
  regular 153, 160, *187*, *209*
  source indicators 200
sterenes 175, 177, **186**, *187*, 208
  isomerization 187, 186, 187
  rearrangement 186–7, *187*, 208, *209*
stereochemistry *see* stereoisomerism
stereogenic centre 32, 34, *34*, 36, 37, 40, 52, 175, *175*, 176, 186, 188, 192, 196, 207, 209, 212
stereoisomerism 32, 45, *46*, **55**, 146, 188, 196; *see also* isomerism
  configurational 32, *33*, 33–4, 38, 43, 44, 52, **55**, 56, 176, 171–2, 175, 185–6, 188, 191–2, 196, 207–10
    absolute 32, 34, *34*, 38, 40, 100, 186, 188, 207–10, 212
    cis and *trans* 32, *34*, 34, 44, 45, 60, 169–70
    D and L *see* stereoisomerism, configurational, absolute
    E and *Z* 45, 220
    R and *S* *see* stereoisomerism, configurational, absolute
  conformational 32, **35**, *35*, 37, *37*, 42, 54, *54*, 55, 64, 100, *160*
  geometric *see cis* and *trans*
  optical 8, 32, 34
steric
  hindrance 175, 176, 210
  interaction 35, 211, 211–12
steroids 34, 44, 50, 52–6, 141, 205, 305, 306, 307
  aromatic 153, *160*, 160, 186, *189*, 192; *see also* steroids, C-ring monoaromatic; steroids, triaromatic
  aromatization 186–8, *210*, 224, *225*, 227, *227*, *233*, 234
  biosynthesis 56
  C-ring monoaromatic 153, 175, 186, *187*, 210, *210*, 224

diagenesis 175, 186–8, *187*, 196
occurrence and function 53
rearranged *see* diasteroids
reduction 186, *187*
regular 186, *189*, 210
ring-A degraded 186
short-chain enrichment 211, 224, *225*
stereoisomerism 186, *187*
structural notation scheme *54*, 54–5
triaromatic 153, *187*, *210*, 210–11, 224
sterols 8, 20, 47, 53, **55**, 57, *134*, 166, 167, 171–2, 175–7, 205
  algal 57, 177, 186, 242; *see also* sterols, phytoplanktonic
  animal 56–7
  bound 171, 177, 186
  diagenesis 177, 179, *187*
  dinoflagellate 57
  free 171, 177, 186
  fungal 57
  higher plant 57, *171*
  phytoplanktonic 57, *171*, 186; *see also* sterols, algal
  plant 55, 56–7, 204, 305–6
  regular 171
  source/environmental indicators 169, *171*, 171, 177, 200
  zooplanktonic *171*
stigmasterol 55, 55, 57, 171–2, 200, 305
stomata 72, 271, *271*
  activity 237
  pores 18
STP *see* standard temperature and pressure
stratigraphy 226, 230
  seismic 165
stratopause 296, *296*
stratosphere 265, 291, 295, **296**, *296*, 297
streams 73
stromatolites 7, 18, 25, 70, 112, 116, 256
strontium *80*, 258
  isotopic ratio 256, 258, 261–2, *262*, 268, 270, 275–6, 274
styrylpyrone 64
Subbotina 282
subduction 6, *23*, 98, *107*, 115, 228, 251, 265, 268
suberan **49**, *133*, 133, 139, 145–6, 151
suberin **49**, 170, 180
suberinite *122*, 133
submarine fans 159
subsidence 203, 232; *see also* thermal subsidence
succinic acid 303
sucrose 37, *39*
sugars 34, **38**, 47, *48*, 67–8, *69*, 94, 94, 101, 106, 117–18, *118*, 120–1, *134*, 172
  stereoisomerism 32
sulphate 19–20, 22, *23*, 47, 94, 94, 97, 97–8, *124*, *134*, 208
  aerosols 265–6, 291
  freshwater 22
  isotopic composition *see* sulphur, isotopic composition
  marine 21, 22, *72*, 97

sulphate (cont.)
  reducers see bacteria, sulphate reducing
  reduction 23, 94, 97–8, 96, 99, 108, 116,
    116, 138, 182–3, 205–6
    assimilatory 97
    bacterial (BSR) 22, 97, 121, 243, 161
    thermochemical (TSR) 97, 161
    zone 94, 98–9
  supply 98–9, 138, 161, 203–4, 243, 269
sulphide 5, 19–20, 22, 31, 94, 97, 97, 116, 121,
    124, 134, 135, 139, 207, 208
  dimethyl (DMS) 97, 97, 266, 278, 278
  flux 266
  oxidation 266
  hydrogen 5, 7, 16, 18, 96, 97, 99, 108, 116,
    123, 127, 135–6, 136, 144, 149, 161, 164,
    165, 197, 208
  incorporation into organic matter see
    sulphur, incorporation into organic
    matter
  iron see pyrite
  isotopic composition see sulphur, isotopic
    composition
  nickel 13, 208
  oxidation 20, 22, 23, 97, 98
    bacterial 15, 21, 22, 23, 97–8, 99, 243
  poly- 135–6, 197
  supply 275
sulphur 3, 16, 74, 80, 97, 98, 99, 161, 205, 242
  bond strength 155, 231
  cycle 13, 94, 97, 97–8, 99
    reservoirs 97, 97–8, 263
  dioxide 5, 5, 16, 19, 23, 98, 265–6
    anthropogenic 291
  forms 97, 124, 208, 242
  gases 260, 266
  hexafluoride 290, 291, 297–8, 298
  incorporation into organic matter 97, 134,
    135–6, 136, 138, 149, 197, 242–3; see also
    biomarkers, sulphurization
  isotopes 4
  isotopic composition 19–21, 22, 161, 239,
    242–3, 263
  organic 30, 31, 41, 97, 119, 121, 123, 137,
    139, 148–9, 151, 161, 242, 266
  oxidizers see bacteria, sulphur oxidizing
  reduction 97
  in volcanic emissions 5, 265–6
sulphuric acid 42; see also sulphate
Sun 2, 3, 8
  light see energy, light; insolation; solar
    luminosity
  -spot cycle 263
  T-Tauri stage 5; see also solar evolution
supercontinents 29, 163, 256, 258, 261, 262,
  267, 270, 271
  fragmentation 29, 70, 267, 272
supercritical fluids 156, 157, 157, 158
supportive tissue 2, 27, 30, 42, 62, 117, 172, 248
supernovae 3, 29
surface
  area of particles 104, 106, 108, 109, 127, 250,
    304

area:volume ratio 69, 86, 89, 93, 238, 259,
    304
  mixed layer (SML) 74, 220, 315, 316
  tension 156
  water see water, surface
suspension feeders see filter feeders
swamp 114, 248, 270; see also mires
  cypress 114
  mangrove see mangroves
  Okefenokee 114
symbiosis 15, 95
synergism 319–20
syntrophy 112
syringaldehyde 173, 174
syringic acid 173
syringyl lignin units 64, 66, 101–2, 129, 173–4,
    173–4

2,4,5-T 313, 313
TAI see thermal alteration index
tannins 64–5, 67, 121, 133
taphonomy 25
tar 123, 301
  sand 163
    Athabasca 164
tardigrade 2
taraxerane 190
taraxerene 190, 191
taraxerol 53, 189–90, 191
Tasmania 298
tasmanites 59, 138, 142, 231
Tasmanites 138
Taxodium distichum see swamp cypress
taxonomy 12, 12, 28, 28, 213
  phylogenetic 14–15, 18
TBT see tin, tributyl
TCDD 308, 309, 313, 315–17, 319–20
  antioestrogenic equivalents (TEQ) 320, 320
tectonic
  activity 21, 70, 107, 109, 159, 162, 164, 228,
    250, 252, 254–6, 258, 261, 263, 265, 270,
    273, 275, 283
  plates 2, 109, 203, 228, 250, 273, 276
    destructive margins see subduction
    constructive margins see spreading ridges
  uplift 107, 144, 148, 227, 230, 233, 234,
    250–1, 258, 261, 265, 276, 284
teichoic acids 47, 48
telinite 122, 122
temperature
  atmospheric see atmosphere, temperature
  bottom-hole 229, 230
  and burial see geothermal gradient
  critical 157, 157
  homogenization 230
  hydrospheric see water, temperature; sea
    surface temperature
  lithospheric 2, 147, 157, 158, 164, 229–30,
    229–30, 233
  surface see Earth, surface temperature
  and phase change 141, 157, 157, 223, 308
  proxies 212–21, 218–19, 218, 220, 222, 265,
    269, 272, 275–6, 276–7, 278, 278, 283

range for life 6, 14, 98, 100, 141, 266, 273,
    279–80
  and reactions 117, 125, 131, 146, 228, 262,
    315; see also reaction rates, temperature
    dependence
  solar 2–3
  and solubility 249–50
  units 322
teratogens 295
termites 287, 287
terpanes 49, 160, 202, 204, 301
  bicyclic 188, 190, 192, 193, 196, 201, 202;
    see also carotane
  pentacyclic 190, 191, 199, 203, 205; see also
    diahopanes; hopanes; methylhopanes;
    neohopanes; norhopanes
  tetracyclic 153, 190, 192, 200-1, 202; see also
    diasteranes; 4-methylsteranes;
    secohopanes; secotriterpanes; steranes
  tricyclic 160, 200, 201, 202; see also
    cheilanthanes
terpenes 49
terpenoids 10, 40, 40, 49–60, 141, 188, 305
  biosynthesis 49, 50
  classification 49, 49
  diagenesis 188, 195
  source indicators 167, 201–3
  in woody plants 49, 51, 52, 53, 56–60, 58,
    61–2, 167, 188–9, 192, 194–5, 199,
    201–3, 202
terpineol 51
Tertiary 26, 27, 28, 114, 137, 138, 162, 163,
    201, 262, 276, 280, 324
testosterone 55, 305, 306, 306, 310
tests 25, 100, 112, 118, 122, 212–13, 240
  calcareous/carbonate 1, 25–6, 79, 98, 166,
    213, 216, 218–19, 235, 240, 240, 249–50,
    255, 258, 280
  siliceous 1, 25–6, 79, 166
tetrachlorodibenzo-p-dioxin
  1,2,3,4- 318
  2,3,7,8- see TCDD
tetradecanoic acid 44
Tetraedron 133
  minimum 133
tetrahydroretene 190
2,6,10,14-tetramethylhexadecane see phytane
tetrahymanol 199, 205
tetramethylbutane 223
1,2,5,6-tetramethylnaphthalene 195, 196
2,6,10,14-tetramethylpentadecane see pristane
tetramethyl-1,2,3,4-tetrahydronaphthalenes
  195, 196
tetrapyrroles 59, 179–80, 182; see also
  chlorophylls; pigments, tetrapyrrole
  oxidative cleavage 177
tetrasaccharides 38
tetraterpenoids 49, 56; see also carotenoids
tetrose 35, 37
Thalassia see grass, sea
Thalassiosora weissflogii 176
thermal
  alteration index (TAI) 221, 225

blanket effect   230
conductivity   145, 230
expansion of surface water   107, 265
history   161, 212, 214, 227–8, 229, 229–30
lag   229
maturation   *see* maturation
maturity   *see* maturity
maximum   *see* Holocene, thermal maximum
relaxation   107, 230
stability   *see* thermodynamic stability
subsidence/contraction   107, 107
thermoacidophiles   *see* archaebacteria, thermoacidophilic
thermochemical sulphate reduction   *see* sulphate reduction, thermochemical
thermocline   74, **74**, 77–8, 84, 109, 205, 281, 283, 315
diurnal   74
permanent/main   74, 75, 85, 79, 81, 83, 248
seasonal   74, 79, 81, 83, 109
thermodynamic
equilibrium   *see* equilibrium, chemical/thermodynamic
stability   14, 35, 146, 188, 193, 201, 203, 207, 209–3, 222, 224–6, 244, 313, 320
thermohaline circulation   *see* ocean circulation, thermohaline
thermophiles   *see* bacteria, thermophilic
Thermoproteus   98
thermosphere   296
Thermotoga   14
Thermus aquaticus   14
thiacycloalkanes   136
Thiobacillus   96, 98
   denitrificans   96, 98
thiols   31, 97, 136, 136
thiophene   31
thiophenic unit   123, 139, 149, 152, 207
Thiothrix   98
thorium   4, 6, 145, 230
threonine   8, 41, 239
thymine   68, 69
tides   115, 159
   red   56
time-temperature index (TTI)   225, 228, 230, 230
tin   311
   organo-   310
   tributyl-(TBT)   308, 309, **309**–11, 310
      chloride   309, 310
      oxide   310
Tm   *see* 22,29,30-trinorhopane
T$_{max}$ *see* Rock-Eval, T$_{max}$
Toba super-volcano   265, 282
tocopherols   199, 206, 207
toluene   43, 140, 152, 203, 203
Tonian   324
torbanites   **122**, 142, 133, 138, 199; *see also* coal, boghead
toxicants   301, 311, 317, 321
toxicity   97, 295, 301, 304, 308, 310, 313–15, 317–20

toxins   292, 294, 319
transamination   41, 65
transfer efficiency   *see* trophic transfer
transformation ratio
transmittance (of light)   123
transportation   139, 299, 308, 317
aeolian/atmospheric   1, 107, 241, 265, 270, 292, 295, 299–300, 311, 315–17
aqueous   1, 19, 70, 107, 109–10, 112, 116, 162, 176, 241, 248–9, 268, 280, 286, 292, 300, 308, 321
glacial   1
by humic material   121
long-range   317
Traps
Deccan   273, 283
Siberian   284
traps
sediment   *see* sediment, traps
petroleum   157–8, 162, 197
stratigraphic   **159**
structural   **159**
trees   22, 27–8, 89, 114, 248; *see also* under individual types
deciduous   89, 172, 270, 273
of life   13, *14*, 14, 15
pine   89
succession   89
rings   217
Triassic   26, 28, 28–9, 262, 324
tricarboxylic acid (TCA) cycle   *see* citric acid cycle
trichlorobiphenyl   *see* PCB-28
1,1,1-trichloro-2,2-di(*p*-chlorophenyl)ethane   *see* DDT
2,4,5-trichlorophenoxyacetic acid   *see* 2,4,5-T
Trichodesmium   86
triglycerides   44–6, 69, 168–9
1,3,5-trihydroxybenzene   *see* phloroglucinol
2,3,6-trimethylalkylbenzenes   184
1,1,3-trimethyl-2-alkylcyclohexanes   183, 184
trimethylbutane   152, 203
5,7,8-trimethylchroman   205
3,6,7-trimethylchrysene   194
trimethylcyclopentane   223
trimethylnaphthalenes   202
1,2,5-   *see* agathalene
1,2,7-   195, 196, 202
trimethylpentane   223
1,2,9-trimethylpicene   194, 202
22,29,30-trinorhopane   54, 55
22,29,30-trinorneohopane   54, 55
triose   37
phosphate   17, 17, 35, 40
triphenylene   299–300
triple point   157
trisaccharides   38
triterpanes   192, 201–3; *see also* hopanes; steranes
A-ring degraded   202
C-ring degraded   *see* 8,14-secohopanes
isomerization   191, 209

triterpenes   189–91, *192*
triterpenoids   49, *50*, 52–3, 60
acyclic   52, 199, *199*; *see also* squalane; squalene
aromatic/aromatization   192–3, *194*, 196, 202, 210
bacterial   *see* hopanoids
diagenesis   188–92, *191*, 194–5
higher plant   54, 190–2, *191*, *199*,   201–3
isomerization   191
oxygen-containing groups   192
pentacyclic   52, 188–92, 202; *see also* hopanoids; triterpenoids, higher plant
structural notation scheme   54
reduction/rearrangement   188–92
ring-A degraded   *194*
ring-C degraded   *see* 8,14-secotriterpenoids
ring cleavage   193–5
tetracyclic   52, 192
triterpenols   190–1, 205
diagenesis   188, *191*–*2*, 192
trophic
level   24, *308*, 312, 317, 319
relationship   98, 179, 234, 240–1, 317
status   13
transfer   24, 319
tropics   *see* environment, tropical
tropopause   296, *296*
troposphere   265, *286*, 289–90, *291*, 295, **296**, *296*, 297, 298
tryptophane   41
Ts   *see* 22,29,30-trinorneohopane
tundra   73, 89, 90–1
turbidites   **108**, 110, 115, 207, *268*; *see also* sediment, density flows
Turonian   272, 274
turtles   304, **306**
tyrosine   41

ubiquinones   67, *67*
UCM   *see* unresolved complex mixture
U$^{k'}_{37}$ 219, 220, 220–1, *222*
Ulva   59
universal
ancestor   *see* common ancestor
gas constant (R)   20, 214
unresolved complex mixture (UCM)   152, 160, 302–4, *304*
unsaturated compounds   **30**, 32, 45, 55–6, 60, 135–6, 149, 177, 179, 188, 220
and melting point   44, 168, 212, 219–20
upwelling   19, 73, 75–6, 83, 85, 85, 86, 92, 104, 115, 250, 275, 279
Antarctic   279
equatorial   85, 250
Namibian   85, 115
Peru   85, 105
uracil   68
uraninites   19, 19–20
uranium   4, 6, 20, 115, 145, 229
uronic acids   35
ursane, A-ring degraded   *see* de-A-ursane
ursanoids   52

vaccenic acid   45, 169
Valanginian   274
valine   8, 41, 168
vanadium   149, 181–2, 182, 208, 225
    ratio to nickel   182, 208
van Krevelen diagram   123, 124, 125, 127, 138, 139, 140, 142, 144, 223
vanillic acid   173
vanillin   173
vanillyl lignin units   64, 66, 101, 128–9, 129, 173–4, 173–4
varves   19, **110**
vascular system   27, 62
vegans/vegetarians   239, 239
vegetables   306, 320
vegetation   204
    combustion   22, 123, 251, 253–4, 287, 287–8, 299–300
    Quaternary changes   241
    zones   see biomes
Vendian   25, 28, 29, 262, 324
Venus   280
veratryl alcohol   102, 104
vertebrates   22, 305, 311–12
Vibrio   96, 101
vinyl group   174, 180
p-vinylphenol lignin units   174
violaxanthin   57, 61
viruses   11, 12, 91, 91
vitamin   86
    A   see retinol
    E   see tocopherols
vitellogenin   306
vitrain   123
vitrinite   122–3, **122**–4, 126, 127, 133, 136, 139, 142, 145, 148
    chemical composition   124, 124, 127
    reflectance   123, 125, 125–6, **126**, 140, 212, 221, 225–6, 226–7, 229, 230, 232, 243–4
        calculated   226, 226, 230, 231
        maturation kinetics (EASY %R$_o$)   227, 229, 230, 231
        suppression   124
    reworked   126, 126
    structure   126, 131, 132
    thermal evolution   123–5, 125, 126, 127, 131–2, 132, 148
vitrite   123
vitrodetrinite   122
volatiles   49, 135, 226, 255, 266, 301
    in coal   see coal, volatiles
    juvenile   5
    reduced   23, 23
    semi-   315
    volcanic   23; see also volcanic emissions
volatility   308, 310, 313–15, 317
volcanic emissions   5, 18, 23, 98, 260–1, 263, 265–6, 268, 273, 273, 275, 279–80, 282–4, 300
volcanism   23, 29, 97, 247, 250–1, 252, 261–2, 262, 289
    circum-Caribbean   282
    eruption style   265, 284

vulcanization   134; see also sulphur, incorporation into organic matter

warming potential   see global warming potential
wash-out   see deposition, wet
Water
    Antarctic Bottom (AABW)   75, 250
    Antarctic Intermediate (AAIW)   75, 75, 85
    Atlantic Intermediate (AIW)   75, 279
    Mediterranean   75, 75
    North Atlantic Deep (NADW)   75, 250, 276, 279, 292, 293
    shut-down   292
    Warm Saline Deep (WSDW)   275, 282–3
water
    atmospheric   see water, vapour
    bodies   246, 248, 252, 292, 307
        still   292
    bottom   20, 75, 102, 105–6, 110, 115–16, 165, 179, 219, 274, 283
    brackish   85, 115, 138, 167, 204
    chemistry   109
    in coal   see coal, water content
    column   18, 21, 69, 71, 78, 79, 80, 96–7, 91, 91–4, 109, 116, 175–80, 219, 259, 269, 280, 282, 291, 315, 320
        stratification   74, 74, 77–8, 77, 80, 83, 86, 104, 105, 109–10, 112, 116, 205, 275, 292
        transformations   see diagenesis, water column
        turn-over/mixing   74, 77–8, 80–1, 83, 116, 248, 265, 283, 288, 292
        zones   79
    continental run-off   see rivers
    deep   19, 21, 74, 75, 75, 85, 86, 92, 105, 110, 116, 217, 248–9, 248–50, 259, 265, 278, 280–1, 281, 283, 286–8
        circulation   see ocean, circulation, deep
        temperature   219, 277
        ventilation   279
    density   27, 74, 75–7, 77–8, 80, 83, 116, 147, 158, 274, 292
    depth   219, 230, 292
    energy   71, 78, 105–6, 110, 115, 300–1
    extraterrestrial   5, 5, 8
    flood-   112–13
    fresh   72, 73, 77–8, 116, 119, 242, 248, 275, 287, 292, 293
    disinfection   321
    ground-   20, 90, 112, 114, 308
    heat capacity   291
    as hydrogen source   153, 232
    hypersaline   see hypersalinity
    ice-melt   72, 293
    ionization   81
    isotopic composition   218–19, 236, 242–3, 277
    lacustrine   77
    and life   2, 6, 10, 16–17, 17, 18, 71–3, 79, 134, 260, 279
    lilies   114–15
    logging   104, 113, 287

marine   see seawater
mean ocean   4
meteoric   161
nutrient content   19, 27, 80, 83, 85, 86, 105, 115–16, 125, 162, 220–1, 249, 256, 275, 278–80
oceanic   see seawater
oxygenation   6, 15, 16, 18, 19, 19–21, 22, 23, 74, 77, 87–8, 87, 93–4, 95, 97–8, 104–5, 106, 109, 110, 112, 115, 116, 206, 242, 259, 268–9, 274–5, 282, 284, 292, 301
phase diagram   157, 164
photolysis   6, 266, 269, 280
pore   6, 93, 96, 98, 104, 108, 125, 145, 156, 158–9, 161, 208, 230, 249–50
potable   311, 321
rain   see precipitation, water/ice
saline   98, 116, 138, 276, 283
salinity   see salinity
sea   see seawater
–sediment interface   117
self-ionization   17
solubility of components   in 19–20, 23, 42–3, 72, 100, 115, 117–19, 129, 132–4, 156, 159, 161–2, 165, 219, 227, 249–50, 278, 280, 282–3, 287, 299, 301, 307, 310–11, 313–14, 316–17, 319, 321
super-heated   10, 228
surface   21, 21, 72, 72, 74, 75, 76, 78, 80–1, 84, 85, 86, 88, 93, 105, 116, 127, 177, 217, 219, 237, 240, 248–9, 248–50, 259, 260, 262, 265–6, 275–6, 280–1, 281, 283, 286–7, 290, 292
table   112–14
temperature   73–4, 78, 165, 218–19, 216, 220, 220, 249, 250, 253, 259, 262, 265, 267, 272, 275, 277, 282–3, 291–2, 311
thermal
    capacity   72
    conductivity   145
    expansion   see thermal expansion, surface water
vapour   5, 7–8, 253, 253, 260, 265, 280, 291–2
    and greenhouse warming   265, 268, 277, 280, 286, 291
washing   see oil, water washing
waste-   292, 300, 307, 311, 321
waves   71, 106, 110, 112, 115, 159, 301
wax   44, 47, 49, 101, 115, 124, 127, 130, 134, 139, 151, 167, 197, 224
    cuticular   47, 121, 122, 124, 152, 179, 241
weather   296
weathering   246, 248, 247, 250, 251–2, 253, 256, 261–2, 270, 276, 279, 280, 285, 301
    of carbonates   see carbonate, weathering
    chemical   250, 262, 276, 279
    continental   80, 86, 87, 258, 270, 275
    of kerogen/organic matter   see organic matter, weathering
    physical   250
    by plants   261–2
    of pyrite   see pyrites, weathering

of silicates *see* silicates, weathering
wetlands 279, *287*, 287
wild fires 22–3, 89, 114, 123, 281, 300
wind 74, 75, 77, 77–8, 83, *85*, 109, 254, 275, 277–8, 292, 300, *308*, 310
  belts 115
  polar easterlies 76
  prevailing 75, 76, 85, *90*, 278
  Trade 76, 76, *90*, 278, 292, 300
  westerlies 75, 76, *90*, 279
wood 64, 67, 101–2, 113–14, *122*, 122–3, *124*, 127–31, 172, 174, 241, 247, 253, 288, 299–300
woodland 73, *91*, 92–3
Woodside Creek 281
work 322

Xanthophyceae/xanthophytes *see* algae, yellow-green

xanthophylls **56**, 57
  cycle **57**, *61*
  diagenesis 185
  oxidation *61*
xenobiotics **295**, 297, *298*, 306, 307–8, 320
  environmental behaviour 308–17
xylans 38, 101
Xylaria 101
xylem 62
xylose *36*, 38, 40, 172, *173*, 175
X-ray crystallography 34

yeasts 27, 93, 179
Yellowstone Park 14
yellow substances *see* humic material
Younger Dryas 217, 292, 293, *293*

zearalenone 306, *307*
zeaxanthin 57, *61*, *185*, 185

zinc *182*
zircon 6
zooplankton **21**, 24, 26, 81, 83, 86, 89, *91*, 91, 176, 280, 287
  biomass *82*
  chemical composition 57, 167, 171–2, 176–7, 179, *206*, 207
  contribution to sedimentary organic matter 24–7, 69, 139
  faecal pellets 27, 80, 91, 176–7, 238, 281
    sinking rate 91
  grazing 26, 81, 83, 86, 89, 92, 175–7, *178*, 179–80, *180*, 200, 238, 248, 266, 281
  efficiency 177
  migration 21, 176
Zostera *see* grass, eel
zwitterion 40, *41*, 213